SATYAM S. NAMPALLI
MAY 2003; NEW JERSEY

Control of Particulate Matter Contamination in Healthcare Manufacturing

Thomas A. Barber

Interpharm Press
Denver, Colorado

INTERPHARM
PRESS

Library of Congress Cataloging-in-Publication Data

Barber, Thomas A.
Control of particulate matter contamination in healthcare manufacturing /
Thomas A. Barber. v. <1> p. cm.
Includes bibliographical references and index.
ISBN 1-57491-072-8
1. Parenteral solutions—Contamination. 2. Particles—Analysis. 3. Parenteral solutions—Quality control. I. Title.
RS201.P37B367 2000
615'.19—dc21 99 18307
CIP

10 9 8 7 6 5 4 3 2 1

ISBN: 1-57491-072-8

Interpharm Press
15 Inverness Way E.
Englewood, CO 80112-5776, USA

Phone: +1-303-662-9101
Fax: +1-303-754-3953
Orders/on-line catalog:
www.interpharm.com

CONTENTS

The color micrographs found near the middle of this text are described in the appendix to Chapter 10 on pages 402 to 408.

PREFACE

It is of value to occasionally reflect on where one has been and where one is going. This is true both with regard to walks in the woods and one's career. This book is a result of an interest in particles that spans more years than I care to remember. This interest began when, as a second grader, my father allowed me to experiment with a simple biconvex 2× glass lens approximately 4 in. in diameter that he had among the miscellany in his writing desk. After cautions regarding the focusing of sunlight and other bright light sources, my father left me to begin my explorations of the unseen world. After some obligatory initial investigation relative to the effects of focusing the summer sunlight on ant hills and flammables, I did actually use the glass for the purpose of observation. While the inversion of images by the magnifier was perplexing, interesting observations could still be made. For my 10th birthday, I received an A.C. Gilbert light microscope made in Japan. For an instrument only 9 in. tall, the images at 50× and 100× were quite acceptable, and I was able to pursue my earlier studies at higher levels of resolution. *Daphnia* species were my first animate subject; then I progressed to the evaluation of blood cells; hair; and particle populations such as salt, sugar, and dust, and insects. This began a lifelong fascination with microscopy, the microscopic, and particles of all sorts.

This book, like my earlier book, is dedicated to the applications of particle analysis in the pharmaceutical industry. Specifically, it deals with the control of particles in "clean" environments and preventing particles from gaining access to areas where they are unwanted. I undertook the writing of this book in the belief that a need existed for a text that would provide basic information regarding the control and analysis of particulate contamination specific to the pharmaceutical industry. Although a large body of literature exists that deals with the principles and theoretical aspects of contamination control, there is no single text available on this subject that addresses the issues specific to our industry; the considerable body of practical knowledge in this area is not readily accessible to the pharmaceutical particle analyst. This book represents an effort on my part to present some of this information in a single sourcebook.

My previous book on this subject, *Pharmaceutical Particulate Matter: Analysis and Control,* was published by Interpharm in 1993. As I gathered information for this second book, I was struck by the logarithmic increase in the amount of information regarding contamination control in general and specifically in the areas of pharmaceutical and medical device manufacturing. The levels of concern and interest in particulate contamination and its control are, at present, significantly increased over those of just five years ago. A good part of this must be assessed as due to enhanced regulatory and compendial interest. This interest is further expanded by the drive to create international standards and achieve harmonization of requirements at the international level. There are perceived (on the part of regulatory agencies) needs for enhanced levels of cleanliness in aseptic processing of drug products and for containment of potent compounds. This perception has been mirrored in an increased focus on particulate matter in regulatory inspections, both here and abroad.

A most exciting and useful change in the past five years involves the explosive expansion of the application of electronic data storage, reduction, and transfer. Information regarding specific interests and activities in the field is available from more than 100 Web sites. This figure does not include the numerous and extremely valuable search services, which allow the individual knowledgeable in the art to unearth information from the professional literature. Books, journals, and compendia are available in CD-ROM format, which allows keyboard access to information that was previously available only to those relatively few who knew where to look. Along these lines, the U.S. FDA has made available a comprehensive and easy-to-use database at their Web site (http://www.fda.gov) that gives the user access to guidelines, regulations, and current initiative information that could not be matched a few years ago by the most comprehensive regulatory affairs libraries in the most affluent companies. Similarly, standards organizations worldwide (e.g., ISO, ASTM, BSI, and DIN) have useful and informative Web sites.

The downside of this increase in available information is that it threatens to bury us in avalanche fashion and prevent effective use simply by virtue of volume. There is one answer to this dilemma—organization. Those who are able to attack and assimilate the great mass of available information will survive and prosper. Less-effective users may fall behind, and no one can afford to be isolated from information in today's competitive environment. In the current environment, based on the volume and value of information, the internal information resources of each manufacturer become extremely valuable resources. The same holds true for those who manage the information resources and serve as gatekeepers for the knowledge indispensable to those working in the area of contamination control.

While the business of contamination control has in some ways changed dramatically for the device or pharmaceutical manufacturer, some basic principles remain the same. The ultimate goal of pharmaceutical manufacturers is still to best serve the needs patients and healthcare professionals who depend on their products. The best interests of the healthcare consumer are served by the cooperative activity between manufacturers, regulatory agencies, and compendial groups, which is essential to the development and production of high-quality pharmaceutical and medical products at minimal cost. Thus, I have also included in this book what I believe to be a useful discussion regarding patient issues with regard to particulate matter, standards-setting activities, and cost-effectiveness of contamination control measures.

As was the case when the previous book was written, the great majority of pharmaceutical products manufactured in the world today are produced in total compliance with compendial requirements and contain extremely low levels of particulate matter. The high level of concern that pharmaceutical manufacturers have for the quality and efficacy of their products has resulted in significant decreases in product particle burden over the past decade. The marketplace becomes more competitive with each passing day. Because of this competitive environment in which pharmaceutical products are developed, manufactured, and marketed, only the manufacturer with a serious commitment to customer satisfaction will be successful. The maintenance of extremely low levels of particulate matter in products remains a necessity with regard to the customer's perception of product quality and safety. In summary, the analysis of particulate matter in general and contamination control in particular is a constantly evolving field, particularly with regard to pharmaceutical and medical products. I have tried to provide the most current information, but the reader is forewarned that eternal vigilance is the price of current knowledge. Due to the extent that GMP, compendial requirements, and regulatory issues impact the interpretation of some types of particulate matter data, I have included what I believe to be an appropriate discussion on these nontechnical subjects.

Of necessity, much of the contents of this book come from my experience and hence reflect my interests and views. I have made a concerted effort to discuss all of my chosen topics objectively. Should others disagree with my views on any of the subjects discussed, I promise to respect their conflicting viewpoints and ask that they respond in kind.

Particle analysis and the control and elimination of particulate matter as a contaminant are application-oriented technologies rather than sciences; they involve the combination of knowledge from a number of areas with a logical, often intuitive, approach to problem solving rather than comprehensive knowledge in a single field such as chemistry or physics. In pharmaceutical particle analysis, the individuals who become most effective in contamination control are generalists rather than specialists. With this, I will wish my readers well and hope that the information they find in this book will recompense them for the time spent in reading.

Tom Barber
September 1999

Any and all presentations of specific technical material, subjective judgments, opinions, and viewpoints included in this book are my own and not necessarily those of Baxter International or any organization representing manufacturers of medical devices or injectable products.

ACKNOWLEDGMENTS

I hasten to confess to the reader that my own personal expertise in the various subject areas dealt with in this book was far from a sufficient basis for the discussions in the text. I am thus deeply grateful to those individuals who have unselfishly provided me with encouragement, support, technical resources, manuscript review, information, and other forms of assistance. Many of these people also have assisted me with other publications, including chapters for books, journal publications, and technical communications of various sorts. I deem myself extremely fortunate to have had the invaluable assistance of other particle analysts, quality and production managers, statisticians, and chemists and engineers of various disciplines, who made it possible for me to extend my knowledge in various areas of technology. To the extent that my attack on problems and progress toward my various goals is characterized not by brilliance but rather by an extreme level of persistence, these individuals have been both patient and persevering.

The first of these individuals should doubtless be Dr. Michael Groves (University of Illinois), a versatile scientist and friend. Mike and I began to discuss what is "allowable" in the way of particulate matter in injectables some 20 years ago and have at this point finally come to some level of (incomplete) agreement. Similarly, I would like to recognize and thank Mr. Alvin Lieberman (Particle Measuring Systems), dean emeritus and true authority regarding the subject of environmental monitoring, for his endless help and enduring good nature. Walter McCrone, John Delly, and Lucy McCrone of the McCrone Research Institute have provided invaluable help with manuscript reviews and have trained numerous Baxter analysts in the microscopic art. Two others should be mentioned here. Mr. Julius Knapp (R & D Associates) has spent what must seem like, to him, a very significant amount of time in tutoring, consulting with, and generally helping the author to gain a basic understanding of the principles of visual inspection and coincidence theory. Finally, I will say a profound thank you to Dr. Damian Neuberger, microscopist (light and electron) of consummate skill who has shown selfless attention over the years to my requests for "just one good picture" and provided the color plates and descriptions for this and my previous book.

I must express my heartfelt gratitude to those at Baxter International who made it possible for me to complete the monumental task of writing this book. I extend my particular thanks in this regard to two very understanding and supportive supervisors: Dr. Kshitij Mohan, Vice President Corporate Research and Technical Services, and Dr. Tom Sutliff, Vice President Medical Materials Technology. Had it not been for the allowances in resources, time, and encouragement that they made available to me, I would doubtless have given up the task as impossible early on.

I will also thank profusely the Administrative and Executive Assistants who dealt with my numerous revisions, changes to drafts, and low level of word processing skill and offered editorial assistance: Ms. Kim Cooper, Ms. Myrna Storm, Ms. Jackie Pardus, and Ms. Jan Simon. I am deeply grateful to my fellow professionals at Baxter, who supported me with discussions, references, critical comment, artwork, and photomicrography: Dr. Damian Neuberger, Dr. Dean Laurin, Ms. Christine Pavek-Hicks, Mr. Neal Zupec, and Mr. Kirk Ashline. Mr. John Williams is the artist and technical illustrator of rare skill who contributed most of the drawings and art for this book. Mr. William Lu also generated numerous drawings, graphics, and data reductions. It would have been impossible for me to present the information in this book as effectively without John's and William's dedicated collaboration. I thank and commend them for their patience and the diligent attention to detail that their drawings and graphics display. Mr. Joseph Barrett deserves thanks and a high level of credit for reviewing the entire manuscript.

I would like to gratefully acknowledge the help given me by the following individuals from organizations outside Baxter: Dr. Chuck Montague and Mr. Peter Rossi (Pacific Scientific), Dr. Benjamin Lu (University of Minnesota), Dr. Holger Somer (TEAM Consulting), Mr. Terry Munson (Kemper-Masterson), Mr. Joseph Belson (USP), Mr. Frank Barletta (USP), Mr. Dale Bares and Dr. James Boylan (Abbott Laboratories), Dr. Ron Wolff (Eli Lilly and Company), Dr. Werner Bergman (Lawrence Livermore National Laboratory), Mr. Dan Berdovich (Micro Measurement Laboratories, Inc.), and Dr. David Nicoli (Particle Sizing Systems). Over the years, I have grown to value critical comments on work that I do and hold those who offer it in high regard. A constructive critic is a true friend.

There are several technical subjects that, when discussed even at the general level that I tried to attain in the chapters of this book, become significantly complex. For their generous support in specific technical areas, I express my gratitude to the following: Dr. John Levchuk of the Food and Drug Administration (FDA) (regulatory perceptions of particulate matter); Mr. Thomas Bzik of Air Products and Chemicals Inc. (airborne particulate matter monitoring statistics); Dr. Theodore Meltzer of Capitola Consulting (process filtration); Mr. Scott Aldrich of Upjohn Pharmacia (compositional analysis and identification); Ms. Anne Marie Dixon of Cleanroom Management Associates (garbing and personnel practice); Mr. Hank Rahe of Contain-Tech (isolation and barrier technology); Dr. Terry Allen of Dupont Corp. (sample collection methods); Dr. Philip Austin of Acorn Industries (cleanroom construction); Mr. Ludwig Huber of Hewlett-Packard (validation); and Mr. Larry Pesko of the St. Vincent Medical Center in Toledo, Ohio (pharmacy practice and historical overview).

In the area of HEPA filtration, I am in the debt of four individuals for their assistance: Mr. Bruce McDonald (Donaldson Filters; Mr. Mark Cutler (Flanders Filters Inc.); Mr. David Crosby (ATI Inc.); and Dr. Rudi Wepfer (LUWA Inc.). I would also like to offer my thanks to the numerous staff persons at the Institute for Environmental Sciences and Technology (IEST), the Parenteral Drug Association (PDA), and the British Standards Institute (BSI) who assisted me in accessing the vast informational network that these organizations comprise.

Finally, I would be remiss if I did not express my deep appreciation to my employer, Baxter International, whose commitment to product quality, customer satisfaction, and the core competency of contamination control technology made it possible for me to complete this book.

Tom Barber

This book is respectfully
dedicated to healthcare
professionals and
their patients.

I

INTRODUCTION AND OVERVIEW

Contamination is often used as a collective term that refers to any materials that occur in some place where they are detrimental or not wanted. An example of contamination is provided by the presence of some soiling material on an object or surface that should rightly be clean, shiny, or bright. Dust on the windshield of a newly washed car, food residue on dishes, grime on the hands or clothes of children, and urban aerosols (smog) are examples of contamination. For centuries, people have applied methods of separating themselves from particulate contaminants. The ancient Egyptians developed sieving to a high level of sophistication for particle separation and collection. Desert dwelling peoples have known for millennia that woven cloth serves to separate sand and dust from air for easier breathing. Settling and decantation are historical methods for removing sediments and flocs from solutions. Soaps were presumably developed following an initial discovery that a mixture of wood ashes and tallow was efficacious for cleaning soiling materials from the hands.

In the industrial practice of pharmaceutical manufacture, there is a significant concern for cleanliness. Contaminants generally fall into three categories:

1. Chemical (such as solvent residues in a mix tank, pyrogens, or traces of a previously processed drug in the one currently being processed)
2. Microbial (replicative) contaminants that may, in a worst-case perspective, be infective
3. Physical (particulate matter)

Thus, particulate matter is only one type of contamination with which the pharmaceutical manufacturer is concerned. It is not as potentially dangerous as microorganisms, nor is it as insidious as an invisible chemical residue. By virtue of the fact that it is now considered an indicator of process or product shortcomings, however, particulate matter has received a great deal of interest in our business. The end user of healthcare products or a healthcare professional generally cannot see microbial or chemical contaminants, but particles above a certain size often are noticed and remarked on by their discoverer so that their presence becomes an issue of concern.

What is a particle? A broad definition is that a particle is any small object having definite physical boundaries in any direction. An electron, a molecule, a blood cell, or even the earth (in the context of the universe) could be considered a particle. Kramer et al. (1970) defined particulate matter as any insoluble foreign material contained in a parenteral solution. This author's definition is that a particle is simply a physical object that is small in relation to the total volume of the system in which it resides. The size and shape of the particle must be thought of in physical three-dimensional terms. Particles may be round, irregular with rough edges, or long and thin such as a fiber. The chemical composition of a particle may, in some cases, be critical. In general, the size of a particle may be simply and uniformly described as the diameter of a circle whose area is equal to the projected area of the particle (Trasen 1969; Barber 1987). Particulate matter present in parenteral solutions and medical devices has been an issue in the pharmaceutical industry since the introduction of injectable preparations (Groves 1973) and remains unavoidable, even with today's well controlled manufacturing processes.

Among the compendial requirements placed on injectable products, particulate matter has historically been an area that receives a significant amount of attention and the single product characteristic that is most likely to involve subjective judgments. Correctly or incorrectly, the particulate matter burden of a product has been taken by some healthcare practitioners, academic investigators, and regulatory personnel as an indicator of overall product quality. This is unfortunate, since particulate matter is, realistically, only a single parameter by which product suitability or conformity may be judged. One reason particulate matter receives so much attention is that it may constitute an obvious defect in product. The pharmacist, physician, nurse, or patient cannot be expected to detect a product nonconformance with regard to pH, potency, drug concentration, or osmolarity without appropriate tests. Particles, however, may be detected by simple visual inspection, or in the case of subvisible particles, quantitated by fairly simple assays. Thus, there is a reasonable (if subjective) inclination to emphasize and be acutely aware of the occurrence of particulate matter.

The material in this chapter is intended to provide a general historical perspective and overview regarding the occurrence of particulate matter in injectable products and devices as well as its detection and control. The author has included a summary of human exposure to particulate matter in aerosols, the origin and philosophy of particulate matter limits, and the role of Good Manufacturing Practice (GMP) in limiting particulate matter. The references to particles in medical products constitute a brief review of the literature. Although many are dated, they relate to the historic, evolving interest in the subject and are in that sense valuable. In a technical or scientific sense the data they contain are still highly relevant to our discussions. Particularly valuable general references are the book by Akers (1985) and the papers by Groves (various).

ENVIRONMENTAL PARTICLES

The environment in which we live contains, literally, thousands of different types of particles. Particles comprise our most intimate (i.e., closest) contact with the rest of the world and are constantly becoming attached to us or our clothing or actually entering our bodies. There is probably no one who has not at some time or other had the unpleasant experience of having a contaminant particle in the eye. Each time we touch a surface, however clean, we take away particles in the film of fatty acids and oils on our hands and fingers, and typically leave behind thousands of skin cells.

The major subject area of this book relates to the occurrence of contaminant particles in drug powders or solutions or medical devices. These particles are, in reality, a second order or indirect effect. They represent contaminants related to the product or manufacturing process, and the patient is exposed to them as a result of intravenous (IV) therapy. In fact, the most direct exposure of humans to particles results from the presence of airborne particles or aerosolized contaminants in the air we breathe. Whether or not inhaled particles should be termed "contaminants" is not a simple question. In fact, all air that an individual breathes, without a filtration device of some sort (i.e., surgical mask, gas mask, filtered respirator) in place, contains large numbers of particles at submicrometer and larger sizes.

If we consider outdoor air in a rural location, our particle counts per cubic foot may well be on the order of 20,000–30,000 particles $\geq 0.5\ \mu m$. The particles counted consist of windblown soils, plant dusts, spores, and pollens, as well as some proportion of man-made dusts, such as those from smokes and agricultural processes. In urban or industrial areas, the counts will be much higher, and the proximate makeup of the population will be skewed toward man-made particles (e.g., auto or diesel emissions, smokes, metal dusts). To describe the higher particle burden of urban air, we tend to use a definition of "contaminant" similar to that applied to pharmaceutical products. In either context, particles become contaminants if they exceed certain numbers or if they are predominantly from a single, known source. It should be noted here that whereas particles in injectable materials or medical devices occur in low numbers and can be generally classified as innocuous, aerosol particles of many types have well defined toxic properties and are a known hazard to human health. Examples are sulfur-containing particles ("acid rain"), asbestiform minerals, metal dusts, tobacco smokes, and some natural materials such as pollens. Airborne particles in urban aerosols may contain trace metals, such as iron, nickel, lead, or calcium, or have bound chemicals such as quinones, phthalates, styrene, napthene, or naphthalene. The poisonous gasses that have been used as weapons of war are often delivered as fine particle aerosols. Incidentally, in the pharmaceutical industry, containment of toxic materials, such as aerosolized chemotherapeutic agents, may be a concern. Particles of specific types can also trigger asthma or bronchitis attacks.

Physiologic Effects of Environmental Particulate Matter

First, let us consider the case with particles in the air that constantly assail the body. Since the face is usually not covered, the body openings (i.e., eyes, nose, ears, and mouth) are particularly attractive routes of entry. The hairs in the nose and ear canal and eyelashes serve as filtration mechanisms that serve to exclude larger particles (including viables such as insects). Secretions (saliva, tears, mucous in the nostrils, and wax in the ears) serve to entrap particles or wash them away. Among all of the organs of the body that serve to "filter out" or collect particles, the lung has the distinction of being the only one exposed to particles from both the circulation (within) and from the outside (i.e., environmental airborne contaminants), and in this regard is subject to special consideration.

The human lung is far from being a helpless target for aerosolized particulate materials. It contains a large population of actively phagocytic cells (alveolar macrophages) that are capable of ingesting hundreds of foreign particles per cell. The lung is also protected by secretory mechanisms that entrap particles in fluids or mucous for subsequent clearance, and by ciliated cells that serve to transport particles from the organ. Small particles (i.e., < 0.5 μm as a ballpark figure) are not deposited in the lung as efficiently as larger particles and may be simply exhaled. Particles larger than 10 μm tend to be deposited or impacted in the nasopharynx or upper respiratory tract and will be cleared in nasal secretions or eventually washed into the esophagus and swallowed. Particles in the range of 0.5–10 μm are those that tend to be deposited at some level in the airways or alveoli in the lung. Based both on historical studies of the deposition of toxic or harmful aerosol particles and more recent research into the delivery of drugs by aerosol particles, the mechanics of particle deposition in the human lung are well understood.

The most useful references include those by Martonen (1991, 1993). The airways of the lungs provide a pathway of low resistance to the bulk flow of air in and out of the "deep" lung, where alveoli perform the essential function of gas exchange. Wiebel (1963) discussed the succeedingly smaller diameter air passages in the lung as "generations" numbered from 0 to 23. Higher numbers refer to airways deeper in the lung with numbers 0 to 5 assigned to the trachea, bronchi, and bronchioles. Based on Wiebel's concept, the lungs are divided into two general compartments or zones—the transport zone and the respiratory zone, as shown in Figure 1.1. The transport zone consists of the first 16 generations of airways composed of the trachea (generation 0), which bifurcates into the two main bronchi, which further subdivide into bronchi that enter the two left and three right lung lobes. The intrapulmonary bronchi continue to subdivide into progressively smaller-diameter bronchi and bronchioles. This zone ends with terminal bronchioles that are devoid of alveoli. The function of the transport zone is to move air by bulk flow in and out of the lungs during each breath. The respiratory zone consists of all the structures that function in gas exchange. These bronchioles subdivide into additional respiratory bronchioles, eventually giving rise to alveolar ducts and finally to alveolar sacs.

In a gross mechanical sense, the lung functions as a complex impaction mechanism, with particles of smaller sizes being impacted at levels deeper in the lung, where airflow velocities

Figure 1.1. Model of airway morphology. Conducting (trachea—terminal bronchioles) and respiratory (respiratory bronchioles—alveolar sacs) zones of the airways.

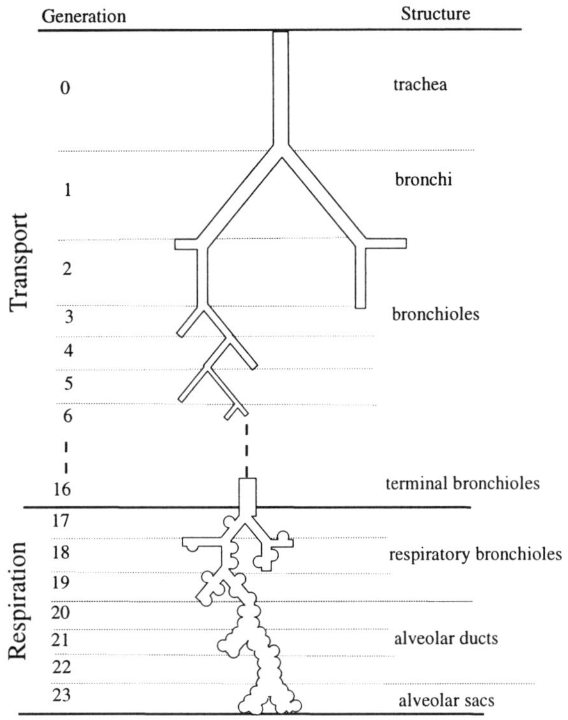

may increase, air passage diameter decreases, and the angles of bifurcation (branching) may be greater. In addition to impaction, two other mechanisms of deposition, sedimentation and diffusion, also occur. The major deposition processes are depicted in Figure 1.2, where particle behavior at a bronchial branching site is illustrated. Inertial impaction occurs when particles of sufficient momentum (product of mass and velocity) are affected by the considerable centrifugal forces generated where the airway network changes direction abruptly. Sedimentation involves particles of sufficient mass deposited by the action of gravity when residence times within airways are large. Particle deposition also occurs as a consequence of random Brownian motion and contact of particles with airway surface (diffusion).

With sedentary breathing rates, deposition by all three mechanisms may be expected to be greatest in the higher-numbered generations (i.e., alveoli) for particles between 0.5 μm and 3 μm. Heavy activity breathing results in a pronouncedly greater deposition in lower-number generations (1–5) for particles > 1 μm in size, due to higher airflow velocities. Soluble particles are dissolved in the fluids that coat the alveolar passages and can thus move through the junctions between the epithelial cells into the capillary lumen. (This is the mechanism of delivery of aerosolized drugs into the circulation.) For more information on this subject, the reader is referred to the references by Altiere and Thompson (1996) and Martonen and Yang (1996). A useful recent review is that of Ding et al. (1997).

Urban aerosols can carry an interesting mixture of particles both in consideration of quantity and chemical composition; the reference by Williams et al. (1995) calls attention to the presence of natural rubber latex allergens in dust

Figure 1.2. Particle deposition mechanisms at an airway branching site.

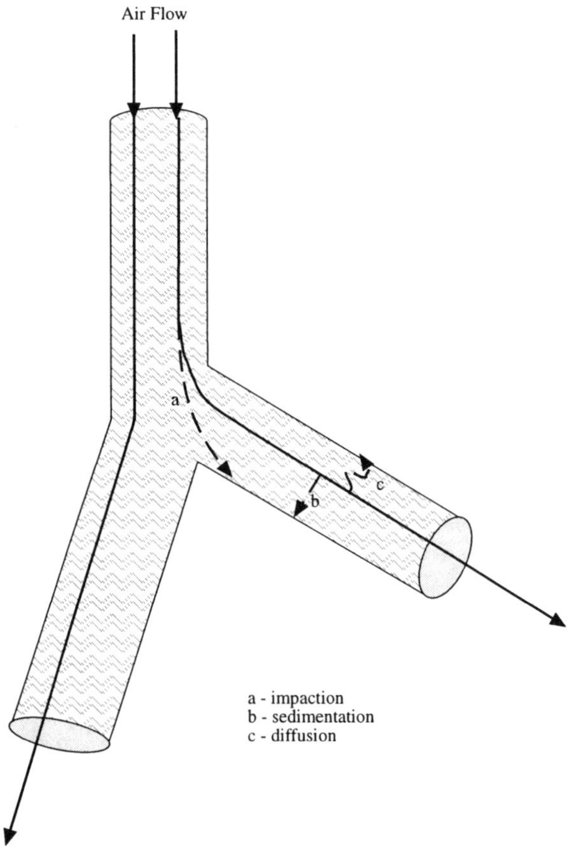

produced by the abrasion of automobiles on highway surfaces (Figure 1.3). These aerosols also contain a high proportional concentration of soil particles composed primarily of aluminum and silicon, which are the most abundant elements on earth.

This discussion serves to establish that the human body is routinely exposed, internally and externally, to large numbers of particles present in the general environment. It seems amazing that there is only a small number of cases in which the people-particle interaction is harmful. In this context, the small numbers of particles to which a patient receiving intravenous therapy is exposed by a circulatory route would seem to be an issue of relatively minor importance. The considerations in the following sections of this chapter and historical perspective bear out this assessment.

CIRCULATORY TRANSPORT AND PARTICLE REMOVAL

Aside from the significant implications of particulate matter with regard to the cleanliness of the manufacturing process, questions regarding the effects of infused particles on patients invariably arise. The review chapter by Pesko (1996) provides the most current and comprehensive source regarding this subject and is recommended reading. The reader is reminded that most of the following discussion primarily relates to particles infused via the route of peripheral veins. There is some confusion in nomenclature applied over the years to injectable dosage forms. An "injectable" is, in the broadest consideration, any material that is intended for injection. The effects of particles in intramuscular (IM) injections have never been a subject of great interest, due to the premise that relatively small amounts of material are injected, and that tissue phagocytosis and local immobilization of the particles would render them harmless. In fact, chronic IM injections, as with insulin, have shown no significant ill effects. Intravenous injectables are administered by a specific route and are often of much larger volume than IM dosage forms. "Parenteral" administration is defined as "other than by the mouth." This literal translation once was a source of some embarassment for the author during a visit to Japanese drug manufacturers. At that point in time my business cards stated that I was an associate director of "other than by the mouth." Of note is the fact that fluids administered intraarterially can, of course, reach organs supplied by that segment of the arterial tree and the greater pressure of the arterial circulation may be a factor in their passing the vessel walls. Most compendial requirements today specify allowable particle numbers for injections or injectables; the U.S. Pharmacopeia (USP) (1995) discusses different limits for large volume injections (LVI) and small volume injections (SVI).

Based on our knowledge of vascular anatomy, we can chart a hypothetical course for an infused particle. If a particle is introduced into the radial vein, it will travel toward the right heart via the systemic venous system, reaching it in as little as 30 sec. Since the diameter of veins increases as they progress to the heart, there is little likelihood of the particle becoming lodged. The particle enters the right atrium, passes the tricuspid valve, enters the right ventricle, and is pumped into the pulmonary artery. The size of arteries, unlike veins, decreases in the direction of flow as they branch. In the case of the pulmonary artery, the vessel finally ends in the alveolar capillary bed of the lung, with the diameter of these capillaries being on the order of 7–12 μm. Theoretically, a particle of greater size will be trapped at some point in the pulmonary vascular bed. While this entrapment of particles $\geq$ 10 μm by the lung generally occurs, larger particles will get through the lung due to arteriovenous "shunts" of larger than capillary diameter.

The body has effective and well-developed mechanisms for dealing with particles to which it is exposed both by external contact (i.e., contact with the environment) and also by routes that circumvent the normal defense mechanisms. Introduced in this latter fashion, the particles evade the body's normal defensive mechanism against materials from the environment, and are then free in an area of the body that is potentially vulnerable. In the circulation, humoral (soluble) factors in the blood and cellular defense mechanisms such as phagocytosis are still effective. Infused particles may be phagocytosed, walled off (sequestered), or, in some cases, simply dissolved. Nonetheless, our efforts to measure, establish limits, and prevent particulate contamination of injectable products and devices are based on our belief that these particles are contaminants, and in some hypothetical way may be detrimental.

Figure 1.3. Particulate matter from an urban aerosol collected on the surface of a glass fiber depth filter. A range of particle types and sizes are present.

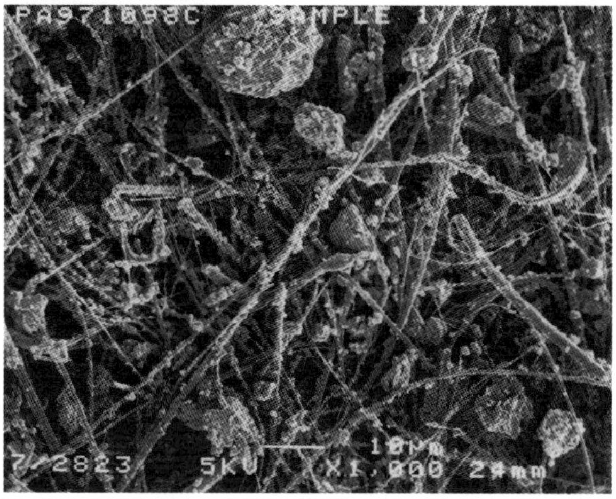

The lung, as discussed above, is constantly exposed to environmental particles of varied chemical composition. It is also prominently exposed to particles from the pulmonary circulation. This includes any particles that might enter the body as a result of IV therapy. Generally, animal studies suggest that the lung constitutes an increasingly effective barrier to infused particulate matter as particles of larger size than 10 μm are considered. Thus, inert particles with diameters larger than 10 μm and many smaller particles may be entrapped in the capillary beds of the lung. In the lung, there is an active mechanism for passing the trapped particles through the capillary walls so that they are excreted into the sputum and mucus and continuously swept from the lung. Further, the alveolar macrophages are highly active phagocytes that can actively pass from the pulmonary circulation into the air space, and particles would be expected to be removed by these cells. Because the lung has an extensive circulation, an extremely high particulate matter load would be required to close down a sufficient area of the lung to cause any physiological effect. Of interest in this regard is that transfusion of blood results in deposition of amounts particulate material in the lung that are far in excess of the particle burden resulting from multiple courses of IV therapy; even so, serious adverse pulmonary effects due to single units of blood or a small number of units are rare today.

While the lung acts as a filter for the venous circulation, not all particles will lodge there. Particles of 50 μm and larger (particularly if administered in large numbers) may reasonably be expected to lodge in the lung simply based on a consideration of their size relative to that of alveolar capillaries, with a consequent decrease in pulmonary function. This would presumably occur with large capillaries being blocked and the function of multiple smaller capillaries thus being restricted by the effect of single, large particles. This situation would not be expected to result with the administration of a parenteral solution containing small numbers of particles at much smaller sizes. The actual mechanics of transport and collection of particles passing through the lungs and lodging in capillary beds in other organs, or being phagocytosed by tissue phagocytes, are not clear.

Comments from early sources pertaining to the size and characteristics of particles are pertinent here. Gardner and Cummings (1931) reported that particles were segregated into different locations according to their size. The larger ones (10–12 μm) were stopped in the pulmonary capillaries, those of intermediate size (3–6 μm) in the spleen and hepatic lymph nodes, and the finest ones (≤ 1 μm) in the liver.

Trasen (1968), summarizing the body of medical knowledge available at that time, found the medical opinion that the greater the irregularity, surface roughness, and length of the particle, the greater was its tendency to block a vessel. Further, harmfulness of particulate matter was believed to be determined by not only the number of particles present but also by their size and ability to become lodged in a blood vessel. This lodging factor was presumably related to the geometric shape and surface characteristics of the particles. It was speculated that a long, thin fiber would be more likely to plug a minor vessel than would a piece of plastic of equivalent mass that might be more nearly spherical in shape and thus progress further to block only a single capillary.

Generally, the vasculature of the lung discussed above is mirrored in other organs. Large vessels, with a diameter significantly larger than that of a particle that might be infused through an IV cannula, transport blood to the organ. Within the organ, the vessels decrease to ever smaller diameters, ending in capillary beds that provide individual cells with nutrients and oxygen and remove waste. This vascular structure has been likened by some authors to that of the system of streets. roads, and interstate highways in the United States (Pesko 1996). In this analogy, the capillary beds are the narrow city streets, alleys, and winding country roads, and a particle might be represented by a truck or other vehicle wider than a single lane. Smaller capillaries have a diameter (with some distension) just sufficient to allow a red cell of 7 μm nominal maximum dimension to pass. Capillary beds with immense filtrative capacity for infused foreign particles are present in the eyes, brain, liver, spleen, and intestine as well as in the lung. The physical effect of any occlusion of a vessel in the lung or other organ will depend upon the degree of collateral circulation available to the affected area and organ function. (One functional definition of "collateral" is "redundant"; see Figure 1.4.) Occlusion of small vessels branching from the pulmonary artery (lung) would have minimal effect since there is more than ample collateral circulation; similar occlusions in the brain, kidney, or retina might be more likely to have a more detrimental result.

PARTICLES AND PATIENTS

Clinicians, scientists, and regulatory officials have long been interested in the possible harmfulness of the particles of foreign matter present in parenteral solutions or introduced in other ways (such as a surgical wound) (Beck 1966). While the historical literature is very interesting, we are generally brought to the conclusion that there is no possibility of patient harm due to the low allowable levels of particles in solutions conforming to the particulate requirements of the world's major compendia today. In examining autopsy specimens from 19 cases, Bruning (1955) described particles of both fibrous and crystalline material in the pulmonary vessels, the vessels being surrounded by a foreign-body reaction. He suggested that IV therapy was the only common factor in these cases. Sarrut and Nezelof (1960) reported 25 cases of pulmonary arterial lesions in autopsy specimens from premature infants, characterized by a macrophage reaction caused by foreign bodies in the pulmonary circulation. The foreign bodies were debris of fine cotton fibers. The only common feature in these cases was the IV infusion of

Figure 1.4. Principle of collateral circulation. (Inset: Latex cast of capillary vasculature of liver.)

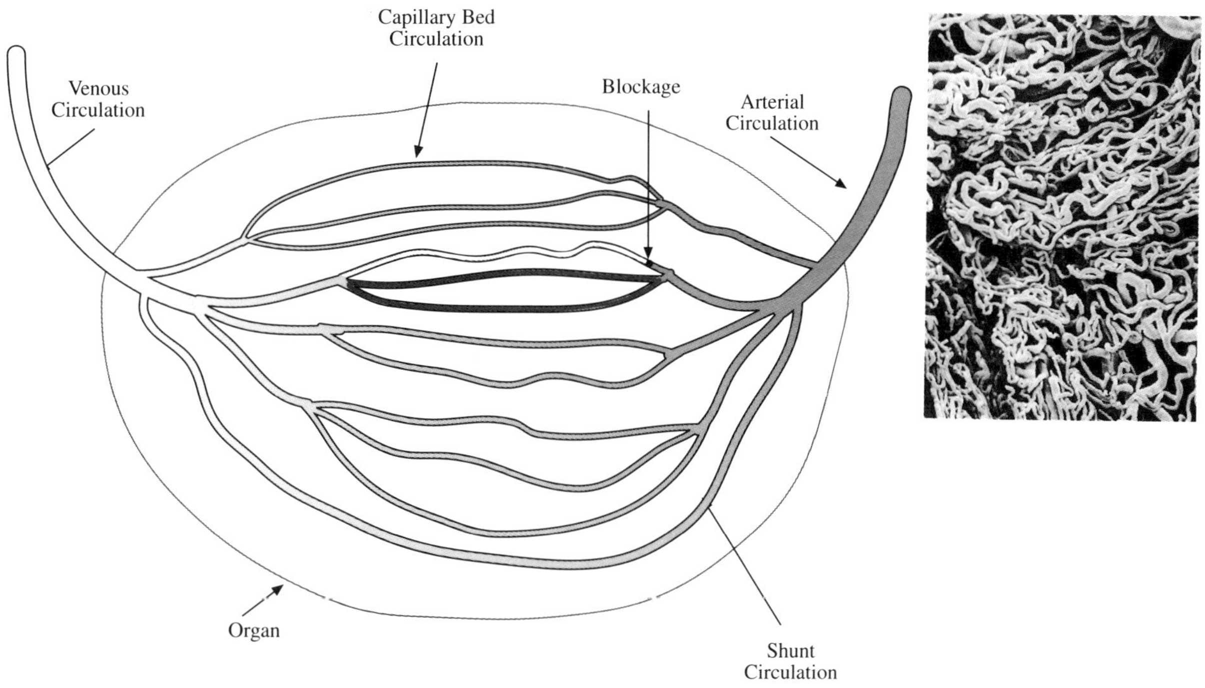

relatively large volumes of fluids. It must be kept in mind that these reports from three decades ago describe circumstances far different from those today in terms of solution particle burdens. In one consideration, these historical reports serve to underscore the dramatic improvements in product quality that parenteral manufacturers have made in the intervening years.

Chason et al. (1963) reported 10 instances of cerebral cotton fiber embolism occurring during carotid angiography discovered during postmortem examination. Although no deleterious effects were noted in the patients, either morphologically or clinically, 2 of the 5 cotton fiber emboli reported by Silberman et al. (1960) were associated with small cerebral infarcts. Garvan and Gunner (1964) discussed 1 case of particles introduced into the cerebral circulation during angiography. Cerebral angiography revealed an angiomatoid malformation in the right occipital lobe of a patient suffering from subarachnoid hemorrhage. A total of 7 granulomas were found in the wall of the aneurysm. Five of these were due to cellulose fibers, and the other 2 contained starch granules.

The study of particulate matter in injectables really began in earnest with the work of Garvan and Gunner in 1963. These authors quantified and classified particles that were present in intravenous solutions manufactured in Australia. In a second paper Garvan and Gunner (1964) gave additional information on the method of detection and the nature of these particles, and presented evidence of the harmful effects of such particles. These observations were based on rabbit experiments and the histological evaluation of human lung tissue taken at autopsy. The histopathologic observations consisted of capillary and arterial granulomas that resulted in a proliferation of histiocytes and contained one or more foreign body giant cells. Each granuloma contained fragments of cellulose fibers. "Vascular bastosis" was suggested by the authors as a descriptive name for these lesions, since bast was the material that made up the filter media used to prepare IV solutions at that time in Australia. The investigators estimated that for every 500 mL of 1965-era Australian intravenous fluid infused into experimental rabbits, 5,000 scattered granulomas had developed throughout both lungs. The lesions observed in humans at autopsy were similar to the granulomas found in experiments on rabbits and contained bast fibers. In those cases in which intravenous fluids were given just prior to death, there were, in some cases, fibrous pneumonitis in which cellulose fibers, crystals, and vascular lesions occurred in the lungs.

As many as 40,000,000 patients may be hospitalized in the United States this year. At an estimated 5 L per hospitalized patient, as many as 200,000,000 L of intravenous fluids might be administered during the course of conventional IV therapy. If we were to additionally consider the administration of intermittent admixtures, directly administered intravenous push medications, and intravenously administered diagnostic agents, we can begin to see the scope of the potential issue, were there any possibility of patient harm from the low number of particles in currently manufactured injectables or devices.

The studies of Garvan and Gunner sparked wide interest in the effect of particles on patients. Goddard (1966) stated that although the data were circumstantial and very limited, they should not cause complacency. Some were of the opinion that particles below 5–10 μm may not be clinically important and that even those up to 20 μm might be tolerated (Krueger and Riggs 1968). Ho (1967) and Ho et al. (1967), pointed out that adverse effects from the injection of particles are not easily demonstrated and that there was little or no conclusive evidence of any significant clinical symptoms. Groves (1968) also stated that there was no real evidence to suppose that particles need be dangerous. Most of these authors conclude, however, that attempts should be made to prevent infusion of particles.

The medical significance of particulate matter may rest essentially on the consideration of the state of patients receiving large amounts of parenteral fluids. Several clinicians (Gross 1967; Gross and Carter 1966; Plein 1969) pointed out early on that particulate matter is more likely to be harmful in hospitalized human patients, who are often recumbent with a sluggish pulmonary circulation and who may be under therapy with corticosteroids. The latter might modify the tissue response so that a localized granulomatous process could become diffuse and extensive throughout the lung or other organs. A summation of the possible effects of particulate contamination was presented by Russell (1970). The major pathological conditions that are produced appeared to be as follows:

- Direct blockage of a blood vessel by foreign matter
- Platelet agglutination, leading to the formation of emboli
- Local inflammatory reactions caused by the impaction of particles in the tissues
- Antigenic reactions with subsequent allergenic consequences

The most recent, and best focused, review of literature on this subject is found in Pesko (1996). A number of conclusions can be drawn from the information summarized in this publication:

- The distribution of injected particulate matter will depend generally on size and to a lesser extent on particle composition.
- Large particles (i.e., ≥ 50 μm) on the basis of the mechanics of circulation (venous infusion → right heart → lung) will be retained in the lung.
- Particles 10 μm and larger pass the pulmonary vascular bed slowly.
- Particles < 10 μm in size are retained in the liver and spleen for long periods of time.
- Phagocytosis by cells of the reticulothelial system is a significant means of clearance only for particles < 10 μm in size.

Pesko reached several conclusions on the physiological effects of particles:

- It appears that the result of introducing particulate matter into the bloodstream is directly related to three major variables: the size of the particles introduced, the number of particles and the rate at that the particles are introduced, and the chemical composition of the particles.
- Pesko suggests that a fourth factor could be the "condition" of the host. This might be defined as the capacity of the host to acclimate to the injurious nature of the foreign insult. This capacity could depend on factors ranging from the general health of the host to the size and ability of various organs to compensate.
- Chemical reactive effects: while polystyrene divinylbenzene microspheres are generally considered to be inert, other types of particles may cause an inflammatory response to be initiated. The seriousness of this insult will depend on the degree of the reaction, its location within vital organs, and the organ's capacity to compensate for the injury. If the reaction is limited, then the injury may be minimal and contained. If the foreign material is antigenic, this may initiate an allergic response by the host that could cause a progressive and much more detrimental reaction.

ANIMAL STUDIES

While the data regarding effects of particles on patients are almost entirely circumstantial rather than experimental, well-designed laboratory studies have been conducted with animal subjects. Animal studies regarding the physiologic consequences of particles began before the interest in human patients sparked by Garvan and Gunner. Animal studies have been characteristically more valuable in understanding the effects of particles than data from humans. Gardner and Cummings (1931) injected rabbits with silica particles 1–12 μm in diameter. Their experiments emphasized the importance of particle size in producing reactions to silica. The largest particles, 10–12 μm in diameter, provoked a foreign body reaction in the lungs that persisted without much further change for nearly 3 years. Smaller ones, from 3 to 6 μm in diameter, caused slowly progressive changes in the spleen and hepatic lymph nodes. Very fine particles of the order of 1 μm in size resulted in progressive proliferation of connective tissue in the liver. Simson (1937) stated that distinctive fibrotic lesions were produced in liver, spleen, and lymph nodes of rabbits following IV injection of siliceous dust of 0.8 μm or less in size.

Brewer and Dunning (1947) reported that occasional particle contamination of ampoule preparations produced no significant pathology in animals. Truly massive doses were required to produce any pathological reaction in rabbits; glass particles were seen in the lungs, liver, and intestine of these rabbits. The spleen was enlarged, the kidney was gorged with blood, and there was evidence of thrombi formation. The authors believed that by careful ampoule processing, the dangers inherent in particle contamination could be largely eliminated.

Van Glahn and Hall (1949) reported that in the routine examination of lung sections over a period of 20 years at Bellevue Hospital and the New York University College of Medicine, 6 cases were noted in which particles were found that were considered to be cotton fibers from saline or glucose solutions. They also conducted experiments in that they produced lesions in animals similar to those they had observed in humans. When cotton fibers were injected in the iliac veins of rats and lodged in the pulmonary arteries, foreign body granulomata were formed. They found no thrombosis or hemorrhage associated with these granulomata, nor did the process lead to infarction of the lungs.

Plugs and fragments of rubber stoppers or diaphragms (also called "cores") are frequently cut out by needles used for withdrawing medications from vials and may be incidentally injected into patients. Magath and McClellan (1950) found that plugs of neoprene or synthetic rubber produced a relatively mild reaction when injected into guinea pigs. They believe it was unlikely that these plugs would produce much damage unless one were injected directly into a blood vessel, or unless one became the nidus for an infection. Injections were given intramuscularly, and in the guinea pig, the fragments of polymer migrated into the popliteal space after being injected into the gluteal muscle.

Wartman et al. (1951) described the results of injecting saline suspensions of filter paper fibers into the ear veins of rabbits. Resulting emboli became impacted in the pulmonary arteries and adhered to the inner arterial walls, causing acute arteritis, and to the right ventricular endocardium. Foreign body granulomas were created in the inner and outer arterial walls. Fibroelastic scarring of the interiors of various blood vessels occurred. The authors thus believed that because the fibers were so effective in producing granulomas that the injection of even minute foreign bodies (cellulose fibers) might be fraught with serious consequences.

Fragments of hair, epidermal scales, and other unidentifiable extraneous matter caused embolic and granulomatous lesions in the lungs of mice as a result of being accidentally introduced by injection into the tail vein (Innes et al. 1956). These foreign particles apparently traveled to the right heart and then were filtered out in the lungs. The final site of lodgement was determined by the size of the particulate matter in the vascular system. Very minute fragments might reach the alveolar capillary bed and produce granulomas. The same lesions were found in the lungs of mice injected repeatedly with normal saline; the more often the mice were injected, the more numerous the lesions. The authors believed most of the lesions to have been embolic in origin, but that none of the embolic lesions were of such severity as to have caused infarction. Stehbens and Florey (1960) found that intravenously administered particulate matter led to agglutination of platelets to which the particles adhered. Thrombi were thus formed that adhered to leukocytes and often entangled some erythrocytes. These thrombi formed strands that "vibrated" in the blood flow and fragmented, forming emboli, which frequently blocked capillaries.

Brewer (1966) reported additional experiments with glass particles involving the injection of several thousand rabbits with these particles. He studied some of them for as long as three years. These injections caused no deaths, and no gross effects were noted. In a rabbit given a large dose over an extended period, macroscopic examination revealed an enlarged liver and spleen. Microscopically, there was a generalized picture of silicosis.

Gross and Carter (1966) carried out a limited preliminary experiment in dogs that indicated that particulate matter of the type found in parenteral solutions intended for IV use may be pathogenic. The choice of particulate matter to be tested narrowed down to two types that were frequently observed as impurities: ground filter paper and a ground plastic material used in closures of containers of parenteral solutions. Pulmonary granulomatous inflammation could be associated with these materials; experimentation indicated that very few, if not actually individual, particles could elicit such lesions.

Twenty young adult beagle dogs were used in further experiments by Gross (1967). Some were injected intravenously with small quantities of fragments of ground filter paper, approximately 74 μm or less in diameter; others were given similar preparations of plastic material used to coat the rubber stoppers in commercially available bottles containing IV injectables. All animals having received as little as 2.5 mg of particulate matter of either type presented numerous pulmonary granulomas, and these were demonstrable as early as 5 days after a single slow injection of 10 mL of saline containing these impurities. The lesions consisted of scattered small nodules throughout the lung (in only one case were they grossly visible) and represented a chronic inflammatory process similar to a foreign-body reaction. It should be noted that the dose used to elicit a response represented millions of paper fibers.

A study conducted by the Parenteral Manufacturers Association (Geisler 1973) involved the IV injection of inert polystyrene spheres into rats over a short period of time. Necropsies were then performed at various intervals over 1 hour to 28 days following injection. The results were summarized as follows:

- Thirteen of 18 rats injected with 8×10^6 particles per kg at a particle size of 40 μm died within 5 minutes.
- Rats showed normal blood studies, organ weights, and pathologic criteria after being injected with either 8×10^5 particles size 0.4 μm to 10 μm, or 4×10^5 particles per kg of particle size 40 μm.
- Particles in the 4 μm size range were found in the lung, liver, and spleen.
- Particles in the 10 μm size range were found primarily in the lung, although some particles were also found in 5 other organs.
- Particles in the 40 μm size range were found in the lungs and myocardial tissue.

All three of these studies involved the injection of large numbers of 40 μm particles into the animals, which showed ill effects. Considering the overwhelming dose, far above that resulting in any course of human IV therapy, the ill effects were not unexpected.

Schroeder et al. (1978a, 1978b) administered varying numbers of polystyrene microspheres to dogs intravenously, and monitored a variety of clinical signs including arterial blood pressure, arterial blood pH, blood gases (oxygen and carbon dioxide), leukocyte count, heart rate, and electrocardiograms. No significant changes were observed during the sphere administration and for 3 hours afterwards; the only adverse effect noted was a slight increase in leukocyte count after an extremely high dose of 2.4×10^9 spheres. After 4 weeks the animals were sacrificed. No evidence of organ damage due to the particles could be demonstrated, indicating that the inert spheres of 3–25 μm infused had no effects physiologically. In the light of the low total numbers of particles contained in currently produced parenteral solutions, it is interesting to consider the lack of any observed effect of the relatively massive experimental doses of particles administered in short periods of time in this and other studies. Whereas Schroeder et al. in this study administered levels of 10^9 particles with no effect, the maximum number of particles ≥ 10 μm received from a 1 L large volume IV solution conforming to USP requirements would be 12,000 ≥ 10 μm. In prolonged administration of parenteral solutions that would reasonably be expected to result in a "worst-case" exposure of a patient to particulate matter, thousands of units would be required to approximate the cumulative level of exposure achieved in this and other animal experiments.

It is interesting to note that some of the few directly demonstrable harmful animal effects have resulted from drug-related materials. A study by Doris et al. (1977) involved the injection of insoluble sodium cephalothin residues and other material into the marginal ear vein of rabbits. Rabbits received injections of D5W residue, D5W filtrate, and the cephalothin filtrate as well. Histologic examination of the veins 24 hours later revealed venous congestion and perivenous hemorrhage in all subjects. The filtrate of D5W had the lowest incidence of identifiable reaction and the mildest reactions. The filtrate and residue of sodium cephalothin caused mild reactions in a high percentage of veins studied, and a substantial number of specimens were found to have severe reactions with damage to the vein wall. Evidence of venous thrombosis was seen only with the injections of the sodium cephalothin filtrate. In the final phase of the study, a dog had a mixture of 2 g/L of cephalothin with 20 mEq/L of KCl in 0.2 percent saline infused for 2 hours in 3 veins, 4 hours in 12 veins, and 6 hours in 12 veins. At the end of the infusion period, each vein was dissected from the point of venipuncture to the saphenofemoral junction and examined histologically. The results of the examination revealed that the 2 hour infusion showed no changes and the 4 hour infusion showed only mild initial edema. The 6 hour infusions, however, showed severe damage to the endothelium as evidenced by focal medial necrosis. This finding is of interest with respect to the use of final in-line filters by many practitioners for administration of cephalosporin drugs to reduce the incidence of phlebitis.

PARTICLE CONTAMINANTS IN MANUFACTURING

Any industry producing articles that must be kept clean must have the means of controlling particle access to the product. Notable examples are the aerospace, food, electronics, and medical industries. The total particulate burden of a medical product is determined by the contribution of particles from five major sources:

1. The environment
2. Packaging materials
3. Solution and formulation components
4. Product-package interactions
5. Process generated particulate matter

Items 2–5 on this list are often categorized as nonenvironmental particle sources, and their control and elimination frequently are more important than control of environmental particulate matter. The control of particles from these sources typically requires more diverse control measures than does the control of dispersed environmental (airborne) particulate matter, and an uncontrolled situation with regard to these particle sources typically will have a more serious effect on product quality than particles from the general environment.

Extraneous particulate material may obviously enter the final product as the result of insufficient cleaning or product assembly issues, such as particle generation by a filler. Formation of particles in the product is a much more complex phenomenon, and can involve degradation, aggregation, agglomeration, precipitation, crystallization, and/or sedimentation reactions. Personnel activity and equipment design and maintenance are critical factors, since point-generated particles arise from these sources. Ideally, nonenvironmental particulate matter is eliminated by process design rather than process troubleshooting. Process design incorporates not only the consideration of process machinery, environmental cleanliness, and filtration, but of container and closure cleanliness as well.

Control of particle formation due to drug degradation or container interactions is much more difficult than control of extraneous particles from the process or environment, requiring studies of container component quality, raw material consistency, and rigorous formulation compatibility/stability testing. On occasion, contaminant particles are readily identified; in other cases, a comprehensive multidisciplinary approach is required. This approach must include understanding the visual appearance of the particles, as well as their physical and chemical characteristics. Efforts to exclude objectionable material from the product do not begin with highly controlled filling environments; R&D efforts that indicate the best drug components, excipients, and packaging components are essential to ensure inherent product quality. Particle identification efforts are an integral part of the quality process; identification of particulate matter often allows location of its source, hence its elimination.

CLASSIFICATION AND SOURCES OF PARTICULATE MATTER

Particulate matter found in parenteral products arises from a number of sources. The content of the reference by Groves (1973) is still of general applicability today. According to this author, intrinsic particulate matter consists of material originally in a solution that is not removed by filtration prior to filling, or particles that occur due to precipitation reactions in the solutions. Extrinsic particles are those that enter the product or its container during the filling operation, such as rubber, metal, or plastic coming from container product contact surfaces. While manufacturing processes today result in levels of particles far lower than those historically present, a review of earlier findings is useful in principle. Later authors agree that the analysis and control of particulate matter contamination in pharmaceutical products is a complex subject (Benjamin 1990; Kalm 1987; Jonas 1966; Skolaut 1966; Vessey and Kendall 1966).

Garvan and Gunner, in their 1964 study, classified particles into the following categories for microscopic identification: whole-rubber particles, chemical particles, cellulose fibers, fungi, and miscellaneous. Davis et al. (1970) attempted to differentiate the deposits found in their studies, and categorized them into particles with a diameter of 1 μm or less, particles of black rubber up to 100 μm, crystals, fibers, brown particles identified as rust, reddish brown particles, starch particles, and diatoms and fungi. The same types of extraneous materials found in parenteral solutions have been reported by a number other authors: Goddard (1966), Gross (1967), Groves (1966), and Jaffe (1970).

CONTAINERS AND CLOSURES

Much of the literature has called attention to the contribution of the rubber closure of solution containers to the particulate matter in the solution. The particle count contributed by the closure and the container is minimized by quality control testing of components and the development and use of efficient wash-rinse cycles to insure cleanliness of containers immediately before the solution is introduced into them; Lachman (1966) pointed out that the development of particulate matter in parenteral solutions in contact with rubber closures may be dependent on one or more of the following factors:

- Surface particles
- Leached materials
- Interaction of leached materials and solutions
- Cleaning techniques
- Reaction between solution and closure
- Roughness of container opening

Perrin (1966) believed all new batches of closures should be checked for the production of particulate matter by reaction with the parenteral solution with which they are to be used. Problems with rubbers and plastics include permeation (oxygen and carbon dioxide in, water out), leaching, sorption, chemical reaction, and change in physical properties. He noted that sulfides used as vulcanizing agents may react with preservatives containing mercury, giving a precipitate of mercuric sulfide.

Preservatives, however, are not permitted in LVI. Zinc often appears on the surface of molded rubber closures, as zinc salts are added to assist the curing process. Reznek (1953) questioned whether the dissolved zinc might reach a level that could be physiologically objectionable or might be the cause of insoluble particle formation by interaction with other constituents of the solution or container. His subsequent experiments confirmed the observation that only a small portion of the total zinc content of the closures is leached under ordinary storage conditions.

Garvan and Gunner (1964) stated that the "ulcers" resulting from the rupture of "blisters" on stoppers and the disintegration of the friable "skin" of the rubber closure were the main mechanisms by which the particulate matter was introduced into the IV fluid during autoclaving. During storage, aging of the rubber increased the friability of the "skin," and the vacuum sealing continued to expand and rupture gas bubbles near the surface of the stopper. This was believed to explain how some bottles of fluid appear to be particle-free on manufacture, but later contain numerous particles. Lacquering of the stoppers was believed by the authors to help inhibit the disintegration of the rubber into the fluid. They believed that the small number of fibers in raw rubber caused no problems in the use of most fabricated rubber articles. The same number of fibers, however, was believed to introduce a possible serious danger to the use of such rubber for stoppers with bottles of IV fluid. An early step to correct this problem involved stoppers coated with a lacquer or with a firmly fitting plastic sleeve.

Results of work by Dungan (1968) indicated that fewer particles are released from lacquered stoppers, and that no further release of particles occurred on repeated autoclaving. The author stated that lacquered stoppers could be made almost particle-free by simple rinsing. The studies indicated that not all lacquered stoppers were better than unlacquered; there were considerable differences in composition of the rubber compound, especially in the types of filling used, finish of mold surfaces, etc., and much depended on the properties of the lacquer used. It was believed that the coating must be strongly adherent and flexible enough not to crack when the stopper is deformed or abraded. In a comparison of lacquered and normal stoppers, the means showed about a threefold decrease in particle counts with lacquered stoppers, although there was appreciable overlap between the individual results. As we will see in a later chapter of this book, lacquered stoppers have been almost totally eliminated, but other coatings are used to attain the same level of protection of the elastomer.

Lachman (1966) found that rubber closures could contribute substantially to particulate matter in LVI solutions, but this was true only for rubber stoppers poorly formulated and inadequately processed. With recent advances in polymer science and technology, it is now possible to manufacture lacquered, epoxy, Teflon®, and polyolefin-coated closures that can effectively prevent the introduction of most particulate matter into the solution. Coring from hypodermic needles, a potential considerable contribution to particulate development in multidose vials, is negligible in other parenteral solutions as the stopper is generally penetrated only once with the needle. The papers by Vessey and Kendall (1966) and Endicott et al. (1966) were in agreement that closures were significant sources of the particles found in injectables.

The particle content of IV fluids in plastic containers has been repeatedly found to be lower than that of glass containers (Davis 1970). Dodd (1965) found it to be a tenth of that of most of the rubber-stoppered infusion solutions. Plastic ampoules contained a statistically significantly lower number of particles, one tenth that of rubber-stopper glass bottles (Groves 1969). The first authors found, however, that fitting a rubber closure to a plastic bag or bottle appeared to raise the count to a level comparable to that of rubber-closed bottles. The latter author stated a concern that the manufacturing process and its associated variables are a third source of particulate matter. Airborne particles and their availability for contamination were believed important in this area. A fundamental approach involves a survey of the entire production facility to identify potential sources of contamination and eliminate them.

Tests by Davis et al. (1970) showed that particle counts varied considerably among different manufacturers, different solutions, and different lots of the same solution of a single manufacturer. Average counts ≥ 25 μm in size for 12 samples ranged from 57 particles per liter for one company's 10 percent dextrose in water, to 2,630 for another company's normal saline. It is now becoming common knowledge, however, that particle counts are often higher in electrolyte solutions. These authors found the average number of particles in 18 samples of normal saline solution in 1000 mL plastic containers to be 137. This represented a significant decrease compared with the number of particles in glass containers.

In 1965, Hurst showed that between 25 and 60 perent of bottles of IV solution from more than one British supplier contained visible particles and were thus considered unfit for use. Garvan and Gunner (1964) found fungal elements in 40 percent of the intravenous solutions investigated. It must be noted, however, that the latter studied solutions of Australian origin, which normally contained higher particulate matter counts than those of U.S. origin at that time. Some samples of commercial origin were found to contain 30,000 particles per mL or 15,000,000 particles per half liter (Groves 1966).

Many authors have discussed the degree of particulate contamination in reference to particle size (see Hurst 1965). Groves (1966) observed that many commercially available solutions of normal saline and dextrose contained considerable numbers of particles larger than the equivalent volume diameter of erythrocytes. He also found physiological saline solution stored in rubber-stoppered glass bottles to contain generally 2,000–15,000 particles per mL above 1.5 μm (Groves 1965). This author also confirmed that many commercial infusion solutions of the time did contain particles of equivalent spherical diameter larger than 10 μm.

Turco and Davis (1971) found that analysis of over 1,000 commercially available solutions from 4 manufacturers revealed a content of foreign particles on the order of 100 particles/L for 1 supplier. This supplier was marketing parenteral solutions in bottles with a screw cap closure at the time of this study. Two other suppliers had average particle counts of approximately 350 particles/L. Over 1,500 particles/L were discovered in solutions of the fourth manufacturer. All counts were on particles of a size greater than 5 μm.

THE IV THERAPY PROCESS

Even at the early dates that these papers were written, it was common knowledge that the particle burden to which patients were exposed came from sources other than the LVI or SVI solutions. Kramer (1970) called attention to the wide variety of particles present in solutions infused, and identified five generic sources of particles:

1. The solution and its components
2. The package and its components
3. The process of manufacture and its variables
4. The sets and devices used in the administration of these preparations
5. The manipulations involving the hospital personnel and the patient

According to this author, items 4 and 5 were believed to be most important. The particle count of the solution itself was believed minimized initially by closely monitoring the water and raw materials used in these preparations. The most important manipulation to minimize the particle count of the solution as it enters the final container was believed to involve the filtration system. Once in the solution container, causes of particulate matter contamination arising from the solution itself would consist of a reaction among the components within the solution; a reaction between the package and the solution; a reaction caused by the air within the package, sunlight, or heat; cold; or some combination of the above.

As previously discussed, the sets and devices used for the administration of these solutions represent another source of particulate contamination. Tests performed on administration sets, syringes, and needles confirmed findings that these devices can frequently be major sources of particles in intravenous infusions (Trasen 1969). Several authors have stated that the use of an administration set, on an average, could cause a 100 percent increase in the number of particles reaching the patient (Davis et al. 1970). Syringes can contribute particulate contamination, especially the reusable type of that time, some of which required boiling for sterilization. Dungan (1968) believed the contamination arising at the point of administration might well be in excess of that in the preparation being administered. Walter (1966) noted that clean, clear parenteral solutions are meaningless to the safety of the patients if they are administered with devices that are heavily contaminated with particulate matter.

Davis et al. (1970) presented compelling data demonstrating that the number of particles increases as the number of manipulations of the infusion solution increases. Maneuvers such as poor handling techniques, mistreatment of the sets and equipment packaging, and subtle items like the gauze used to cleanse the site of the injection were all important. This author believed even a fragment of the patient's own tissue, created by needle coring, can be considered as an undesirable particle in the bloodstream. Historically, most authors have believed that specific education programs at the hospital level will be necessary to minimize such sources of particle contamination.

Ho (1967) believed that in-hospital manipulations that interfere with efforts to produce high-quality solutions included the contamination of solutions by the introduction of drug additives, particularly lyophilized drugs. Data on drug incompatibilities are not always available to the user, and the result can be gross accumulation of precipitates. Kramer et al. (1970) classified pharmaceutical incompatibilities as pharmacologic, physical, or chemical. A chemical incompatibility occurs when an undesirable chemical reaction takes place, thereby altering the original drug composition and activity. For example, if sodium amobarbital solution is combined with promethazine hydrochloride, a precipitate is formed and the original drug composition is lost. Kramer et al. conducted a study with the addition of KCl to 5 percent dextrose in water. When filtered separately, none of the individual solutions of either showed signs of a precipitate on a Millipore® filter. When the KCl, dextrose, and water were mixed and filtered, a visible precipitate was present on the filter with 64 of the 66 bottles tested (96.9 percent). Qualitative evaluation made by diffraction and X-ray spectrographic scans revealed the precipitates to be mostly silica and some alumina. The authors considered it highly probable that the precipitate was material leached from the glass. Davis et al. (1970) found that lyophilized vitamins made a greater contribution to particles when compared to KCl liquid. They believed that as the number of additives to the basic solution was increased, the levels of particulate matter also rose.

The mixing of drugs is particularly likely to result in physical instability of the resulting solution, which may cause formation of particles apparent as a visible haze or precipitate. Such reactions are generally well known to pharmacists, and admixtures are routinely checked for visible particulate matter. An entire book, *Trissel's Tables* (Trissel and Leissing 1996), is devoted to identification of solution incompatibilities. To list but a few incompatibilities, we find that morphine sulfate and pentobarbitol, penicillin and polymixin B, and ascorbic acid and erythromycin may result in precipitation or other manifestations of physical instability.

Ho et al. (1967) made an interesting point by calling attention to the storage of solutions. They pointed out that although particles may eventually come from rubber and glass for various reasons, very little is known concerning the effects of storage conditions. The flocculation of particles may change the particle size distribution with time, especially in solutions containing electrolytes. Garvan and Gunner (1963) also emphasized the contribution of the rubber stoppers to particulate matter in solutions during shelf life. All their studies were conducted on stoppers of natural rubber. More information must be secured concerning the effect of solution aging on particulate matter. If a long shelf life does allow for greater interaction between a fluid and the glass container and/or closure, and this reaction is the cause of most particulate matter, then a short shelf life would tend to eliminate or lessen this potential cause (Archambault 1966). Skolaut (1966) saw a need for a reasonable expiration date for solutions when stored under suitable conditions, established jointly by the FDA and the solution producer.

In addition to particulate contained in LVI solutions that are administered to the patient, particles generated by or contained in SVI (especially reconstituted powders), administration sets, syringes, needles, and ampoules, and those resulting from the procedure of IV therapy itself, are a concern. Thus, although the manufacturer puts forth a great deal of effort to deny particle access to product and remove the particulate matter present in a solution through filtration during production, some low level of particles will be found in all products. The user of a parenteral product, whether a pharmacist, doctor, or nurse, may also inadvertently put particulate matter into the product during intravenous therapy administration. (See sources in Figure 1.5.)

A notable recent tragic instance of patient harm due to particulate matter involved the deaths of several patients due to pulmonary embolism caused by large numbers of calcium phosphate particles (crystals) in improperly prepared total nutrient admixtures (TNA) (Hill et al. 1996). One primary concern in formulating TNAs is the solubility of calcium and phosphorus, especially in pediatric patients who require higher concentrations of both minerals in their nutritional supplements. The calcium and phosphorus solubility in TNA is affected by pH, temperature, amino acid formulation and concentration, type of calcium salt, duration of storage, order of mixing, and contact with lipids. If chemical incompatibility exists, precipitates or

Figure 1.5. Sources of particulate matter in intravenous therapy.

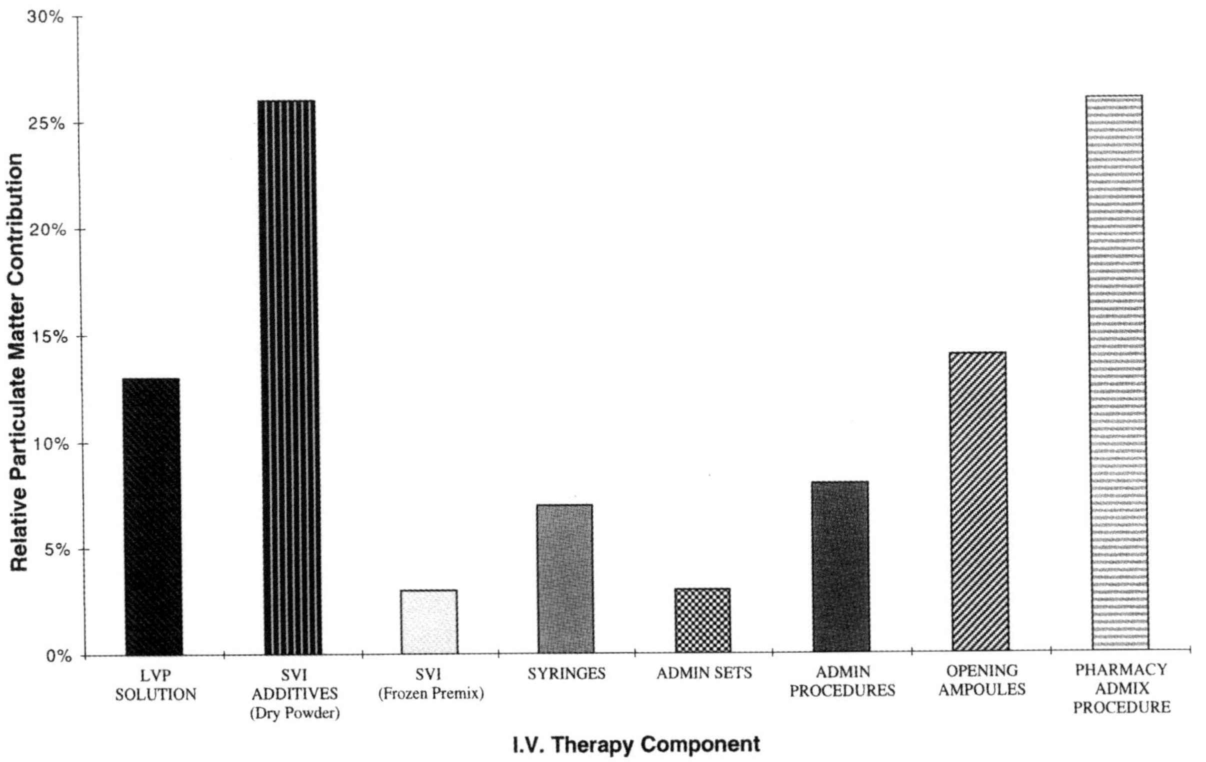

aggregates of particles will form. These mineral precipitates have also been reported to cause catheter occlusions in patients receiving TNAs. In the incident reported by Hill, the pH of the amino acid component, transient elevation of calcium and phosphorus concentrations during mixing, and the lack of agitation during automated preparation of the formulation were identified as the causative factors producing the fatal precipitate.

USE OF IN–LINE FILTERS

Data have been published indicating the efficacy of the elimination of particles by final filtration (Davis et al. 1970; Ernerot and Sandell 1967). Davis et al. conducted studies with final filters and stated that "it appears this device is nearly 100 percent effective in removing particulate matter from IV infusion solutions." The final filter tested contained a Millipore membrane made of cellulose ester(s), which has a controlled-size pore of 0.45 μm placed at the distal end of the intravenous set.

Over the past two decades, investigations related to the influence of final in-line filtrations with 0.2 μm filters on the incidence of infusion phlebitis suggest some relationship of infused particles and phlebitis. A number of the studies performed (e.g., Falchuk et al. 1985) have suggested that the number of particles in the intravenous medications were directly related to an earlier onset of phlebitis in patients receiving these drugs via the IV route. Other authors have questioned the economic viability of in-line filters and asked whether the uncertain danger from particles justifies the increased cost. More recent research (Rypins et al. 1990; Maddox et al. 1983; Gotz et al. 1985) into the causes of infusion phlebitis has lead to the conclusion that in-line filtration is definitely not broadly effective in reducing the incidence of the reaction. Knowledge that has evolved out of studies of phlebitis and particulate matter conspicuously includes the fact that phlebitis is a much more complex reaction than originally supposed. The definition of phlebitis, the nature of the drug infused, the vascular access procedure, the vein used for infusion, the osmolarity of infused solution, the type of catheter, patient sensitivity, and microbiological issues have all been determined to be critical factors that may singly or in combination outweigh the effect of particulate matter. The role of in-line filters in the adsorption of pyrogens that might be responsible to some extent for phlebitis has not been thoroughly investigated.

DETECTION AND COUNTING OF PARTICLES

The most widely used detection or examination methods are visual in nature and are carried out on unopened solution containers (i.e., are nondestructive) (see Chapter 7). These methods are used both to detect the presence of particles and allow estimation of their size, but are not often used for quantitation. Cotton fibers or other particles of visible (macroscopic) size will rarely be present in large enough numbers to be detectable by a particle counter, and so visual examination will always need to be part of routine testing procedures.

The occurrence of visible sized particles in medical products and injectables is a particularly pertinent example of the benefits of process control of particulate matter. It is, unfortunately, very difficult to perform a uniform critical visual inspection of injectable solutions using human inspection. This is due to the fact that visual inspection by a human analyst is critically dependent on a number of factors, including the following:

- Nature of the particle (color, shape, opacity, reflectance)
- Type and intensity of lighting
- Visual acuity of the inspector
- Inspector training
- Container clarity
- Container volume
- Interval of inspection
- Analyst fatigue and psychological factors

In recognition of the physical limitations of the visual inspection process, the major pharmacopoeia have historically described acceptable product as "essentially free" (USP), "practically free" (British Pharmacopoeia [BP]), or "free of particles" visible under specific conditions (Japanese Pharmacopoeia [JP]). These descriptions are generally imprecise, and in the case of the JP, apply only to certain types of product. Suggested improvements to the USP requirement would substitute "no evidence of" (visible particles) for "essentially free." This constitutes no improvement or clarification, and, worse yet, implies an absolute standard. The only control for visible particles in product is obviously a process that eliminates larger particles through process design and control.

One hundred percent inspection of all product by automated means has been suggested as a means of generating product with no particles of visible size (Falchuk et al. 1985). Application of a 100 percent inspection process as a means of eliminating units containing particles ≥ 100 μm in size has been shown to be practical for ampoule product, for which a very critical automated visual inspection can be applied. Although SVI vial units can also be inspected automatically, vial product may have particles adherent to the closure that are released during shipping, after the manufacturing inspection has been applied. Thus, visible particles undetected by the 100 percent inspection at manufacture may be detected by the user. Dry powder vials, which are often found to contain visible particles when reconstituted, cannot be 100 percent inspected at manufacture because visual inspection is a destructive analysis for this product type (Johnson et al. 1970). Ironically, ampoules are not only the product type most readily 100 percent inspected, but are also the product type that most often contains the highest levels of large particles (glass fragments) when used.

It is impossible for the manufacturer to inspect the majority of LVI product critically enough so that assurance of no visible particles can be obtained. This is due not only to the fact that particles not present in the solution when inspected may be released during transport or storage, but also to the inadequacy of both automated and manual inspection of LVI product and the general unsuitability of many plastic LVI container types for inspection (Archambault 1966). Translucent plastic containers (as used in Japan) cannot be inspected critically by either automated or manual means, although machine inspection with intense light is vastly superior to human inspection.

Visual inspection enables larger particles or gross contamination to be readily detected. It was historically believed that the lower size limit of particles that are visible was about 50 μm (Brewer et al. 1947; Groves 1968; Saylor 1966). This technique, conducted with human inspectors, is presently used by many manufacturers for inspection of finished products, but incorporates multiple error sources including lighting intensity, time of inspection, and operator subjectivity, experience, and fatigue.

Garvan and Gunner (1963) used a light scattering apparatus for visual inspection, the light beam being introduced through the bottom of the bottle and the particles viewed from the horizontal position. Particles over about 5 μm in diameter and of moderately high refractive index were seen as twinkling spots of light, often appearing to change color as the angle of viewing changed. Small particles over about 10 μm could be seen as rather faint but definite spots of light with this critical means of inspection.

In fact, control of particulate matter by inspection is not practical for the majority of injectable products. Particulate matter must be controlled by a carefully validated and monitored process operated strictly under the principles of GMP. A large percentage of LVI product in Japan is marketed in translucent polyethylene containers that cannot be critically inspected by either human or automated methodology as mentioned above. Nonetheless, the levels of visible particles in these containers are very low. This low level of particulate matter is achieved through manufacturing processes that are designed and validated to generate product with an extremely low level of visible-sized particles. Thus, a suggestion for compendial groups who struggle with requirements for visible particulate matter is to require control by process validation and monitoring rather than to rely on compendial requirements that include such phrases as "essentially free" or "practically free" or "bear no evidence of." These ambiguous terms are of limited usefulness and are unenforceable.

SUBVISIBLE PARTICLE DETECTION

Techniques of particle counting for quantitation of subvisible contaminant particles in injectables are relatively new technologies (Kirnbauer 1970). The two basic approaches for determining particle size distributions and concentrations in solutions are the microscopic technique and automatic particle count devices. Microscopic membrane counting is the method of preference when information on particle shape, identification, and distribution are desired as well as a numerical analysis (see DiGrado 1970). Measuring the dimensions of a particle is readily accomplished with this technique, and a further advantage is that microscopy can be an extremely useful tool when attempting to trace the source of the particulate contamination (Kirnbauer 1970). The technique consists of concentrating the contaminant by filtering a defined portion of the liquid sample through an analysis membrane and subsequent microscopic examination of the membrane surface for the particulate contaminants. As opposed to electronic instruments, the microscope method allows measurement of the longest axis of a particle. Consequently, a fiber 100 μm in length by 5 μm in width would be scaled at 100 μm by microscopy, whereas the light obscuration (LO) methods (e.g., the "HIAC" method) would consider it as only a 30–50 μm particle, and the electrical zone sensing method (Coulter®) would report it as about 20 μm, based on volume.

The data obtained by microscopic membrane counting are considered by some to more nearly approach the true "absolute" values. Only with this method can a permanent visual record of the sample be retained for particle identification, comparisons, and storage for future reference. It is for these reasons that this method is often considered the standard reference when comparing and reporting results (DiGrado 1970). Unfortunately, this investigator found the reproducibility of this method between two trained technicians not to be good, and greater than the statistical limits considered attainable. Other disadvantages associated with use of this technique were found to be related to the time-intensive aspects: ultracleaning of equipment is time consuming; microscopic counting is time consuming; rupturing of the membranes results in additional work and time; and wrinkling of dried membranes, requiring refocusing of the microscope, further extends the time for analysis.

The Coulter® principle uses resistance modulation of electrical current through a small orifice to detect and size single particles. With this method, the operator determines the number and sizes of particles suspended in an electrically conductive liquid. This is done by forcing the suspension to flow through a small aperture having an immersed electrode on either side. As a particle passes through the aperture, it changes the resistance between the electrodes. A voltage pulse of short duration is produced that has a magnitude proportional to the volume of electrolyte displaced, and hence to the particle volume (size). The series of pulses is then electronically scaled and counted. The counts at thresholds of 2 μm and 5 μm were of interest to the developers of the first BP particle count test, as the numbers of contaminating particles increases as the particle threshold decreases, and statistically the level of counting accuracy is improved (Groves 1969).

For purposes of comparison, it should be noted that Coulter® counter determinations are in terms of particle volume, sizes being expressed as equivalent spherical diameter. This instrument presents certain disadvantages, as listed by DiGrado (1970):

- The reliability of small sample aliquots is questionable.
- The problem of spurious counts has not been remedied.
- Ultracleaning of equipment is extremely time-consuming.
- The liquid stirring system is a problem; since it is not a positive drive, it is difficult to make fine adjustments, and its speed varies with the volume of the solution in a beaker.
- Extraneous electrical interference is still a problem.
- The instrument requires a highly trained technician.

The LO counters (see Chapter 8) operate by a light extinction methodology and show less intralaboratory variation and sensitivity to external influences than the Coulter® counter. In this device, sample fluid passes through a small rectangular tube and past a window. As long as the number of particles does not exceed the specified concentration, the particles will pass the window one by one. A parallel light beam is formed by the window to an exact size and directed through the sample fluid and onto a photodiode. Each particle passing the window interrupts a portion of the light beam according to its size. This causes a momentary reduction (or pulse) in the photodiode output signal that is proportional to the size of the particle.

The advantages of the LO counter were believed by DiGrado (1970) to be as follows:

- Relatively simple to operate
- Does not require as extensive technician training as the Coulter® counter
- Easily cleaned to produce very low background counts
- Countability of the entire volume in a container
- Speed
- Reproducibility of result
- Relative insensitivity to outside influences

The instrumental and microscopic counting methods are both destructive in that they must be conducted by opening the containers. Nondestructive methods (e.g., visual inspection) have the advantage that every container in a production batch can be examined (Groves 1968). Interpretation of the results of destructive methods must necessarily be presumptive for the entire lot by the nature of the analytical method involved and the subsequent sampling design (Innes 1956).

PARTICLE CONTROL—STANDARDS OR GMP?

The controversy regarding which of these two methods is more effective for control of particulate matter is ageless (see Chapter 3). Krueger and Riggs (1968) observed that there was no appreciable data of any kind (biological or analytical) usable to project a standard based on a number-size distribution of particles. They believed data were needed for a total systems approach and should include the following:

- Consideration of available instruments and their adaptation to the proposed uses
- The instruments' capabilities and limitations, including the accuracy, precision, and replication of results
- Methods for sample handling and examination that will minimize differences between analytical laboratories
- The results of collaborative studies that will show the nature and scope of the interlaboratory differences
- Determination by the proposed method of the characteristics of solutions presently available—as produced and after maximum acceptable storage times and conditions—a rather complete market survey (the correlation between these data and that of visual examination is also needed)
- Information on present limits of manufacturing capability
- Information on the contributing sources of the particulate matter present as a solution is being administered to show the proportion attributable to the particulate matter present in a final container, and the proportion that is attributable to the environment, method of administration, or other contributing factors
- Information about the differing biological effects associated with different particle size–number distributions for consideration in developing a meaningful standard

Krueger and Rigg's issues remain to some extent today, but the particulate matter requirements of the world's major pharmacopeia effectively eliminate any issue of patient risk with manufactured solutions.

In any consideration of particles in injectable products, the definition of extraneous particles becomes extremely important. This definition must be made based on a consideration of GMP principles, applicable means of analysis, and patient well-being. Particulate matter in injectable products may be defined as randomly sourced extraneous material of heterogeneous chemical composition that is not amenable to analysis by chemical means due to the extremely low levels present and its diverse composition (see USP [1995], JP [1996], and BP [1993]). This material is "unintentionally" present. There are three very important concepts embodied in the definitions provided by the three compendia.

1. Particulate matter currently exists at extremely low levels in injectable products so that there is no demonstrable evidence of adverse patient effects.
2. The material cannot be monotypic, but rather results from a variety of sources inherent in a GMP–controlled production process.
3. The material is not amenable to chemical analysis due to the small mass that it represents and its heterogeneous composition. Thus, the appropriate analytical methods for

enumeration of this material must be sensitive physical tests that detect size and quantitate the material based on its optical properties.

Of extreme importance to this definition is the concept of GMP control of particulate matter in pharmaceutical products. If a particulate material from a single source becomes dominant in a product, control of the production process is lacking. Monotypic particulate matter is not random in occurrence and frequently may be analyzed at high levels of sensitivity. Specifically, materials such as silicone oil microdroplets from an elastomeric closure or a syringe piston do not fall within this definition; such container-related materials are inherent in the product and can be quantitated at high sensivity by chemical means.

The compendial requirements for particulate matter found in compendia today can be traced back to the 1970s. Kramer (1970) and Kramer et al. (1970) stated that when considering the establishment of specific size limitations for particulate matter, the diameter of the blood vessels must be taken into account. Leukocytes are larger in diameter than either erythrocytes or platelets. Trasen (1969) concluded that standards for particle size ranges should begin at 5 μm maximum length, a minimum figure based on the diameter of the capillaries, that the author believed to be 7 μm. White blood cells range in diameter from 8 to 12 μm. These papers apparently do not allow for the passage of leukocytes, which must be much larger than erythrocytes and which are less flexible. The smallest capillaries, therefore, must expand to a minimum of 12 μm to allow these living cells to pass through. From this, Kramer et al. (1970) concluded that it would seem that if any standards are to be formulated, no value above 12 μm should be established as a maximum diameter size for particulate matter. Furthermore, the author believed that even a particle diameter smaller than 12 μm will not eliminate the possibility of formation of granulomas and pulmonary infarcts.

STANDARDS FOR MEDICAL DEVICES

The desirability of controlling product particle burden by means of process control rather than compendial limits is also true with regard to medical devices (Illum et al. 1978a and 1978b). During the past decade, there have been numerous discussions between industry and the various regulatory and enforcement authorities regarding whether particulate matter requirements for devices are necessary or desirable. In this case, as in others, assurance that a clinical need has been identified should be confirmed prior to determining whether or not a standard should be developed (see Williams and Barnett 1973). Any proposals for particulate limits must be tempered with the knowledge that there is no significant evidence regarding the adverse clinical effect of particles from devices. If a need were to be identified for a specific product, controls should logically be established for that product.

Comparing the particulate matter levels in the medical devices evaluated by manufacturers and independent investigators to the particulate levels indicated as acceptable per the USP General Chapter <788> for LVI for single-dose infusion, it appears that devices generally do not significantly contribute to particulate levels being infused into a patient. Additionally, it appears that products produced under controlled conditions meet customer expectations as evidenced by total absence of customer complaints regarding subvisible particulate matter and the very low number of complaints related to large, visible particles that have been received, based on the millions of devices used annually. The USP test for LVI for single-dose infusion states that a solution passes the test "if it contains not more than 25 particles per mL that are equal to or larger than 10 μm and not more than 3 particles per mL that are equal to or larger than 25 μm in effective linear dimension" (by the LO test). These ranges are significantly above the level of particulate the device industry has been experiencing.

Differences in devices and their use further complicate the establishment of a standard, and it is evident that a single standard could not possibly address the entire range of devices and problems. Complex devices, such as autologous transfusion filters, pharmapheresis sets, blood warmers, and oxygenators, can be expected by nature of their size and construction to have larger particle burdens than administration sets. A Dacron® vascular graft and a winged cannula cannot be expected to shed comparable numbers of particles since a vascular graft has a much greater surface area than the tiny cannula and has the potential to release some number of particles throughout its life span. Other medical devices, such as syringes and blood oxygenators, have material such as silicone or surfactants purposely applied to aid in the product's function. Eliminating these additives to control particulate matter could adversely effect the function of the product. Failure to allow for the particulate matter generated as a result of the additives within the allowable limits range could also result in an unwarranted rejection of the product.

In summary, the main points that underscore the general lack of need for a device particulate matter requirement are as follows:

- The absence of any clinical evidence indicating that particulate matter requirements on medical devices would benefit patients
- The impracticality of requirements for devices due to the many different types of devices and the lack of any uniformly applicable test method
- The fact that only GMP will control device particulate burden (testing to demonstrate compliance with requirements may represent cost without benefits)
- Present availability of detailed GMP plans for device manufacture that have proven capable of controlling device particulate burden

With regard to any proposal for particle limits, an evaluation of the cost-benefit ratio should also be made to ascertain if the costs encountered to comply with a standard are equal to the expected benefit. In a worst-case scenario, small companies might be unable to continue operating due to inordinate cost increases. Increased manufacturing costs as a result of the implementation of an additional standard of unsubstantiated benefit will inevitably result in increased cost to the consumer that would appear to be particularly undesirable at a time when the medical community around the world is attempting to control costs. Process controls and effective implementation of the applicable GMPs have continued to be the medical device industry's methods of choice for reduction of particulate matter levels.

Syringes have also been of historical interest (Illum 1980). A recent USP proposal (*Pharmacopeial Forum,* Jan.–Feb. 1997) of a requirement for disposable syringes and the response received provide an interesting study in the difficulty of generating requirements for devices. Although this method is unsuitable as proposed, there is a favorable consideration with regard to this and other requirements proposals being generated by the USP. In the case of the syringe proposal, the criticism by industry experts of the proposed limits and for the testing of syringes will be given rational consideration, thereby allowing the development of a more suitable, realistic final requirement.

SUMMARY

Particles are simply fragments of matter that are small in comparison to the volume of the system in which they are found. (Grammatically, the plural of "particle" is either "particulate matter" or "particles." In a strictly correct sense, "particulate" is an adjective and should not be used alone without its accompanying noun "matter" [i.e., a particulate].) The human body is constantly assailed by particulate matter originating from a wide variety of sources. If we eliminate the consideration of ingested particles, which seems reasonable, we are left with concerns relating to particles that we are exposed to by contact mechanisms, environmental airborne particles, and particles that enter by unnatural routes, such as wounds, surgery, and injection. The vast majority of airborne particles of respirable size cause no injury and are dealt with effectively by the defense mechanisms of the lung and respiratory tract. The body also has a considerable armamentarium against injected particles that circumvent many of its first-line defenses.

The occurrence of low numbers of heterogeneously sourced particles is inevitable in the manufacture of injectable products and medical devices. The numbers of particles that occur in product manufactured under conscientiously applied current Good Manufacturing Practice (cGMP) are extremely low. There is no practical issue of adverse patient effects due to the particle burden of current injectable products. This low level of particulate matter has been achieved through the efforts of manufacturers toward product improvement rather than through regulatory activity. Based on a rational consideration of particulate matter occurrence, the numbers of particles in product cannot be further reduced by either additional requirements or tighter limits, since these have no effect on the manufacturing process. The manufacturing process itself is our sole available control on particulate matter. Ample GMP regulations are currently in place to provide for enforcement activities against noncompliance manufacturers. GMPs, rather than monitoring or regulatory or compendial requirements, control particles. It is the manufacturer's commitment to product quality that has resulted in the current high-quality levels, cost-effectiveness, and efficacy of parenteral solutions.

The obvious ultimate goal in considering particulate matter in pharmaceutical products is its control at levels far below those at that there is any issue of patient well-being. Although the manufacturers of injectable products, compendial bodies, and regulatory agencies share this goal, there is a distinct divergence in philosophy regarding how this control is to be attained. The charters of the world's various compendia address the assurance of product safety and efficacy, that is, patient well-being. In this regard, it is reasonable that pharmacopeia include tests and limits for extraneous particulate matter. The critical consideration so often overlooked in discussions of particulate matter is the extremely low level of particles present in products produced under current GMP conditions. Interestingly, the products currently produced in Japan, the United Kingdom, Europe, and the United States have an extraneous particle burden so low as to make issues of adverse patient effects practically nonexistent. This is true even with circumstances of chronic exposure, such as with patients who have been on total parenteral nutrition or hemodialysis for long periods of time. (In the latter case up to 25 years.)

With regard to the occurrence of particles in any pharmaceuticals or medical products, there are generally four components of particulate matter control that pertain irrespective of the product type and manufacturer:

1. Process design
2. Process and environmental particulate monitoring
3. Enumeration of particles in product
4. Particulate matter identification

The first item (process design) is of overwhelming importance. Design of a process to eliminate particles results in more positive control than any removal of particles after the process is in place. Currently, the industry's emphasis in all four of these areas has turned from particle control to particulate source elimination. During manufacture, it is necessary to protect the product from particles related to the process by which it is made, for example, fillers, mixers, stoppers, and the product container itself. Frequently, for troubleshooting, it becomes necessary to measure particulate matter added to the product at each step in the process.

Based on the basic investigations conducted up to the present time, and in available literature, conclusions regarding the danger to patients from particles are the following:

- The quality levels of parenteral products, to include particulate matter considerations as well as other measures, have improved dramatically within the past 25 years (the post-Garvan and Gunner era).
- This characteristically extremely low extraneous particle burden has resulted from application of advancing processing technology by the manufacturer rather than compendial limits or enforcement activities.
- There is some evidence, the majority of it somewhat conflicting and confusing, that levels of particulate matter in injections exceeding that allowable by pharmacopeial limits may potentially give rise to clinical issues.
- Even with today's large-scale use of intravenous therapy, there is no documented evidence of serious patient issues resulting from the chronic or short-term administration of intravenous solutions complying with pharmacopeial requirements.

While these conclusions paint an optimistic picture for the patient, constant vigilance on the part of the manufacturer is necessary to maintain the reality of patient exposure that is far below any level of concern. The manufacturer accepts this responsibility and insures the low particle burden of injectables and devices through the conscientious application of GMP; constant monitoring of product environments, processes, and products; and the attention to details of quality management beyond any regulatory scrutiny or compendial requirements.

REFERENCES

Akers, M. J. 1985. *Parenteral quality control: Sterility, pyrogen, particulate, and package integrity testing.* New York: Marcel Dekker, Inc., pp. 143–197.

Altierre, R. J., and D. C. Thompson. 1996. Physiology and pharmacology of the airways. *In: Inhalation aerosols—physical and biological basis for therapy,* edited by C. Lenfant. New York: Marcel Dekker.

Archambault, G. F., and A. W. Dodd. 1966. Macroscopical light-testing procedure for large volume parenterals. *Proceedings of the FDA symposium on safety of large volume parenteral solutions,* pp. 15–17.

Barber, T. A. 1987. Limitations of light blockage particle counting. In *Proceedings of the PDA Conference on Liquid Borne Particle Inspection and Metrology,* pp. 147–175.

Beck, W. C. 1966. Particulate contamination of the surgical wound. *Guthrie Clin. Bull.* 36:64–75.

Brewer, J. H. 1966. Panel discussion: Particulate matter. IV. Toxicology. *Bull. Parenteral Drug Assoc.* 20:35–37.

Brewer, J. H., and J. H. F. Dunning. 1947 An in-vitro and in-vivo study of glass Particles in ampules. *J. Amer. Pharm. Assoc.* 36:289–93.

Benjamin, F. 1990. Particulate matter: A historical review. In *Proceedings of the PDA International Conference on Particle Detection, Metrology and Control,* pp: 28–65.

BP. 1993. *The British Pharmacopoeia.* Addendum (1994). Her Majesty's Stationery Office.

Bruning, E. J. 1955. Origin and significance of intra-arterial foreign body emboli in lungs of children. *Virchow's Arch. Pathol. Anat. Physiol. Klin. Med.* 327:460.

Chason, J. L., J. W. Landers, and R E. Swanson. 1963. Cotton fiber embolism. A frequent complication of cerebral angiography. *Neurology* 13:558–560.

Davis, N. M., S. Turco, and E. Sivelly. 1970 Particulate matter in i.v. infusion fluids. *Bull. Parenteral Drug Assoc.* 24:257–270.

DeLuca, P. P., B. Conti, and J. Z. Knapp. 1988. Particulate matter II. A selected annotated bibliography. *J. Parent. Sci. and Technol.* 42 Supplement.

DiGrado, C. J. 1970. Panel discussion: liquid-borne particle counting in the pharmaceutical industry. III. Method evaluation. *Bull Parenteral Drug Assoc.* 24:62–67.

Ding, J. Y., C. P. Yu, L. Zhang, and Y. K. Chen. 1997. Deposition Modeling of Fibrous Particles in Rats: Comparison with Available Experimental Data. *Aerosol Sci. and Tech.* 26:403–414.

Dodd, H. 1965. Particles in intravenous fluids. *Lancet* 2:241.

Doris, G. G., B. A. Bivins, D. L. Rapp, P. Weiss, P. P. DeLuca, and M. B. Ravin. 1977. Inflammatory potential of foreign particles in parenteral drugs. *Anes. and Anal. Current Research* 56:422–428.

Dungan, D. J. 1968. Particulate contamination in pharmaceutical preparations for injection. *Aust. J. Pharm.* 49:599–564.

Endicott, C. J., R. Giles, and R. Pecina. 1966. Particulate matter—its significance, source, measurement, and elimination. In *Proceedings of the FDA Symposium on Safety of Large Volume Parenteral Solutions,* pp. 62–70.

Ernerot, L., and E. Sandell. 1967. Membrane filtration during administration of infusion fluids for elimination of particulate matter. *Acta Pharm. Suedica* 4:293–296.

Falchuk, K. H., L. Peterson, and B. J. McNeil. 1985. Microparticulate-induced phlebitis: Prevention by in-line filtration. *N. Engl. J. Med.* (Jan 10) 312:78–82.

Gardner, L. U., and D. E. Cummings. 1931. Studies on experimental pneumoconiosis: inhalation of asbestos dust; its effect upon primary tuberculosis infection. *J. Ind. Hyg.* 13:112.

Garvan, J. M., and B. W. Gunner. 1963. Intravenous fluids: a solution containing such particles must not be used. *Med. J. Aust.* 2:140–5.

Garvan, J. M., and B. W. Gunner. 1964. The harmful effects of particles in intravenous fluids. *Med. J. Aust.* 2:1–6.

Geisler, R. M., J. M. Garvan, B. Klamer, R. U. Robinson, C. R. Thompson, W. R. Gibson, F. C. Wheeler, and R. G. Carlson. 1973. The biological effects of polystyrene latex particles administered intravenously to rats: A collaborative study. *Bull. Parenteral Drug. Assoc.* 27 (3):101–117.

Goddard, J. L. 1966. Address to the Parenteral Drug Association. *Bull Parent. Drug Assoc.* 20:183.

Gotz, V. P., K. H. Rand, and B. S. Kramer. 1985. Effect of filtering amphotericin B infusions on the incidence and severity of phlebitis and selected adverse reactions. *Drug Intell. Clin. Pharm.* 19 (June):436–439.

Gross, M. A. 1967. The danger of particulate matter in solutions for intravenous use. *Drug Intel.* 1:12–13.

Gross, M. A., and C. J. Carter. 1966. The pathogenic hazard of particulate matter in solutions for intravenous use. *Proceedings, FDA Symposium on Safety of Large Volume Parenteral Solutions,* pp. 31–37.

Groves, M. J. 1966. Some size distributions of particulate contamination found in commercially available intravenous fluids. *J. Pharm. Pharmacol.* 18:161–167.

Groves, M. J. 1965. Particles in intravenous fluids. *Lancet* 2:344.

Groves, M. J. 1968. Harmful effects of particles in parenteral solutions: Methods for their detection. *J. Hosp. Pharm.* 25:17–21.

Groves, M. J. 1973. *Parenteral Products.* London: Hinemann Medical Books, Ltd., p. 316.

Hill, S. E., L. S. Heldman, D. H. Goo, P. E. Whippo, and J. C. Perkinson. 1996. Fatal microvascular pulmonary emboli from precipitation of a total nutrient admixture solution. *J. Parent. and Enter. Nutrition* 20:81–87.

Ho, N. F. H. 1967. Particulate matter in parenteral solutions. I. A review of the literature. *Drug Intel.* 1(1):7–11.

Ho, N. F. H., R. L. Church, and H. Lee. 1967. Particulate matter in parenteral solutions. II. Particle size distribution in intravenous injections and administration sets. *Drug. Intel.* 1:356–361.

Hurst, D. A. 1965. Particles in intravenous fluids. *Lancet* 2:181–182.

Illum, L. 1980. Characterization of particulate contamination released by application of parenteral solutions. 3. Particulate matter from syringes. *Arch. Pharm. Chem. Sci. Ed.* 8:109.

Illum. L., V. G. Jensen, and N. Moller. 1978a. Characterization of particulate contamination released by application of parenteral solutions. I. Particulate matter from administration sets. *Arch. Pharm. Chem. Sci. Ed.* 6:93.

Illum, L., V. G. Jensen, and N. Moller. 1978b. Characterization of particulate contamination released by application of parenteral solutions. 2. Particulate matter from cannulas. *Arch. Pharm. Chem. Sci. Ed.* 6:169.

Innes, J. R. M., E. J. Donàti, and P. P. Yevich. 1956. Pulmonary lesions in mice due to fragments of hair, epidermis and extraneous matter accidentally injected in toxicity experiments. *Amer. J. Pathol.* 34:161–167.

Jaffe, N. S. 1970. Elimination of particulate contamination from ophthalmic solutions for intraocular surgery. *Bull. Parenteral Drug Assoc.* 24:218–225.

JP. 1996. *The Pharmacopoeia of Japan*, 12th ed. Tokyo: Yakuji Nippo, Ltd.

Johnson, K. T., C. D. Helper, and J. P. B. Gallardo. 1970. Particulate contamination in vials of sterile dry solids. *Am. J. Hosp. Pharm.* 27:968–976.

Jonas, A. M. 1966. Potentially hazardous effects of introducing particulate matter into the vascular system of man and animals. *Proceedings of the FDA Symposium on Safety of Large Volume Parenteral Solutions*, pp. 23–27.

Kalm, M. 1987. Historical review of particles. In *Proceedings of the PDA Conference on Liquid Borne Particle Inspection and Metrology*, pp. 70–74.

Kirnbauer, E. 1970. Panel discussion: liquid-borne particle counting in the pharmaceutical industry. I. Microscopic counting. *Bull. Parenteral Drug Assoc.* 24:53–58.

Kramer, W. 1970. Inspection for particulate matter essential to I.V. additive program. *Drug Intel. Clin. Pharm.* 4:311–313.

Kramer, W., J. J. Tanja, and W. L. Harrison. 1970. Precipitates found in admixtures of potassium chloride and dextrose 5percent in water. *Amer. J. Hosp. Pharm.* 27:548–553.

Krueger, E. O., and T. H. Riggs. 1968. Objectives: PMA parenteral particulate matter committee. *Bull. Parenteral Drug Assoc.* 22:99–103.

Lachman, L. 1996. The contribution of rubber closures to particulate matter in large volume parenteral solutions. In *Proceedings of the FDA Symposium on Safety of Large Volume Parenteral Solutions*, pp. 51–55.

Maddox, R. R., J. R. John, L. L. Brown, and C. E. Smith. 1983. Effect of in-line filtration on post infusion phlebitis. *Clin. Pharm.* 2:58–61.

Magath, T. B., and J. T. McClellan. 1950. Reaction to accidentally injected rubber plugs. Amer. *J. Clin. Pathol.* 20:829–833.

Martonen, T. B. 1991. Aerosol therapy implications of particle deposition patterns in simulated human airways. *J. Aerosol Med.* 4:25–40.

Martonen, T. B. 1993. The behavior of cigarette smoke in human airways. *Am. Ind. Hyg. Assoc. J.* 53: 6–15.

Martonen, T., and Y. Yang. 1996. Deposition Mechanics of Pharmaceutial Particles in Human Airways. In *Inhalation aerosols—physical and biological basis for therapy,* edited by C. Lenfant. New York: Marcel Dekker.

Perrin, J. H. 1966. The reactivity of solvent in parenteral solutions. In *Proceedings of the FDA Symposium of Safety of Large Volume Parenteral Solutions,* pp. 73–78.

Pesko, L. J. 1996. Physiological consequences of injected particles. In *Liquid and surface borne particle measurement handbook,* edited by J. Z. Knapp, T. A. Barber, and A. W. Lieberman. New York: Marcel Dekker.

Plein, E. M. 1969. Preparation and use of intravenous solutions: a review of incompatibilities and related problems. *Hosp. Pharm.* 4:5–13.

Reznek, S. 1953. Rubber closures for containers of parenteral solutions. I. The effect of temperature and pH on the rate of leaching of zinc salts from rubber closures in contact with (acid) solutions. *J. Amer. Pharm. Assoc.* 42:288–291.

Russell, J. H. 1970. Pharmaceutical applications of filtration. *J. Hosp. Pharm.* 28:125–126.

Rypins, E. B., B. H. Johnson, I. Reder, J. Sarfeh, and K. Shimoda. 1990. Three-phase study of phlebitis in patients receiving peripheral intravenous hyperalimentation. *Am. J. Surg.* 159:222–225.

Sarrut, S., and C. Nezelof, C. 1960. A complication of intravenous therapy: Giant cellular macrophetic pulmonary artenitis. *Presse Med.* 11:375–377.

Saylor, H. M. 1966. Panel discussion: Particulate matter. II. Visual inspection. *Bull. Parenteral Drug Assoc.* 20:31–33.

Schroeder, H. G., B. A. Bivins, G. P. Sherman, and P. P. DeLuca. 1978a. Physiological effects of subvisible microspheres administered intravenously to beagle dogs. *J. Pharm. Sci.* 67 (4):501–507.

Schroeder, H. G., B. A. Bivins, G. P. Sherman, and P. P. DeLuca. 1978b. Distribution of radiolabeled subvisible microspheres after intravenous administration to beagle dogs. *J. Pharm. Sci.* 67 (4): 508–513.

Skolaut, M. W. 1966. Particulate matter in large volume solutions as viewed by the hospital pharmacist. In *Proceedings of the FDA Symposium on Safety of Large Volume Parenteral Solutions*, pp. 18–21.

Silberman, J., H. Cravioto, and J. Feigen. 1960. Foreign body emboli following cerebral angiography. *Arch. Neurol.* 3:711.

Simson, F. W. 1937. Experimental cirrhosis of liver produced by intravenous injection of sterile suspensions of silicious dust. *J. Pathol. Bacteriol.* 44:549–557.

Stehbens, W. E., and H. W. Florey. 1960. The behaviour of intravenously injected particles observed in chambers in rabbits' ears. *Quart. J. Exp. Physiol.* 45:252–264.

Taylor, S. A. 1982. Particulate contamination of sterile syringes and needles. *J. Pharm. Pharmacol.* 34:493–495.

USP. 1995. *The United States Pharmacopeia*, 23rd edition. Easton, Pa., USA: Mack Printing Company.

Trasen, B. 1968. Membrane filtration technique in analysis for particulate matter. *Bull. Parenteral Drug Assoc.* 22:1–8.

Trasen, B. 1969. Analytical techniques for particulate matter in intravenous solutions using membrane filters. *Ann. N. Y. Acad. Sci.* 158:665–673.

Trissel, L. A., and N. C. Leissing. 1996. *Trissel's tables.* Lake Forest, Ill., USA: Multimatrix, Inc.

Turco, S. J., and N. M. Davis. 1971. Detrimental effects of particulate matter on the pulmonary circulation. *J. Amer. Med. Assoc.* 217:81–82.

Van Glahn, W. C., and J. W. Hall. 1949. The reaction produced in the pulmonary arteries by emboli of cotton fibers. *Amer. J. Pathol.* 25:575–584.

Vessey, I., and C. E. Kendall. 1966. Determination of particulate matter in intravenous fluids. *Analyst* 91:273–279.

Wartman, W. B., B. Hudson, and R. B. Jennings. 1951. Experimental arterial disease. II. The reaction of the pulmonary artery to emboli of filter paper fibers. *Circulation* 4:756–763.

Weibel, E. R. 1963. *Morphometry of the Human Lung.* Berlin: Springer Verlag.

Williams, A., and M. I. Barnett. 1973. Particulate contamination in intravenous fluids, administration sets and cannulae. *Pharm J.* 211:190.

Williams, P. B., M. P. Buhr., R. W. Weber, M. A. Volz, J. W. Koepke, and J. C. Seiner. 1995. Latex allergens in respirable particulate air pollution. *J. Allergy Clin. Immunol.* 95:88–95.

II

SOURCES AND CHARACTERISTICS OF CONTAMINANT PARTICLES

This chapter provides some basic information on sources of particles and particle technology and nomenclature that will equip the reader to better understand material presented later in this book. Too often, the technologist interested in particulate matter in a pharmaceutical context proceeds unaware of much of the large body of general information on the subject. Numerous standards relating to particulate matter are in place worldwide (see Mielke 1990, 1991; Yu 1992), and there is sufficient information and dedicated technology to constitute this field as a "science" in its own right. General information also serves to guide the thought processes of the reader regarding sources of contaminant particles that can enter the process or product, which is often the single most important consideration in the manufacturing context. Useful general references are those by Austin (1995) and Benjamin (1990). The former reference is extremely valuable in that it is a detailed and comprehensive compendium of information on not only cleanroom use and construction but on particle sources and control as well.

In pharmaceutical manufacturing, we can define antibiotic drug powders, emulsions, and aerosols for respiratory therapy as beneficial to people's existence. Other particles, such as those responsible for the blue color of the sky at midday and particles present in "clean" air that we routinely inhale in and exhale, are really neither harmful or beneficial. Similarly, fibers of paper generated by writing on paper or particles released by the operation of machinery are a consequence of everyday processes or activities. It is a third category of particulate matter, particles detrimental in some fashion to people or their activities, which is the subject of this book.

Detrimental particles are most commonly defined as contaminants. Just as a weed is a plant that grows in some location where it is not wanted, a contaminant particle exists in an environment where it is unwelcome. Examples are provided by the single 1 μm particle that can "kill" a very large-scale integrated circuit, causing a significant loss to the manufacturer, or the visible fiber that renders a vial of injectable product a reject according to compendial regulations. Microbes, which are contaminant biological particles with significant potential for harm to human life, are beyond the scope of our discussion here. In summary, success in a number of economic ventures, of which pharmaceutical manufacturing is one, critically depends on limiting or controlling the numbers of particles that gain access to product.

DEFINITIONS

Adopting the definition that a particle is a discrete fragment of matter that is small relative to the system in which it resides is a good starting point. It follows from this definition that a single, individual particle is a small unit of matter with physical properties approaching those of the material from which it is formed.

The following (essentially universal) definitions apply to groupings of particles*:

- *Bulk material:* A material of any type that is of nonparticulate form or that consists of a population or mass of particles in a collective context. Examples might be a block of concrete or a kilogram of drug raw material powder.
- *Primary particle:* The individual unit under consideration in a population or grouping of particles. The primary particle, although it may adhere to other particles to form a larger group, retains a unique identity. In a sheet of paper, these are defined as individual paper fibers; in a crystallized drug powder, individual crystal fragments are the primary particles.
- *Aggregate:* A group particles held together by strong atomic or molecular forces.
- *Agglomerate:* A group of particles held together by the weaker forces of adhesion or cohesion.
- *Flocculate (floc):* A still weaker grouping of particles easily broken up by gentle shaking or stirring.
- *Aerosol:* A solid or liquid particle dispersed in a gas.
- *Hydrosol:* Solid or liquid particles dispersed in a liquid.
- *Foam:* A gas dispersed in a solid or liquid.
- *Comminution:* Any process that serves to reduce a bulk material into smaller fragments of that material. Examples of man-made processes based on abrasion are sanding, grinding, and finishing.
- *Condensation:* A gas-to-particle conversion in the atmosphere, such as the condensation of combustion vapors or the fusion of oil droplets in an aqueous system.
- *Evaporation:* The loss of liquid by a droplet, leaving behind only solid material contained in the droplet. Evaporation generates salt particles over oceans through aerosolized drying of seawater droplets.
- *Aggregation or agglomeration:* Combining processes that do not produce particles in a strict sense, but change particle size as small particles form associations and adhere together, forming a larger particle. These processes influence particle size distribution by decreasing the number of small particles and increasing the numbers of larger particles. By common convention, an aggregate refers to a group of particles held together by persistent, tenacious bonding mechanism(s); agglomerations are looser associations.
- *Biologicals:* A broad term that includes particles produced by or from living (or deceased) organisms. Such particles are formed by any process that can produce a fragment or small piece of tissue that subsequently becomes a part of a particle population in a gas, liquid, or solid system or on a surface. Spores, pollens, bacteria, and fungi fall into this category.

SIZE AND CLASSIFICATION OF PARTICLES

The conventional unit of measurement for particles is the micron (historical) or micrometer (current) and is abbreviated as μm. A micrometer is one-millionth of a meter (10^{-6} m), or approximately 0.000040 in. Airborne particles range in size from 0.001 μm to 1,000 μm, the latter settling from the air of a cleanroom very quickly. Particles of 0.001 μm border on molecular size. The unaided eye viewing particles in a strong light beam can see airborne contaminants as small as 30 μm due to the diffuse scattering of light. Human hair varies in diameter from 60 μm to 170 μm and can be visually detected. Most individual bacterial cells are smaller than 5.0 μm.

A wide variety of particles in terms of size, shape, and composition are found in pharmaceutical manufacturing environments. They vary in size from 0.001 μm to 100 μm or more and possess different shapes, chemical

*See Military Standard 124C and Federal Standard 209E for other useful definations.

compositions, refractive indexes, electrical properties, and densities. The terms *coarse* and *fine* particles are sometimes used to divide particles of all types into two large classes. Coarse particles are usually taken to be those with sizes greater than 1.0 μm, and fine particles are smaller than that size. In addition to the variety of particle characteristics, there is also a wide variation in the quantity or the numbers of particles in a given location. The concentration of particles ranges from less than 10^{-2} μg/m^3 in the stratosphere to more than 10^9 μg/m^3 in smokestacks and other industrial situations. As the pharmaceutical and device industry moves toward more complex products, it becomes necessary to control particles of ever decreasing size in cleanrooms.

Fine particles fall into the size range of materials that chemists refer to as colloids. Colloid chemistry is a subdivision of physical chemistry that involves the study of particles that have one or more dimensions in the range between 1 nm (nanometer, 10^{-9} m) and 1 μm. This science includes not only finely divided particles but also films, fibers, foams, pores, and surface irregularities of bulk material. Colloidal particles may be gaseous, liquid, or solid and occur in varieties of suspensions (often imprecisely called solutions) (e.g., solid/gas [aerosol], solid/solid, liquid/liquid [emulsion], gas/liquid [foam]). In this size range, the surface area of the particle is so great relative to its volume that unusual phenomena occur. As one example, such particles do not settle out of a suspended system by gravity and may readily pass through filter membranes that trap *coarse* particles. Macromolecules (proteins and other high polymers) are at the lower limit of the colloidal size range. Historically, the upper limit of the colloidal size range was taken as the point at which the particles can be resolved under a light (optical) microscope. Natural colloidal systems include rubber latex, milk, blood, egg white, and so on.

Particles can also be classified as either natural or man-made. However, a dissociation based on the specific processes producing the particles aids in understanding particle size distributions and chemical composition. Tables 2.1 and 2.2 (Murphy 1984) and Figures 2.1 and 2.2 depict particle population size/composition in different environments. These tables and figures are useful for general knowledge and because each of the sizes and compositions is represented in pharmaceutical manufacturing facilities.

Table 2.1 summarizes the major processes of particle production, including the range of particle sizes that results from a given process, and provides examples of natural and man-made particle sources for different particle size ranges. A single general particle source (e.g.,

Table 2.1. Mechanisms of Particle Production

Mechanism	Size Range (μm)	Naturally Occurring Particles	Man-Made Particles
Condensation	< 1.0	Smoke, normal gas-particle conversion products, water droplets	Smoke, combustion nuclei, gas-particle conversion products from man-made raw materials
Breakup	> 10	Dust, erosion products	Industrial dusts, dust from demolition, construction, and agricultural activities
Evaporation	0.1–10	Sea salt nuclei, dusts from hydrologic cycle	Dried agricultural sprays, dried paint sprays
Biological	0.01–10	Bacteria, viruses, plant spores, pollen, fibers	Environmental sources
Hydrologic	0.01–0.1	Bubbles bursting on ocean surface, forest fires	Irrigation, hydroelectric power dams, boating, application of insecticide and herbicide sprays
Combustion	0.1–1.0	Forest fires, bubbles bursting, volcanic activity	Smelting, metallurgical and manufacturing processes
Comminution	1.0–10	Large-scale erosion of the earth's surface, volcanic activity, tornadoes	Milling, sanding, crushing, grinding, construction and agriculture activities, combusting (fly ash)

smog) may result from more than one particle production process, so that the particle size distribution and chemical and physical characteristics of the particle population become complex considerations.

Aerosol dispersions are populations rather than individual particles; they may be produced by both natural forces and as a result of human activity. Table 2.2 provides further general information on particle relative sizes in aerosols.

Given the intense interest often generated regarding the composition of particles and where they come from, Tables 2.1 and 2.2 also cause the reader to think in terms of categories of particles and their possible sources. For instance, smokes and vapors in the atmosphere have their counterparts in chemical vapors generated by plastics extrusion, blow molding, or injection molding; water vapors with small particle size distributions may be found in solution filling areas. Such vapors, consisting of liquid microdroplets, can give spurious counts on optical particle counters and can result in artifactual failures to meet room limits at the ≥ 0.5 μm size. Dusts, usually the next largest size category, are represented by drug powders inadvertently aerosolized during mixing. Such aerosols can be pervasive in plants where drug powders are manufactured or filled and are the reason for extensive containment measures in some facilities where powders (e.g., chemotherapy agents), if released, may constitute a distinct hazard.

Figure 2.1 is a representation of the dimensions of some particle populations; Figure 2.2 shows the size of familiar objects and common methods of analysis. These distributions present interesting general information, but they also have practical and academic value. Diesel engines can produce millions of spherical carbon particles of submicrometer size during each minute of their operation. Some years ago, qualification of a newly built aseptic fill facility of a major domestic manufacturer of injectable products was delayed until a consultant could determine that the cause of airborne particle count excursions between 9:00 A.M. and 11:00 A.M. each weekday was the exhaust of switching locomotives in a nearby railyard penetrating HEPA (high efficiency particulate air) filters.

Particle Size Standards

From a dimensional requirement, the numbers of ≥ 0.5 μm particles are specified in many pharmaceutical environments as the cleanliness descriptor; smaller particles are typically counted in microelectronics controlled environment areas (CEAs). Using this value as a point of reference allows the practical use of documents already existing in the cleanroom and contamination control field. These documents allow a rational approach to controlling both airborne and surface-positioned particles. However, this does not consider the size of microorganisms. Some organisms, such as bacteria, are as small as 0.2 μm in size. In the United States, air cleanliness has historically been based on Federal Standard 209A and its subsequent revisions, the latest being 209E. This document discusses cleanliness levels from Class 1 to Class 100,000, as well as correlation with international levels called M Levels. Fortunately or unfortunately, the cleanliness classifications consider only the total particles per volume (cubic foot or liter),

Table 2.2. Sources and Mechanisms of the Formation of Common Aerosols

Aerosol	Mechanisms of Formation	Agents of Formation	Examples
fog, mist, smoke	condensation from a gas or vapor	condensation	soot, smoke, fog, clouds, condensation nuclei, artificial aerosols
dusts	dispersion of powder or disintegration ion of matter	nature, wind, human activity	pollen, sandstorm, road dust, fly ash
smog, haze	condensation of photo-chemical reaction products	sunlight	smog, haze, condensation nuclei
mist, spray	atomization of a solution, evaporation	atomization of a liquid	sea spray, salt nuclei, dust from humidifiers, road spray

Figure 2.1. Size ranges of common particle populations in comparison to wavelengths of electromagnetic radiation. (Courtesy of SRI, Inc.)

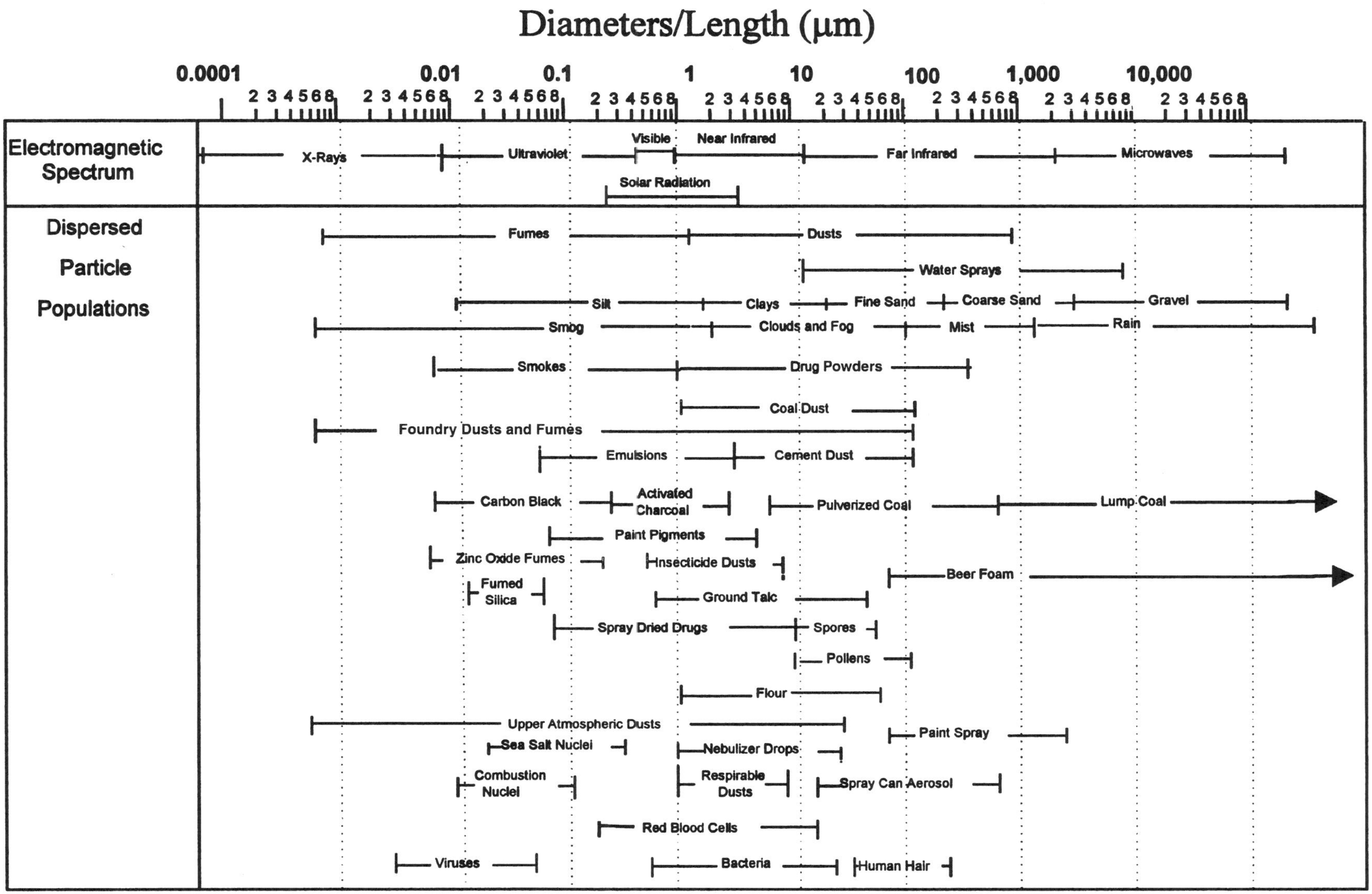

Figure 2.2. Scale of size ranges of particle populations, dimensions of familiar objects, and applicable modes of analysis. (Courtesy of Gelman Sciences)

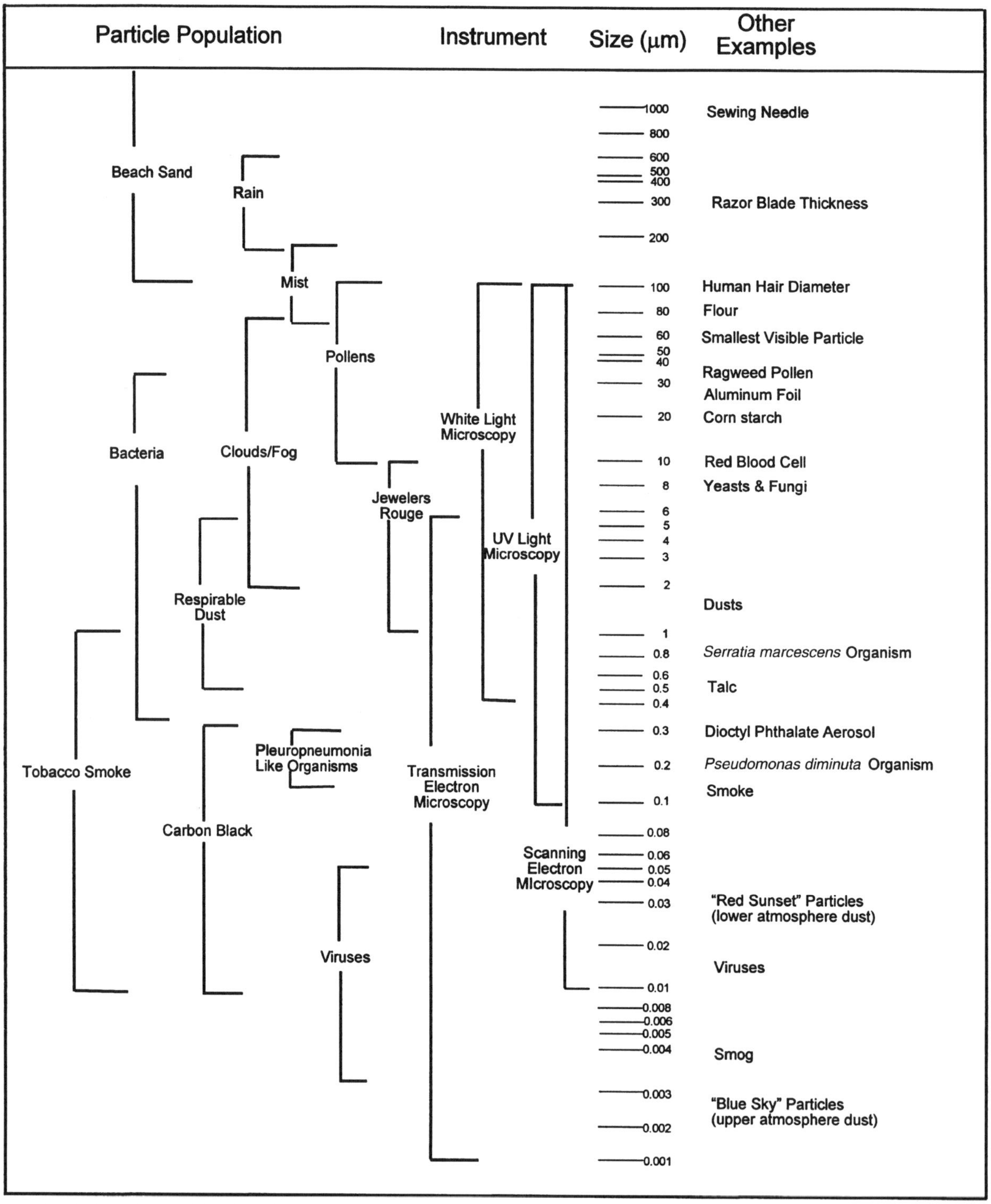

regardless of whether the particles are living or nonliving material.

In consideration of the environmental cleanliness of a CEA, the number of small particles is most often (rightly or wrongly) emphasized. This is largely because of the nature of environmental standards and the way that airborne particles are counted (i.e., particles $\geq$ 0.5 μm/ft^3 [m^3] of air). This is unfortunate, since these small particle counts lead to an underestimation of total airborne counts. The assumed distribution shown in tabular form in various standards is also misleading. For instance, according to Federal Standard 209E, an area with 100,000 counts at $\geq$ 0.5 μm/ft^3 can be expected to have only 700 particles $\geq$ 5 μm in the same volume.

Several considerations are in order here. First, the distributions shown in this and other standards are entirely hypothetical and are unlikely to occur in reality. The line log-log plot on which such distributions are based can be viewed in the real world as pivoting (seesawing) about the 1 μm size point or the 5 μm size point to give smaller numbers of 0.5 μm particles and greater numbers of 5 μm particles. Second, such distributions represent the ambient air of a process room, but not the air near a process point or a person. In these critical areas, the numbers of both small and large particles increase, with a disproportionate increase in larger particles. Likewise, airborne particle counts of the general environmental air in a room under in-use conditions do not reflect fibers that may be of considerable size (length) but, because of their shape and low mass, float for some distance and remain suspended on air currents. Airborne counts do not reflect large particles that may be generated at a process point as a result of activities (e.g., fill line intervention).

SOURCES OF PARTICLES

Given this general information, the reader's philosophical outlook on what particles are, their size ranges (or distributions), and where they come from should already have been broadened somewhat. At this point, we can now begin to consider the sources of particles that can occur in pharmaceutical products and the nature of particle populations and sources in manufacturing environments. The logic that has prevailed for many years is that there are three sources of particles in the environment. (For purposes of this discussion, the environment comprises everything except the product and the components of its container.) These sources are the air, people, and process points.

Particles less than 0.5 μm in size in air can readily gain access to a manufacturing facility despite HEPA filtration (Schicht 1988). Examples are soot from diesel engines and airborne road dusts. One route of entry is via the ventilation system, another is the street clothes of employees. HEPA filters in high classification areas (e.g., Class 100,000) are of little value in removing layers of fine dust that are deposited by sedimentation on floors, walls, machinery, and packing cartons. In regard to the latter, an astounding revelation for many is that the polyethylene bags used so widely for packaging closure components are typically thickly coated with particles of polyethylene waxes (extrusion aids) and other detritus added by the air used in the blown film extrusion process by which they are formed.

With regard to production process points, the buildup of residue from the activity carried out is unavoidable. Surface areas adjacent to powder-filling tubes, for instance, must be carefully designed or they can become coated with drug powder. Some manufacturing plants use robots to transfer cartons of finished goods or drugs in production areas. What happens to the rubber and metal fragments generated by the wheels of the robots or wear particles (grease, metal, rubber) generated by other machinery?

Two things are overwhelmingly important in the design of a production process: minimizing particle generation and protecting the product from particles. Particles are everywhere, and those who design and operate pharmaceutical production facilities must do so with a basic awareness of this fact and constant vigilance. While being adept at problem solving and particle source elimination are valuable when problems arise, the better course is to design and build particle elimination into a process so that the process protects the products.

Many of the particles with which we are concerned in the manufacture of pharmaceuticals or devices are materials generated by the process itself. Although such materials will be considered in greater detail in later chapters (9, 10, 11, and 13), an introduction here will serve to begin the thought processes. Consider packaging materials: when any frictional, abrasive, tearing, or rubbing force is exerted on paper packaging materials or fibrous materials like paper, paper fibers and fiber fragments are generated. Immediately, the advantages of plastic wrapping materials come to mind. Yet, most of the commonly used plastic sheeting materials contain extrusion aids that can segregate as fine particles on surfaces. The sealing or cutting of plastics with heat results in char, or fine, black, carbonized particles that can become attached to or fall into product. The search for the ideal packaging material is ongoing.

Many of the materials that are either optimal or absolutely necessary in the manufacture of a product are finely divided so that uniformity of mixing or dispersion is facilitated. Drug raw materials are excellent examples of this; in fact, solutes for solution manufacture, such as crystallized dextrose, contain large numbers of "fines" as well as the crystals of the bulk material. The same may be said of granular and powdered plastics used for the fabrication of plastic containers.

There is a distinct class of materials that the author categorizes as "nonparticles." These cause problems for test methods but have no consequence for the product. One example might be water vapor that is often present in areas where large-volume injections (LVIs) are produced by high-speed automated fillers. The water droplets are counted by particle counters. The same effect is noted with liquid vapors produced by the forming or molding of plastics. These are often low molecular weight organic materials, such as flexibilizers added to the polymer. The dioctyl phthalate (DOP) vapor that is used to challenge HEPA filters also falls into this category. Solid "smokes" may also be produced, but these are rare in production environments.

Drug powders escaping from an area in which they should be contained become contaminants. These materials can, in the case of some drugs, create a toxic hazard or serve as sensitizing agents. Filling operations and other powder handling operations are potential sources. Containment measures typically reverse the normal pressure gradients that flow downward in air pressure as one moves from clean to less clean areas. The negative pressures used for containment serve not only to contain powders but also to concentrate them in critical areas.

Personnel-generated contaminants are an unavoidable, invariably present, and overwhelmingly important component of the total particle population in cleanrooms or CEAs. Personnel are significant sources of contamination, especially the generation of skin and hair. The skin continuously sheds particles in the form of dead cells and cell fragments. The cells are rounded flakes, 10–20 μm in diameter and 2–4 μm thick. (It is possible to find fragments as small as 5 μm.) Tests have shown the reality of viable bacteria dispersion by overall body emissions; normal activities can release several hundred colony-forming units (CFU) per minute per person, even when clean clothing is worn. The particle release or emission rate increases with activity, indicating that a combination of higher breathing rates and bodily movements will raise the rate of release of bacteria. Oral and nasal emissions present serious problems when these vapor droplets with biological matter are deposited on sterile items in clean facilities (Austin 1995).

Activities of employees during breaks or at home can also carry over contamination into the cleanroom or CEA environment. Tobacco smoke has been detected in exhalations hours after smoking. Vapors from some foods persist in the breath long after ingestion. Cosmetics may not be completely removed from the skin and hair during washing and gowning. Such emissions should be prevented from reaching sensitive items.

Contaminants can be divided into inorganic and organic materials. Inorganic materials can be natural or man-made solids or liquids. They can consist of one compound or complex mixtures. Organic substances released as particles may be in the form of solids such as synthetic fibers, liquids such as oils, or gases. The word *clean* is, in itself, only a relative statement. The definition of clean as it applies to one CEA operation may mean dirty or contaminated in another operation. The term *clean* is really a strong function of the product requirements.

SOURCES OF PARTICLES IN THE CONTROLLED ENVIRONMENT AREA

The intent of the preceding sections of this chapter was to introduce the reader to particle sources and types of particles occurring in the general environment. The following discussion will serve as a further introduction to some of the particles found in a controlled environment.

Air sampling in a "typical" pharmaceutical CEA using a microporous filter membrane and a vacuum pump to draw air through the filter will normally result in collecting a range of particle types. The number and types of particles will depend both on the use of the area and its numerical classification. Measures to eliminate particles are targeted primarily at the source of the contaminant material. In areas where any type of packaging is applied to the product or removed, synthetic and plant (paper) fibers abound. Filling lines yield glass, plastic, and elastomer particles as well as aerosolized drug powders or liquids. Form-fill-seal processes may be designed to minimize particle release, but one particle characteristic of these operations often seen is the micrometer or submicrometer-sized spheres of polymers released by forming, cutting, or sealing operations. In fact, the operation of any type of machine in a CEA is suspected as a particle source, and many instruments can generate or release particles. In this regard, it was determined years ago that the very particle counting devices used to obtain data for CEA classification required filtration of the air that had been passed through the sensor, as well as cooling air that had been swept across the internal electronic components and was then exhausted into the room. The vacuum pumps that may be used to collect particles from air onto a membrane filter for microscopic analysis are also particle sources and are best kept outside. (A length of tubing may be used to provide a vacuum source in the clean area, or better, a clean house vacuum system may be used).

Speaking in general terms, the sources of contaminant particles in CEAs and cleanrooms used in the manufacture of pharmaceutical products and devices may be sorted into the same general categories as for other industries:

- Air handling and filtration system
- Particles shed by the room construction materials
- Personnel activity
- Production materials (including packaging)
- Equipment and instrumentation

We may begin by considering the air handling system (also termed the heating, ventilation, and air-conditioning or HVAC system) by which clean air is supplied to the CEA. Air handling systems are designed so that the particle burden of supplied air is effectively controlled at some low level by filters. Some fractional portion of the air volume is recirculated, and the remainder may be made up of outside air or air from other areas of the facility. The function of air filters is never absolute, and small numbers of particles will escape filtration and be released into the air of the CEA. Higher numbers of particles in the air supply to the final filters protecting the clean area will simply result in the higher numbers of particles penetrating the filters. Thus, the quality of supply air to the final filters and the prefiltration system is worthy of careful consideration.

What about the materials from which the walls, ceilings, and floors are constructed and the furniture and equipment? As requirements for cleanliness have become more critical over the years, the range of materials suitable for the construction of walls, floors, and other surfaces and the range of coatings and construction methods has increased apace. Austin (1995) provides an encyclopedic consideration of all aspects of construction. The evolution of CEA construction technology has been in the direction of nonshedding, durable base materials that are coated with tough, nonreactive, nonshedding coatings, such as epoxy or other resin paints. Square corners and seams are avoided, as these constitute areas that are difficult to clean and will trap dust, dirt, and soils. Floor-to-wall and wall-to-ceiling joints, as well as corners of the room are most often "coved" or rounded with a radius of 2–5 in. Ceilings are typically fabricated from seamless, nonshedding materials or suspension systems specifically designed to be nonshedding and not prone to flexing or breathing with pressure differential changes. Furniture, counters, benches, and cabinets especially designed for cleanrooms are also available. Floors, importantly, are constructed of materials resistant to the wearing or ablative forces caused by constant contact with the feet of personnel and the wheels of carts, chairs, stools, or other furniture or equipment.

Material control, both in the manufacture of devices and injectable products, is a key tenet of Good Manufacturing Practice (GMP). In some instances, the component parts of which a product is assembled may enter the process CEA in a package from which they must be removed before use; finished product units invariably leave the production facility in packages of some type. As a worst-case example, one can consider cardboard cartons. Opening these involves either tearing or the separation of layers that are glued together; both of these actions will result in large numbers of coarse paper fibers being propelled violently into the air. Vinyl blister packages closed with one of the spun-bonded polyolefin fabrics are also very popular. This packaging envelope is much superior to paper but still generates a significant number of particles when opened. Polyethylene bags of one type or another are frequently used as "clean" containers, but extrusion aids such as waxes used in their manufacturing and polyethylene flakes may contaminate components transported or stored in such containers. If not blow molded with clean air, the bags will also contain (and release) environmental contaminants. Packages and packaging materials have been in contact with a variety of surfaces during shipment and, thus, also have an adherent particle burden.

The product and process must be diligently protected from personnel. The high inherent particle burden of personnel is another factor in addition to the particle burden generated by the process itself. Product protection is accomplished directly through protective garb and indirectly through training, monitoring, and procedural measures. Not only must we consider the particles generated by people by virtue of their existence as life-forms (exhaled moisture, perspiration, shed skin cells, microbes), but other adherent debris, such as fibers of synthetic and vegetative origin, and contact contamination (also called dry transfer contamination or touch contamination) from a multitude of sources. As well as constituting stand-alone sources of contaminants, people interact with and contact the process and the product. Personnel are the subject of the following section of this chapter.

The production process itself, including the operation of equipment and communication of the CEA environment with other areas of a facility, is another consideration. In the filling of injectable products into glass or plastic containers, particles are generated by handling or friction of containers against transport surfaces and by fillers or other equipment. Larger particles (> 20 μm) rapidly settle out on work surfaces or on the floor. Particles that come to rest on surfaces are often white or gray, difficult to see, and typically remain undetected. These may be picked up by personnel (gloves, shoe covers), on casters, or on other parts of equipment and carried about. They may also be resuspended in the air and moved for some distance. Abrasion of the feet of personnel or other contact mechanisms on the floor constantly generate particles. Corrosion products are also generated in a CEA or cleanroom. Despite the prevalent use of stainless steel, rusting occurs due to the use of electrolyte solutions, and the pressure vessels used to hold solvents serve as reservoirs for contaminants. The movement of bearing surfaces such as those in the wheels of carts generates wear particles as do hinges and the sliding doors of cabinets. Iron rust is a brittle, friable material, and deposits of iron oxide may be shed and dispersed; stainless steel is resistant but not immune to corrosion.

Despite our best efforts to isolate the cleanroom from the rest of the faculty, material from adjacent, less clean areas will also gain access to the process area. Despite overpressures, turbulence generated when doors are opened and personnel enter or leave enable particles to move into the cleaner area. The principle that pertains here is a gradient or probability effect; the higher the number of particles in an adjacent, less clean area, the more likely that particles will drift or be carried "up" the concentration gradient into the cleanroom. Construction dust particles of various sorts and particles from carpet, writing materials, clothes, and abrasion or machinery operations are important in this regard. Such particles often have a high static charge and cling tenaciously to surfaces, equipment, packages, and garments. Even the most durable paints and surface coatings can be detached, and paint fragments sometimes may be the entity responsible for large visible particles in LVI containers. Wear particles often constitute a source for the "smudges" of dark particles that are readily visible against white packaging materials. These are typically composed of agglomerates of metal, grease, and polymers, such as Nylon® or Teflon®, and serve as a fingerprint of the contact surfaces involved.

In light of these considerations, it becomes obvious that an envelope of protective factors must counter the potential sources of particles that may enter the CEA itself. The envelope includes the following:

- Appropriate design and construction
- Workstations that minimize particle generation and buildup

- Air handling (HVAC) systems that eliminate particles from the air supply
- Equipment selected for minimal particle shedding
- Materials control
- Process flow and separation of clean from less clean
- Personnel protection (actually protecting the product from the people)
- Minimizing personnel activity in process areas

A number of authors (e.g., Austin 1995, Thorton 1990) have identified proportional contributions of different factors to the total particle burden in clean areas or the CEA. These proportional contributions from different sources will vary extensively based on the process and the sizes and sources of particles considered. Based on this author's experience, hypothetical contributions from various sources in clean labs and production areas might reasonably be as shown in Table 2.3.

Table 2.3. CEA Particle Sources

Area	Factor	Contribution ≥ 10 μm	Contribution ≥ 50 μm
Clean Lab	Workstation	10%	5%
	Equipment	15%	25%
	Materials*	5%	25%
	Process Flow**	1%	5%
	Air Handling	5%	1%
	Personnel***	70%	> 30%
Aseptic Filling	Workstation	5%	2–3%
	Equipment	5%	25%
	Materials	5%	5%
	Process Flow	< 5%	10%
	Air Handling	< 5%	1%
	Personnel	85%	40%
Device Production	Workstation	25%	25%
	Equipment	15%	5%
	Materials	15%	25%
	Process Flow	5%	5%
	Air Handling	5%	< 5%
	Personnel	25%	10%

*Includes production or analytical disposable, packaging

**Includes adjacent area effects

***The worker and activity

Based on this table, it can been seen that each type of clean area may be expected to give a different type of contamination "fingerprint" based on its usage and the process it contains. Personnel are a prevalent source, but not always the most significant one when particles of specific sizes or types are considered. The characteristic particulate burden varies with regard both to the source and composition of particles. Particularly with regard to medical device production, the greatest source of large particles is often in the workstation where extensive cutting, welding, sealing, and manipulation may occur. Figure 2.3 shows an example of proportional particle concentrations in a "spider" chart. This format of data presentation allows particle count contributions from different sources in different CEAs to be presented in a form that facilitates comparison.

In terms of patterns of particle generation and the likelihood of particles entering the product or its package, a number of systematic considerations have been suggested. Perhaps that most easily understood and applied is the TSC concept of Greiner (1994). This author dealt with the probability of product contamination (*P*) as the product of time of exposure (*T*), susceptibility (*S*), and particle concentration (*C*) to which the product is exposed (e.g., $P = T \times S \times C$). In this simplistic and effective scheme, if all three factors are unity (1), then the probability of exposure is 100 percent; If one factor is reduced to 0.3, the probability becomes 30 percent; if any factor is reduced to zero, the overall probability of contamination is zero. This concept is further discussed in Chapter 11.

Sources of particles, which come to reside in a product, have been classified by authors and technologists in a number of ways. Perhaps the earliest classification and facts still having the widest currency was that of Groves (1964), who referred to "extraneous" and "intrinsic" sources. Others (Whyte 1981) have referred to particles or formulations as container or solution related. Any such categorizations are generally appropriate and useful as long as they lead to systematic considerations of all the sources of particles that exist in a product.

A general consideration of types of particle sources is also useful.‡ Going from the least to most significant risk for product, these sources may be ranked in four classes:

‡This classification is one that the author has found useful; it is by no means widely accepted or used.

Figure 2.3. Spider chart of proportional particle contributions in a clean area.

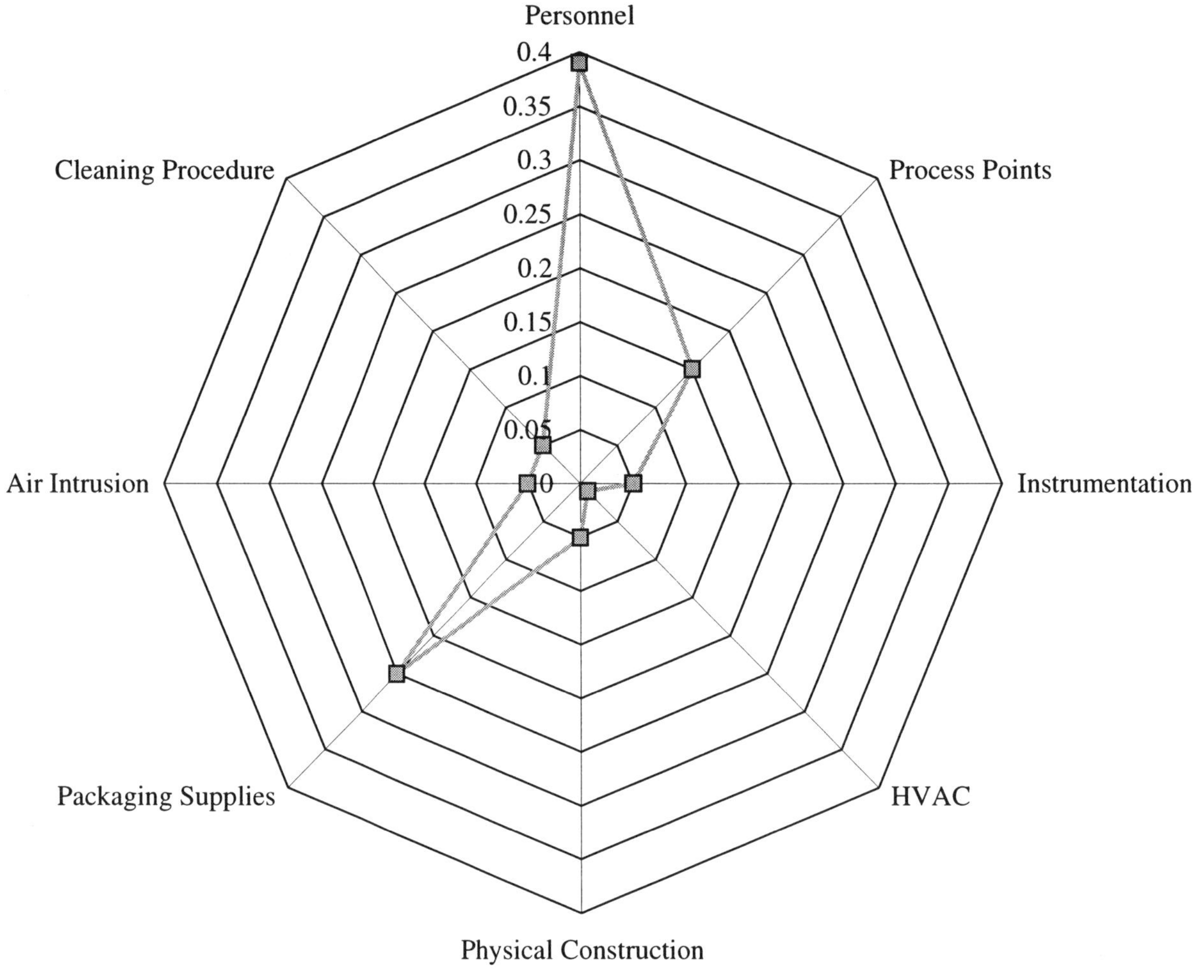

Source	Proportional Contribution	Source	Proportional Contribution
Personnel	0.39	Physical Construction	0.05
Process Points	0.15	Packaging Supplies	0.25
Instrumentation	0.05	Air Intrusion	0.05
HVAC	0.01	Cleaning Procedure	0.05

1. Diffuse or (environmental)
2. Localized
3. Point related
4. Product related

A diffuse source is illustrated by the ambient air of the process room. The airborne, heterogeneous particle population has a number of miscellaneous component sources, such as air entering at doors, particles passing filters, or particles from other sources (including people) that are of sufficiently small size to be suspended, dispersed, and well mixed in the air. These particles are amenable to control through airflow patterns, overpressure, and HEPA filtration and are least likely to enter the product. Many types of particles may be dispersed in the CEA environment.

Cleanrooms that enclose heat-sealing or molding operations may employ special exhaust systems to contain the particles produced by these activities. If this consideration is not made, air currents may diffuse fine plastic spheres and vapors throughout the cleanroom. This is especially true of turbulent (nonunidirectional) flow cleanrooms. Fiberglass particles from HEPA filters, insulation, and ceiling tiles and other dust particles caused by maintenance activity are prevalent in a start-up of a new facility and may be components of the ambient particle population for some time thereafter. Biological materials from outdoors and the surrounding areas may be transported into the cleanroom. In some areas of the world, plant cell structures, such as pollen and spores, are much more numerous than in other areas. In colder climates, plant growth is slowed in fall and winter, and the source decreases in importance. Plants may even be covered with snow so that this problem is reduced. In subtropical areas such as Puerto Rico, the problem may be severe year-round.

In contrast to diffuse sources, localized sources represent particle generation that is typically delimited to a smaller area at higher concentrations. Examples might be a vial unscrambler, turntable, or accumulator that generates suspended glass fragments (small numbers of which may enter the empty vials), or a stopper feeder bowl that generates debris as stoppers are rubbed against one another. Both of these may be significant particle generators, but both are usually located at sites some distance from where the filled product will be exposed. In the external environment, an uncontrolled plume of smoke from a factory stack or the column of ash from a volcano are also localized sources. As distance from the source increases, the particle numbers are diluted, and the concentration of particles per unit volume of air is decreased. With distance from the source, as particle number concentration decreases, so does the probability of contamination.

The point source designation incorporates a consideration of both the source of particles and its proximity to the product. The point source creates a risk for the product at a specific point in the process. When particle sources involved move close to the point where the product is exposed and susceptible, or a means of direct transfer is present, local sources become a point source. Examples of point sources include touch contamination by workers, ruptured filling head filters (which release both filter fragments and retained particulate matter), buildup of particles from stoppers on a stoppering punch, and misaligned fill tubes that touch the bottle mouth as they feed in. Air or compressed gases used for various purposes can also comprise point sources, particularly if these are in inert gases used as an overblanket on a product. In the assembly of medical devices, buildup of debris at assembly stations constitutes a point source of particles, which may be directly transferred to the product being assembled.

As the reader begins to think about the possibilities, the range of point and localized sources increases. A damaged HEPA filter escaping detection can be a localized source, or if in a fill line enclosure, perhaps a point source. Backstreaming of hydrocarbon or silicone oils from vacuum pumps in a lyophilization oven may be point sources for a drug product. Grinding, milling, or spray drying processes may incorporate particulate matter into the drug particles and thus qualify as point sources. Time-motion considerations suggest that storage and shipment factors are often additional, serious considerations with particulate matter that occurs in or on medical devices. The blister pack packaging configurations so widely used for devices (e.g., Tyvek® over a rigid vinyl bubble) may be subject in some cases to internal abrasion by the hard, plastic parts of devices during the vibration of shipment. Some containers for devices contribute paper fibers. In these cases, the "point" involved in particle generation involves time as well as location.

The last of the four particle source types mentioned above is the product-related or "intrinsic" type of Groves (1964). In this classification, Groves prefers to emphasize those particles that result from the physical instability or chemical reactions of the product itself, rather than those such as solids extruded from stoppers or formed through reaction with glass. Drug reactions and instabilities create a wide range of possibilities for the production of reactant precipitant, impurity, and degradant moieties. It is greatly to the credit of the drug manufacturer that these are almost totally avoided in a wide range of products.

Personnel as Particle Sources

Personnel and their activities are generally the most significant source of nonviable particles in controlled environments; most experts identify personnel as the single most significant contributor of viables. Based on the rationale that people contribute high numbers of particles and interact with materials with equipment and other sources, the highest particle levels in the cleanroom will occur when the process is operating with people present (i.e., the operational state). The rate of release of particles by work in the cleanroom will depend heavily on two factors: the potential numerical level of the particle source relating to personnel and the levels of activity.

An individual can depend on mobility, activity, and other variables to be a carrier of a multitude of particle types. Robots or automated machinery do not replace the human worker, and moving parts that roll, slide, impact, and rotate do so at the cost of particle generation, perhaps in close proximity to the product. Importantly, the human worker is not only prolific in regard to particle generation, but can also be trained to minimize the number of particles released. Humans, unlike robots, are trainable to change their actions sufficiently and, more importantly, can think. The human worker can be trained not only to recognize the more obvious particle sources but also to think in terms of contamination control. Evidence of this is seen in the aseptic production processes of some companies, which emphasize employee training in particulate matter control as a means of microbe control.

The elements of control of particulate contaminants from this source include the following:

- In-depth education regarding sources of particles
- Training in appropriate procedures for particle elimination and control
- Use of appropriate protective garb
- CEA access control
- Monitoring of particles generated by individuals
- Establishing a "partnership" between managers and CEA workers to minimize contamination

The last item is extremely important. In one large, Japanese injectable manufacturing facility visited by the author several years ago, he became curious about the text of a sign posted in almost every area of the plant. When asked, a company representative stated that the sign carried the message, "Every person must be a quality inspector for particles." The intent here is obvious; if *every* worker is trained regarding the sources of particulate matter and is diligent in eliminating particles from products, a powerful force for minimizing the occurrence of contaminants is brought into play.

A significant portion of both the fibers and particles greater than 5.0 μm in size found in cleanrooms typically originate from humans or human activity. Personnel moving about in an otherwise clean atmosphere release particulate matter and gaseous contamination behind them (Munson and Sorenson 1990); personnel activity also results in the reentrainment and emission of particles (Kim and Flynn 1991). The outer layers of the skin shed continuously, the shedding rate and particle size depending generally on the amount of abrasion to which the skin is exposed and its moisture content. Drier skin equals more shedding; hence, most cleanroom managers avoid cleaning products that dry the skin). The human body is notorious for releasing skin cells and skin fragments that can serve to transport large numbers of bacteria. This fact is an overwhelming consideration of regulatory agencies in their scrutiny of airborne particle counts in aseptic filling areas.

Many of the particles coming from people can be effectively subjected to control measures. Particles that may be temporarily trapped in the lungs from air outside the CEA are carried into the clean area and may be exhaled and released. In the past, when cigarette smoking was much more common than at present, "second-hand" smoke deposited in the cleanroom was a widespread problem in the microelectronics industry. The wearing of surgical masks minimizes particle release from this source. For obvious reasons, cosmetics must be exhaustively removed before entering the clean area. Fragments of fibers from street clothing are common CEA contaminants, as are fragments of paper fibers. Paper and its shed particle burden are a pervasive contaminant.

One major component of the personnel-sourced CEA contaminant particle population is fragments of plant cell material. Humans use paper extensively for packing material, writing, cleaning, and many other more specific tasks, but paper particles are nonviable. In fact, if paper fiber fragments or skin cells were viables, it would probably be impossible to perform a successful media fill. Using a tissue to blow one's nose can result in the generation of tens of thousands of paper fibers and fiber fragments. The author estimates that 90 percent of all fibrous particles isolated in CEAs are paper fibers or fiber fragments. The general category into which this material is placed in a microscopic analysis is usually "vegetative fibers."

The effect of clean garments and HEPA filtration is to generally decrease the number of particles present in cleanroom air to levels far lower than those in the home, office, or factory environment, but the approximate makeup of the particle populations in the different locations that originate from humans may be similar. Some of particles in CEAs come from contaminated street clothes and the skin beneath the clean garments (avoidable); others are present simply because of human bodily functions such as perspiration or breathing (unavoidable, but controllable). Many times, the particles collected and analyzed from cleanroom air samples will tell much about the living habits and homes of employees (Austin 1995; Clements 1990). People collect contaminants on their clothing, skin, and hair from various locations that they visit; after an employee's personal street clothes are washed in the home laundry, they are placed in a clothes dryer that draws unfiltered air from the home with its airborne load of fine particles (dust, animal hair, skin cells) through the drying clothes. Thus, fibers are not only related to the garment material but also to debris from the washing process.

Changing clothes invariably generates large numbers of fibers, yet change rooms with their skin cells, hair, and fibers from papers and clothing are typically located adjacent to clean areas for convenience in access. Street clothes have the potential of releasing millions of particles. If worn underneath cleanroom garments, particles that shed can pass through the clean garment and into the cleanroom. The shedding through protective garb depends on the openness of the weave of the protective garment worn and the levels of particles shed by the street clothes underneath. As a secondary consideration, the movement of personnel resuspends people-generated contaminants that settle on the floor. Hence, the use of tacky mats, air showers, shoe cleaners, and HEPA–filtered vacuum cleaners.

In many companies, the garbing policy understandably involves removing street clothes before donning clean overgarments. Instead, synthetic plant uniforms of synthetic fiber or specifically designed undergarments are worn beneath the outer garb. Leaving street clothes and the billions of particles on their surfaces outside the cleanroom complex will invariably have a demonstrable beneficial effect in decreasing the number of particles emitted. Simple procedures such as using two hairnets instead of one and "double-glove" procedures in higher classification areas are also often effective in reducing particle contaminant levels in the cleanroom or CEA.

Cosmetics are a well-known source of particles in CEAs and are insidious in that they are likely to be deposited close to critical process points or workstations. Ground talc (magnesium aluminum silicate) is a finely divided mineral that is widely used as a base for cosmetics. The talc material combined with fragments of the skin may be shed with bacteria attached in the cleanroom. Other cosmetics are produced by using a variety of metal oxides and other compounds for color and texture. Commonly used cosmetics contain such elements as aluminum, titanium, barium, calcium, magnesium, and iron. On another note, finely milled white flour and powdered sugar are ingredients of many baked products and can follow the employee into the cleanroom. Each breath of air taken by a human worker is exhaled with moisture, and biological debris adds to the cleanroom population of airborne particles. According to Austin (1995), the speaking of consonants such as B, D, G, J, K, T, and S produces a jet of air, carrying with it contaminants from the respiratory tract and mouth.

The proper use of protective garments is overwhelmingly important in controlling particle release by personnel. In an effort to better understand the contamination level in cleanrooms, Austin (1995) began, over 30 years ago, collecting and correlating emission data on cleanroom personnel. These data are summarized in tabular form as the Austin Contamination Index (Table 2.4). This information and other found in the literature (notably, Whyte 1984) suggest the relationship shown in the Figure 2.4. Disregarding for a moment the mental effects of personnel training and motivation and considering only physical factors, we see that there are a number of key determinants of the rate of shedding of particles from beneath the garment and from the surface of the clean garment itself. These include garment coverage,

Table 2.4. Personnel Particle Generation per the Austin Contamination Index

Personnel Activity	Snap Smock	Standard Coverall	2-Piece Coverall	Tyvek® Coverall	Membrane Coverall
No movement	100,000	10,000	4,000	1,000	10
Light movement	500,000	50,000	20,000	5,000	50
Heavy movement	1,000,000	100,000	40,000	10,000	100
Change position	2,500,000	250,000	100,000	25,000	250
Walk 2.0 mph	5,000,000	500,000	200,000	50,000	500
Walk 3.5 mph	7,500,000	750,000	300,000	75,000	750
Walk 5.0 mph	10,000,000	1,000,000	400,000	100,000	1,000

material and pore size of the garment fabric, and the rate of movement and/or type of activity.

The most significant contribution of particles from personnel will be particles in the 10–25 μm size range, consisting of skin cells, droplets of moisture, and extraneous debris from the small areas of exposed skin and clothing worn under protected garb (Busnaina 1987; Austin 1995). In supplying personnel with protective garments, we effectively enclose them in a filter. How the garb is worn and cleaned, as well as what is under it will be primary factors in particle release. Like other filters, garb will allow fewer particles to pass if it is challenged with lower particle numbers. Rapid movement or violation of good cleanroom practice (such as wiping one's hands on the surface of the garment or touching the face, then touching the product) increases particle release. Thus, training is of great importance. The openings in the weave of conventional polyester fabrics range from about 25 μm to 50 μm, thus skin cells, fiber fragments, and/or other debris can be pushed through by the bellows effect when workers move.

Two factors considered by Austin (1995) are of interest here: the nature of "street" clothing and the surface area exposed if it is worn under clean garb. The cotton or cotton blend fabrics of which street garments are predominantly composed are characterized by the relatively short length of individual fibers and their tendency to release "lint" or fiber fragments on flexion. With movement of the worker, the fiber fragments set free can be pushed through the weave of the overgarment. Further, since the total surface area of street clothes is in excess of 30 ft^2, a significant number of fibers can be expected to be available for release under the protective garb.

A final consideration in the contribution of particles by personnel regards the adherence of particles and their transportation on the surface of garments. Due to the fact that polymeric garments can generate and hold a significant static charge, they tend to attract and adhere particles that may be released in critical areas. Current generation garb with conductive carbon fibers interwoven into the fabric at a level of < 1 percent minimizes the effect of this factor. In donning garb, contact with surfaces, street clothes, or the ungloved hands is always to be avoided.

In a summary consideration, establishing an appropriate garbing program includes addressing the following factors:

- Type of garment (smocks vs. coveralls, etc.)
- Disposables versus reuse
- Type of fabric
- Static charge factors (carbon fiber fabric?)
- Washing (vendor vs. in-house)
- Garb particle testing
- Training (donning, etc.)
- Personnel monitoring for particle emissions

Contaminants in Device and Injectable Manufacture

Cleanroom or CEAs for the pharmaceutical and device industry offer a somewhat different

Figure 2.4. Effects of various parameters on the release of particles by personnel.

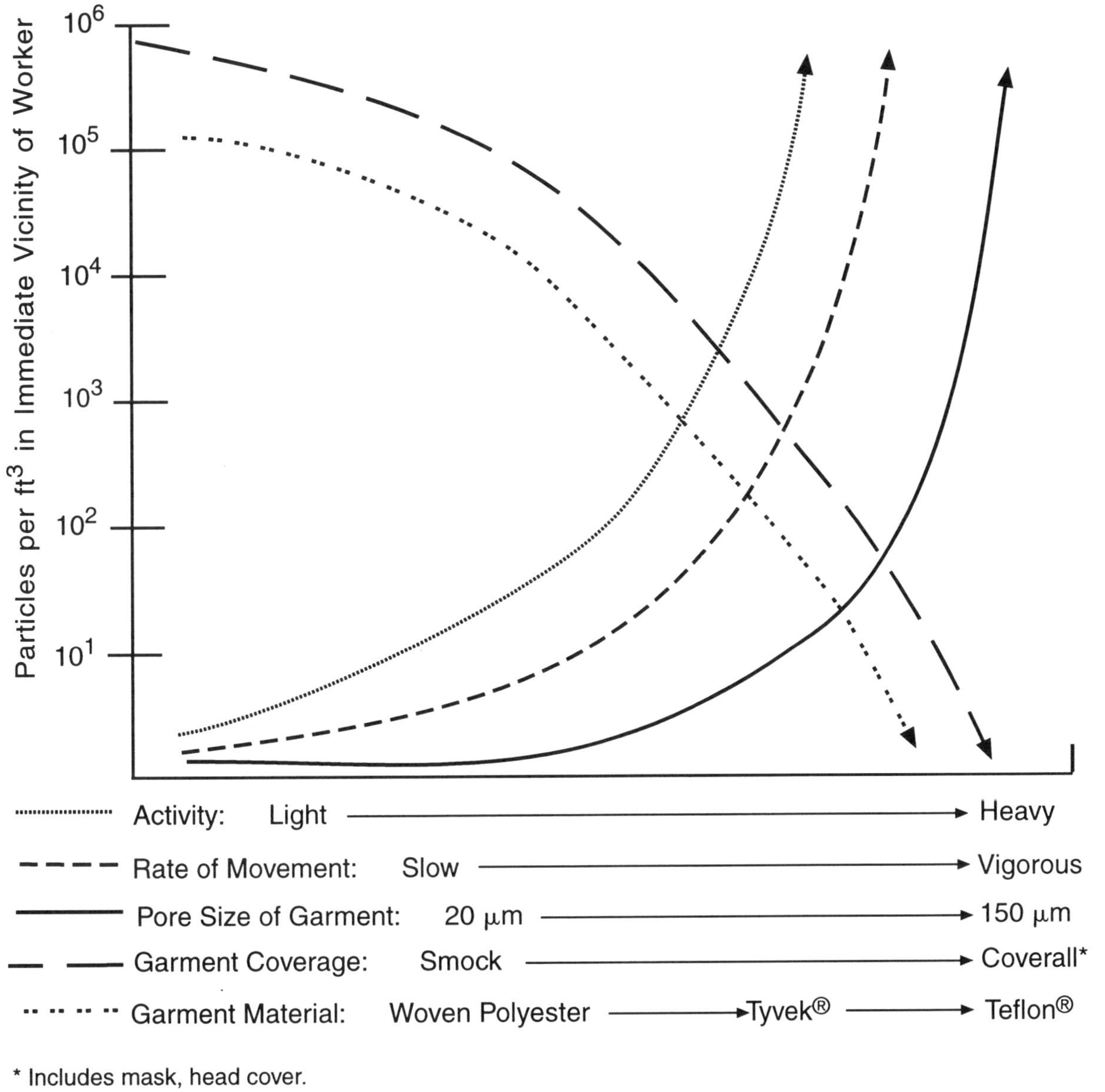

operational challenge than those used in the medical device, electronics, semiconductor, or aerospace industries (Lieberman 1986; Dixon 1990). The cleanliness requirements of pharmaceutical products and devices are relatively fixed and constant, relating to the dimensions of vasculature and formed blood elements in the human body. In contrast to this stable scenario, the ongoing development of integrated circuits by the electronics manufacturer dictates that the integrated circuits become ever more highly miniaturized and, hence, more sensitive to smaller particles. In the contemporary production of medical devices and injectable products, the number and sizes of environmental contaminant particles are not sufficient to degrade either product function or affect patient well-being. In other industries/technologies, however, particles ranging from 0.1 µm to 10 µm in size can cause operational failures.

The author has been chided by one of his friends in the hydraulics industry on a number of occasions regarding the lower number of allowable particles in hydraulic fluids in comparison to injectables. This is a somewhat superficial assessment, but dust, grit, dirt, metal particles, and fibers do cause seal and surface wear in hydraulic devices and accelerate failure. (A well-known supersonic transport aircraft at one time incorporated on-board particle counters with remote sensors to quantify particles in its hydraulic fluids). In microcircuits, particles 0.1 times the size of the finest conducting path (critical design dimension) can cause chip failure. Here the unseen particles are actually responsible for damage; in injectables and medical devices, it is most often the particles that are seen that raise concerns. In the application of high vacuum technology necessary for space exploration, seal damage is also accelerated by wear particles. In regard to optical surfaces, coatings of fine particles can degrade both the absorptive and reflective properties of mirrors and lenses, telescopes, and spectrophotometers.

The solution product cleanliness requirements of the pharmaceutical industry (U.S. Pharmacopeia [USP] <788>) were historically derived from the human body and the physical dimensions of its vasculature. One of the primary assumptions was that the smallest capillary of the human body had a diameter of approximately 7 μm (the diameter of a red blood cell). In fact, capillaries are tremendously distensible, and a fixed dimensional consideration is somewhat impractical. If one reviews the basic inert particulate matter size allowances in requirements of all the world's pharmacopeias for particulate matter, it seems that controlling particles of 10 μm and larger in size has become the accepted standard. Accounting for possible variations and adding a safety factor, possibly the best size control level could be established at 5 μm as per the current British Pharmacopoeia (BP) and European Pharmacopoeia (EP). The 10 μm lower size of the USP is also eminently reasonable on the basis that capillaries as well as cells are distensible, that red blood cells are not the smallest circulating cells, and monocytes as large as 14–15 μm in size circulate through the capillaries as well.

Today, the elimination of particles from injectable solutions is heavily dependent on the use of filters. Filters currently used in the manufacture of injectable products are extremely efficient in removing both nonviable and viable particles from the solutions being processed. There are little published data about the levels of particulate matter found prior to filtration. There is no one type of particulate matter common to all injectables. In the past 20 years, a wide variety of chemical materials have been found in parenteral solutions. Groves presented early data in his text (1973) that is given in tabular form in Table 2.5 (see also Groves 1993).

Groves made some simple calculations regarding a hypothetical, heavily contaminated LVI solution. These calculations may be revisited in terms of the current USP compendial limits. Dosage forms, even with a particle burden approaching the USP or BP limits, would in fact contain only very small amounts of material. Consider for a moment an LVI solution within the current USP limits. A 1 L solution, by microscopic test, would be allowed to contain 12,000 particles > 10 μm in size and 2,000 particles > 25 μm in size. If a density of 1 is assumed,

Table 2.5. Approximate Sizes of Some Contaminants Reported in Intravenous Solutions (Groves 1973)

Contaminant	Size Range (μm)
Insect parts	20–1,000
Glass fragments	1–1000
Rubber fragments	1–500
Trichomes	10–100
Metal particles	1–100
Cellulose fibers	1–100
Lubricating and machine oil	1–100
Plastic fragments	1–100
Starch	5–50
Fungi	5–10
Zinc oxide	1–10
Calcium carbonate	1–10
Plasticizer droplets	1–10
Silicone oil droplets	0.01–10
Carbon black	1–5
Clay	1–5
Diatomaceous earth	1–5
Talc	1–5
Bacterial fragments	0.1–5
Viruses	0.05–0.1

this equates to .053 μg of material in 1 L or a concentration of 0.000053 ppm or 0.053 ppb in 1 L. Groves also called attention to two other interesting factors bearing on the contamination level to which a patient was exposed; these are as true today as at the time this text was written:

1. Over a period of weeks, a patient may receive much as 100 L of parenteral fluid following surgery or a serious trauma.
2. It is virtually physically impossible to remove every vestige of unwanted particulate matter from an injectable solution, and it becomes extremely expensive to remove most particulate matter below certain size thresholds; the smaller the size, the more prohibitive the cost.

Groves concluded that some unwanted contamination would invariably be administered to a patient. He went on to define the problem of determining a balance between the necessity for cost-effective measures to remove particulate matter and the issues of possible patient harm.

The considerations of particulate matter in medical devices differ fundamentally from those pertaining to injectable products. In medical device manufacture, many contaminant sources are present in the manufacture and assembly of components. Normally, single component items are manufactured and then built stepwise into subassemblies and then the finished device. Contaminant sources are present in all operations. Because devices for the infusion of intravenous (IV) solutions or for the administration or filtration of blood are made of multiple components that are subject to numerous fabrication and assembly steps, a great potential for the inclusion of particles exists. As examples, filter fragments often result from the assembly of fragile membrane filters into their housings, and char and other inclusions can occur in tubing.

Elastomeric components of either solution containers or medical devices may be significant sources of particles and are the target of significant efforts to remove particles before their use. A consideration of the manufacture of these components provides a general case study relative to safeguards and concerns with regard to device manufacturing factors that govern contaminant particle generation. There are two very basic considerations pertaining to the device manufacturing process: (1) Elastomeric stoppers, injection sites, and so on are, at some stage in their production, very sticky and tend to adhere particles. (2) The production of such items involves blending of the bulk polymer and addition of finely powdered additives such as fillers, curing agents, and catalysts. In the production of stopper closures, for instance, initial steps involve mixing of the basic polymer (e.g., butyl rubber) with kaolin (aluminum silicate), titanium dioxide (TiO_2), and powdered chemicals such as peroxides or thiazoles. This additive material may be poorly controlled in the manufacturing facility; material is transported by air currents and pervades all unprotected areas of the plant. The stoppers are molded in large sheets that are exposed to mold release agents, typically a sprayed silicone emulsion that also is aerially dispersed. They are cut from the sheets in which they are molded by punch presses, perhaps with the aid of a lubricant soap solution to aid cutting. This step has the potential to add metal fragments as well as point-generated particles of other types.

The end result is that it is very difficult to protect rubber items from the contaminating materials associated with their manufacture; despite rigorous washing, the parts may be recontaminated from the environment before they can be packaged and shipped. One plan used by some manufacturers involves a second, separate site where stoppers are not manufactured or are manufactured only from premixed materials. In this second "clean" site, stoppers produced at a first facility may be thoroughly cleaned before "packing out" in cleaned containers. A key philosophy observed here is that containment measures must be justified by product cost. The heroic measures necessary to clean stoppers and contain powdered material in initial production may be unjustified based on increased costs. Another useful observation to be made is that some users of elastomeric parts would rather pay less and apply their own in-house additional cleaning measures prior to use. This plan allows the device or drug dosage form manufacturer to have the ultimate control over adherent particles on the part in question.

In fact, sources and origins of particles with which we are concerned in pharmaceutical manufacture are almost as diverse as those shown in Tables 2.1 and 2.2 and Figures 2.1 and 2.2 earlier in this chapter for particles in the general environment. Glass vials, ampoules, and bottles may contain cleaning materials or "slag" from the forming process. Glass of out-of-specification composition can either release ionic materials that will react with product or shed glass fragments through delamination or breakdown on solution exposure. Vial or bottle washers may remove one type of particulate matter and deposit another. Stoppers, as mentioned earlier, can both shed particles of filler, accelerator, catalyst, or stabilizer, and, because of their sticky nature, may bear a surface coating of adherent or embedded particles.

With regard to injectables, control of the process and a uniform process that eliminates or minimizes potential sources of particles controls the particle burden of the solution in the vial—not testing of the product or elimination of particle problems after they occur. A simple summary consideration of the potential sources of contaminant particles that might find their way into a 25 mL stopper vial of a liquid dosage form in a glass stopper vial suggests the scope of the GMP control measures that a manufacturer must have in place. First, the formulation itself must be properly designated, or else physical instability will result in the formation of crystalline or amorphous particles independent of the environmental particle burden. The stopper must be chosen for its nonreactive properties, cleanliness, and its lack of any tendency to release modifiers or low molecular weight materials into the formulation.

The stopper must be treated with minimal amounts of silicone. The glass vial must be exhaustively cleaned of any extraneous particles and, as well as the stopper, must be kept clean until assembled together. The stopper feeder mechanism must be kept clean. The air that contacts the vial and stopper should be Class 100 or better. The liquid formulation going into the vial must be filtered at a 0.2 μm (aseptic processing) or 0.45 μm retention rating to remove particles and microbes. The filling machine fluid path downstream of the fill head filter must be clean and free of any particle-shedding gaskets or seals, and the filling tubes must be properly adjusted to avoid chipping the mouth of the vial. The glass of the vial must be nonreactive with the product and properly annealed to prevent weak areas. In short, anything that touches or contacts the container-closure system or the formulation must have the lowest possible particle burden.

Filling machines, bottle washers, stoppering machines, and various types of machines that feed or handle components in assembly incorporate hydraulic, systematic, or solenoid valves or actuators. The latter all generate particles (wear debris) at some level; typically, the rate of particle generation is dependent both on design of the device and wear. In the years prior to the mid-1980s, piston-based filling pumps were notorious for a high shed particle burden, typically composed of metal and gasket fragments. The wide current use of other types of fill mechanisms, particularly rolling diaphragm fillers in which a minimal number and area of moving parts contact the solution, have almost eliminated these sources of particles; thus the use of filling head filters has become less critical. High-shear pumps, blenders, mixers, and homogenizers are always suspect of shedding metal or polymer fragments; these may be problematic in emulsion or suspension products that cannot be final filtered.

Medical devices entail some different considerations. With regard to the potential for the contamination of finished items by particulate matter, the manufacturing processes used for medical devices are exceedingly complex. This is to a great degree due to the wide range of objects classified as medical devices. If one reviews the Food and Drug Administration (FDA) classification of devices, one finds that device items such as powered wheelchairs and examination gloves are classified as devices, as well as pacemakers, oxygen regulators, some types of vessel grafts, and solution administration or transfusion sets. These latter types of devices fall into the FDA's Type III grouping, that is, devices that support or sustain human life, and which, should failure occur, present a "potential unreasonable risk of illness or injury."

Not being knowledgeable in the manufacture of pacemakers or other electronic devices, the author has the impression that the U.S. FDA expectation for particulate control in this type of device may involve stringent control measures consistent with those applied in the microelectronics industry. This would seem to be reasonable, given the electronic principle of operation of these devices. The situation with devices with which the author is more familiar (i.e., solution administration sets and blood transfusion or separation devices) is simpler in some respects but still relatively complex. An example is provided by a simple IV solution administration set. These devices consist of tubing that serve to transport IV solution from the container to the catheter inserted into the radial (or other) vein of the patient. Similar devices serve for the purpose of blood transfusion or access to central catheters. In addition to a 1–2 m total length of vinyl (or other polymer) tubing of 1–3 mm ID, components of this type of set may include

- a luer or other type connecting fitting to connect to the patient IV catheter,
- a transparent drip chamber for visualizing the drip rate,
- a pump segment or section of tubing that facilitates use of an electronic infusion device,
- "Y" sites or additive sites at which medications may be injected,
- elastomeric septa for needle insertion,
- needleless access devices,
- terminal filters, and
- a plastic spike or steel connector used to access the solution being infused.

Components for devices such as administration sets are often made of hard, rigid, dimensionally stable polymers, such as acrylics or polycarbonate. Flash, whiskers, and other particles attendant to the molding or forming process of these materials may be present. With typical devices of a wide range of types, however, assembly operations are the invariant predominant source. This holds true regardless of whether these operations are executed by humans or by automated or robotic processes. The latter have their own inherent particle-generating capabilities, and metal fragments and wear debris may predominate. Such is also the case with automated packaging machinery.

The assembly of solution administration devices may be used as an illustration of potential ways of particle generation. First, the majority of the device by weight consists of tubing that forms a flow path for the infused solution. The tubing is extruded, and the polymers used for this type of tubing require the use of extrusion aids and modifiers, typically lower molecular weight polymers, waxes, or oils. These materials can segregate at the tubing solution contact surface, and their distribution is controlled by manufacturing conditions and the plastic formulation. In many circumstances, each fitting or subcomponent of the device is added by cutting the tubing and inserting the subcomponent. Particles (crescentic fragments or flakes) result from cutting, especially if corroded or dull blades are inadvertently used. Subcomponents such as rigid plastic "Y" sites may be added with either a solvent or radio frequency sealing.

The insertion of the tubing into the subcomponent or vice versa is often a butt joint with an interference fit, which has the potential for scraping particles of polymer from the softer polymer (usually the tubing). Subcomponents can bear their own intrinsic particle burden, such as fragments of reinforcing material, modifiers, or contact contaminants such as paper fibers. Final filters or burette chambers that necessarily have been subject to multiple assembly operations bear their own particle burdens. Components may also have been exposed to cleaning operations, which simply comprise a means of decreasing the particle burden or, worst case, effecting its redistribution. Dogma here tells us that it is much easier to manufacture components under clean conditions and keep them clean than it is to clean them; washing procedures actually may add particles if not carefully designed and controlled; drying ovens have historically been problematic in this regard.

Other sources of particles that contribute to the particle burden of the finished set include the following:

- Pneumatic manipulators or handling devices
- Pressure test devices
- Storage containers
- Polyethylene (or other) storage bags
- Transport and storage
- Manual inspection procedures
- Solvents or adhesive used in sealing

Thus, the finished device may be expected to contain some numbers of particles from multiple sources. One further note of interest here regards the particle testing of devices. Devices are often tested by flush or wash procedures, whereby infusion solutions or filtered water is used to wash particles from the device for enumeration. The washout of particles, i.e., the number of particles released per volume of solution used, almost invariably follows a log-order decrease with increasing flush solution volume. Thus, if a volume of 1 L is sufficient to remove all particles from the device, 0.9 of the total particles will be released with the first 100 mL of wash solution. Hence, serious errors in estimating particle numbers may be made based on the particle numbers in the initial volumes of flushed solution.

Obviously, based on only this short discussion, particles deposited on or adherent to devices and injectable containers and closures in manufacturing can be expected to vary significantly in physical characteristics. Key physical properties of the particles with which the pharmaceutical particle analyst is concerned are as follows:

- Shape: Spherical → Angular, sharp-edged
- Texture: Smooth → Rough
- Hardness: Soft, deformable → Hard, brittle
- Size: 1 μm → 1,000 μm
- Reactivity: Inert → Reactive

With regard to physical and chemical properties, two other points regarding the occurrence of particles in injectables and devices differ from those in the microelectronics and aerospace industries. These involve concerns over mechanical damage to the device or product container-closure system and to chemical reactivity. In the aerospace industry, many mechanical and optical devices will only function properly based on the presence of highly polished surfaces or surfaces that are micropolished. Particles that can deform or degrade such surfaces are detrimental and may cause accelerated failure rates. Particles that are hard and/or rough textured may cause damage through abrasion. In microcircuits, the presence of a single conductive or insulating particle of submicrometer size can alter the electrical properties of a minute area of a very large-scale, integrated circuit and destroy a chip worth thousands of dollars. Such detrimental mechanical effects on the function of a medical device are rare, and the presence of contaminant particles usually does not affect the physical function of a container-closure or a drug delivery system.

Another important consideration in regard to contaminant particles generated by dosage forms present in the environment and shed by the components of devices or injectable containers is the inert or benign nature of the contaminant particles. The materials from which medical devices and solution containers are made are rigorously screened for toxicological properties. The materials used must invariably be shown to have extremely low potential for causing ill effects on the patient. Thus, at the low levels of any reactive materials in particulate form to which a patient might be exposed, the possibility of adverse chemical effects is extremely remote.

PARTICLE TRANSPORT, DEPOSITION, AND ADHERENCE

Based on our discussions up to this point, surfaces in the CEA, work areas, process equipment, machines, and instrumentation all are sources of particles. The mechanisms by which particles may be removed from cleanroom surfaces (including cleaning) are complex. Particles, once adhered to surfaces, are difficult to remove. Particles are unlikely to be dislodged by air currents from the surfaces to which they have been attached, due to the strength of the electrical attraction between the particle and the surface. This force required to remove the particle depends on the physical properties of the surface and of the particle, but it is disproportionately great in comparison to the small mass of a settled particle, perhaps as if a bulldozer were required to remove a golf ball!

With some particles of lightweight materials, such as paper fibers, that are not uniformly in contact with the surface, the flowing airstream may effect removal of the particle, but this is in a rare case. Consider a smooth, long paper fiber that, because of its twisted, irregular morphology and extreme length, has only a small proportional area of contact with the surface. When contacted by the moving airstream, the fiber is first displaced by sliding or rolling in the direction of the airflow. As the air velocity increases, aerodynamic lift may develop that twists and raises the particle from its points of attachment. Eventually, the fiber may be carried out of the boundary layer close to the surface and into the general airspace where air currents can move it about. This process is responsible for the removal of particles when high velocity, ionized air "knives" are used to clean films and other flat surfaces, but it is extremely unlikely to occur with particles that are exposed only to the nominal 90 ft/min velocity of directional airflow from a HEPA filter. As a matter of record, air knives are generally ineffectual without an angularly opposed vacuum pickup for the particles dislodged and are also of marginal utility on irregularly contoured surfaces.

Airborne contaminant materials are contained in the air that a properly designed HVAC system removes from the CEA process area through return plenums. This airborne particle load must be removed from the volume of air that will be recirculated into the CEA through HEPA filters. The quantities and sizes of the particles that comprise the airborne particle population depend on the type of filtration system used. If little or no filtration is provided for some areas of the building from which the recirculated air is drawn, the challenge to the HEPA filters feeding the CEA may be significant.

The directional transport of particles generated in the CEA is an important factor in the design and placement of clean zones and clean workstations. If horizontal unidirectional airflow is used, the direction of transport of particles, whether generated in the room or entering through air filters, is from the vertical wall filter banks toward the return plenum that is typically at the opposite side of the room. In this scenario, each process point has the potential for contaminating those downstream of it, and each workstation impedes the effectiveness of the airflow in particle removal. The air mass through (or across) the room tends to be characterized by complex eddy currents and has many zones of stagnation in which particles (sometimes seen as "dust bunnies" or fiber agglomerates) can collect.

In contrast, vertical unidirectional airflow typically maintains its directional characteristics until it comes in close proximity with the work surface; particles of all sizes tend to be carried generally downward and may be effectively removed, particularly if floor-mounted return plenums are used. According to the data in Figure 2.5, particles larger than about 10 μm settle rapidly from the air, whereas particles smaller than this size may be expected to settle less rapidly with decreasing size. The aerial transport and deposition of particles, once generated, is effectively a complex resolution of the vectors related to airflow direction, velocity, and gravitational settling. For particles of approximate unit density and smaller than 0.5 μm in size, random movement due to diffusion also becomes a significant consideration.

The original particle generation process will do much to determine the types of contaminants that can be reentrained. Different surfaces and different physical processes produce different types of particles. For example, wall, ceiling, or floor surfaces subject to mechanical or wear trauma may produce particles that are highly charged. Metal or metal-coated walls do not produce particles, but the solvent and cleaners used to keep such surfaces "clean" in aseptic fill areas can generate particles as a result of evaporation that leaves nonvolatile particles and residues behind. The deposits from the cleaning process will contain residues of dissolved material and contaminants that are left on the surface when the solvent evaporates. Thus, tap water is generally not used as a solvent in cleaning aseptic filling or critical device assembly areas, since it will leave behind a residue of insoluble precipitates that become a potential source of airborne contaminant particles.

Ambient aerosols are composed of charged particles. The particles carry a charge dependent on their generation process, or a charge may be gained through exposure to ionizing radiation, through contact with charged surfaces, or agglomeration with other charged particles. Once particles have been charged, they will be attracted to grounded or oppositely charged surfaces. After deposition, the effects of the initial electrical charge will effect the establishment of adhesive forces. Thermal energy gradients cause thermal precipitation to take place when the substrate (surface) is at a lower temperature than the surrounding air. Convincing evidence of the charge on polymer films or other forms of polymeric material can be seen simply by peeling a length of cellophane tape from a roll and holding it close to some black pepper. Particles of the seasoning material will be violently drawn to the tape from some appreciable distance. This same phenomenon can be observed in cleanrooms and CEAs due to employee use of polyester garments or decreased relative humidity and resulting electrostatic attractive forces.

The phenomenon known as triboelectric charging is responsible for the strong static charge observed when tape is unrolled. This charge generation mechanism is notably present in the manufacture of medical devices. When two polymer surfaces in contact are separated,

Figure 2.5. Particle settling rates based on size (courtesy of SRI International, Menlo Park, Calif.).

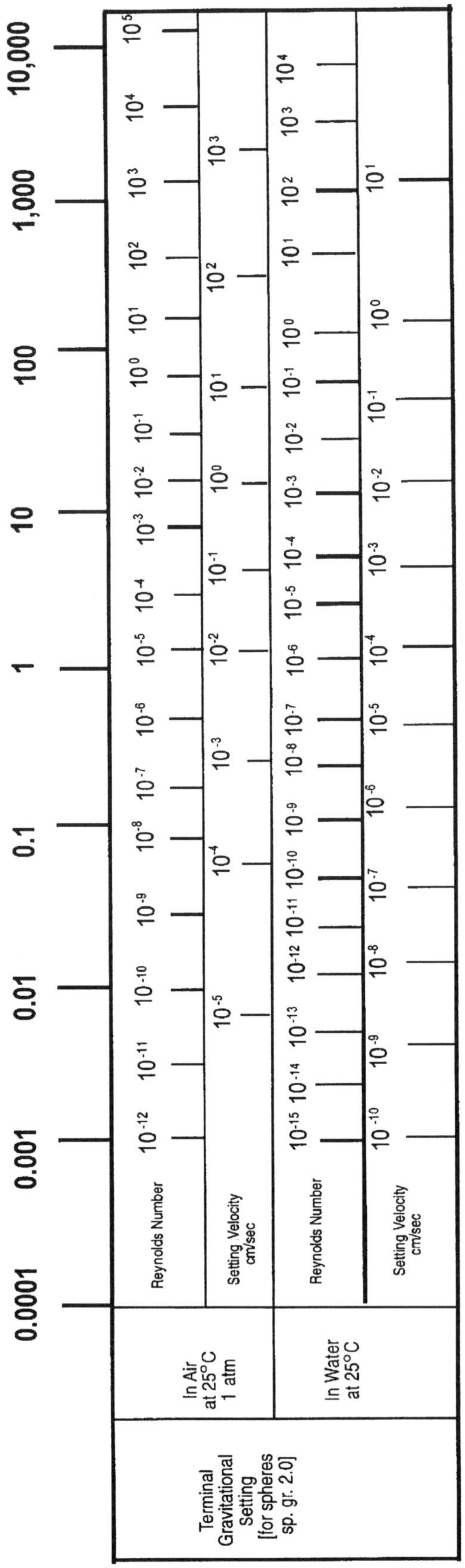

one surface will gain electrons and become negatively charged, and the other will lose electrons and be characterized by a positive potential. Closer, more intimate contact or adhesion increases the extent of the triboelectric effect. The level of charge developed by unrolling plastic sheeting can be thousands of volts, and particles may be attracted and held tenaciously by the charged surfaces. Charge neutralization is a key step in keeping the surfaces of polymeric materials clean.

VIABLE PARTICLE CONCERNS

What about replicative (viable) particles? What is the origin of viable particles of such significant concern in aseptic filling operations? The constant release of skin by personnel in the form of dead cells and cell fragments forms airborne particles that are "rafts" on which bacteria float and drift in the ambient air of the aseptic fill area (Munson and Sorenson 1985). Skin cells are ovoid flakes 20–40 μm in diameter and 2–4 μm thick. (It is possible to generate skin fragments as small as 2–3 μm.) Tests have shown the extent of viable bacteria dispersion by overall body emissions, normal activities can release several hundred colony-forming units (CFU) per minute per person, even when protective clothing is worn. The emission rate increases with activity, indicating that a combination of higher breathing rates and bodily movements generates elevated bacteria emission rates. (See the classic studies of Whyte [1981] and Whyte and Bailey [1986], and the later publications of Whyte [1996] and Whyte et al. [1998].)

While no one has ever been able to demonstrate a single, unique proportional relationship between viable and nonviable counts, the fact is that in dry, clean areas (e.g., < Class 10,000), where human activity is a significant consideration, the viable count levels will show some correlation with overall ≥ 0.5 μm particle counts; a "spike" elevation in nonviable counts may often result in higher than average viable counts. Viable count increases will, of course, not track such short-term perturbations as the addition of vials to a turntable (glass fragments) or restarting a powder filling line after a stoppage.

Bacteria are also brought into the cleanroom or CEA on skin surfaces or released by personnel respiration and carried on the surfaces of airborne dust particles, or released from surfaces bearing biofilms with adsorbed organic material nutrients and bacterial colonies. If the deposition of organisms is in an area where the bacteria can find nutrients and are not disturbed for some period of time, then the organisms will multiply and may form "hot spots" from which bacteria may be emitted for long periods of time. These have been the reason for the failure of more than one media fill. Important here is the ability of bacteria to form diffuse colonies in areas of residual moisture and scant nutrients.

In aseptic filling areas, it should be kept in mind that cleaning by poorly suited methods may raise the bioburden of the air and surfaces rather than decreasing it, thus raising the bioburden of the product. Bacteria rarely exist as individual entities in most environments. They are normally found on or in a larger "dirt" particle, water droplet, or skin fragment on which they may be aerially transported (they may be present on skin cells because of their ability to proliferate there in the first place). Most individual bacteria in cleanrooms or CEAs are in the size range from 1 μm to 3 μm in diameter, whereas skin, plant fiber, or dust particles that carry the organisms may be as large as 15–20 μm or larger. The number of organisms released depends on a first approximation of personnel activity; thousands of organisms can theoretically be emitted per minute with only moderate activity if personnel protection is inadequate.

Contamination from viable particles in the cleanroom or CEA HVAC supply is an additional issue (Calder 1984). Of note is that this latter source of organisms was responsible for the original outbreak of "Legionnaire's Disease." In secluded regions of the HVAC system, the deposited bacteria have the capability of growing in extensive colonies and then can spread throughout the process. Because viable particles are deposited on floors, walls, and work surfaces or in vessels or piping, these surfaces must be sanitized or sterilized before significant growth occurs; this will occur only if systematic, appropriate cleaning regimens exist. The presence of viable particles in liquid handling process areas for injectable products is a problem of potential severity in any cleanroom or CEA situation. Bacteria can multiply in any environment where the proper temperature, adequate moisture, and nutrient materials are present.

A consideration of this summary scenario regarding the prolific nature of the cleanroom worker as a source of potentially viable particles gives some interesting insight into the concern that FDA auditors have regarding airborne particle counts. While, in the view of the manufacturer, airborne particle counts are low and represent a manufacturing operation that is both validated and well controlled with respect to its release of viable organisms, the auditor knows the high potential for the release of potentially virulent organisms that is inherent as a process contribution from each human worker. To ensure patient safety, this must be viewed as an ever-present threat to the compromised patient. The presence of organisms such as *Staphylococcus aureus* on the skin constitutes a clear and present danger that can only be neutralized through high sterility assurance levels (SALs) and consistent adherence to procedures that probabilistically ensure that the release of viables from the human worker is vanishingly low. The safeguards against potentially infective organisms gaining access to a product closely involve employee training, monitoring, and attention to the protocols and procedures of cleanroom operations.

The critical and variable relationship between total instrumental counts and viable counts of airborne microbes is complex. Considerations include the following:

- Means of microbial enumeration (e.g., liquid vs. solid media)
- Physiological status of organisms (starved vs. log phase, disinfectant effects)
- Sources of nonviable particles in the cleanroom
- Restriction of human activity in the cleanroom
- Design of the cleanroom (especially airflow patterns)
- Sampling methods and means of airborne counting applied

Thus, in a reasonable regulatory perspective, the maintenance of the total number of particles in environmental air and in solutions and devices at appropriate low levels is a descriptor of process control. More importantly, in some areas, the number of particles enumerated by various means is scrutinized in assessing the potential for the access of viables to the process. Because of these concerns, it is useful to consider some of the factors relating to microorganisms as particles.

As with nonviables, the predominant source of viable particles in clean manufacturing areas or CEAs must also be considered to be the human worker. The skin of humans is colonized with strains of bacteria that are potentially pathogenic to patients (and an increasing number to patients who are immunocompromised). The aerobic diphtheroids (gram-positive, nonsporulating bacilli) form a large part of this population; examples are organisms of the genera *Actinomyces, Corynebacterium,* and *Listeria. Staphylococci* may be present on the skin in significant numbers even after showering and garbing; thus, they are the subject to release or transfer by the hands, gloved or ungloved. They are also capable of passing or being transported through protective garb. More than one FDA 483 observation in aseptic processing areas has involved a worker touching the skin in the area of their face, then touching processing equipment or product containers. Organisms are also exhaled in the breath. While the surgical masks worn are rigorously challenged for the removal of bacteria, they only constitute a barrier when properly worn.

The bacterial flora of the human respiratory tract is composed of high numbers of diverse genera. These include relatively hardy species of *Streptococci* and *Staphylococci*. The contamination of cleanrooms from this source will, of course, occur more readily if personnel are suffering from bronchial infections, when secretions may bypass masks and contaminate surfaces. These organisms may include pathogens such as the hemolytic *Streptococci* and *Staphylococci*. Gloves constitute an effective barrier that keeps organisms from the hands (the most dangerous surface) from escaping, but touch contamination of the glove surfaces when they are donned or used is a real risk.

The presence of low amounts of humidity or water allowed to condense on surfaces may lead to growth. Deionized water supplies have been known to harbor large populations of *Pseudomonads* that can contaminate and grow freely in water systems that have been dechlorinated. There are some species of organisms (e.g., the widely dispersed *Serratia marcescens*) that have been isolated when infections have followed therapy with injectable solutions. Control of moisture (humidity or surface moisture) is a key to the control of microbial growth.

Bacterial cells exist in a range of sizes and stages. In response to low nutrient conditions or other adversity, some organisms form spores that are resistant to heat and drying. The airborne spore, encountering favorable conditions will germinate to form a vegetative cell capable of growth and multiplication. Cocci (spherical) forms are often predominant in air; these include *Staphylococci* and *Streptococci*, both potentially infective. *Pseudomonads* are also present in unfiltered air. One specific genus, *Pseudomonas diminuta*, is used to challenge solution filters with a 0.2 μm retention rating because of its small size. These organisms are uncomfortably close to the particle size believed to be most penetrating to HEPA filters (0.2–0.3 μm). Thus, a risk of microbial access through air handling systems must be assumed to exist at some level.

Once access to the clean area is gained by a microbe, it has potential to form a colony of millions of organisms. In viable sampling of CEAs, organisms are not enumerated as individuals but as CFUs. While it is possible that a single organism can multiply to form millions, it is unlikely that CFUs are individuals; more than likely, these represent "clumps" or groups of organisms. The larger the number of organisms in such a microbial agglomerate, the greater the probability that some organisms in a replicative state will be present, and the greater the probability of grow-out to form a colony or biofilm. Three factors are required for growth: moisture, nutrients, and suitable temperature. Individuals in clumps of organisms are also more likely to be protected against desiccation. Organisms have a wide range of dietary preferences and/or capabilities, and only trace amounts of nutrients are necessary to support a low level of growth. Thus, it is easy to see how the moist air of some filling areas in combination with aerosolized solutions might support growth; the frequent cleaning of aseptic filling areas, if not properly carried out, may actually result in local conditions favorable to growth.

PARTICLE DETECTION AND ANALYSIS

The use of cleanrooms and clean areas does not ensure clean products; testing and sampling are essential to monitoring both the environment and product quality (Francis 1983; Gerbig 1985). Controlling contamination levels in cleanrooms today is based largely on identification of the contaminant material and empirical knowledge of the possible sources of specific contaminants rather than a thorough overview before production commences. (Particle identification is dealt with in Chapter 10 also.) Those responsible for CEA or production area operations must sometimes rapidly identify and verify the location and effects of specific sources of contaminants and control their emissions. The characterization of contaminants characteristic to production area inlet and outlet areas and those related to specific cleanliness classes often aid in locating problem sources and show the value of component isolation. Principal component analysis allows the positive definition of major sources in the cleanroom, thus allowing rapid specification of corrective action should problems occur.

The detection of particles is the first step. The indication of the presence of particles in numbers that exceed allowable levels can come from a number of sources. In CEAs, the routine monitoring of airborne particle counts may indicate excursions above normal levels or class limits. Detection and enumeration of particles in CEAs may be carried out by a number of technologies (see Chapters 8, 9, 10, and 12 of this book). The most commonly used method, and that on which CEA classifications are established, involves the use of an optical particle counter (OPC) that uses the light scattered by particles in the range of 0.1 μm to 5.0 μm to size and count particles. The basic information required is the number of particles > 0.5 μm per ft^3, or M3. To collect these data, a relatively simple and inexpensive OPC will suffice. This type of measurement, although widely used, is subject to the deficiency that only airborne particles are counted and that particles of larger sizes (i.e., > 3–5 μm) may not be adequately sampled both due to their relatively low numbers and difficulties in getting them into the counter (see Chapter 7).

Given the potential adverse effects produced by airborne particles and their differing properties, it becomes apparent that it is essential to define and eliminate particle sources. The detection and counting of particles and the collection of particulate matter samples are discussed in detail in Chapters 7, 8, 9, and 12, but some discussion of philosophy is useful at this point. In addition to airborne particle counting, the monitoring of fluids in process lines or the release testing of final product per USP or other compendial requirements may also serve to indicate elevated count levels. Indications of the presence of contaminant, subvisible particles may come from either instrumental or microscopic tests. The importance of detecting visible particles in the process and on workstations, assembly points, or machinery is often neglected to the extent that the first indication of high particle numbers comes from the final inspection of the solution or device or from the customer. Both of these end-stage detection situations have possible negative economic implications for a product batch or for uninterrupted production.

Direct visual detection of particles on CEA surfaces is generally not a critical technique. Such surfaces are often white or light colored and may convey a specious impression of cleanliness. Most particles are white or light colored due to diffuse scattering of white light from their surfaces. Low-angle observation with a magnifier may be more effective in their detection but still must be considered noncritical. Significant enhancements to visual observation may be made by reducing the intensity of ambient lighting and applying long wavelength ultraviolet (UV) irradiation to fluoresce the deposited particles. The effectiveness of the techniques is due to a simple increase of signal to noise. White particles (signal) against a white background (noise), when both are illuminated by the same wavelengths and intensity of light, are a very difficult visual target. When the background illumination is diminished and the particles become white light sources due to fluorescence, the detection is powerfully enhanced. This technique is particularly effective for fibers and filter fragments that serve as "markers" for other types of particles that may not fluoresce as strongly. Figure 2.6 shows black particles on a white background and white particles on a black background.

Figure 2.6. Effect of background contrast on particle visualization.

A handheld UV light of specially selected low-energy frequency will increase the detection capability in this fashion by a factor of 100 times. The Contam-a-Light (Acorn Industries, Livonia, Mich.) is ideal. In one device manufacturing company, every workstation is provided with such an inspection device to ensure the cleanliness of work surfaces, tools, microscopes, fixtures, racks, manipulators, product, and anything else that will be in contact or in near proximity to the product. The procedure for using this inspection device is to hold it approximately 4 in. away from the surface at an angle. This will illuminate the area to be inspected and allow the inspector to examine the surfaces under the light.

The multiple sources of particles in clean manufacturing areas in the pharmaceutical industry almost invariably dictate that multiple particle size distributions exist. There are invariably more particles at smaller sizes to be sure, but several discrete distributions may exist over the range from largest to smallest, so that a distribution with "lumps" results. Interestingly, International Organization for Standardization (ISO) standard 14644-3 provides descriptions of methods for quantifying both ultrafine particles (≤ 0.1 μm in size) and macroparticles (≥ 5 μm in size), which are useful in quantifying particles over the range of sizes that may exist in a CEA.

While the number of particles present in the ambient air is important, these data do not provide an indication of the number of particles that settle on and adhere to surfaces. They also do not indicate the number of particles that may be present at a critical process point, such as the tabletop at an assembly position in the manufacturing process for a medical device. Subvisible particles deposited on surfaces or those that have the potential for such deposition may be studied using a number of techniques and involve microscopy or magnification, including

- the use of filtration techniques,
- adhesive pickup followed by microscopy,
- application of witness plate techniques, and
- surface scanning devices.

All of these methods are dealt with in Chapter 12.

The ultimate goals of the microscopic study of particle morphology are identifying particles and determining their origin. This latter activity is commonly referred to as "sourcing" the contaminant particle or particle population. Once the source is determined, it may be eliminated through process changes, facility modification, or personnel training. The following properties of particles are used as morphological descriptors in characterizing, identifying, and sourcing a particle using light microscopes:

- General appearance
- Size
- Shape
- Color by transmitted and/or reflected light
- Transparency, translucency, or opacity
- Anisotropy and birefringence
- Refractive indices and dispersion

Microscopic techniques in general have the advantage of allowing the analyst to see the particles and thus both characterize and accurately size them. The most universally applied method for microscopy of airborne particles, described in Federal Standard 209E, involves drawing some specified volume of air through a microporous membrane filter and then subjecting the filter to microscopic evaluation. Particles > 5 μm in size are amenable to analysis by this method; ISO 14644-1 describes particles of this size as macroparticles. Adhesive tape "lifts," whereby particles are removed from the surface using the sticky surface of a transparent tape, are commonly applied in forensic investigation and are often useful in the CEA. Particles may be examined in situ on the tape or removed from the tape with solvent and concentrated by centrifugation prior to analysis (see Chapter 12).

Probably the most used method for collecting samples of deposited particles involves the use of witness plates. These are simply sampling plates that can be placed at specific locations in the CEA, allowed to collect particles on their surfaces, and then examined microscopically or by other means. A wide range of materials may be used, including glass microscope slides, black vinyl tape (adhesive side up), sheets of photographic film, bacteriological agar dyed black and poured into disposable plastic dishes, and so on. The summary criterion for the sample surface is that it be smooth and free of particles before sampling begins. Relation of the time of exposure to the area of the plate allows calculation of the particle deposition rate on the basis of surface area. In addition to microscopes, automated surface scanning devices may be used to evaluate witness plates.

Today's particle analyst is often confronted by the fact that many of the easy problems have been solved, but there are still opportunities to be found. Paper wipes and towels have been replaced with nonshedding cloth ones, but the edges of some "particle free" wipes are still prone to shed fibers. Fill line operators may wear overalls and head covers and double glove, but if these protective clothes are not donned and worn properly, they may not be wholly effective. Aseptic process areas are rigorously cleaned by wiping and swabbing; if these operations are not carefully carried out, sources not only of particles but of microbes may be left behind. In order to solve the more complex problems of today's cleanroom operations, it is necessary for the particle analyst to be intimately familiar with all process aspects if identification of potential particle sources is to be successful.

PHILOSOPHY OF CONTAMINANT PARTICLE CONTROL

This chapter has dealt with particle sources, particle generation, and control measures in an introductory fashion and begins a consideration of factors in contamination control that are presented in greater detail in the chapters that follow. Based on the foregoing discussion, contaminant particles are defined as those that are unacceptable in or on product or within the product package due to number, chemical composition, size, or source. In the pharmaceutical industry, significant efforts are directed toward minimizing the risk of airborne microbiological contamination. Generally, for the control of viable or nonviable particles, the implementation of various types of manufacturing constraints and well-directed local controls will be far more cost-effective than enormous expenditures for ultraclean (Class 10 or better facilities) when one wants to improve the SAL for aseptically manufactured products. Contamination is an integrated effort dependent on the severity or level of product exposure, time, product susceptibility, and particle concentration; the crux of contamination control is to manage these four critical factors. Product protection is afforded by minimizing each factor and, at the same time, protecting the product, process, or function against them.

USP general chapter <788> (and other compendial requirements for particulate matter in injections) is intended to allow for the low numbers of randomly sourced particles of heterogeneous origin that are to be expected under GMP. These limits reflect a realization that injectable products cannot be manufactured without some low number of particles from random (miscellaneous) sources entering the product. The discussions by Dr. Michael Groves that expound on this philosophy (Groves 1965, 1966, 1973; Groves and DeMalka 1986) are valuable in understanding both sources of particles and their control. Compendial limits are not intended to apply to particles that are predominantly from a single source, such as glass fragments from a filling operation or degradant particles resulting from lyophilization. Thus, while particle numbers are explicitly limited, sources are also an implicit concern. On this basis, regulatory auditors worldwide may use particulate matter levels as an indication of the extent to which GMPs are followed. Elevated particle count data may point to possible non-GMP conditions and indicate that further investigation is necessary. This approach is not without basis in fact. For example, investigations of visible fibers in device packages (blister packs) led investigators to determine that garb was being sent home with personnel to be laundered. On the other side of the coin, the philosophy of "allowable" versus "contaminant" particles (Groves 1986) states that it is not possible to eliminate particulate matter completely from injectable products or medical devices, and that low numbers of particles of heterogeneous origin will occur in product despite GMP conditions.

Broadly speaking, it is possible to consider contaminant particles and plot their elimination in two ways. Too often, the approach used is a reactive one, whereby the presence of a contaminant is observed due to some negative effect, such as customer complaints or batches that cannot be released. At this point, the manufacturer takes heroic measures to isolate, identify, and eliminate the offending material. This "firefighting" approach typically proves effective, but high costs in terms of production delays, lost product, consultant fees, facilities modification, and regulatory issues are often entailed. Far more appropriate in the pharmaceutical industry is a predictive, prospective, or proactive approach that particle sources are anticipated beforehand (i.e., in process development) and eliminated or brought under control prior to damage being caused. It is, of course, impossible to carry out any human activity in totally predictive fashion, especially the complex production processes for some of today's novel, high-technology drug and device products. The best course in this regard will be to anticipate particulate matter issues to the maximum extent possible during process development so as to avoid problems later on. Time-to-market and competitive pressures are often a negative in this approach.

Based on the foregoing discussions, particles in the device or injectable manufacturing area result from the environment, the process, people, and product components. Specific control concepts and control measures have historically been directed at all of these sources. Environmental particle control is based on a barrier concept, whereby a virtual barrier (e.g., unidirectional airflow) or an absolute physical barrier (e.g., a mini-environment enclosure) is used to protect product or prevent dispersal of a potent material (Leary 1992; Loest 1991). Control mechanisms employed include the following:

- Air filtration
- Directional airflow
- Pressure differentials
- Access control
- Material control
- Area classifications
- Physical enclosures

As we have discussed, product components include widely diverse materials (e.g., the tubing of solution administration sets, bulk drug raw material powders, stoppers). Washing is generally a last resort, except for glass solution containers and stoppers. Personnel control measures may be divided into those dictated by GMP and ancillary measures.

Solution filtration (i.e., the use of solution, e.g., line or "production") filters with a given retention rating will only ensure that solution is free of particles larger than a specified size at the downstream surface of the filter. It has no beneficial effect regarding particles released by process lines or equipment between the filter and the product, or on particles from other sources, e.g., containers. The only process step in which washing is usually effective in particle removal is for glass containers. The washing of closures is less effective and generally more difficult to conduct effectively. Particles removed by washing may be replaced by others added in drying or depyrogenation ovens. With regard to plastic device components, clean molding followed by clean handling and packaging is invariably far more cost-effective than washing. In general, for all components of injectables or devices, it is far less expensive to fabricate components cleanly and protect them from contamination than to attempt to remove particles once they become attached.

HEPA air filtration simply serves to provide a source of clean air to process areas. Efficiency is 99.97 percent at 0.3 μm. Some fallacies regarding HEPA filtration are as follows (Flanders Filters Inc. 1989):

- HEPA airflow can be used to "wash away" particles that have adhered to a product or surface.
- More HEPA coverage is always better.
- The use of HEPA filtration and unidirectional airflow will ensure that the product is protected from extraneous particles.
- The application of HEPA filters in process areas will excuse shortcomings in process design, personnel protection, and so on.
- HEPA filtration is necessary in all CEAs.

Another concept should be briefly mentioned here. In some cases, the contaminant particles to be controlled are those of a powder or liquid originating from the product itself. This area of contaminant control is termed *containment.*

Particle containment involves the protection of personnel from particles generated by a manufacturing process rather than protecting a process from heterogeneously sourced contaminants. Most often, the concern is to protect personnel from toxic, aerosolized drug powders (e.g., chemotherapeutic agents) or potential sensitizing agents (e.g., penicillin, cephalosporins). The mixing of two types of drugs is also a concern. "Potent" compounds are pharmaceutical chemicals that are effective in very low doses in causing a positive or negative response in a human subject. Drugs of the future will almost universally demonstrate activity at far lower doses than in the past. Such highly active compounds will present significant challenges in personnel protection.

The so-called "barrier" isolators interpose an actual physical barrier between the material(s) to be contained or protected and the operator. The barrier walls may be made of plastic, glass, metal, or combinations of these. Barrier isolators in medical product manufacture provide maximum product protection/containment with a high degree of assurance through the selective physical containment of critical process steps. An example is a barrier enclosure for aseptic filling processes. Barrier isolators and other mini-environments are clean spaces designed to be of minimal size to enclose, contain, and/or protect a specific process or process stage.

As demands for product cleanliness increase, we must realize that it is impossible to (within the bounds of practical economics) make a large space as clean as a small space; a consideration of all economic factors (e.g., utilities, capital outlay, personnel) points to a favorable consideration of isolator-type mini-environments.

There is an interesting parallel between the predictive approach to particulate matter contamination control and the study of ecology. The reader may be familiar with ecology as a life science, which deals with the organism in the total context of its existence in and interactions with its surroundings. No organism can be considered in isolated fashion; it must be studied as a component of the whole system, as a member of groups and subsets of similar and dissimilar creatures and as a member of a population. Similarly, particles exist as populations originating from specific and general sources; the populations and sources are best understood and controlled if a detailed analysis of the whole process is made, including complex interactions relating to materials control, personnel, drug and container interactions, airflow patterns and filtration air pressure gradients, and other factors. In any case, particle sources generate particle populations.

SUMMARY AND CONCLUSION

A wide variety of particle types and particle populations occur in nature and in human-controlled environments, such as those in which injectables and medical devices are manufactured. Similarly, particles arise from a multitude of processes and sources. Contaminant particles are those that are undesirable by virtue of their source, number, chemical composition, or physical characteristics.

Of all the principles regarding contamination control, the most basic and valuable is this: Particulate matter contamination control must begin at the design stage of the product, process, or manufacturing facility. If this proactive (i.e., forward-thinking) procedure is in place, the process control measures under which manufacturing is conducted will control particles in the product with minimal expenditure and panic, or firefighting exercises can be avoided, to say nothing of regulatory concerns.

The lowest levels of contaminant particles are achieved by using clean product components that are kept clean throughout the assembly of the final product. Reliance on the removal of contaminants, by any means, will ultimately result in a higher product particle burden. This philosophy must be followed in process design. Elimination of particulate matter through product protection must be achieved in all areas where the product is not protected by the final container.

All personnel working in manufacturing areas critical to the particle burden of the product must be effectively trained with regard to particle sources and particulate matter control and elimination; the lowest levels of particulate matter will be achieved only if each individual involved in production accepts a responsible role.

It is impossible to keep larger process areas as clean as smaller process areas, regardless of the costs for control measures that a manufacturer is willing to accept. Thus, mini-environments, microenvironments, and isolators must be considered for application across a wide range of production conditions. Process design to eliminate particle sources is critical to contaminant control. Cost-effectiveness will be achieved by a comprehensive knowledge of particle sources and the application of least-cost approaches for particle elimination.

REFERENCES

ASTM. 1989. *Standard practice for processing aerospace liquid samples for particulate contamination analysis using membrane filters.* Document No. 311. Philadelphia: American Society for Testing and Materials.

Austin, P. R. 1995. *Encyclopedia of cleanrooms, clean manufacturing areas and aseptic areas.* Livonia, Mich., USA: Contamination Control Seminars.

Benjamin, F. 1990. Particulate matter: A historical review. *Proceedings of the PDA International Conference on Particle Detection, Metrology and Control,* pp. 28–65.

Bradley, A., S. Probert, C. S. Sinclair, and A. Tallentire. 1991. Airborne microbial challenges of blow/fill/seal equipment: A case study. *J. Paren. Sci. Technol.* 45:86–101.

Burris, N., and A. Elfe. 1992. Biocleanroom validation for terminally sterilized medical devices. *Proceedings of the 38th Annual Technical Meeting of the Institute of Environmental Sciences,* pp. 405–413.

Busnaina, A. A. 1987. Modeling of cleanrooms on the IBM personal computer. *Proceedings of the 33rd Annual Technical Meeting of the Institute of Environmental Sciences,* pp. 187–201.

Calberg, D. M. 1984. The microbiological assessment of biocleanroom quality. *Pharm. Manuf.* 1:14–17.

Clemens, R. W. 1990. International methods for sizing and counting detachable particulate contaminants on cleanroom garments. *ICCCS* 90: 117–120.

Cooper, D. W. 1989. Towards Federal Standard 209E: Partial versus complete inspection of clean air zones. *J. Env. Sci.* 32:31–33.

Dixon, M. 1990. Human contamination issues in cleanroom manufacturing. *Proceedings of the PDA International Conference on Particle Detection, Metrology and Control,* pp. 65–81.

Federal Standard 209E. 1992. *Federal standards for clean room and work station requirements, controlled environment.*

Flanders. 1997. *HEPA filters and filter testing,* 4th ed., Technical Bulletin No. 581 D. Washington, D.C.: Flanders Filters, Inc.

Francis, T. 1983. Clean rooms and contamination control. *ASTM Std. News* (May): 16–18.

Gerbig, F. T. 1985. Recommended practice for testing cleanrooms. *Med. Dev. Diag. Ind.* 7:58–62.

Groves, M. J. 1965. Particles in intravenous fluids. *Lancet* 2:344.

Groves, M. J. 1966. Some size distributions of particulate contamination found in commercially available intravenous fluids. *J. Pharm. Pharmacol.* 18:161–167.

Groves, M. J. 1968. Harmful effects of particles in parenteral solutions: Methods for their detection. *J. Hosp. Pharm.* 25:17–21.

Groves, M. J. 1973. *Parenteral products.* London: Hinemann Medical Books, Ltd., p. 316.

Groves, M. J. 1993. *Particulate matter sources and resources for healthcare manufacturers.* Buffalo Grove, Ill., USA: Interpharm Press, Inc.

Groves, M. J., and S. R. DeMalka. 1986. The relevance of pharmacopeial particulate matter limit tests. *Drug Dec. Commun.* 2:285–324.

Kim, T., and M. Flynn. 1991. Airflow pattern around a worker in a uniform freestream. *Am. Ind. Hyg. Assoc. Jour.* 52 (July):91–97.

Leary, H. R. 1992. Designing automated high speed packaging lines for cleanroom operations. *Pharm. Eng.* 12:26–34.

Lhoest, W. J. 1991. Design and selection of pharmaceutical production equipment in the scope of modern automated plants—Part 1. *Pharm. Eng.* 11:181–198.

Lieberman, A. 1986. Particle control and air cleanliness for state-of-the-art microelectronics manufacturing. *Micro Contam.* 4:29–32.

Liu, B. Y. H. 1995. Class notes, cleanroom aerosol technology short course. Minneapolis, Minn., USA. University of Minnesota, Dept. of Mechanical Engineering.

Mielke, R. L. 1990. Revision of FED-STD-209D and HIL-STD-1246B and development of IES contamination control practices in the USA. *Swiss Contam. Control* 3:289–292.

Mielke, R. L. 1991. Testing and certifying cleanrooms: A status report on the applicable standard and recommended practice. *Micro Contam.* 9:52–56.

Military Standard 1246C. *Product cleanliness levels and contamination control program.* Washington, D.C.: Department of Defense, Army Missile Command.

Munson, T., and R. L. Sorenson. 1990. Environmental monitoring: Regulatory issues. In *Sterile pharmaceutical manufacturing—Applications for the 1990s,* vol. 2, edited by M. J. Groves, W. P. Olson, and M. H. Anisfeld. Buffalo Grove, Ill., USA: Interpharm Press Inc., pp. 163–184.

Murphy, C. H. 1984. *Handbook of particle sampling and analysis methods.* Deerfield Beach, Fla., USA: Verlag Chemie.

Schicht, H. H. 1988. Engineering of clean room systems: General design principles. I: Standards, guidelines, fundamental concepts. *Swiss Contam. Control* 1:15–20.

Thornton, R. M. 1990. Pharmaceutical sterile cleanrooms. *Pharm. Tech.* 14:44–48.

Whyte, W. 1981. Setting and impaction of particles into containers in manufacturing pharmacies. *J. Paren. Sci. Technol.* 36:255–268.

Whyte, W. 1996. In support of settle plates. *J. Pharm. Sci. Tech.* 50:201–210.

Whyte, W., and P. Bailey. 1985. Reduction of microbial dispersion by clotting. *J. Paren. Sci. Technol.* 39:51–60.

Whyte, W., W. Matheis, A. Dean, M. Netcher, and M. Edwards. 1998. Airborne contamination during blow-fill-seal pharmaceutical production. *J. Pharm. Sci. Tech.* 52:89–99.

Yu, A. 1992. New Chinese aerospace industrial standard for clean rooms/clean zones. *ICCCS* 92: 357–360.

III

REQUIREMENTS AND TEST METHODS FOR INJECTABLE SOLUTIONS AND MEDICAL DEVICES

In this chapter, the author has attempted to provide a general discussion of requirements and test methods applicable to devices, injectable products, and container components. The present situation, particularly with regard to requirements for particulate matter levels in medical devices, is moving very rapidly. Each manufacturer should have in place a systematic means of "keeping a finger on the pulse" of world compendia and standards-setting organizations (e.g., ISO [International Organization for Standardization], AAMI [Association for the Advancement of Medical Instrumentation], DIN [Deutsches Institut fur Normung) and tracking new requirements initiatives. The situation is to some extent a competitive one, with the world's compendia and standards organizations vying with one another to be the first to implement specific standards and establish precedents.

The generation of the widely applicable and technically sound requirements needed in today's pharmaceutical manufacturing environment is a difficult and complex undertaking. Significant levels of technical input, product data, knowledge of best demonstrated practices, and consideration of economic and political factors are required. (The author has been fortunate in being able to furnish input to this process through the design of test methods and considerations of industry data.) Up to the present, there has been an apparent (unfortunate) tendency of both compendial and standard organizations to work *in vacuo,* with the input of only a small number of local technical authorities and with limited recognition of broader national or international expertise. This approach makes eventual harmonization more difficult and results in standards that are technically flawed and of narrow applicability. The primary practical concern of requirements for particulate matter in both injectables and devices is the maintenance of manufacturing standards rather than patient issues. In this light, standards should be based on round-robin studies, as is the frequent practice of the U.S. Pharmacopoeia (USP). Such studies serve three purposes:

1. A large number of industry experts in a given area are involved (the greater body of expertise invariably rests with the manufacturer).
2. The data collected from a broad product base may be subject to a comprehensive review by industry, compendial, and regulatory authorities.

3. Reasonable limits and sound technical methods may be achieved by an appropriate consensus.

The proposed standards discussed later in this chapter for medical devices are the most recent issues. Because of the current rapid evolution of device standards, the reader is cautioned that the versions of some standards included in this chapter may have been revised by the time this book goes to press; caution is necessary to be sure that the version being evaluated is the most current.

As outlined in Chapter 2, the age of awareness of the presence and undesirability of large amounts of particulate matter in injectables dawned in the early to mid 1960s. The papers by Garvan and Gunner (1963, 1964) called attention to the potential for patient harm from this source. Since that time, the world situation in regard to particles in injectables has changed markedly, primarily due to a response by manufacturers and to a lesser extent to the response of regulatory agencies and compendia. While limits and regulations (requirements) specify maximum amounts of particles that are allowed, only the attention of the manufacturer to Good Manufacturing Practice (GMP) and specific items such as container cleanliness and filtration procedures can actually effect particle control and reduction.

The current situation is far different than that in the 1960s. In those earlier years, injectables and devices might have contained significant numbers of contaminating particles, often from a single source; today's injectables and devices contain only minimal numbers of particles, originating from a number of miscellaneous sources in the production process. Significant progress in particle reduction has been due to advances in the following areas:

- Container and closure cleanliness
- Optimization of stopper composition
- Drug purity
- Filtration technology
- Process design for particle elimination (most important)

A particle is not invariably a contaminant. A product becomes "contaminated" if one of three conditions is met: (1) the particle burden is above applicable limits; (2) particles are monotypic and thus related to a known source; or (3) particles are of a reactive rather than an inert chemical composition and are potentially damaging on this basis. In recognition of the differences between the acceptable low particle burden present in devices and injectables manufactured under cGMP (current Good Manufacturing Practice) and the higher level of particles (often of a single type) present in product made by less–well-controlled processes, international compendial bodies have drifted toward a harmonization of definitions for particulate matter.

The terms *mobile undissolved substances, unintentionally present, randomly sourced,* and *of heterogeneous composition* all apply to the low levels of allowable particles of visible or subvisible size for which compendial allowances are made. With the level of manufacturing control applied by the world's major pharmaceutical and device producers and the resulting strict compliance with appropriate limits, any patient risk from pharmaceuticals and devices as produced is a nonissue. The same consideration does not, unfortunately, apply to admixtures or specific circumstances of use. Thus, it seems that particle numbers to which a patient is exposed may be best further controlled not by limits on manufacturers, but by attention to use of injectables and devices and the control of admixture practice. It is imperative that further requirements proposed be technically well founded, considerate of best demonstrated practice in manufacture, and directed at areas in which there is a need for further regulation.

A significant number of standards are in place in various countries of the world that apply directly or indirectly to contamination control in the pharmaceutical industry. Others apply specifically to pharmacy operations. Generally, these standards may be divided into solution-borne and airborne particulate areas. A number of standards-setting organizations are involved, including the ISO, the AAMI, the USP, the Japanese Pharmacopoeia (JP), the European Pharmacopoeia (EP), and the ASTM (American Society of Testing and Materials). This chapter and the next will serve to identify some of the many standards already in force and discuss some of the technical issues in generating new requirements for medical device and product manufacture.

The application of GMP is as important to the control of particulate matter in product as it is in assuring consistent quality with regard to other product parameters. Applicable GMPs for various countries and economic unions are spelled out either in governmental (regulatory) documents or in compendia (pharmacopeia). It is also common practice to refer to the "c"GMPs. The *c* stands for current, indicating applications of GMP that not only comply with the principles and provisions spelled out in the current issue of the appropriate regulations, but also incorporate best demonstrated manufacturing and quality assurance practices. Examples of best demonstrated practice in regard to particulate matter control can be found in several areas.

Let us consider the control of adherent particles on elastomeric closures by washing. The adherence to GMP might be effected simply by developing a validated distilled water wash cycle with a final Water for Injection (WFI) rinse prior to siliconizing and sterilization. Best demonstrated practice might also involve monitoring the particle burden of stoppers as received from the vendor, applying a wash cycle specifically designed to remove types of particles known to be present, and monitoring the effectiveness of particle removal. With regard to bulk drug raw materials for liquid filling, adherence to GMP might be represented by the tracking of key degradant profiles by a light obscuration counter as well as by chemical means. In addition, the manufacturer might monitor levels of extraneous particles in a raw material drug powder by microscopic testing and quantitate amorphous particles related to degradation by the same means. In any consideration, adherence to GMP and innovative application of appropriate technology to ensure product purity, efficacy, and low particle burden by manufacturers has decreased particle levels in injectables for the past 25 years.

A most important criterion is that standards development addresses a need. While at the time of Garvan and Gunner's first paper, particulate matter in Australia and some other countries was a serious issue, such is not currently the case. Several rationales have been presented to justify guidelines, limits, and concerns regarding particulate matter (see Groves and DeMalka 1976). Some reasonably argue that realistic particle standards should be consistent with the capabilities of existing technology and, in this sense, a measure of adherence to GMP; others believe that particle limits are necessary to control the cumulative particulate matter "insult" the patient receives. In the United States, the USP LVI and SVI requirements have been rationalized on both accounts. Numerous articles have been published concerning the size and numbers of particles in large-volume injections (LVIs) and small-volume injections (SVIs) (see Groves [1969] and DeLuca et al. [1986]). Various studies have utilized a variety of methods for counting particles, including the microscope and light obscuration, light scattering, and electrical zone-sensing instrumental particle counting techniques.

The technology and philosophy of the enumeration of particles in product and the assessment of product particle burden in comparison to the applicable limits has been a subject of great interest to manufacturers, regulatory agencies, and compendial groups. Historically, many of those concerned with particulate matter in pharmaceutical products have believed that the particle burden of LVI solutions is the principal source of particles to which patients are exposed. This belief is erroneous, but a series of compendial particulate matter standards for LVI products have been proposed and implemented in different countries. Some have been adopted and are currently enforced.

The summary history of the development of the present USP requirement for particulate matter presented below is interesting (see also Anderson and Higby 1995). The reader should note that any excerpted sections from compendia or requirements may not conform exactly to the original text. With particular regard to compendial test methods, the reader should not rely on the excerpted sections for any purpose other than a basis for discussion. Reference must be to the compendia or other published standards directly for any purposes relating to compliance.

> USP XIII (1905): Diphtheria antitoxin—clear or slightly turbid.
>
> NF VI (1936): Ampoule solutions . . . must be substantially free from precipitate, cloudiness, specks, fibers, cotton, hairs, or undissolved material.
>
> NF VII (1942): . . . *substantially free* shall be construed to mean a preparation that is free from *foreign bodies* that would be readily discernible with the unaided eye when viewed through a light reflected from a 100-watt mazda

lamp using as a medium a ground glass and a background of black and white.

USP XII (1942): Appearance of Solution or Suspension Injections which are solutions of soluble medicaments must be clear and *free of any* turbidity or undissolved material that can be detected readily without magnification when the solution is examined against black and white backgrounds with a bright light reflected from a 100-watt mazda lamp or its equivalent.

USP XIII (1947): Clarity of Solutions: Injections that are solutions of soluble medicaments must be clear and *substantially* free of any turbidity or undissolved material that can be detected readily without magnification when the solution is examined against black and white backgrounds with a bright light reflected from a 100-watt mazda lamp or its equivalent.

USP XV (1955): "Injections" <1>: Every care should be exercised in the preparation of all products intended for injection, to prevent contamination with microorganisms and foreign material. Good pharmaceutical practice requires also that each final container of injection be subjected individually to a physical inspection, whenever the nature of the container permits, and that every container whose contents show evidence of contamination with visible foreign material be rejected.

USP XIX (1975): Original Microscopic Test (<788>) Limits: The large-volume injection for single-dose infusion meets the requirements of the test if it contains not more than 50 particles per mL that are equal to or larger than 10 μm, and not more than 5 particles per mL that are equal to or larger than 25 μm in effective linear dimension.

Supplement No. 3, USP XX (1980): Note: for dextrose-containing solutions, do not enumerate morphologically indistinct material showing little or no surface relief and presenting a gelatinous or film-like appearance. Since in solution this material consists of units of the order of 1 μm or less and is liable to be counted only after aggregation and/or deformation on the membrane, interpretation of enumeration may be aided by testing a specimen of the solution with a suitable electronic particle counter.

USP XXI (1985): SVI requirement added to <788> for Light Extinction Counting—Interpretation: The small-volume injection meets the requirements of the test if it contains not more than 10,000 particles per container that are equal to a greater than 10 μm in effective spherical diameter, and/or 1000 particles per container equal to or greater than 25 μm in effective spherical diameter.

USP XXIII (1995): Summary total revision of USP XXII Version of <788>. New microscopic and light obscuration methods. Stratified test with light obscuration assay performed first. Limits decreased to lower level than those of the JP (see Table 3.1).

RATIONALES FOR STANDARDS

Regulatory agencies and the compendia have historically placed their reliance on limiting allowable numbers of particles as a means of achieving control. This approach is useful, and particulate matter limits worldwide have had some effect on increasing product quality with regard to particle content. However, this

Table 3.1. USP 23 <788> Limits for Particulate Matter in Injections

LVI (per mL basis)	≥ 10 μm	≥ 25 μm
Light Obscuration	25	3
Microscopic	12	2
SVI (per container basis)	≥ 10 μm	≥ 25 μm
Light Obscuration	6,000	600
Microscopic	3,000	300

approach is limited in that it does not represent the ultimate measure of control. Dependent on the degree to which a manufacturer incorporates uniformity as a feature of its process, units tested for compliance with limits may or may not accurately represent untested units of a batch.

Although more requirements or tighter limits may be attractive from a regulatory standpoint, the desired result with regard to an assurance of patient protection cannot be achieved by this means. The only sensible, practical, and totally effective control of particles in product is achieved by GMP (Akers 1985). GMPs, as specified by a large number of compendial and regulatory documents worldwide involve (1) the definition and operation of a manufacturing process that does not generate nonrandom particle populations and is thoroughly validated in that regard and (2) monitoring of the crucial parameters of that process to ensure that it continually operates within the established operating range. Well controlled, functionally validated processes allow the greatest degree of control over particles in product.

This philosophy of GMP control of particulate matter has significant implications regarding enforcement activities. One need only review historical and current FDA 483 observations to confirm that ample grounds for enforcement activity are provided by GMP regulations presently in force. If a process is validated to produce units with low levels of particles and is strictly controlled and adequately monitored, the product generated will, by definition, have a well-controlled particle burden. It is interesting to consider what value particulate limits really have. While limits may have the beneficial effect of restricting the activities of a few manufacturers operating poorly controlled processes, regulatory activities against such vendors will invariably be based on GMP issues rather than the actual particle burden of product. There is an unfortunate tendency on the part of some compendial groups to conceptualize tighter particulate limits as an assurance of higher product quality. Even more unfortunate is the tendency of some to view higher numbers of requirements and more restrictive limits as a measure of compendial stature. These views in no way promote patient well-being. In the contemporary pharmaceutical manufacturing environment, the manufacturer and its process protects the patient rather than compendial requirements and regulatory activity.

The standards proposed by the Department of Health in Australia following Garvan and Gunner's findings (1963, 1964) were as follows (Vessey and Kendall 1966):

- Not more than 1,000 particles/cm^3 at 2.0 μm
- Not more than 250 particles/cm^3 at 3.5 μm
- Not more than 100 particles/cm^3 at 5.0 μm
- Not more than 25 particles/cm^3 at 10.0 μm

In a visit to the United States, Dr. Gunner discussed the particle limit proposed in Australia as less than 250 particles/mL that are 3.5 μm and larger (Endicott 1965). Dr. Gunner verified that this figure was selected on the following basis:

- 3.5 μm particles are representative of the range of particles in any intravenous (IV) solution.
- A 3.5 μm particle is easy to count with presently available equipment.
- Controlled-size latex particles in this range can be obtained as a standard for this specification.
- "Good" manufacturers can attain this figure with proper care and technique.

Dr. Gunner emphasized that there was no clinical significance to the figure selected. The fourth item incorporated one of the first allusions to the principle of "bench-marking" or best demonstrated practice with regard to particles.

Ho (1967) found that large-volume parenteral (LVP) solutions could be made with far fewer particles than the proposed Australian standard with the current state of technology, production methods, and controls. He believed the Australian standard to be arbitrary because the answer to a seemingly fundamental question—How many particles and what size particles can be tolerated with safety in a parenteral solution?—is complicated by the following factors:

- The extremely low probability of eliminating particles perfectly on a commercial scale
- The state of the technology in pharmaceutical manufacturing and quality control
- The present uncertainty of the clinical effects of extraneous particles of all kinds

- The medical problem of long-term therapy involving a depleted patient receiving a voluminous supply of parenteral fluids

With regard to the simplest case, LVI solutions, the USP, British Pharmacopoeia (BP), and JP each specify allowable levels of extraneous particles that may be present on a per mL or per unit basis. This allowance is made in recognition of the fact that it is impossible to manufacture injectable products totally free of particles. Products manufactured under the strict conditions of GMP control in the United States, Europe, and Japan will invariably contain some amount of extraneous particles. Implicit in the compendial requirements is the philosophy that particles in excess of those levels resulting from the conscientious application of GMP are not allowable. Thus, although the compendial limits may differ significantly between countries, the concept of control is quite similar (DeLuca et al. 1988).

Particle size distributions have been reported for particles in a wide range of injectable products. It has been observed that sometimes there is a log-log relationship of particle size and number, and some workers have been able to summarize their data using the following equation (Groves and Muhlen 1987; Muhlen 1986):

$$\ln N = \ln N_{1.0} - M \ln D$$

where N is the cumulative number of particles at the threshold of diameter (D), $N_{1.0}$ is the value of N, D is 1.0 μm, and M is the slope of the log-log plot. Based on the result of these size distributions, a variety of limits have been suggested for both LVI and SVI solutions. While the premise of a log-log distribution of particles in injectables has received considerable support, it is unclear whether or not the extraneous particle population in injectables actually conforms to any single distribution. The data obtained with any method of particle enumeration will be inextricably linked to the method used. Thus, a given unit of injectable product will contain one distribution of particles by light obscuration, another by a Coulter® counter, and yet a third by microscopy. Thus, to make any general statements regarding particle distribution in an unqualified fashion is, at best, tenuous.

Further, light obscuration particle counters do not see all types of particles equally well, and the size assigned a particle will critically depend on shape and refractive index (Trasen 1969). Thus, the distribution of particles detected will depend on the particle type being counted. In fact, the historical data on which a log-log distribution is based shows a considerable variability of distribution and could also be shown to fit Poisson, negative binomial, log-normal, or other distributions. These data have, in some cases, been compared without regard to the precision or accuracy of individual data points or the types of counters used (Barber 1987). In some cases, light obscuration data have been "lumped" together with Coulter® data (Groves 1969). In short, it would appear that data have sometimes been interpreted in favor of theory in disregard of data interpretation and technical and scientific principles.

Unfortunately, most published material purporting to summarize the occurrence of particles in parenteral solutions has never shown enough data points to be widely indicative of marketed product. Some of the instruments used, especially light obscuration counters, show less than 50 percent sensitivity at the particle sizes counted in early (pre-1960) studies. Also, historical sampling techniques for instrumental counting (small aliquots and inadequate particle suspension) have been shown to alter particle distributions (Knapp and Abramson 1990). Counting injectable solutions with the multichannel light obscuration instruments currently available shows that significant discontinuities can exist between the wide-set thresholds used in previous studies. Given the widely known tendency of light obscuration counters to undersize large particles, it would be particularly unfortunate to base limits on the hypothetical log-log distribution, since these particles typically are not correctly represented in the database. Based on all of these considerations, limits precluding numbers above a certain value at a specific size appear to be a better choice for the present than any requirement based on an assumed distribution.

The limits sets of the world's most widely used standards are shown in Table 3.2. The 1966 proposed Australian standard is included for historical perspective. This standard emphasized smaller-sized particles, which, unfortunately, are detected less well than larger sizes. A happy circumstance at present is that worldwide efforts at harmonization of standards is bringing standards together, both in terms of numerical limits for particles and methodologies.

Table 3.2. Various Solution Particulate Matter Limits

		Limit Counts (per mL)					
Standard	**Counting Method**	**>2 μm**	**>3.5 μm**	**>5 μm**	**>10 μm**	**>20 μm**	**>25 μm**
US (USP 23)							
(LVI)	Microscopic	—	—	—	12	—	2
	LO*				25		3
(SVI)**	Microscopic				3,000		300
	LO*				6,000		600
Australia (proposed 1966)	Coulter	1,000	250	100	—	5	—
	LO*	—	—	—	—	—	—
United Kingdom (BP 1993)	LO*		—	100	50	—	—
Japan (JP XIII)							
(LVI)	Microscopic	—	—	—	20	—	2
European Pharmacopoeia (3rd ed.)	LO***	—	—	100	50	—	—
Pharmeuropa (1997 proposal)	LO*						
LVI					25	3	
SVI (liquid)**					6,000	600	
SVI (powder)					10,000	1,000	

*Light obscuration particle counting

**USP 23 SVI limits are on a per container basis

***Limits for dialysis solutions only

The latest issue of the USP (USP 23, 1995) and subsequent supplements resulted in the world's most comprehensive compendial particle test methodology (General Chapter <788>). It incorporates a two-stage test method with light obscuration testing followed by microscopic analysis, if necessary, to resolve failures. It also decreased allowable limits and revised the method of microscopic counting.

Happily, the procedure for establishing compendial requirements has been the subject of significant changes around the world. Most recently, the approach of obtaining technical input from the manufacturers of the products involved and instrument vendors in advance of the publication of the requirements proposals has been taken. The result has been requirements proposals that are technically sound and more appropriate to the product to which they apply. It has proven particularly critical to obtain the consensus of manufacturers, since the producers of pharmaceutical products invariably have a thorough understanding of the principles of analyses related to their products. In the United States, cooperative activities between the pharmaceutical industry and the USP, such as those involved in the development of the improved <788> tests for subvisible particles in solutions, suggest that this course will become a Standard Operating Procedure (SOP) in the future.

REQUIREMENTS FOR INJECTABLE PRODUCTS

The four most widely used requirements for injectables (USP, JP, BP, and EP) are summarized in this section. A detailed discussion of visual inspection is not included here, since this is the subject of a separate chapter (Chapter 7). A tabular comparison of the methods is provided at the end of this section (Table 3.3). An interesting function of the drive toward harmonization of drug and device regulatory requirements is that particulate matter requirements for injections worldwide seem to move closer together with each passing year.

The following subsections of this chapter include a number of test methodologies that are either in force presently or have been proposed for inclusion in compendia or in the standards of organizations such as the ISO. A discussion of the need for a solid technical basis in any requirement or standard is provided. The intent of including these test methods and discussions is twofold. First, the reader will be able to gain some understanding of the types of test methods that are currently in force and must be complied with as appropriate. Second, a comparison may be made between the structure and conduct of current methods and those that have been proposed. With regard to the latter, initial proposals often are technically flawed, containing both test methods and limits sets that are not appropriate. Thus, in the case of proposed tests and requirements, a critique is provided.

A caution that must again be emphasized here is that the pharmacopeial test methods excerpted (USP, JP, BP, and EP) are only partial (i.e., are incomplete). In some cases, the author has paraphrased or excerpted some portion of the full text. Thus, the complete method from the individual compendia (not these excerpts, which may be in error and are certainly out of context) must be used as a basis for SOPs and in-company standards.

British Pharmacopoeia, Appendix XIII Addendum 1996 (Figures 3.1 and 3.2)

Particulate Contamination

At present, the BP (1994, Addendum 1996) and the third edition of the EP contain test methods for both subvisible and visible particulate matter. Limits are provided only for the light obscuration test. The following commentary is made in the BP.

> As foreshadowed in the Introduction to the Addendum, 1994, the European

Figure 3.1. BP Light Obscuration Test Method.

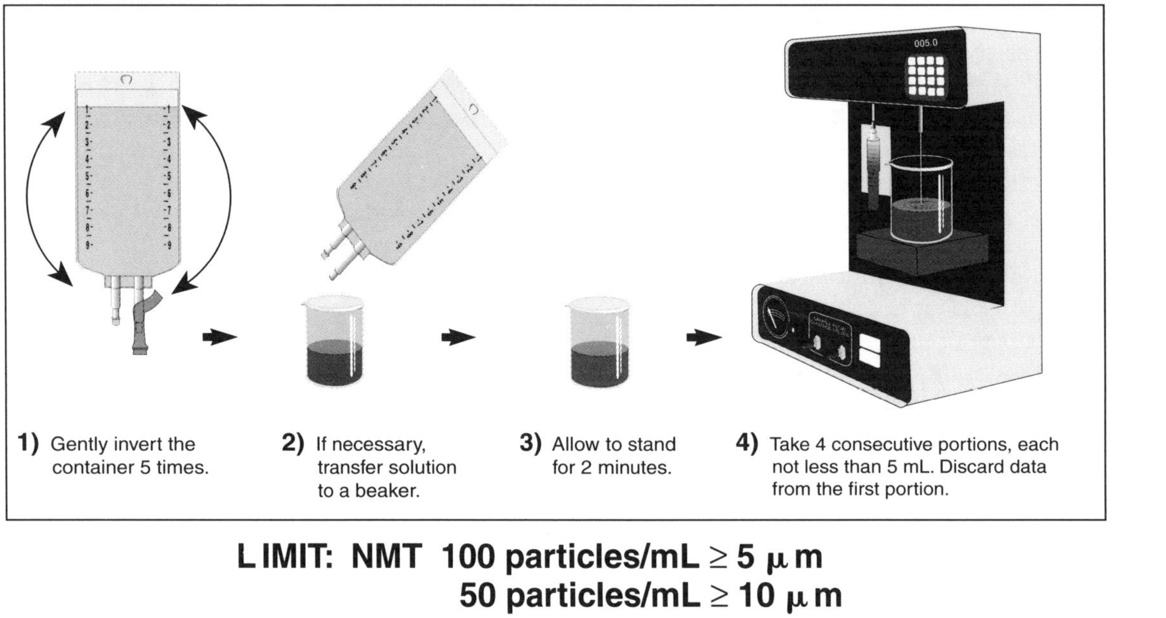
005.0
1) Gently invert the container 5 times.
2) If necessary, transfer solution to a beaker.
3) Allow to stand for 2 minutes.
4) Take 4 consecutive portions, each not less than 5 mL. Discard data from the first portion.
LIMIT: NMT 100 particles/mL ≥ 5 μm
50 particles/mL ≥ 10 μm

Figure 3.2. Apparatus for visible particles.

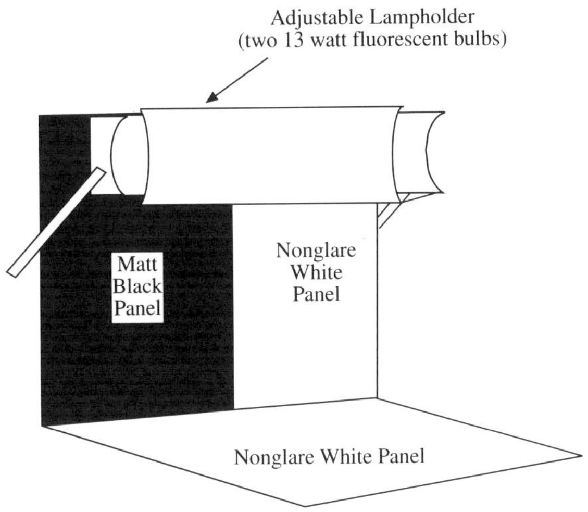

Pharmacopoeia text for particulate contamination has been expanded. In addition to the test for sub-visible particles (reproduced as Appendix XIII A in the addendum 1994 and invoked in monographs for certain preparations for parenteral use), the text now also includes a test for visible particles and a microscopic test. These additional tests have been included in this Addendum as Appendices XIII B and C respectively.

The test for visible particles describes standardized viewing conditions but, as yet, sets no criteria of acceptance. Work is continuing to reach agreement on criteria suitable for application as a pharmacopoeial "check" test that would provide a control analyst with a simple procedure for the assessment of the quality of parenteral solutions. It is emphasized that the test is not intended for use by a manufacturer for batch-release purposes. It is expected that a manufacturer would obtain assurance of the quality of his product with respect to visible particulate matter by 100% inspection and rejection of unsatisfactory items before release or by other appropriate means in accordance with good pharmaceutical manufacturing practice (GMP).

In contrast, it should be noted that it is not intended to invoke Appendix XIII C in monographs of the Pharmacopoeia; the microscopic test is intended to provide a qualitative method for identifying and characterizing any particles that might be present in a solution. Such methodology may be useful to a manufacturer, for example, when investigating the possible source of contamination of a product with a view to developing the means to avoid it.

A. SUBVISIBLE PARTICLES

Particulate contamination of injections and parenteral infusion consists of extraneous, mobile undissolved particles, other than gas bubbles, unintentionally present in the solutions. The type of preparation for which this test is required and the requirements to be applied are stated in the relevant monograph.

Apparatus. Use a suitable apparatus based on the principle of light blockage which allows an automatic determination of the size of particles and the number of particles according to size. Calibrate the apparatus using dispersions of spherical particles (EPCRS) of known size between 5 μm and 25 μm. These standard particles are dispersed in particle-free water. Care must be taken to avoid aggregation of particles during dispersion.

Author's note: The EPCRS standards may be unsuitable for the intended purpose due to not being available in a sufficient range of sizes, not having a "monosize" distribution, and limited availability.

General Precautions. The test is carried out under conditions limiting particulate contamination, preferably in a laminar-flow cabinet.

Clean, particle-free glassware may be obtained by very carefully washing the glassware and filtration equipment used, except for the membrane filters, with a warm detergent solution and rinsing with abundant amount of water to remove all traces of detergent. Immediately before use, rinse the equipment from top to bottom, outside and then inside, with particle-free water.

Take care not to introduce air bubbles into the preparation to be examined, especially when a fraction of the preparation is being transferred to the container in which the determination is to be carried out.

In order to check that the environment is suitable for the test and that the glassware is properly cleaned and to verify that the water to be used is particle-free, the following test is carried out. Determine the particle contamination of five samples of particle-free water, each of 5 ml, according to the method described below. If the number of particles of 10 μm or greater size exceeds 25 for the combined 25 ml, the precautions taken for the test are not sufficient. The preparatory steps must be repeated until environment, glassware and water are suitable for the test.

Method. Mix the containers of the sample by slowly inverting the container five times successively. If necessary, cautiously remove the sealing closure. Clean the outer surface of the container opening using a jet of particle-free water and remove the closure, avoiding any contamination of the contents. Eliminate gas bubbles by allowing to stand for 2 minutes.

Remove four portions, each of not less than 5 ml, and count the numbers of particles with sizes equal to or greater than the limits specified below or in the individual monograph, as appropriate. Disregard the result obtained for the first portion, and calculate the mean number of particles for the preparation being examined.

For preparations that are required to comply with this test, the mean number of particles does not exceed 100 per ml greater than 5 μm and does not exceed 50 per ml greater than 10 μm, unless otherwise stated in the monograph.

B. VISIBLE PARTICLES

Particulate contamination of injections and parenteral infusions consists of extraneous, mobile undissolved particles, other than gas bubbles, unintentionally present in the solutions.

The test produces a simple method for the detection of visible particles. It is performed in accordance with the provisions of good manufacturing practice.

Apparatus. The apparatus (see Fig. 59 [Figure 3.2 of this chapter]) consists of a viewing station comprising:

(a) a matte black panel of appropriate size held in a vertical position,

(b) a non-glare white panel of appropriate size held in a vertical position next to the black panel,

(c) an adjustable lampholder fitted with a suitable shaded, white-light source and with a suitable light diffuser (a viewing illuminator containing two 13-watt fluorescent tubes, each 525 mm in length is suitable). The intensity of illumination at the viewing point is maintained between 2000 lux and 3750 lux, although higher values are preferable for colored glass and plastic containers.

Method. Gently swirl or invert each individual container, ensuring that air bubbles are not introduced, and observe for about 5 seconds in front of the white panel. Repeat the procedure in front of the black panel.

C. MICROSCOPE METHOD

Particulate contamination of injections and parenteral infusions consists of extraneous, mobile undissolved particles, other than gas bubbles, unintentionally present in the solutions.

The test is intended to provide a qualitative method for identifying any particles that may be present in a solution and for determining their characteristics. This may provide an indication of the possible origin of the contamination. From such information it may be possible for a manufacturer to develop means to avoid the contamination.

General Precautions. Carry out manipulative procedures in a laminar-flow cabinet.

Preparation of Materials. Very carefully wash the glassware and filtration equipment used, except for the membrane filters, with a warm detergent solution and rinse with abundant amounts of water to remove all traces of detergent. Immediately before use, rinse the equipment from top to bottom, outside and then inside with particle-free water.

Apparatus. Use a stainless-steel or glass vacuum-filtration system equipped with a grid membrane filter having a suitable porosity and color.

A binocular microscope equipped as follows may be used:

(a) an achromatic objective with a magnifying power of 10,

(b) eyepieces with a magnifying power of 10, of which at least one is equipped with a reticle allowing the accurate measurement of particles 10 μm or larger in size,

(c) an illumination system for observation with incident or reflected light,

(d) an object micrometer to calibrate the eyepiece reticle.

Method. Assembly of the membrane filter and the filtration apparatus: Rinse the funnel and the filter holder base with particle-free water; pick up a grid membrane filter with forceps that have previously been washed; rinse both sides of the filter with particle-free water by holding the filter in vertical position and sweeping the stream slowly back and forth from top to bottom to eliminate all particles. Place the rinsed filter, grid upward, on the filter holder base of the vacuum-filtration apparatus, making sure that the filter is well centered on the holder base. Install the filtering funnel on the base without sliding the funnel over the grid side of the filter. Invert the assembled unit and rinse the inside of the funnel and the filter surface with a jet of particle-free water for about 15 seconds. Place the unit on the filter flask.

Filtration of the sample: Transfer 25 ml of the homogenized preparation to be examined to the filtration apparatus. Allow to stand for about 1 minute. Apply the vacuum and filter. Release the vacuum gently and rinse the inner walls of the funnel with a jet of particle-free water. Avoid directing the jet onto the filter surface. After turbulence has dissipated, vacuum-filter the rinsing. Maintain the vacuum for a moment to dry the filter. Release the vacuum. Carefully remove the funnel; pick up the filter with steel forceps previously washed with a jet of particle-free water and place the filter in a clean Petri dish. Place the dish in the laminar-flow cabinet to dry the filter slightly, avoiding the formation of folds, and then place the Petri dish on the microscope stage and count the particles present on the filter as described below.

Preparation of the blank: Prepare a blank under the same conditions as the sample. The blank consists of 25 ml of particle-free water passed directly through the funnel of the vacuum-filtration apparatus.

Calibration under the microscope: Calibrate the reticle of the microscope eyepiece using an object micrometer.

Determination: Either visual examination or an electronic recording device can be used. Examine the entire surface of the filter at a magnification of 100 and illuminated with incident light or reflected light. Count and classify the particles according to the sizes previously chosen, ≥10 μm. For each class, subtract the blank count from that of the sample. If the blank contains more than five particles 25 μm in size or larger, the operational conditions are unsatisfactory and the test must be repeated.

Interpretation: Where possible, identify the type of particles detected and determine their characteristics. Where relevant, the number of particles detected can also be recorded.

Japanese Pharmacopoeia XIII (Figure 3.3)

Foreign Insoluble Matter Test for Injections

Method 1. This method is applied to injectable solutions, or solvents for drugs to be dissolved before use. When the outer surface of the container is cleaned, injectable solutions or solvents for drugs to be dissolved before use must be clear and free from foreign insoluble matter that is readily detectable when inspected with the unaided eye at a position of luminous intensity of about 1000 luxes, right under an incandescent electric bulb. As for injections contained in plastic containers, the inspection is performed with the unaided eye at a position of luminous intensity of 8000 to 10,000 luxes, with incandescent electric bulb placed at appropriate distances above and below the container.

Method 2. This method is applied to the preparations to be dissolved before use. Clean the outer surface of the container, and dissolve the contents in the attached solvent or in water for injection carefully, avoiding contamination of extraneous substances. The solution must be clear and free from foreign insoluble matter that is clearly detectable when inspected with the unaided eye at a position of luminous intensity of about

Figure 3.3. JP XIII Test Method.

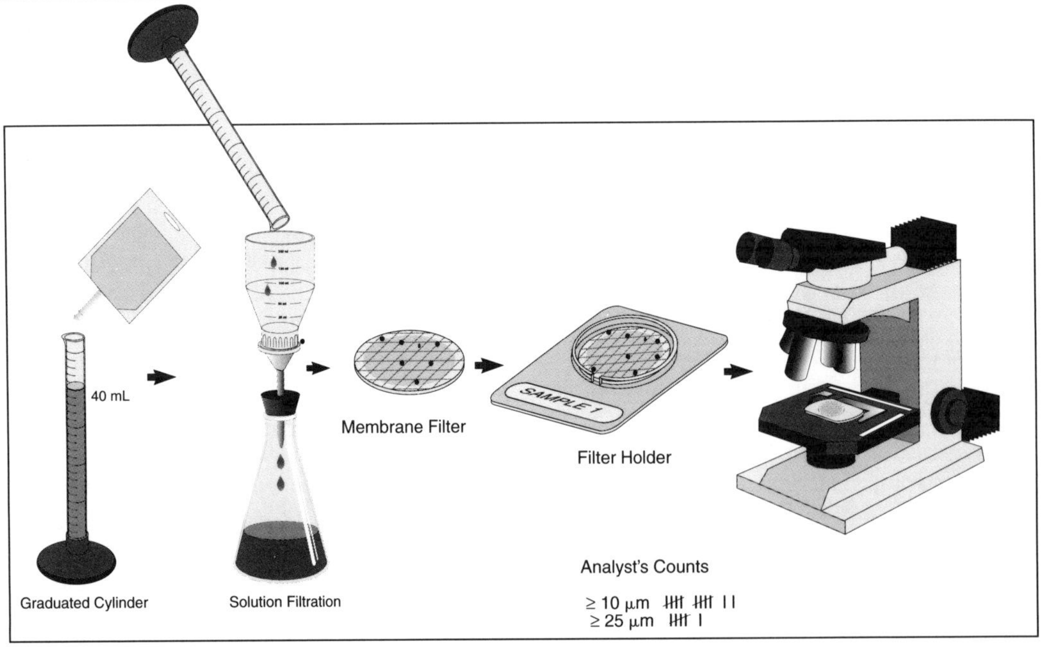

LIMIT: NMT 20 Particles/mL ≥ 10 μm
2 Particles/mL ≥ 25 μm

1000 luxes, right under an incandescent electric bulb.

Insoluble Particulate Matter Test for Injections

The test prescribes limits for the size and the number of insoluble particulate matter. The limits are not more than 20 particles per ml that are equal to or larger than 10 μm and not more than 2 particles per ml the are equal to or larger than 25 μm.

Apparatus. Use a microscope, filter assembly for trapping insoluble particulate matter and membrane filter for determination.

Microscope: The microscope is equipped with a micrometer system, a mobile stage and an illuminator and is adjusted to 100 magnifications.

Filter assembly for trapping insoluble particulate matter: The filter assembly for trapping insoluble particulate matter consists of a filter holder and a clip that are made of glass or other materials that do not affect the test, can be fitted with 25 mm diameter membrane filter for determination, and can be used under reduced pressure.

Membrane filter for determination: The membrane filter for determination is white in color, 25 mm in diameter, 0.45 or 0.5 μm in nominal pore diameters and is imprinted with about 3 mm grid marks. When examined in advance, the insoluble particulate matter on the filter is not more than 5 particles that are equal to or larger than 10 μm, and particles that are equal to or larger than 25 μm should not be found. When necessary, wash the filter with purified water for particulate matter test.

Reagent. *Purified water for particulate matter test:* Purified water that contains not more than 10 particles that are equal to or larger than 10 μm in 100 ml. Prepare before use by filtering through a membrane filter with a nominal pore size of 0.5 μm or less.

Procedure. Carry out all operations carefully in clean equipment and facilities that are low in dust. Fit the membrane filter for determination to the filter holder, and fit them with the clip. After washing the inside of the filter holder with purified water for particulate matter test, filter by suction 200 ml of purified water for particulate matter test at a rate of 20 to 30 ml per minute. Apply the vacuum until the surface of the membrane filter is free from water, and remove the membrane filter. Place the filter in a flat-bottomed Petri dish with the cover slightly ajar, and dry the filter fully at a temperature not exceeding 50°. After the filter has been dried, place the Petri dish on the stage of the microscope.

With the vertical incident light produced by a suitable illuminator, adjust the grid of the membrane filter to the coordinate axes of the microscope, adjust the microscope so as to get the best view of the insoluble particulate matter, then count the number of particles that are equal to or larger than 10 μm and equal to or larger than 25 μm within the effective filtering area of the filter, moving the mobile stage, and ascertain that the number is not more than 20. In this case the particle is sized on the longest axis.

Fit another membrane filter for determination to the filter holder, and fit them with the clip, then wet the inside of the filter holder with several ml of purified water for particulate matter test. Clean the outer surface of the container, and mix the sample solution gently by inverting the container several times. Remove the closure carefully and pour out about 50 ml of the solution with gentle mixing in such manner as to wash the opening of the container. Immediately after that, measure 40 ml of the solution using a measuring cylinder, that has been rinsed well with purified water for particulate matter test, and pour it into the filter holder along the inner walls. Apply the vacuum and filler mildly so as to keep the solution always on the filter. As for viscous sample solutions, dilute the solution about three-fold with purified water for particulate matter test, and then filter as described above. When the amount of the solution on the filter becomes small, add 30 ml of purified water for particulate matter test in such manner as to wash the inner walls of the filter holder. Repeat the process 3 times with 30 ml of the water. Apply the vacuum gently until the surface of the membrane filter is free from water. Place the eater in a Petri dish, and dry the filter fully at a temperature below 50° with the cover slightly ajar.

After the filter has been dried, place the Petri dish on the stage of the microscope, and count the number of particles that are equal to or larger than 10 μm and equal to or larger than 25 μm within the effective filtering area of the filter according to the same procedure of the microscope as described above. In this case the particle is sized on the longest axis.

U.S. Pharmacopeia 23 (1995)

The USP 23 version of General Chapter <788> includes a revised presentation of the light obscuration assay and an entirely new method of microscopic particle counting. This method is described in detail in chapter 4 and will not be dealt with at any length here.

The improved microscopic assay represents an entirely new technical procedure with respect to the method in earlier versions of the USP. It is carefully designed to provide greater precision in sizing and enumeration of particles and to detect types of particles not visualized under the microscopic conditions of the previous test. While many advantages accrue to the use of the new test, it is fairly complex in execution, and the initial attempts of an inexperienced analyst may lead to many questions and uncertainties.

There are four elements of the new test that represent dramatic departures from the old USP microscopic test:

1. The use of a combination of incident oblique and episcopic illumination
2. Filtration of the entire volume of a unit being tested
3. The use of equivalent circular area particle sizing
4. Application of either total counting of the particles collected on a membrane filter or statistical (partial) counting in the event that numbers of particles are too numerous to count are present

A summary outline of the BP, JP, and USP tests and requirements is provided in Table 3.3.

OPHTHALMIC SOLUTIONS

In the United States, a proposal has also been put forth for requirements on the particle burden of ophthalmic solutions (*Pharmacopeial Forum* 22:2525–2527, July–August 1996). The test method is analogous to the USP XXII <788> microscopic test method; a suggestion to use the newer USP 23 microscopic method has been made by industry representatives.

Test Preparation—After mixing by inverting the containers 25 times within 10 seconds, or as needed to suspend any particles, open and combine the entire contents of not less than 10 containers to obtain a volume of not less than 25 mL. Transfer the combined contents to the funnel. Allow to stand for 1 minute, and apply the vacuum and filter.

Determination—Examine the entire membrane filter in a suitable microscope under 100× magnification with the incident light at an angle of 10° to 20° with the horizontal. Count the number of particles having effective linear dimensions equal to or larger than 10 μm and equal to or larger than 25 μm. Also observe any particles having an effective linear dimension larger than 50 μm. Using the Membrane Filter and Assembly, perform a blank determination as directed under Test Preparation, beginning with "wash the inner walls of the funnel with a jet." Subtract the total counts obtained in the blank determination from the uncorrected total counts obtained for the Test Preparation.

Table 3.3. A Comparison of USP, JP, EP, and BP Particulate Matter Requirements

Requirements	USP 23 and Supplements	EP 3rd Ed. 1997, BP 1993, Addendum 1996	JP XIII 1996
Particulate Matter	<788> Particulate Matter in Injections	Particulate Contamination: Subvisible Particles; Visible Particulate; Microscope Method	Part I. 11. Injections, p. 7 (JP Forum: Vol. 5, No. 2 [1996], pp. 3 and 7–11)
Sections	Light Obscuration Particle Count Test, Microscopic Particle Counting Test	Subvisible Particles, Visible Particles, Microscope Method I	Foreign Insoluble Matter Test for Injections, Insoluble Particulate Matter Test for Injections
Presentation	Light obscuration particle count test is described in greater detail compared to others, includes provisions for calibration, resolution tests. Microscopic test incorporates a novel method of particle sizing using circular diameters similar to British Standard graticule. Importance of a statistically sound sampling plan based upon known operational factors and analyst training are emphasized.	Minimal method descriptions sufficient to perform the tests	Greater detail in test description than EP, BP.
Light Obscuration Test	The test applies to large-volume injections labeled to indicate a content for use in excess of 100 mL. Applies also to single-dose or multiple-dose small-volume injections labeled as containing 100 mL or less that are either in solution or in solution reconstituted from sterile solids. Injections packaged in syringes and cartridges are exempt from this requirement. Test apparatus: An electronic, liquid-borne particle counting system is used. It is not possible to specify exact methods. Instrument standardization: The following standardizations should be performed, including sample volume accuracy, sample flow rate, calibration (with a minimum of 3 calibrators), sensor resolution, and particle counting accuracy.	Subvisible particles: Automatic particle counting method (light obscuration) is described for determination of size and number of particles according to size by light blockage. Calibration with dispersions of spherical particles CRS with known particle size between 5 and and 25 µm in particle free water. The particulate contamination of five samples of particle-free water is determined. Remove four portions of the test sample (each of not less than 5 mL) for size determination, the result of the first portion is disregarded.	No comment re: light obscuration test for particles in injections; light obscuration test is prescribed as test for empty plastic infusion bags that will count immiscible plasticizer microdroplets.

Continued on next page.

Continued from previous page.

Requirements	USP 23 and Supplements	EP 3rd Ed. 1997, BP 1993, Addendum 1996	JP XIII 1996
Limits for Subvisible Particles	The injection meets the requirements of the test if the average numbers of particles present in the units tested do not exceed the appropriate value listed in Table 1. If the average number of particles exceeds the limit, test the article by the Microscopic Particle Count Test (Table 2) (pp. 1816 and 1819). Table 1: Light Obscuration Assay Count Limits SVI (per container) ≥ 10 μm = 6,000 ≥ 25 μm = 600 LVI (per mL) ≥ 10 μm = 25 ≥ 25 μm = 3 Table 2: Microscopic Method Particle Count Limits SVI (per container) ≥ 10 μm = 3,000 ≥ 25 μm = 300 LVI (per mL) ≥ 10 μm = 12 ≥ 25 μm = 2	BP limits for injectable solutions; EP limits for dialysis solns only Limits: Light Obscuration LVI (per mL) ≥ 5 μm = 100 ≥ 10 μm = 50	Unless otherwise specified, injections meet the requirements of the Foreign Insoluble Matter Test for Injections. Furthermore, aqueous infusions having a capacity 100 mL or more meet the requirements of Insoluble Particulate Matter Test for Injections (p. 8). Foreign Insoluble Matter Test for Injections Insoluble Particulate Matter Test for Injections (LVI) ≥ 10 μm = NMT 20/mL ≥ 25 μm = NMT 2/mL
Visible Particles	General descriptions only (in General chapter <1>). All units must be visually inspected and any showing evidence of contamination with visible particles must be rejected. Reconstituted dosage forms must be "essentially free" of visible particles. Conditions of inspection not specified.	The unit is first tested before a white "panel" to visualize the particles. The procedure is repeated before a black panel. The arrangement of the background panels is described in the method writeup in the compendium and a drawing of an inspection "booth" is provided. Intensity of illumination is between 2000 and 3750 lux.	Foreign Insoluble Matter Test for Injections. Two methods are described: Method 1 is applied to liquid dosage forms; method 2 is applied to reconstituted solutions. Both use an intensity of illumination of 1000 lux.

> **Interpretation**—Duplicate Test Preparations and blanks may be examined as directed. If the blank determination yields more than 5 particles having effective linear dimensions of 25 μm or greater, the operational environment is unsatisfactory and the test is invalid.
>
> The ophthalmic solution meets the requirements of the test if it contains not more than 50 particles per mL that are equal to or larger than 10 μm, not more than 5 particles per mL that are equal to or larger than 25 μm in effective linear dimension, and not more than 2 particles per mL larger than 50 μm.

Author's note: These limits and the linear dimension measurement are from USP XXII <788>; limits at the ≥ 50 μm have been added. Also, it seems that the implementation of this requirement will be delayed to allow pharmaceutical laboratories to perform a round-robin study on current and expired product.

STOPPERS—DRAFT AMENDMENT TO ISO 8871

Stoppers have historically been a component of injectable product that was suspected of contributing the majority of extraneous particles found in many vial or bottle products. At present, all major stopper manufacturers and many users perform particulate matter testing on stoppers, both in-process and for final product. Stoppers are offered for purchase at various cleanliness levels. Importantly, the test methods applied by manufacturers are technically sound and are specifically suited to the process and elastomeric closure type of interest.

The generalized tests in the proposed annexes of the ISO standard are at present, in contrast, technically lacking and would be less well suited for application to closures than the specific tests that manufacturers already have in place. The mechanics of the tests will hopefully be improved based on further working group discussions. Below are excerpted portions of the proposed annexes of the identified ISO proposed standard that are deserving of specific critical comment:

> **(1) Introduction**
>
> The pharmaceutical industry requires to an increasing extent concrete details from the rubber manufacturer about the presence of particles the closures may release to the injectable in case that closures are used as primary packaging materials in direct contact with pharmaceutical preparations. This request has been taken into account by elaborating annex N and annex O.

As indicated previously, such "details" can already be obtained from stopper manufacturers who apply appropriate tests to their product. The author is not aware of any "requests" for tests such as those specified in annexes N and O; the generation of the requirement seems to have been prompted by the erroneous presumption on the part of individuals unfamiliar with pharmaceutical manufacture that there is a need for these requirements.

> **ANNEX N**
>
> (2) N.2.1. Preparations
>
> N.2.1.1. Carry out all operations in such an environment that no extraneous particles can interfere. This involves wearing suitable garments and using a suitable clean air work station, e.g. providing laminar flow to Class 100 (Federal Standard 209D of the USA), as well as suitable decontaminated tools and handling means.

The word *garments* does not specify gloves—the single most important part of analyst protection. FS-209D should not be referenced; it has been superseded by FS-209E. The "preparation" section totally disregards precautions to avoid contamination of the stoppers by handling. As anyone familiar with the testing of closures knows, improper handling, transport, or sample collection will render the test result invalid.

> (3) N.2.1.2. Prepare a rinse fluid by dissolving 3 g of commercially available highly concentrated sodium N-methyl-N-oleyltaurate powder in 10 litres . . .

Selection of detergent should depend on the specific material of which the closure is made.

A single detergent most likely will not provide optimal removal of particles for all types of elastomers.

> (4) N.2.2. Test

> N.2.2.1. In a cleaned Erlenmeyer flask place a number of intact parts to be tested, with a total surface of approximately 100 cm^3.

Any meaningful assessment of particle burden on closures must be made on the basis of particles per part. A measurement of particles per surface area is approximate only if the material tested is used on a surface area basis (e.g., sheeting for plastic containers). The conversion to stopper surface area is unnecessary, provides a source of error, and implies a nonexistent commonality between types and sizes of stoppers.

> Add 50 ml of the filtered rinse fluid according to N.2.1.2. Shake for 20 seconds by means of a machine imparting motion in the form of a horizontal circle of 12 ± mm diameter at 300 to 350 min^{-1}.

First, 50 mL of rinse fluid will not have a uniform cleaning effect on varying numbers of stoppers of different sizes. Stoppers should be tested in a system specifically designed for a given stopper size. Different container sizes and rinse volumes are necessary for each stopper size. Second, circular agitation of the type described will result in the stoppers being entrained in a circular orbit with the solution flow in the flask. The appropriate agitation consists of a horizontal back-and-forth motion that results in a rapid reversal of the direction of lid and stopper movement, which is highly effective in dislodging particles.

> Filter the rinse fluid over a light gray membrane of maximal pore size of 0.8 μm printed with a grid of 3 mm squares.

The filter used must be as dark as possible to enhance contrast of light-colored particles, which are predominant on closures. Black filters (that appear gray when viewed with a microscope) are available and preferable.

Note: The inclusion of a "coordinate" system as described in the draft for filter color serves no useful purpose and implies a precision that is impossible to attain with the test as described. This is particularly true when "between parties" arrangements are allowed, which can change the whole outcome of the test and make any meaningful comparisons of data impossible.

> (5) N.2.2.2. Prior to making the test prepare a blank filter . . .

This reflects poor organization of the test description; this section should appear before—not after—the analyses section.

> (6) N.2.2.3. Count the particles on the filters using a suitable microscope . . .

Uniform conditions of microscopic observation must be specified if uniform particle detection and sizing is desired. The requirement as written (i.e., a "suitable" microscope) would allow microscopes with a wide range of magnification and resolution to be applied. This variation in conducting the test would make the collection of meaningful data impossible. Calibration of the microscopes used in the test, another essential component, is omitted.

> Divide the particles into the following categories, using the longest visible dimension as the classifying parameter:

> Class I: larger than 25 μm and smaller than or equal to 50 μm

> Class II: larger than 50 μm and smaller than or equal to 100 μm

> Class III: larger than 100 μm

The threshold of "visible" size range in liquid pharmaceutical products and on closures is not universally agreed on but is between 50 and 100 μm depending on critical variables such as lighting, time of inspection, and training level of inspectors. Class I size particles cannot be reproducibly detected by visual inspection. "The longest visible dimension" does not adequately specify what is to be measured. Means of measurement are also not specified.

ANNEX O

(7) O.2.1. Preparations

O.2.1.1. Use an automated electronic particle counter based on light blockage or electrical resistance measurements, which is capable to discriminate between the following particle size categories, compared to spherical particles as a standard:

The use of an electronic particle counter to enumerate particles from medical devices and pharmaceutical container components has been repeatedly rejected by compendial groups, regulatory agencies, and manufacturers. The instrument not only undersizes particles but is subject to a multitude of count errors due to air bubbles, refractive index effects, and immiscibles. Undersizing is most pronounced for fibers, which is the single most important class of particles on stoppers. For electrical resistance instruments, an ionic electrolyte solution must be used, resulting in additional labor and another blank procedure. Also, serious objections must be raised to the redundant performance of the Annex O test when performance of the Annex N test would result in a membrane sample on which articles 5 μm and larger could be counted microscopically. The additional 2 μm count data that would be collected using electronic counters is generally of little value, due to the high level of count artifact occurring at this size when stoppers are tested.

This is an example of a standard section in early stages of development. The tests described in Annex N and Annex O of the proposed amendment to ISO 8871 will not provide the information they are designed to collect and are not necessary, having been preempted by tests that manufacturers of elastomeric components and pharmaceutical products already have in place. The methods specified would result in highly variable and sometimes invalid data, which would bear no definable relationship to a manufacturing process. Tests such as those proposed in the annexes would fail to detect materials leached from stoppers, since leaching occurs over time and is not typically present during or immediately after manufacture; the generation of solid particles thus occurs during the shelf life of a product. In like manner, particulate matter added during shipping and handling subsequent to manufacturing would not be detected by the proposed test. Sample handling and sample selection (i.e., how a valid sample is collected from a batch of closures), which are the two most important parameters for this type of test, are not even addressed. In view of the facts that (1) closure manufacturers already implement tests for particulate matter, (2) GMP controls for both finished product and component particle burden are in place, and (3) the tests described in Annex N and Annex O are technically invalid, another effort at this standard seems justified.

REQUIREMENTS AND TESTING FOR MEDICAL DEVICES

During the past decade, considerable interest has been expressed by regulatory and compendial groups in the United States, Canada, the United Kingdom, and Europe regarding the need for particulate matter requirements for medical devices. Most recently, the USP has proposed a requirement for disposable blood bags used for the collection and separation of blood components. A number of widely accepted requirements for devices (e.g., those of AAMI and ISO) are already in place, and it seems unlikely that additional requirements are needed. The majority of the few documented occurrences of high levels of particulate matter in specific devices are based on large (visible-sized) particles. The most appropriate test method would thus appear to be a microscopic assay performed on the total effluent volume of a flush solution passed through a device. The principles outlined in Chapters 1 and 2 regarding monitoring and GMP control of environmental and process-related particles also apply to medical device manufacture. Similarly, the sections regarding particulate matter analysis and enumeration by various means are applicable to devices.

The generic categorization of "medical devices" includes an extremely wide range of injectable and noninjectable medical materials. Some are electrical, electronic, and/or electromechanical and have little or no association with injectable products. Particulate matter is most certainly not an equal concern for all device types. The discussion earlier in this chapter has been directed almost entirely toward injectable materials. This is reasonable, since the majority of published information on the subject of particulate matter in medical products pertains to LVI and SVI. The fact that less has been published regarding particles in medical devices is reflective of the healthcare professional's generally decreased level of concern over the infused particle burden arising from this component of IV therapy. There are a number of suggested test methods (DIN, BSI, AAMI) but no compendial requirements. In the United States, the FDA expects that medical devices will be monitored for subvisible particulate matter as a principle of GMP.

The FDA, through the Center for Devices and Radiological Health (CDRH), groups devices in three general classes (I–III) based generally on their complexity and degree of risk and benefits. Class I devices require the least control, being simplest in design and having minimal potential for user harm. This category includes exam gloves, elastic bandages, and hand-held surgical instruments. Class II devices include powered wheelchairs, infusion pumps, and surgical drapes; they are subject to the general controls applied to Class I, as well as special controls such as labeling requirements and postmarket surveillance. Class II devices are subject to premarket notification and GMP. Class III devices are usually those that sustain human life and may have an inherent high risk level with regard to patient injury. These are subject to premarket approval or the 510(k) process, as well as general and special controls. Pacemakers, < 6 mm vascular grafts, and some implants are in this class, as well as many IV solution and blood contact devices.

Despite its status as a general world standards organization, rather than a specialized one, the ISO at this point has a number of standards for medical devices (see the reference listing). Many of these do not include particulate matter requirements. The key compliance criterion regards whether a specific requirement is "normative" or "informative"; normative sections of standards, at least in theory, specify a test method that is, in fact, a referee test, with limits that must be met to demonstrate compliance. An ISO standard may be adopted as normative by a national medical, compendial, or regulatory body so that it has the same status as compendial requirements; this possibility is a good reason for manufacturers of medical devices and pharmaceuticals to track ISO standards development and become an active part of the development process. The current ISO standards development process affords significant opportunities for the healthcare manufacturer to become involved.

The variety of medical products categorized as "devices" is so wide as to cause difficulties in any general discussion, or with regard to any uniform application of regulatory requirements. The author refers to the general device category as parenteral-type devices or solution contact devices. With regard to particulate matter, this broad grouping may be usefully subdivided into several subcategories. The most obvious subcategory includes solution transfer sets used in a pharmacy, solution and blood administration sets, syringes, and other drug delivery devices. A second subcategory includes oxygenators, autologous transfusion devices, blood warmers, and similar devices applied in surgical procedures. The third subcategory consists of chronic use devices, such as central venous catheters, transfer sets used with continuous ambulatory peritoneal dialysis (CAPD) therapy, and hemodialyzers.

Just as there is no evidence to indicate that the low levels of particles present in injectable products manufactured under GMP are a hazard to patients, there is also no indication that any significant patient health issues are related to particulate matter from parenteral-type devices. While a very limited number of studies have suggested that occasionally specific devices may have high particle burdens (Dimmick et al. 1976), recent comprehensive studies of Gour (1987) and Di Paolo et al. (1990) show that a wide range of parenteral-type devices have low solution contact path particle burdens.

In the Gour study, the majority of a wide range of tested devices met the acceptability criteria imposed by the investigator (a maximum of 25 particles ≥ 10 μm and 5 particles ≥ 25 μm per mL of eluent). Di Paolo et al. commented that IV administration sets are potential sources of substantially lower particulate contamination than the infused solutions themselves, and that an inline filter in a set functions effectively not only in removing particles from devices but also from the infusate solutions.

Since the aim of the latter study was to test relatively large numbers of various types of administration sets, an electronic particle counter was chosen. This enabled the authors to examine a large number of sets in a relatively short time. It should be pointed out that this type of equipment is not generally well suited for qualitative examination of devices for particulate matter, or for reliable counting of larger (≥ 25 μm) particles. Indeed, this consideration prompted the German Standards Institute (DIN) at one time to recommend the use of a light microscope to analyze particles in intravenous administration sets.

Parenteral-Type Devices

Presently, the primary concern of both regulatory agencies and manufacturers are devices that have a contact path for blood or solutions that will be infused directly into the venous circulation of a patient. Examples of devices of this type are as follows:

- Administration, recipient, blood sets
- Dialyzer, oxygenator inlet, and outlet sets
- Hollow fiber dialyzers
- Membrane oxygenators
- Pump cassettes
- Cell elimination filters
- Heat exchangers
- Empty plastic containers for IV solutions
- IV catheters
- Needles
- Central catheters

Many of the particles associated with medical devices originate at process assembly points; typically, a much smaller number comes from the environment. Each component part assembled into devices typically contributes some variable level of unique particles. Interestingly, because of the distinct sources of particles found in devices, mixed particle populations composed of subvisible 1–50 μm particles (representing the inherent burden of components), and small numbers of larger particles of visible size (predominantly process related) may occur. Currently manufactured parenteral-type devices span a range of complexity from single administration sets and extension sets to blood oxygenators, extracorporeal circuits, blood warmers, and autotransfusion devices. Large numbers of individual component parts may be included in devices of the latter types. Even simple devices may incorporate parts made from several different types of polymeric material; many manufacturers test individual components as a means of qualifying processes or vendors.

With parenteral-type devices, the primary concern is particulate matter that can be washed from the solution contact path and thus be infused into a patient along with the particulate burdens derived from LVIs or SVIs, admixtures, and so on. Importantly, the particulate burden to which a patient receiving parenteral therapy is exposed is a function of the total process of solution administration, including the techniques of admixture, addition of drug, and vein access. Thus, limiting the patient particle loading cannot be reliably achieved by limiting particulate matter related to individual components of the system, such as parenteral solutions or administration sets. We cannot consider the particle burden delivered to the patient as dependent on any single, physical component of the system that delivers a parenteral solution.

The only completely effective approach to minimizing the exposure of patients to particles involves the use of in-line disposable filters for IV infusion systems. Terminal filters are effective in reducing patient exposure to all types of particulate matter resulting from the infusion process, not merely the components present in parenteral devices, solutions, or infusion sets. Filters have proven useful in some cases in reducing the incidence of phlebitis related to infusate particle burden (Falchuk et al. 1985; Barnett et al. 1983; Rypings et al. 1990) discussed in Chapter 1. However, these few cases of patient benefit do not justify the wide-scale use of in-line filters. Current IV therapy practice, based on more recent research and in-house studies conducted in large hospitals, dictates the use of terminal filters only for some specific drugs and for total parenteral nutrition (TPN) or total nutrient admixtures (TNA).

With regard to the standards for injectable products, the reader will find a common thread between the JP, the USP, and the BP/EP in terms of the principles of the methods. The situation is not the same with device requirements, many of which are in the proposal stage. The user (i.e., medical device manufacturer) is, at present, forced to devise his or her own method, most of which are based on simple flushing of a device with a volume of clean water or saline consistent with a use volume and performing a microscopic light obscuration test on the effluent. Manufacturers, based on intimate knowledge of their product, are best qualified to develop such tests. The BSI test method proposed in 1989 but never officially implemented, the AAMI methods, and the HIMA–PMA (Health Industry Manufacturers Association–Pharmaceutical Manufacturers Association) test methods all provide useful starting points for those developing tests.

British Standards Institution Method (1986)

The BSI proposal, never implemented, describes a sound test method applicable to simple administration sets.

I. Sample

Test 5 sets of the product.

II. Apparatus

The following apparatus and materials are required:

1. Straight-sided borosilicate beakers of 200 ml graduated capacity
2. A top-loading balance, measuring in divisions of 0.1 g or less
3. A sodium chloride intravenous infusion BP containing 9 g of sodium chloride per liter
4. An instrument capable of counting the number of particles suspended in a known volume of electrolyte solution having equivalent sphere diameters equal to or greater than 2 μm and equal to or greater than 5 μm.

III. Method

A. Run approximately 100 ml of sodium chloride infusion into the beaker in small aliquots, rinsing and discarding the infusion.

B. Collect 150 ml of the infusion in the beaker and determine particle counts at 2.0 μm and 5.0 μm. Note: The test should not proceed until consistent conditions have been obtained, and periodic checks should be made to ensure that these criteria are maintained.

C. Empty the beaker and drain in the inverted position or use a fresh, clean beaker.

D. Handling the set as little as possible, remove the connector protector and squeeze the filter chamber before inserting the connector into the infusion supply. Invert the set before releasing the filter chamber so that air can escape down the set and is not trapped. It is essential that the flow control be opened to release air, and closed before reinserting so as not to lose fluid. The filter chamber should be full of sodium chloride infusion and the drip chamber half full. If more infusion is needed in the set, remove the connector, gently squeeze the filter chamber to remove air, replace the connector, and release the filter chamber.

E. Arrange the set so that the connector is 1.0 m to 1.1 m above bench height, and the beaker on the balance is 700 mm to 900 mm from the infusion source and drip stand.

F. Collect 200 g of infusion in the beaker at a drip rate of 9 g/min to 11 g/min, taking care to collect the first few drops issuing from the set.

G. Immediately determine particle counts at 2.0 μm and 5.0 μm on five replicate samples from the solution to the beaker using Coulter counter or light extinction (i.e., obscuration) counter.

HIMA–PMA Method (1980) (Figure 3.4)

Author's note: This method was originally used in an industry round-robin study in 1980; it has since been modified by a number of companies for use in-house in monitoring.

1. Test Preparation

A. Test Environment

All procedures should be performed in a laminar flow hood equipped with HEPA filters that are certified to meet Class 100 conditions. The laminar hood should operate 24 hours a day for reliable results.

Figure 3.4. HIMA–PMA test method for administration sets.

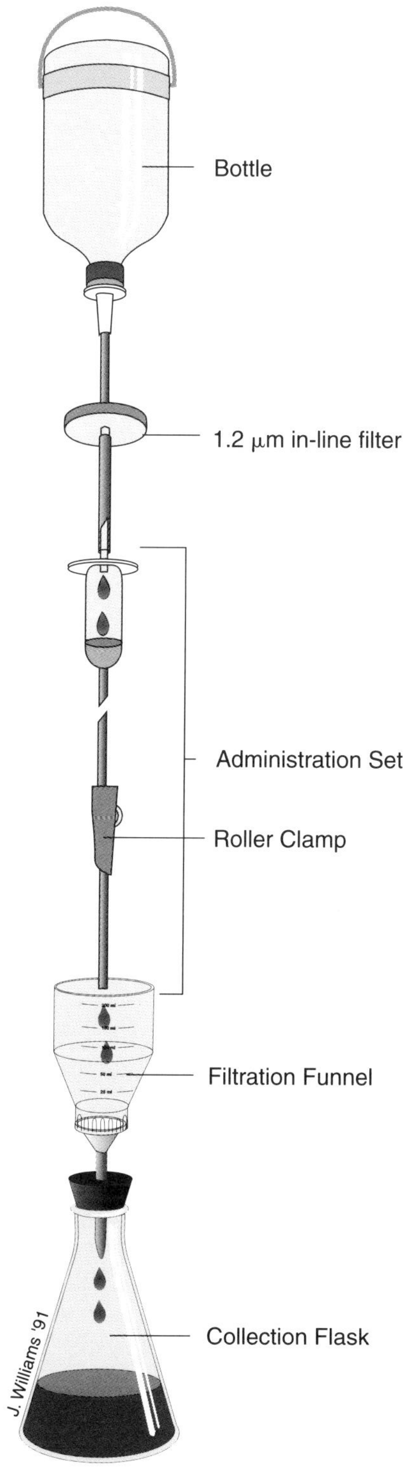

B. Apparatus Preparation

1. All solvents (water and isopropyl alcohol) used to clean the apparatus must be filtered through a membrane filter (with a maximum pore size of 2.0 µm) into the precleaned, pressure-dispensing vessels. The solvent dispensers are fitted with membrane filters as specified.

   a. Water—25 mm, 1.0 µm, white, plain

   b. Isopropyl Alcohol—25 mm, 1.0 or 2.0 µm, white, plain

C. Filter Holder Assembly

Clean the filtration apparatus and accessory items before use, as follows:

1. Wash filtration apparatus with a warm solution of liquid detergent dilute as per package instructions).
2. Rinse in 0.45 mm filtered water.
3. Before each use, ultraclean the filtration apparatus utilizing pressure vessels and a "gun" with a 0.45 mm filter.
4. Rinse with filtered isopropyl alcohol.
5. Follow with a final rinse of 0.45 mm filtered water.

D. Membrane Filter Assembly

1. Clean the inside top and bottom of the petri slide with filtered purified water. Set aside to dry in the laminar flow hood with the top of the petri slide ajar;
2. Rinse the blades of the unserrated forceps with a forceful stream of filtered purified water;

3. Using the cleaned forceps, remove a 0.8 μm black-gridded filter from the package. Holding the filter in a vertical position and starting at the top of the nongridded side, sweep a stream of filtered water back and forth across the surface, working slowly from top to bottom, so that the particles will be rinsed downward off the filter. Repeat the process on the gridded side;
4. Place the cleaned membrane filter (grid side up) on the support screen of the base. Place the ultracleaned funnel on the base and secure with the spring clamp;
5. Repeat the above steps for each membrane filter and assembly.

E. Reservoir Assembly

1. Attach an ultraclean piece (approximately 2″ in length) of clear plastic tubing to the flat end of an ultra-clean, long or short point connector;
2. Place a 1.2 μm (or smaller retention rating) filter into the 47 mm in-line filter holder and assemble the filter holder as required:
3. Insert the inlet of the in-line filter holder into the plastic tubing that is attached to the long or short point connector;
4. Attach an ultraclean piece (from 2–6″ in length) of clear plastic tubing into the outlet of the in-line filter holder;
5. Using a hemostat, clamp the plastic tubing that is attached to the long or short point connector (note: a roller clamp will also serve this purpose);
6. Insert the exposed end of the long or short point connector into the stopper of a 1000 ml IV bottle of Sterile Water for Injection, USP.

II. Method

A. Blank Analysis

1. A blank analysis should be performed on the apparatus rinse water at the beginning of the test period, and after each solvent-dispensing filter change. A blank analysis should be performed on the eluate water from the IV bottle setup at the beginning and end of every test day;
2. After assembling the test apparatus and filtration equipment, discharge and discard 100 ml of apparatus rinse water from the filter jet solvent dispenser;
3. Fill the filter holder with 50 ml of rinse water and vacuum through an 0.8 μm porosity gridded membrane filter.
4. Examine the membrane filter microscopically at 100× magnification for particulate matter. The particulate should not exceed 5 particles ≥ 10 μm, or 2 particles ≥ 25 μm, and no fibers.

B. Test Procedure

1. After assembling the test apparatus and filtration equipment, attach the test sample to the plastic tubing that is attached to the outlet of the in-line filter assembly.
2. Flush the lumen of the test sample with 50 ml of eluate water, and collect the solution directly into the filtration funnel that houses a 0.8 μm black-gridded membrane filter;

3. Turn on the vacuum and filter the solution until approximately 10 ml of eluate remain in the funnel;
4. Flush an additional 50 ml of water through the test sample and vacuum filter as above;
5. Repeat this procedure until 100 ml of filtered water has been flushed through the test sample;
6. After all 1000 ml of eluate have been filtered, rinse the funnel walls with 25 ml of apparatus rinse water, using care not to direct the stream of water onto the filter surface;
7. After turbulence has dissipated, vacuum filter the rinse water;
8. Turn off the vacuum and remove the filter carefully with ultraclean unserrated forceps, and place onto a petri slide using a piece of double sided tape;
9. Allow the filter membrane to dry in the laminar flow hood with the cover slightly ajar;
10. Upon drying, count the particulate contained on the membrane filters as follows:
    a. position the incident light illuminator at an angle of 10°–20° to ensure maximum definition for counting and sizing;
    b. using 100× magnification, count the particles for counting and sizing in the following size ranges: 10–<25 μm and ≥25 μm. Additionally, count and size fibers in a "fiber" category.

AAMI Method for Blood Filters (1989; reapproved 1993)

The AAMI has two standards that apply to the levels of particulate matter in devices. These pertain to blood filters and autologous transfusion filters. The appropriate parts of the former are excerpted below. The caution to a user of these methods to obtain complete copies from AAMI as mentioned earlier holds.

> **4.2.3.1 Filter Cleanliness.** All operations shall be performed in an operating, certified laminar flow hood equipped with high-efficiency particulate air (HEPA) filters. Install a dummy filter (i.e., a filter housing not containing any filter medium) in a fixture containing a test sample holder, fluid circulation apparatus, and an analysis membrane holder downstream of the test sample holder. The fluid shall be USP Water-for-Injection filtered through a 0.8 μm filter. Wash the test fixture and both sides of a black-gridded 0.8 μm pore size, blank analysis membrane using this fluid. Install the analysis membrane using smooth-tip forceps, and allow the fluid to flow through the dummy filter sample at a rate of 500 cc/min, or at the maximum pressure drop of 450 mm Hg (8.7 psi, gauge), for 5 min. Remove the analysis membrane, place it in a petri dish container with the cover slightly ajar, dry in a laminar flow hood, and count microscopically the particles collected on the membrane surface (as per Applicable Document 2.4 and/or 2.5). Repeat the procedure from the beginning using a test filter sample. The blank count may not exceed 10 percent of the maximum acceptable counts given in 3.2.3.1. Calculate the difference between the test and blank counts for all particles larger than 10 μm in diameter, larger than 25 μm in diameter, and for all fibers. All counts shall be less than or equal to those given in 3.2.3.1.

> **3.2.3.1 Filter Cleanliness.** The filter shall provide no more than the following effluent particle levels when subjected to the water or other solvent flush test of 4.2.3.1:

(1) 0.90 particle per milliliter of solvent flush larger than 10 micrometers.

(2) 0.35 particle per milliliter of solvent flush larger than 25 micrometers.

(3) 0.65 fiber per milliliter of solvent flush.

The test membrane will be placed at the effluent port, and the total volume of the system will be drained through the test membrane by gravity. A vacuum may be used to assist in filtration through the test membrane. After the analysis membrane is removed, it is placed in a petri dish container with the cover slightly ajar, and dried in the laminar flow hood. The particles collected on the membrane surface are counted microscopically. The particle/fiber counts for all particles larger than 10 μm in diameter, for those larger than 25 μm in diameter, and for all fibers is calculated as:

$$\text{Particle/Fiber Count} = \frac{\text{Count Test} - \text{Count Control}}{\text{Total Test Fluid Volume}}$$

Author's note: This method was originally developed for the testing of blood microaggregate filters containing a nonwoven mat of polyethylene terephthalate fibers. Thus, relatively high fiber limits are allowed—0.65 fibers per mL × 4,500 mL flush volume.

DEVICE STANDARDS PROPOSALS

In addition to test methods that are already in place, proposals for new requirements are being made. Many of these proposals initially reflect generally inappropriate considerations of the technical methods proposed and are unrealistic with regard to product currently being produced, which is proven to be safe and effective. It is extremely important that manufacturers respond to new method proposals so that the elements of any such monograph are pertinent and correct. They should submit their comments, which may support, oppose, or recommend modifications, to the proposed standards in writing. It is equally important the standards-setting groups receive and incorporate information into the standards. This feedback procedure will ensure that standards are reasonable and serve the intended purpose. Manufacturer input ensures that the requirements are relevant and commensurate with the needs for the proposed article. The definition of "requirement" should be clearly understood. A requirement specifies both a test method and a limits set for the test; it typically constitutes an evaluation that must be passed to gain approval of some standards-setting or compendial group. The ISO terminology for different tests is aptly descriptive; "normative" tests are those that must be passed to prove adherence to a norm or standard. "Informative" tests are merely specified as a uniform means of collecting data.

In the generation of new requirements, regulatory and compendial bodies should be concerned that a specific need is being addressed. Such a need might generally be defined as a situation in which particle numbers are elevated due to a lack of control in manufacture of a specific type of device or injectable. The requirements proposed should also have a sound technical basis. Ideally, manufacturers, who are the experts with regard to particle burden of their product, should be consulted and asked to collaborate in the process. While the situation for requirements proposed has improved, these considerations are not always made in some cases. One of the unfortunate aspects of harmonization and world requirements has been the tendency of standards organizations to compete with regard to getting specific requirements "on the books" so that their standard may (for whatever motive) be adapted as an international requirement. In the following subsections, the author reviews some current proposals and some from the recent past that fall somewhat short in regard to either the method of analysis specified, the perception of need for the proposed standard, or the particle count limits chosen.

In the generation of any requirement for particulate matter in "infusion" or solution administration sets, it is important to take the following into consideration:

- The absence of any clinical evidence indicating that particulate matter requirements for medical devices would benefit patients.
- The difficulty of generating requirements for devices because of the great diversity of

types of devices, and the lack of any uniformly applicable test method.

- Only cGMP will control device particle burden (requirements may represent cost without benefits).
- The wide use of cGMP in device manufacture has resulted in the particulate matter burden of marketed devices of this type being controlled at very low levels.

Syringes and Needles (Proposed USP Method)

A recent USP proposal (*Pharmacopeial Forum,* Jan.–Feb. 1997) of a requirement for disposable syringes and the response received provides an interesting study in the difficulty of generating requirements for devices. Specific comments on excerpted sections of the proposed syringe test method are given below.

> Particulate matter <788>—Proceed as directed for large-volume injections under Particulate Matter in Injections <788>. Unless a needle is an integral part of the unit, attach a needle that has been flushed thoroughly with membrane-filtered water. For sample preparation, select 10 syringes or cartridges having a capacity of less than 50-mL, or 3 syringes or cartridges having a capacity of 50-mL or greater.

The needle itself, even with the cursory step of "flushing thoroughly" with clean water, may be expected to add particulate matter to the test. Attaching the needle to the syringe will likewise add plastic shavings. The number of syringes is poorly considered. Filtration of a 10 mL volume of water from ten 1 mL syringes will result in a very high background contribution due to extraneous particles and the high surface-to-volume ratio of the volumetric flask used to collect the particles. Further, all of the particles from smaller volumes of test fluid can be expected to "pile up" on one small area of the microscopic test membrane so that statistical or total counting will not be possible.

> Fill each previously unfilled syringe to its rated capacity with membrane-filtered water, then expel the contents into a suitable, cleaned volumetric flask. Use the same flask to collect the contents from all the syringes or cartridges.

A blank criterion for the water should be given. The procedure of expelling the contents of a syringe into a volumetric flask is difficult to conduct clearly, and the contribution of particles from the component of the test itself cannot be measured. Possible particle sources are both the hands of the analyst and extraneous particles from the surface of the syringe. Obtaining a satisfactory blank for a volumetric flask is very difficult, and this type of container (especially those with a ground glass mouth) is very difficult to clean thoroughly.

> Examine the sample visually for foreign matter.

This direction is insufficient to specify uniform inspection and is consequently unsatisfactory. How is foreign matter to be discriminated from material related to the syringe or its manipulation? Consider the test article. The disposable syringe is often packed in a blister pack. When opened, the fiber portion of the pack is torn and generates large numbers of particles. Further, when the pack is torn open, static electricity is generated that attracts the fibers generated to the outer surface of the syringe. There is no way to ensure that these do not enter the test as contamination via contact of the syringe with the collection vessel, or by transfer on the analyst's hands. The blanks cannot account for these extraneous particles. Additionally, the proposal states that the test articles must be "essentially free from particles that can be observed on visual inspection." Industry testing indicates that syringes of current production contain small numbers of difficult to detect particles, hence critical variable inspection could result in one company failing a batch, while a less critical inspection at another company would pass the same batch. Thus, even though the current production of disposable syringes represents state-of-the-art production methods and cGMP, this standard could remove them from the market on the grounds of visible particles.

Visual inspection results are critically dependent on a number of key factors, including lighting, time of inspection, training of inspectors, particle optical characteristics, and the size of particles present in the tested suspension.

The volume of solution inspected also critically affects the result, and the proposed test would result in a variable volume from 10 mL to 750 mL being inspected, based on the volume of the syringes tested. The fact that the suspension must be "essentially free" of visible particles is the only information on this assay that suggests that those authoring the proposal have, in fact, given a somewhat superficial consideration to this aspect of the test.

> Using the same sample, proceed as directed for large-volume injections in Particulate Matter in Injections <788>, except to filter the entire sample. Count and size the total number of particles collected on the upper surfaces of the filter. Calculate the average number of particles per syringe or cartridge.

If the limits are to be on a per mL basis, why are we calculating the number of particles per syringe? USP 23 <788> instructs the user to filter the whole volume, so this step is not an exception.

> Interpretation—The syringe or cartridge meets the requirements of the test if it is essentially free of visible particles of foreign matter and the sample contains not more than 10 particles equal to or greater than 10 µm in effective linear dimension and not more than 1 particle equal to or greater than 25 µm in effective linear dimension per 1 ml nominal volume of the tested syringe or cartridge.

(Effective linear dimension measurement is not the method in <788>.)

> If the sample does not conform at this stage, retest using pooled material from a second sample of twice the number of specimens (20 or 6) from the same batch. The requirements are met if the results conform to the criteria stated above.

This step of doubling the amount of material filtered for a retest can well be expected to result in a second membrane with particles too numerous to count. A screening of several different types of disposable syringes by the proposed method shows that a significant number of lots (e.g., up to 25 percent) may be expected to fail the test for visible particles, subvisible particles, or both. Thus, the proposed limits and test method are inconsistent with the particle burden of present product.

The subvisible particulate limit for empty syringes is much more strict than that allowed for SVI. To illustrate more fully, a sealed glass ampoule with no moving parts or internal lubricant can contain 6,000 particles > 10 µm, while a 1.0 mL syringe, consisting of moving parts lubricated by silicone may contribute only 10 particles > 10 µm per mL based on flushings of the syringe with filtered distilled water. Subvisible particulate limits should vary with syringe size. According to the proposal, a 1.0 mL syringe may contain only 0.17 percent of the total number of particles contributed by an SVI in an ampoule. The test would allow 100 particles/syringe > 10 µm for a 10 mL syringe, or 1.7 percent of the USP limit for SVI, and 500 particles/syringe > 10 µm for a 50 mL syringe, or 8.33 percent of the limit. This is not consistent with SVI requirements, where there is a constant limit per container, irrespective of size. According to the proposal, larger, cleaner syringes have higher limits. In fact, smaller syringes will be found to have a much higher particle burden on either a per syringe or per mL basis.

Per this description, a 1 mL syringe, a difficult-to-inspect mechanical device that must be actuated 3 times using the test, may contain only 10 particles ≥ 10 µm and 1 particle ≥ 25 µm. A 1 mL ampoule (SVI), a clear, nonshedding container with no moving parts, may contain 10,000 particles ≥ 10 µm, and 1,000 ≥ 25 µm. In this case, the proposed syringe particulate standard is 1/1,000 of that allowable for an SVI. A comparison of particles allowed from these different product groups is shown in Table 3.4 for purposes of comparison.

Data from syringe manufacturers indicate that a number of syringes produced by domestic and foreign manufacturers will not meet the proposed limits. In looking at the results of testing of plastic syringes, most of the failures occur in the 1–3 mL size range. Converting these data to a per-syringe basis reveals that there are more than twice as many particles per syringe in the 1–3 mL volume range as in the 5–10 mL range. This again indicates the unsuitability of a single limit. Glass syringes incorporate a ground glass plunger bearing on a ground glass barrel surface. The number of actuations becomes very critical,

Table 3.4. Comparison of LVI and SVI Limits to Those Proposed for Syringes

USP Product	# of Particles ≥10 mm	# of Particles ≥25 mm
LVI (1,000 mL)	12,000	2,000
SVI (per container)	6,000	600
Empty Syringe—1 mL (proposed)	10	1
Empty Syringe—20 mL (proposed)	200	20

since more actuations simply generate more glass fragments or release more particles of other types.

The subject proposal for particulate matter requirements for disposable syringes and cartridges appears to be based on an initial cursory overview of the situation regarding particulate matter in these devices. The many technical issues indicate that only a very brief consideration was given the write-up of the method. This situation is not unusual in initial drafts of requirements proposals for which industry input is not requested. In the case of this proposal for syringes, and in other USP initiatives, input regarding the testing of syringes from industry representatives should be given rational consideration by the USP, thereby allowing the development of a more suitable, realistic limits proposal. There appears to be no present, demonstrable need for a requirement pertaining to the number of subvisible or visible-sized particles in disposable syringes or cartridges. The ISO requirements for syringes (ISO 7886-1,2) do not involve particulate matter testing but rather deal with physical, chemical, and functional parameters.

Solution Administration Sets

ISO/CD 8536-4 (Infusion Equipment for Medical Use), Annex F, contains a test for particulate contamination that reads as follows:

Annex F (informative)

Test for particulate contamination

F.1. Determination of visible particles

F.1.1. Principle

The inner fluid pathway surfaces of infusion sets may superficially be contaminated with particles visible to the eye. Such particles may be transferred to infusion solutions administered through the set and deteriorate the quality of such preparations. The present method purports to evaluate contamination of this kind of collecting and counting the particles detached by rinsing from the inner fluid pathway surfaces of an infusion set.

F.1.2. Procedure

F.1.2.1. Carry out all operations in such an environment that no extraneous particles can interfere. This involves wearing suitable garments, non-powdered gloves and using a suitable clean air work station, e.g. providing laminar air flow to e.g. class 100 (US Federal Standard 209 E) as well as suitably decontaminated tools and handling means.

F.1.2.1.2. Prepare a rinse fluid by dissolving 3 g of highly concentrated sodium N-methyl-N-oleyltaurate powder in 10 l of water conforming to ISO 3696 grade 1 or grade 2. Make provisions for supplying the rinse fluid under pressure using a final membrane filter with maximum pore size of 1.2 μm.

F.1.2.2. Test

F.1.2.2.1. Fill a clean 50 ml glass syringe with 50 ml of the rinse fluid. Connect the syringe to the closure-piercing device by appropriate means and empty the 50 ml of rinse fluid through the infusion set at such a flow rate that the injection pressure obtained reaches 3 bar. Collect the rinse fluid in a clean Erlenmeyer flask. Filter the rinse fluid over a light gray membrane filter with a pore size of 0.8 μm provided with green grid lines at 3 mm distance.

Repeat this operation with the same syringe with a second 50 ml portion of the rinse fluid and filter the same way.

Store the filter suitably.

NOTE—The color of the filter may significantly affect the test results. If no specific details have been agreed on between parties, the colour should be medium gray and meet the following coordinate ranges in the CIE system:

L between 60 and 70%

a between -4.7 and -3.7%

b between -4.7 and -3.7%

This specification is recommended for measurements with a membrane filter with a 3 mm square green grid.

F.1.2.2.2. Prepare a blank filter following the procedure as described in F.1.2.2.1 by emptying the rinse fluid directly from the syringe into the Erlenmeyer flask. The blank counts shall satisfy the following criteria when performing total counts as indicated in clause F.1.2.3.

Class I: ≤ 5 particles;

Class II: ≤ 1 particle;

Class III: 0 particles.

Results for infusion sets tested are only acceptable if calculated with a blank determination that meets these criteria.

F.1.2.2.3. Count the particles on the filter using a suitable microscope under a magnification of about 50× and appropriate illumination, incident angle with the slide stage between 0 and 10 degrees.

Classify the particles into the following categories, using the longest visible dimension as the classifying parameter:

Class I: 25 μm < d ≤ 50 μm

Class II: 50 μm < d ≤ 100 μm

Class III: 100 μm < d

F.1.2.3. Proposed acceptance criteria

Class I: ≤ 100 particles;

Class II: ≤ 20 particles;

Class III: 0 particles.

F.1.3. Expression of results

For each test report

- average flow rate obtained during the injection
- estimated surface of the inner fluid pathway
- total count of particles found in each of the three classes
- counts of particles found in each of the three classes with a minimum of one count of the blank tests performed
- average count of particles for 10 cm^2 of tested surface found in each of the three classes, rounded to one decimal.

The test described in this standard proposal is also technically flawed and would not provide for accurate enumeration of particles that might be present in the solution contact path of devices so tested. Further, it is doubtful that a standard of this type will be of any practical utility, due to the wide range of "infusion" sets of various types on the market and the fact that manufacturers of devices currently have more appropriate tests in place. The fact that an identical test method has been proposed for two general categories of sets, which vary significantly in construction and use, is particularly disturbing.

With specific references to the technical aspects of the proposed standard, there are a number of critical problems with the proposed test method, which would seriously limit its usefulness. In F.1.1, the word *contaminated* is used.

Administration and infusion sets inevitably contain, as a result of manufacture, small numbers of randomly sourced particles that occur despite the application of cGMP in manufacture and are allowable under the definition of "particulate matter" provided by the USP. These particles cannot be classified as contaminants, since they are an inevitable consequence of the manufacturing process.

In F.1.2.1.1, the environment for the test is not sufficiently well described to allow reproducible conduct of the test between laboratories. The control of the conditions of the test is also not sufficient (i.e., "under cleanroom conditions"). Most contamination impacting the test will result from analyst manipulations rather than from environmental air. There are no specified controls regarding contamination of the analytical membrane when it is drying.

In F.1.2.1.2, "highly concentrated" detergent is not the infusate with which sets will be used. A more appropriate rinse solution is normal saline or distilled water.

In F.1.2.2, the 100 mL volume of flush solution prescribed is less than the infusate volume used with most sets. It is insufficient to flush such sets exhaustively; it will result in an overestimation of numbers of particles in the test item when the number of particles per mL of flush solution is considered. Different types of devices will require different volumes. Also, the flow rate of the rinse should be selected based on the use and volume of the set. Use of a uniform pressure will result in a more or less thorough rinsing of particles from a set depending on its volume. A very important consideration is the range of device types that would have to be covered by a test of the type described. There are many types of "infusion sets" for which the test as specified will not work, such as burette, filter, pump, and capillary lumen sets.

In F.1.2.2.2, the allowance is almost surely too high. Only one count of a blank is required, and the normal variability expected with an allowable count of 5 particles 25–50 μm in size would dictate that some data collected would contain an unacceptable blank contribution. A blank should be taken before each test.

In F.1.2.2.3, the microscopic test is not described in sufficient detail to allow for reproducible conduct between laboratories. What is "about 50×?" Is a range of 40× to 60× acceptable? What about 30× to 70×? The magnification will critically impact the accuracy and precision of sizing, with the danger that particles < 50 μm in size could be oversized and placed in the next higher size category. Lighting at an angle of 0° to 10° is not useful for critical particle sizing due to the poor definition of particle boundaries. How the measurement of particles is to be performed is not specified, and no accounting is provided for fibers, which are a unique and distinct class of particles that may be found in sets produced under non-GMP conditions. Lastly, the description of the analysis to be performed is insufficiently detailed to allow consistency of testing between laboratories.

In F.1.2.3, the JP and USP both allow 2 particles ≥ 25 μm per mL of solution for a microscopic test. Thus, the limit for a device tested as described would be twice as restrictive as these standards. Further, as stated above, the flush volume and flow rate for that test would have to be matched to a wide variety of sets.

In the handling of blank counts, the procedure specified is totally unacceptable. Blank subtraction as described, in theory, will allow a blank of any level to be accepted as long as it is subtracted for the counts obtained from the set. A high blank count will have an attendant high variability, and since the blank contribution to a specific test is unknown, a significant error source is incorporated.

Infusion Containers—DIN 58363-15 Proposal (12/24/97)

(Proposal for the addition of a requirement for infusion bags and bottles made of plastic and adoption of the DIN standard as an ISO norm)

Introduction

In some countries, national or regional pharmacopeia or other government regulations are legally binding and these requirements take precedence over this International Standard.

Field of Application

The standard shall be applicable to plastic containers, having one or more chambers for parenterals from 50 ml to 5000 ml, such as, for example, film bags or blow-molded plastic bottles for direct infusion *(injections)* solutions. It contains requirements related to safe handling and on physical, chemical, and biological testing of *these containers*.

(5) Test Method

5.1.6. Particle contamination

Infusion containers shall be manufactured so that contamination with particles is avoided.

Maximum permissible levels

- 200 particles per ml nominal capacity with a diameter ≥ 2.0 μm and
- 20 particles per ml nominal capacity with a diameter ≥ 5.0 μm.

Testing according to 5.1.7.

5.1.7. Particle contamination

Empty containers are filled under clean room conditions with the volume corresponding to the nominal capacity with water for injection that has been filtered previously through a membrane filter with a pore width of 0.2 μm. After being closed, the containers are sterilized and stored for approximately 12 h.

The particle content of the container contents is then determined with a particle counting device working according to the light blockade method. The values of the blank sample shall be taken into account.

The subtraction of blank counts is not a sound principle, given that the contribution of blank counts for a specific analysis will never be known. A higher blank count will reflect not only a higher contribution of extraneous particles to an analysis but also a higher variability of that contribution.

The intent is stated to revolve around limiting "the maximum quantity of particle contamination." Although well intended, the proposed test and test method will not serve to give the user any assessment of the level of "particle contamination" in a plastic infusion container. Contamination is defined by the USP as extraneously sourced (e.g., from the manufacturing process) materials that are unintentionally present in an infusion unit. These particles have been shown to exist in a distribution with the highest number at about 10 μm and smaller numbers both higher and lower than this size. An estimate of these extraneous particles cannot be made by counting at ≥ 2 μm. All of the world's major compendia have dropped the ≥ 2 μm size (EP, BP) due to it not being meaningful. Another key factor in not using this size is the unsuitability of light blockage counters for testing in this range.

The test that the proposed document intends to apply in not specified in sufficient detail to allow reproducibility between laboratories, even if it were performed at the ≥ 10 μm size. Performance of the assay must be the same in different laboratories. What blank level is allowable? Filling under unspecified "clean room conditions" as specified will not protect the test article from contamination. How are different types of containers to be opened and closed so that particles generated by the process itself are not counted? How is "previously filtered" water transferred into the container?

When an empty bag is filled as described and sterilized, minute amounts of organic material present in a hydrophobic film on the plastic surface are removed and can enter the fluid to form micelles, immiscible microdroplets or persistent air bubbles in the range of 1 μm to 5 μm. Some of these materials have surfactant properties, and can form agglomerates that appear as particles to an instrument.

This release of material from the surface occurs with all plastic bags to some extent and can result in erroneous light blockage counts centered at about 2 μm, which have no relation to either contamination or container suitability. This population of counts will be totally distinct from "contaminate" particles and will be counted to different extents by different types of particle counters. This phenomenon makes the proposed test generally unsuitable for its intended purpose. The JP, in performance of almost exactly the same test, acknowledges that there are difficulties and instructs that counts be performed at ≥ 5 μm (100 particle limit), > 10 μm, and > 25 μm.

The materials addressed in (5) above that will be counted at the 2 μm size are really chemical materials, not extraneous particles. If total counts allowable (≥ 2 μm, 200; ≥ 5 μm, 20) were obtained from a solution based on these chemical materials, the total amount of chemicals involved would be approximately 1 ppb. This figure is far below the limits allowed for chemical entities (Table 2 of the draft) that are tested after the bag has been rinsed twice then sterilized. The foam allowed that must disappear in 0 seconds is visible foam—the microbubbles produced by the foam at 2 μm size will be much more persistent.

Also totally inappropriate is that Table 2 of the draft allows "weak opalescence" in the turbidity test. This "weak opalescence" will be composed of a number of fine particles far in excess of those allowed. This is to say that a solution with "weak opalescence" will certainly fail the test for particles.

ISO Draft Standard 3826—Plastic Collapsible Containers for Human Blood and Blood Components

ISO Draft Standard 3826 has been in the working group stage for several years. The present version (1997) has a number of points of commonality with the proposal for infusion containers reviewed above, including a test for "opalescence," which recognizes that containers made of some polymers will release small amounts of plasticizers when "steam extracted." The requirements for particulate matter testing in this standard draft involve only visible particulate matter. Section 5.2.8 requires that

> Plastic containers shall be manufactured such that contamination with particles is avoided. When tested as in B.4 the plastic container fluid path should be free of visible particles.

The latter section instructs the analyst to fill the test unit with "purified water" (i.e., 0.2 μm filtered). Section B.4.3 requires that

> The fluid in the plastic container shall be inspected by an appropriate method that will readily detect visible particles.

This instruction, of course, without specification of a test method, is of little value insofar as describing a test and results in the same sort of confusion that can result from "essentially free" criteria.

One very interesting point regarding the parallel path (or competitive) development of standards is provided by the simultaneous development of a standard for plastic blood bags by the USP. Standards of this type, in general, would seem to be more appropriate for inclusion in a compendium than among the standards of a general standards organization such as the ISO. The USP proposal is reviewed below.

USP Proposal for Blood Bags (Blood Storage Containers)

The somewhat similar proposal published by the USP targeting blood storage containers is excerpted in part below (*Pharmacopeial Forum*, 23 (5):563, Sept.–Oct. 1997):

> **Sterile, Single-Use Plastic Large-Volume Containers for Human Blood and Blood Components**
>
> Particulate matter <788>—Rinse the outside of a container with water to be used in the test, and fill the container to the nominal capacity volume with sterile 0.9% sodium chloride solution. Hermetically close the container when the amount of air is about 50 mL per 500 mL of the nominal capacity. Wipe clean the outside of the container, mix the contents by inverting the container 5 to 6 times, and immediately insert a sterile needle of an infusion assembly that has been flushed thoroughly with water that has been filtered through a membrane filter. Discard about 50 mL of the first spontaneous effluent, and collect the next 300 mL of effluent in a suitable cleaned volumetric flask. Examine this effluent visually, note any visible particles, and proceed as directed for large-volume injections for single-dose infusion.
>
> Interpretation—The requirements of the test are met if the effluent is free of visible particles and if not more than 10 particles present in the effluent are equal to or greater than 10 μm in effective linear

> dimension and not more than 1 particle is equal to or greater than 25 μm in effective linear dimension. If the requirements are not met, retest, using the pooled effluents from not less than three containers from the same batch. Determine the average number of particles per container tested; the requirements of the test are met if the number of particles does not exceed the values indicated above.

There are several issues here that the reader should be able to identify without any clues given. Using a syringe needle to withdraw the contents of the container is unnecessary and is a step prone to add contaminants to the analysis. Even a syringe needle flushed with filtered water cannot be relied on not to add particulate matter in countable sizes to the analysis. Filtration of the saline flush solution is not specified. The collection vessel has not been blanked for visible particles; the visible particles, if present, are noted, but no mention is made of what should be done with the data observations. The limits are too low to be worthy of practical consideration; as written, the limit is 10 particles ≥ 10 μm in size and 1 particle ≥ 25 μm in size per 300 mL sample volume. This is much lower (by more than an order of magnitude) than the glassware blank specified in USP 23 <788> that allows 20 particles ≥ 10 μm in size and 5 particles ≥ 25 μm in size per 50 mL sample volume tested by microscopy. Presumably, the microscopic test would be used, since this would be the only means of testing the entire 300 mL aliquot of the sample.

SUMMARY AND CONCLUSION

The world's major compendia include test methods and numerical limits for subvisible particles and qualitative requirements for visible particulate matter in injectables. Thus, the great majority of parenteral products manufactured are in conformance with standards that specify a particle burden far below levels of patient concern. While the existence of requirements and limits provides a superficially reassuring impression for the healthcare professional, they have little or no effect on the actual numbers of particles found in product.

The present quality level of injectables has been attained through advances in the technology for the production of parenterals and the conscientious application of this technology in combination with GMP by the pharmaceutical manufacturer. Limits do not serve to control the numbers of particles in product; they do serve as useful criteria for the critical monitoring of the production process, since the internal particle count limits of most manufacturers are significantly below the compendial limits.

Particles in injectable dosage forms originate from a wide variety of sources, and the particle burden of an injectable is typically low in comparison to the potential total particle contribution from other IV therapy components (such as admixing and administration procedures). The only reasonable control of particles in injectables is effected by GMP and process monitoring—not additional regulatory requirements. Requirements should be developed only in cases where a need can be clearly established based on clinical evidence. The same is generally true for devices.

Since GMP (rather than level of testing or limits) determines the particle burden of devices, the only effective control of particulate matter in devices will result from the conscientious implementation of GMP by all device manufacturers. For manufacturers that presently have well-defined GMP programs of proven effectiveness, it seems likely that compendial limits (and additional testing) will only result in higher production costs, which will produce no benefits in the form of reduced particle burden.

In general, a consideration of the particulate matter levels in medical devices and the particle levels acceptable (per the USP and other compendial requirements) for LVI or SVI suggests that neither devices nor injectables significantly contribute to the particulate levels to which the patient is exposed. Additionally, it would seem that both types of products meet customer expectations, as evidenced by the total absence of customer complaints regarding subvisible particulate matter and the very low number of complaints received related to large (visible) particles. This is an impressive commentary based on the fact that millions of devices and in excess of 200 million units of injectable products are used annually.

REFERENCES

Akers, M. J. 1985. *Parenteral quality control: Sterility, pyrogen. particulate, and package integrity testing.* New York: Marcel Dekker, Inc., pp. 143–197.

Anderson, L., and G. J. Higby. 1995. *The spirit of voluntarism: A legacy of commitment and contribution—The United States Pharmacopeia 1820–1995.* Rockville, Md., USA: United States Pharmacopeial Convention, Inc.

Barber, T. A. 1987. Limitations of light blockage particle counting. Paper presented at the meeting on Liquid Borne Particle Inspection and Metrology, 11–13 May, in Washington, D.C.

Barnett, M. I., N. A. Armstrong, D. C. James, and B. K. Evans. 1983. Particle contamination from administration sets. *Brit. J. Parent. Ther.* 4:8–17.

DeLuca, P. P., S. Bodapatti, D. Haack, and H. Schroeder. 1986. An approach to setting particulate matter standards for small volume parenterals. *J. Parenter. Sci. Technol.* 40:2–13.

DeLuca, P. P., B. Conti, and J. Z. Knapp. 1988. Particulate matter II. A selected annotated bibliography. *J. Parent. Sci. and Technol.* 42 Supplement.

Di Paolo, E. R., B. Hirsch, and A. Pannatier. 1990. Quantitative determination of particulate contamination in intravenous administration sets. *Pharm. Weekbl.* 12:190–195.

Dimmick, J. E., K. E. Bove, J. McAdams, and G. Benzing. 1976. Fiber embolization, a hazard of cardiac surgery and catheterization. *New Engl. J. Med.* 292:685–687.

Endicott, C. J., R. Giles, and R. Pecina. 1966. Particulate matter—its significance, source, measurement, and elimination. In *Proceedings of the FDA Symposium on Safety of Large Volume Parenteral Solutions,* pp. 62–70.

Falchuk, K. H., L. Peterson, and B. J. McNeil. 1985. Microparticulate-induced phlebitis: Its prevention by in-line filtration. *N. Engl. J. Med.* 312: 78–82.

Garvan, J. M., and B. W. Gunner. 1963. Intravenous fluids: A solution containing such particles must not be used. *Med. J. Australia* 2:140–145.

Garvan, J. M., and B. W. Gunner. 1964. The harmful effects of particles in intravenous fluids. *Med. J. Australia* 2:1–6.

Gour, L. 1987. Particulate matter in parenteral type medical devices. *Pharmacopeial Forum* (May/June):2506–2522.

Groves, M. J. 1969. The size distribution of particles contaminating parenteral solutions. *Analyst* 94:992–999.

Groves, M. J., and S. R. DeMalka. 1976. The relevance of pharmacopeial particulate matter limit tests. *Drug Dec. Commun.* 2:285–324.

Groves, M. J., and E. Muhlen. 1987. The parenterals numbers game—Newer ways of looking at particulate contamination. *J. Parent. Sci. Technol.* 41: 116–120.

Ho, N. F. H. 1967. Particulate matter in parenteral solutions. I. A review of the literature. *Drug Intel.* 1:7–11.

Knapp, J. Z., and L. R. Abramson. 1990. A systems analysis of light extinction particle detection systems. In *Proceedings of the PDA International Conference on Particle Detection, Metrology and Control,* pp. 283–297.

Muhlen, E. 1986. An index for the particulate contamination in parenterals—Its fundamentals and its application to quantitative determination by photometric control and fully automated image analysis. Paper presented at the Congress of the F.I.P., 9–13 September in Helsinki.

Rypings, E. B., B. H. Johnson, B. Reder, I. J. Sarfeh, and K. Shimoda. 1990. Three-phase study of phlebitis in patients receiving peripheral intravenous hyperalimentation. *Am. J. Surg.* 159: 222–225.

Trasen, B. 1968. Membrane filtration technique in analysis for particulate matter. *Bull. Parenteral Drug Assoc.* 22:1–8.

Trasen, B. 1969. Analytical techniques for particulate matter in intravenous solutions using membrane filters. *Ann. N. Y. Acad. Sci.* 158:665–673.

Trautman, K. A. 1997. *The FDA and worldwide quality systems requirements guidebook for medical devices.* Milwaukee: ASQC Quality Press.

Turco, S. J., and N. M. Davis. 1971. Detrimental effects of particulate matter on the pulmonary circulation. *J. Amer. Med. Ass.* 217:81–82.

Vessey, I., and C. E. Kendall. 1966. Determination of particulate matter in intravenous fluids. *Analyst* 91:273–279.

STANDARDS, COMPENDIA, AND REGULATORY REFERENCES

AAMI. 1989. *American national standard for autotransfusion devices.* Arlington, Va., USA: Association for the Advancement of Medical Instrumentation.

AAMI. 1989. *American national standard for blood transfusion microfilters.* Arlington, Va., USA: Association for the Advancement of Medical Instrumentation.

BP. 1993. *British Pharmacopoeia.* 1993. London: Her Majesty's Stationery Office.

BSI. 1986. BS 2463 (Draft). *Transfusion equipment for medical use,* Part 2, Specification for giving sets. London: British Standards Institution.

DIN. 1996. DIN 58363-15. *Infusion containers and accessories* Part 15: Infusion bags and bottles made of plastic. Berlin: Deutsches Institut fur Normung.

European Commission. 1995. *Guide to Good Manufacturing Practice, Annex on the Manufacture of Sterile Medicinal Products.* Brussels.

European Pharmacopoeia, 3rd ed., Sections 2.2.1. and 2.2.2.

FDA. 1996. *Medical Devices; Current Good Manufacturing Practice (CGMP) Final Rule; Quality System Regulation.* Rockville, Md., USA: Food and Drug Administration.

HIMA/PMA Task Force. 1980. *Final report on particulate in intravenous equipment.* Washington, D.C.: Health Industry Manufacturers Association.

Intravenous therapy guideline. 1985. Ottawa, Ontario: Health Services Directorate, Health Service and Protection Branch.

ISO 8536-4. *Infusion devices for medical use*—Infusion set for single use, gravity feed. Geneva: International Organization for Standardization.

ISO 10993-4. *Biological evaluation of medical devices*—Part 4: Selection of tests for interaction with blood. Geneva: International Organization for Standardization.

ISO 10993-5. *Biological assessment of medical products*—Part 5: Tests for cytotoxicity: in vitro methods. Geneva: International Organization for Standardization.

ISO 10993-12. *Biological evaluation of medical devices*—Part 12: Sample preparation and reference materials. Geneva: International Organization for Standardization.

ISO/CD 13485. *Quality systems for medical devices.* Supplementary requirements for ISO 9001. Geneva: International Organization for Standardization.

ISO/CD 13488. Quality systems for medical devices. Supplementary requirements for ISO 9002. Geneva: International Organization for Standardization.

JP. 1996. *The Pharmacopoeia of Japan,* 13th ed. Tokyo: Yakuji Nippo, Ltd.

Pharmeuropa, 19 (3):444–445 (Sept. 1997).

USP. 1985. *The United States Pharmacopeia,* 22nd ed. Easton, Penn., USA: Mack Printing Company.

USP. 1995. *The United States Pharmacopeia,* 23rd ed. Easton, Penn., USA: Mack Printing Company.

IV

USP 23 <788> PARTICULATE MATTER IN INJECTIONS

The test method for subvisible particulate matter in injections that is presented in the U.S. Pharmacopeia (USP) General Chapter <788> is the most comprehensive analysis of this type in any of the world's compendia. It is applicable not only to injections bearing the letters "USP" following the product name on the label copy but also other injections for which the test is specified. Exemptions include chemotherapy drugs, radiopharmaceuticals, diagnostic agents such as X-ray contrast media, and prefilled syringes. Importantly, solutions for which the monograph directs that a final filter be used in administration are also exempt.

The test is performed not only by domestic (U.S.) pharmaceutical manufacturers, but by overseas manufacturers who make product for sale in the United States, or product that must comply with the USP. Due to the wide use of this method, its relative complexity, and the several rounds of revisions that have been made since the requirement went into effect in January 1996, the following discussion of the method should be useful to the user in understanding and applying the current version of the requirement. The intent of this chapter is to provide the reader with considerations regarding specific sections of the method and to assist in understanding the principles and philosophy of the tests prescribed. At this point, the reader is again warned that the interpretive information included in this chapter represents the opinion of the author rather than any USP policy or position. In this regard, the best course of action is to contact the USP with interpretive questions.

Two basic principles regarding this method and the USP itself should be clear to the reader. First, the purpose of the USP is to protect the patient by ensuring that drugs administered are of the highest quality with regard to purity, efficacy, and safety. The <788> requirement is one of a series of limits tests, which taken together serve to achieve this goal. The intent of <788> is to limit the number of heterogeneously sourced extraneous particles that an injection can contain. In that regard, it indirectly serves to ensure that adequate Good Manufacturing Practice (GMP) safeguards as specified in the USP apply to product as manufactured and over its shelf life. The wide use of limits tests as manufacturing standards or over the course of stability studies is by the choice of the manufacturer, not the USP.

The origin of the <788> test method is interesting. In 1990, the USP announced the intent to replace the USP XXII version of <788> with a new requirement, which incorporated only the light obscuration assay. This would have meant that the microscopic assay, long used as an arbiter in resolving artifacts (i.e., spurious counts) obtained using the light obscuration assay, would no longer be available for this purpose. The manufacturer unfortunate enough to collect counts due to air bubbles or immiscible fluids such as silicone with a light obscuration counter, in the absence of a microscopic test, would have no means of referee analysis left. Rather than see this situation arise, the Health Industry Manufacturer's Association (HIMA) committed to develop an improved microscopic test suitable for use in the requirement and update the light obscuration test as well. This task was completed, with input and participation from more than 30 manufacturers of parenteral products in the United States and abroad in early 1994.

Because of the complexity of the method and the multiple comments and inputs received, a number of minor points in the method have been the subject of revision since the method was published. Revisions appear in the November–December 1994, March–April 1995, and July–August 1995 issues of *Pharmacopeial Forum* and in the First, Second, and Fifth Supplements to USP 23.

While the revisions are interesting as an historical perspective, the user should make absolutely certain that the Fifth Supplement or later version be used as a basis for tests applied in a specific laboratory. The Fifth Supplement version is included in its entirety below with comments and explanations. The author's purpose is to provide information on the intent of the method and its technical conduct. Interpretations of the method by the user, particularly with regard to sampling plans and technical changes to suit a specific product type, are solely the responsibility of the user and must be defensible in regard to regulatory agency scrutiny.

The USP <788> method, as published, is directly applicable to a wide variety of injection products. Those using the method are reminded, however, that the principle of assay equivalency must apply when USP assays are modified or when an alternate assay is chosen. Auditors (Food and Drug Administration [FDA] investigators) may require proof that assay modifications are justified and that a prediction of batch suitability obtained through use of a modified method is equivalent to that provided by the USP assay.

CHAPTER FORMAT

The USP 23 <788> requirement from the Fifth Supplement to USP 23 is excerpted almost in its entirety in the following sections of this chapter. Following each excerpted section are the author's comments pertaining to that section. Figures and line drawings are added as the author believes necessary to illustrate specific points or principles in greater detail than is possible in the USP text.

Note: On pages 3356 and 3357 of USP 23 <788> is an explanation of the various symbols that are used in the excerpts provided in this chapter.

EXCERPTED REVIEW

<788> PARTICULATE MATTER IN INJECTIONS*

Change to read:

> Particulate matter consists of mobile, randomly-sourced, extraneous substances, other than gas bubbles, that cannot be quantitated by chemical analysis due to the small amount of material that it represents and to its heterogeneous composition. Injectable solutions, including solutions constituted from sterile solids intended for parenteral use, should be essentially free from particles that can be observed on visual inspection. The tests described herein are physical tests performed for the purpose of enumerating subvisible extraneous particles within specific size ranges.

This short paragraph is extremely important! The definition of particulate matter in the *Introduction* to USP <788> is clearly intended to exclude homogeneous monotypic materials that exist in the form of a precipitate or a suspension. The "randomly-sourced extraneous substances" addressed in the definition applies to the low numbers of particles of heterogeneous composition resulting from even the most closely monitored and validated cGMP process. Examples are fragments of filter material and small pieces of

*U.S. Pharmacopeia/National Formulary, Fifth Supplement, USP 23/NF 18.

paper fibers. Such materials are acknowledged by manufacturers, the USP, and the FDA to occur and are allowable at very low levels in injectable product; the intent of the requirement is simply to place limits on the numbers of these particles. The type of particulate matter at which the requirement is directed relates directly to the manufacturing process, not to the solution or solution-container interactions.

The requirement makes a clear distinction between materials of this type and materials present in solutions in small quantities that can be analyzed by chemical means (e.g., silicone oil). The clear intent of <788> is to use instrumental and microscopic particle counting only for the quantitation of particulate material that cannot be analyzed by other methods. Any particulate material that is homogeneous in composition (such as a drug degradant) is a distinct chemical entity and can be identified and/or precisely quantitated by sensitive contemporary chemical techniques.

A very important descriptor in the USP definition of particulate matter is its random nature. Particulate matter, by the USP definition, is randomly sourced and, hence, cannot be clearly defined as to origin. Materials such as precipitates or drug degradants in the form of amorphous or crystalline materials not only originate from a defined source (i.e., the product itself), but they also can be characterized in terms of the chemical reaction by which they are produced.

> Microscopic and light obscuration procedures for the determination of particulate matter are given herein. ■It is expected that most articles will meet the requirements on the basis of the light obscuration test alone; however, it may be necessary to test some articles by the light obscuration test followed by the microscopic test to reach a conclusion on conformance to requirements.■1 ■ ■5

The <788> requirement is based on the stratified test principle, with the microscopic test performed only as a second stage test to support or assist in the interpretation of the first pass light obscuration test results. If the light obscuration test data exceed limits and the microscopic test is within limits, the product passes. It may behoove the manufacturer to explain such results.

> All large-volume injections for single-dose infusion and those small-volume injections for which the monographs specify such requirements are subject to the particulate matter limits set forth for the test being applied, unless otherwise specified in the individual monograph.

The intent here is only that products tested under the USP <788> requirement for particulate matter in injections must meet the limits specified for the product volume in question. Stated another way, all large-volume injections (LVIs) (product labeled to contain greater than 100 mL) must meet the limits specified for LVI (for light obscuration—not more than (NMT) 25 particles/mL ≥ 10 μm and 3 particles/mL ≥ 25 μm). Small-volume injection (SVI) product (≤ 100 mL) must meet the limits specified for SVI on a per container basis (for light obscuration—NMT 6,000 particles/mL ≥ 10 μm and 600 particles/mL ≥ 25 μm). The only determinant of which limits set applies is product volume. The verbiage in the excerpt above is in the process of being changed to "<788> meets the requirements," a direction that is more easily understood.

Not all SVI product is required to be tested per <788>. The USP, in looking for a way to specify which SVI product should be tested, chose to do this by adding a sentence at the end of each monograph if testing of a specific product in SVI form was required. The wording chosen was "meets the requirements under <788> for small volume injections." This wording intends to indicate that a SVI dosage form of the monographed item must be tested for particulate matter and must meet the SVI limits. Unfortunately, the sentence structure has led many users of the <788> assay to believe that the LVI dosage form of a monographed product that has this sentence must also meet the SVI limits. The author has received in excess of 100 questions on this subject since 1990, and the USP has probably received many more.

> Not all injection formulations can be examined for particles by one or both of these tests. Any product that is not a pure solution having a clarity and a viscosity approximating those of water may provide erroneous data when analyzed by the light obscuration counting method. Such materials may be analyzed

> by the microscopic method. Emulsions, colloids, and liposomal preparations are examples. Refer to the specific monographs when a question of test applicability occurs. Higher limits are appropriate for certain articles and are specified in the individual monographs.

The intent here is to allow microscopic testing alone if the viscosity is too great for a light obscuration counter sampler to draw at a satisfactory flow rate. Should the viscosity be too great to allow filtration for the microscopic test, dilution (next excerpt) is allowed. Dilution to decrease the number of particles to be counted in a given solution volume is not mentioned; one reason for this omission is the concern that some particles may be dissolved when diluent is added. In the future, further dispensations may be added for "problem" product that is difficult or impossible to test on light obscuration counters, such as bicarbonate-containing solutions that generate bubbles when drawn into a sensor under negative pressure.

> ■In some instances, the viscosity of a material to be tested may be sufficiently high so as to preclude its analysis by either test method. In this event, a quantitative dilution with an appropriate diluent may be made to decrease viscosity, as necessary, to allow the analysis to be performed.■5
>
> In the tests described below for large-volume and small-volume injections, the results obtained in examining a discrete unit or group of units for particulate matter cannot be extrapolated with certainty to other units that remain untested. Thus, statistically sound sampling plans based upon known operational factors must be developed if valid inferences are to be drawn from observed data to characterize the level of particulate matter in a large group of units. Sampling plans should be based on consideration of product volume, numbers of particles historically found to be present in comparison to limits, particle size distribution of particles present, and variability of particle counts between units.
>
> Sampling plans appropriate for the prediction of batch suitability are prescribed. It is essential that statistical input be obtained regarding the appropriate numbers of units to be tested per batch of specific product. A discussion of sampling plans is provided in Appendix III of this chapter.
>
> ***Change to read:***
>
> LIGHT OBSCURATION PARTICLE COUNT TEST
>
> **USP Reference Standards** [11]—*USP Particle Count RS.*

The *Change to read* notation indicates that a change to the section on the USP particle count reference standard is included in this section.

> The test applies to large-volume injections labeled as containing more than 100 mL, unless otherwise specified in the individual monograph. It counts suspended particles that are solid or liquid. This test applies also to single-dose or multiple-dose small-volume injections labeled as containing 100 mL or less that are either in solution or in solution constituted from sterile solids, where a test for particulate matter is specified in the individual monograph. Injections packaged in prefilled syringes and cartridges are exempt from these requirements, as are products for which the individual monograph specifies that the label states that the product is to be used with a final filter.

The intent of the last sentence of this paragraph is to provide monographed exemptions for certain products if necessary. An example might be a unique emulsion product of high viscosity, which would be destabilized by dilution and, thus, could not readily be tested by either the microscopic or light obscuration method. Exemptions may be applied for (at the time of this writing) by contacting Mr. Frank Barletta at the USP and will be considered by the Water and Parenterals Subcommittee of the Committee for Revision and the Division of Standards Development. No exemptions have yet been applied for.

The reader is cautioned against applying for an exemption simply because a product has inherently high counts or is expensive (e.g., a monoclonal antibody [MAb]). This is not a sound technical approach and would result in the product not being tested. Such a request is unlikely to draw regulatory attention to the product, since FDA representatives who have been elected to committee positions at the USP do not officially represent the FDA, but rather contribute specific expertise as do other committee members from the pharmaceutical industry and academia. Avenues other than exemption should also be considered, such as use of unique diluents for problem products or the generation of in-house test methods that will comply with the principles of <788> through a technically sound and validated modification of the test. Steps should always be taken to address any source of elevated counts. For expensive products, smaller amounts of material might be tested if validation of the exception is performed.

Test Apparatus

> The apparatus is an electronic, liquid-borne particle counting system that uses a light-obscuration sensor with a suitable sample-feeding device. A variety of suitable devices of this type are commercially available. It is the responsibility of those performing the test to ensure that the operating parameters of the instrumentation are appropriate to the required accuracy and precision of the test result, and that adequate training is provided for those responsible for the technical performance of the test.
>
> It is important to note that for Pharmacopeial applications the ultimate goal is that the particle counter reproducibly size and count particles present in the injectable material under investigation. The instruments available range from systems where calibration and other components of standardization must be carried out by manual procedures to sophisticated systems incorporating hardware- or software-based functions for the standardization procedures. Thus, it is not possible to specify exact methods to be followed for standardization of the instrument, and it is necessary to emphasize the required end result of a standardization procedure rather than a specific method for obtaining this result. This section is intended to emphasize the criteria that must be met by a system rather than specific methods to be used in their determination. It is the responsibility of user to apply the various methods of standardization applicable to a specific instrument. Critical operational criteria consist of the following.
>
> *Sensor Concentration Limits*—Use an instrument that has a concentration limit (the maximum number of particles per mL) identified by the manufacturer that is greater than the concentration of particles in the test specimen to be counted. The vendor-certified concentration limit for a sensor is specified as that count level at which coincidence counts due to simultaneous presence of two or more particles in the sensor view volume comprise less than 10% of the counts collected for 10-μm particles.
>
> *Sensor Dynamic Range*—The dynamic range of the instrument used (range of sizes of particles that can be accurately sized and counted) must include the smallest particle size to be enumerated in the test articles.

The user is given license here to buy any model of counter deemed appropriate (see Appendix I of this chapter). The user should assess several key items prior to purchase. Foremost is that the instrument is capable of meeting key criteria specified in <788>, and that the user will not be the first to apply this instrument in a pharmaceutical lab or compendial testing. Training is specified and should be conducted for or attended by users of the instrument. This training, as for other types of training, in the use and operation of the laboratory instruments should be documented. Test methods published in the USP are considered to be validated methods. This aegis of validation applies to the basic assay, but the user should be warned that regulatory auditors may expect that a specific instrument used be validated further with regard to its specific operation in performance of the test (see Chapter 11).

The extent of such validation typically increases in a direct proportion to the extent to which the instrument is software driven in standardization and operation. The sensor concentration limit and dynamic range may be accepted on vendor certification, depending on individual company policy and the amount of vendor information available.

A particularly important component of training regards teaching an analyst to recognize artifactual or erroneous data. The term *artifactual* applies to counts that may include air bubbles, spurious counts from external disturbances, or problems related to the mixing of solutions in the sensor or contamination on sensor windows. Importantly, reasons for rejecting data should be spelled out and familiar to the analyst.

> **Instrument Standardization**
>
> The following discussion of instrument standardization emphasizes performance criteria rather than specific methods for calibrating or standardizing a given instrument system. This approach is particularly evident in the description of calibration, where allowance must be made for manual methods as well as those based on firmware, software, or the use of electronic testing instruments. Appropriate ■instrument qualification■$_2$ is essential to performance of the test according to requirements. Since different brands of instruments may be used in the test, the user is responsible for ensuring that the counter used is operated according to the manufacturer's specific instructions; the principles to be followed to ensure that instruments operatewithin acceptable ranges are defined below.
>
> The following information for instrument standardization helps ensure that the sample volume accuracy, sample flow rate, particle size response curve, sensor resolution, and count accuracy are appropriate to performance of the test. Conduct these procedures at intervals of not more than six months.

The implicit instruction in these two paragraphs is more important than what appears in print. The term *instrument qualification* was originally *validation* in drafts of the method. Some FDA representatives who had input into the method felt that total validation of a given commercial instrument by the user was not called for; thus, *qualification* was used in place of *validation* (again, see Chapter 11).

The policy at some companies will probably be to consider that instrument standardization test specifications in the following portions of <788> will suffice for instrument qualification, but that software-driven instruments will require some level of user software validation. Whatever term is used to describe this procedure, it consists of obtaining assurance that the instrument functions as intended. Software may be subjected to the validation process, which is rapidly becoming standardized across the industry, involving requirements definitions, test plans, change control, and other items. Additionally, assessment should be made of the "ruggedness" of the assay as performed with a specific instrument system. Particularly important are items such as drift of threshold settings, susceptibility to vibration and electrical disturbances, and cleanliness requirements.

> SAMPLE VOLUME ACCURACY
>
> Since the particle count from a sample aliquot varies directly with the volume of fluid sampled, it is important that the sampling accuracy is known to be within a certain range. For a sample volume determination, determine the dead (tare) volume in the sample feeder with ■filtered■$_2$ distilled ■or deionized■$_2$ water that has been passed through a filter having a porosity of 1.2 µm or finer. Transfer a volume of ■filtered distilled or deionized water■$_2$ that is greater than the sample volume to a container, and weigh. Withdraw through the sample feeding device a volume that is appropriate for the specific sampler, and again weigh the container. Determine the sample volume by subtracting the tare volume from the combined sample plus tare volumes. Verify that the value obtained is within ■■$_2$ 5% of the appropriate sample volume for the test. Alternatively, the sample volume may be determined using a suitable Class A graduated cylinder (see *Volumetric Apparatus* [31]). [NOTE—instruments of this type require a variable tare volume. This is the amount of sample withdrawn

prior to counting. This volume may be determined for syringe-operated samplers by setting the sample volume to zero and initiating sampling, so that the only volume of solution drawn is the tare. Subtract the tare volume from the total volume of solution drawn in the sampling cycle to determine the sample volume.]

The reasons for needing to know the sample volume are intuitively obvious. The principle is to separate the tare and sample volumes, then ensure that the sample volume corresponds to that specified. The procedure for doing this may be highly instrument specific. It may be necessary to set various sample volumes on the instrument, then measuring the total amount of sample drawn to get that volume. Once the tare is determined, the sample volume is easily derived. A critical point is that counts should be collected only from the sample volume (i.e., counting should commence coincident with completion of the tare volume and cease when the sample volume is completed.)

SAMPLE FLOW RATE

Verify that the flow rate is within the manufacturer's specifications for the sensor used. This may be accomplished by using a calibrated stop watch to measure the time required for the instrument to withdraw and count a specific sample volume (i.e., the time between beginning and ending of the count cycle as denoted by instrument indicator lights or other means). Sensors may be operated accurately over a range of flow rates. Perform the *Test Procedure* at the same flow rate as that selected for calibration of the instrument.

The sample flow rate selected should generally be in the middle of the vendor's specified range. However, the user sampling more viscous solutions may opt for lower flow rates. Slower flow rates (outside the vendor range) may result in higher counts at specific sizes; higher flow rates can also yield lower counts. Calibration should be performed at the flow rate that will be used for the test, even if there will be a range of flow rates different from that used in calibration over which minimal or no change in counts may be expected.

CALIBRATION

Use one of the following methods.

Manual Method—Calibrate the instrument with a minimum of three calibrators, each consisting of near-monosize polystyrene spheres ■having■1 diameters of about 10, 15, and 25 μm, in an aqueous vehicle.* The calibrator spheres must have a mean diameter of within 5% of the 10-, 15-, and 25-μm nominal diameters and be standardized against materials traceable to NIST standard reference materials. The number of spheres counted must be within the sensor's concentration limit. Prepare suspensions of the calibrator spheres in water at a concentration of 1000 to 5000 particles per mL, determine the channel setting that corresponds to the highest count setting for the sphere distribution. This is determined by using the highest count threshold setting to split the distribution into two bins containing equal numbers of counts, with the instrument set in the differential count mode (moving window half-count method). Use only the central portion of the distribution in this calculation to avoid including asymmetrical portions of the peak. The portion of the distribution, which must be divided equally, is the count window. The window is bounded by threshold settings that will define a threshold voltage window of ± 20% around the mean diameter of the test spheres. The window is intended to include all single spheres, taking into account the standard deviation of the spheres and the sensor resolution, while excluding noise and aggregates of spheres. The value of 20% was chosen based on the worst-case sensor resolution of 10% and the worst-case standard deviation of the spheres of 10%. Since

*ASTM standard F658-87 provides useful discussions pertaining to calibration procedures applying near-monosize latex spheres.

the thresholds are proportional to the area of the spheres rather than the diameter, the lower and upper voltage settings are determined by the equations:

$$V_L = 0.64 V_s,$$

where V_L is the lower voltage setting and V_s is the voltage at the peak center, and

$$V_U = 1.44 V_s,$$

where V_U is the upper voltage setting.

Once the center peak thresholds are determined, use these thresholds for the standards to create a regression of log voltage versus log particle size, from which the instrument settings for the 10- and 25-μm sizes can be determined.

Automated Method—The calibration (size response) curve may be determined for the instrument-sensor system by the use of validated software routines offered by instrument vendors; these may be included as part of the instrument software or used in conjunction with a microcomputer interfaced to the counter. The use of these automated methods is appropriate if the vendor supplies written certification that the software provides a response curve equivalent to that attained by the manual method and if the automated calibration is validated as necessary by the user.

Electronic Method—Using a multichannel peak height analyzer, determine the center channel of the particle counter pulse response for each standard suspension. This peak voltage setting becomes the threshold used for calculation of the voltage response curve for the instrument. The standard suspensions to be used for the calibration are run in order, and median pulse voltages for each are determined. These thresholds are then used to generate the size response curve manually or via software routines. The thresholds determined from the multichannel analyzer data are then transferred to the counter to complete the calibration. If this procedure is used with a comparator-based instrument, the comparators of the counter must be adjusted accurately beforehand.

Three methods of calibration were described above. All follow the same general principle, that of the "moving window half count" shown in Figure 4.1. This consists of moving a "window" of three threshold settings through the instrument response curve for a "monosize" standard. The figure shows the window as defined by sphere diameter, whereas sensor output is proportional to area. The standard sizes specified are an important change from USP XXII. The 10 μm and 25 μm spheres serve to anchor the curve points; the 15 μm size is a necessary reference point to ensure accurate sizing when the USP particle count reference standard is used with laser diode sensors.

Note the requirement for validation of the software calibration. Current and future instruments that are suitable for analysis of pharmaceuticals are software based, and at least one major vendor of particle counters has notified users that older manual models in the hands of customers will only be supported with regard to maintenance and repair as long as present stocks of spare parts last. Thus, the user will be forced to deal with software validation to some extent. Counters automated to some extent are presently, in fact, the only type available for purchase; the major difference between the vendors is the degree to which function of the instrument is under software control.

In this situation, the user is best advised to become familiar with validation. Two texts are highly recommended: Chamberlain (1991) and Mallory (1994). Both provide summaries as well as detail, and this subject is discussed in Chapter 11. Many larger companies have validation "cultures" and will require that the particle counting systems be validated per existing plans and SOPs. Smaller companies may choose to perform minimal validation in-house and rely on vendor information, which is organized into a "package" for the instrument and managed in ongoing fashion. This latter approach puts one at the mercy of the vendor, and in the author's experience, defects both in vendor-supplied information and problems with instrument function have been encountered as a result of more

Figure 4.1. Moving window half count calibration.

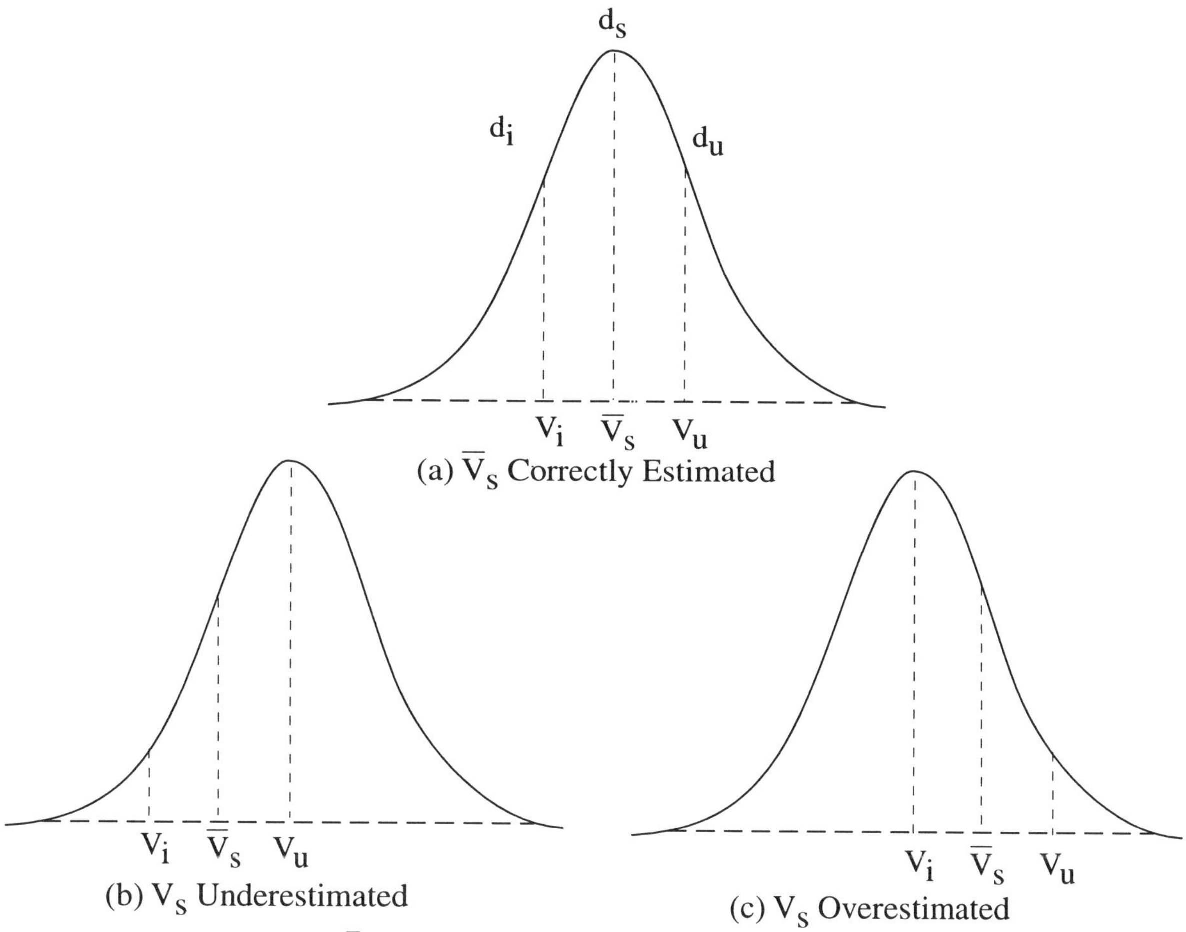

d_s: median diameter: sensor output is $\overline{V}_s$

d_i: deviates from median diameter by -20%: sensor output is $\overline{V}_i$ equal to $(0.8)^2\,\overline{V}_s$

d_u: deviates from median diameter by +20%: sensor output is $\overline{V}_u$ equal to $(1.2)^2\,\overline{V}_s$

rigorous validation procedures performed in-house.

SENSOR RESOLUTION

The particle size resolution of the instrumental particle counter is dependent upon the sensor used and may vary with individual sensors of the same model. Determine the resolution of the particle counter for 10-μm particles using the monosized 10-μm calibrator spheres. The relative standard deviation of the size distribution of the standard particles used is not more than 5%. Acceptable methods of determining particle size resolution are (1) manual determination of the amount of peak broadening due to instrument response; (2) using an electronic method of measuring and sorting particle sensor voltage output with a multichannel analyzer; and (3) automated methods.

Manual Method—Adjust the particle counter to operate in the cumulative mode or total count mode. Refer to the calibration curve obtained earlier, and determine the threshold voltage for the

10-μm monosize spheres. Adjust 3 channels of the counter to be used in the calibration procedure as follows:

> *Channel 1* is set for 90% of the threshold voltage.
>
> *Channel 2* is set for the threshold voltage.
>
> *Channel 3* is set for 110% of the threshold voltage.

Draw a sample through the sensor, observing the count in Channel 2. When the particle count in that channel has reached approximately 1000, stop counting, and observe the counts in Channels 1 and 3. Check to see if the Channel 1 count and the Channel 3 count are 168 ± 10% and 32 ± 10%, respectively, of the count in Channel 2. If not, adjust Channel 1 and Channel 3 thresholds to meet these criteria. When these criteria have been satisfied, draw a sample of suspension through the counter until the counts in Channel 2 have reached approximately 10,000, or until an appropriate volume (e.g., 10 mL) of the sphere suspension has been counted. Verify that Channel 1 and Channel 3 counts are 168 ± 3% and 32 ± 3%, respectively, of the count in Channel 2.

Record the particle size for the thresholds just determined for Channels 1, 2, and 3. Subtract the particle size for Channel 2 from the size for Channel 3. Subtract the particle size for Channel 1 from the size for Channel 2. The values so determined are the observed standard deviations on the positive and negative side of the mean count for the 10-μm standard. Calculate the percentage of resolution of the sensor by the formula:

$$100\left[\left(\sqrt{S_O^2 - S_S^2}\right)/D\right] < 1$$

in which S_O is the highest observed standard deviation determined for the sphere, S_S is the supplier's reported standard deviation for the spheres, and D is the diameter, in μm, of the spheres as specified by the supplier. The resolution is not more than 10%.

See Figure 4.2 for an outline of resolution determination. Resolution is a key determinant of counting accuracy, since the width of the instrument response to particles of a given size determines to what extent particles smaller than that size will be counted above the threshold in question. This can be a serious issue in the counting of SVI units, which can have high numbers of particles per mL slightly less than the 10 μm threshold. Another comment is that the statement "Verify that Channel 1 and Channel 3 counts are 168 ± 3 percent and 32 ± 3 percent of the count in Channel 2" incorporates an editorial error; the numerical values should be 1.68 ±3 percent and 0.32 ± 3 percent instead of the whole numbers presented.

> **Automated Method**—Software is available for some counters that allows for the automated determination of sensor resolution. This software may be included in the instrument or used in conjunction with a microcomputer interfaced to the counter. The use of these automated methods is appropriate if the vendor supplies written certification that the software provides a resolution determination equivalent to the manual method and if the automated resolution determination is validated as necessary by the user.

Figure 4.2. Sensor resolution determination.

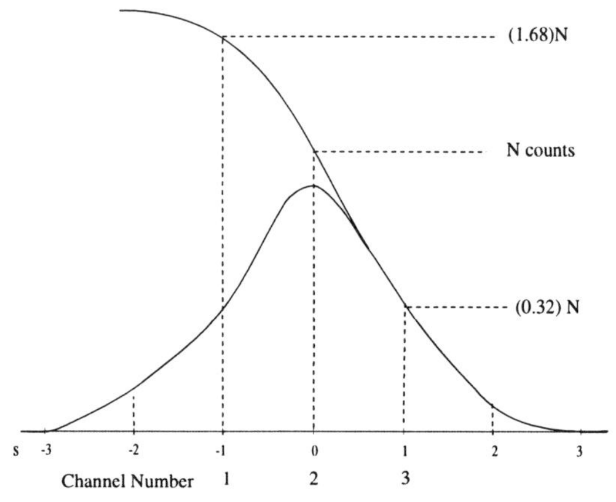

Electronic Method—Record the voltage output distribution of the particle sensor, using a multichannel analyzer while sampling a suspension of the 10-µm particle size standard. To determine resolution, move the cursor of the multichannel analyzer up and down the electric potential scale from the median pulse voltage to identify a channel on each side of the 10-µm peak that has approximately 61% of the counts observed in the center channel. Use of the counter size response curve to convert the mV values of these two channels to particle sizes provides the particle size at within 1 standard deviation of the 10-µm standard. Use these values to calculate the resolution as described under *Manual Method.*

In the Sensor Resolution section, note that under "Automated Method" validation is again spelled out.

PARTICLE COUNTING ACCURACY

Determine the particle counting accuracy of the instrument using *Method 1* (for small-volume injections) or *Method 2* (for large-volume injections).

Method 1—

Procedure—Prepare the suspension and blank using the USP Particle Count RS. ■With the instrument set to count in the cumulative (total) mode, collect counts at settings of not less than 10 µm and not less than 15 µm.■$_5$ Mix the blank by inverting 25 times within 10 seconds, and degas the mixture by sonicating ■(at ■80 to 120 watts■$_5$) for about■$_2$ 30 seconds or by allowing to stand. Remove the closure from the container, and gently stir the contents by hand-swirling or by mechanical means, taking care not to introduce air bubbles or contamination. Stir continuously throughout the analysis. Withdraw directly from the container three consecutive volumes of not less than 5 mL each, obtain the particle counts, and discard the data from the first portion. [NOTE—Complete the procedure within five minutes.] Repeat the procedure, using the suspension in place of the blank. From the averages of the counts resulting from the analysis of the two portions of the suspension at ■not less than■$_5$ 10 µm and from the analysis of the two portions of the blank at ■not less than■$_5$ 10 µm, calculate the number of particles in each ml■■$_2$ by the formula:

$$(P_S - P_b)/V,$$

in which P_s is the average particle count obtained from the suspension, P_b is the average particle count obtained from the blank, and V is the average volume, in mL, of the 4 portions tested. Repeat the calculations, using the results obtained at ■the setting of not less than■$_5$ 15 µm.

Interpretation—The instrument meets the requirements for *Particle Counting Accuracy* if the count obtained at ■not less than■$_5$ 10 µm and the ratio of the counts obtained at ■not less than■$_5$ 10 µm to those obtained at ■not less than■$_5$ 15 µm conform to the values that accompany the USP Particle Count RS. If the instrument does not meet the requirements for *Particle Counting Accuracy*, recalibrate with the remaining suspension and blank. ■■$_1$ If the results of the second test are within the limits given above, the instrument meets the requirements of the test for *Particle Counting Accuracy*. If on the second attempt the system does not meet the requirements of the test, determine and correct the source of the failures, and retest the instrument.

Method ■2■$_5$—

Procedure—Using standard calibrator spheres having a nominal diameter of 15 to 30 µm, prepare a suspension containing between 50 and 200 particles per mL. Degas the suspension by ■sonicating■$_2$ ■(at 80 to 120 watts)■$_5$ for ■about■$_2$ 30 seconds or by allowing to stand. Properly suspend the particles by stirring gently, and perform five counts on 5-mL volumes of the suspension, using the particle counter 10-µm size threshold. Obtain the mean cumulative particle count per mL. Pipe a volume of this

> suspension containing 250 to 500 particles into a filter funnel prepared as described for *Filtration Apparatus* under *Microscopic Particle Count*. After drying the membrane, count the total number of standard spheres collected on the membrane filter. This count should be within 20% of the mean instrumental count per mL for the suspension.

One notable thing about the particle counting accuracy test is that two methods are specified. The USP PCRS simply provides a 15 μm sphere suspension of narrow distribution and known number for use as a test standard. Counts at the ≥ 10 μm threshold should be approximately 2 times those at the ≥ 15 μm threshold, which splits the 15 μm peak. A second method is specified for samples used in the testing of LVI to allow a larger volume of calibrate suspension, equal to larger sample volumes to be tested. The PCRS is fairly expensive ($460.00 per set), and its use should not be undertaken until the instrument is believed to be in calibration. One way to avoid wasting a kit is simply to first use a suspension of 15 μm standard spheres of narrow distribution as a check on the calibration before the standard is run. This suspension should behave just as the standard does. While a number of spheres will not be known, this is not the critical factor; any counter with a reasonable setting of thresholds should count all of the 15 μm spheres at the ≥ 10 um threshold. The more critical test component is the splitting of the 15 μm peak, which is readily checked using the sphere suspension made up in the lab.

A word here is in order regarding whether instrument standardization should be performed in the laboratory by users or by the vendor's service personnel. Generally, it is the author's belief that the user is better off to perform the calibration based on the degree of control afforded and the level of knowledge about the instrument that is attained as a result. If the vendor performs the standardization, the user should ensure that all components of the USP procedure are addressed. Vendor or service lab calibrations commonly involve shipping the sensor to a service facility. User-performed calibrations have the advantage of allowing the particle counting instrument to be standardized as a system (i.e., a specific sensor in combination with a specific counter and sampler). This latter approach is generally more desirable for GLP applications.

Early on, some service labs performed calibration by their service specifications, then certified to the user that a "USP" procedure had been performed. Regulatory auditors have shown some tendency to be less critical of the vendor calibrations than those performed by the user.

What if a user performs both LVI and SVI testing employing the same instrument? Must they perform both count accuracy tests? There is no easy answer, but the implicit intent of the USP is that the count accuracy test be applied using a volume of test material equal to that used in performance of the test. Thus, if LVI are tested with 5 mL samples, this rationale would tend to justify the use of the PCRS alone.

Another editorial note: Under **Method 2** in the third to the last sentence . . . *Filtration Apparatus* under *Microscopic Particle Count* the word *Test* should be added (*Microscopic Particle Count Test*). With regard to the entire excerpt, the particle size ranges for the tests are incorrectly described. The current version reads: "From the averages of the counts at not less than 10 μm and from the analysis of the two portions of the blank at not less than 10 μm, calculate . . . ." The suggested change is as follows: Change the phrase "not less than" wherever it appears in this section (a total of five times) to "greater than or equal to." Reason for the change: In editing of the USP 23 version of <788> the editors changed the ≥ symbol to "not less than." The correct verbal description of ≥ is "greater than or equal to."

Test Environment

> Perform the test in an environment that does not contribute any significant amount of particulate matter. Specimens must be cleaned to the extent that any level of extraneous particles added has a negligible effect on the outcome of the test. The test specimen, glassware, closures, and other required equipment preferably are prepared in an environment protected by high-efficiency particulate air (HEPA) filters. Nonshedding garments and powder-free gloves preferably are worn throughout the preparation of samples.
>
> Cleanse glassware, closures, and other required equipment preferably by immersing and scrubbing in warm, nonionic

> detergent solution. Rinse in flowing tap water, and then rinse again in flowing filtered ■distilled or deionized■$_{2}$ water. Organic solvents may also be used to facilitate cleaning. [NOTE—These steps describe one way to clean equipment; alternatively, particulate-free equipment may be obtained from a suitable vendor.] Finally, rinse the equipment in filtered ■distilled or de-ionized■$_{2}$ water, using a hand-held pressure nozzle with final filter or other appropriate filtered water source, such as distilled ■or deionized■$_{2}$ water passed through a capsule filter. The filter used should have a porosity of 1.2 μm or finer.

With regard to the test environment, the word *preferably* is used three times. In a practical consideration, it is not necessary to have the light obscuration counter in a HEPA–protected environment, but care must be exercised in preparing samples (particularly pooling) and in cleaning glassware. Gloves and garments are definitely preferable. The user who takes advantage of "preferably" to avoid using safeguards may encounter problems with elevated background (blank) counts. A number of schemes for protecting the sample will result in satisfactory blanks.

> To collect ■blank■$_{2}$ counts, use a cleaned vessel of the type and volume representative of that to be used in the test. Place a ■50-mL■$_{2}$ volume of filtered ■distilled or deionized■$_{2}$ water in the vessel, and agitate the sample in the cleaned glassware by inversion or swirling. Degas by sonicating ■(at■80 to 120 watts■$_{5}$ for about■$_{2}$ 30 seconds or by allowing to stand. Swirl the vessel containing the water sample by hand or agitate by mechanical means to suspend particles. Withdraw and obtain the particle counts for three consecutive samples of not less than 5 mL each, disregarding the first count. If more than 10 particles of 10 μm or greater size, or more than 2 particles of 25 μm or greater size are observed in the combined 10-mL sample, the environment is not suitable for particulate analysis: the filtered ■distilled or deionized■$_{2}$ water and glassware have not been properly prepared or the counter is generating spurious counts. In this case, repeat the preparatory steps until conditions of analysis are suitable for the test.

The procedure for obtaining "blank" or background counts is intended to ensure that the contribution of extraneous particles by the environment and test articles is held below a specific level. The intent is to use a cleaned vessel of the type employed in the test. If a 30 mL beaker is used for the posting of samples, obviously a 50 mL volume of water is too great; this semantic error will be addressed in future changes to the method to specify "50 mL or smaller volume as appropriate." "Disregard" means "discard" or otherwise disallow the first count.

One problem that arises in this procedure for "blanking" the test is that a 50 mL blank volume may be greater than the pool volume tested. As an example, ten 2.5 mL units of monoclonal antibody give a 25 mL pool volume. This pool may be tested easily in a 30 mL beaker, but if this volume is placed in a 50 mL beaker, air may be drawn into the sensor due to the decreased solution depth. The overriding consideration here is that the blank test is intended to give an uniform baseline of background counts. Thus, the user would probably be best advised to perform the blank as directed, then test in a 30 mL beaker. The careful individual would probably choose to verify that a blank performed on a 30 mL beaker would give the same result as with a 50 mL beaker; the result should, of course, be documented.

> **Test Procedure**
>
> TEST PREPARATION
>
> ■Prepare the test specimens in the following sequence. Outside of the laminar enclosure, remove outer closures, sealing bands, and any loose or shedding paper labels. Rinse the exteriors of the containers with filtered distilled or deionized water as directed under *Test Environment*. Protect the containers from environmental contamination until analyzed. Withdraw the contents of the containers under test in a manner least likely to generate particles that could enter the sample. Contents of containers with removable stoppers may be withdrawn directly by removing the closures. Sampling devices having a needle

to penetrate the unit closure may also be employed. Products packaged in flexible plastic containers may be sampled by cutting the medication or administration port tube or a corner from the unit with a suitably cleaned razor blade or scissors.

The sample material must be removed in the manner least likely to generate contamination. This is the key consideration to the entire procedure. Some steps may be problematical in this regard. The removal of rubber stoppers can free or generate debris between the stopper and vial neck. Importantly, units of 25 mL or greater may be tested singly (number of samples is important as stated above). The 10-unit pool is statistically a single sample. One consideration is that sufficient volume for the test must be pooled. A volume of nearly 20 mL will be required for 3–5 mL samples and tare volumes.

Dry or lyophilized products may be constituted either by removing the closure to add diluent or by injecting diluent with a hypodermic syringe having a 1.2-µm or finer syringe filter. If test specimens are to be pooled, remove the closure and empty the contents into a clean container.

The number of test specimens must be adequate to provide a statistically sound assessment of whether a batch or other large group of units represented by the test specimens meets or exceeds the limits. If the volume in the container is less than 25 mL, test a solution pool of 10 or more units. Single small-volume injection units may be tested if the individual unit volume is 25 mL or more. For large-volume injections, single units are tested. For large-volume injections or for small-volume injections where the individual unit volume is 25 mL or more, fewer than 10 single units may be tested, based on the definition of an appropriate sampling plan.■5

■PRODUCT■5 DETERMINATION

■Depending upon the dosage form being tested, proceed as directed under the appropriate category below.

Liquid Preparations—*Where the volume in the container is less than 25 mL*—Prepare the containers as directed under *Test Preparation*. Mix and suspend the particulate matter in each unit by inverting the unit 20 times. [NOTE—Because of the small volume of some products, it may be necessary to agitate the solution more vigorously to suspend the particles properly.] In a cleaned container, open and combine the contents of 10 or more units to obtain a volume of not less than 20 mL. Degas the pooled solution by sonicating for about 30 seconds or by allowing the solution to stand undisturbed until it is free from air bubbles. Gently stir the contents of the container by hand-swirling or by mechanical means, taking care not to introduce air bubbles or contamination. Withdraw a minimum of three aliquots, each not less than 5 mL in volume, into the light obscuration counter sensor. Discard the data from the first portion.

Where the volume in the container is 25 mL or more—Prepare the containers as directed under *Test Preparation*. Mix and suspend the particulate matter in each unit by inverting the unit 20 times. Degas the solution by sonicating for about 30 seconds or by allowing the solution to stand undisturbed until it is free from air bubbles. Remove the closure of the unit or effect entry by other means so that the counter probe can be inserted into the middle of the solution. Gently stir the contents of the unit by hand-swirling or by mechanical means. Withdraw not less than three aliquots, each not less than 5 mL in volume, into the light obscuration counter sensor.

Discard the data from the first portion.

Dry or Lyophilized Preparations—Prepare the containers as directed under *Test Preparation*. Open each container, taking care not to contaminate the opening or cover. Constitute as directed under *Test Preparation*, using the specified volume of filtered water or an appropriate filtered diluent if water is not suitable. Replace the closure, and manually agitate the container sufficiently to ensure dissolution of the drug. Alternatively, the product may be reconstituted with filtered diluent using a hypodermic needle attached to a syringe to pierce the closure. [NOTE—For some dry or lyophilized products, it may be necessary to let the units stand for a suitable interval, and then agitate again to effect complete dissolution.] After the drug in the constituted sample is completely dissolved, mix and suspend the particulate matter present in each unit by inverting it 20 times prior to analysis. Proceed as directed for the appropriate unit volume under *Liquid Preparations*, and analyze by withdrawing a minimum of three aliquots, each not less than 5 mL in volume, into the light obscuration counter sensor. Discard the data from the first portion.

Products Packaged with Dual Compartments Constructed to Hold the Drug Product and a Solvent in Separate Compartments—Prepare the units to be tested as directed under *Test Preparation*. Mix each unit as directed in the labeling, activating and agitating so as to ensure thorough mixing of the separate components and drug dissolution. Degas the units to be tested by sonicating or by allowing the solution to stand undisturbed until it is free from any air bubbles. Proceed as directed for the appropriate unit volume under *Liquid Preparations*, and analyze by withdrawing a minimum of three aliquots, each not less than 5 mL in volume, into the light obscuration counter sensor. Discard the data from the first portion.

Products Labeled "Pharmacy Bulk Package Not for Direct Infusion"—Proceed as directed for *Liquid Preparations or Dry or Lyophilized Preparations* where the volume is 25 mL or more. Calculate the test result on a portion that is equivalent to the maximum dose given in the labeling. For example, if the total bulk package volume is 100 mL and the maximum dose volume is 10 mL, then the average light obscuration particle count per mL would be multiplied by 10 to obtain the test result based on the 10-mL maximum dose. [NOTE—For the calculations of test results, consider this maximum dose portion to be the equivalent of the contents of one full container.]

Calculations

Pooled Samples (Small-volume Injections)—Average the counts from the ■two■1 or more aliquot portions analyzed. Calculate the number of particles in each container by the formula:

$$PV_1/V_a n,$$

in which P is the average particle count obtained from the portions analyzed, V_1, is the volume of pooled sample, in mL, V_a is the volume, in mL, of each portion analyzed, and n is the number of containers pooled.

Individual Samples (Small-volume Injections)—Average the counts obtained for the 5-mL or greater aliquot portions from each separate unit analyzed, and calculate the number of particles in each container by the formula:

$$PV/V_a,$$

in which P is the average particle count obtained from the portions analyzed, V is the volume, in mL, of the tested unit, and V_a is the volume, in mL, of each portion analyzed.

Individual Unit Samples (Large-volume Injections)—Average the counts obtained for the two or more 5-mL aliquot portions taken from the solution unit. Calculate the number of particles in each mL of injection taken by the formula:

$$P/V,$$

in which P is the average particle count for an individual 5 mL or greater sample volume, and V is the volume, in mL, of the portion taken.

■For all types of product, if the tested material has been diluted to effect a decrease in viscosity, the dilution factor must be accounted for in the calculation of the final test result.■5

The calculations are straightforward, with the possible exception of a pharmacy bulk pack. Here, the maximum dose is taken as the volume of a single SVI container and compared to limits in this fashion. For example, a count of 20 particles at the ≥ 10 μm size per mL would equate to 200 particles at this size in a 10 mL dose, and thus be well within the limits of 6,000 particles/container ≥ 10 μm.

Interpretation

The injection meets the requirements of the test if the average number of particles present in the units tested does not exceed the appropriate value listed in Table 1. If the average number of particles exceeds the limit, test the article by the *Microscopic Particle Count Test.*

	≥ 10 μm	≥ 25 μm
Small-volume Injections	6000	600 per container
Large-volume Injections	25	3 per mL

NOTE: The limits apply to the average count for each unit tested. Obviously, a manufacturer will not expect to obtain a result almost at the limits and will take pains to ensure that any such result can be statistically supported. Given a normal distribution of counts, an average value near the limits (e.g., 10 percent below) means that some units in the batch will exceed the limits.

LIGHT OBSCURATION TEST SUMMARY

A summary of changes from the USP XXII <788> *Light Obscuration Test* is as follows:

- More latitude is allowed for automated calibration methods (per vendor methodology).
- More latitude given in counter choice.
- Allowance for testing of single units (25 mL or greater) instead of pooling.
- Stratified test plan allowed with light obscuration test as a first pass method, followed by the microscopic test to resolve any light obscuration failures.
- Use of 10 μm, 15 μm, and 25 μm calibrated particles allowed.
- Limits decreased for both LVI and SVI.
- Counter validation is required if an automated counter is used.
- A training requirement is specifically called out.

The light obscuration method is also outlined in Figures 4.3 to 4.5.

Figure 4.3. The USP 23 <788> Stage 1 Light Obscuration Test

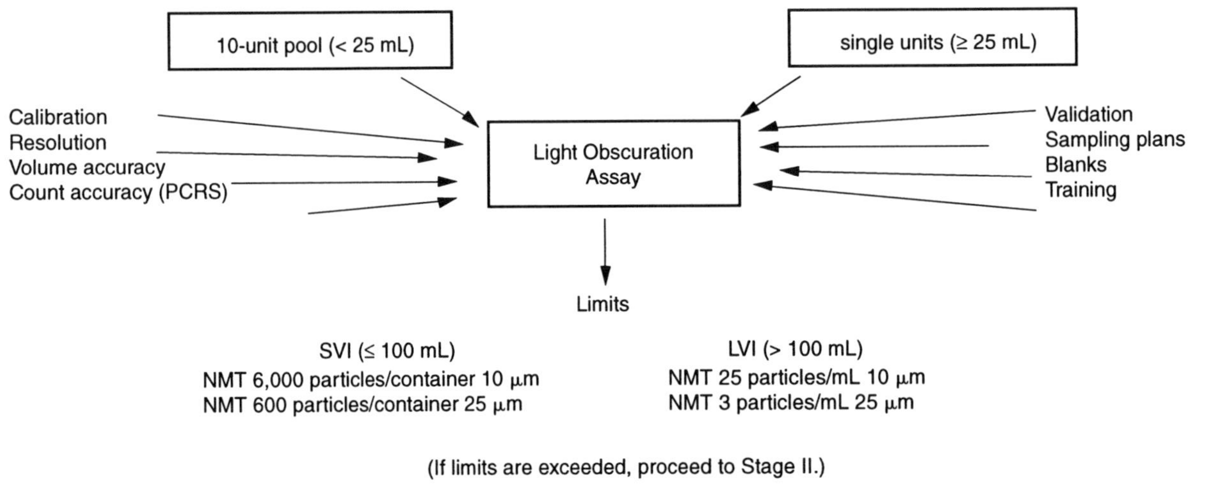

Figure 4.4. The USP 23 <788> Light Obscuration Test (LVI)

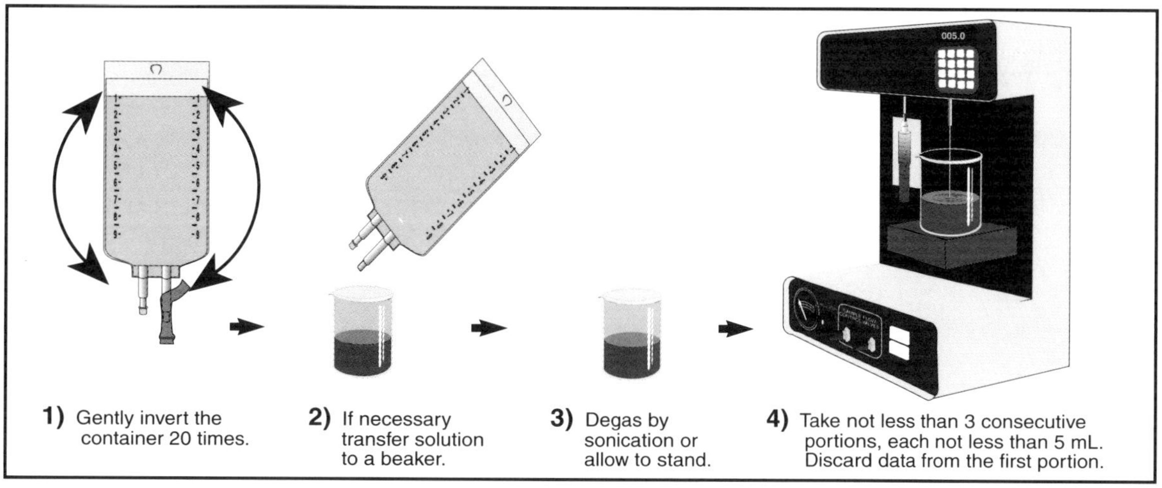
005.0
1) Gently invert the container 20 times.
2) If necessary transfer solution to a beaker.
3) Degas by sonication or allow to stand.
4) Take not less than 3 consecutive portions, each not less than 5 mL. Discard data from the first portion.

USP LVI Limit: NMT 25 particles/mL ≥ 10 μm
3 particles/mL ≥ 25 μm

USP SVI Limit: NMT 6,000 particles/mL ≥ 10 μm
600 particles/mL ≥ 25 μm

Figure 4.5. The USP 23 <788> Light Obscuration Test (SVI)

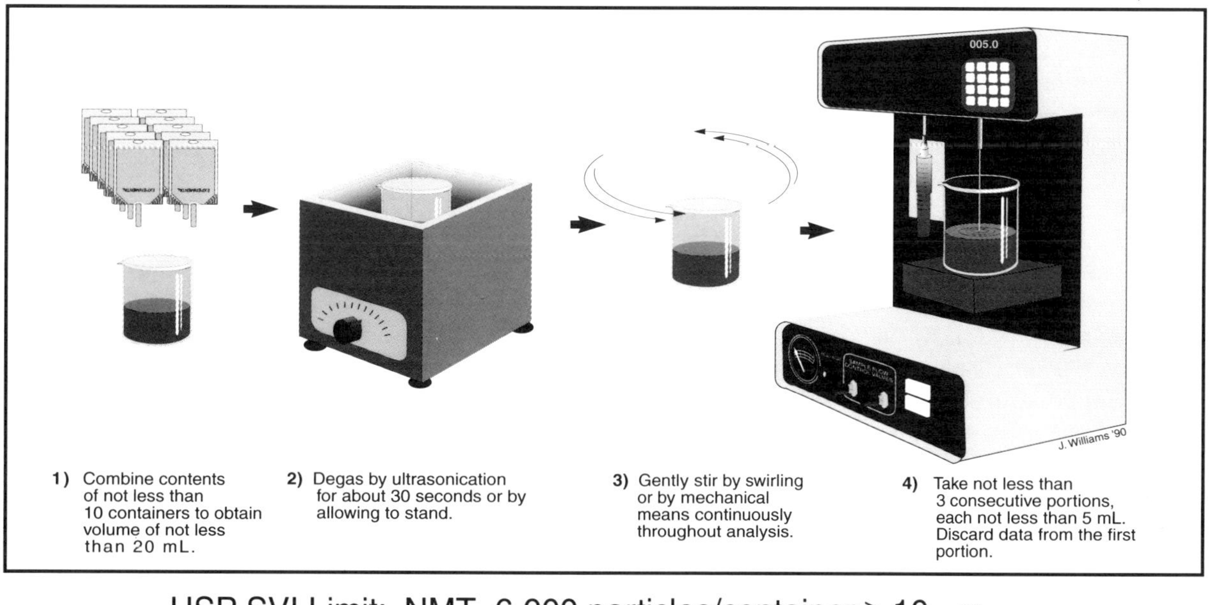

USP SVI Limit: NMT 6,000 particles/container ≥ 10 μm
600 particles/container ≥ 25 μm

THE USP <788> MICROSCOPIC TEST

The microscopic assay is not amenable to intuitive execution, even for an experienced analyst. Hence, significant space is allotted for text and figures in explanation. Training sessions are available through the Parenteral Drug Association and are very useful for those performing the test for the first time.

> ***Change to read:***
>
> MICROSCOPIC PARTICLE COUNT TEST
>
> The microscopic particulate matter test may be applied to both large-volume and small-volume injections. This test enumerates subvisible, essentially solid, particulate matter in these products on a per-volume or per-container basis, after collection on a microporous membrane filter. Some articles cannot be tested meaningfully by light obscuration. In such cases, individual monographs specify only this microscopic assay. Solutions exempted from analysis using the microscopic assay are identified on a monograph basis. Examples are solutions of viscosity too high to filter readily (e.g., concentrated dextrose, starch solutions, or dextrans). ■In performance of the microscopic assay do not attempt to size or enumerate amorphous, semiliquid, or otherwise morphologically indistinct materials that have the appearance of a stain or discoloration on the membrane surface.■$_1$ These materials show little or no surface relief and present a gelatinous or film-like appearance. Since in solution this material consists of units on the order of 1 μm or less, which may be counted only after aggregation or deformation on an analytical membrane, interpretation of enumeration may be aided by testing a sample of the solution by the light obscuration particle count method.

Like the previous <788> test, the new test is filtration based and only particles that are filterable (i.e., can be collected on a microporous membrane) will be enumerated. Exemptions mentioned on a monograph basis do not exist as yet. Again, high viscosity may cause a problem for both light obscuration and microscopic assays; a product with a problematical viscosity can be diluted, based on information provided earlier for the light obscuration assay. Such dilution must be quantitative. In many cases, the method may be modified to test suspensions or emulsions not amenable to light obscuration; color or refractive index are not considerations.

The microscopic assay is of specific and critical importance regarding two types of materials: (1) liquids, such as silicone compounds that are immiscible in aqueous systems, which can cause spurious light obscuration counts if dispersed in microdroplet form, and (2) amorphous (semiliquid) materials that can result from degradative processes with proteins, amino acids, or drugs. The latter will be collected on a filter but cannot be accurately counted due to their indistinct morphology. The microscopic assay obviously will not count immiscible liquids, so what is to prevent high levels of silicone in product from being discounted based on the microscopic assay? The microscopic assay is only performed as a second stage assay. A high percentage of failed light obscuration tests with a microscopic "pass" will warn both the manufacturer and auditors that something is amiss. Similarly, high levels of amorphous material responsible for light obscuration failures would be detected on microscopic testing.

In designing the improved test, careful attention was given to factors increasing the sensitivity of detection and count precision. Two illumination types are combined to effect sizing of particles by equivalent circular diameter, which results in higher measurement precision (i.e., decreased variability between two analysts, or for multiple counts by a single analyst). Also, total unit/pool volume filtration contributes to decreased assay variability regarding intersample variation. Finally, provision for statistical counting is made for heavily particulated membranes as might result from a 10-unit SVI pool.

> **Test Apparatus**
>
> **Microscope**—Use a compound binocular microscope that corrects for changes in interpupillary distance by maintaining a constant tube length. The objective and eyepiece combination of lenses must give a magnification of 100 ± 10×.

The objective must be of 10× nominal magnification, a planar achromat or better in quality, with a minimum numerical aperture of 0.25. In addition, the objective must be compatible with an episcopic illuminator attachment. The eyepieces must be of 10× magnification.■ ■5 In addition, one eyepiece must be designed to accept and focus an eyepiece graticule. The microscope must have a mechanical stage capable of holding and traversing the entire filtration area of a 25-mm or 47-mm membrane filter.

Regarding microscope selection, a philosophical principle is of great importance: Critical judgments must be made based on the magnified image that will be obtained from particulate matter on a filtration membrane. A higher quality image will facilitate the analysis significantly, and the best available or affordable microscope should be used. Changes in interpupillary distance must be made without changing the optical path length, which will change magnification.

Illuminators—Two illuminators are required. One is an external, focusable auxiliary illuminator adjustable to give incident oblique reflected illumination at an angle of 10 degrees to 20 degrees. The other is an episcopic brightfield illuminator internal to the microscope. Both illuminators must be of a wattage sufficient to provide a bright, even source of illumination and may be equipped with blue daylight filters to decrease operator fatigue during use.

The episcopic brightfield illuminator is essential to the performance of the revised test (see Figure 4.6). This accessory is not available for many older microscopes, and may necessitate the purchase of a new microscope. Note that the path of both illumination and imaging reduction are through the objective lens. This has the effect of generating specular reflection from any smooth-surfaced particle with a surface oriented normal to the viewing axis. Thus, particles that would be invisible or poorly seen with lateral oblique illumination can be visualized.

Figure 4.6. Types of microscopic incident illumination.

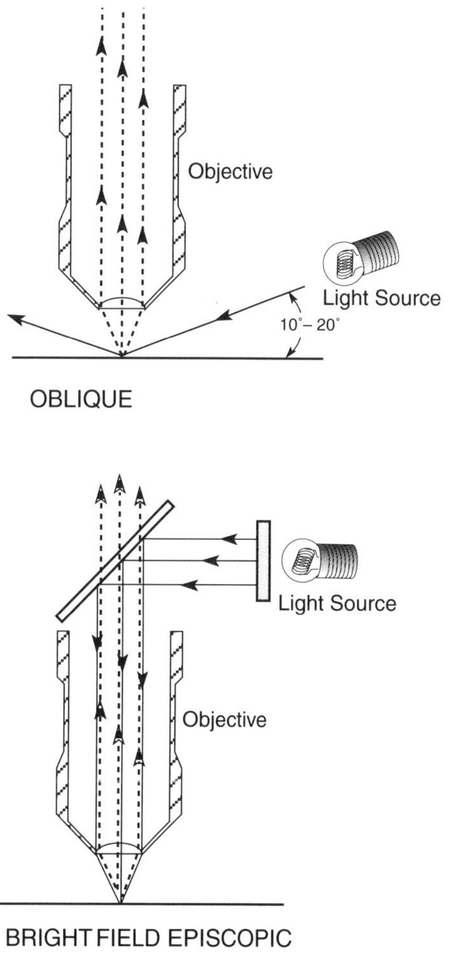

Circular Diameter Graticule—Use a circular diameter graticule (see Figure 1) matched to the microscope model objective and eyepiece such that the sizing circles are within 2% of the stated size at the plane of the stage.

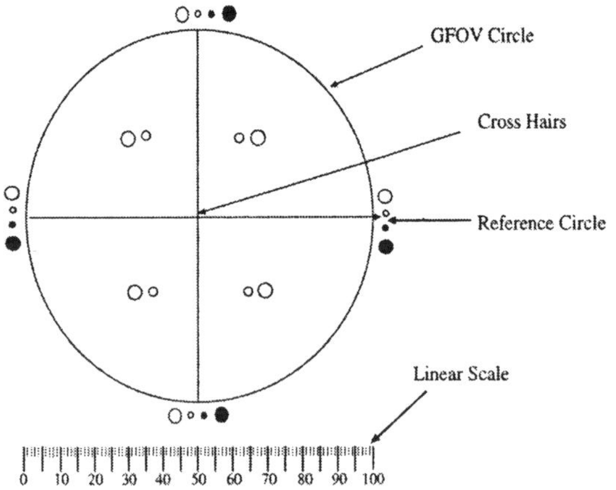

Fig. 1. Circular diameter graticule. The large circle divided by crosshairs into quadrants is designated the graticule field of view (GFOV). Transparent and black circles having 10-μm and 25-μm diameters at 100× are provided as comparison scales for particle sizing.

Micrometer—Use a stage micrometer, graduated in 10-μm increments, that is certified by NIST.

The circular diameter graticule (see Appendix II of this chapter for vendors) represents the most significant change in the USP 23 version of <788>. The graticule is shown in Figure 1 of the USP text above. The new graticule provides a means of sizing particles based on their equivalent circular area diameter rather than a length (linear) measurement. This mental comparison can be made more quickly than a linear measurement per the old <788> method, which was accomplished by the analyst mentally referencing the particle to a fixed linear scale. The operator of the circular diameter graticule is described in a following section. A NIST–traceable stage micrometer remains a necessity for microscope calibration and determining the precision of the circular diameter graticule.

Filtration Apparatus—Use a filter funnel suitable for the volume to be tested, having a minimum diameter of about 21 mm. The funnel is made of plastic, glass, or stainless steel. Use a filter support made of stainless steel screen or sintered glass as the filtration diffuser. The filtration apparatus is equipped with a vacuum source, a *solvent dispenser* capable of delivering solvents filtered at 1.2 μm or finer retention rating at a range of pressures from 10 psi to 80 psi, and *membrane filters* (25 mm or 47 mm non-gridded or gridded, black or dark gray or of suitable material compatible with the product,■2 with a porosity of 1.0 μm or finer). Use suitably cleaned blunt forceps to handle membrane filters.

A suitable filtration funnel is shown in Figure 4.7. The most widely used filtration apparatus for the test is made of polysulfone or polycarbonate and is transparent, with easily

Figure 4.7. Plastic filtration funnel.

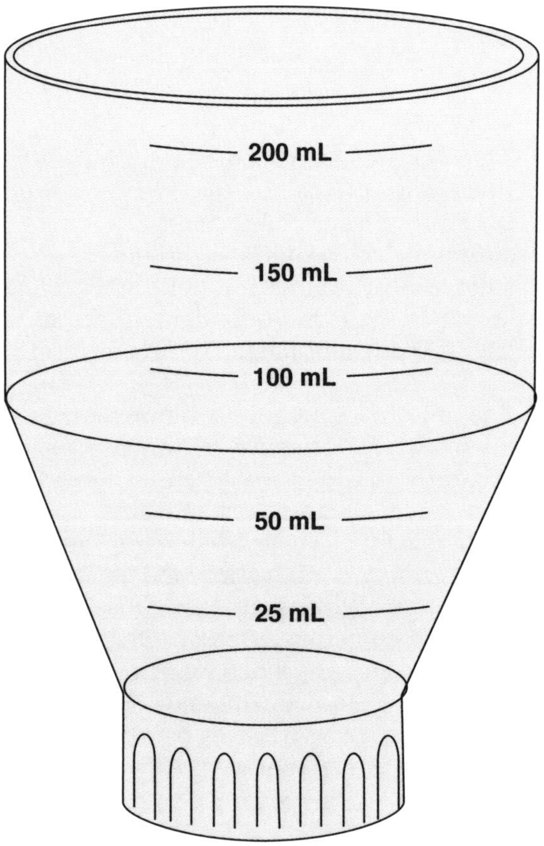

readable gradations. Glass and stainless steel are allowed but are not popular because of their weight, the possibility of breakage (glass), and their being more difficult to handle and assemble. The complete filtration apparatus consists of a funnel and base. Use of a screen support filtration diffuser is prescribed in order to improve the flow characteristics through the filter and achieve a more even particle distribution. Gridded or ungridded dark-colored filters may be used. While many analysts prefer gridded membranes to assist them in keeping track of the location on the membrane, the grid lines can interfere with counting, and particles falling on grid lines are more difficult to observe. The larger diameter (47 mm) filter has the advantage of providing a 4× greater surface area for particle retention and will facilitate counting of higher numbers of particles.

Test Environment

A laminar flow hood or other laminar airflow enclosure, having a capacity sufficient to envelope the area in which the analysis is prepared, with HEPA-filtered air having not more than 100 particles (0.5 µm or ■larger)■$_5$ per cubic foot. For the blank determination, deliver ■a 50-mL■$_2$ volume of filtered ■distilled or deionized■$_2$ water. ■■$_2$ Apply vacuum, and draw the entire volume of water through the membrane filter. Remove the membrane from the filter funnel base, and place atop a strip of doublesided tape in a petri slide or petri dish. After allowing the membrane to dry, examine it microscopically at a magnification of 100×. If not more than ■20■$_2$ particles 10 µm or larger in size and ■5■$_2$ particles 25 µm or larger are present within the filtration area, the background particle level is sufficiently low for performance of the microscopic assay.

■Throughout this procedure, it is preferable to use powder-free gloves and thoroughly clean glassware and equipment. Prior to conducting the test, clean the work surfaces of the laminar flow enclosure with an appropriate solvent. Glassware and equipment should be rinsed successively with a warm, residue-free solution of detergent, hot water, filtered distilled or deionized water, and isopropyl alcohol. [NOTE—Prior to use, filter the distilled or deionized water and the isopropyl alcohol with filters having a porosity of 1.2 µm or finer.] Perform the rinsing under the laminar flow enclosure equipped with HEPA filters. Allow the glassware and filtration apparatus to dry under the hood, upstream of all other operations. Preferably, the hood is located in a separate room that is supplied with filtered air-conditioned air and maintained under positive pressure with respect to the surrounding areas.■$_5$

The test environment is of relatively greater importance for the microscopic assay than for the light obscuration test. There are several reasons for this. Particles that settle from the air onto the surface of the solution being filtered will be collected on the analytical membrane and will be counted. Further, there are more manual manipulations in the microscopic test, resulting in more opportunities to transfer contaminants to the test sample or membrane. Particles settling on the surface of a liquid to be analyzed by light obscuration counting do not pass through the surface into the solution. Since the instrument probe draws solution from beneath the surface, these particles do not affect the analysis. Powder-free gloves and nonshedding smocks should definitely be used, although the method description states "preferably." Empirically, the analysts performing the test will learn what steps are essential to obtaining an appropriate blank and will build their individual techniques around those steps.

MICROSCOPE PREPARATION

Place the auxiliary illuminator close to the microscope stage, focusing the illuminator to give a concentrated area of illumination on a filter membrane positioned on the microscope stage. Adjust the illuminator height so that the angle of incidence of the light is 10° to 20° with the horizontal. Using the internal episcopic brightfield illuminator, fully open the field and aperture diaphragms. Center the lamp filament, and focus the microscope on a filter containing particles. Adjust the intensity of reflected illumination until particles are clearly visible and show pronounced shadows.

> Adjust the intensity of episcopic illumination to the lowest setting, then increase the intensity of episcopic illumination until shadows cast by particles show the least perceptible decrease in contrast.

The steps in achieving proper illumination are critical to the conduct of the method. A key problem with the previous version of <788> was that the simple source of lateral-reflected illumination did not allow critical visualization of reflective flakelike particles (often crystalline), which might represent degradation of a drug or precipitate. Similarly, the presence or absence of amorphous material could not be assessed. Massive (equant) particles that cast a shadow on the side of the particle away from the light source were, however, readily visualized and sized. The combination of bright field episcopic illumination with lateral-reflected illumination allows both thin, reflective, crystalline particles and equant particles to be visualized critically. At the same time, amorphous particles, which are thin, transparent, and often resemble a stain on the membrane surface, are not well visualized, and are unlikely to interfere with the counting of other types of particles.

To achieve the proper conditions of illumination, it is essential to follow the procedure outlined in the last sentence of the excerpted paragraph. Episcopic illumination must be used to image thin, transparent, reflective particles. If too much episcopic illumination is used, however, very thin amorphous material will be visualized and may confuse the analysis.

> OPERATION OF CIRCULAR DIAMETER GRATICULE
>
> The relative error of the graticule used must initially be measured with an NIST-certified stage micrometer. To accomplish this, align the graticule micrometer scale with the stage micrometer so that they are parallel. (Compare the scales, using as large a number of graduations on each as possible). Read the number of graticule scale divisions, *GSD*, compared to stage micrometer divisions, *SMD*. Calculate the relative error by the formula:
>
> $$\blacksquare 100[(GSD - SMD)/SMD]_{\blacksquare 2}$$
>
> A relative error of ± 2% is acceptable. The basic technique of measurement applied with the use of the circular diameter graticule is to transform mentally the image of each particle into a circle and then compare it to the 10- and 25-μm graticule reference circles. The sizing process is carried out without superimposing the particle on the reference circles; particles are not moved from their locations within the graticule field of view (the large circle) for comparison to the reference circles. Use the inner diameter of the clear graticule reference circles to size white and transparent particles. Use the outer diameter of the black opaque graticule reference circles to size dark particles.
>
> Rotate the graticule in the right microscope eyepiece so that the linear scale is located at the bottom of the field of view, bringing the graticule into sharp focus by adjusting the right eyepiece diopter ring while viewing an out-of-focus specimen. Focus the microscope on a specimen, looking through the right eyepiece only. Then, looking through the left eyepiece, adjust the left eyepiece diopter to bring the specimen into sharp focus.

The circular diameter graticule is obtainable from a number of sources (see Appendix II of this chapter). These graticules are specifically made for individual microscopes, and the maker will request sufficient information from the user regarding the microscope to ensure that the graticule is of the appropriate size. On receipt of the graticule, the relative error must, nonetheless, be measured to assure proper dimensions. This procedure is shown in Figure 4.8. In the figure, 100 GSD = 98 SMD. Thus, according to the formula provided in <788>,

$$\begin{aligned} RSD &= 100[(GSD - SMD)/SMD] \\ &= 100[(100 - 98)/98] \\ &= 2 \text{ percent} \end{aligned}$$

The current verbiage, "Use the inner diameter of the clear graticule reference circles to size white and transparent particles. Use the outer

Figure 4.8. Determining graticule relative error.

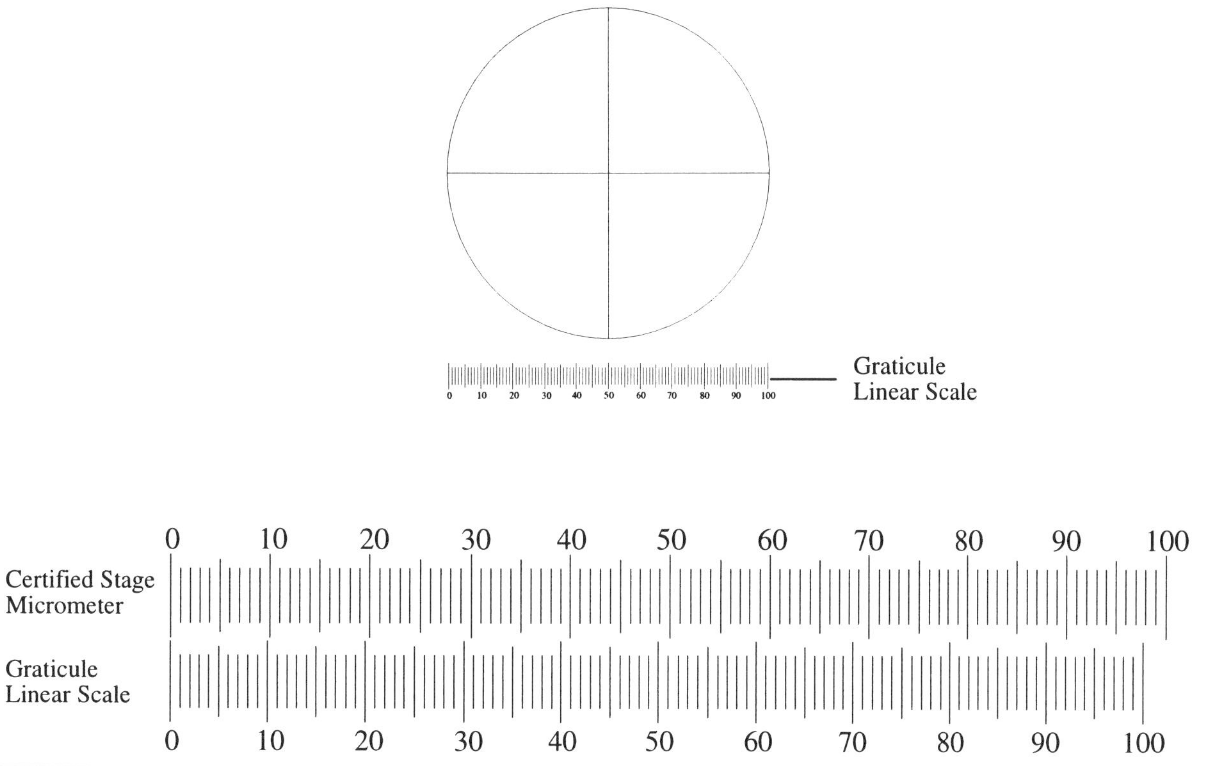

diameter of the black opaque graticule reference circles to size dark particles." is somewhat misleading. The origin of this description is obscure, but it probably dates from the developmental stage of the method when experimentation with the size of the circular scales was occurring. At any rate, the future wording at some time will read "Compare the area of the particle being sized to that of the black or transparent circles. Use the area of the clear graticule reference circle to size white and transparent particles. Use the area of the black circle to size dark particles."

The microscopic particle sizing process is shown in Figures 4.9 and 4.10. The basic mechanism of measurement is to assign the particle a diameter based on a metal conversion of its shape to that of a circle and compare the resulting area to the graticule. The particles are then sized by simply comparing their area to the area of the appropriate reference circle. Because dark and light particles of identical size will appear different sizes (white particles appear larger), black and white (clear) reference circles are provided. In Figures 4.9 and 4.10, the particle being recovered has an area of the same as the reference circle; in Figure 4.11, only one particle has an area greater than the 25 μm circle, 7 particles are between 10 μm and 25 μm in equivalent circular diameter, and 2 are < 10 μm in equivalent circular diameter.

Stepwise, the sizing procedure is as follows:

- Mentally rearrange particle area into a circle.
- Compare the particle circular area to the reference circle without superimposing the particle on the reference circle.
- Count particles ≥ reference circles.

With practice, an analyst can rapidly make this comparison with much greater precision than a comparison of the particle maximum dimension to a linear scale. Elongate particles are not oversized. Focusing the microscope eyepieces individually per the last paragraph in the excerpt above and bringing the graticule into sharp focus are critical to accurate sizing without

Figure 4.9. Circular area diameter.

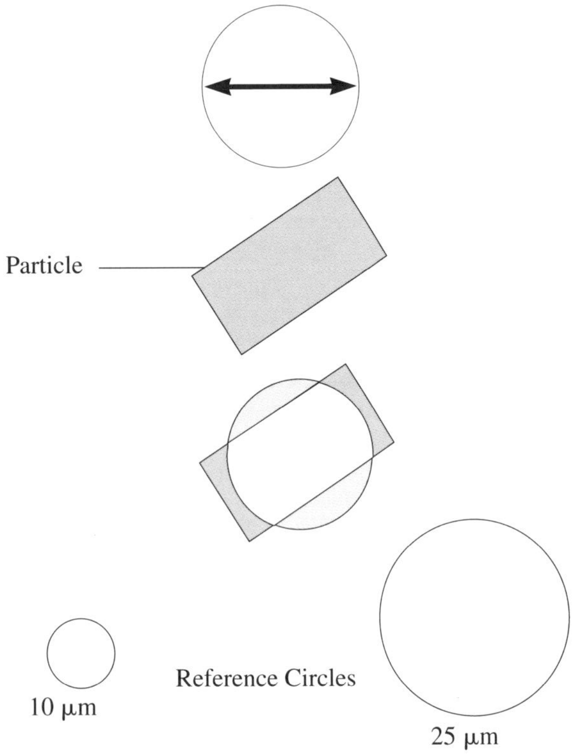

Figure 4.10. Equivalent circular diameter.

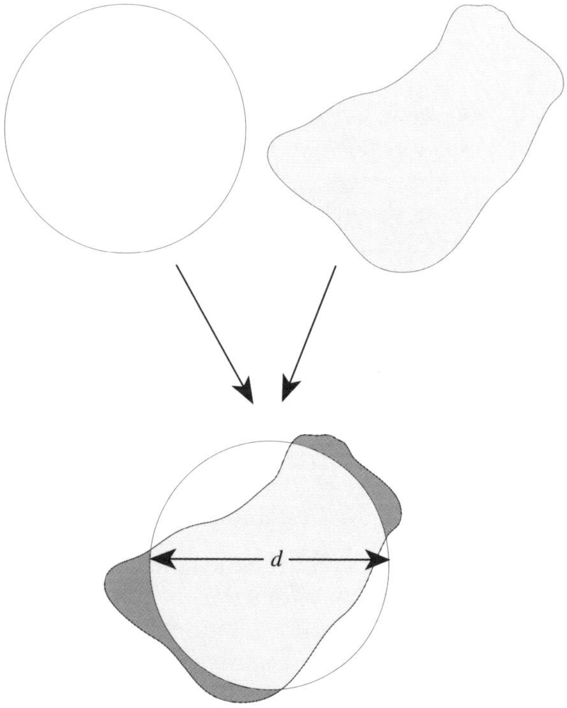

Figure 4.11. Circular area diameter of mixed colored particles.

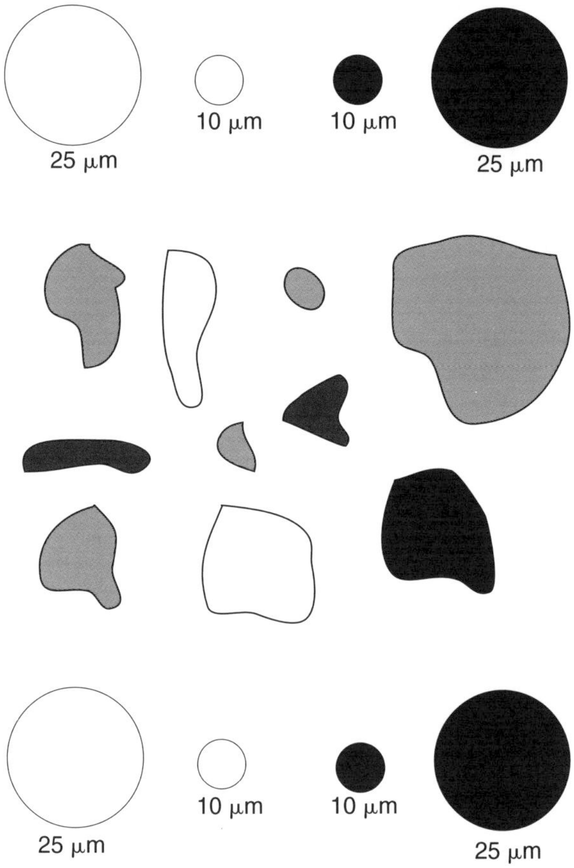

eyestrain. The right eye is typically the master eye and the graticule must be in the right eyepiece.

PREPARATION OF FILTRATION APPARATUS

Preferably, wash the filtration funnel, base, and diffuser in a solution of liquid detergent and hot water. Rinse with hot water. Following the hot water rinse, apply a second rinse with filtered distilled or deionized water, using a pressurized jet of water over the entire exterior and interior surfaces of the filtration apparatus. Repeat the pressurized rinse procedure using a filtered isopropyl alcohol. Finally, using the pressurized rinser, rinse the apparatus

> with filtered ■distilled or deionized water.■2
>
> ■Remove a membrane filter from its container, using ultra cleaned blunt forceps. Use a low pressurized stream of filtered purified water to wash both sides of the filter thoroughly by starting at the top and sweeping back and forth to the bottom. Assemble the cleaned filtration apparatus with the diffuser on top of the filtration base, placing the clean membrane filter on top of the diffuser. Place the funnel assembly on top of the filtration base, and lock it into place.

One comment to be made here is that extraneous particles must be exhaustively removed from both the filtration apparatus and filter so that acceptable blanks are obtained. The instructions are intended to be general so that the analyst can develop a specific technique based on the steps given. The sampling and testing of units is intended to conform to that earlier described for light obscuration counting. Other comments for that assay also apply here.

A comment is also in order regarding the size of the filtration funnel. A large funnel (e.g., 250 mL) is more useful for LVI, since refilling is minimized. For SVI pool volumes, smaller funnels should be used. This applies whether or not a partial count is performed. The stream of water from the pressure rinse should never be directed onto the filter surface. The goal is to collect particles in an even distribution on the filtration area (Figure 4.12) of the membrane. Examples of even and uneven particle distributions on a membrane are shown in Figure 4.13.

TEST PREPARATIONS

> ■Proceed as directed for *Test Preparation* under the *Light Obscuration Particle Count Test*, beginning with "Prepare the test specimens in the following sequence," and ending with "For large-volume injections, single units are tested." For small-volume injection units containing a volume of 25 mL or more and tested singly and for large-volume injections, the entire unit volume is tested. For large-volume injections or for small-volume injections where the individual unit volume is 25 mL or more, fewer than 10 single units may be tested, based on the definition of an appropriate sampling plan.

Figure 4.12. Areas on a membrane filter.

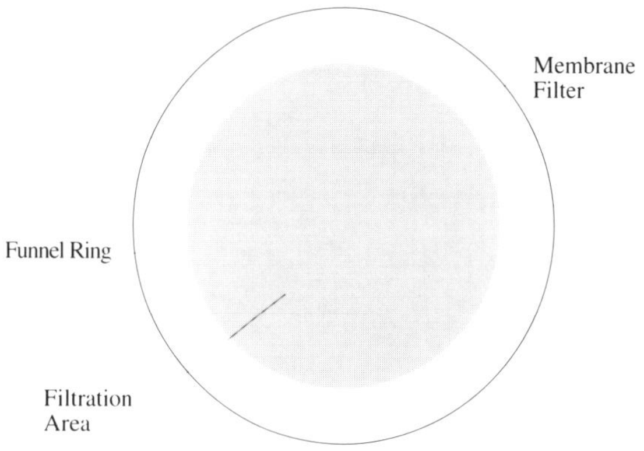

The limits apply to the average count for each unit tested. Obviously, a manufacturer will not expect to obtain a result almost at the limits and will take pains to ensure that any such result can be statistically supported. Given a normal distribution of counts, an average value near the limits (e.g., 10 percent below limits values) may mean that some units in the batch will exceed the limits. Two changes suggested here are as follows:

1. *Current:* "If the volume of the container is less, test a solution pool of 10 or more units." *Suggested change:* "If the volume of the container is less than 25 mL, test a solution pool of 10 or more units." *Reason for change:* To identify the 25 mL or greater volume for a single unit test.
2. *Current:* "For small-volume injections containing a volume of 25 mL or more and tested singly for large-volume injections, the entire unit volume is tested." *Suggested change:* "For small-volume injections containing a volume of 25 mL or more and for large-volume injections, the entire unit volume of single units is tested." *Reason for change:* The current wording is in error, as can be seen by reading it.

Figure 4.13. Examples of even and uneven particle distribution.

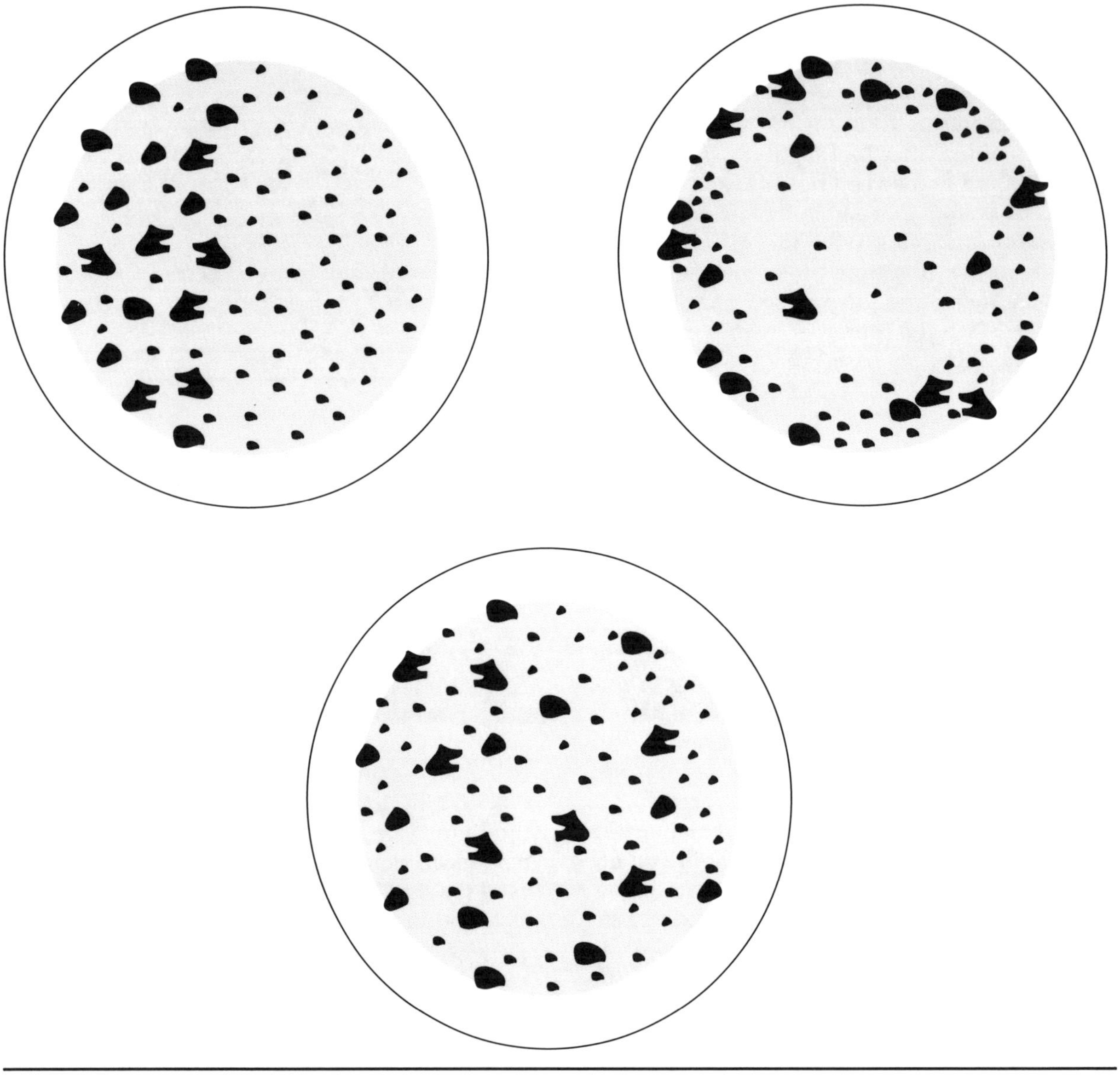

PRODUCT DETERMINATION

Depending upon the dosage form being tested, proceed as directed under the appropriate category below.

Liquid Preparations—Thoroughly mix the units to be tested by inverting 20 times. Open the units in a manner consistent with the generation of the lowest possible numbers of background particles. For products less than 25-mL in volume, open and combine the contents of 10 or more units in a cleaned container. Filter large-volume injection units individually. Small-volume injection units having a volume of 25-mL or more may be filtered individually.

Transfer to the filtration funnel the total volume of a solution pool or of a single unit, and apply vacuum. If the volume of solution to be filtered exceeds the volume of the filtration funnel, add, stepwise, a portion of the solution until the entire volume is filtered. If the partial count procedure is to be used (see *Partial Count Procedure* under *Enumeration of Particles*), do not allow the liquid volume in the filtration funnel to drop below approximately one-half of the funnel volume between refills. [NOTE—Use a filter funnel appropriate to the volume of solution if a partial count procedure is to be employed. This is necessary to ensure even distribution of particles on the analytical membrane.]

After the last addition of solution, begin rinsing the walls of the funnel by directing a low-pressure stream of filtered, distilled, or deionized water in a circular pattern along the walls of the funnel, and stop rinsing the funnel before the volume falls below about one-fourth of the fill level. Maintain the vacuum until all the liquid in the funnel is gone.

Remove the filtration funnel from the filtration base while maintaining vacuum, then turn the vacuum off, and remove the filter membrane with blunt forceps. Place the filter in a petri dish or similar container, secure in place with double-sided tape, and label with sample identification. Allow the filter to air-dry in the laminar-flow enclosure with the cover ajar.

Dry or Lyophilized Preparations—To test a dry powder vial or similar container of drug powder, constitute the material with an appropriate diluent, using the method least likely to introduce extraneous contamination, as directed for *Test Preparation* under *Light Obscuration Particle Count Test.* Using a solution pool of 10 or more units, or the desired number of single units, proceed as directed for *Liquid Preparations.*

Products Packaged with Dual Compartments Constructed to Hold the Drug Product and a Solvent in Separate Compartments—Activate each unit as directed in the labeling, agitating the contents sufficiently to ensure thorough mixing of the separate components, and then proceed as directed for *Liquid Preparations.*

Products Labeled Pharmacy Bulk Purchase Not for Direct Infusion—For *Products Labeled "Pharmacy Bulk Package—Not for Direct Infusion,"* or for *Multiple-dose Containers*, proceed as directed for *Liquid Preparations*, filtering the total unit volume.

Calculate the test result on a portion that is equal to the maximum dose given in the labeling. Consider this portion to be the equivalent of the contents of one full container. For example, if the total bulk package volume is 100 mL and the maximum dose listed is 10-mL, the microscopic total unit volume count test result would be multiplied by 0.1 to obtain the test result for the 10 mL dose volume. [NOTE—For the calculation of test result, consider this portion to be the equivalent of the contents of one full container.] ■$_5$

Enumeration of Particles

The microscopic test described in this section is flexible in that it can count, in particles per mL, specimens containing 1 particle per mL as well as those containing significantly higher numbers of particles per mL. This method may be used where all particles on an analysis membrane surface are counted or where only those particles on some fractional area of a membrane surface are counted.

TOTAL COUNT PROCEDURE

In performance of a total count, the graticule field of view (GFOV) defined by the large circle of the graticule is ignored, and the vertical cross hair is used. Scan the entire membrane from right to left in a path that adjoins but does not overlap the first scan path. Repeat this procedure, moving from left to right to left until all particles on the membrane are counted. Record the total number of particles that are 10 µm or larger and the number that are 25 µm or larger. For large-volume injections, calculate the particle count, in particles per mL, for the unit tested by the formula:

$$P/V,$$

where P is the total number of particles counted, and V is the volume, in mL, of the solution. For small-volume injections, calculate the particle count, in particles per container, by the formula:

$$P/n,$$

in which P is the total number of particles counted, and n is the number of units pooled (1 in the case of a single unit).

The specifics of counting are shown in Figures 4.14 and 4.15. The GFOV–verified diameter is used to determine the scan line width. To avoid over- or undercounting, each particle must be counted only once. The technique most frequently used involves the analyst continuously using the field of view from right-to-left and left-to-right across the filtration area of the membrane (Figure 4.13), as in "mowing the lawn." At the end of each horizontal scan, the filter is indexed vertically by one graticule diameter. This may be done using visual reference or by the microscope venue. Alternatively, the GFOV may be moved in steps of one graticule diameter. If this is done, the following counting rules (Figure 4.15) should be followed:

Figure 4.15. Counting within the graticule field of view.

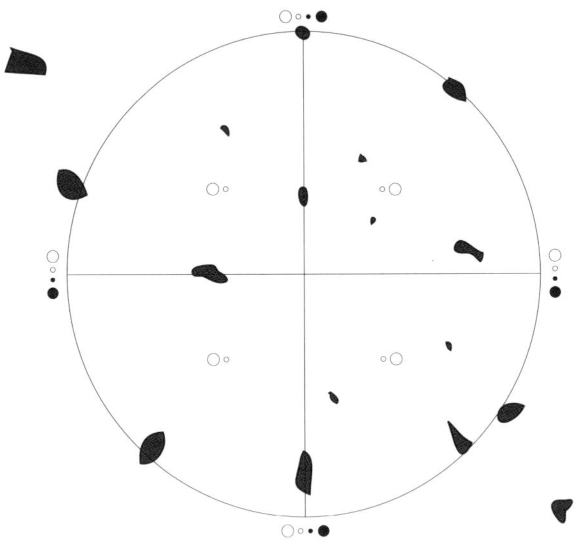

Figure 4.14. Filtration area scan using graticule field of view.

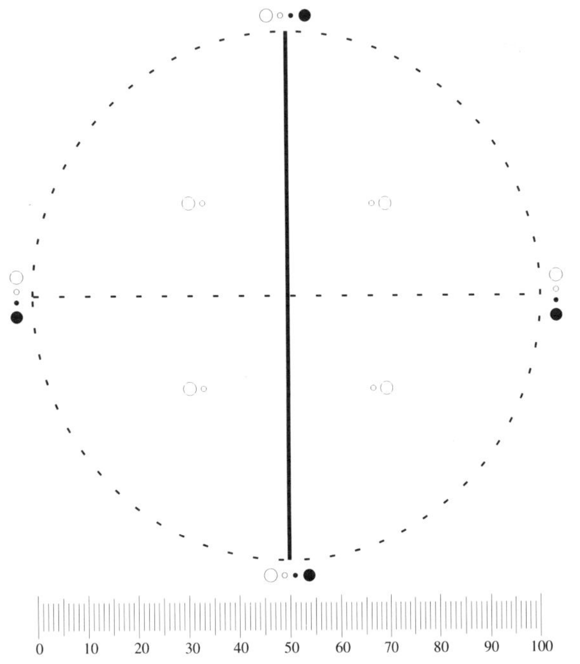

- Do not count particles outside the GFOV. (The count obtained will be cumulative (i.e., a 30 μm particle will be counted in both the ≥ 10 μm and ≥ 25 μm size ranges.)
- Count particles inside GFOV ≥ 10 μm and ≥ 25 μm.
- For particles touching the GFOV: Count particles touching right half of the GFOV circle. Do not count particles touching the left half of the GFOV circle. (The latter procedure is essential when a partial count is performed, as described in the next section.)

The filtration of some types of product preparations such as 1 L or larger volume containers or SVI sample pools may result in the collection of large numbers of particles on the analytical membrane so that total counts of particles present cannot be performed. In such cases, a partial count of particles on the membrane may be performed as described below. Use of a larger (e.g., 47 mm) filter membrane, or filtration of less than the total unit or pool volume, may be used, if necessary, to combine these procedures with the performance of a partial count.

One of the more critical problems that has arisen with the microscopic method regards the collection of large numbers of particles on filters when some test preparations are filtered. An example might be a 10-unit pool with 6,000 particles allowable at ≥ 10 µm/unit (60,000 particle total) or a 1 L LVI unit with 25 particles ≥ 10 µm/mL (25,000 total). When this occurs, partial counting alone may not be sufficient to complete the analysis. A large filter may be used to effect an approximately 4-fold reduction in particle sample material. The analyst must take care to keep the particulate evenly dispersed in the volume to be filtered if a partial volume is used. The objective for either partial or total counting is to reduce particle numbers to the point where individual particles are discernible. If particles touch or are so closely adjacent that multiple particles will be sized as a single particle, serious count errors will result. Such membranes should be designated as "too numerous to count" and additional sample preparations made as necessary.

PARTIAL COUNT PROCEDURE

To perform a partial count of the particles on a membrane, start at the right center edge of the filtration area and begin counting adjacent GFOVs. When the left edge of the filtration area is reached, move one GFOV toward the top of the filter and continue counting GFOVs by moving in the opposite direction. Moving from one GFOV to the next can be accomplished by one of two methods. One method is to define a landmark (particle or surface irregularity in the filter) and move over one GFOV in relation to the landmark. A second method is to use the vernier on the microscope stage to move one millimeter between GFOVs. To facilitate the latter, adjust the microscope x and y stage positioning controls to a whole number at the starting position at the center right edge of the filtration area, then each GFOV will be one whole division of movement of the x stage positioning control. If the top of the filtration area is reached before the desired number of GFOVs is reached, begin again at the right center edge of the filtration area one GFOV lower than the first time. This time move downward on the membrane when the end of a row of GFOVs is reached. Continue as before until the number of GFOVs is complete.

For large-volume injections, if a partial count procedure for the ≥10-µm and ≥25-µm size ranges is used, calculate the particles per mL by the formula:

$$Pa_t/A_pV,$$

in which P is the number of particles counted, a_t is the filtration area, in mm^2, of the membrane, A_p is the partial area counted, in mm^2, based on the number of graticule fields counted, and V is the volume, in mL, of solution filtered. For a solution pool (for small-volume injection units containing less than 25 mL) or for a single unit of a small-volume injection, calculate the number of particles per unit by the formula:

$$Pa_t/A_pn,$$

in which n is the number of units counted (1 in the case of a single unit), and the other terms are as defined above.

■For all types of product, if the tested material has been diluted to effect a decrease in viscosity, the dilution factor must be accounted for in the calculation of the final test result.■5

Note: If a larger filter is used or if a fractional volume of a unit's contents or a sample pool is filtered, this decreased volume must be accounted for in the final calculations.

The partial count is well described here (fields are placed as in Figure 4.16). An example of a calculation is as follows:

Figure 4.16. Placement of fields on membrane for partial count.

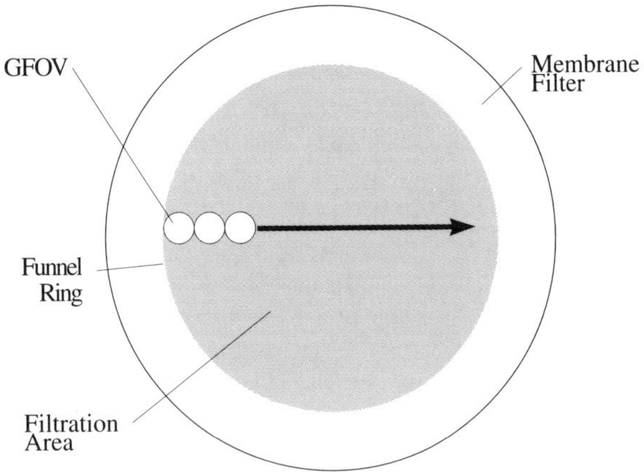

$$\text{particles per container} = P \times Af/N \times Ag$$

where P is the number of particles, $P10$ is the number particles $\geq$ 10 µm, $P25$ is the number particles $\geq$ 25 µm, Af is the filtration area in mm^2 ($A_{Circle} = \pi r^2$), N is the number of GFOVs counted, and Ag is the graticule area in mm^2 (0.785). For example,

$$N = 20$$

$$P10 = 100$$

$$P25 = 25$$

$$Af = 346.18 \text{ mm}^2$$
(for 21 mm diameter filtration area)

$$Ag = 0.785 \text{ mm}^2$$
(1 mm diameter GFOV)

thus,

$$\text{particles/container} = P \times Af = P\,(10, 25) \times 346.18/N \times Ag = 20 \times 0.785$$

$$\text{particles on membrane} \geq 10\ \mu\text{m} = 2{,}205$$

$$\text{particles on membrane} \geq 25\ \mu\text{m} = 551$$

Interpretation

The injection meets the requirements of the test if the average number of particles present in the units tested does not exceed the ■appropriate■5 value listed in Table 2.

Change to read:

Table 2. Microscopic Method Particle Count.

	10µm	25µm
Small-volume Injections:	3000	300 per container
Large-volume Injections:	12	2 per mL

With regard to the limits set, the average count of units tested is compared to the limit. Again, adequate sample sizes should be calculated to ensure a valid assessment of batch suitability.

SUMMARY OF MICROSCOPIC TEST

The USP 23 <788> Microscopic Test changes may be summarized as follows:

- A combination of reflected and brightfield episcopic illumination is used to optimize the detection of all particle types.
- Particles are sized by equivalent circular area diameter using a specially designed graticule.
- The total volume of individual units (25 mL or greater) is filtered, or the total volume of a solution pool for smaller units.
- Provision is made for partial (statistical) counting of heavily particulated membrane (as from an SVI pool).
- The blank is changed.
- Allowable particle numbers are decreased 4-fold ($\geq$ 10 µm) and halved ($\geq$ 25 µm).

Summaries of the microscopic test method are shown in Figures 4.17 and 4.18.

Figure 4.17. USP 23 <788> Microscopic Test outline.

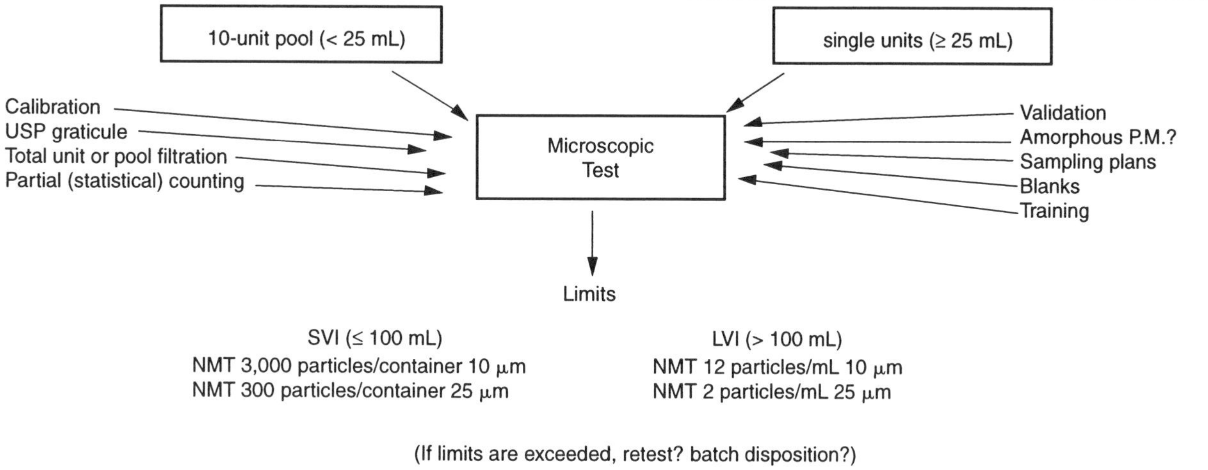

Figure 4.18. USP 23 <788> Microscopic Test.

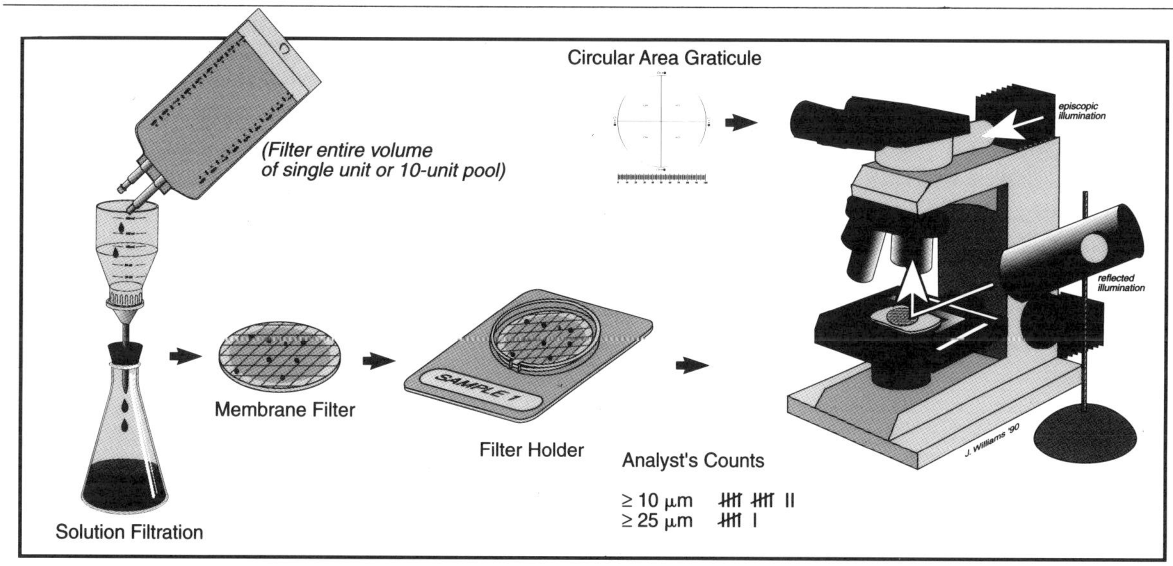

USP SVI Microscopic Limit: NMT 3,000 particles/container ≥ 10 µm
300 particles/container ≥ 25 µm

USP LVI Microscopic Limit: NMT 12 particles/mL ≥ 10 µm
2 particles/mL ≥ 25 µm

SUMMARY

The USP <788> requirement is complex and provides the most complete methodological description in any of the world's pharmacopeia. The user must become familiar with the entire method to ensure appropriate application to different product types. Validation of the instrumental test method as applied in a specific laboratory and thorough training of analysts performing the microscopic assay is necessary. The user should keep in mind that modifications of the test are possible if they are carefully validated. The question regarding modifications to the test are best answered in the USP General Notices, Procedures section. Specifically, it states,

> Every compendial article in commerce shall be so constituted that when

> examined in accordance with these assay and test procedures, it meets all of the requirements in the monograph defining it. However, it is not to be inferred that application of every analytical procedure in the monograph to samples from every production batch is necessarily a prerequisite for assuring compliance with pharmacopeial standards before the batch is released for distribution.

In the author's opinion, the USP is stating that all products purporting to meet the compendial standards, when tested by these official methods, must pass. However, the USP does not specifically require that the manufacturer do each and every USP test for the in-process control of a production batch.

Later in this same section, the General Notices state,

> Compliance may be determined also by use of alternative methods, chosen for advantages in accuracy, sensitivity, precision, selectivity, or adaptability to automation or computerized data reduction or in other special circumstances.

This language speaks for itself. Insofar as the USP is concerned, the user has the option of selecting alternative methods that are believed to best suit one's purposes of assuring compliance. Of course, alternate methods and their application are also subject to GMP and GLP requirements. Therefore, it is necessary to have data that demonstrate that products meeting such in-house tests do, in fact, comply with the USP requirements.

A most important fact follows in the next sentence of the General Notices.

> However, pharmacopeial standards and procedures are interrelated; therefore, when a difference appears or in the event of a dispute, only the result obtained by the procedure given in this pharmacopeia is conclusive.

In conclusion then, the General Notices, under Procedures, give manufacturers and enforcement agencies (i.e., the FDA) the option of examining official products by alternative methods that have been validated as suitable for determining compliance. However, in the event of a dispute over a product's compliance, the official method is controlling. This principle has been well accepted for many years.

A specific word of warning is thus in order for those who modify the <788> method to make it more suitable for a specific product type or for a given set of operational circumstances (e.g., the sampling of small volume, high-tech, expensive biopharmaceuticals). Inspection principles, particularly those of the FDA can be expected to vary to some extent based on the knowledge base, training and experience of inspectors. Compliance with a limit set is enforceable by law. Particulate matter limits are included. Individual inspectors, although generally guided by inspection guidelines for different product types, may question departures from strict literal application of the <788> method differently.

The best advice for the user is to validate all exceptions (e.g., volume of sample tested or blank volume) vigorously to show equivalence to the component of the method as published. One way to consider this is to look at <788> as one would a chemical assay. Better safe than sorry. The validation may be of the component of the method that differs or of the whole assay with regard to equivalence. Principles include the following:

- Make a conscious decision regarding method modification; don't just "do it."
- Don't make the decision alone—involve others as logic and/or company policy directs.
- Document test results thoroughly.
- Keep reports of validation and supporting data handy.

USP <788> is an evolving document. Despite a continuing need for minor editorial or semantic changes in some sections, the methodology described is a workable, uniform, and reproducible test for particulate matter in injections. The standard describes a reasonable standard of a level of sophistication consistent with today's high-quality injectable products.

BIBLIOGRAPHY

Akers, M. 1985. *Parenteral quality control: Sterility, pyrogens, particulate, and package integrity testing.* New York: Marcel Dekker, Inc.

Aldrich, D. S. 1997. Membrane-based counting of the particulate matter load in parenteral products. *Microscope* 45:373–383.

Barber, T. A. 1987. Limitations of light blockage particle counting. Paper presented at the meeting on Liquid Borne Particle Inspection and Metrology, 11–13 May in Washington, D.C.

BP. 1994. *British Pharmacopeia,* vol. II. London: Her Majesty's Stationery Office.

Chamberlain, R. 1991. *Computer systems validation for the pharmaceutical and medical device industries.* Libertyville, Ill., USA: Alaren Press.

DeLuca, P. P., S. Bodapatti, D. Haack, and H. Schroeder. 1986. An approach to setting particulate matter standards for small volume parenterals. *J. Parenter. Sci. Technol.* 40:2–13.

DeLuca, P. P., B. Conti, and J. Z. Knapp. 1988. Particulate matter II. A selected annotated bibliography. *J. Parent. Sci. and Technol.* 42. Supplement.

Endicott, C. J., R. Giles, and R. Pecina. 1966. Particulate matter: Its significance, source, measurement and elimination. In *Proceedings of the Symposium on Safety of Large Volume Parenteral Solutions.* Rockville Md., USA: Food and Drug Administration, pp. 62–75.

Groves, M. J. 1969. The size distribution of particles contaminating parenteral solutions. *Analyst* 94:992–999.

Groves, M. J. 1973. *Parenteral products.* London: Hinemann Medical Books, Ltd.

Groves, M. J. 1992. An improved method for calculating data from particulate matter limit tests of small volume injection solutions. *Pharmacopeial Forum* 16:4100–4101.

Groves, M. J. 1993. *Particulate matter: Sources and resources for healthcare manufacturers.* Buffalo Grove, Ill., USA: Interpharm Press, Inc.

Groves, M. J., and S. R. DeMalka. 1976. The relevance of pharmacopeial particulate matter limit tests. *Drug Dec. Commun.* 2 (3):285–324.

Groves, M. J., and E. Muhlen. 1987. The parenterals numbers game: Newer ways of looking at particulate contamination. *J. Parent. Sci. Technol.* 41:116–120.

Mallory, S. R. 1994. *Software development and quality assurance for the healthcare manufacturing industries.* Buffalo Grove, Ill., USA. Interpharm Press Inc.

USP. 1995. *The United States Pharmacopeia,* 23rd ed. Easton, Penn., USA: Mack Publishing Co.

APPENDIX I. VENDOR AND EQUIPMENT INFORMATION

This vendor listing is not intended to be comprehensive; rather, it serves as a starting point in a search for lab equipment useful in the analysis of particulate matter, supplies for the "clean" laboratory, clean room garments, and other essential materials and instruments. More detailed lists of where to purchase specific items can be found in pharmaceutical or clean room trade publications and journals, such as *Clean Rooms, Pharmaceutical Technology, Medical Device and Diagnostic Industry,* and the *Journal of Parenteral Science and Technology.*

Particle Counters

Climet Instruments Company
1320 W. Colton Ave.
Redlands, CA 92374
725-793-2788

HIAC-Met One
World Headquarters
11801 Tech Road
Silver Spring, MD 20904
800-638-2790 or 301-680-7000

Malvern Instruments Inc.
10 Southville Road
Southborough, MA 01772
508-480-0200

Particle Measuring Systems, Inc.
1855 S. 57th Court
Boulder, CO 80301
910-940-5891

Rion Co., Ltd.
20-41, Higashimotomachi 3-chome
Kokubunji, Tokyo 185, Japan
0423-22-1133

Calibration Materials

Bangs Laboratories, Inc.
979 Keystone Way
Carmel, IN 46032
317-844-7176

Coulter Electronics, Inc.
601 W. 20th Street
Hialeah, FL 33010
800-327-6531

Duke Scientific Corp.
1135-D San Antonio Rd.
Palo Alto, CA 94303
415-962-1100

Specialty Materials Dept.
2-11-24
Tsukjii, Chuo-ku, Tokyo
03-5565-6610

National Institute of Standards and Technology
Office of Standard Reference Materials
Chemistry Bldg, Room B311
Gaithersburg, MD 20899
301-975-2000

Light Microscopes

Nikon Inc. Instrument Group
623 Stewart Avenue
Garden City, NY 11530
516-222-0200

Olympus Corporation
4 Nevada Drive
Lake Success, NY 11042-1179
516-488-3880

Reichert Scientific Instruments
Microscopical Optical Consulting, Inc.
P.O. Box 586
Valley Cottage, NY 10989
914-268-6450

Wild-Leitz
Wild Heerbrugg Ltd.
9435 Heerbrugg Switzerland
071-703131

Carl Zeiss, Inc.
One Zeiss Drive
Thornwood, NY 10594
914-747-1800

Microfiltration Supplies

Gelman Sciences
600 S. Wagner Road
Ann Arbor, MI 48106
800-521-1520 or 313-665-0651

Micro Filteration Systems
6800 Sierra Court
Dublin, CA 94568
800-344-7132 or 415-828-6010
MFS sintered glass support cat. #311200

Millipore Corp.
80 Ashby Rd.
P.O. Box 9125
Bedford, Massachusetts 01730-9125
800-225-1380 or 617-275-9200
Millipore stainless steel support cat. #XX 30 025 10

Pall Filtration
2200 Northern Blvd.
East Hills, NY 11548
800-645-6532 or 516-484-5400

Sartorius Corp.
140 Wilbur Place
Bohemia, Long Island, NY 11716
516-563-5122

Cleanroom Supplies

Allegiance Healthcare Corp.
1425 Lake Cook Road
Deerfield IL 60015
847-940-5000

Berkshire Corporation
River Street
Great Barrington, MA 01230
800-242-7000 or 413-528-2602

Clean Room Products, Inc.
1800 Ocean Avenue
Ronkonkoma, NY 11779
516-588-7000

Clestra Cleanroom Technology
4003 Eastbourne Drive
Syracuse, NY 13206
315-437-2152

E.I. Du Pont De Nemours & Company
Fibers
Laurel Run Building
P.O. Box 80, 705
Wilmington, DE 19880-0705
302-774-1000

Teijin Shoji America, Inc. (garments)
42 W. 39th St., 6th Floor
New York, NY 10018
212-840-6900

Texwipe
650 E. Crescent Ave.
P.O. Box 575
Upper Saddle River, NJ 07458
201-327-9100

APPENDIX II. MANUFACTURERS OF "USP"* TYPE CIRCULAR AREA GRATICULE

KRI (Klarmann Rulings, Inc.)
P.O. Box 4795
Manchester, NH 03108
Phone: 603-424-2401
Fax: 603-424-0970
Estimated cost: Master Already Made

Applied Image, Inc.
653 East Main St.
Rochester, NY 14609
Phone: 716-482-0300
Fax: 716-288-5989
Estimated cost: $600/Master
Tolerances

1. Theoretical perfection ± 0.8 μm
2. Line ± 0.75 μm

Gage-Line Technology, Inc.
121 LaGrange Avenue
Rochester, NY 14613-1511
Phone: 716-458-5310
Fax: 716-458-0524
Estimated cost: $400/Master
Tolerances

1. Line width and location 0.0001 in. (2.5 μm) on the reticle or 10–20 μin. (2.5–5.0 μm) at the stage

*This is not an official USP item.

Leco Corporation
3000 Lakeview Avenue
St. Joseph, MI 490852396
Phone: 616-983-5531

Graticules, Ltd.
England 0732-359061

NOTE: Prices and tolerances are based on 1998 information.

APPENDIX III. SAMPLING PLANS

"Statistically valid" sampling plans are mentioned in the text of <788>. The inputs necessary to define an adequate sampling plan include statistics, historical data, and a consideration of the method itself. It is unusual to encounter an individual in a technical, quality management, or scientific position in a company manufacturing injectables or devices who does not have a basic knowledge of statistical principles. The data for particle counting tests require special handling in the sense that data distributions encountered are of non-Gaussian, and the tests themselves are subject to significant variability.

Statistical validity may be defined simply as the "strength" of the number representing the average particle counts obtained from the units tested per <788>; the strength of this number depends on how closely our test methods and sample selection have allowed us to approximate the "true" average counts for the batch or other group tested.

The assurance required is that our numbers reasonably approximate the real average values that would be obtained by testing all of the units in the batch, and that we have some measure of how much our assay result may vary from that value. Statisticians are adept at applying the appropriate tests to data sets to obtain measures of central tendency (e.g., means) and set confidence limits on the averages determined. Too often, however, no account is taken of the technique or method-based variability of this method. This variability source can be significant with all types of particle counting tests. The importance of this variability source may be based on the consideration that while multiple tests of the same unit by a chemical method such as chromatography will yield closely similar results, multiple particle count assays may be expected to show much more variation. This result is based both on the random nature of the occurrence of extraneous particles versus the uniformity of dissolved chemical entities and the imprecision of the tests themselves.

"Valid" sampling plans thus have two components that must be considered in order to develop an assessment of the strength of the assay result derived: assay variability and the sampling plan itself. Variation related to the test itself must be minimized through adherence to the requirement as described and further considerations related to conduct of the test in individual labs; once the test is optimized, appropriate sampling is based on statistical principles. To define a valid sampling plan, a consideration of the variability of the test method is a good place to begin.

Assay Variability

The light extinction and microscopic particle counting assays described in chapter <788> of USP 23 are limit tests or pass-fail assays. The desired result is a simple prediction of whether a batch or other sample group tested meets or exceeds the established compendial limits. Given the variability of the light extinction and microscopic tests, even in optimized form, it is impractical to test a sufficient number of units to determine actual numbers of particles present with any reasonable degree of confidence. As mentioned above, this is not a chemical assay (such as a titration) for which standards of accuracy and precision are established and readily available. The instrumental particle test incorporates considerable variability simply because of the randomness of particle distribution in a tested container. Specific human factors also have an effect, such as differences in the manner in which analysts agitate or enter a solution unit.

Typically, when two identical light obscuration instruments with the same sensor and counting electronics and calibrated by the same method are used to count an ideal standard suspension of mixed latex beads, a variation of as much as ±10 percent or greater results between counts from the two instruments. In counting nonideal particles, such as extraneous particles in parenteral solutions, this between-instrument variability increases to more than 20 percent. With regard to the variability of test articles, between-unit variability of particle counts also increases as the unit size decreases, thus introducing a considerable variability component into the assay result. This instrumental variability is also compounded by slight variations in technique between human analysts. Subtle factors such as extent and method of agitation, method of entering a unit, and depth of probe tube insertion into the solution volume of a unit being tested can all make a difference in counts. Although a great deal of effort has been expended in individual laboratories to minimize the effect of these variables in the manual test, their impact is still significant.

The microscopic test as described in <788> is estimated to incorporate a coefficient of variation, based on round-robin studies conducted during its development. As with other statistical tests, the extent to which the sample approximates the true number or value depends on the number of samples. Thus, the more particles counted on a membrane, the better the estimate of the real number present. In the case of heavily particulated membranes, the process of counting is subject to the limitations of diminishing returns or diminishing marginal utility; the counting of more particles gives an increasingly better estimate of the total number present, but the rate at which that estimate increases in strength declines as greater numbers are counted. A statistical confidence in counts of the number of particles on a membrane may be assessed by a simple procedure, such as the one provided in Section A70 of Federal Standard 209E.

> *A70.3 Counting particles.* Estimate the total number of particles, 5 μm and larger, present on the membrane filter by examining one or two of the selected field. If this estimate is greater than 500, use the procedure for counting particles described in A70.4.

(Author's note: for USP counting, ≥ 10 μm would be substituted for ≥ 5 μm.)

> If the estimate is less than 500, count all of the particles on the entire effective filtering area of the membrane. Scan the membrane by manipulating the stage so that particles pass under the calibrated ocular scale. The size of a particle is determined by its longest dimension. The eyepiece with its calibrated ocular scale may be rotated if necessary. Using a manual counter, tally all particles with sizes in the range of interest. Record the number of particles counted in each field.
>
> *A70.4 Statistical particle counting.* When the estimate of the number of particles, 5 μm and larger, on the membrane filter exceeds 500, a statistical counting method should be used. After a unit field has been selected, particles are counted in a number of fields of that size until the following statistical requirement is met:
>
> $$FAN > 500$$
>
> Where: F = number of unit fields counted, and N = total number of particles counted in F unit fields. The total number of particles on the membrane is then calculated from the following equation:
>
> $$P = N \times A/(F \times a)$$
>
> Where: P = total number of particles in a given size range on the membrane, N = total number of particles counted in F unit fields, F = number of unit fields counted, a = area of one unit field, and A = total effective filtering area of the membrane.

Consider the following case describing how the increase in confidence with higher *P* value works. On the assumption that the number of particles is governed by the principles of Poisson statistics, the standard deviation of the count will be equal to the square root of the number of particles counted. If a membrane is counted and 100 particles ≥ 25 μm in size are present, the 90 percent confidence limit for the count is 100 particles ± 10 and the 90 percent confidence limit for the per mL count based on a 90 mL sample is 90/90 and 110/90 or 1.0 to 1.2 particles/mL. A higher confidence may be tested by the use of the appropriate multiplier for the confidence level defined by *P*. For example, for the count of the membrane, a multiplier of 3 (99 percent of the values will lie within three standard deviations of the mean) would give a count range of 130 to 70 for a 99 percent confidence that the true ≥ 25 μm count for the unit tested is in the range of 130/90 to 70/90 or 1.4 to 0.8 particles/mL, which is also below the limits value.

Sampling Plans

In the context of the USP, a "statistically valid" sampling plan must perform but one simple function—provide an assessment of the probability that the mean counts for units within a batch that remain untested will exceed the USP limits. For release of product, this probability must be sufficiently low that the batch may be judged acceptable based on the mean particle count from units tested. Given the test result, this judgment can be put in the form of a question. What is the probability that the batch average exceeds limits? The answer will differ for different types of product. In some cases, the probability that an average count for a sample of the size tested will exceed limits may be $\leq 10^{-6}$. In this case, the decision that the batch is acceptable is readily made. As the average test result draws closer to the limits value, the probability that the actual true mean exceeds the limits increases, and the decision will become more difficult. Sampling plans may receive regulatory scrutiny and must be devised with the input of professional statisticians. The number of samples chosen must be adequate to provide a statistically sound assessment of whether a batch or other large group of units represented by the samples meets or exceeds the limits.

In the case of any batch of product, there are three key parameters that will define necessary numbers of samples to obtain a valid statistical assessment of batch suitability or suitable confidence that the batch does not exceed limits. All are based on historical data for that product.

1. The distribution of particle counts at ≥ 10 μm and ≥ 25 μm for individual units. This is specified as numbers of units with specific count results, and is sometimes a Poisson distribution.
2. Average count for units of the product and the relationship of this value to limits.
3. Variability of counts between units.

Once the distribution of count data (Poisson, log normal, negative binomial, etc.) is known and the mean, variance, and standard deviation of counts are determined, the probability of exceeding limits based on samples of different sizes may be calculated from the statistical tables available to statisticians. At this point, the probability of a result exceeding the limits when the true mean count is at some given level below the USP limit (false rejection) and the probability of a result that is below the limits when the true mean is above the limit (false acceptance) may be determined. An example of a sampling plan devised in this fashion is as follows:

Consider the case of a manufacturer who produces SVI units (< 100 mL) of saline and dextrose product in glass bottles. The appropriate USP limit for such units is 6,000 particles per container ≥ 10 μm, and 600 particles per container ≥ 25 μm. This equates, for a 60 mL container, to 60 particles/mL and 6 particles/mL at the ≥ 10 μm and ≥ 25 μm sizes, respectively. The manufacturer knows that the product typically evidences particle counts far below these limits (typically less than 10 particles/mL ≥ 10 μm, and less than 1 particle/mL ≥ 25 μm). Further, the manufacturer is convinced that the process by which these units are made is stable and well controlled. In this situation, considerable latitude is available to the manufacturer in defining a batch release assay other than the 10-unit sample pool specified by the USP for SVI product, and the sampling plan will likely involve testing only a small number of units, even from relatively large batches.

In the initial statistical consideration by the manufacturer in our example, it becomes obvious that the testing of individual units provides a better estimate of the particle burden of individual units and the variability of particle burden between units than does a pooled sample. This may be explained as follows: If X represents the particle count per 5 mL of sampled volume, there will be some overall rate of particle detection (I). Thus, the count distribution may be modeled as a Poisson distribution with rate I. If we pool ten 50 mL units and then test 3 sample aliquots, we are in effect creating 3 independent Poisson random variables—X_1, X_2, and X_3. Hence, the average determination has a true variance of I/3. If we take 3 independent samples from each of 5 units, we are in essence generating 15 variables—$X_{1,1}$, $X_{1,2}$, $X_{1,3}$, $X_{2,1}$, . . . , $X_{5,3}$—that are independent and distributed as Poisson random variables. The average determination of these 15 samples has a true variance of I/15. Hence, the 5-unit sample plan provides a more precise estimate of the particle count than a 10-unit pool.

Similar considerations may be made based on the degree of confidence required in an answer versus the number of samples tested. With a known distribution of particle counts from product, assume that the historical process average is X and that a 95 or 99 percent confidence level in accepting sampled product as being at or below this value is desired. Sample sizes for 95 and 99 percent confidence statements determined from tabular statistical data are shown in Table 4.1, along with the accept criteria.

Example: Suppose the process average is 160, and 99 percent confidence is desired. Then the required sample size is 2 and the accept criterion is 175. If the average of the two samples is less than or equal to the accept criterion, the test is passed. If the test passes, one can state that with 99 percent confidence, the average particle count for the batch is below 200.

The concept of a warning or alert limit is critical to any other sampling scheme in which a number of samples much lower than the total number of units in a batch will be tested. An alert limit consists of an assay result at some point below the USP limits at which action will be taken by the manufacturer prior to actually exceeding the limits for the compedial test. This typically consists of testing a much larger number of samples selected from across the production interval of the batch (i.e., first, middle, and end of production) or chosen to reflect any other key factors that might affect particle burden, such as filter changes or filler differences. With an alert limit established, a manufacturer may be able to justify testing smaller members of units since, in effect, a more stringent count limit is being applied. For example, based on the distribution of a historical data set, an appropriate alert limit may be established such that if the mean count of a sample group exceeds the alert limit, $P > 0.01$, the true mean count exceeds the product limit. The smaller the sample size used, the lower the alert limit at which probability of failure of product limits is greater than 0.01.

The internal alert limits usually used by manufacturers, however, are generally much less than the USP limits. The manufacturer in our example above might select an alert limit of 35 particles per mL (≥ 10 μm) or 3 particles per mL (≥ 25 μm). Since these alert limits are so much lower than the actual compendial limit, sampling done with a smaller number of units might be shown not to affect the performance of the test adversely.

Here we will consider two sample sizes, 10 units and 5 units, and compare their respective error rates in determining particulate level. Table 4.2 presents the percentage of lots that would be accepted under the decreased sample scheme given some true average particle count. (This table only serves to provide a generic example and is not appropriate for direct application in sampling a specific product.) For each value of *t*, 3 sample sizes are reported: 1 unit, 5 units, and 10 units. As can be seen when *t* is close to the limit (33 to 39 for ≥ 10 μm and 3 to 4.5 for 25 μm), the difference between the sampling plans is evident. For example, if the true mean particle count were 39 for > 10 μm particles, then a 10-unit sampling plan would be preferred. Such a plan would result in false acceptance of a lot no more than 3 out of 100 tests. A 5-unit test would incorrectly accept these lots 1 out of 12 times.

The apparent superiority of the 10-unit test is thus true only for a narrow range of *t* values near the limit. If the true mean number of particles is much greater than the limits, it does not really matter which test is used, as the change of incorrectly accepting a lot falls to zero quite rapidly. This is more true of the 10 μm particles than the 25 μm particles. For either size of particle, note that the USP limits are far above the

Table 4.1. Samples Sizes and Accept Criteria to Demonstrate Below 200 particles per mL in 50 mL Bag with Specified Process Average (sample number, maximum count allowable to accept)

Process average (X)	90	150	160	170	180	190
95 percent confidence	—	(1,170)	(2,175)	(3,182)	(6,189)	(22,194)
99 percent confidence	(1,106)	—	(2,175)	(4,180)	(8,187)	(31,194)

Table 4.2. Probability of Accepting[a] a Lot Under Various Sampling Schemes (True mean [t] is real number of particles per mL)

Sample Size	True Mean (> 10 μm)	Acceptance[b] Probability (%)	True Mean (> 25 μm)	Acceptance Probability (%)[b]
1	5	100.00	0.5	99.825
5	5	100.00	0.5	100.00
10	5	100.00	0.5	100.00
1	10	100.00	1.0	98.101
5	10	100.00	1.0	99.993
10	10	100.00	1.0	100.00
—	—	—	2.0	79.403
—	—	—	2.0	82.106
—	—	—	2.0	97.243
1	29	88.409	—	79.403
5	29	99.314	—	82.106
10	29	99.972	—	97.423
1	31	79.364	—	—
5	31	94.795	—	—
10	31	98.815	—	—
1	33	67.675	—	—
5	33	79.441	—	—
10	33	86.996	—	—
1	35	54.479	3.0	64.723
5	35	52.009	3.0	56.809
10	35	51.421	3.0	54.835
1	39	29.392	—	—
5	39	7.948	—	—
10	39	2.134	—	—
1	50	1.621	5.0	26.503
5	50	0.000	5.0	2.229
10	50	0.000	5.0	0.159
1	100	0.000	10.0	1.034
5	100	0.000	10.0	0.000
10	100	0.000	10.0	0.000
1	200	0.000	20.0	0.000
5	200	0.000	20.0	0.000
10	200	0.000	20.0	0.000

[a]with alert limit of: NMT 35 particles/mL > 10 μm; NMT 3 particles/mL > 25 μm

[b]calculated based on a Poisson distribution of count data

true means limits in Table 4.2. Any lot that had mean particulate levels near the USP limits would surely fail the test whether 5 or 10 were sampled. The added power of a 10-unit test is only useful for those lots whose mean falls in the questionable range near the 35 and 3 count limits. Smaller sample size lots that are just over the alert limits would be more likely to be accepted, but it is very unlikely that any of these "bad" lots would have particle counts near the USP limits.

The author will provide one other example of setting alert limits and sample size based on tabularized statistical data and the input of a professional statistician. As an illustration to determine a sample size, let X and S represent the average and standard deviation of historical counts and let L represent the corresponding particulate matter limit. Then the sample size (n) can be calculated as follows:

$$\left(\frac{3S}{L-X}\right)^2$$

The sample size n should be rounded up. The resulting sample size ensures that the upper control limit for a control chart of the average counts is below the corresponding particulate matter limit. The upper control limit is

$$\frac{X+3S}{\sqrt{N}}$$

Example 1: For an LVI, suppose that historically ≥ 10 μm counts average 2.0 particles/mL with a standard deviation of 1.0 particles/mL. Based on these results, the resulting sample size is 0.017. Therefore, the minimum sample size of 2 is sufficient. The upper control limit for a control chart using a sample size of 2 is

$$\frac{2.0+3(1)}{\sqrt{2}}=3.53$$

This could be used as an alert limit for the average.

Example 2: Suppose instead that the historical LVI average is 15.0 particles/mL with a standard deviation of 6.0 particles/mL. Based on these results, the resulting sample size is 3.24. Therefore, a larger sample of 4 should be tested. The upper control limit for a control chart using a sample size of 4 is

$$\frac{15.0+3(6.0)}{\sqrt{4}}=16.5$$

This could be used as an alert limit for the average.

The author will add one word in closing regarding the distribution of particle counts from injectables and devices. Contaminant particle populations may obligingly follow theoretical distributions or be ill-behaved. Data from single units is often distributed as a log-normal or Poisson distribution, while means or data from sample pools is distributed normally. This must be taken into account in statistical manipulations.

A word regarding the intent of the verbiage in USP 23 <788> about sampling plans is also in order. The compendium is specifying not only a specific test method that must be followed, but also that the number of units tested in relation to batch size and historical data need to be taken into account.

> Sampling plans should be based on consideration of product volume, numbers of particles historically found to be present in comparison to limits, particle size distribution of particles found to be present, and variability of particle counts between units.

A wide variety of appropriate sampling plans may be devised based on statistical input from professionals. The FDA will expect a consideration of the four critical factors listed in the USP in defining whatever plan is chosen.

V

ENVIRONMENTAL STANDARDS, MANUFACTURING OPERATIONS, AND GOOD MANUFACTURING PRACTICE

This chapter will provide a general background discussion of the application of various environmental monitoring standards, guidelines, and provisions of the current Good Manufacturing Practices (cGMPs) to pharmaceutical and device manufacturing operations. In the near future, a new International Organization for Standardization standard (ISO 14644-1) and its companion documents will provide a unified standard for environmental monitoring that should simplify present compliance activities that are based on standards around the world.

Worldwide, there are presently a significant number of standards in place governing the operation, construction, and monitoring of clean facilities (e.g., controlled environment areas [CEAs]). Similarly, there are a number of Good Manufacturing Practice (GMP) regulations in force for various countries or national unions, such as the European Economic Community (EEC), that impact environmental monitoring and control. A small number of these documents are listed in Appendix II (excerpted from IES-RP-CC009). While not intended to be all-inclusive, the listing does provide a clearer picture of the scope of compliance-related literature in various countries. There are also specific regulatory guidelines.

A list of references specific to environmental monitoring is also included with the chapter on airborne particle counting (see Chapter 9). The references for this chapter are intended only to support the general discussion of environmental standards. The papers by Cooper (1988, 1989), Cooper and Milholland (1990), Grotzinger and Cooper (1989), Cooper et al. (1991), and Bzik (1986) provide basic discussions of the statistics of particle counting that are as applicable today as when they were written. The same is true for the paper by Borden et al. (1989); Wen and Kasper (1986) provide introductory discussions of the mechanics of monitoring that remain valid. Finally, those by Sommer (1989), Montague and Sommer (1990), and Sommer and Harrison (1991) will allow those generally interested in the operation of airborne particle counters to find operational reference.

INTERNATIONAL STANDARDS FOR CLEANROOM CLASSIFICATION

The nomenclature used by different nations and/or industry groups to refer to CEAs is somewhat confusing in and of itself. The term *clean* specifies that contamination control measures (minimally for particles) are in place in an area. In the United States, it is fairly common to refer to clean areas at Class 100 or cleaner as "cleanrooms." The "class" designation comes from U.S. Federal Standard 209E (FS-209E). In the business of pharmaceutical manufacturing, we often speak of "aseptic cores" or "aseptic processing areas" without the word *clean* being used. These areas are required to be Class 100 or cleaner, based on the Food and Drug Administration (FDA) guidelines for aseptic processing.

The ISO 14644-1 standard, which will in some measure yet to be determined supersede national standards such as FS-209E and British Standard (BS) 5295, speaks of "cleanrooms" and clean zones," the difference being that a cleanroom has walls, while a clean zone may not. Some standards documents of other nations refer to all classified areas as cleanrooms. Happily, while the nomenclature is different, the principles of operation, monitoring, and classification are essentially identical.

To further understand the present situation, which is characterized by a number of different environmental standards being used in different areas of the world, it is useful to look briefly at the evolution of these standards. With the development of technology during World War II and the years after, principles evolved for operating, classifying, and monitoring clean areas. In similar fashion, a vocabulary was developed. The common bases of the evolving standards were found in an exchange of technical information between workers in different countries and in the scientific and technical literature. The extent to which standards were developed depended on the available or required technology and the industrial developmental status of each country. For example, CEA technology is closely tied to needs in the nuclear, aerospace, pharmaceutical, and electronics industries. Thus, for Europe, the Americas, and Asia, the early leaders in this area were the United Kingdom, the United States, and Japan. There was also considerable activity in Australia regarding contamination control documents.

The common technical roots of such documents resulted in close similarities between documents, test methodologies, and CEA classification schemes. At present, the most widely used documents in the world are FS-209E, BS 5295, and Japanese International Standard (JIS) 9920. The close commonality of these documents is shown in Tables 5.1–5.3, which present each respective classification scheme. Table 5.4 represents the CEN 243 (1994) document that, along with some of the others, will be superseded by ISO 14644-1. The conversion factors between the metric and English systems of volume measurement are based on the fact that 1 ft^3 is equal to 28.3 L and 1 m^3 is equal to 35.3 ft^3.

The intent of such general specifications is to provide a technical commonality between user groups; there is no intent or effort made to provide operational descriptors for specific manufacturing processes. Such specifications are generated with the realization that specific points may need to be modified in lower level documents due to operational necessity. Any such modifications should be consistent with the requirements of the parent document. Modifications must typically be shown to be adequate for compliance; that is, those procedures done differently must be carried out such that the principle of equivalence can be established.

ISO Document Levels

In a broad assessment, three grades of documents are required for ISO compliance within a specific company. These are often addressed as Levels 1, 2, and 3. The company umbrella document is Level 1, specifying basic principles to be followed. An example would be a company document relating to ISO 9000 principles. Level 2 contains general compliance documents that specify policies within a given division of a company. Examples of Level 2 documents might be pharmaceutical documents specifying procedures to be followed for compliance with the GMP regulations of a specific country or region (i.e., 21 CFR in the United States). The most detailed documents are Level 3, typically pertaining to specific processes (e.g., aseptic fill or barrier isolator operation), production of a specific product, or operations of a work group. Obviously, in this context, the situation for a pharmaceutical manufacturer remains complex. For example, considerations later in this chapter for ISO 14644-1 will show that the document governs only basic principles to follow and does not significantly impact the more detailed procedures already in place for regulatory compliance.

Implementation of ISO 14644-1 should benefit the manufacturer by providing a basic foundation for CEA classifications and a methodology for testing critical (normative)

Table 5.1. Airborne Particulate Matter Cleanliness Classes per FS-209E

Class Name**		0.1 μm Volume Units		0.2 μm Volume Units		0.3 μm Volume Units		0.5 μm Volume Units		5 μm Volume Units	
SI	English***	(m^3)	(ft^3)	(m^3)	(ft^3)	(m^3)	(ft^3)	(m^3)	(ft^3)	(m^3)	(ft^3)
M 1		350	9.91	75.7	2.14	30.9	0.875	10.0	0.283	—	—
M 1.5	1	1,240	35.0	265	7.50	106	3.00	35.3	1.00	—	—
M 2		3,500	99.1	757	21.4	309	8.75	100	2.83	—	—
M 2.5	10	12,400	350	2,650	75.0	1,060	30.0	353	10.0	—	—
M 3		35,000	991	7,570	214	3,090	87.5	1,000	28.3	—	—
M 3.5	100	—	—	26,500	750	10,600	300	3,530	100	—	—
M 4		—	—	75,700	2,140	30,900	875	10,000	283	—	—
M 4.5	1,000	—	—	—	—	—	—	35,300	1,000	247	7.00
M 5		—	—	—	—	—	—	100,000	2,830	618	17.5
M 5.5	10,000	—	—	—	—	—	—	353,000	10,000	2,470	70.0
M 6		—	—	—	—	—	—	1,000,000	28,300	6,180	175
M 6.5	100,000	—	—	—	—	—	—	3,530,000	100,000	24,700	700
M 7		—	—	—	—	—	—	10,000,000	283,000	61,800	1,750

CLASS LIMITS*

* The class limits shown in this table are defined for classification purposes only and do not necessarily represent the size distribution to be found in any particular situation.

** Concentration limits for intermediate classes can be calculated, approximately, from the following equations:

$$\text{Particles/m}^3 = 10M\,(0.5/d)^{22}$$

where M is the numerical designation of the class based on SI units, and d is the particle size in μm, or

$$\text{Particles/ft}^3 = N_c(0.5/d)^{22}$$

where N_c is the numerical designation of the class based on English (U.S. customary) units and d is the particle size in μm.

*** For naming and describing the classes, SI names and units are preferred; however, English (U.S. customary) units may be used.

Table 5.2. Airborne Particle Count Cleanliness Classes of JIS 9920

Cleanliness Class

Particle Size (μm)	Class 1	Class 2	Class 3	Class 4	Class 5	Class 6	Class 7	Class 8
0.1	10^1	10^2	10^3	10^4	10^5	(10^6)	(10^7)	(10^8)
0.2	2	24	236	2,360	23,600	—	—	—
0.3	1	10	101	1,010	10,100	101,000	1,010,000	10,100,000
0.5	(0.35)	(3.5)	35	350	3,500	35,000	350,000	3,500,000
5.0	—	—	—	—	29	290	2,900	29,000
Particle size range of cleanliness	0.1–0.3		0.1–0.5		0.1–0.5	0.3–5.0		

Remarks: Class 3, Class 4, Class 5, Class 6, Class 7, and Class 8 are equivalent to Class 1, Class 10, Class 100, Class 1,000, Class 10,000, and Class 100,000 of FS-209E, respectively.

Table 5.3. BS 5295 (1989) Requirements for Controlled Environment Installations

Class of Environmental Cleanliness	Maximum Permitted Number of Particles per m^3 (equal or greater than stated size)					Max. Floor Area (m^3) per Sampling Position for Cleanrooms	Classified and Unclassified Areas	**Classified Area Adjacent to Lower Classification
	0.3 μm	0.5 μm	5 μm	10 μm	25 μm			
C	100	35	0	NA*	NA*	10	15	10
D	1,000	350	0	NA*	NA*	10	15	10
E	10,000	3,500 (100)	0 (0)	NA*	NA*	10	15	10
F	NA	3,500 (100)	0 (0)	NA	NA	25	15	10
G	100,000	35,000	200	0	NA*	25	15	10
H	NA*	35,000	200	0	NA*	25	15	10
J	NA*	350,000 (10,000)	2,000 (57)	450 (13)	0 (0)	25	15	10
K	NA*	3,500,00 (100,000)	20,000 (570)	4,500 (130)	500 (14)	50	15	10
L	NA*	NA*	200,000	45,000	5,000	50	10	10
M	NA*	NA*	NA*	450,00	50,00	50	10	NA

* No specified limit.

** This applies only to cleanrooms and totally enclosed devices.

Table 5.4. Airborne Particle Count Allowances of Proposed CEN 243*

Classification Number	Maximum Permitted Number of Particles/m^3 of a Size Equal to or Greater Than the Considered Size						
	0.1 μm	0.2 μm	0.3 μm	0.5 μm	1 μm	5 μm	10 μm
0	25	6**	NA	(1)	NA	NA	NA
1	250	63**	23**	10	NA	NA	NA
2	2,500	625	278**	100	25	NA	NA
3	25,000	6,250	2,778**	1,000	250	10	NA
4	NS	62,500	27,778**	10,000	2,500	100	25
5	NS	NS	NS	100,000	25,000	1,000	250
6	NS	NS	NS	1,000,000	250,000	10,000	2,500
7	NS	NS	NS	(10,000,000)	2,500,000	100,000	25,000

*Withdrawn in accord with ISO 14644-1.

NA = not applicable; NS = no specified limit; ( ) = for reference only; ** = rounded values parameters; however, it will not remove the manufacturer's requirement to follow regional or national standards that give more detailed instructions of how pharmaceutical manufacture is carried out. This general scheme is presented in Figure 5.1. The ISO document will presumably supersede the various national umbrella documents (e.g., FS-209E, BS 5295, JIS 9920, and CEN 243), which will benefit some companies that have international operations. Note that ISO 14644-1 compliance may eventually be required for ISO 9000 certification. While Figure 5.1 illustrates the general nature of the system, it falls far short of illustrating the intricacies of some local situations.

PHILOSOPHY OF COMPLIANCE

The following portions of this chapter comprise a discussion of the philosophy of compliance with both umbrella type documents (e.g., ISO 14644) and national, regional, or local standards that apply more directly to the manufacture of medical products and pharmaceuticals. Of utmost importance here are two principles:

1. The manufacturer must have a comprehensive knowledge of *all* standards that apply in a specific manufacturing and/or marketing location.

2. Specifications in written format must be in place as Level 1 or 2 documents, which will serve to ensure general compliance. Level 3 documents must be developed as necessary to describe specific processes.

The tie-in between control of environmental particulate matter and GMP for pharmaceuticals and medical devices is spelled out without equivocation in all of the major GMPs of the world. For instance, in those of the United Kingdom, the requirement is called out in the GMPs as a requirement for "adequate premises and space." More specifically, it is specified that for aseptic processing, rooms with conventional filtered airflow and with contained workstations (in the form of filtered air hoods or laminar airflow protection at working points) are more appropriate than laminar airflow rooms. Rooms for aseptic processing should, in the unmanned state, comply with the conditions specified for Grade 1B in the *Table of Basic Environmental Standards* (Table 5.5 in this chapter).

Figure 5.1. Hierarchy of requirements for environmental control in pharmaceutical manufacture.

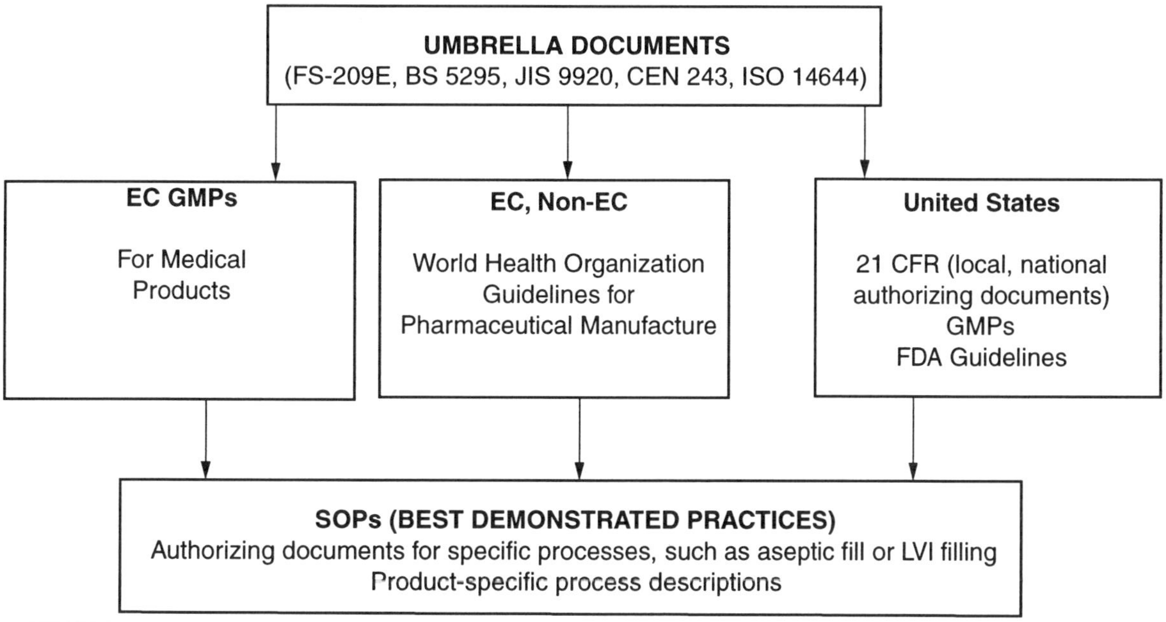

Table 5.5. Various Standards Applicable in the United Kingdom (1991)

		Particles/m^3 per (ft^3 in parentheses) of Size Equal to or Greater Than in Micrometers				
Authority	**Class**	**0.5**	**1**	**5**	**10**	**25**
MEDICAL DEVICES						
BS 5295 (1976)	Class 3	NA	1,000,000) (28,317)	20,000 (566)	4,000 (113)	300 (6)
BS 5295 (1989)	Class K	3,500,000 (99,111)	NA	20,000 (566)	4,500 (127)	500 (14)
FS-209E	Class 100,000	3,531,400 (100,000)	NA	24,720 (700)	NA	NA
EEC GMP 1987 (Draft)	Grade D	3,500,000 (99,111)	NA	20,000 (566)	NA	NA
Orange Guide 1983	Grade 3	3,500,000 (99,111)	NA	20,000 (566)	NA	NA
TERMINALLY STERILIZED SOLUTION						
BS 5295 (1976)	Class 2	300,000 (8495)	NA	2,000 (57)	30 (1)	NA
BS 5295 (1989)	Class J	350,000 (9,911)	NA	2,000 (57)	450 (13)	0
FS-209E	Class 10,000	353,140 (10,000)	NA	2,472 (70)	NA	NA
EEC GMP 1987 (Draft)	Grade C	350,000 (9,911)	NA	2,000 (57)	NA	NA
Orange Guide 1983	Grade 2	300,000 (8,495)	NA	2,000 (57)	NA	N
ASEPTIC FILLING						
BS 5295 (1976)	Class 1	3,000 (65)	NA	0	NA	NA
BS 5295 (1989)	Class F	3,500 (99)	NA	0	NA	NA
FS-209E	Class 100	3,531 (100)	NA	NA	NA	NA
EEC GMP 1987 (Draft)	Grade B	3,500 (99)	NA	0	NA	NA
Orange Guide 1983	Grades 1A and 1B	3,000 (65)	NA	0	NA	NA

With people present and with work in progress, Grade 1/A conditions should be maintained under the contained workstations where products are exposed and aseptic manipulations are carried out. Solutions intended for use as large-volume injections (LVIs) or small-volume injections (SVIs), eye drops, contact lens solutions, peritoneal dialysis solutions, and for irrigation (including nonintravenous water for irrigation) should be prepared in a room complying with the conditions specified for Grade 2. The objective should be to prepare a pyrogen-free solution with low microbial and particle counts suitable for later sterilization. The filling of products to be terminally sterilized should be carried out in rooms complying with the conditions specified for Grade 2. Extra precautions in the form of contained workstations and/or laminar airflow protection are deemed necessary when solutions intended for intravenous use are filled into wide-necked containers.

The various GMP regulations not only specify that product be manufactured under specific conditions of appropriate cleanliness, but they often indicate what level of cleanliness must be used in the production of different types of pharmaceuticals. As regulations and compendia of different countries are considered, the matrix of requirements that applies to a given product made in a specific country of the world becomes increasingly complex.

Pharmaceutical/biopharmaceutical cleanroom requirements depend on manufacturing procedures and on the final use of the manufactured products. For example, the manufacturing requirements for an oral drug are not the same as those for a parenteral drug. Oral drugs are ingested, and by the time they get into the bloodstream, many body defense mechanisms have been activated (starting with saliva). It is most likely that viable organisms present in the ingested product will be destroyed. When a parenteral drug is injected into the bloodstream, the administration route bypasses the body's natural defenses. A parenteral product must, therefore, be free of viable organisms that are capable of multiplying to cause an infection (i.e., sterile). This is why cleanrooms used in oral drug pharmaceutical plants may in some cases be Class 100,000, while cleanroom classifications for parenteral products are more stringent, ranging from Class 100 to Class 10,000. Biopharmaceutical drugs are mostly injectable drugs for which special processes are employed to prepare and/or alter biological molecules in product destined to become diagnostic or therapeutic.

As discussed earlier, a manufacturer of healthcare products in any given geographic or national location (or for a specific international market) has the burden of determining that all appropriate requirements are met. A key consideration of our discussion here regards how the basic standards for environmental monitoring are woven into the application of GMP through compendial and/or regulatory requirements. The ultimate goal of this process for both the manufacturer and the regulator is a simple and effective mechanism of compliance and product protection. Several examples of this process given are presented in the following portions of this chapter. First, as an example of complex compliance situations that may hopefully be avoided in the future, the situation in the United Kingdom at the beginning of this decade is reviewed. In this case, multiple sets of regulatory requirements pertained to the product manufactured, and several sets of standards differing in minor respects were enforced. Secondly, the relatively simple situations in the United States are explored, where FS-209E and U.S. Pharmacopeia (USP)/FDA requirements apply. Then the somewhat more involved present situations in the United Kingdom and EEC are considered based on the British and EEC standards. Finally, the ISO 14644 series of documents are discussed.

The United Kingdom—Complexities in Compliance

Compliance with appropriate standards is relatively simple with regard to products to be sold in a single, national marketplace (i.e., the domestic market). Production for international marketing, however, requires diligence on the part of the manufacturer to comply with what may be a complex array of standards.

The situation at the beginning of this decade in the United Kingdom provides an illustration of the difficulties of compliance in the presence of multiple requirements and standards. This is summarized in Table 5.5; the first group of data refers to Class 3 areas. All dry production areas producing finished dry product for terminal sterilization were liable for inspection by the Procurement section of the DHSS. This department audited against the *Blue Guide*, which referred to rooms in terms of BS 5295 (1976). The second group of data pertains to areas where terminally sterilized products are mixed and filled. These areas were subject to inspection from the Medicines branch of the DHSS, which audited against the *Orange Guide* (1983). The *Orange Guide* refers to rooms in terms of grades (e.g., Grade 1, Grade 2). The third grouping in the table pertains to aseptically filled units (e.g., compounding). This endeavor was subject to the Medicines branch and the *Orange Guide* Grade 1A or 1B requirements. (Grade 1B is differentiated from 1A in terms of the permitted number of viable organisms per m^3.) In this undeniably complex situation, the requirements differed to various degrees not only with regard to the use of different standards for enforcement but also on the dates of the standard involved (Tables 5.6, 5.7, 5.8, and 5.9).

Environmental Monitoring and GMP in the United States

FS-209E is a U.S. regulation that applies to most clean spaces. *Airborne Particulate Cleanliness Classes in Cleanrooms and Clean Zones* defines and classifies cleanrooms and also describes methods of measuring space cleanliness. Section 3.4 of FS-209E states that a clean zone is

> A defined space in which the concentration of airborne particles is controlled to meet a specified airborne particulate cleanliness class.

Table 5.6. Summary of Requirements for BS 5295 (1976)

Controlled Environment (cleanroom, workstation, or device)	Recommended Airflow Configurations	Recommended Periodicity for Air Sampling and Particle Counting	Maximum Permitted Number of Particles per m^3(equal to or greater than stated size)					Percent Final Filter Efficiency
			0.5 µm	1 µm	5 µm	10 µm	25 µm	
Class 1	Unidirectional	Daily or continuous by automatic equipment	3,000*	NA	Nil	Nil	Nil	99.995
Class 2	Unidirectional	Weekly	300,000	NA	2,000	30	Nil	99.95
Class 3	Unidirectional/ conventional	Monthly	—	1,000,000	20,000	4,000	300	95.00
Class 4	Conventional	3 per month	—	—	200,000	40,000	4,000	70.0
Controlled Area	Normal ventilation	—	—	—	—	—	—	—
Contained Work Station	Unidirectional	To suit required class and application	—	—	—	—	—	99.997
Portable Tent	As selected	To suit required class and application	To suit required class*					

*Subject to maximum particle size of 5 µm (see clause 4).

Table 5.7. Orange Guide 1983—Appendix 1: Basic Environmental Standards for the Manufacture of Sterile Products

Grade	Final Filter Efficiency (determined by BS 3928)[1]	Recommended Minimum Air Changes/h	Max. permitted # particles per m^3 equal to or above:[2]		Max. permitted # of visible organisms per cm^3[(2,3)]	Nearest Equivalent Standard Classification		
			0.5 μm	5 μm		BS 5295[4]	FS-209B[5]	VDI 2083.P.1.[6]
1/A (Unidirectional airflow work-station	99.997%	Flow of 0.3 m/s (vertical) or 0.45 m/s (horizontal)	3,000 85/ft^3	0	less than 1	1	100	—
1/B	99.995%	20	3,000	0	5	1	100	3
2	99.95%	20	300,000 8,500/ft^2	2,000 57/ft^2	100	2	10,000	5
3	95.0%	20	3,500,000	20 000	500	3	100,000	6

Air pressure should always be highest in the area of greatest risk to product. The air pressure differentials between rooms of successively higher to lower risk should be at least 1.5 μm (0.06 in. WG).

[1]BS 3928: Method for Sodium Flame Test for Air Filters, British Standards Institution, London, 1969.

[2]This condition should be achieved throughout the room when unmanned, and recovered within a short "cleanup" period after personnel leave. Conditions should be maintained in the zone immediately surrounding product whenever product is exposed. (Note: It is accepted that it may not always be possible to demonstrate conformity with particle standards at the point of fill with filling in progress, due to generation of particles or droplets from the product itself.)

[3]Mean values obtained by air sampling methods.

[4]BS 5295: Environmental Cleanliness in Enclosed Spaces, British Standards Institution, London, 1976.

[5]U.S. Federal Standard FS-209B, 1973.

[6]Verein Deutscher Ingenieure 2083, P.1.

Table 5.8. Blue Guide—Air Classification System for Manufacture of Sterile Products (Also see EC GMPs)

Grade	Maximum permitted number of particles/m^3 equal to or above:		Maximum permitted # of viable microorganisms per m^3
	0.5 µm	5 µm	
A laminar airflow work station	3,500	0*	Less than 1*
B	3,500	0*	5*
C	350,000	2,000	100
D	3,500,000	20,000	500

Notes:

- Laminar airflow systems should provide a homogeneous air speed of about 0.30 m/s for vertical flow and 0.45 m/s for horizontal flow.
- In order to reach the B, C, and D air grades, the number of air changes should generally be higher than 20 per hour in a room with a good aerobic pattern and appropriate HEPA filters.
- Low values involved here* are only reliable when a large number of air samples are taken.
- The guidance given for the maximum permitted number of particles corresponds approximately to FS-209E as follows: Class 100 (grades A and B), Class 10,000 (grade C), and Class 100,000 (grade D).

Table 5.9. Blue Guide—Class Limits in Particles per Cubic Feet of Size Equal to or Greater Than Particle Sizes Shown (µm)[1]

	Measured Particle Size (µm)				
Class	0.1	0.2	0.3	0.5	5.0
1	35	7.5	3	1	NA
10	350	75	30	10	NA
100	NA	750	300	100	NA
1,000	NA	NA	NA	1,000	7
10,000	NA	NA	NA	10,000	70
100,000	NA	NA	NA	100,000	700

NA = Not Applicable

[1]The class limit particle concentrations shown in Table 4 and Figure 1 are defined for class purposes only and do not necessarily represent the size distribution to be found in any particular situation.

Section 3.5 defines a cleanroom as

> A room in which the concentration of airborne particles is controlled and which contains one or more clean zones.

This standard classifies cleanrooms in the United States from Class 1 through Class 100,000 in terms of the number of particles of size 0.5 µm and larger allowed per cubic foot.

The testing of cleanrooms and clean zones is covered in some detail in FS-209E. Methods for verifying the air cleanliness class and monitoring of airborne particles also are included. The document does not apply to particles deposited on equipment or supplies within the cleanroom, and it is not intended to separate viables and nonviables. It provides a specification for building, commissioning, and maintaining cleanrooms that are generally applicable to pharmaceutical manufacture. The criticality of the number of airborne particles for processes in the electronic industry makes the application of FS-209E a necessity. Exposure of critical process areas is directly related to particle concentration, and one particle deposited on a microcircuit during fabrication has some defined statistical probability of becoming a defect. The variable criticality of airborne microorganisms in pharmaceutical industry cleanrooms makes the standard less well applicable in our industry. The reasoning that the fewer particles present in a cleanroom the less likely that airborne microorganisms will be present is accepted broadly, but the presence of small numbers of organisms has little significance for a terminally sterilized product. Thus, in the manufacture of injectables and devices, one typically sees modifications of the principles specified in this standard to suit the specialized technical and regulatory concerns that pertain.

GMP in the United States: Laws and Regulations

The manufacturer of pharmaceuticals and medical devices in the United States is governed by a complex matrix of different types of requirements, including laws, guidelines, and GMP.

Law. Law 501 of the Food, Drug, and Cosmetic (FD&C) Act states that a drug, device, diagnostic, or bulk pharmaceutical chemical is considered to be adulterated if not manufactured in accordance with cGMP. The purpose of cGMP is to ensure process control on a repeatable basis. This is a cornerstone requirement that ties pharmaceutical and device manufacture to GMP.

The cGMPs are found in parts 210, 211, 212, 606–680, and 820 of Title 21 of the *Code of Federal Regulations* (21 CFR) as well as in the USP. Of the cGMPs, those that affect heating, ventilation, and air-conditioning system (HVAC) design are found in *Part C—Buildings and Facilities*, Sections 211.42 and 211.46.

> 21 CFR 211.42—DESIGN & CONSTRUCTION FEATURES
>
> a. Any building . . . shall be of suitable size, construction and location to facilitate cleaning, maintenance and proper operation.
>
> b. . . . flow of components, drug products . . . shall be designed to prevent contamination.
>
> c. . . . There shall be separate or defined areas . . . to prevent contamination or mix-ups as follows:

Aseptic processing is a specific focus:

> c.10. Aseptic processing, which includes as appropriate:
>
> (i) Floors, walls and ceilings of smooth, hard surfaces that are easily cleanable;
>
> (ii) Temperature and humidity controls;
>
> (iii) An air supply filtered through HEPA . . .; and
>
> (iv) A system for monitoring environmental conditions . . .

Guidelines. The sterilization of drugs is accomplished by terminal sterilization or aseptic processing. In terminal sterilization (the procedure preferred by the FDA), the final product in its final container is sterilized with heat or radiation. In aseptic processing, the drug product, container, and closure are separately sterilized and then brought together. Because of their sensitivity to heat, most biological (biopharmaceutical) products are produced using aseptic processing.

The FDA guidelines for various processes define in more detail what type of CEA is required, depending on product/container exposure at different stages of manufacture. These guidelines are binding unless the alternative solution proves to be equivalent to or better than the guidelines. The most specific instruction on clean space requirements is in the *Guideline on Sterile Drug Products Produced by Aseptic Processing* (FDA 1987). It specifically addresses two types of exposure areas requiring separation and control: critical areas and controlled areas.

Critical areas are sectors in which the sterilized dosage forms, containers, and closures are exposed to the environment. Air must be Class 100 when measured not more than 1 ft from the work site, upstream of filling and closing operations. Air has to be supplied at the point of use through high efficiency particulate air (HEPA) filters, with an air velocity of 90 FPM (feet per minute) ± 20 percent, the principle being that airflow rates higher than 110 FPM may be more likely to induce turbulence. The microbial count of the room cannot exceed 0.1 CFU/ft^3 of space. The room pressure should be at least 0.05 in. water gauge (WG) positive, relative to adjacent less clean areas.

Controlled areas are those in which the unsterilized product, in-process materials, and containers and closures are prepared. Included are areas where components are compounded and where components, in-process materials, and drug products contact surfaces of equipment, containers, and closures after the final rinse of such surfaces and are thus exposed to the plant environment. The room may be no less than Class 100,000 with a minimum of 20 air changes per hour (AC/h). The room microbial count cannot exceed 2.5 CFU/ft^3 of space. The room pressure should be 0.05 in. WG or more positive in relation to adjacent, less clean areas. The FDA guidelines for aseptic filling specify Class 100 conditions and laminar flow. Particularly confusing situations can arise in the national arena, depending on what guidelines are used as a basis for the national GMPs. For example, Canadian GMPs specify Class 100 laminar flow conditions for the filling of terminally sterilized LVI product based on World Health Organization (WHO) guidelines; this level of air cleanliness is really not necessary for terminally sterilized LVI, but can be required and enforced on a national basis.

USP Requirements for Compounding

Another example of informative instruction is found in General Chapter <1074> of USP 23. This is an informative (i.e., for informational purposes) rather than a normative (enforceable by law) chapter, but could be interpreted by a field inspector as describing the best demonstrated practice. In this section, an environmentally controlled workspace suitable for the aseptic processing of a drug product is defined as a suitably constructed, properly functioning, and regularly certified device that sweeps the workspace with HEPA–filtered, laminar-flowing air at a velocity of 90 FPM ± 20 percent, such as a laminar airflow workbench (LAW). An LAW is required for both low-risk and high-risk operations. Since the airflow velocity is relatively low, the LAW must be located in an environmentally controlled room or a space otherwise separated from less controlled work areas (such as the main pharmacy) by partitions, plastic curtains, or similar solid dividers. Such a room is called a "buffer" room. As is the case in other pharmaceutical applications, FS-209E has been adapted and modified to meet pharmacy compounding needs. It is generally accepted that if fewer particles are present in an operational cleanroom or in other controlled environments, the microbial count under operational conditions will be less.

The environmental quality within a buffer room must be demonstrably better than that of adjacent areas to reduce the risk of contaminants being blown, dragged, or otherwise introduced into the LAW. The level of cleanliness of the air in the buffer room, in conjunction with the expertise of the operator, is critical to maintaining Class 100 conditions within the LAW. Air entering the buffer room should be HEPA filtered, temperature and humidity conditioned, and meet the requirements for at least a Class 100,000 cleanroom for low-risk operations and Class 10,000 for high-risk operations (see Table 5.1). In addition to cleaning the air flowing in and providing at least 10 AC/h, cooling is essential because of the continual buildup of heat from the circulating air flowing through the blower and HEPA filter of the LAW. It should be noted that the circulation of air from the buffer room through the HEPA filter of the LAW will enhance the cleanliness of the air, particularly during nonuse periods.

INTERNATIONAL STANDARDS

In the context of its use, a standard may be generally defined as a document that provides a guideline, set of rules, or description of a process that allows for reproducible conduct of a given activity between groups. The author will admit, early on, that he does not have a comprehensive knowledge of neither the various standards organizations nor their activities. The reader who goes beyond this chapter for a better knowledge of the world standards bodies and their work will find this area complex and not amenable to cursory review.

For the American manufacturer, the British Standards Institute (BSI) constitutes an invaluable resource for the standards of the United Kingdom and the EEC (as well as ISO standards).

These are well indexed in a thousand-page hard copy catalog, which includes the CD-ROM version of the same information. The BSI maintains an office in the United States (Herndon, Va.) for the convenience of users in this country. Standards of other nations that have not been adopted as British standards are also available through BSI.

Perhaps the best presentation of the world's standards organizations, their aims, and their interactions is provided by British Standard 0-1:1997, aptly entitled *A Standard for Standards—Part 1: Guide to the Context, Aims and General Principles.* The scope of this document is stated to be a description of the overall context of standardization and an explanation of how standards are used in trade, contracts, and how they may be involved in legislation.

This explanation of these subject areas provided has proven invaluable to the author, and it is highly recommended. Included in BS 0-1 are discussions of principles of standards development, beginning with the criteria that standards should be

- needed and wanted,
- used,
- agreed to at the widest level,
- impartial, and
- planned.

International GMP Documents

The first WHO draft text on GMP was prepared at the request of the 20th World Health Assembly (resolution WHA 20.34) in 1967 by a group of consultants.

The WHO Expert Committee document on pharmaceutical manufacture has key similarities to the EC and United Kingdom documents, since it was derived from the EC GMPs, which in turn were significantly based on established British standards. Importantly, It states that to obtain air of the required characteristics, methods specified by the national authorities should be used. The following should be noted:

- Classes A, B, C, and D represent progressively less clean or less well-controlled areas.
- Laminar airflow systems should provide a homogeneous air speed of about 0.30 m/s for vertical flow and about 0.45 m/s for horizontal flow; however, precise air speeds will depend on the type of equipment. (This reflects the poorer clearance efficiency of horizontal flow rooms).
- In order to reach the B, C, and D air grades, the number of air changes should generally be higher than 20 AC/h in a room with a good airflow pattern and appropriate HEPA filters.
- Low values for containment are reliable only when a large number of air samples is taken.
- The guidance given for the maximum permitted number of particles corresponds approximately to FS-209E (1992) as follows: Class 100 (Grade A and B), Class 10,000 (Grade C), and Class 100,000 (Grade D).

A caveat is added to the effect that it may not always be possible to demonstrate conformity with a particulate matter air standard where filling is in progress, owing to the generation of particles or droplets from the product itself.

A basic principle of the WHO document is that each manufacturing operation requires an appropriate air cleanliness level in order to minimize the risks of particle or microbial contamination of the product or materials being handled. The document also recognizes that utilizing isolator technology and automated systems to minimize human interventions in processing areas can produce significant advantages in ensuring the sterility of manufactured products. When such techniques are used, the recommendations in these supplementary guidelines, particularly those relating to air quality and monitoring still apply, with appropriate interpretation of the terms *workstation* and *environment.*

> Terminally sterilized solutions should generally be prepared in a grade C environment in order to give low microbial and particulate counts, suitable for immediate filtration and sterilization. Solution preparation could be allowed in a grade D environment if additional measures were taken to minimize contamination, such as the use of closed vessels. For parenterals, filling should be done in a laminar airflow workstation (grade A) in a grade C environment. The preparation of other sterile products, e.g., ointments, creams, suspensions,

and emulsions, and filling of containers should generally be done in a grade C environment before terminal sterilization.

The handling of starting materials and the preparation of solutions to be aseptically filled should be done in a grade C environment. These activities could be allowed in a grade D environment if additional measures are taken to minimize contamination, such as the use of closed vessels prior to filtration. After sterile filtration, the product must be handled and dispensed into containers under aseptic conditions in a grade A or B area with a grade B or C background, respectively.

Other sterile products prepared from sterile starting materials should be handled in an aseptic way. The handling of starting materials and all further processing should be done in a grade A or B area with a grade B or C background, respectively.

High standards of personal hygiene and cleanliness are essential, and personnel involved in the manufacture of sterile preparations should be instructed to report any condition that may cause the shedding of abnormal numbers or types of contaminants; periodic health checks for such conditions are desirable. Actions to be taken for personnel who could be introduced under a microbiological hazard should be decided by a designated competent person. Outdoor clothing should not be brought into clean areas, and personnel entering the changing rooms should already be clad in standard factory protection garments. Changing and washing should follow written procedure. The clothing and its quality has to be adapted to the process and the workplace and worn in such a way as to protect the product from contamination. Wrist-watches and jewelry should not be worn in clean areas, and cosmetics that can shed particles should not be used. Clothing should be appropriate to the air grade of the area where the personnel will be working. The description of clothing required for each grade is specified as follows:

Grade D: The hair and (where appropriate) beard should be covered. Protective clothing and appropriate shoes or overshoes should be worn to avoid any contamination coming from outside the clean area.

Grade C: The hair and (where appropriate) beard should be covered. A single or two-piece trouser suit, gathered at the wrist and with a high neck, and appropriate shoes or overshoes should be worn. The clothing should shed virtually no fibers or particulate matter.

Grade B and A: Headgear should totally enclose the head and, where appropriate, beard; It should be tucked into neck of the suit. A face mask should be worn to prevent the shedding of droplets; sterilized, non-powdered rubber or plastic gloves and sterilized or disinfected footwear should be worn; trouser-bottoms should be tucked inside the footwear, and garment sleeves into the gloves. The protective clothing should shed virtually no fiber or particulate matter and should retain particles shed by the body.

For every worker in a grade B or A area, clean, sterilized protective garments should be provided at each work session, or at least once a day, if monitoring results justify. Gloves should be regularly disinfected during operations, and masks and gloves should be changed at least at every working session. The use of disposable clothing may be necessary. Clothing used in clean areas should be laundered or cleaned.

Particular attention should be paid to the protection of the zone of greatest risk (that is, the environment immediate to the product and cleaned components, air supplies, and pressure differentials may need to be modified if it becomes necessary, to contain materials such as pathogenic, highly toxic, radioactive, or live viral or bacterial materials). Decontamination facilities and the treatment of air leaving a clean area may be necessary for some operations.

With regard to the cleanliness of the air itself, it should be demonstrated that airflow patterns do not present a contamination risk. For example, care should be taken to ensure that airflow does not distribute particles from a particle-generating person, operation, or machine to a zone of higher product risk. A warning system should be included to indicate failure in the air supply. An indicator of pressure difference should be fitted between areas where this difference is important, and the pressure difference should be regularly recorded. Consideration should also be given to restricting unnecessary access to critical filling areas (e.g., grade A filling zones), by the use of a physical barrier. A conveyor belt should not pass through a partition between a clean area B and a processing area of lower air cleanliness, unless the belt itself is continuously sterilized (e.g., in a sterilizing tunnel).

Whenever possible, equipment used for processing sterile products should be chosen so that it can be effectively sterilized by steam or dry heat or other methods. It should be designed and installed so that operations, maintenance, and repair can be achieved outside the clean area. Equipment that has to be taken apart for maintenance should be resterilized after completion wherever possible when equipment maintenance is carried out within the clean area. Clean instruments and tools should be used, and the area should be clean and disinfected where appropriate before processing recommences, if the required standard of cleanliness and/or sepsis have not been maintained during the maintenance procedure.

All equipment including sterilizers, air-filtration systems, and water treatment system (including stills) should be subject to planned maintenance, validation, and monitoring; its approved use following maintenance work should be documented. Water treatment plants should be designed, constructed, and maintained so as to ensure the reliable production of water of an appropriate quality. These should not be operated beyond their designed capacity.

Many developments occurred in GMP in the intervening years since the first version of the WHO document was introduced, and important national and international documents (including new revisions) have appeared.

- *Guide to Good Pharmaceutical Manufacturing Practice (Orange Guide)*, Her Majesty's Stationery Office. 1983. [Superseded by the 1992 EEC guide, which took much of its material from the UK document.]
- *Direction de la Pharmacie et du Medicament,* Bonnes Pharma Paris, Ministere des Affaires Sociales et de la Solidarite Nationale, Secretariat d'Etat charge de la Sante, 1985. [Superseded by the 1992 EEC guide.]
- *ASEAN Good Manufacturing Practices Guidelines*, 2nd ed., Association of South-East Asian Nations, 1988.
- *Good Manufacturing Practice for Medicinal Products in the European Community.* Commission of the European Communities, 1992.
- *Guide to Good Manufacturing Practice for Pharmaceutical Products.* Convention for the Mutual Recognition of Inspection in Respect of the Manufacture of Pharmaceutical Products (PIC), 1992.

Other guidelines that have appeared in recent years include GMP texts applicable to the manufacture of bulk pharmaceutical chemicals as opposed to the manufacture of formulations of dosage forms (Pharmaceutical Inspection Convention [PIC] guidelines 1987; various national documents). Another important development in the industry at large is the appearance of the guidelines of ISO, specifically the ISO 9000 to 9004 standards for quality systems (1987, revised 1990). These developments, together with plans to expand and revise the certification scheme, call for a revision of the WHO GMP text.

The European Community (EC, EEC, EN) GMPs

The EC Guide for GMP was first published in 1989, and became compulsory for EC nations on

1 January 1992. This Guide was approved by the representatives of the pharmaceutical inspection services of the EC member states and replaced national guidelines or other relevant GMP requirements. It was intended to provide the EC pharmaceutical industry and national inspection services with an understanding of what compliance with GMPs entails. The Pharmaceutical Inspection Convention (PIC), an authority of the European Free Trade Association (EFTA), has a similar document; the two are generally harmonized to the extent that compliance with the current version of the EC GMPs entails compliance with the PIC guidelines.

The European GMP Guide was patterned after the Orange Guide (UK GMPs). It originally contained nine chapters, each containing a statement of principle and a glossary; it also contained a separate supplementary guideline on the manufacture of sterile medicinal products. This latter (Annex) was similar to Section 9 of the 1983 Orange Guide. In 1992, a reprint of the EC Guide with 11 additional annexes besides the one on the manufacture of sterile medicinal products was published. The most recent revision of this Annex was made in 1997 and has important implications for the manufacture of injectable products. It also applies to products manufactured in other countries (e.g., the United States) for the EC market.

The revised Annex 1 provides additional information on:

- the classification of CEAs by particle counts;
- the monitoring of controlled environments;
- barrier isolators;
- blow-fill-seal processing; and
- sterilization, including media fills.

It is imperative that U.S. manufacturers whose products will be affected by the EC GMP document have a good grasp of how its principles apply to their operations. For an American manufacturer unused to compliance with the guidelines, directives, requirements, and GMP regulations of nations other than the United States, the playing field can be extremely confusing at first. Interpharm Press (Buffalo Grove, Ill.) is a source for various compliance documents of other nations, including the European Commission literature of the European Community. Standards may also be obtained from the BSI as mentioned earlier and domestically from the Technical Standards Services (Alliston Park, Penn.). One specific volume the author believes very useful is the *EU Directive Handbook* (Barley 1997) from St. Lucie Press (Boca Raton, Fla.) on the subject of EU Directives.

Two excellent summary articles are included in the references of this chapter: (1) Del Valle (1997), which appeared in the *Journal of Pharmaceutical Engineering*. This author discusses the specific differences between the U.S. and EC GMPs. (2) Delattin (1998), which was published in the *Journal of Parenternal Science and Technology*.

The current Annex on sterile medicinal products incorporates a number of changes to address criticisms of the original EC Guide of 1992. The EC document continues to reasonably specify that each manufacturing operation requires an appropriate air cleanliness level in order to minimize the risk of particle or microbial contamination of the product or materials being handled. Table 5.10 gives the minimum air grades required for different manufacturing operations. The particulate matter levels shown in the table are intended to be maintained in zones immediately surrounding the product whenever the product is exposed to the environment.

The most significant change to Annex 1 in the 1997 revision on sterile manufacturing is to the airborne particle classification, as shown in Table 5.10. The "Grades" table is divided into "at-rest" conditions and "in-operation" conditions. At rest is defined as "the condition where the installation is complete with production equipment installed and operating but with no operating personnel present." The "in-operation" state is defined as "the condition where the installation is functioning in the defined operating mode with the specified number of personnel working."

The maximum values for particles equal to or above 0.5 μm and 5 μm (two sizes must be counted) from the earlier classification system have been retained for all grades in the "at-rest" condition and for the Grade A in the "in-operation" condition. Requirements have been lowered for Grades B and C in the "in-operation" state to a value that corresponds to the next less clean grade in the "at-rest" condition. As an example, during production, the particle count requirements for Grade B correspond to the "at-rest" requirements for Grade C, and the requirements for Grade C "in operation" are those of the requirements for Grade D "at rest." The

Table 5.10. Airborne Classification System for the Manufacture of Sterile Products (EC GMPs for Medicinal Products) (Table 1 of Annex)

	Air Classification System			
	At Rest[b]		In Operation	
Grade[a]	Maximum Permitted Number of Particles per m³ Equal to or Above 0.5 μm	Maximum Permitted Number of Particles per m³ Equal to or Above 5 μm	Maximum Permitted Number of Particles per m³ Equal to or Above 0.5 μm	Maximum Permitted Number of Particles per m³ Equal to or Above 5 μm
A	3,500	0	3,500	0
B	3,500	0	350,000	2,000
C	350,000	2,000	3,500,000	20,000
D	3,500,000	20,000	not defined	not defined

a. In order to reach the B, C, and D air grades, the number of air changes should be related to the size of the room and the equipment and personnel present in the room. The air system should be provided with appropriate filters such as HEPA for grades A, B, and C.

b. The guidance given for the maximum permitted number of particles in the at-rest condition corresponds approximately to U.S. Federal Standard 209E and ISO classification as follows: grades A and B corresponds with Class 100; M* 3.5, ISO 5; grade C with Class 10.000; M 5.5, ISO 7; and grade D with Class 100.000; M 6.5, ISO 8.

*M class designations are based on the International System of Units (SI).

maximum permitted number of particles for Grade D "in operation" is not given. A variance of ±20 percent has been allowed for the air velocity of laminar flow systems, and a single value of 0.45 m/sec at work level is applicable in both horizontal and vertical flow areas. This is generally in agreement with U.S. FDA guidelines.

Table 5.10 also shows that a minimum of 20 AC/h in areas of Grades B, C, and D is not necessarily an appropriate measure, and that particle levels (viable or nonviable) in the air are determined by airflow rates and patterns and particles entering the area. It also states that "the number of air changes should be related to the size of the room and the equipment and personnel present." This bases the numbers of air changes on the activity in the room, which is ultimately the most practical consideration. Room grades are also cross-referenced to ISO 14644-1 and U.S. Federal Standard 209E. Table 5.11 shows these different classification systems.

Table 5.11. Various Airborne Particulate Classification Levels of Pharmaceutical Controlled Environment Areas

Grade*	0.5 μm**	5 μm**	FS-209E	U.S. Customary	ISO 14644-1
A	3,500	0	M 3.5	100	5
B	350,000	2,000	M 5.5	10,000	7
C	3,500,000	20,000	M 6.5	100,000	8
D	Not defined	Not defined	M 6.5	—	—

*In operation.

**Maximum permitted number of particles per m³ equal to or above.

Appropriate Air Cleanliness

Examples of operations to be carried out in areas of the various grades are shown in Table 5.12, which has been divided into conditions for terminally sterilized products and those for aseptic preparations (see Delattin 1998). The principle of unusual risk for terminally sterilized products is emphasized; a cleaner environment is required for processing and filling a terminally sterilized product that is determined to be unusually at risk during preparation. Examples are a product that supports microbial growth (i.e., deltrose), one which must be held for prolonged periods before sterilization, one filled into wide-mouthed containers, or one filled by a process that requires longer than usual time.

This important principle is worthy of a moment's further discussion. The filling of LVI to be terminally sterilized in a Class 100 environment is foreign to the principles of U.S. manufacture. The 1992 EC Guide left this as a stumbling block for U.S. firms, many of which ran high-speed lines, sterilized product immediately after filling, and filled product into the narrow ports of plastic bags on automated machines. This type of process, per the 1997 revision, which clearly does not require a Class A enclosure, should be done in a Grade A environment with a Grade C background when the product is exposed and not subsequently filtered.

The handling of sterile drug materials should be done in a Grade C environment if they will be filter sterilized later in the process; if not, the process should be conducted in a Grade A zone with a Grade B background. The handling and filling of aseptically prepared products, either both LVI and SVI, should be done in a Grade A environment with a Grade B background. The preparation and filling of ointments, creams, suspensions, and emulsions into containers allows for contamination. Exposure of the filled container for longer periods of time prior to stoppering is also considered an unusual risk. The preparation of such products is directed to be done in a Grade C environment (instead of Grade D), and filling is to be carried out in a Grade A laminar flow Class 100 environment (instead of Grade C).

Importantly, the EC guidelines no longer mandate air change rates. In the United States, engineering practice calls for 50–60 AC/h for Class 10,000 rooms (Del Valle 1997). There is also the GMP Guidance that specifies a minimum of 20 AC/h for "controlled" areas and 90 FPM ± 20 percent for "critical" areas. Some pharmaceutical companies may use as low as 25 AC/h for Class 10,000 rooms if no significant sources of particles are present.

For critical areas, the airflow rate has changed from 0.3 m/sec (60 FPM) vertical and

Table 5.12. Application List of Grades (Table III of Annex)

Grade	Terminally Sterilized	Aseptically Prepared
A	Filling of products, when unusually at risk	Aseptic preparation and filling Handling of sterile starting materials and components (unless sterile filtered later) Transfer of partially closed containers (in open trays)
B		Background for grade A Transfer of partially closed containers (in sealed trays)
C	Preparation of solutions, when unusually at risk Filling of products	Preparation of solutions to be sterile filtered
D	Preparation of solutions and components for subsequent filling	Handling of components after washing

Unusually at risk (preparation of components): if the product actively supports microbial growth or must be held for a long period before sterilization or is not necessarily mainly in closed vessels.

Unusually at risk (filling): if the filling operation is slow, or the containers are wide-necked or are necessarily exposed for more than a few seconds before sealing.

0.45 m/sec (90 FPM) horizontal flows to the latter figure for both flow regimens. For controlled areas in the United States, the airflow rate still remains at 90 FPM ± 20 percent for critical areas and a minimum of 20 AC/h in controlled areas. The pressurization level has changed from positive to 15 Pa or 10–15 Pa (0.04 to 0.06 in. WG). In the United States, pressurization still remains at 0.05 in. WG. U.S. guidelines require a positive pressure of 0.05 in. WG between adjacent rooms of different air cleanliness. The EC requires positive pressure, with a cleaner room required to be at a higher static pressure than an adjacent, less clean space. The UK Orange Guide mention of 1.5 mm (15 Pa) WG overpressure carries over to the EC GMPs that are based on the UK GMPs. Thus, 1.5 mm WG is normally accepted as sufficient. In the fill room of the EC, a Class B background is to be achieved during "static" (nonoperational) conditions. In the United States, room classifications are to be achieved during room operation with personnel present.

In summary, the U.S. guidelines require controlled areas to be a minimum of cleanliness Class 100,000 (ISO 8), with a minimum airflow rate of 20 AC/h. The EC guidelines require a Class 10,000 (ISO 7) area with an appropriate airflow rate to maintain that condition. U.S. guidelines require critical areas to be Class 100 (ISO 5), with a minimum background of Class 100,000. (ISO 8). The Class 100 vertical airflow rate should be 90 FPM ± 20 percent. The EC guidelines require both the critical area and the room where the critical area is located for aseptic processing to be Class 100. A critical area, (Class A) requires an airflow rate 0.45 m/sec (89 FPM). The Grade B background requires an airflow rate sufficient to maintain the appropriate cleanliness level in operations.

Of significant importance is the consideration that the filling of terminally sterilized LVI does not require laminar flow Class 100 protection unless the product is unusually at risk. It has long been believed by manufacturers that a Class 100 laminar airflow workstation for the filling of terminally sterilized parenterals is an unrealistic requirement. One reason for this line of thinking is that the number of airborne nonviable particles that may enter containers is so low compared to the numbers allowed in injectable solutions by compendial standards; in this context, adequate environmental cleanliness is demonstrated by compliance with the compendial standards for solution-borne particulate matter. For terminally sterilized products, filling of products not "unnecessarily at risk" may thus be conducted in a Grade C area. The Grade C area is equivalent to a U.S. customary Class 10,000 at rest and a Class 100,000 area in operation. These are generally the conditions applied by U.S. manufacturers at present for this purpose: The status quo thus is compliant.

Standards for General Application: FS-209E

With regard to the fate of FS-209E in view of the development of the comprehensive ISO 14644 document series, the author makes the following comments: The IES documents will become mandatory in Europe and the United Kingdom six months after they are published in final form. Their actual adoption by appropriate certifying authorities may take much longer. While the U.S. General Services Administration (GSA) and the Institute of Environmental Sciences and Technology (IEST) have discussed "sunsetting" FS-209E, no formal action in this regard will be taken until the ISO documents are issued in final form. Thus, FS-209E can be expected to serve as a basic standard in the United States for the immediate future.

FS-209E is simply intended to provide the U.S. user of a CEA with a uniform plan to ensure that airborne particulate matter is controlled at a level below that which adversely affects product or clean processes. It falls short of defining sampling plans applicable to the wide ranging use of CEAs. Thus, the pharmaceutical user is still forced to interpret how the standard applies in specific situations and to decide how to implement testing as necessary to protect manufacturing product and process control. The standard provides the user latitude in interpretation, so that specific sampling plans for situations not clearly spelled out in the standard may be developed. The provisions of FS-209E pose some difficulties for both pharmaceutical manufacturers and the FDA.

Factors that made previous FS-209 series documents generally unsuitable for application in the medical products industry are still present. Most notable is the emphasis on air cleanliness in cleanroom operations rather than point-generated or personnel-generated particles that can have severe effects on pharmaceutical product quality. Others problems are the failure to provide for smaller numbers of sample points in high classification areas with nonunidirectional airflow, where air mixing may allow the assessment of air cleanliness with less sampling, and a provision for counting at any of a large number of different particle sizes that could encourage a "count until it passes" approach, which is totally unacceptable to the FDA. Specific considerations relating to FS-209E that complicate its implementation by the pharmaceutical manufacturer include the following:

- A somewhat confusing presentation of a combined classification based on both metric and English systems
- A more complex statistical consideration involving sample size, number, location, and reduction of count data
- Introduction of concepts, such as ultrafine particle measurement and sequential sampling, generally not applicable to pharmaceutical manufacture
- Failure to include a provision for verifying clean zones and cleanrooms based on historical data and product particle burden
- Inclusion of a statistical approach directed primarily at decreasing count variability through sampling plans at a given test point rather than eliminating the cause of the variability
- Allowance of remote sampling based on tubing transport with only a brief discussion of tubing materials and other technical factors
- Inclusion of only summary discussion of implications of counts taken in clean areas under "as built," "at rest," or "operational" conditions

U.S. pharmaceutical manufacturers have characteristically complied with the intent and principle of FS-209E without necessarily following each of its specific directions. The FS-209E document primarily reflects the input of vendors of cleanroom equipment and individuals closely associated with the microelectronics and aerospace industries; representatives from these industries accounted for most of the members of the IES committee that drafted FS-209E. Only 4 individuals from the medical industry were among the original 34 members of the committee. As a result, the document does not have the optimal makeup to make it generally applicable to the manufacture of medical products.

The terms *cleanroom* and *clean zone* that appear in the title of FS-209E are lifted directly from the terminology of the microelectronics and aerospace industries. In these industries, the terms imply areas in which particles are often strictly controlled at levels of not more than 10–100 ≥ 0.5 μm particles/ft^3 of air. In contrast, areas in which medical devices and parenteral solutions are manufactured, with the exception of aseptic or sterile fill complexes, do not require this degree of particle control. Typically, manufacturing areas in pharmaceutical manufacturing facilities are more correctly termed CEAs or clean zones than cleanrooms. In these areas, airborne particles are controlled at levels below the point where particles have any demonstrable negative effect on finished product. The control of particles in such areas is closely tied into the GMP approach, whereby well-characterized and carefully controlled production processes and environments ensure consistent product quality.

With regard to airborne particulate matter monitoring plans for pharmaceutical manufacture, the four most critical sections of the standard are as follows:

1. Class definitions presented in Table 5.1 (again, Class 100,000 is the highest classification area described)
2. Concepts of "verification" and "monitoring" and associated sample plans
3. The statistical handling of data
4. Definition of alternate cleanliness classes

A commentary on all four items follows, as well as some discussion of areas of less critical importance. The publication by Munson and Sorenson (1991) (who were at the time of writing in FDA management), specifically points out defects in FS-209E and sites its weakness with regard to pharmaceutical processing.

Class Limits

The presentation of class limits made in Table 1 of FS-209E is reproduced as Table 5.1 of this chapter. While it is stated that classes as specified in FS-209D remain acceptable, confusion results from the use of both the metric and

English systems and suggests issues to be dealt with in the future. For the U.S. pharmaceutical manufacturer, the change to metric designation is inevitable; the metric system is the basis for the ISO 14644 documents that will probably supersede FS-209E. The reader is again cautioned to obtain complete original copies of these and other documents that are quoted in the text, so that any information used will be free of possible transcription errors and may be read in the total context of the document.

A word of caution is in order regarding the size of particles to be counted for classification purposes. The data in Table 5.1 indicate that, depending on the classification of the area considered, particles from 0.1 μm to 5 μm in size are specified for monitoring. The use of ≥ 0.5 μm particles is generally satisfactory if sampling times are to be kept within reason (except in the case of a low classification area where counts at the smaller sizes provide a distinct statistical advantage). Counts at the ≥ 5 μm size should generally be avoided, due to the low numbers of particles encountered and the unfavorable sampling dynamics for particles of that size with an optical particle counter (OPC). Microscopic testing should be avoided except where absolutely necessary, such as in areas where water vapor or a similar interference exists. If counting at 5 μm is resorted to by either microscopic or instrumental measures, the reason for the variance in technique should be documented.

Any indication that a monitoring plan allows for a "count until it passes" or a sliding scale with regard to the size chosen for counting can be expected to raise a flag for a regulatory investigator—and rightly so. The critical consideration in aseptic filling is the maintenance of cleanliness at critical process points rather than in the room environment away from the process. Further, single counts at a sample point that exceed the class limit in aseptic filling must be considered to indicate inappropriate conditions, even if the average count at a sample point is within limits. A useful consideration of sampling statistics for low classification areas is provided in the paper by Cooper and Milholland (1990).

Although the text of FS-209E states that the class limit particle concentrations (shown in Table 5.1) are not intended to represent particle size distributions in a particular situation, the entire system of class limits is, nonetheless, based on a presumed relationship of particles at different sizes. Unfortunately, this basic flaw in logic seriously questions the value of the sampling plans and statistical data handling provided in other sections of the standard.

For example, the Class 100,000 area is defined by 100,000 particles at ≥ 0.5 μm and 700 particles at ≥ 5 μm. If either of these two sizes is used to define a class limit, they *must* relate to a real distribution or a real ratio relationship between particle numbers at the two sizes. Otherwise, the user's count measurement at one of these sizes will have no relationship to the count at the other size; a measurement at the ≥ 5 μm size allowed by the standard will have no predictive value for the count obtained at ≥ 0.5 μm if the ratio of counts between particles at both sizes cannot be specified. In fact, any user collecting counts at sizes other than ≥ 0.5 μm has no idea whether or not his or her area meets the appropriate 0.5 μm class limit. The current sampling plans of most pharmaceutical manufacturers avoid this problem by providing for instrumental counting only at the ≥ 0.5 μm size to define the classification.

The particle size distributions described in Table 1 of FS-209E may also generate problems for pharmaceutical manufacturers. Theoretically, one could "pick and choose" to find a size at which an area consistently meets the class limits and then count for record at that size (such as 0.2 μm, 0.5 μm, or 5 μm). Understandably, this method is unacceptable to the FDA and untenable on the basis of the technical principles involved, since the counts at the different particle sizes specified cannot be shown to have any definable relationships. The FDA aseptic processing guidelines state that

> air in the immediate proximity of exposed sterilized containers/closures and filling/closing operations is of acceptable particulate matter quality when it has a per-cubic-foot particle count of no more than 100 in a size range of 0.5 μm and larger (Class 100) when measured not more than one foot away from the work site during filling/closing operations.

Any monitoring plan that measures at particle sizes other than 0.5 μm to circumvent the requirement to supply air as clean as possible to the area where aseptic drug products are assembled would most certainly be the subject of an adverse inspectional observation (a 483).

Verification and Monitoring Procedures

FS-209E correctly separates the criteria for monitoring and verification activities. Verification is intended to ensure that the cleanroom functions as intended with regard to the assigned classification. Importantly, the number of sample sites specified for verification activities does not apply to monitoring. The requirement for monitoring or ongoing testing to ensure that the environment remains within class limits simply specifies that a monitoring plan must be in place that is based both on the degree of cleanliness control necessary for product protection within the room(s) in question and the airborne particulate matter cleanliness class. The monitoring requirement allows manufacturers of medical products considerable latitude in establishing monitoring plans, as long as the following are specified:

- Frequency of testing
- Operational conditions during which testing is performed
- Method of counting
- Sample location
- Sample number
- Sample volume
- Method of data interpretation

The standard allows both microscopic and instrumental counting to be performed for particles ≥ 5 μm in size, but results from the two methods cannot be combined or interchanged.

Verification is described as follows:

> 5.1. *Verification of airborne particulate cleanliness.* Verification, the procedure for determining the compliance of air in a cleanroom or clean zone to an airborne particulate cleanliness class limit or a U descriptor, or both, as defined in section 4, shall be performed by measuring the concentrations of airborne particles under the conditions set forth in 5.1.1 through 5.1.4. The particle size or sizes at that the measurements are to be made for verification shall be specified, using the appropriate format as described in 4.4.
>
> 5.1.1 *Frequency.* After initial verification, tests shall be performed at periodic intervals, or as otherwise specified.[1]
>
> 5.1.2 *Environmental test conditions.* Verification of air cleanliness shall be accomplished by measuring particle concentrations under specified[1] operating conditions, including the following.
>
> 5.1.2.1 *Status of cleanroom or clean zone during verification.* The status of the cleanroom or clean zone during verification shall be reported as "as-built," "at-rest," "operational," or as otherwise specified.[1]

Verification tests are to be performed initially and at periodic intervals as specified. In fact, verification activities mentioned in FS-209E (i.e., determination of particle counts, air velocity, air volume change, room air pressure, airflow patterns, temperature, humidity, vibration, and personnel effects) are measured by pharmaceutical manufacturers before accepting a newly built room and monitored on an ongoing basis. This ongoing monitoring would seem to eliminate any need for periodic validation testing; monitoring activities, in fact, constitute verification conducted on an ongoing basis, precluding the requirement for process shutdown or periodic intensive testing.

Probably the most critical point in explaining compliance with FS-209E to an inspector is the footnote (see below) defining use of the term *as specified.* This seemingly innocent footnote is also extremely important to user interpretation of the standard since it allows the pharmaceutical manufacturer to specify the sampling plan (i.e., the degree of control) necessary for compliance.

---

[1]When terms such as "shall be specified," "as specified," etc. are used without further reference, the degree of control needed to meet the requirements will be specified by the user or contracting agency.

Frequency of the verification test is *as specified*. Many pharmaceutical CEAs were initially verified years before FS-209E was available. This initial verification has since been borne out by product tested by USP methodology found to have a particle burden far below compendial limits. Thus, the degree of control is, by definition, sufficient in older CEAs. Under these conditions the user might choose to specify that reverification will not be required, and that continuous monitoring using the test methods described in the standard will be the most effective measure to assess airborne particle levels. This explanation of why a manufacturer may not perform periodic reverification of a CEA also relates to present sample plans that contain fairly low numbers of sample points; the FS-209E verification procedure spells out a sample number much higher than manufacturers presently use in higher class rooms with nonunidirectional airflow. The frequent monitoring performed by pharmaceutical manufacturers offsets to some extent the smaller number of sample points. In fact, the location of sample points in critical areas is far more important than the number of sample points.

Monitoring

The monitoring procedure per FS-209E is defined in section 5.2 of the standard.

> 5.2 *Monitoring of airborne particulate cleanliness.* After verification, airborne particulate cleanliness shall be monitored while the cleanroom or clean zone is operational, or as otherwise specified. Other environmental factors (e.g., air velocity, air change rate, room pressurization, airflow parallelism, airflow turbulence, temperature, and humidity) such as those listed in 5.1.2.2 may also be monitored as specified[1] to indicate trends in variables that may be related to airborne particulate cleanliness.
>
> 5.2.1 *Monitoring plan.* A monitoring plan shall be established based on the airborne particulate cleanliness and the degree to which contamination must be controlled for protection of process and product, as specified.

In the case of 5.2.1, the qualifying term *as specified* is used to indicate that monitoring after initial verification is necessary only when the user believes it is required as a production control. The pharmaceutical manufacturer must ensure, through monitoring and GMP process control, that environmental particulate matter levels are consistently within control limits on each day of production. Thus, in the manufacturing of pharmaceuticals, monitoring and the attendant assurance of GMP conditions are overwhelmingly important. The pharmaceutical manufacturer's approach is based on a level of monitoring activity that ensures that the environmental particulate matter counts are maintained at a level below any product concern (i.e., at a level where there is no product effect with regard to the USP particulate matter limits).

Two things become apparent after reading the description of verification operations. First, the verification process is more appropriate to ensure that a newly built facility meets specifications, rather than for determining that production is being carried out under appropriate conditions of cleanliness in an established facility. Prior to acceptance, any new facility constructed by a pharmaceutical manufacturer will be tested by the verification plan specified in the standard. Monitoring is also critical to compliance with FS-209E. In this context, as in other GMP applications, monitoring is simply ongoing testing at a level of frequency necessary to verify process control. The basis of our compliance with FS-209E is found in the monitoring concept.

Interestingly, though a number of sample locations are prescribed in FS-209E for verification, the conditions of the test (e.g., operational, at rest, etc.) are not prescribed. This is a shortcoming, since the user may verify the clean area under any conditions he or she deems fit. In contrast, a pharmaceutical manufacturer's emphasis in testing (monitoring), and that of the FDA, is always on the operational condition.

[1]When terms such as "shall be specified," "as specified," etc. are used without further reference, the degree of control needed to meet the requirements will be specified by the user or contracting agency.

The best philosophy is simply to test at intervals sufficient to ensure that airborne particulate matter is below levels at which there is any issue of product quality. In an objective comparison of the results of monitoring versus the verification required by FS-209E, one must conclude that monitoring procedures result in more frequent testing than the standard requires and are, thus, a better procedure for assessing product exposure to environmental particulate matter.

Required Number of Sampling Sites

For the manufacturer of pharmaceuticals or medical services, there are a number of technical issues related to the sampling methods proposed by FS-209E. Some of the most significant relate to the instrumental monitoring method itself. The difference in counts obtained by two different instruments calibrated with the same test particles can vary by ±50 percent. While instrumental counting of airborne particles constitutes a rapid and cost-effective means of assessing general particle levels, this method is susceptible to critical error effects. Coincidence is one source of error. Another and more severe error is the counting of water microdroplets or other dispersed fluids as particles (e.g., liquid vapors resulting from heat seal operations). These count errors will lead to count data that are erroneously high with respect to the true level of particles present in an area. Fortunately, FS-209E provides for microscopic testing to resolve artifactual counting by instrumental counters.

It is important to keep in mind that the number of sample sites described in the standard relate to verification, not monitoring. The number of sample sites is determined as follows:

5.1.3.1 *Sample locations and number:* Unidirectional airflow. For unidirectional airflow, the sample locations shall be uniformly spaced throughout the clean zone at the entrance plane, unless otherwise specified, except as limited by equipment in the clean zone.

The minimum number of sample locations required for verification in a clean zone with unidirectional airflow shall be the lesser of (a) or (b):

(a) *SI units:*
$A/2.32$, where A is the area of the entrance plane in m^2

English (U.S. customary) units:
$A/25$, where A is the area of the entrance plane in ft^2

(b) *SI units:*
$A \times 64/(10^M)^{0.5}$, where A is the area of the entrance plane in m^2, and M is the SI numerical designation of the class listed in Table 1.

English (U.S. customary) units:
$A/(N_C)^{0.5}$, where A is the area of the entrance plane in ft^2, and N_C is the numerical designation of the class, in English (United States customary) units, listed in Table 1 (Table 5.1 of this chapter).

The number of locations shall always be rounded to the next higher integer.

5.1.3.2 *Sample locations and number:* Nonunidirectional airflow. For nonunidirectional airflow, the sample locations shall be uniformly spaced horizontally, and as specified vertically, throughout the clean zone, except as limited by equipment within the clean zone.

The minimum number of sample locations required for verification in a clean zone with nonunidirectional airflow shall be equal to:

> *SI units:*
> $A \times 64/(10^M)^{0.5}$, where A is the floor area of the clean zone in m^2, and M is the SI numerical designation of the class listed in Table 1.
>
> *English (U.S. customary) units:*
> $A/(N_C)^{0.5}$, where A is the floor area of the clean zone in ft^2, and N_C is the numerical designation of the class in English (United States customary) units listed in Table 1 (Table 5.1 of this chapter).
>
> The number of locations shall always be rounded to the next higher integer.

The intent of FS-209E is to reduce variation and inaccuracies in count data by increasing the number of samples. However, the sample requirements specified by section 5.1.3.2 do not address instrument bias or alter error effects. It is also interesting to note that FS-209E allows the use of 0.1 ft^3/min (CFM) instrumental sampling rates to collect data that are then multiplied 10-fold to determine the particles/ft^3. This procedure results in a significant increase in data variability above that occurring when 1 ft^3 samples are taken. To some extent, this sampling error counteracts any advantages of using greater numbers of sample points. Thus, the use of more sample points will not necessarily result in a more accurate representation of the particle levels present in a given area.

A consideration of a typical pharmaceutical manufacturer's sampling plan indicates that a smaller number of sample points taken at frequent intervals and then compared to historical data provide a more representative assessment. The reliability of our product with regard to particle levels being below the USP limits is also far higher than that of electronic components manufactured in Class 100 cleanrooms. In the latter situation, extremely small numbers of particles can cause product failure. In a final analysis, historical particulate matter data obtained from product release testing indicate that our present monitoring is sufficient to ensure that airborne particulate matter is maintained at a level so low as to have no effect on product quality.

There is one other error in logic involved in the verification of sampling plans specified in FS-209E. In some situations, more samples may be required for nonunidirectional airflow than for directional. Since the airflow is turbulent in nonunidirectional airflow areas, airborne particulate matter tends to be more uniform within the room volume; in fact, it can often be adequately represented by smaller numbers of samples. The key factor in this consideration is the variability of the air volume that the count represents. In the laminar airflow of a Class 100 room, the air is not mixed, and each cubic foot of the volume represented by a sample may have a significantly different count level. Levels of counts per cubic foot are also low (typically 10–30). In fact, in CEAs with mixed airflow and higher counts, the random distribution of the higher numbers of particles in mixed air makes each cubic foot sample far better than a similar sample volume taken in a Class 100 room. In these areas, adequate sampling can be based on a 0.1 ft^3 sample volume rather than 1.0 ft^3. Again, the actual location of sample sites is critical.

Statistical Data Treatment

FS-209E prescribes methods for data handling to assess whether a CEA meets or fails the assigned class requirements (Section 5.4, Appendix E). Plans for the statistical reduction of airborne particle count data are available from the literature, where that given in the standard assumes a normal distribution of count data. The user should be warned that this assumption is not necessarily valid for counts below 100 particles/ft^3. Most statisticians agree that with counts below this level, count data are distributed according to a non-Gaussian distribution. Thus, the data reduction plan specified by FS-209E may be inappropriate for critical areas. The limits for counts per the standard are as follows:

> 5.4.1 *Acceptance criteria for verification.* The air in a cleanroom or clean zone shall have met the acceptance criteria for an airborne particulate cleanliness class (see Table 1 for standard limits or U descriptor) when the averages of the particle concentrations measured at each of the locations fall at or below

the class limit or U descriptor. Additionally, if the total number of locations sampled is less than ten, the mean of these averages must fall at or below the class limit or U descriptor with a 95% UCL.

The calculation presented in this section is a straightforward calculation of mean, standard deviation (SD), and upper confidence limit (UCL). The rationale for this criterion of acceptance based on average counts is, however, diametrically opposed to the principles of GMP and FDA enforcement policies. The FS-209E rationale is described in Appendix E.

E20. *The statistical rules.* The first rule requires that, for each location sampled, the average particle concentration not exceed the class limit or U descriptor. The second rule requires that an upper 95% confidence limit, constructed from all of the location averages, not exceed the class limit or U descriptor; this rule applies only when fewer than ten locations are sampled.

The rationale for the first rule is that the cleanliness of the air must be checked at multiple locations in a cleanroom or clean area. The number of locations is a function of the size of the area to be checked. The average is of all measurements of performance rather than an absolute one; thus, the average particle concentration at each sampling location is the base level of statistical summarization found in the standard. All variation among the measurements taken at each location is ignored, except to the extent it affects the average at that location and the two statistical rules. It may be possible, therefore, to average out sampling variability, spikes, and time trends in data collected at a given location. Minimizing the impact of sampling variability is a desirable consequence of using an average; the potential for obscuring real changes in particle concentration (spikes, cyclicity, or other time-related trends) is an undesirable consequence of using an average. While individual count data are distributed in nonnormal fashion, averages of multiple counts are distributed normally. The principle of ignoring variation in counts at a location has been specifically objected to by the FDA, which is further evidence of the lack of applicability of FS-209E to the pharmaceutical industry.

The statistical data treatment provided in FS-209E serves to correct for smaller numbers of sample points by assigning a higher "t" value. The UCL at the 95% level is calculated as:

$$\text{UCL} = (\text{mean count}) + [(\text{SD of mean}) \times (\text{“t” value})]$$

This statistical treatment in itself tends to make a greater number of sample points less necessary.

Count data for a specific CEA may better be handled according to appropriate statistical methods in a plant or company SOP. If the distribution of historical count data is defined, alert and action limits may be correctly specified for use in monitoring activities. In this context, as with the monitoring of other parameters, the alert and action limits are defined by their probabilistic separation from the mean count levels based on historical data.

The statistical provisions of FS-209E require the 95 percent confidence limit of the cleanroom count to be within the class limit. The testing methods on which the calculations are based have a degree of accuracy much lower than this statistical treatment implies. In order to make a valid statistical inference from any measurement, it must not contain bias or nonrandom errors. Yet discrepancies of as high as 50 percent can be found between particle counting instruments from different manufacturers and in some cases between instruments of the same manufacturer. To an extent, this renders the application of a precise statistical treatment and the drawing of statistical inferences inappropriate.

This becomes an extremely important consideration with regard to data reduction counts in Class 100 areas. Because of the day-to-day variability of single particle counters based on light scattering and the variation between different models of counters, counts of < 100 particles/ft^3 incorporate a level of variability unrelated to the actual numbers of particles present. Further, at the low count levels typical of Class 100 areas, the count distribution most frequently conforms to a Poisson distribution. The SD of the mean for a Poisson distribution is divided by *N*, where *N* is the average count. The confidence interval defined by this approximation invariably results in a large relative standard deviation (RSD). For example, if the mean count for a Class 100 CEA is 75 particles/ft^3, the SD rounds to 9 particles/ft^3, and a 2σ confidence interval is 75 ± 18 particles. The RSD equals 12 percent, which is unacceptable precision for any critical measurement. The best means of handling Class 100 CEA data would seem to be simply to trend the mean counts obtained and use the class limit as both a cutoff and alert limit; i.e., if all counts are below the level of 100 counts/ft^3, the area is considered to be within limits. A second reasonable approach is to use the CEA class limit as an action or shutdown limit, and a value separated from the mean by 2 SD as the alert limit.

Regulatory Shortcomings of FS-209E (see Munson and Sorenson 1990)

The impression among some regulators is also that FS-209E is poorly suited for the pharmaceutical industry. First, there is a concern with the emphasis placed on verification of the count levels in an as-built controlled CEA or cleanroom. The in-use condition is of overwhelming importance in pharmaceutical manufacture, and monitoring rather than verification is, thus, of greatest importance.

Consider a sample point where counts of 5, 7, 9, 6, 8, 54, 11, 17, 14, and 109 are collected at successive sample intervals. This extreme spread of values at the sample point is obviously not in keeping with the aseptic fill guidelines and indicates an out-of-control condition. A pharmaceutical manufacturer operating under these conditions would, in fact, be out of compliance with the regulatory guidelines, although the average counts is within the guidelines. The FDA would find this situation unacceptable. This example demonstrates the inadequacy of applying the 95 percent confidence limits to the count averages in an area. It might be more appropriate to apply these statistics to the readings at each sampling point or to all of the count values obtained to demonstrate that the clean area is, in fact, under control and uniformly meets the appropriate class limits.

Another problem with the FS-209E method of particle count calculation is that an area that is under control and operating within class limits can fail because of statistical manipulation. Some areas in an aseptic fill room will be much cleaner than others, and this may be entirely in keeping with the principles of product protection. For instance, samples collected in an area just below HEPA filters could be quite clean, with counts of 0 to 10/ft^3. Other areas sampled near the filling nozzles might be in the range of 40 to 50 counts. Areas near vials that contact each other on moving conveyor belts may have counts in the 50 to 100 range. Although the averages at these sample points might vary only slightly, the room may still fail based on FS-209E criteria when statistical treatment is applied to the sample point average. Ironically, if some critical areas have higher counts, the FS-209E statistical test may result in the area passing.

It must be emphasized that the number of sampling locations for verification specified in FS-209E, $A/25$, or A/the square root of the class, whichever is less, is intended to be a minimal number. If a fraction remains, the number derived from this calculation must be rounded up. Special equipment/protection in some pharmaceutical CEA configurations could dictate the use of more sample points. For example, although the area under a laminar flow module over a linear fill process may be less than 50 ft^2, the length of the module, equipment placement, and multiple points of entry may dictate that more than the minimum of two sampling locations should be used in the area. Sampling locations should be indicative of worst-case conditions at the point of fill. All values at each sampling location should meet the class limit, and all readings taken must demonstrate consistent control of particulate matter within class limits. If aseptic filling of powder material precludes sampling in some areas during filling, particle counts in these areas must be rigorously validated under "in-use" conditions with no powder present.

For Class 100, measuring at 0.5 μm, the minimum volume per sample is 0.2 ft^3. In the FDA view, this sample volume is not adequate enough to assess aseptic processing facilities. For critical areas, the FDA viewpoint is that, based on a minimum of two sampling locations, and using an OPC with a flow rate of 1 CFM, the sampling time should be 5–10 min at each sampling location, to give a total sample of 10–20 ft^3. This is not, obviously, an absolute criterion, and other valid sample plans may be devised (see Whyte 1983). The intent is to collect a large enough sample so that valid assessments may be made based on the small numbers of counts that will be collected and the variability between repeated samples.

The statistical handling of data prescribed in FS-209E also causes problems for the FDA. Remember that the FDA philosophy and the principles of aseptic fill stress product protection and minimal levels of particles in areas where product is unprotected. The acceptance criteria in FS-209E requires that the average concentration at each sample point be within the class limit and that the mean of these averages meet the class limit with a 95 percent confidence limit. Varying numbers of samples can be collected at each sampling point, and an area can meet Class 100 even though some sample point readings exceed the class limit.

Generally speaking, two types of process areas exist within pharmaceutical manufacturing for which airborne particulate matter monitoring is applied: sterile or aseptic fill areas and larger controlled environments. As stated above, the cleanliness requirements for these two types of areas differ dramatically. In fact, the operation of the former type of facility is governed by the FDA guidelines for aseptic processing, and adherence to these guidelines constitutes regulatory compliance irrespective of the requirements of FS-209E. In CEAs that have higher limits for airborne particles, such as 10,000 to 100,000, the practical requirement is that environmental particles be controlled at levels below those that have any demonstrable effect on product particle burden.

The higher numbers of test points and the addition of Class 10 and Class 1 cleanrooms in FS-209E are directed at ensuring air cleanliness for processing a product highly vulnerable to particles at many stages of manufacture, such as electronic circuit wafers. In these processes, the presence of one particle on a silicon chip can result in a total defect. This problem is entirely different from that in controlled pharmaceutical manufacturing areas; our internal monitoring is directed toward controlling particles at levels consistently shown to result in an increase in product particle burden. With regard to solution-borne particles in filled LVI product, particle content is almost solely dependent on container-related sources, process equipment, and particle shedding by production filters. Thus, the primary control of particulate matter in solution is exerted by solution fill line filters, not by air filtration. Historical data from manufacturers of injectables and devices often illustrate the independence of solution particle burden to airborne particle levels. A review of these data suggests that airborne counts show no relationship to the low particle burden of the final product, and the coefficient of correlation for the data is, in fact, poor. Particles generated at process points have no such effect (Whyte 1983).

The inherent variability of the test methodology used in counting airborne particulate matter is also important (Lieberman 1988; Liu and Szymanski 1987). This is true when instrumental counting is applied, and particularly if microscopic and instrumental counts are compared. Instrumental counts in a 100,000 particles/ft^3 device manufacturing room compared to microscopic counts ≥ 5 μm provides an example of this. During manufacturing operations, molding, heat sealing, and extrusion operations are carried out. Instrumental counts at ≥ 0.5 μm are frequently out of limits, while microscopic ≥ 5 μm counts invariably pass. Disregarding test method variability, a large component of the variation in counts between the two methods is due to the fact that plasticizer vapor produced by heat sealing, solvent vapors, or water vapor is counted as particles by instrumental counters. This vapor must be considered as a significant component of all instrumental counts collected in this area. Thus, by definition, any instrumental count in these rooms must be assumed to represent some lower count of real particles. Thus, instrumental counts collected represent the "worst case." Since a spurious count component exists, the application of precise statistics to these instrumental data serves no useful end, and the use of more sample points will not give a more accurate assessment of particle levels present.

The concepts of alert limits and data trending, well known in pharmaceutical manufacture, are omitted from FS-209E. Trending or trend analysis is a useful method of achieving confidence in the airborne particle level in a CEA. Trending compares the current day's measurement against an historical database containing months of measurements. The pharmaceutical manufacturer aspires to a steady state of predictable compliance with regard to particle counts in a CEA or cleanroom. Any upward drift of data must be detected. Trending of these data is an important component of process control. In order to trend data, confidence limits must be established around the average counts at a sample point or for a room.

The alert limit concept and its relation to class limits may be defined as follows. Class limits may be class levels as specified by FS-209E or a manufacturer's internal limits. When class limits are exceeded, this signals drift from normal operating conditions and requires prompt action. Alert limits are airborne count levels or ranges that, when exceeded, signal a potential drift from normal operating conditions. A situation where alert limits have been exceeded may not signal specific action, but require only that the area receive more than standard monitoring. In other instances, one level may be used with a single excursion eliciting an alert response, and multiple or sequential deviations requiring corrective action.

Thus, an alert limit simply flags a change in the particles/ft^3 in a CEA so that some action can be taken before the shutdown limit is exceeded. For a given CEA, there is a specific class limit (C_L). Historical count data will allow an historical mean count level (C_M) for this area to be established. Once the distribution of counts in the historical database from which C_M is derived is known, an alert limit C_A can be established at some level such that $C_L > C_A > C_M$. Depending on the C_A chosen, the probability that any mean for a CEA that is within C_A actually represents a population that exceeds C_L may be calculated.

For example, assume that for a Class 100,000 area the historical C_M is 50,000 particles/ft^3 and that C_M + 2 SD is equal to 70,000 counts. Based on the distribution of historical data, any mean count obtained from 10 samples taken in the CEA during monitoring activities that falls within the alert limits will have some probability of representing a count population with some values exceeding C_L. In the room under consideration, this probability may equal 1.0×10^{-4}. Thus, there is one chance in 10,000 that any 10-count sample (for which the mean is below the C_A of 70,000) actually represents a population with out-of-limits counts.

On initial consideration, this summary approach to setting alert limits may seem to be rather imprecise and nonuniform in nature. The reader is referred to Bzik (1986), Cooper (1988, 1989, 1990), Cooper and Milholland (1991), Grotzinger and Cooper (1989), and Chapter 9 for more detailed analyses. Keep in mind, however, that the idea of the alert limit is simply to define a level of counts at which investigation will be undertaken without exceeding the class limits. Thus, a number of alert limits may be chosen that serve this end equally well. As discussed above, however, the closer the alert limit is to the class limit, the higher is the probability that a population of counts within the alert limits contains counts that exceed the class limit.

ISO International Standard 14644-1

ISO 14644-1 (*Cleanrooms and Associated Controlled Environments: Part I Classification of Airborne Particulates*) is mentioned in various contexts with regard to "umbrella" requirements. It is the ISO international counterpart to FS-209E issued in 1998. Some changes on the part of U.S. pharmaceutical and device manufacturers will be in order to ensure compliance with the new standard, which presumably will be essential to ISO certification at some time in the future. Immediate compliance would not be possible due to the fact that the necessary companion documents have not been published. It will become obvious to the reader that ISO 14644-1 will impact pharmaceutical manufacturing operations much in the same way FS-209E did. Methods of sampling to verify or establish certain particle count levels are specified, but the pharmaceutical application of different classes of cleanliness for specific processes will remain unchanged.

One important step for a pharmaceutical or device producer considering compliance with ISO 14644-1 will be the development of a position document describing the in-company philosophy and rationale used in assessing the steps needed for compliance. If, as may be the case, a manufacturer chooses not to comply to the letter with some sections or points, the manufacturer must address why these points are not appropriate for a particular operation.

The eventual worldwide adoption of the ISO 14644 series requirements will probably have a number of significant implications for U.S. pharmaceutical procedures. These are, for the most part, presently in some draft stage. Nomenclature indicating the developmental status of these standards is as follows:

- CD = Committee Draft (followed by number and date to include revision)
- DIS = Draft International Standard (also number and date)
- FDIS = Final Draft International Standard (legal for contract use)
- FD = Final Draft (issue date)

Given that the ISO document series is currently incomplete, and that uncertainties regarding the extent to which future harmonization of requirements for devices and injectables will occur exist, the author can only speculate. Of pivotal consequence to U.S. manufacturers is that the FDA considers the documents in this series at this time to be totally voluntary standards that have no implications regarding domestic drug or device manufacturing operations. If a company chooses to adopt them, this will have no impact with regard to regulatory assessments of compliance with FDA directives, guidelines, and cGMP (e.g., independent of the ISO documents); the manufacturer must consistently meet FDA expectations.

At some time in the future, the 14644 document series may be adapted by governmental agencies in the United States as a replacement for FS-209E in the interest of harmonization. The IES reportedly has FS-209F in-process, and this document will "mesh" or be consistent with the ISO 14644-1 documents, although less comprehensive. The number of sample points required for compliance with the ISO document are less than for FS-209E. In some other ways, too, 14644-1 is less stringent. The FDA will likely not accept the allowance in either document to average counts to pass a classification test. The FDA view is that all test points must pass class limits, especially Class 100.

The adoption of ISO standards to replace regional or national standards in Europe may be a protracted process, and assessment of compliance seems sometimes to be less rigorous than in the United States. The adoption can be much more rapid if their use is mandated by the EC (based in Brussels) or if their development is requested by the European Commission on the grounds of the standard being necessary to serve European interests. There is also a process called a CEN parallel inquiry, whereby the CEN (EC) can identify an ISO standard in process as desirable for EC use and request that development of the standard go forth under the aegis of both ISO and CEN.

The status of a standard requested to be developed by the EC is that of a "mandated" standard. A mandated standard becomes a normative requirement for EC member nations and, if mandated for the medical products industry, will be enforceable by law by regulatory agencies such as the Medicines branch of the Department of Health and Social Services (DHSS) in the United Kingdom. If not mandated, the standard can be used or not (i.e., it is a voluntary standard). As is the case in the United States, individual European EC member nations may adopt ISO standards even if they are not adopted by the EC as an "umbrella" requirement. If ISO standards or provisions thereof are not mandated by the EC, they may still become enforceable as law in the individual EC member nations based on their designation by the regulatory agency of the country concerned. At present, ISO 14644-1 is not mandated by the EC, but indications are that some provisions (e.g., a particle count for CEN classification) may be either adopted by the EC for the EC GMPs or referred to or referenced so that compliance will be required. This is logical since many of the requirements in the EC GMPs are the same as ISO 14644-1.

The EC GMPs are legally enforceable in the United Kingdom based on their EC commitments; there are gaps between the ISO 14644-1 document and the BS 5295 document that was superseded. In Germany, DIN standards are almost invariably enforceable as law. When a national standard is withdrawn in favor of an ISO document, as BS 5295 has been, some "flushing out" is necessary to remove vestiges of compliance with the old document and complete adoption of the new. While BS 5295 has been withdrawn, it will take UK manufacturers and regulators some time to weave ISO 14644-1 into compliance structures completely. The fundamental difference between the two is size of particles counted, ISO 14144-1 requiring one size (of the choice of the tester) and the EC GMPs requiring two sizes. So much of both the FS-209E and BS 5295 documents have been incorporated into ISO 14644-1 that neither the United States nor the United Kingdom should have substantial issues with it or changes to be made in specific or overall compliance procedures.

The ISO 14644-2 document (monitoring to prove continued compliance) is, in contrast to ISO 14644-1, a problem. The issues relate primarily to a failure to recognize the value of monitoring as proof that verification or requalification is not needed. Indications are, however, that those responsible for ISO 14644-2 are willing to incorporate changes that will facilitate compliance in multiple user nations. Monitoring activities required by ISO 14664-2 are less rigorous (i.e., testing is less frequent than those required by BS 5295). The quality of the air—the criteria against which air cleanliness classes are determined—is essentially the same for ISO 14644-1 and FS-209E. The most significant (and probably beneficial) change in the ISO document is the allowance of less frequent "strategic requalification" or complete reverification of a facility. If the company in question applies computer-based facility monitoring systems (FMS) to conduct "continuous" monitoring (user input to the 14644-2 working group in this regard emphasized the money spent for these systems), the loss on investment through periodic manual requalification, and the continuous database acquisition make frequent requalification unnecessary.

The only general inspection mechanism currently in place in the United States is the TÜV inspections for compliance with ISO 9000 series documents (i.e., ISO 9000, 9001, 9002, 9004). These are general "umbrella" documents that are directed at ensuring that quality systems are in place. TÜV auditors are experts in "quality systems," which are based on 11 key components. TÜV auditor judgment on the existence or lack of a quality system is based on the evidence of a methodical, structured means of ensuring that key steps (e.g., a means of establishing customer requirements) are in place. The TÜV auditors do not (yet) audit at the level of ensuring the use of specific standards to satisfy a specific technical requirement (e.g., ISO 14644-1 for airborne particle monitoring). At best (or worst), they will look for a systematic approach to airborne particle monitoring that would comply generally with the ISO document (e.g., monitoring for FS-209E). They may want assurance that the requirement used is complimentary to ISO 14644-1. Many national documents meet these general expectations.

Historically, TÜV auditors who assess ISO compliance have proven to be eminently reasonable with regard to the substitution of test plans and methods that do not comply to the letter with ISO requirements but are adequate for the intended purpose. All such rationales should, of course, be written out. Auditors may rightfully expect a firm being audited to be familiar with the principles being applied in the use ISO 14644-1. As harmonization of requirements for medical product manufacturers proceed, it behooves pharmaceutical manufacturers to be reasonably aggressive in assessing whether or not their procedures comply with the ISO and other umbrella requirements for environmental controls and classification. This type of overview of operations will almost invariably indicate that principles of compliance are in place due to GMP and the conscientious care taken by the manufacturer with regard to patient well-being.

The ISO 14644-1 and 14698 documents being produced by ISO/TC 209 are basically only generic, recommended standards for voluntary compliance. Even in the United States, FS-209E is only mandatory for U.S. government agencies or by vendors doing business with the federal government. It is voluntary for commercial trade, including the pharmaceutical industry. While ISO standards are voluntary for U.S. trade and commercial concerns, ISO standards may become mandatory for EC member nations six months after their publication date. Regulatory authorities in various nations of the EC determine and/or specify levels of cleanliness, frequency of testing, and microbial limits in accord with the appropriate cGMPs. ISO/TC 209 and 14644-1 only offer suggestions on the first two of these—except, of course, now for EC nations.

At present, the biggest risk to multinational compliance by U.S. pharmaceutical companies is the inspections conducted when product made in the United States is cleared for importation into an EC member nation. For example, a shipment from Puerto Rico to France might be inspected by the French Ministry of Health to ensure that appropriate testing of manufacturing facilities had been carried out. Specific questions might involve how air classification was performed in the manufacturing facility. A "how we comply" rationale previously drafted will be invaluable in this situation. Compliance with EC standards in some few cases may be specifically questioned by the inspectors due to national pride or chauvinism. Remember, the EC requires two particle sizes ($\geq$ 0.5 μm and $\geq$ 5 μm) to be counted, and EP limits are based on less than 1 ft^3 of air due to the metric to English system conversion.

Organization

ISO 14644-1 is the first in a series of documents that will deal with the classification, construction, and operation of cleanrooms and CEAs. The documents will include the following:

- ISO 14644-2: Monitoring CEAs
- ISO 14644-3: Metrology and Test Methods
- ISO 14644-4: Design and Construction
- ISO 14644-5: Cleanroom Operations
- ISO 14644-6: Terms, Definitions, and Units
- ISO 14644-7,8: Mini-environments and Isolators

ISO 14644-2 is of extreme importance in terms of CEA operations. ISO 14644-3 governs the test methods applicable for compliance; notably particle counting, measurement of airflow patterns and pressure differentials, and HEPA filter testing. ISO 14644-5 spells out mandatory activities, ISO 14644-7 is generally applicable to barrier isolation processes. An important consideration is the unknown extent to which the FDA will embrace ISO 14644 series. While the FDA is in favor of harmonization, they will look closely at the requirements of the series in terms of how well they maintain the status quo in aseptic filling operations. Another issue regards the extent to which pharmaceutical manufacturer quality management will decide to adopt the tenets of the new ISO documents. Documents relative to biocontamination control, classification, and monitoring are being developed as ISO 14698-1,2,3 on a parallel path.

An evaluation of a manufacturer's environmental test plans versus the matrix set forth in ISO 14644-1 should methodically compare the key components of the two. These criteria are as follows:

- *Primary Criteria:* test method, sample sites (number & location), sample intervals, particle size on which classification is based, data reduction rules
- *Secondary Criteria:* room air change rates, overpressures, testing to obtain proof of HEPA filter integrity

The primary criteria involve data collection; the secondary criteria relate to measures to ensure the control of particles in circulating air. Again, ISO 10,000 series documents, like ISO 14644, are "umbrella" documents, general in nature and pertaining to a wide variety of manufacturing operations. They do not attempt to provide specific instructions for various manufacturing activities; instead, they list general criteria that provide a basic standardized framework to be built upon as necessary to suit specific operations.

There are two primary factors that tend to restrict the use of general authorizing documents such as ISO 14644 and FS-209E in the highly controlled manufacturing of medical devices and injectables. First, there are no specific instructions given regarding the types of activities carried out in different classifications. Secondly, in the heavily regulated and controlled medical products industry, ISO 14644 does not take into account regulatory standards that apply to device and solution manufacture that may differ from country to country. The key consideration is that ISO 14644-1 specifies how CEAs are classified on the basis of particle counts, without mention of how these levels (classes) pertain to product. Points of the new standards that are being developed that represent specific issues are outlined below.

CEA Classification per ISO 14644-1

ISO 14644-1 covers the classification of air cleanliness in cleanrooms and associated controlled environments. Classification is specified and accomplished exclusively in terms of the concentration of airborne particles. The only particle populations considered for classification purposes are those having cumulative distributions based on threshold (lower limit) sizes ranging from 0.1 μm to 5 μm. Classification may be accomplished in any of three occupancy states.

While ISO 14644-1 does not allow for microscopic testing or retesting for the purposes of classification (see Working Group 3 document), latitude is allowed in terms of the particle size used for classification, such that Class ISO 8 (equivalent to current U.S. Class 100,000) may be verified or monitored over any particle size from 0.1 μm to 5 μm. Thus, the manufacturer may be able to save a considerable amount of time and effort by simply testing and classifying rooms where vapor is an issue with particle counts at 0.5 μm on the basis of the 5 μm instrumental count. ISO classes are based on the metric system (counts/m^3), rather than the English system (particles/ft^3) (see Table 5.13, Figure 5.2).

Table 5.13. ISO 14644-1 Particle Count Classes—Cleanroom/CEA

	Max. Concentration (particles/m³ of air) for Particles Equal to and Larger Than the Considered Sizes Shown Below (Values rounded to the nearest whole number)					
Classification #	**0.1 µm**	**0.2 µm**	**0.3 µm**	**0.5 µm**	**1 µm**	**5 µm**
ISO 1	10	2	—	—	—	—
ISO 2	10	24	10	4	—	—
ISO 3	1,000	237	102	35	8	—
ISO 4	10,000	2,370	1,020	352	83	—
ISO 5 (IES Class 100)	100,000	23,700	10,200	3,520	832	29
ISO 6 (IES Class 1,000)	1,000,000	237,000	102,000	35,200	8,320	293
ISO 7 (IES Class 10,000)	—	—	—	352,000	83,200	2,930
ISO 8 (IES Class 100,000)	—	—	—	3,520,000	832,000	29,300
ISO 9	—	—		35,200,000	8,320,000	293,000

Class Designation

The designation of airborne particulate matter cleanliness for cleanrooms and clean zones per the document include:

a) The classification number, expressed as ISO class N's;

b) The occupancy state to which the classification applies; and

c) The considered particle size(s), and the related concentration(s), as determined by the classification formula, where each considered particle size is between 0.1 µm and 5 µm, inclusive.

Particle Size and Classification Number

If measurements are to be made at more than a considered particle size, each larger particle diameter (e.g., *D*2) shall be at least 1.5 times the next smaller particle diameter (e.g., *D*1).

$$D2 \geq 1.5 \times D1$$

Airborne particulate cleanliness shall be designated by a classification number, *N*. The maximum permitted concentration of particles, C_n, for each considered particle size, D, is determined from the formula:

$$C_n = 10^N \times \left(\frac{0.1}{D}\right)^{2.08}$$

where C_n represents the maximum permitted concentration (in particles/m³ of air) of airborne particles that are equal to or larger than the considered particle size, C_n is rounded to the nearest whole number; *N* is the ISO classification number, which shall not exceed a value of 9. Intermediate ISO classification numbers may be specified, with 0.1 the smallest permitted increment of *N; D* is the considered particle size in µm; and 0.1 is a constant with a dimension of µm.

Establishment of Sampling Locations

Average Particle Concentration at a Location (*P*)

Note: This procedure is applicable to any number of sampling locations.

$$P = \frac{(T_1 + T_2 + \ldots T_x)}{X}$$

where *P* is the average particle concentration at one location. T_1 to T_x are the particle concentrations of the individual

Figure 5.2. Particle size distribution for ISO 14644 (particles/m^3). $C_n = 10^N \times (0.1/D)^{2.08}$

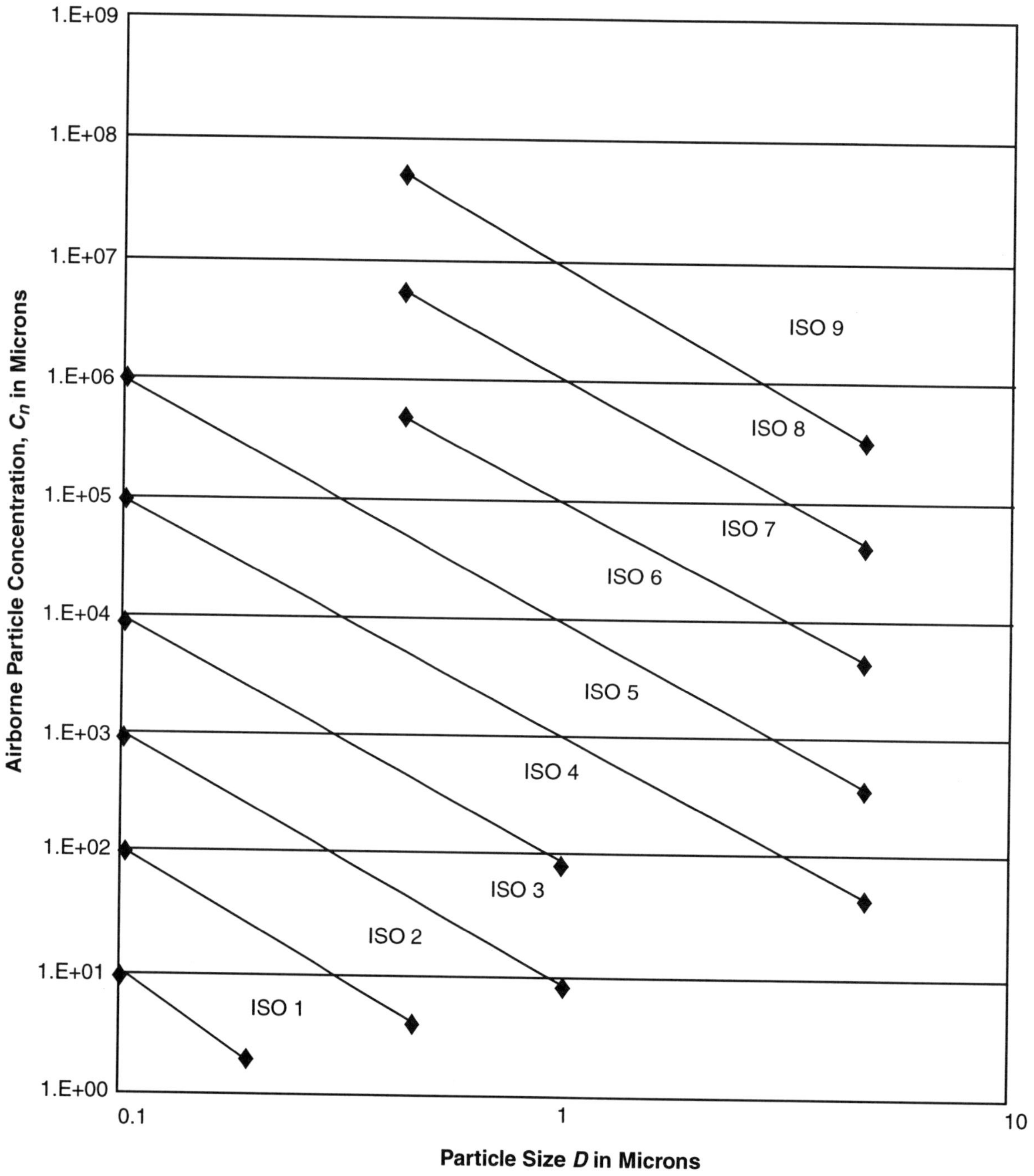

samples. X is the number of samples taken at the location.

Computation of 95 Percent Upper Confidence Limit (UCL)

Note: This procedure is applicable only to situations where the number of sampling locations is at least two but not more than nine. In such circumstances, this procedure shall be used in addition to the algorithm:

$$M = \frac{(P_1 + P_2 + \ldots P_x)}{Y}$$

where M is the mean of the averages. P_1 to P_X are individual location averages. Y is the number of individual location averages.

All individual location averages are equally weighted, regardless of the number of samples taken at any given location.

Calculation of Statistical Parameters

The *SD* of the mean is calculated as follows:

$$SD = \sqrt{\frac{(P_1 - M)^2 + (P_2 + M)^2 + \ldots + (P_y - M)^2}{(Y-1)}}$$

Standard error of the mean (*SE*) is:

$$SE = \frac{SD}{\sqrt{Y}}$$

Computation of 95 Percent UCL (III)

UCL = $M + (t \times SE)$

Where: t is student's t factor that is dependent on the number of individual averages (Y), and guarantees with 95% confidence that all individual averages are within the computed UCL. The values of the t factor for 95% UCL are as shown in Table 17.

Table 17. t factors.

# of Individual Averages (Y)	2	3	4	5–6	7–9
t	6.3	2.9	2.4	2.1	1.9

Annex B (Sampling Plans)

Annex B is a normative section. The minimum number of sample points is based on the area of the CEA in *square meters*. This will result in a lower number of samples than FS-209E, which bases the sample number on an area measurement; the ISO sample number is irrespective of classification. The number of sample points arrived at by this computation is similar to the test presently used by manufacturers. But the new document (2–9 sample points) requires that the 95 percent UCL of the mean be determined to establish compliance. The statistical calculations required are brutally straightforward.

> The minimum number of sample points is based on the area of the CEA in square meters.
>
> $$N_L = \sqrt{A}$$
>
> where N_L is the minimum number of sampling locations (rounded up to a whole number). A is the area of the cleanroom or clean air controlled space in m^2.
>
> **Computing the 95 Percent UCL**
>
> When the number of locations sampled is more than one and less than ten, compute the mean averages, standard deviation, standard error, and 95% upper confidence limit from the average particle concentrations for all locations. When there is only one or more than nine locations sampled, computing of the 95% upper confidence limit will not be applicable.
>
> **Documentation**
>
> a. The name and address of the testing organization, and the date on which the test was performed.

Author's note: Table 17 and the table numbers in the following discussion are those of the ISO document rather than the chapter.

b. The number and date of this part of the ISO standard, i.e., ISO 14644-1:199X.

c. A clear identification of the physical parameters for the cleanroom or clean zone (including the ISO classification reference to adjacent areas if necessary), and specific designations for coordinate of all sampling locations.

d. The specified designation criteria for the cleanroom or clean zone, including the ISO classification, the relevant occupancy state(s), and the considered particle size(s).

e. Details of the test method used, with any special conditions relating to the test or departures from the test method, and identification of the test instrument and its current calibration certificate.

f. The test results, including particle concentration data for all sampling location coordinates.

The requirement for detail of the data to be recorded indicates that data information sufficient for GMP documentation would be collected.

Compliance

Many U.S. biomedical companies have stated an intent to incorporate the requirements of the new ISO document into their manufacturing and laboratory operations as necessary to secure ISO certifications. The best course for the pharmaceutical manufacturer may be to develop a rationale for compliance (position paper) based on a minimal change of their present procedures.

From the perspective of the pharmaceutical manufacturer, issues with the current version of ISO 14644-1 include the following:

- There is no clear provision for "monitoring" versus "verification." Thus, all provisions of the ISO document are applicable only to initial and periodic verification and qualification or certification testing.
- As with FS-209E, calculation of the test results (based on average counts) will be unacceptable to the FDA.
- No provision is made for using the microscopic test method.
- Class designations differ from those presently in use.
- A performance specification for OPCs will be included (ISO 14644-3).
- The number of sample locations will increase in some areas, depending on user interpretation.

Application of the Standard

ISO 14644-1 is a guidance document with a general methodology. The testing plan, if followed, would require a substantial increase in time and effort for testing with no measurable benefit for some users. After an evaluation of ISO 14644-1 and pharmaceutical manufacturer's specifications, the latter often seem superior to ISO 14644-1 from the standpoint of time, cost, continuity of data collection, and sample site placement. Historical data will typically show that pharmaceutical manufacturer cleanrooms and clean areas meet the ISO cleanliness classes without the need for increased sample numbers.

The issue of regulatory compliance is central to any comparison of ISO 14644-1 to the pharmaceutical manufacturer's current monitoring plans. Strictly speaking, as stated above, FS-209E and its revision is binding only upon federal government agencies. FS-209E has, however, been used widely in private industry and has been considered a default standard. Because ISO 14644-1 will be the future original source document defining air cleanliness classifications (Class 100, etc.), private industries' use of ISO cleanliness classes will also imply adherence to the ISO document. This may not be a problem, since the burden of developing the specifications and procedures for compliance is left with each user. Using ISO 14644-1 as an industry standard would be a problem, since all users would presumably be forced to follow the same specific procedures regardless of differences between user applications.

Addressing the issue of regulatory compliance by deciding to follow company specifications and not the ISO standard places the burden of proof on the pharmaceutical manufacturer to prove that their specifications measure air cleanliness appropriately. This can readily be demonstrated by showing how pharmaceutical manufacturer specifications are similar to ISO 14644-1 and in fact may be superior, and how historical data from pharmaceutical manufacturing facilities meet ISO air cleanliness standards. Based on the proceeding discussion as to similarities between particle testing per pharmaceutical manufacturer specifications and the ISO standard, the similarities outweigh the differences. Both pharmaceutical manufacturer specifications and ISO 14644-1 incorporate validation and monitoring plans; both use the same instrumentation and general methodology. Pharmaceutical manufacturer specifications are generally more comprehensive, specifying frequent periodic challenges of HEPA filters and monitoring of airflow patterns, room air changes, and room pressure differentials.

One significant difference between ISO 14644-1 and most pharmaceutical manufacturer specifications is the sample number. ISO 14644-1 stipulates that the minimum sample number be based on room size (m^3) divided by the square root of the air cleanliness class. In comparison, manufacturer's specifications typically specify a smaller number of samples based on square footage. Pharmaceutical manufacturers may justify their different approach on two counts. There exists a large and well-characterized historic database on the air cleanliness of cleanrooms; thus, fewer samples are needed to detect a change in a room under these conditions. Further, the manufacturer will typically opt to focus higher numbers of sample sites or more frequent sampling on critical areas rather than the general environment.

Clearly, for reasons of compliance and historical perspective, it may be in the best interest of pharmaceutical or device manufacturers to continue to use their present airborne particle monitoring procedures. Demonstrably, these procedures effectively control airborne particle levels and detect departures from the historic data for the CEA in question. As was the case with FS-209E, the differences between the ISO 14644-1 requirements and most manufacturer's current procedures are periodic verification testing, sample size, arbitrary class definition, and statistical data handling; these are not of overwhelming importance for achieving the central goal of adequate cleanliness levels necessary for product protection. Present pharmaceutical manufacturer's airborne particle monitoring procedures fully meet this goal.

Monitoring Activity for Compliance with ISO 14644-1 (ISO 14644-2)

Partly at the insistence of pharmaceutical industry representatives on the ISO Working Groups contributing to ISO 14644-1, a companion standard (ISO 14644-2) relating to the monitoring of cleanrooms and CEAs was developed. The reader is reminded that only ISO 14644-1 has been issued, and that ISO 14644-2 and the others of the series remain at various draft stages and are subject to change from the form discussed here. As explained earlier, the concept of monitoring critical parameters of CEA operation is essential to obtaining proof of ongoing compliance in a cost-effective fashion. The concept is widely understood and universally applied in the U.S. pharmaceutical industry as well as those of most other nations. Because of this extreme importance and the necessity of developing a valid approach to prove continued compliance, the majority of the text of the current ISO 14644-2 document is excerpted below with a discussion of major points of interest. The intent of this document is to prescribe contiguous testing (on relatively frequent intervals) at a smaller number of sites than would be required for verification activity. This approach is, in fact, what manufacturers of medical products do now to prove continued compliance to regulatory agencies.

> Introduction
>
> This International Standard provides a mechanismn to prove continued compliance with ISO 14644-1 and specifies minimum requirements for testing and monitoring. In any testing plan, consideration should also be given to the particular operational requirements, risk assessment of the installation and its use.
>
> Cleanrooms and associated controlled environments provide for the control of airborne particulate contamination to levels appropriate for accomplishing contamination-sensitive activities. Products and processes that benefit from the

control of airborne contamination include those in such industries as aerospace, microelectronics, pharmaceuticals, medical devices, and healthcare.

The intent of ISO 14644-2 is specified to provide a means of proving continued compliance with the parent document. Minimum requirements are spelled out to that end. Also surfaced here is the key issue of product impact (risk assessment) of environmental airborne particles on the product being manufactured. This was a principle of the earlier FS-209 documents (e.g., FS-209B), in which the operator of a facility was directed to elaborate test plans and specific frequencies of testing that would ensure protection of product at the necessary level of assurance.

The consideration of protection of product is the overriding concern in any industry applying contamination control measures (paragraph 2). With regard to pharmaceutical manufacture, the production of LVI product with particle burdens below the USP limits shows, by definition, that the impact of environmental contamination is below the level at which product is detrimentally affected. Most LVI manufacturers will have years of monitoring into both for governmental aid and solution product to show that this is the case.

The situation with SVIs is more complex. Particularly as aseptic filling is being carried out, monitoring of airborne particulate matter, pressure differentials, and room air changes is carried out at a high level. In this activity, the presence of liquid microdroplets or powder in the area of fill head nozzles may decrease the value of airborne counts. Counts in areas surrounding the process must invariably reflect less than 100 particles/ft^3 ≥ 0.5 μm in size. Microbial monitoring will be of overwhelming importance. With medical devices, the situation is more akin to that with LVI.

Specification

1. Scope

This part of this International Standard specifies requirements for periodic testing of cleanrooms and associated controlled environments to prove their continued compliance with ISO 14644-1 for the designated class of environmental cleanliness.

These requirements invoke the test described in ISO 14644-1 for the classification of the cleanroom or clean zone. Additional tests are also specified, to be carried out based on the requirements of this International Standard. Optional tests, to be applied at the user's discretion, are also identified.

This part of ISO 14644 also specifies requirements for the monitoring of a cleanroom or clean zone (hereafter referred to as an installation) to provide evidence of its continued compliance with ISO 14644-1 for the designated classification of airborne particulate cleanliness.

Semantics are important here, as in other parts of the standard; with an international document is the fact that different readers, based on their first language background, may make different interpretations of specific text. The statement made here might better be "only periodic checking (monitoring) will serve as practical proof of compliance."

In the preceding section, mention was made of the historical particle count data that are available to most pharmaceutical manufacturers. These data have been compared to internal and regulatory limits and found appropriate to continued operation, or dealt with as an out-of-limits condition. Classes specified for areas in which different activities are conducted are adequate using the same empirical definition. A very important component of the value of historical data in verifying ongoing compliance is that particle counts or other information, such as pressure differentials, on which suitability judgments are made are taken under in-use (worst case) conditions. In fact, the in-use data are extremely valuable; other data (i.e., as-built or at rest) is of academic value alone if we are concerned about environmental suitability for the manufacturing process in question or about product protection.

Based on this eminently logical rationale, the basis for explaining any company's system for compliance with ISO 14644-1 must emphasize monitoring rather than initial certifications (qualification) as the only real proof of environmental suitability.

2. Definitions: For the purpose of this standard the following definitions apply. (See also 14644-1.)

3.6. Frequency Intervals

3.6.1. *Continuous:* updating that occurs constantly

3.6.2. *Frequent:* updating that occurs at specified intervals not exceeding 60 minutes during operation.

3.6.3. *6 months:* updating that occurs at an average interval not exceeding 183 days throughout periods of operational use, subject to no interval exceeding 190 days.

3.6.4. *12 months:* updating that occurs at an average interval not exceeding 366 days throughout periods of operational use, subject to no interval exceeding 400 days.

3.6.5. *24 months:* updating that occurs at an average interval not exceeding 731 days throughout periods of operational use, subject to no interval exceeding 800 days.

3.7. Requalification

Execution of the test sequence specified for the installation to demonstrate compliance with ISO 14644-1 according to the classification of the installation, including the verification of the selected pre-test conditions.

3.8 Test

Procedure undertaken in accordance with a defined method to determine the performance of an installation or element thereof.

3.9 Monitoring

Observations made by measurement in accordance with a defined method and plan to provide evidence of the performance of an installation.

NOTE: This information may be used to detect trends in operational state and to provide process support.

There are some problems with the definitions section, which may be related to semantics and translation. First, there is no clear demarcation between "continuous" and "frequent." "Constantly updated" will depend on timing of the interval of sampling. In low count areas, a sampling interval of 30 min could be chosen; this would satisfy criteria for "frequent" but would involve only 2 "updates" in 60 min. The frequency interval then jumps from 60 min to 6 months and makes no provision for weekly, biweekly, or monthly testing, which has proven entirely adequate for many pharmaceutical manufacturing conditions. We trust this defect will be corrected as ISO 14644-2 develops. Consensus procedures for the development of this document will hopefully result in an elimination of this deficiency. In fact, the only drug manufacturing operation realistically requiring short interval (60 min) or continuous updates would be an aseptic filling operation.

4. Demonstration of Continued Compliance

4.1 Principle

Continued compliance with air cleanliness (ISO class) requirements specified for the installation is verified by performing specified tests and by documenting the results. Monitoring data are used as an indication of installation status and may determine the frequency with which tests are carried out.

4.2 Testing for Continued Compliance

The reference test method and the maximum time intervals between such tests to prove continued compliance with the designated ISO class is given in Table 1.

Note: Where the installation is equipped with facilities for continuous or frequent monitoring of the airborne particulate concentration and the differential pressure, the maximum time interval for the test specified in Table 1 may be extended to 24 months provided that the results of continuous or frequent monitoring remain within the specified limits.

Table 1. Schedule of testing to demonstrate particle count compliance[1)]

Class	Maximum time interval	Test Method
≤ ISO 5	6 months	ISO 14644-1 Annex B
> ISO 5	12 months	ISO 14644-1 Annex B

1)This test will normally be performed in the operational state, but may also be performed in the at-rest state in accordance with the designated ISO classification.

Testing for continued compliance represents an evaluation of parameters listed in Table 1 by the test plans specified in the ISO 14644-1 and ISO 14644-3. The issue is that the test plans specified in these two documents are more rigorous than necessary to ensure compliance. The logical question is that if monitoring tests specifically designed to ensure compliance are being conducted, and if product protection requirements are specified, why it is necessary to construct verification level testing?

This leads to an even more interesting question. Considering the current high costs of manhours, why would it be necessary to perform verification level testing on a CEA process area for which years of in-limits historical monitoring data are available? It would seem that a "grandfather" clause might be added to save the user from having to perform the test to implement ISO 14644-1. As in the earlier section, remedial action and record retention as specified. Requalification is described in Section 3.4.

4.2.2. Where the application requires them, tests as given in table 2 shall be carried out to demonstrate compliance. The requirement for each of these tests to be carried out shall be determined by agreement between the Customer and supplier.

Table 2. Schedule of additional tests for all classes[1)]

Test Parameter	Maximum Time Interval	Test Procedure
Airflow velocity	12 months	ISO 14644-1 Annex B4
Airflow volume[2)]	12 months	ISO 14644-1 Annex B4
Air Pressure[3)] Difference	12 months	ISO 14644-3 Annex B

1)This test will normally be performed in the operational state, but may also be performed in the at-rest state in accordance with the designated ISO classification.

2)Airflow volume may be determined by either airflow velocity or airflow volume measurement techniques.

3)This test will not apply to clean zones that are not totally enclosed.

4.2.3. In addition to the normative tests given in tables 1 and 2, other tests may be included by agreement between customer and supplier as considered appropriate to the installation, such as those listed in the informative Annex A.

4.2.4. Where instruments are used for testing, they shall be calibrated in accordance with current industry practice.

4.2.5. If the test results are within the limits specified, then the installation is in a condition of continued compliance.

If any of the test results exceeds the limits specified, the installation is not in compliance and appropriate remedial action shall be taken.

Following remedial action, requalification shall be undertaken.

4.2.6. Requalification of the installation shall be undertaken after any of the following:

(a) Completion of remedial action implemented to rectify an out of compliance condition (see 3.2.5.).

(b) A significant change from the current performance specification, such as a change of operational use.

Note: The significance of a change should be determined by agreement between the customer and the supplier.

(c) Any interruption of air movement that significantly affects the operation of the installation.

(d) Special maintenance, such as change of final filters, that significantly affects the operation of the installation.

Requalification involves the verification level test matrices. The key would seem to be whether or not the manufacturer deems that a "risk to the operation" has occurred. An appropriate approach to this consideration might also be based on historical data. With regard to shutdown of the air handling system, for instance, if years of monitoring data show that a 2 h clean down is sufficient to return to class level, why perform verification tests?

4.3. Monitoring

4.3.1. Routine monitoring of airborne particle concentration and other parameters shall be performed according to a written plan. Note: This monitoring will normally be performed with the installation or part thereof in the operational state.

4.3.2. The airborne particle monitoring plan shall be based on risk assessment (see Annex B) related to the application of the installation. The plan shall include, as a minimum: predetermined sample locations, minimum volume of air per sample, duration of measurements, replicate measurements as required, time interval between measurements, particle size or sizes to be counted, count acceptance limits, count alert, action and excursion limits if appropriate.

Note 1: Monitoring of other attributes, e.g. differential pressure, temperature and humidity may also be undertaken in the same way as above.

Note 2: If continuous or frequent monitoring is specified in the plan for airborne particle counting and differential pressure monitoring, the schedule for the particle count test may be modified by extending the time between tests (see 3.2.1. note).

This section is a concession to those with input into the development of the standard who routinely perform remote monitoring with an FMS.

4.3.3. If the monitoring gives results that exceed specified action limits, the installation shall be considered non-compliant and appropriate remedial action shall be taken. Following remedial action, appropriate tests (see 4.2 and Annex A) shall be performed to determine if the installation is in compliance. If compliance has been achieved, the monitoring may be resumed.

4.3.4. Where instruments are used for monitoring, they shall be calibrated in accordance with current industry practice.

Annex A (Informative)—
Optional further tests

In addition to the normative tests specified in tables 1 and 2, optional tests, such as those listed in table 3, may be included within the testing schedule.

Table 3. Schedule of optional tests

Test Parameter	Class	Interval	Standard
Installed Filter Leakage	All Classes	24 months	ISO 14644-3 Annex B6
Airflow Visualization	All Classes	24 months	ISO 14644-3 Annex B7
Recovery	All Classes	24 months	ISO 14644-3 Annex B13
Containment Leakage	All Classes	24 months	ISO 14644-3 Annex B14

In summary, ISO 14644-2, in combination with ISO 14644-1 and ISO 14644-3, will provide a reasonable framework by which pharmaceutical manufacturers may operate with minimal changes to existing GMP–based environmental monitoring. The biggest question would appear to be the necessity for periodically applying the time-consuming verification level tests.

Annex A shows tests that are informational rather than compulsory. Interestingly, in-place testing of HEPA filters is not a normative test, although specified in ISO 14644-1 as normative. If this test remains informative, the manufacturer may be allowed latitude and cost savings. Again, it will be interesting to see to what extent the FDA accepts the ISO requirements. Annex B deals with the principle of risk assessment, which most producers of medical products are familiar with as a point of GMP.

Annex B (Informative) Guidance on the influence of risk assessment on cleanroom tests and monitoring.

The risk assessment pertaining to a particular cleanroom application will affect the following:

i) Monitoring plan.

ii) The interpretation of monitoring data.

iii) The actions to be taken as a result of the monitoring data obtained.

iv) The selection of parameters to be measured from table 2.

v) The selection of parameters to be measured from table 3.

The remainder of ISO 14644-2 should, similarly, not cause concern to the pharmaceutical manufacturer. Monitoring of testing is presently conducted per GMP by protocols or specifications. Risk analysis is specified as the basis of choice for tests to be performed. Similarly, a plan is required for actions to be taken in the event that data values are not within specification. The only problematical portion of this section (and section 4) regards the requirement of testing per the ISO 14644-1 method or per ISO 14644-3 following corrective (remedial) action.

Metrology (ISO 14644-3)

ISO 14644-3 is being developed by a Working Group convened in Japan. This is probably the single most important peripheral document in terms of impact on plant operations, since it specifies the types of instrumentation and test methods to be used for measuring pressure differentials, airflow patterns, and HEPA filter testing.

There are few "surprises" in this document for pharmaceutical manufacturing; the means of conducting the tests are specified in ISO 14644-1 (e.g., enumeration of airborne particles) and other tests are performed for the characterization of CEAs and cleanrooms. The latter include testing of humidity, temperature, particle fallout, particle intrusion, recovery rate, pressure differentials, airflow direction and flow pattern, and challenging of HEPA filters. The last subject has proven to be somewhat problematical for this Working Group, of which the author is a member.

The challenging of HEPA filters, as described in Chapter 13 of this book, is widely practiced in the pharmaceutical industry, using an oil aerosol as a challenge medium and an aerosol photometer to detect leaks. This is performed generally according to IES standard procedures but is poorly standardized in the industry. ISO 14644-3 calls for determining the extent of mixing of the upstream aerosol and precise calculation of the scan rate. These two additions will probably necessitate the change of some company SOPs.

Monitoring of Large Particles (ISO 14644-3 Annex B12—Particle Deposition Test)

An area of specialized but critical interest for the manufacturer of both injectable products and devices is the nonstoring and control of large extraneous particles. These particles are "ignored" by conventional particle counters with their 1 CFM flow rates and restricted sampling area. Specialized methods are required. A section of the future ISO 14644-3 document will deal with this area, giving, in the author's opinion, a useful description of the process and method of large particle monitoring.

B12.1 Application

This annex describes procedures and equipment for sizing and counting particles which are or may be deposited from the air onto product or work surfaces in the installation. Deposited particles are collected upon witness plates with appropriate surface characteristics similar to those of the at risk surface under consideration, and are sized and counted using optical microscopes, electron microscopes, or wafer scanning instruments. A particle fallout photometer may be used to obtain particle deposition data. Data for deposited particles should be reported in terms of mass or number of particles per unit surface area per unit time.

B12.2 Procedure

B12.2.1 Test method for definition of particles on witness plates

The witness plate should be placed in the same plane as the at risk surface, and should be retained in position by either vacuum or adhesive fixing to its back surface. The witness plate should be at the same electrical potential as the test surface. Where possible, the witness plate should be recessed to ensure its surface at the same level as the test surface (tabletop, floor, tool exterior, etc.) rather than situated on to of that surface. The following procedures and methods should be considered when manipulating witness plates:

- Verify that all cleanroom systems are functioning correctly, in accordance with operational requirements.
- Identify each witness plate uniquely and clean it as required to reduce the surface particle concentration to the lowest possible level. Determine the background concentration of particles on each witness plate.
- Maintain 20% of the witness plates as controls. These must be handled in exactly the same manner as the test witness plates until the test witness plates are placed in their test position.
- Transport all witness plates to the te4st locations in such a manner as to prevent airborne particles from contaminating their surfaces.
- Expose the test witness plates for time intervals ranging up to 48 hours, depending upon the cleanroom type, itís mode of operation, and the particle counting equipment that will be used. The exposure time should be adjusted, if necessary, to obtain sufficient particle deposition upon the witness plate surface to provide statistically valid data that satisfies user requirements.
- Collect the exposed witness plates by reversing the exposure procedure and store the exposed witness plates in their closed containers so that they are protected from further contamination.

B12.2.2 Counting and sizing collected particles

Counting and sizing of particles collected on witness plates is carried out to obtain reproducible data that can be used to access and categorized the cleanliness of the area being tested.

When using an optical light microscope, calibrated linear or circular graticules may be used for the particle sizing measurements. With an electron microscope, calibrated gratings with know line spacings may be used to relate the image dimensions to actual size. When using a surface scanner, size calibration information supplied by the vendor can be sued. Data from counts over a partial area of the witness plate may be extrapolation to the entire surface area (statistical counting). Extrapolation may be made as described in Annex B3 section 3.1.1, section B3.3.1.2(a).

- Count and size the particles on all witness plates, including the control plates. Enumerate particles on the total area of all witness plates and categorize them in appropriate particle size ranges, based on the particle diameters.
- Determine the surface concentration of deposited particles for each witness plate:

  $D = (Nt - Nb)/Aw$

  where,

  D: deposited surface concentration of particles,

  Nt: Total surface concentration particles,

  Nb: Number of particles larger or equal to the defined minimum size on the witness plate surface after cleaning, but before exposure to the cleanroom environment, and

  AW: witness plate area in cm^2

- Average the values of *D* for the control witness plates.
- Determine the net increase in surface concentration for each witness plate by subtracting the average control witness plate concentration from the average test witness plate concentration. Divide the net concentration by the test witness plate exposure time. This calculation yields a particle deposition rate (PDR) in terms of particles deposited per square centimeter per unit time.
- Record the mean PDR value and its standard deviation.

B12.3 Apparatus

B12.3.1 Selection of witness plate material

Depending upon particle size to be detected and means of measurement the following may be used.

- Microporous membrane filters
- Double-sided adhesive tape
- Petri dishes
- Petri dishes containing a contrasting color (black) polymer, such as polyester resin
- Photographic film (sheet)
- Microscope slides (plain or with evaporated metal film coating)
- Glass or metal mirror plates
- Semiconductor wafer blanks
- Glass photomask substrates

The surface smoothness of the witness plate must be appropriate to the size of the particles which will be counted to ensure, so that the particles are easily visible. The means of measurement employed must be capable of resolving and measuring the smallest particle size to be enumerated.

B12.3.2 Apparatus

A variety of instruments may be used in counting and sizing particles which have settled onto the witness plate surface. These fall into three general categories, depending upon the size of the particles of concern:

- Light microscopes (particles larger than or equal to 2 μm)
- Electron microscopes (particles larger than or equal to 0.20 μm)
- Surface analysis scanners (particles larger than or equal to 0.1 μm)
- Particle fallout photometer (up to 10,000 ppm)

When choosing the counting and sizing instrument to be used, consideration must be given to detection of particles in the relevant size range. Other factors to be considered include the time required for sample collection and analysis, the time required for characterization of the method. The instrument used should have a valid calibration certificate.

B12.4 Data reporting

Record the following data, as specified by agreement between customer and supplier, for specification of the installation:

1. Identification of instrument operator or data collection facility.
2. Identification of instruments and its calibration status.
3. Type of tests and measurements, and measuring conditions.
4. Clean zone specification level of the installation, measured or target value.
5. Measured point locations
6. Raw data for each measurement, if specified.

The only significant area for negotiation in the 14644-3 document remains the means of specifying and testing the HEPA filters used for removing particles from the air supplied to CEAs and cleanrooms. The widely used method of leak testing with an oil aerosol results in some (unacceptable to microelectronics manufacturers) level of carryover into tested areas and results in adsorption of the test material by the glass fibers of the filters. Thus, it seems to be necessary to specify two methods of testing using both particle counters and photometers, one for the microelectronics industry and the other for pharmaceutical manufacturers. The final version of this document will include fairly rigorous criteria for the performance and calibration of OPCs, the basic particle counting instrument used in classifying CEAs and cleanrooms.

SUMMARY AND CONCLUSION

For manufacturers of pharmaceuticals and medical devices, a complex hierarchy of rules, regulations, laws, and standards govern operation. With the pressure to establish international standards and, ultimately, standards for the world community, manufacturers are pressed to comply not only with local, national, and regional rules but also with world standards documents.

In the present situation, until there are international documents at all levels, the manufacturer will be forced to maintain a working knowledge not only of local documents, but others, such as the ISO 14644 series. A saving factor is that umbrella documents, such as ISO 14644, are only intended to specify sound general rules that manufacturers may use to assess their present procedures and, on objective assessment, be assured their method is adequate to fulfill the standard's principles.

The author's work with the ISO as a U.S. delegate has left him impressed with several points of the standards development process. First, there are the many points of technical commonality between standards documents and requirements for many nations. It seems as though technology, in this sense, provides a basis of commonality and harmonization between nations just as other intellectual components of our different cultures. Another source of surprise is how well individuals from different countries, with different native languages, can communicate when their expertise and technical interest are topics of discussion. The author's counterparts from other nations have proven to be very generous with regard to furnishing input for shared methodologies. This spirit of free exchange of information and compromise evident in ISO Working Groups has also been impressive. Harmonization would seem, in large part, to represent simplification and, more importantly, a means of removing technical barriers, which can separate people just as effectively as cultural barriers. Thus, although the process of harmonization is complex and time-consuming, it is necessary and desirable for the common good. The process has the potential not only of ensuring a key component of product quality but also reducing costs of compliance.

REFERENCES

Barley, A. R. 1997. *EU directive handbook.* Boca Raton, Fla., USA: St. Lucie Press.

Borden, P., J. Munson, D. Bartelson, and M. McClellan. 1989. Real time monitoring of large particle fallout for aerospace applications. In *Proc. Ann. Meeting Inst. Env. Sci.,* pp. 394–396.

BS 5295. 1989. *Environmental cleanliness in enclosed spaces:* Part 0. Terms and definitions for cleanrooms and clean air devices; Part 1. Specifications for cleanrooms and clean air devices; Part 2. Method for specifying design, construction, and commissioning of cleanrooms and clean air devices; Part 3. Guide to operational procedures and disciplines applicable to cleanrooms and clean air devices; Part 4. Specifications for monitoring cleanrooms and clean air devices to prove compliance with BS-5295 Part 1. London: British Standards Institute.

Bzik, T. J. 1986. Statistical management and analysis of particle count data in ultraclean environments, Part I. *Microcontamination* (May): 89–90.

CEN/TC 243 N31. 1994. Cleanroom technology part I. Classification of cleanrooms and clean air controlled spaces by airborne particulate content. Brussels: European Committee for Standardization.

Cooper, D. W. 1989. Towards Federal Standard 209E: Partial versus complete inspection of clean air zones. *J. Environ. Sci.* 33:31–33.

Cooper, D. W. 1988. Statistical analysis relating to recent Federal Standard 209 (cleanrooms) revisions. *J. Environ. Sci.* 31:48–52.

Cooper, D. W., and D. C. Milholland. 1990. Sequential sampling for Federal Standard 209 for cleanrooms. *J. of the IES* 33:28–32.

Cooper, D. W. 1990. Particle statistics for contamination control: An introduction. In *Proceedings of the PDA International Conference on Particle Detection, Metrology and Control,* pp. 183–208.

Cooper, D. W., R. J. Miller, and J. J. Wu. 1991. Measurements with condensation nucleus counters and an optical particle counter in a cleanroom. *J. of the IES* 32:702–711.

Delattin, R. 1998. EU status of GMP for sterile products. *J. Paren. Sci. Technol.* 52:82–88.

Del Valle, M. A. 1997. Effects of the 1997 revision of the European Community GMPs on HVAC design for fill/finish facilities. *Pharm. Eng.* (Nov.–Dec.): 34–41.

Good Manufacturing Practice for Medicinal Products in the European Community. 1992. Brussels: Commission of the European Communities.

Grotzinger, S. J., and D. W. Cooper. 1989. Cost-effective allocation of samples to preselected locations. *Eng. Technol.* 21:1–20.

FDA. 1987. *Guideline on sterile drug products produced by aseptic processing.* Rockville, Md., USA: Food and Drug Administration, Center for Drugs and Biologics.

FS-209E. 1992. Airborne particulate cleanliness classes in cleanrooms and clean zones, Federal Standard 209-E. Washington, D.C.: General Services Administration.

ISO 14644-1. 1998. Cleanrooms and associated controlled environments. Part 1: Classification of airborne particulate cleanliness for cleanrooms and clean zones. Geneva: International Organization for Standardization.

JIS B 9920.1995. Measuring methods for airborne particles in cleanrooms and evaluating methods for air cleanliness of clean rooms. Tokyo: Japanese Standards Association.

Lieberman, A. 1988. Requirements, instrument capabilities and potentials for particle monitoring in pharmaceutical manufacturing areas. *Pharm. Eng.* (July–Aug.): 65–78.

Liu, B. Y. H., and W. W. Szymanski. 1987. Counting efficiency, lower detection limit and noise level of optical particle counters. *Proceedings of the Meeting of Inst. of Env. Sci.,* pp 417–421. Available from the Institute of Environmental Science, Mt. Prospect, IL 60056.

Munson, T. E., and R. L. Sorensen. 1990. Environmental monitoring: Regulatory issues. *In Sterile pharmaceutical manufacturing: Applications for the 1990s*, edited by M. J. Groves and W. Olsen. Buffalo Grove Ill., USA: Interpharm Press Inc., pp. 163–184.

Montague, W., and H. Sommer. 1990. Performance parameters of optical aerosol particle counters. *Filt. News* (Nov./Dec): 26–30.

Rules Governing Medicinal Products in The European Community, Volumes IV. 1989. Brussels: European Committee for Standardization. .

Sommer, H. T. 1989. Resolution, sensitivity, counting efficiency and coincidence limit of optical aerosol particle counters. In *Proceedings of the Fine Particle Society Conference* (21–25 August, Boston) (per HIAC, Silver Springs, Md., USA).

Sommer, H. T., and C. F. Harrison. 1991. Aerosol size concentration measurements from scattered light signals. In *Proceedings Conference Sensor '91* (14–16 May, Nuremburg,Germany) (per HIAC, Silver Springs, MA).

USP. 1995. *United States Pharmacopeia,* 23rd ed., Chapter <1077>, *Current good manufacturing practices in manufacturing, processing, packing or holding of drugs.* Rockville, Md., USA: The United States Pharmacopeial Convention, pp. 1907–1921.

Wen, H. Y., and G. Kasper. 1986. Counting efficiencies of six commercial particle counters. *J. Aerosol Sci.* 187 (6):947–961.

WHO. 1992. *Expert committee technical committee report on specifications for pharmaceutical preparations.* Geneva: World Health Organization.

Whyte, W. 1983. A multicentered investigation of cleanroom requirements for terminally sterilized pharmaceuticals. *J. Parent. Sci. Tech.* 37:184–197.

APPENDIX I. GOOD MANUFACTURING PRACTICE–RELATED DOCUMENTS

Major Authorizing Agencies:
- World Health Organization (WHO)
- Association of South-East Asian Nations (ASEAN)
- European Community (EC)
- Pharmaceutical Inspection Convention (PIC)

Countries Having GMP Requirements for Pharmaceutical Manufacture:
- Australia
- Canada
- Chile
- Denmark
- Egypt
- France
- Greece
- Guyana
- Hong Kong
- India
- Israel
- Italy
- Japan
- The Netherlands
- Malaysia
- New Zealand
- Switzerland
- South Africa
- South Korea
- Turkey
- United Kingdom
- United States

APPENDIX II. INTERNATIONAL CONTAMINATION CONTROL STANDARDS

These examples were excerpted from the Institute of Environmental Sciences *Glossary of Contamination Control Standards Documents Related to Construction, Operation, and Monitoring of Controlled Environment Areas.*

Contamination Control Documents from the United States

Recommended Practice for HEPA and ULPA Filters

Document number: IES-RP-CC001
Latest date: 1993
Reference number: 1084
Country: United States
Language: English
Categories: 1400, 1410, 1440
Source: IES
Number of pages: 24
Abstract: This recommended practice covers basic provisions for HEPA and ULPA (ultralow penetration air) filter units as a basis for agreement between buyers and sellers. Six levels of performance and six grades of construction are included.

Recommended Practice for Laminar Flow Clean Air Devices

Document number: IES-RP-CC002
Latest date: 1983
Reference number: 1085
Country: United States
Language: English
Categories: 1010, 1020, 1030
Source: IES
Number of pages: 12
Abstract: This recommended practice covers basic provisions for laminar flow clean air devices, including definitions, performance evaluation, and major requirements of the devices.

Recommended Practice for Testing Cleanrooms

Document number: IES-RP-CC003
Latest date: 1987
Reference number: 1086
Country: United States
Language: English
Categories: 1200, 1210, 1220
Source: IES
Number of pages: 12
Abstract: This recommended practice covers testing methods for characterizing the performance of cleanrooms. It is intended to assist planners, designers, manufacturers, and buyers in preparing detailed specifications for cleanroom procurement and for ensuring cleanroom operational compliance. Performance tests are recommended for three types of cleanrooms at three operational phases. The tests evaluate and characterize the overall performance of the cleanroom or clean zone system. Airflow, filters, particle levels, air pressure, airflow parallelism, room integrity, particle fallout, temperature and relative humidity, light, noise, and vibration are tested.

A Glossary of Terms and Definitions Related to Contamination Control

Document number: IES-RD-CC011
Latest date: 1985
Reference number: 1093
Country: United States
Language: English
Categories: 2500
Source: IES
Number of pages: 16
Abstract: This document includes common terms and their definitions as they relate to contamination control.

Recommended Practice for Testing HEPA and ULPA Filter Media

Document number: IES-RP-CC021
Latest date: 1993
Reference number: 1103
Country: United States
Language: English
Categories: 1400, 1410, 1440
Source: IES
Abstract: Test methods are given to determine physical and filtration properties of HEPA and ULPA filtration media.

Recommended Practice for Cleanroom Operation

Document number: IES-RP-CC026
Latest date: 1994
Reference number: 1107
Country: United States
Language: English
Categories: 1010, 1020
Source: IES
Abstract: This practice provides guidance for maintaining the integrity of cleanrooms during ancillary operations, such as the preparation of materials, modification of the facility, and installation and repair of equipment. Methods are given for verifying cleanliness.

Airborne Particulate Cleanliness Classes in Cleanrooms and Clean Zones

Document number: Federal Standard 209E
Latest date: 1992
Reference number: 1156
Country: United States
Language: English
Categories: 1110, 1120, 1130
Source: GSA
Number of pages: 48
Abstract: This document establishes standard cleanliness classes for cleanrooms and clean zones, based on specified concentrations of airborne particles. It prescribes methods for verifying air cleanliness.

Particulate Matter in Injections

Document number: USP 23, <788>
Latest date: 1992
Reference number: 1167
Country: United States
Language: English
Categories: 1600, 1900
Source: USP
Number of pages: 2
Abstract: Limits for particulate matter are prescribed here for individual articles in containers for both LVIs and SVIs. Both measurement methods and quantities are stated.

Cleanrooms and Clean Workstations: Laminar Flow Cleanrooms

Document number: MIL-STD-1695 (AS)
Latest date: 1977
Reference number: 1157
Country: United States
Language: English
Categories: 1020, 1120
Source: NAEC
Number of pages: 59
Abstract: This document sets out specific requirements for laminar flow cleanrooms.

Contamination Control Documents from Australian Agencies

Cleanrooms and Clean Workstations: Non-Laminar Flow Cleanrooms— Class 350 and Cleaner

Document number: AS 1386.3-1989
Latest date: 1989
Reference number: 2003
Country: Australia
Language: English
Categories: 1010, 1020, 1030
Source: SAA
Number of pages: 6
Abstract: This document sets out specific requirements for nonlaminar flow cleanrooms with an air cleanliness of at least Class 350 in accordance with AS 1386.1-1989.

Cleanrooms and Clean Workstations: Operation and Inspection of Cleanrooms

Document number: AS 1386.6-1989
Latest date: 1989
Reference number: 2006
Country: Australia
Language: English
Categories: 1020, 1120
Source: SAA
Number of pages: 6
Abstract: Recommendations and requirements are given for the operational procedures and inspection of laminar flow cleanrooms and nonlaminar flow cleanrooms. It complements AS 1386.2, AS 1386.3, and AS 1386-4.

Cleanrooms, Workstations and Safety Cabinets—Methods of Test: Determination of Air Velocity and Uniformity of Air Velocity in Clean Workstations and Laminar Flow Safety Cabinets

Document number: AS 1807.1-1989
Latest date: 1989
Reference number: 2008
Country: Australia
Language: English
Categories: 1030, 1310
Source: SAA
Abstract: This standard sets out a series of methods, including specifications for all test equipment. Test methods for testing safety cabinets and determining recovery times of cleanrooms are included.

Cleanrooms, Workstations and Safety Cabinets—Methods of Test: Determination of Performance of Clean Workstations and Laminar Flow Safety Cabinets Under Loaded Filter Conditions

Document number: AS 1807.2-1989
Latest date: 1989
Reference number: 2009
Country: Australia
Language: English
Categories: 1020, 1030
Source: SAA
Number of pages: 4
Abstract: This standard sets out a series of methods, including specifications for all test equipment. Test methods for testing safety cabinets and determining recovery times of cleanrooms are included. An installation integrity test of nonterminally mounted HEPA filters is included.

Cleanrooms, Workstations and Safety Cabinets—Methods of Test: Determination of Performance of Laminar Flow Cleanrooms Under Loaded Filter Conditions

Document number: AS 1807.4-1989
Latest date: 1989
Reference number: 2011
Country: Australia
Language: English
Categories: 1120, 1130
Source: SAA
Abstract: This standard sets out a series of methods, including specifications for all test equipment.

Cleanrooms, Workstations and Safety Cabinets—Methods of Test: Determination of Integrity of HEPA Filter Installations Not Terminally Mounted

Document number: AS 1807-7-1989
Latest date: 1989
Reference number: 2014
Country: Australia
Language: English
Categories: 1440
Source: SAA
Abstract: This standard sets out a series of methods, including specifications for all test equipment. Test methods for testing safety cabinets and determining recovery times of cleanrooms are included. An installation integrity test of nonterminally mounted HEPA filters is included.

Cleanrooms, Workstations and Safety Cabinets—Methods of Test: Particle Counting Cleanrooms by Microscopic Sizing and Counting

Document number: AS 1807-9-1989
Latest date: 1989
Reference number: 2016
Country: Australia
Language: English
Categories: 1910, 1920
Source: SAA
Abstract: This standard sets out a series of methods, including specifications for all test equipment. Test methods are given for testing airborne particle concentrations by collecting airborne particles on a substrate that is then examined microscopically for particle concentration and size.

Cleanrooms, Workstations and Safety Cabinets—Methods of Test: Determination of Airflow Parallelism in Laminar Flow Cleanroom

Document number: AS 1807-11-1989
Latest date: 1989
Reference number: 2018
Country: Australia

Language: English
Categories: 1130, 1310
Source: SAA
Abstract: This standard sets out a series of methods, including specifications for all test equipment. Test methods are given for determining the parallelism of airflow stream lines.

Contamination Control Documents from Belgian Agencies

Contamination Curves

Document number: BH77/1
Latest date: 1992
Reference number: 2040
Country: Belgium
Language: French, English
Categories: 1130, 1910, 1920
Source: ICCE BRUSSELS
Number of pages: 9
Abstract: This standard gives procedures for monitoring and particle counting in different working conditions, in order to identify the origin of particles and their influence on area classification. This aids in identifying the origin of contamination that may arise from working in the critical area.

Decontamination Time

Document number: BH 77/2
Latest date: 1992
Reference number: 2041
Country: Belgium
Language: French, English
Categories: 1130, 1920
Source: ICCE BRUSSELS
Number of pages: 10
Abstract: This standard, similar to the determination of recovery time, consists of controlling the efficiency of the air-conditioning and air distribution systems, overpressure, and airflow rate in a cleanroom by decreasing particulate levels artificially introduced into the room. The method defines the particle generation procedure and the number and location of sampling points.

Contamination Control Documents from Canadian Agencies

Good Manufacturing Practices

Document number: H42-2/1-1989
Latest date: 1989
Reference number: 2042
Country: Canada
Language: French, English
Categories: 1800, 2500
Source: CGSB
Number of pages: 162
Abstract: The scope of this publication includes all GMP–related activities in the production and import of drugs for human or veterinary use. It states applicable principles and practices. Basic environmental standards, including final air filter efficiency, air change rates, air cleanliness levels, and viable organism concentrations are stated.

Contamination Control Documents from Chinese Agencies

Cleanliness Classification and Verification for Cleanrooms and Clean Zones

Document number: QJ2214-91
Latest date: 1991
Reference number: 2044
Country: Peoples Republic of China
Language: English
Categories: 1120, 1130
Source: CMAI
Abstract: This standard defines cleanroom classes. It also includes measurement at 0.1 g and requires filter leak testing. It recommends class levels for various production processes in the Chinese aerospace industry.

Contamination Control Documents from English Agencies

Determination of Particle Size Distribution Part 7: Recommendations for Single Particle Light Interaction Methods

Document number: BS 34 06 Part 7
Latest date: 1988
Reference number: 2047
Country: United Kingdom
Language: English
Categories: 1510, 1900

Source: British Standards Institute
Number of pages: 20
Abstract: This document gives recommendations on methods of determining particle size distribution and concentration by single particle counting and sizing instruments using light interaction methods. Information is also provided to assist in making comparisons between interlaboratory measurements and those made by different instruments.

Environmental Cleanliness in Enclosed Spaces 0: Terms and Definitions for Cleanrooms and Clean Air Devices

Document number: BS 5295 Part 0
Latest date: 1989
Reference number: 2048
Country: United Kingdom
Language: English
Categories: 1110
Source: British Standards Institute
Number of pages: 4
Abstract: Terminology is described to define clean area performance and testing methods.

Environmental Cleanliness in Enclosed Spaces 1: Specifications for Cleanrooms and Clean Air Devices

Document number: BS 5295 Part 1
Latest date: 1989
Reference number: 2049
Country: United Kingdom
Language: English
Categories: 1010, 1110
Source: British Standards Institute
Number of pages: 17
Abstract: This standard specifies the designated classification and performance requirements for nine classes of cleanliness for cleanrooms and clean air devices. Environmental, construction, and design requirements and methods of test to demonstrate compliance are included.

Environmental Cleanliness in Enclosed Spaces 2: Method for Specifying the Design, Construction and Commissioning of Cleanrooms and Clean Air Devices

Document number: BS 5295 Part 2
Latest date: 1989
Reference number: 2050
Country: United Kingdom
Language: English
Categories: 1010, 1110, 2500
Source: British Standards Institute
Number of pages: 14
Abstract: A method is given for specifying requirements for clean areas. It provides a checklist of design, construction, furniture, and commissioning requirements that may or may not be relevant.

Environmental Cleanliness in Enclosed Spaces 3: Guide to Operational Procedures and Disciplines Applicable to Cleanrooms and Clean Air Devices

Document number: BS 5295 Part 3
Latest date: 1989
Reference number: 2051
Country: United Kingdom
Language: English
Categories: 1020, 1120
Source: British Standards Institute
Number of pages: 8
Abstract: This standard lays out some guidelines for procedures and personnel activity requirements applicable for implementation in cleanrooms and clean air device operations.

Environmental Cleanliness in Enclosed Spaces 4: Specification for Monitoring Clean Rooms and Clean Air Devices to Prove Continued Compliance with BS 5295: Part 1

Document number: BS 5295 Part 4
Latest date: 1989
Reference number: 2052
Country: United Kingdom
Language: English
Categories: 1030, 1130
Source: British Standards Institute
Number of pages: 10
Abstract: This standard specifies the procedures to be followed in order to prove continued compliance during the life of a cleanroom or clean air device with the requirements of the designated class to which it is required to be operated. The procedures may also be applied to monitoring of controlled spaces.

Contamination Control Documents from French Agencies

Limiting Enclosures: Classification of Enclosures According to Their Scaling Capacity

Document number: NF M 52-200
Latest date: 1982
Reference number: 2054
Country: France
Language: French
Categories: 1010
Source: AFNOR

Contamination Control Documents from German Agencies

Type Testing of High Efficiency Submicron Particulate Air Filters

Document number: DIN 24184
Latest date: 1974
Reference number: 2069
Country: Germany
Language: German
Categories: 1440
Source: DIN

Test Method for Particle Analysis in Liquids: Determination of Particles with Optical Particle Counter

Document number: DIN 50452 Part 2
Latest date: 1991
Reference number: 2075
Country: Germany
Language: German
Categories: 1510, 1910, 1920
Source: DIN
Abstract: Requirements for liquid-borne particle counter specifications are stated for correlation when measuring submicrometer particles. Procedures for measuring interlaboratory correlation are given.

Cleanroom Technology—Fundamentals, Definitions, and Determination of Cleanliness Categories

Document number: VDI 2083
Reference number: 2076
Country: Germany
Language: German
Categories: 1100, 1110, 1120, 1130
Source: VDI
Abstract: Cleanroom technology is covered in 10 segments: (1) fundamentals, definitions, and standards; (2) construction, operation, and maintenance; (3) measurement methods; (4) surface cleanliness; (5) comfort ventilation; (6) personnel in the working place; (7) process media cleanliness; (8) usefulness of equipment for cleanrooms; (9) quality, production, and distribution of clean water; (10) media supply systems.

Contamination Control Documents from Japanese Agencies

Standard for the Design of Air Cleaning Apparatus for Air Conditioning Systems

Document number: JACA no. 3B
Latest date: 1978
Reference number: 2077
Country: Japan
Language: Japanese
Categories: 1310, 1400
Source: JACA
Number of pages: 32

Standard of Test Method for Air Cleaning Devices

Document number: JACA no. 10C
Latest date: 1979
Reference number: 2078
Country: Japan
Language: Japanese
Categories: 1030, 1440
Source: JACA
Number of pages: 32

Standard of Test Method for Composition of Air Filter Media

Document number: JACA no. 11
Latest date: 1978
Reference number: 2079
Country: Japan
Language: Japanese
Categories: 1410
Source: JACA
Number of pages: 5

Guidance for Operation of Cleanrooms

Document number: JACA no. 14C
Latest date: 1992
Reference number: 2082
Country: Japan
Language: Japanese
Categories: 1120
Source: JACA
Number of pages: 34

Standardization and Evaluation of Cleanrooms

Document number: JACA no. 24A
Latest date: 1992
Reference number: 2092
Country: Japan
Language: Japanese
Categories: 1130
Source: JACA
Number of pages: 57

Measuring Methods for Airborne Particles in Clean Room and Evaluation Methods for Air Cleanliness of Clean Room

Document number: JIS B 9920
Latest date: 1989
Reference number: 2097
Country: Japan
Language: Japanese, English
Categories: 1130, 1920
Source: JIS
Number of pages: 35
Abstract: This standard specifies the measuring methods for concentrations of airborne particles in cleanrooms and the methods of evaluating air cleanliness of cleanrooms.

Contamination Control Documents from Swedish Agencies

Standard for Open LAF Units

Document number: Open LAF Units Standard
Latest date: 1985
Reference number: 2108
Country: Sweden
Language: Swedish
Categories: 1010, 1310
Source: R3-NORDIC
Number of pages: 35
Abstract: This standard provides information on goals and application for open LAF units, dimensional and material specifications, airflow control, and power requirements. Operational procedures and control demands for airflow, filtration, and resistance against expected chemical attack are also given.

Contamination Control Documents from Swiss Agencies

Classification and Utilization of Air Filters

Country: Switzerland
Abstract: This standard provides terms, definitions, symbols, and units for the testing of air filter cells (including HEPA and ULPA filters). It provides classification of air filter cells, defines testing and evaluation of air filter systems, and defines realization and operation of air filter systems.

VI

REGULATORY PERSPECTIVES RELATED TO THE CONTROL OF PARTICULATE MATTER CONTAMINATION

Governmental regulation of the manufacture of pharmaceuticals and medical devices is a critical concern in most countries of the world. The subject of this chapter is thus of significant practical importance. The author deliberated for some time over the title of this chapter. The term *perspective* was eventually chosen over *perception* for the reason that "perspective" conveys an image of a specific, objective viewpoint. It also suggests the reasonable context of differing viewpoints. Our perspective is pivotal regarding the context in which all of us, including regulators, the compendial body, and the pharmaceutical industry, view particulate matter in medical products.

As the author set out to discuss the impressions of the Food and Drug Administration (FDA) and regulatory activities regarding particulate matter in this chapter, it rapidly became obvious that it was not possible to characterize and summarize the viewpoints of individual field inspectors, reviewers, or FDA management personnel; all that can be achieved is a series of generalizations. Nonetheless, the author believes that the following discussion will be valuable to the reader in terms of an increased general understanding of the ways inspectors and reviewers use and assess particulate matter data.

Additionally, with regard to perceptions and perspectives, all of the opinions and subjective impressions expressed in this chapter are wholly those of the author, not those of Baxter International Inc. or any other manufacturer of healthcare products. At the outset, the author apologizes to the FDA for using the terms *inspector, auditor,* and *investigator* somewhat interchangeably. Although these terms have specific meanings within the FDA, the general meaning is well understood to the manufacturer and serves to convey the intended information.

In the perspective of individual auditors, the importance of particulate matter and the implication of the levels present may vary widely. As discussed earlier in this book, compendial particulate matter limits are not intended to apply to particles that are predominantly from a single source, such as numerous glass fragments from a filling operation or degradant particles resulting from lyophilization. The intent is, instead, to limit the number of particles from random sources occurring in products made under the appropriate application of Good Manufacturing Practice (GMP). These particles from random sources and of heterogeneous composition are impossible to exclude from product but are "acceptable" due to the source and low numbers in which they occur. Interpretation of the compendial verbiage in an enforcement context may be expected to vary based on the specific situation and the experience base of the inspector.

The references for this chapter generally fall into two categories, both of which contain useful information with regard to understanding regulatory perspectives on and approaches to particulate matter in injectables and medical devices. First, there is an author-indexed section that references articles published in professional journals and, in some cases, trade magazines. The *Journal of Parenteral Science and Technology* is the single most useful source in this regard and contains articles that are both of current interest and well written and edited. In line with the subject of this chapter (i.e., regulatory perspectives), the majority of these articles are authored by individuals who were formerly employed by the FDA in a policy-making position with regard to injectable or device manufacture. Their teachings and doctrines at present (and probably for years to come) will continue to influence field inspector perspectives and audit strategy to some extent. These individuals are, in alphabetical order, H. L. Avalone, E. M. Fry, D. L. Michaels, T. E. Munson, R. L. Sorenson, and Ronald Tetzlaff. These persons have the distinction of laying much of the groundwork for the principles of inspection, investigation, and enforcement practiced today by the FDA. The author has selected references based on their content rather than their date of publication; the reader may wish to search the literature for more the recent references that are available.

Mr. Avalone is, perhaps, the most prolific author among high-level individuals who have worked at the FDA and has published on a wide range of subjects relating to regulatory and inspectional issues (1980; 1983a; 1984a; 1988a,b; 1989a,b; 1990) and policy and expectations (1982; 1983b; 1984b; and 1985). Of particular interest to the reader who has a role in quality assurance are the inspectional findings discussed in the first category of references.

Mr. Fry published most of his journal articles (1982; 1984a,b; 1984a,b) on the subjects of FDA doctrine and principles. Mr. Michaels' references deal with the principles of drug regulation and enforcement (1982; 1984; 1985). Mr. Sorenson (1984; 1985; 1986) was the FDA expert on environmental monitoring and authored the aseptic processing guidelines, a document that has stood the test of time for more than a decade (see also Munson and Sorenson 1991). Mr. Tetzlaff (1983; 1984; 1987; 1988; 1990; 1993) needs no introduction for those involved in the aseptic processing of drug products. While at the FDA, he was widely known and respected for his detailed and methodical investigation management; the section of this chapter dealing with audits of environmental monitoring is based on the 1990 reference.

The other section of the references contains specific FDA, or governmental publications, the majority of which are guidelines for various aspects of injectable or device manufacture. Of specific note are those regarding the aseptic processing of drugs (1987), quality control laboratories (1993), bulk pharmaceutical chemicals (BPCs) (1991), the manufacture of sterile drug substances (1994), and *U.S. v. Barr Laboratories, Inc.* (1993). The FDA *Investigations Operations Manual* (1994) is the basic guide for investigators and provides valuable insight into the principles of inspection and into the highly organized nature of the enforcement process. Those readers who take the time to review just those references that appear to be of particular interest will gain an invaluable insight into inspectional principles. Of particular interest to most readers will be the highly structured, methodical approach that expert investigators employ.

For those with interests regarding regulatory perspectives, the FDA home page on the Internet is a very useful tool (http://www.fda.gov). This home page can be accessed from computers with Internet access software and allows rapid and efficient access to more than a dozen general subject headings by icon, ranging from toxicology research and medical products reporting through biologics, field operations, human drugs, and tobacco. In the author's experience (I am, at best, semiliterate in the use of computers), some web sites operate much more quickly and efficiently than others. However, if the reader selects the "Field Operations/Imports" icon from the FDA home page, you are connected to a wealth of valuable information, such as inspectional references (which were not easily obtained in the not-too-distant past).

Compliance policy guides are also easy to access, and there is an industry assistance category as well.

One final note: In gathering information for this chapter, the author found that it was difficult to discuss regulatory activity in the United States without some level of understanding of the statutory and organizational framework involved. In a cursory overview, the U.S. Pharmacopeia (USP) sets rules regarding medicinal product safety, purity, and efficacy and the FDA enforces them. Thus, the following sections of this chapter discuss how and why the FDA operates, as well as specific consideration of enforcement activities and philosophies related to particulate matter.

THE FDA

The FDA was established by the Food and Drug Act of 1906, primarily to deal with drug quality. In 1938, the Food, Drug, and Cosmetic (FD&C) Act focused on the safety aspect of drugs; this was amended in 1962 to deal with drug efficacy. More recently, the FDA has been empowered to deal with biological products and medical devices. There is a legal link to the compendium of drug standards established by the USP. The FD&C Act recognized the USP, NF (National Formulary), the AOAC (Association of Official Analytical Chemists), and the Homeopathic Pharmacopeia. There is a close functional relationship between the USP and the FDA. The USP, in essence, sets standards but cannot enforce them. The FDA recognizes the USP–set standard and is empowered by the FD&C Act to enforce them. In the evolution of its functions, the FDA has come to issue regulations concerning New Drug Applications (NDAs) and current Good Manufacturing Practices (cGMPs). The addition of the word *current* to the term philosophically indicates the evolving, dynamic nature of these practice and suggests the basis of cGMP on best demonstrated practice. More will be said shortly regarding the cGMPs. Suffice it for the present that their purpose is to identify and to control critical process factors with the intention of building in quality, rather than relying solely on final product testing to select it (FDA Human Drug GMP Notes 1993; various FDA Policy and Procedure Guides).

FDA inspectional authority is found in the following documents:

- FD&C Act, 704 (21 USC 374), 510(h)
- GMP (21 CFR 211, 600, and 802)
- GLP (21 CFR 58)
- Public Health Service Act, 351 [42 USC 262(c) and (d) (1)]

The administrative procedure for the establishment of regulations by the FDA involves printing of the proposed regulation in the *Federal Register*. A period of time is provided for comments. If drastic changes result from the comments, a revised proposal may be reissued. Otherwise, the FDA resolves the comments and issues a final regulation. At present, a final regulation must be signed by the Commissioner of Food and Drugs or by his or her designee. The process can be lengthy; it may take several years before the FDA formulates and approves a proposal internally, a year or so to solicit and resolve comments, and at least another year or more to secure the Commissioner's signature to establish the regulation.

As a field function of the regulatory process, FDA inspectors often visit pharmaceutical manufacturing facilities on general inspections to see whether compliance with current Good Manufacturing Practice (cGMP) is in place. Other occasions for a visit may be a follow-up to a cGMP inspection, in response to an New Drug Application (NDA) complaint, or to investigate the circumstances surrounding a product recall. The inspector will list his or her critical observations, if any, on a standard form, called an "FDA 483" for the number of the form. The official nomenclature for this form is a "notice of inspectional observations." A copy of the 483 is given by the inspector to the company individual most responsible for the referenced activity. This may be the plant production or quality manager. Thus, company management becomes directly and immediately involved. Generally, pressure is brought to bear at all levels of a company for rapid correction of a cited condition, regardless of whether it represents a serious violation or not. Indeed, the implication of a 483 observation is such that the manufacturer often simply acts to accommodate the inspector's judgment in order to avoid further involvement with the regulatory process.

The 483 typically receives a review for appropriate action at the FDA district headquarters. Several courses of action are possible. The FDA district or central headquarters may issue a warning letter, which indicates the significant GMP violations and asks the company to respond with corrective actions. The firms usually has 15 days to respond. Failure to respond will leave the FDA with the impression that the firm does not intend to correct the GMP violations. If the violations represent a health hazard or have been found in previous inspections, the FDA may request product seizure and/or injunction against the company. If the FDA can prove that the company intended to break the law, company officials may be prosecuted. U.S. federal marshals are sometimes called to enforce cessation of operations or seize "violative" product.

Importantly, each regulatory auditor has the courage of conviction to enforce the various regulations based on a knowledge of specific historical instances in which significant patient harm has been done or could be done were the regulatory agent not present to protect the patient. Notable among these are the following:

- 1906 (Purity)—Contaminated meat products from the Chicago stockyards.
- 1937 (Safety)—Sulfanilamide contaminated with antifreeze resulted in patient deaths.
- 1962 (Efficacy)—Thalidomide resulted in deformed babies in Europe.
- 1976 (Devices)—Dalkon Shield intrauterine device (IUD) proved harmful to patients.

FDA Concerns with Regard to Particulate Matter

The manufacturer will always be interested, understandably, in compliance as a philosophy. Philosophy is the basis of practice. In order to understand more about the FDA consideration of particulate matter, the manufacturer of devices or injectables should have some understanding of the magnitude of the task that faces regulatory and auditory personnel involved in enforcement activities. The system in the United States, including the U.S. Code of Federal Regulations (CFR), as well as numerous USP and FDA requirements, guidelines, and directives serves as an example. In general, enforcement activities are intended to ensure the purity, safety, and efficacy of medical products. In this context, both the CFR requirements for GMP and USP assay result specifications (e.g., the <788> particulate matter limits) are enforceable as law. In the reasonable and understandable perception of a regulatory auditor, these requirements relate to safety, purity, and efficacy, and their enforcement is one (most direct) means of ensuring product suitability.

As was discussed earlier, it is not possible to define a single perspective or philosophy for investigators with regard to enforcement activity relating to particulate matter contamination. Significantly less information is available in the compendial and regulatory literature directly relating to particulate matter than is the case for sterility assurance, pyrogens, facility suitability, and quality assurance. Much of the information on FDA perspectives is thus indirect and comes from discussions, presentations, and audit findings on subjects such as high efficiency particulate air (HEPA) filtration, airflow patterns, and personnel training and control. While some (few) inspectors appear not to address particulate matter data directly in inspections unless specific complaints have been received, others are very interested in particle count data collected from the product or the environment in routine testing. One perception on the part of many inspectors would seem to be that relative particulate matter levels provide a valuable indicator as to how well cGMP is observed in the company being inspected. Stated another way, this philosophy suggests that high product particle burdens, particularly if accompanied by high variability between batches, may indicate that other production areas may be worthy of investigation. This appears particularly true if customer complaints or customer perception are involved.

Inspectors are almost invariably conscious of the fact that particulate matter from extraneous sources gains access to the product via multiple routes. There also seems to be a realization on the part of some investigators that particulate matter control is a probabilistic procedure, with measures such as filtration (air or liquid) or washing being more likely to attain the highest levels of cleanliness if the particle burden to be removed or excluded (i.e., challenge) is minimized. While inspectors are aware that the number of viables does not relate in a specifically predictable fashion to nonviables, the route that allowed nonviable particles access to the product may also allow viables egress.

Control mechanisms should be designed and applied to minimize the likelihood of contamination. Even though the inside of an isolator is "sterile" and supplied with HEPA–filtered air, the principles of directional airflow should still be observed to achieve the lowest probability of particle access to vials being filled.

Thus, for better or worse, the contaminant particle content of a product is believed by both the majority of manufacturers and regulatory agencies to relate in some way to the overall "quality" of the product. Here again, one of the few nonsubjective definitions that we can assign to "quality" is the freedom from defects or degree of conformance to requirements. While the USP limits for particulate matter in solutions provides numerical limits for allowable numbers of particles in injectable products, there are no such well-defined limits for particles in or on other types of products. It is in these areas of pharmaceutical and medical device manufacture where there are no clearly defined requirements that regulatory issues and questions over whether or not a manufacturer is applying GMP most often occur. Notable examples of this are provided by the concerns in recent years over particles in the containers (packages, blister packs) in which medical devices are marketed, and over visible particles in syringes and injectable products.

While the apparent concern in some cases may be particulate matter as a process indicator (i.e., in line with the concept of statistical process control), actual perspectives of investigators would appear to go well beyond this consideration. In the author's experience, FDA training on visible particulate matter appears to be well structured, and visible particle data may be routinely viewed as a process or product quality indicator to a greater extent than subvisible particle data.

The inspector may have a rather complex expectation as far as what levels of particulate matter may reasonably be expected to be present in product, based on training, experience, and the literature, both historical (see, for example, Turco and Davis 1970) and current (Avis and Levchuk 1995).

In the various editions of *Remington's Pharmaceutical Sciences* (or "Remington's" as it is commonly called), a chapter historically authored by Dr. Kenneth Avis has dealt with the subject of "parenteral preparations." This text serves not only as a valuable reference for teaching in universities and colleges, but it is useful to the FDA as a source of general information on pharmaceutical topics. In the most recent (1995) edition of this text, Dr. Avis and coauthor Dr. John Levchuk discuss in some detail the different types of injections (i.e., solution products, dry powder dosage forms, suspensions, and emulsions) and the cGMP and good pharmacy practice–related "expectations" for these products. An example of an expectation for an injectable solution is the property of clarity. This property is also described in chapter <1> of USP 23 and preceding volumes. Visible particulate matter in a solution means that one of the required properties has not been attained, hence an interest on the part of inspectors in the visual inspection result. Such expectations are finished product attributes. In this consideration, particulate matter as a process indicator may be less important than its impact as an expectation or quality attribute.

Another judgment may be made by inspectors is based on the expectation for controlling particulate matter. 21 CFR Part 211, Subpart D, Section 211.72 (filters) and Sections 211.80–211.94 (control of components and drug product containers and closures) serves as the basis of a perspective or expectation on how the generation of particles by filters and container components should be controlled. The context in which or the standard against which inspectors may view product also critically involves the control measures that are in place to deny particle access to the products (e.g., HEPA filtration, directional airflow) or to control particles on components (stoppers, bottles, bags). This is to some extent understandable based on the effectiveness that process filtration controls subvisible particles.

At this point, then, we can come up with a several hypothetical perspectives regarding the implications of particulate matter contamination:

- Visible particulate matter represents a desirable property or quality expectation not attained in specific units in which it occurs.
- Particulate matter is a direct factor, indicating that a quality attribute is lacking.
- Particulate matter is a general process indicator.
- While the occurrence of particulate matter at some low level in product is unavoidable, the occurrence of miscellaneous particles at

levels in excess of compendial limits or particles due to degradation constitutes contamination.

On another note, the status of investigator training and expertise has evolved along with the technology of drug and device manufacture. Today's FDA field auditor is a trained professional who may have a broad general knowledge base in science, GMP, and enforcement. In audits, he or she will persistently attempt to determine whether production conditions comply with GMP and whether product uniformly meets USP requirements. The best auditors will display the same level of diligence and competence in investigation that the manufacturer devotes to product quality. Given the competence of today's inspector, the belief on the part of any manufacturer that systematic defects in compliance will be "overlooked" or "won't come up" in an inspection is often an error in logic.

One present FDA emphasis in both laboratory and manufacturing inspections that was not present in inspections during the 1980s appears to be the training of personnel in both GMP and GLP (Good Laboratory Practice) principles and in the interaction between the two (see Levchuk 1991). Audit results within the last three years reflect an increased scrutiny of this compliance component. This is understandable, in that the performance of a company (or "firm") will be critically dependent on how well the individual can perform his or her job, which is critically dependent on training and attitude.

Training quality and effectiveness, or, rather, a lack of these qualities, may be found in a number of recent observational reports. Not only that employees have received training but also that the training program has met its goals in terms of acquired competencies is being scrutinized. This emphasis has been evidenced in a range of areas, from visual inspection to the filling room floor. Training is obviously critical in areas such as aseptic processing, where personnel actions bear critically on the process outcome. With regard to the specific subject of this book, USP 23 <788> calls out the responsibility of the firm to ensure that "those performing the test" are adequately trained. The USP 23 <788> procedure is a particularly good example of the need for training, due to its relative complexity versus earlier versions (see Chapter 4). Review of training, evolution of training programs, and adequate training documentation are integral parts of the training program.

CURRENT GOOD MANUFACTURING PRACTICES

The FDA has developed the cGMPs with the intention of identifying and controlling key steps in pharmaceutical processing. These are generally accepted as defined by 21 CFR Parts 210 and 211. These regulations are supplemented from time to time by guidelines on specific processes published by the FDA. Found in the cGMP regulations are the minimum practices and methods to be used in, and the facilities and controls to be used for, the manufacturing, holding, processing, or packaging of a drug to ensure that the requirements for purity, safety, and efficacy claimed by the maker are met. By means of the cGMPs (often abbreviated as GMPs), it becomes possible to validate that equipment, components, and containers do indeed function as they are intended to. The result, as stated, is a shared reliance on building quality into the product, rather than a dependence on final product testing alone. Required, in the words of the FDA, is

> The attaining and documenting of sufficient evidence to give reasonable assurance, given the current state of science and the art of drug manufacturing, that the process under consideration does or will do what it purports to do or is expected to do.

The so-called umbrella GMPs, finalized in 1978, are broad guidelines covering many processing aspects. Subsequently, more pointed GMPs were proposed for large-volume parenterals (LVPs); their consideration was urged for small-volume parenterals (SVPs) as well. (Note: The current term is "injectables" rather than "parenterals.") Detailed GMP compliance programs for both types of parenterals were put forth by the FDA in early 1983; the proposed GMPs for LVPs were never finalized and were eventually withdrawn. Nevertheless, their strictures are often used by FDA inspectors as the basis for critical reviews of process practices. Indeed, the willingness on the part of pharmaceutical manufacturers to conform to the tenets of the GMPs for LVPs served to establish their applicability and to extend it to SVPs as well. However, with regard to filtration practices, the relevant GMPs elicit their authority not from the GMPs for LVPs, but rather from their derivation under the equipment section of the umbrella GMPs. One way or another, the proposed GMPs for LVPs do serve as guidelines for FDA inspectors.

The umbrella GMPs contain two definitions pertinent to particulate matter. A fiber is defined as a particle having an aspect ratio of 3:1 or greater; a nonfiber-releasing filter is described as one that does not release fibers after a flushing. Particles whose lengths do not exceed their widths by multiples of three or more are not called fibers, but the umbrella GMPs suggest that particles be monitored along with organisms, and the proposed GMPs for LVPs established acceptable particle count levels. The umbrella GMPs stipulate that nonfiber-releasing filters be used in liquid filtrations. This provision relates in principle, back to 1975, when asbestos-containing filters were still in use. If a fiber-releasing filter is required in a process, its use necessitates a 0.2 μm-rated membrane filter downstream to remove the fibers released from the filter. Special attention was shown to asbestos fibers. The continuing use of asbestos-containing filters was permitted only on a temporary basis, for 30-day periods, upon petition, subject to a demonstration of earnest endeavors to replace them. Consequently, the use of asbestos in pharmaceutical manufacturing had virtually ceased within 1 year (21 CFR, 211.40, 1978).

The GMPs proposed for LVIs were more specific than the umbrella. They required initial and final integrity testing of filters, although the type of test was not specified. They also stipulated the use of hydrophobic vent filters for tanks that contain processing solutions, such as Water for Injection. Filters were required on compressed air lines and on air lines to sterilizers. The intent of these GMPs was clearly the filtrative control of all fluids constituting or in contact with the pharmaceutical product during processing. Process validation as indicated was required as well. With regard to LVIs, the proposed GMPs specified the use of 0.45 μm-rated filters to reduce both the particle count and the bioburden prior to terminal sterilization of the product. Mixing, filtering, filling, and sterilization were to be completed within an 8 h period. In the actual manufacturing situation, many LVI manufacturers are validating or have validated 12 h and 16 h durations for their operations, and find FDA acceptance based on adequate validation.

Since the purpose of GMP is, in part, to reduce particles in pharmaceutical preparations, it follows that the environment wherein the operation takes place should be as clean as possible. USP 23 (1995) describes the physical requirements of the aseptic processing area. Central to its features is the maintenance of virtual microbiological barriers. The imposition of differential positive air pressure serves to prevent the inward leakage of air; the air supplied to the aseptic areas is sterilized by HEPA filtration; air locks are provided at the entrances to rooms. These items are certain to receive scrutiny in inspections. As shown in Table 6.1, the GMP concepts relate both to devices and drugs, with separate GMP regulations in the CFR.

Some of the requirements for drug particulate matter testing relate to stability studies conducted by the manufacturer to comply with the FDA Stability Guidelines (Table 6.2). Stability study data receive attention both in GMP audits and in NDA submission review. The GMPs, or, in some cases, the guidelines of the International Conference on Harmonisation (ICH) and the FDA 1993 Draft Guideline Covering Stability Testing Requirements for New Dosage Forms and for Variations and Changes to New Drug Substances and New Drug Products.

CURRENT REGULATORY INTERESTS—DRUGS AND INJECTABLE PRODUCTS

Based on audit findings, publications and presentations in local and national meetings of manufacturer's organizations such as the Parenteral Drug Association (PDA), FDA perceptions of particulate matter contamination are closely similar to those of the pharmaceutical industry. In drug manufacture, contamination is an integrated effect dependent on the severity or level of product exposure, time, product susceptibility, and particle concentration; the crux of contamination control is to manage these critical factors. Product protection is afforded by minimizing each factor and, at the same time,

Table 6.1. Overview of GMP Concepts

Element	Drug	Device
Quality Policy	O	X
Responsibility	X	X
Resources	X	X
Management Review	O	X
Design Control	O	X
Document Control	X	X
Purchasing	O	X
Traceability	X	X
Process Control	X	X
Special Processes	X	X
Inspection	X	X
Laboratory Control	X	X
Corrective Action	O	X
Distribution	X	X
Records	X	X
Quality Audits	O	X
Annual Reviews	X	O
Training	X	X
Service	O	X
Statistics	O	X

X = included; O = not included

Table 6.2. Stability Study Requirements (explicit and implicit) That Involve Particulate Matter Analysis*

Requirement	Reference
SVP's should be evaluated at least for strength, appearance, color, particulate matter, pH, sterility, and pyrogenicity (at reasonable intervals)	SG III.B.6.h GMP 211.167.a
For LVP's a minimum study should include: strength, appearance, color, clarity, particulate matter (USP or equivalent), pH, volume (plastic containers), extractables (plastic containers), sterility and pyrogenicity (at reasonable intervals).	SG III.B.6.i GMP 211.167.a
Testing should cover not only chemical and biological stability but also loss of preservative, physical properties and microbiological attributes.	ICH DP 48756.TP
Stability data includes interaction between drug and container-closure, leachables into drug formulations during storage should be assessed by sensitive procedures.	SG III.B.1 GMP 211.166.a.4
Sample times should be chosen so that any degradation can be adequately characterized, sufficient to establish stability characteristics of the drug product.	SG III.C.1.c ICH DP 48757.TF
For significant changes of products known to be relatively unstable, 6 months' data at the normal recommended storage temperature, as well as the data from accelerated conditions, may also be required.	SG V.D.1
Parenterals (except ampoules) should be stored inverted or on side to determine whether contact of drug product or solvent with the closure system affects product integrity or results in leaching from the closure.	SG III.B.6.h, i

*Excerpted from *Draft Guidelines Covering Stability Testing Requirements for New Dosage Forms and for Variations and Changes to New Drug Substances and New Drug Products.*

protecting the product, process, or function against them. In the production of sterile injectables, the highest levels of FDA and manufacturer attention are being directed toward minimizing the risk of airborne microbiological contamination. The present impression of both regulators and manufacturers is that the implementation of various types of manufacturing constraints and local controls will be far more cost-effective than enormous expenditures for Class 10 or better facilities when one wants to improve the sterility assurance level for aseptically manufactured products.

In FDA and industry concepts, particle and microbial control are based on a barrier concept, whereby a virtual barrier (e.g., unidirectional airflow) or an "absolute" physical barrier (e.g., a mini-environment or other enclosure) is used to protect product or prevent dispersal of a potent material. Barrier mechanisms include the following:

- Air filtration
- Directional airflow
- Pressure differentials

- Access control
- Material control
- Area classifications
- Physical enclosures

The FDA also appears to expect the principle of process control, which is based on the elimination or removal of particles originating from specific steps in the manufacturing process, to be applied. Some proactive means of doing this include process design, robotic procedures, the use of "clean" product components, and assembly steps designed for low particle generation. Product components include widely diverse materials, for example, the tubing of solution administration sets, bulk drug raw material powders, and stoppers. Factors in particulate matter exclusion that have received scrutiny include vendor certification audits, the monitoring of component particle levels, washing validation, filter qualification, and container-closure selection.

Personnel and their activities are universally regarded as the most significant source of viable and nonviable particles in controlled environments. Elements of the control of particulate contaminants from this source include education regarding the sources of particles, training in appropriate procedures for particle elimination and control, protective garment selection, aseptic core area access control, and, most importantly, monitoring of particles generated by individuals. The only process step in which washing is almost invariably effective in particle removal is for glass containers. The washing of closures is less effective and generally more difficult to conduct effectively. The greatest problem with stoppers in this regard may be due to mechanical effects—friction, sloughing during the washing process or handling. Particles removed by washing may be replaced by others added in drying or depyrogenation ovens. With regard to plastic device components, clean molding followed by clean handling and packaging is far more cost-effective than washing. The degradation of drug formulations to form particulate matter during sterilization is related to the manufacturing operation; longer-term physical instability during the shelf life is a "stability" issue.

With regard to the mechanisms of process control, statistical process control (SPC) is a widespread industry practice; this includes the trending of data related to product release, such as particle counts. In this context, SPC constitutes a structured method by which particle monitoring data can be used to assess and increase process capability with regard to the control of particulate contamination. The assessment of whether a process is "in control" will always be made during inspections; a 483 observation relative to an uncontrolled process is very serious. The use of clean zones in higher classifications has been given attention to show how well cleanliness is maintained. The use of physical barrier enclosures around sensitive process areas (e.g., aseptic filling lines) is a cost-effective method for achieving a high level of control over microbes and particulate contaminants. The validation of these devices is required.

Perennial key points of cGMP that can impact particulate matter control and may receive audit scrutiny during investigations of drug products are as follows:

- Control and/or classification of manufacturing areas
- Positive pressure environments (airflow from clean to less clean areas)
- Airflow patterns and airflow velocity (particularly for aseptic filling)
- Filtered air (HEPA or other)
- Particle count monitoring
- Corrective action program
- Suitable premises, equipment, and materials
- Trained personnel
- Adequate transport and storage

Cleaning processes used by a manufacturer impact bioburden, chemical residues, and particulate matter. Equipment cleaning procedures are being reviewed more closely by FDA investigators for proof that they have been adequately validated and are being applied in practice. The FDA allows considerable latitude for firms to determine which sampling procedures, analytical methods, and specification limits are appropriate for specific equipment and products. However, FDA auditors will expect to see evidence that the choices are scientifically sound. Both bulk and dosage form manufacturers are under pressure from FDA investigators to produce proof that their cleaning procedures work effectively and are being applied in practice.

The FDA wants to see written procedures, i.e., protocols, describing how cleaning processes were validated.

The containment of drug powders generally involves the protection of personnel from particles generated by a manufacturing process, rather than protecting a process from heterogeneously sourced contaminants. Most often, the concern is to protect personnel from aerosolized drug powder particles that are toxic (e.g., chemotherapeutic agents) or potential sensitizing agents (e.g., penicillin, cephalosporins). The mixing of two types of drugs is also a concern. Containment measures should be well documented and monitored. Cross-contamination by chemicals or particles is not permitted under any circumstances. However, the fact that a plant is, or can be, used for manufacturing multiple drugs, even simultaneously, is not in itself objectionable, with only a few exceptions. The cGMPs require that there be separate facilities and/or completely separate air handling systems equipment and facilities within a building for the production of penicillin-type drug products. The FDA also encourages that separate facilities and air handling systems be used for the production of potent steroids, alkaloids, cytotoxics, and/or cephalosporins.*

BPCs (bulk pharmaceutical chemicals or bulk drug raw materials) are currently coming under a higher level of inspectional scrutiny than in previous years; much of the material used in the United States to manufacture sterile drug products is produced abroad, which calls for an increased level of inspectional effort and enhanced vigilance on the part of the FDA. The point at which the final BPC product is initially recovered (usually as a moist cake from a centrifuge or filter press) should be in a clean environment that is not exposed to airborne contaminants such as dust from other drugs or industrial chemicals. Typically, the damp product will be unloaded into clean, covered containers and transported elsewhere for drying and other manipulations. These subsequent operations should be performed in separate areas because, once dry, the BPC is more likely to contaminate its environment; this in turn makes it likely that other products in the same area might become contaminated. The primary consideration is that the building and facilities should not contribute to an actual or potential contamination of the BPC.

The USP defines an impurity as any component of a drug substance (excluding water) that is not the chemical entity defined as the drug substance. Impurities may exist in the form of soluble chemical compounds, extraneous contaminants, or insoluble drug-related particulate matter.

From the perspective of both FDA inspectors and review chemists, impurities in drug substances present a series of problems that an applicant should have extensively addressed in the NDA submission: finding, quantitating, isolating, and identifying impurities. This sequence leads to proposals for impurity limits by an NDA applicant, and FDA decisions on the specifications—both methods and limits—for impurities in the drug substance. FDA decisions on limits are reached jointly with the pharmacology reviewer and may involve the medical review staff as well. Impurities may be present as particles as well as in a soluble chemical form.

A specific example of particulate impurities is provided by the presence of amorphous particles. These materials are specifically called out in <788> as showing an indistinct, gelatinous, or semisolid appearance. In a bulk drug material, these materials may represent an impurity such as a low polymer of the drug indicating that degradation has occurred due to the presence of moisture or residual solvent. Worse yet, they may be indicative of ongoing degradation that might continue after filling or mixing of the final product. This material may be detected by microscopy at levels below the sensitivity of chemical analytical technique. Hence, the microscopic assay can serve as an important descriptor of drug quality with regard to both degradants and extraneous particles. In the latter case, the presence of large numbers of extraneous contaminants may lead to questions regarding the chemical purity of the drug. There is also the possibility of impurities in particulate form that are chemical variants of the drug substance (e.g., enantiomers, diastereoisomers, geometric isomers, etc.).

Based on the inspection of BPC facilities, current FDA bulk drug expectations with regard to particulate matter would seem to be increased over those in this area in the past. Auditor questions posed to users of BPCs (i.e., manufacturers of the final dosage form) have included the following:

*See proposed GMP revisions, 3 May 1996.

- Are particle test results available?
- Have degradants and amorphous particulate matter been quantified?
- What levels of contaminant particles are present?
- What is the composition of the extraneous particle population?
- Is the particulate matter level monitored as a component of vendor qualification?
- Does the particle burden indicate non-GMP conditions at the vendor?

Based on these considerations, the bulk drug used in developing the new dosage forms or products should be tested microscopically at some point prior to signing the contract and "locking in" a vendor. A vendor should not be approved without a favorable outcome in this testing. It is desirable to determine both the levels of degradant particles and extraneous particles. The former provide an indication of process quality with regard to drug purity; the latter give a indication of process cleanliness. Either type of particle can become the subject of FDA scrutiny. High levels of particles can cause blinding of production filters, and amorphous particles may indicate that degradation pathways are in place, which could result in field issues later on.

CURRENT REGULATORY INTERESTS—DEVICES

In the *Federal Register* of October 7, 1996 (Volume 61, Number 195) and in 21 CFR Parts 808, 812, and 820, the FDA published the "Final Rule, Medical Devices; Current Good Manufacturing Practice (cGMP)". The document bore the following introduction:

> SUMMARY: The Food and Drug Administration (FDA) is revising the current good manufacturing practice (cGMP) requirements for medical devices and incorporating them into a quality system regulation. The quality system regulation includes requirements related to the methods used in, and the facilities and controls used for, designing, manufacturing, packaging, labeling, storing, installing, and servicing of medical devices intended for human use. This action is necessary to add preproduction design controls and to achieve consistency with quality system requirements worldwide. This regulation sets forth the framework for device manufacturers to follow and gives them greater flexibility in achieving quality requirements.

This document is detailed and comprehensive; it addresses "contamination" in the manufacture of devices of different types and will doubtless serve as a broad basis for interpretive enforcement activity for many years in the future. In the light of this most recent directive on the application of cGMP to device manufacture, it is interesting to review the results of recent FDA inspections of device manufacturing facilities. They provide some additional indication of the concerns of the FDA and its thinking regarding device particulate issues. First of all, there seems to be a general acceptance on the part of investigators that subvisible particulate matter is not a serious concern in devices. There is, however, considerable regulatory interest in the occurrence of visible particles both within the solution contact path of devices and in the packing of administration sets and other items.

It seems likely that the FDA's general perception regarding particulate matter in devices is not dramatically different from that of most device manufacturers. Broadly speaking, the role of the investigator who visits a device manufacturing plant is to ensure that devices are made, processed, stored, and handled according to approved procedures, and that GMP is followed by the manufacturer. While particulate matter (particularly of visible size) may be used as a "first pass" indicator of difficulties with injectables, its most significant impact point for the investigator is that if process control is lacking, GMP may have been violated. The occurrence of particulate matter may provide the investigator with clues that lead to specific process difficulties.

Specific examples of this approach are provided by the occurrence of material entirely foreign to the process within the package of a device or in the device itself. The occurrence of human hairs or fibers from street clothing (e.g., wool) implicates gowning practices. The presence of insect parts or plant material suggests that air classification between clean and less clean production areas has not been maintained, or that particulate matter from areas outside the plant has not been denied access. The presence of animal hair may point to inappropriate laundering of smocks or protective garments. Fragments of packaging material may indicate inadequate housekeeping at critical process points. All of these materials provide indications of potential process difficulties that may be the basis of inspectional observations or regulatory actions.

There have been a number of indications that the perception of FDA management in regard to particulate matter also bears key similarities to that of device manufacturers. The FDA realizes that there is no "magic number" with regard to levels of particles in or on devices or within packaging. The occurrence of particulate matter must be judged on a case-by-case basis with

- the use of the device,
- the location of particles,
- the nature of production process,
- the amount of particulate matter present,
- the type of particulate matter occurring, and
- GMP implications of particulate matter.

All of the above need to be considered by the manufacturer. There is an apparent realization on the part of auditors that the importance of particulate matter differs dramatically depending on the use of a device (i.e., an implantable device, a blood contact device, or a device that enters a sterile field is likely to be judged far more critically than a simple disposable solution administration set). FDA policy in this regard appears sufficiently flexible so that enforcement action may be taken at a low level of particulate matter for one type of device, while another type of device with a higher level of inherent particulate matter burden (such as an autologous transfusion device) may be deemed acceptable with significantly higher levels of particles.

Some recent inspection results reflect a high level of scrutiny of visible particulate matter, with the emphasis placed on the presence of particles in packaging and on the external surface of device equal to or greater than that placed on particles in the devices themselves. Not only are devices such as administration sets being closely scrutinized, but also devices such as heat exchangers, cooling jackets for surgical laser systems, procedure trays, nebulizers, and the like. The basic rationale involved in this enhanced concern over visible particles seems to be the investigators' perception that the presence of visible particles in device packaging is reflective of undesirably high levels of process-related particles, which in turn indicates non-GMP compliant manufacturing conditions.

Attendant to perceived problems with visible particulate matter, investigators have been demonstrating a new level of initiative in inspecting device manufacturing facilities. Conditions of manufacture, judged acceptable on previous inspections, have during later inspections been judged unacceptable. Investigators have, on some occasions, used magnifying glasses in inspections. There has been some suggestion by investigators that the presence of any visible particles (either within a device or its package) constitutes grounds for rejection of the product (i.e., that an unwritten requirement for zero visible particles is being enforced). The terms *adulterated* or *contaminated* have been applied to product when visible particles have been detected. This increased scrutiny of particulate matter is no doubt at least partly the result of the Safe Medical Devices Act of 1996, with its more stringent requirements for regulation, reporting, and documentation related to devices.

The issue of particles in packaging materials, such as blister packs or pliable film coatings, represents another problem for the pharmaceutical industry. The presence of a single visible particle exterior to the solution path of a device represents no risk to a patient. The difficulty of the total elimination of particles from packaging materials is so great as to represent an impossibility for any production method that will be economically practical. Thus, the cost versus benefits issue is raised once again. Importantly, the applications of GMP to the packaging of devices is far less specific than with the actual manufacture of devices themselves. A high level of technical refinement is often used in the creation of a package that is a durable sterile barrier for a device and applied by automated machinery. Total particle elimination from high-speed packaging operations represents an extremely difficult task. High-speed packaging machinery is extremely cost-effective, but it is almost impossible to protect against static electricity and the consequent attraction and adherence of small numbers of visibly sized particles. Particle control in this regard is critically dependent on the control of point-generated particles in cutting, drawing, and sealing operations, and in the control of static electricity through humidity, ionized air, and other measures.

The FDA appears to realize that the manufacturer is faced with complex issues in controlling the occurrence of particulate matter in devices. The detection of visible particles in devices and their packaging materials is subject to an even higher number of critical variables than is the case with injectable products. It is, in every sense of the word, impossible to create a uniform requirement for the visual inspection of devices due to the wide variation in physical construction, size, and application. It is quite conceivable that a complex device with a high solution path particle burden might easily pass a visual inspection due to the opacity of its components, while a simple device that could be more critically inspected because of its transparency would be deemed "unsuitable" based on the detection of a single visible particle. While the application of GMP in device manufacture results in extremely low levels of visible particles, the manufacture of devices with no visible particles is beyond the scope of GMP and beyond the capabilities of contemporary manufacturing processes.

Despite the considerable difficulties, any manufacturer of medical devices is well advised to have a monitoring plan for visible particulate matter in place. This plan should include sampling methods, test methods, and accept-reject criteria based on appropriate attribute testing standards (e.g., MIL-STD 105-E). The concept of cGMP is extremely important and is a matter that manufacturing organizations must take seriously. An investigator who sees a high level of monitoring of critical and environmental control at Plant A on Monday will be favorably impressed; this favorable impression will rapidly give way to dissatisfaction during a visit to Plant B on Tuesday, if they do not have similar controls in place. Thus, consistency between processes and inspection measures and industry standards regarding particulate control will be critical with regard to the outcome of inspections.

Procedure or specialty trays, which contain all of the devices and related articles necessary to perform a specific procedure in a single "kit," have recently come under regulatory scrutiny. Many contain injectable products that have variable expiry dating and storage conditions. An example would be a tray containing the articles necessary for a spinal tap procedure. This would include anesthetic, disinfectant sponges, syringe(s), cannulae, a sample collection vial, gauze pads, disposable forceps, and other items. Particulate matter levels of the drug and solution contact articles is not the only concern here; levels in the kit packaging material, i.e., blister pack, are also critical.

Despite the absence of USP requirements for particulate matter in devices, auditors have found ample grounds for enforcement activities based simply on whether or not device production adheres to cGMP. Some issues that have arisen are the following:

- Failure to monitor particulate matter levels of finished product
- Failure to establish alert and action limits for particle counts resulting from monitoring
- Failure to maintain environmental controls on particulate matter
- Inadequate housekeeping procedures
- Failure to monitor and trend airborne particulate matter
- Presence of cardboard in production areas
- Lack of directional air movement
- Lack of specific dress codes and hygiene procedures
- Failure to implement and maintain GMP–required training procedures

Section 820.46, Environmental Control, of the Device GMP regulations is considered by the FDA to be a "discretionary" requirement; that is, the degree of environmental control to be maintained must be consistent with the intended use of the device. Details of how to achieve this control are left to the manufacturer to decide.

ENVIRONMENTAL MONITORING

FDA perceptions and procedures relating to airborne particulate matter are closely tied to those for microbes. In fact, with regard to aseptic procedures, particle counts seem to be frequently used as a readily obtainable "handle" on the potential microbial levels in a room or process area. Given the criticality of cleanliness in aseptic fill processes, this is probably as it should be.

This section is based on or excerpted from the 1993 presentation made by then FDA

International Parenteral Drug Specialist Ronald Tetzlaff. In this position, Tetzlaff devoted considerable attention to developing computer-aided techniques for auditing microbiological environmental programs. During the past decade, there has been an increased awareness of the importance of particulate matter monitoring as an adjunct to microbiological environmental monitoring in parenteral drug processes. This heightened awareness has been reflected in increased investigational scrutiny and investigator interest in environmental monitoring and airborne particle counts.

The FDA position with regard to monitoring airborne particulate matter is generally straightforward and allows considerable latitude of interpretation. The identification of acceptable test frequencies and suitable quality levels for environmental or airborne particulate matter is the responsibility of each manufacturer. There are various monitoring methods adequate to demonstrate that equipment, facilities, and personnel are maintained within acceptable airborne particulate matter quality levels. The objectives of such a program include establishing at least the following:

- Identifying quality level requirements in each manufacturing environment.
- Establishing suitable test limits/frequencies.
- Conducting routine monitoring of air quality to determine the need for filter testing and maintenance; heating, ventilation, and air-conditioning (HVAC) adjustments; and so on in each area.
- Scheduling periodic evaluations to assess the overall effectiveness in maintaining suitable conditions of cleanliness (i.e., to determine the need for changes in manufacturing and quality assurance procedures).

Beyond the general language contained in the GMPs and the FDA aseptic guideline, there are few "official" documents that provide specific FDA requirements for air quality in terms of particle counts. This overall lack of specific regulatory requirements for airborne monitoring is undoubtedly one of the reasons that controversies have persisted. While the broad language in the CFR intentionally permits and encourages flexibility, firms need to be careful not to misinterpret FDA expectations. The FDA expects firms to use sound scientific principles for establishing monitoring programs (i.e., appropriate for their operations and consistent with current best demonstrated industry practices).

FDA personnel have for years recognized the importance of environmental testing, and FDA investigators frequently find objectionable conditions while reviewing environmental monitoring programs. Most investigators spend a significant amount of time in this area during inspections, devoting additional attention to known or suspected problems. Some FDA investigators and administrators have published papers describing inspection techniques or policy issues about environmental monitoring. Notable examples are listed in the references.

The FDA, as policy, does not give information on how to construct specific types of device assembly areas. FDA expectations, however, for environmental cleanliness in a given process are critical to how a process is developed and how Premarket Approval (PMA) applications are written and reviewed. An FDA inspector will evaluate the physical layout and process description based on the PMA, a facility tour, blueprints, knowledge of applicable standards, and best demonstrated practice and cGMP. The combination of these considerations will lead to judgments based both on objective and subjective processes and is interesting to review. The inspectors are individuals, and some (generally minor) differences in opinion are to be expected.

Neither the cGMPs nor the FDA will specify how a controlled environment area (CEA) must be built for a specific product. While some concise instruction is available for aseptically filled product, terminally sterilized product and devices are less well covered. Standards and practices have been, in most cases, written to allow considerable latitude on the part of the user. The user is allowed to determine the most efficient and cost-effective ways to achieve the desired operational characteristics and product protection. The basic principle that the inspector uses as a yardstick is whether adequate protection of the product is attained. What is "adequate" has to be interpreted and defined by the user of the CEA in question. A key consideration regarding adequacy of protection seems to be that the FDA expects a generally even maintenance of class conditions over the entire work area in the room. The FDA will expect all readings within a room to be within the specified CEA class.

Inspectors well know that uniform conditions of air cleanliness in a room are achieved most readily with a vertical flow design, whereby clean air is furnished from ceiling-mounted HEPA filters to the level of the bench or workstation without obstructions. Horizontal airflow is disrupted by the first obstruction that it encounters; for the remainder of its pass through the room, it becomes subject to eddies, vertical flow, turbulence, and particle buildup. Since this design is somewhat dated and inferior in function to a vertical flow system, it may be considered as not representative of best demonstrated practice or FDA/cGMP expectations.

The FDA appears to audit new CEAs with a reasonable expectation of seeing current technology and best demonstrated practice applied. Investigators are invariably conscious of the effects of overpressures and other factors. Some manufacturers have been cited because videotapes purporting to show proper airflow patterns were not convincing, and because unidirectional airflow did not reach the work surfaces. Remember that airflow from a general room environment can intrude into the Class 100 hoods via vortices around the operators. In a horizontal flow room, this intruding air will be less clean than if vertical flow is used. FDA auditors will often look for unidirectional air protection around clean workstations. If the room has horizontal flow, there will always be turbulent (nonunidirectional) flow as the distance from the filter wall increases; thus workstations further from the wall are less well protected.

USP general chapter <1074> deals with good pharmacy practice in regard to compounding operations, and the ISO document addresses the construction of cleanrooms and controlled environments. In <1074>, the necessity of protecting Class 100 laminar flow workstations from less controlled (i.e., less clean operations) to reduce the risks of contaminants being dragged, blown, or otherwise introduced into the clean workstations is discussed.

In a PMA inspection, investigators can be expected to look carefully for the application of best demonstrated practice and adherence to FDA guidelines for aseptic processing. Engineering drawings and HVAC specifications may be requested. The number of air changes per hour, recovery capability, and pressure differentials between rooms will be questioned. An interesting circumstance that has turned up occasionally on FDA regulatory audits involves a manufacturer who has two in-house processes, each with different criteria for air cleanliness, physical and virtual barriers, personnel and industrial controls, and so on for making the same product. This can be expected to be the subject of FDA scrutiny and rescrutiny. This situation is most likely to become a focus when a manufacturer is modernizing or expands manufacturing facilities or if there is a product problem that appears to be tied to the old process. Questions may also arise when an older process area is repaired rather than being rebuilt to match current practices. On the other side of the coin, one of the FDA rationales here might be that the old process is capable of turning out acceptable product today just as when it was new.

FDA investigators may use a wide variety of inspection approaches and strategies to evaluate environmental monitoring programs. Tetzlaff (1993) described the use of a portable computer (and software applications) to assess the extent of GMP compliance, giving particular emphasis to areas where some firms continue to experience difficulties in satisfying FDA expectations. Key to internal audits of such systems (and FDA inspections) is a methodical detection of trends and patterns. Environmental monitoring programs are especially suited for computer-aided inspection techniques. Large amounts of data are continually being collected. Firms are expected to have systems in place to recognize and detect problems in a timely and systematic fashion.

A principal objective of FDA inspections of environmental monitoring programs is to determine if patterns or trends may exist that may adversely affect quality assurance decisions. Computer-aided inspection techniques permit the detection of problems that may not be readily apparent from manually reviewing individual test records. One objective of computerized audits is to extract from the firm's data those portions that may reflect objectionable patterns or trends (problems). The extracted information is then used as the basis for response or corrective action.

A most important consideration is that systematic assessments of the facility and environmental conditions be used to predict likely deficiencies in environmental monitoring programs. This method initially involves a careful, systematic, visual inspection (tour) of the manufacturing areas and controlled environments for the purpose of assessing maintenance and housekeeping (negatives are inadequate cleaning, poorly designed equipment, or areas that are relatively inaccessible for monitoring.) Such a tour may be relatively limited (a few minutes to a few hours) or may be more extensive (sometimes requiring a day or more to complete), but it is extremely important to the outcome of the inspection. The appearance of the premises should reflect uniform and continuous cleanliness and monitoring procedures that are effective in verifying this state of affairs.

It is not unusual for the investigator(s) to request a schematic of the facility and its services. By marking up this diagram, it is possible for the investigator to "organize" the firm's operations into a logical format that is readily used in evaluating a complex process and facility. This approach simplifies complex interrelated systems by combining operations into logical groupings that the auditor can effectively and efficiently manage. An important benefit is the initial tour to note conditions and practices that seem likely to have an impact on environmental quality.

As the tour of the facility proceeds, the blueprint is used as guide to ensure that all important areas are visited, which allows the auditor to keep his or her perspective even if there are unexpected "diversions" (such as investigating interesting issues found during the preliminary tour). Examples of conditions that may be observed during a tour that may elicit a more detailed review include the following:

- Do product and product contact surfaces appear to be maintained under control necessary for aseptic processes?
- Are visible residues present on surfaces that do not appear to be designed for effective cleaning or sanitization or on noncritical surfaces?
- In the area of personnel activities, such factors as gowning practices, improper personnel actions, or excessive or inappropriate entry to aseptic or controlled environments. (By watching the employees over a short time period, the auditor will be able to judge the degree of environmental particulate matter control practiced by the employees.)
- The auditor will be interested in employee actions or activities that are not reflected by normal sampling and/or validation programs. For example, employees may be observed doing tasks, functions, or controls that were not done during validation but are routine components of the process.
- Evidence of inactivity in rooms within controlled or critical areas may also be a "hot button." If the inactive rooms are in close proximity or connected to active areas or have been inactive for significant periods of time, they may adversely affect air quality in the adjacent operating clean areas.

Most auditors make a systematic assessment of environmental monitoring frequency and limits for each manufacturing area. They will also review available Standard Operating Procedures (SOPs) for content and identify specific areas for which limits would be expected and then verify that they exist and/or are suitable. Based on the intended use of each area, a preliminary assessment is made to determine if specifications (frequencies and limits) seem to be "reasonable." Particular attention is given to verifying that limits have, in fact, been defined for each selected area and of their suitability.

The investigator will also assess how well the firm's management has reviewed the monitoring data. If the auditor finds that the firm has well-documented reviews that seem to be reasonable in scope and content, then no further scrutiny may be given. Many investigators will review the adequacy of documentation required by GMP regulations and subjectively assess how well responsible personnel are familiar with their own data. Part 211.108(e) of the GMP regulations requires that periodic evaluations be made to determine the need for changes in drug product specifications or manufacturing or quality control procedures. A firm's periodic evaluation reports show the type of reviews and the degree (depth) of evaluations made by responsible management.

The overall assessment of whether to go into further detail in a given area of interest will depend on the answers to the following questions:

- Do periodic evaluation reports exists?
- Have responsible management reviewed/evaluated and signed-off the reports?
- Do documentation systems exist for demonstrating that problems/errors have been followed-up, investigated, and appropriate corrective action taken?
- Are routine written evaluations performed at a frequency that seems reasonable?

- Do the reports reflect both "good" and "bad" data? For example, do data look "too good," "too often"?
- Do the reactions seem reasonable when "bad" data (out-of-limit) occur? (e.g., How well have they been evaluated and do corrective actions seem appropriate?)
- Do annual reviews of the process reflect an adequate detail of scrutiny or merely a perfunctory overview that confirms the "status quo"?

Based on the answers to these questions, the investigator will be able to determine whether specific areas of the plant/process are "target-rich environments" for observations or whether additional time is unlikely to result in any significant observations. If the firm's reports appear realistic and corrective actions seem reasonable, it may be decided that further scrutiny is not warranted.

Veteran auditors (some trained by Ron Tetzlaff) will systematically examine the firm's recordkeeping system(s) for environmental monitoring data collection, storage, and retrieval. This review has several important purposes:

- To determine record medium (electronic format vs. manual hard copy paper)
- To observe the record style and format (graphs, charts, databases, tables, etc.)
- To identify record organization (such as by room, test, location, date or other)
- To determine storage/retrieval capabilities (records on-site vs. remote locations)

The auditor's assessment in most cases (as with drug or device audits) will include a judgment of how rapidly the firm is able to provide requested records. This line of investigation may be subtle or more obvious. FDA investigators realize that the amount of time needed will vary with each request. However, excessive delays frequently have reasons that are related to GMP compliance issues. For example, firms that are unable to readily retrieve records for the FDA may very well not have routinely retrieved them for their own management reviews or internal audits. Firms that quickly provide records (even if the data are questionable) may be viewed far more favorably than those who seem to withhold information.

Investigators have historically been interested in outlier airborne particle count data for two important reasons. First, abnormally high (or low) values may be indicative of unusual conditions or problems, especially if there is a prevalence of such data. For example, monitoring data showing excessive counts may signal the need for follow-up (such as retesting, corrective action, or investigations). FDA investigators would want to make certain that the firm had documented appropriate correction of the problem(s) and taken steps to prevent recurrence. Another objective in identifying outlier data is to determine if unusual conditions have been detected and properly reacted to by responsible management. An effective audit technique is to pick one or more outlier values (exceeded limits) from a test report. Using the out-of-limit sample as an actual point of reference, the audit may follow the firm's documentation for this test failure. Missing data may become a rapid inspection focus.

Missing data that receive attention are those that are expected or required to be available but are not. For example, SOPs may specify samples to be collected at a certain frequency, but for some reason sample reports cannot be located (e.g., samples were not collected, test reports were not prepared, laboratory accidents, lost records, etc.).

VISUAL INSPECTION

Based on the author's level of familiarity with the subject, FDA training on visible particulate matter is well structured, and visible particle data may be reviewed as a process or product quality indicator to a greater extent than subvisible particle data. Visual inspection of product will indicate some low number of units manufactured with a GMP compliant process to contain visible particles. The inspection process and the reduction of data from the visual inspection may be considered by an investigator to have specific implications regarding process control and product quality. If product cannot be 100 percent inspected (e.g., powder or lyophilized material, solution in brown bottles or translucent plastic containers), the manufacturer is well advised to have a valid sampling plan for testing sample groups by destructive analysis, such as the USP microscopic test or visual inspection of reconstituted solutions in glass containers.

For visual inspection, both the defect rate observed and actions taken are important. One of the most important procedures considered by some inspectors is an audit or reinspection of previously inspected units (for example, 100 units from a batch of 5,000) after rejected units from the initial 100 percent inspection have been discarded.

Reinspection of the smaller sample group pulled will provide information both on the rate of occurrence of particles in released product (the theoretical reject rate on the reinspection is "0") and the rate of false rejections, which reflects the visual inspection process. If the second inspection rate approaches that of the first, detection at the first 100 percent inspection is random and ineffectual. Both the first and second inspections should have action levels. For instance, if the historical average rate of rejection for a code is 4 percent, statistical analysis will show some higher defect rate at which investigative action should be taken. Lower than normal defect rates should also not occur with a consistent process and inspection procedure. As well as initiating investigation of the reason for an increase in particles (process and inspection procedure) investigative action may include pulling additional sample groups for reinspection and 100 percent reinspection to ensure that the visual inspection is effective at rejecting a higher number of units at the higher rate of occurrence. If the cost of rejects in a concern, rejected units ("culls") may be reinspected and units free of particles returned to the batch.

The rationale and conduct of the visual inspection process should be clearly documented in SOPs; personnel should be trained in these SOPs. Investigations may be conducted based on either microscopic/filtration procedures including analysis by electron microscopy, X-ray and identification or on a more critical visual inspection that will allow the types of particles to be determined. For instance, paper fibers may be a "miscellaneous" type of particulate matter, and glass or metal may be responsible for the increase. These latter are indicative of process problems.

Inspectors may discriminate between particulate matter sources based on when the particulate matter is detected. Process-related particle numbers (i.e., extraneous particles) do not normally increase in number over the course of a stability study; increasing visible particle numbers during the course of a stability study may reflect formulation problems or product-container interactions. Retain samples as well as stability samples should be inspected; retain samples may be exposed to different conditions of storage, handling, and transportation conditions than stability samples.

With visual inspection as well as with filtration and environmental controls, the principle of minimal challenge should be observed. If an inspector is challenged with higher numbers of particles, he or she may be unable to reject as effectively as with the normal lower particle burden. The human element of the visual inspection process is always a potential issue for an auditor. The investigation of reject rates that are higher or lower than the historical average should include investigation of the process for human inspector variability; humans are subject to influence by manager or supervisor concern over increased defect percentages or by anecdotal information regarding process problems.

QUALITY CONTROL LAB PRACTICE IN PARTICULATE MATTER ANALYSIS

The pharmaceutical quality control laboratory is considered by the FDA to serve one of the most important functions in pharmaceutical production and control. General considerations applicable to chemical testing also pertain to particulate matter testing, and the latter has become an item of increasing interest in recent years. A significant portion of the cGMP regulations (21 CFR 211) pertain to the quality control laboratory and product testing. Similar concepts apply to bulk drugs. The FDA *Guide to Inspections of Pharmaceutical Quality Control Laboratories* (July 1993) supplements other inspectional information contained in other inspectional guidance documents. (See also *U.S. v. Barr Laboratories, Inc.* 1993.)

The interests of auditors in particulate matter in drug products is not limited to manufacturing operations. Increased emphasis is currently placed on laboratory particle testing results that appear in submissions or are applied as batch release criteria for final product. Recent preapproval inspections have extended to the laboratory procedures used to collect data according to the USP <788> test method, and indications are that the same criteria regarding the handling out-of-limit test values, documentation data review, and other GLP components as for chemical labs are being applied. While in the not-too-distant past there was evidence that many inspectors believed that little numerical emphasis could be placed on particle count data, data are today in some cases being evaluated for variability, and questions regarding reasons for count variation have been asked.

Inspectors today seem to be more knowledgeable regarding the interpretation of count data. The difference between extraneous particle counts (which typically remain constant over time), counts due to container components such as silicone oil, and counts due to degradants (which may trend upward with time) appear to be well understood. Outlier data and count excursions have been questioned. As mentioned earlier, training is being scrutinized more closely than in earlier years, and the training of analysts performing particulate matter testing is no exception. Whereas method validation for particle counting tests had not been a consideration prior to 1993 (to the best of the author's knowledge), by 1998 there were numerous instances of investigator questions that indicated an expectation that particle counter performance qualifications should be performed (see Chapter 11).

Auditors may be expected to evaluate a firm's system for investigating laboratory test failures. These investigations represent a key item in deciding whether a product may be released or rejected and form the basis for retesting and resampling. Laboratory records and logs represent a vital source of information, which allows a complete overview of the technical ability of the staff and of overall quality control procedures. All SOPs should be complete and adequate, and the operations of the laboratories should conform to the written procedures. Specifications and analytical procedures should be suitable and, as applicable, in conformance with application commitments and compendial requirements.

In USP 23 <788>, the light obscuration test is described as the first stage test to be performed for particulate matter. One of the oft-cited weaknesses of this test is the tendency to generate spurious counts due to various sources of interference. These artifactually elevated counts may constitute out-of-limit data that must be addressed appropriately, which provides a challenge for the lab manager. In the recent *Barr* decision, the judge used the term *out-of-specification* (OOS) laboratory result rather than the term *product failure,* which is more common to investigators and analysts. This decision understandably figures in the current thinking of FDA investigators. The judge ruled that an OOS result identified as a laboratory error by a failure investigation or an outlier test and overcome by retesting is not a product failure.

Laboratory errors occur when analysts make mistakes in following the method of analysis, use incorrect standards, and/or simply miscalculate the data. Laboratory errors must be determined through a failure investigation to identify the cause of the OOS. Once the nature of the OOS result has been identified, it can be classified into one of the three categories above. The inquiry may vary with the object under investigation. The exact cause of analyst error or mistake can be difficult to determine specifically, and it is unrealistic to expect that a source for analyst error will always be determined and documented. Nevertheless, a laboratory investigation consists of more than a retest. The inability to identify an error's cause with confidence affects retesting procedures, not the investigation inquiry required for the initial OOS result. Typically, if the cause of an OOS result cannot be identified, the result in question must be reported to quality management (QM) as a failing result. QM will expand the investigation to include manufacturing information and will determine further testing to be performed.

An almost universal expectation seems to be that a firm's analysts should follow a written procedure, checking off each step as it is completed during the analytical procedure, with expected laboratory test data to be recorded directly in notebooks or on a worksheet. These commonsense measures enhance the accuracy and integrity of data. Specific procedures must be followed when single and multiple OOS results are investigated. For the OOS result, the investigation and inquiries must be conducted before there is a retest of the sample. An analyst conducting the test should report the OOS result to the supervisor, and the supervisor should conduct an informal laboratory investigation/interview process that addresses the following areas:

- Discussion of the testing procedure
- Discussion of the calculation
- Examination of the instruments
- Review of the notebooks containing the OOS result

An alternative means to invalidate an initial OOS result, provided the failure investigation proves inconclusive, is the "outlier" test. However, specific restrictions must be placed on the use of this test:

- Firms cannot frequently reject results on this basis.
- The USP standards govern its use in specific cases only.
- The test cannot be used for chemical testing results.

It is not deemed appropriate to utilize outlier tests for statistically based tests, i.e., content uniformity and dissolution.

When the laboratory investigation is inconclusive (the reason for the error is not identified), the firm

- cannot conduct two retests and base release on the average of three tests,
- cannot use the outlier test in chemical tests, and
- cannot use a resample to assume a sampling or preparation error.

Laboratory auditors are specifically directed at this point in time to evaluate a company's retesting SOP for compliance with scientifically sound and appropriate procedures. The *Barr* district court ruling provides an excellent guide for evaluating some aspects of a pharmaceutical laboratory, but it should not be considered as law, regulation, or binding legal precedent. The court simply ruled that a firm should have a predetermined testing procedure, and it should consider a point at which testing ends and the product is evaluated. If the results are not satisfactory, the product is rejected.

Retesting following an OOS result was ruled appropriate only after the failure investigation is underway and the failure investigation determines whether retesting is appropriate. It is appropriate when analyst error is documented (i.e., the original result is invalid) or the review of analysts work is "inconclusive," but it is not appropriate for known and undisputed non-process or process-related errors.

The court ruled that retesting must be done on the same, not different, samples; it may also be done on a second aliquot from the same portion of the sample that was the source of the first aliquot or on a portion of the same larger sample previously collected for laboratory purposes. Care must also be taken in identifying and documenting the rationale for resampling. Firms cannot rely on resampling to release a product that has failed testing and retesting unless the failure investigation discloses evidence that the original sample is not representative of the batch or was improperly prepared.

USP <788> provides for averaging particle count data. Data averaging was also identified by the court as a potential source of nonrepresentative test results. The opinion of the court was that as a general rule this practice should be avoided because averages hide variability among individual test results. This phenomenon is particularly troubling if particulate matter testing generates both OOS and passing individual results that, when averaged, are within specification. Here, relying on the average figure without examining and explaining the individual OOS results may be misleading and unacceptable.

Fortunately, most laboratory auditors who are familiar with particle count data appreciate the between-unit variability that is encountered on occasion, especially with dry powder SVI dosage forms. These products in < 25 mL volume containers must be pooled to give sufficient material for the <788> test. This procedure constitutes a physical averaging, but since individual unit results are not known, between-unit variability is not an issue. Variability becomes obvious as single units of ≥ 25 mL volume are tested and may be pronounced with multiple dose vials that contain powder-filled or lyophilized product. The potential high level of count variability for some product was the basis of the first <788> test for SVI requirement for pooling. In a multistage test like <788>, dissolution, pyrogen, or content uniformity, the "test" is not considered "OOS" until the last stage test result fails. The test method has specific instructions to be followed if the stage does not meet certain criteria.

Laboratory equipment usage, maintenance, and calibration logs; repair records; and maintenance SOPs will also be examined by most lab auditors. In this context, particle counters and microscopes are the same as other instruments. The existence of the equipment specified in the analytical methods will be confirmed, and its condition noted. Auditors are directed by the inspectional guide to determine whether equipment is being calibrated and used properly and, in addition, to verify that the equipment in any application was in good working order when it was listed as being used to test product. Thus, frequently, the auditor will investigate whether equipment was present and in good working order at the time specific batches were tested.

PREPARATION FOR FDA INSPECTIONS

An investigator is bound by training and philosophy of enforcement to assess the extent to which the entire product output of a manufacturer meets requirements. Medical devices or injectable products will, in many cases, be applied in the administration of therapy to patients who are, at best, unknowing of the consequences of the use of substandard devices and drugs and, at worst, may be incapacitated, unconscious, immunologically suppressed, or otherwise compromised with regard to any decision regarding the use of a device or drug for their treatment. Since the use of a given unit of drug or a specific administration set cannot be predicted at manufacture, all products must measure up to certain high standards. In laboratory audits, the correctness of analytical data is a prime concern in the assurance of safe and efficacious products; the auditor will thus attempt to ensure that proper care and attention to detail are evident in the generation of analytical data relating to specific products on which decisions are based. The decision-making process through which critical judgments are made may also come under scrutiny. Critical decisions may involve batch retention or release, investigation of out-of-limits or OOS results, or extent of method validation. All critical decisions should be documented. The documentation should be written with the inspector in mind. It should clearly guide the inspector from the problem statement to the conclusion.

Experienced investigators frequently appear to use a "snapshot" approach to GMP inspections (see Nilsen 1997). This involves looking at key items in a hierarchy of importance in order to obtain a general impression of overall GMP or GLP compliance. If this initial overview is good, then the inspection may be relatively brief. Conversely, if the overview raises basic issues, then a more detailed inspection may result, lasting for weeks or even months and involving additional inspectors. For a laboratory inspection, the overview will typically include a review of equipment calibration, qualification, and SOPs; an examination of sample logs, various data, and training records; and a review of one or more protocol studies from beginning to end to review the logic of the analytical path, documentation, and decision process.

In a preapproval inspection, the occurrence of deficiencies in data may cast doubt on the overall suitability of the laboratory system. Minor laboratory deficiencies usually result only in 483 observations that can be easily addressed; major deficiencies, such as suitable method validation, lack of an instrument calibration program, missing or suspect documentation, or superficial handling of OOS results without appropriate investigation, will result in a deficiency letter, which stops review of the file until deficiencies are corrected.

Established firms that are experienced in dealing with the FDA realize and accept that striving for a perfect inspectional record (i.e., no observational items) is impractical both in terms of product quality and economics. The pivotal concern in economics is that there is a point at which no further assurance of product safety, efficacy, purity, and patient protection is gained through additional efforts to anticipate and address all possible auditor or reviewer questions (i.e., the principle of diminishing returns). The laboratory manager or manufacturer must define the point at which substantial or adequate compliance has been established; while compliance is an operational umbrella under which activities are carried out, it does not constitute the primary reason for existence of the lab or manufacturing operation. The achievement of this realistic, attainable, and practical goal is the intent of both the FDA and the enlightened manufacturer. This principle is Nilsen's (1997) "continuous state of substantial compliance," with substantial compliance being defined as having all of the required elements of laboratory and manufacturing practice in place, coupled with management actions and attitudes that demonstrate adherence to the intent of the cGMP regulations and the incorporation of sound technology.

The preinspection consideration given to areas involved in the control of particulate matter contamination does not differ in any significant way from the basic plan that has evolved at most firms for other areas. The list of items below, although emphasizing laboratory practice and not all-inclusive, includes major considerations and serves to illustrate the principles of a preinspection self-assessment. Nilsen (1997) suggests the concerns may be put in the context of questions that those responsible for the management of a given area or function may ask themselves.

- What impression will be made on an inspector by the overall appearance of the laboratory (or manufacturing) area? This initial visual impression is extremely important and may establish in a few minutes the nature of the inspector's impression of the site and allow formulation of an inspection strategy. Does the area look clean and well organized? Don't neglect, by any means, environmental controls and housekeeping in particle analysis areas.
- Do SOPs exist for each laboratory operation, including analytical methods, equipment calibration, change control, failure investigations, method validation, equipment qualification, training, safety and housekeeping, documentation practices, management of standards, evaluation and auditing of data, archiving, sample processing, equipment identification, outside lab audits, labeling and dating of reagents and solutions, management responsibility, document review, and the generation of standard procedures and authorizing documents?
- Are in-house procedures available for all analytical test methods? (Even USP procedures should be detailed as in-house methods.)
- Are training records current, cataloged or indexed, and readily accessible in a central location? Are responsible individuals identified? (Remember, every analyst must be certified as being trained in every procedure he or she performs.)
- Does a sample tracking system exist that provides for a consistent, unique identification assignment to every sample and a concise description of the nature of the analysis to be performed? Is chain of custody and reasonable sample security established? Are sample logbooks are current?
- Is a systematic and comprehensive approach to metrology, calibration, and use/control of analytical standards in place? Are particle size/number standards within their expiration dates? Do current calibration/standardization tags or stickers appear on all instrumentation?
- Is it demonstrable that SOPs are followed as written and that technicians have read and understood SOPs? (Individual analysts may be queried on this during tours or walk-throughs.)
- Do documentation practices themselves applied comply with those described in the CFR and are they described in an SOP or other authorizing document? Are safeguards for original data and lab notebooks sufficient?
- Can the chain of original data be readily traced within the analytical process and related to specific samples?
- Have all analytical methods been validated, including USP method modifications and the in-house counterparts of methods such as the <788> particulate matter test?
- Are OOS results properly dealt with? (Remember the *Barr Labs* decision.) Not only should an SOP for OOS result handling exist, but management responsibility for approval for rejection or acceptance of these results should be spelled out. Is there a mechanism for capturing OOS data at the moment of collection by an analyst and ensuring that it receives the attention of the manager or supervisor? What is the rationale for the allowance of retesting?
- What are specific concerns in preparation for focused inspections related to approval of NDAs? (Note: These inspections are generally performed on the basis of the manufacturer's time frame; failure to prepare properly may be hard for investigators to understand; also, pervasive problems found in a preapproval inspection can result in a general investigation.) Steps in preparation may include anticipating what documents the investigator(s) may want to review and, prior to the inspection, reviewing these documents to include supporting data. Becoming familiar with the work and anticipating questions regarding outliers and deviations from protocols are also important. It is also generally helpful to review all notebooks, documents, and protocols pertaining to specific testing involved in the inspection.

Interfacing with the investigator(s) is an important consideration in inspection preparation. Not only personnel responsible for directly interacting with the investigators but also other (laboratory) personnel should be familiar with inspection etiquette. The focus in this training should be on performing the manufacturer's role smoothly and effectively and objectively providing the requested information concisely and accurately.

FOCUS OF REGULATORY INSPECTIONS (SUBJECTS FOR QUESTIONS)

The subject matter here relates generally to the focus of inspections by the FDA. The author has no detailed knowledge of inspection results and/or inspectional strategies in other countries, and such a broad-scale consideration would be beyond the scope of this book. Based on limited information, however, the author generalizes that key inspection items in the United Kingdom, EEC, Canada, and Australia parallel to some extent the concerns of the FDA in the United States. A comment that has been heard in a number of other countries, including Japan, is that the detail of FDA inspections performed on foreign manufacturers who seek to market in the United States is greater than that of the regulatory agencies in each country. It is important to remember that because of the size of the U.S. market for drugs and devices, many countries overseas may expect FDA audits at any time.

It is likewise extremely important to be mindful of the fact that FDA auditors and inspectors are individuals, and their concerns in an inspection will be governed by individual perceptions and experience as well as by FDA guidelines, the USP, cGMP, and formal training received. The author has attempted to outline below some inspectional perspectives based on historical perspectives, audit findings, and information presented in different contexts by FDA administrative, scientific, and enforcement personnel. The reader is forewarned that these summary items are merely example and are in no way to be considered a comprehensive listing of items on which inspectional observations may be made.

Particulate Matter as a Process Indicator

As observed earlier, a key principle that may be observed in historical and current audit findings is the use of particulate matter as a process indicator. In this context, particulate matter as an audit item offers two unique advantages:

1. Particulate matter burdens can be judged against very simple requirements:
   - Presence or absence (visual inspection)
   - USP numerical limits
2. The simple requirements for particulate matter tie into
   - cGMP,
   - formulation suitability,
   - container-closure systems,
   - degradants, and
   - process cleanliness.

In a somewhat similar fashion, for reviewers of product application files, particle count data may be used as a first pass indication of whether or not all is well with regard to the product formulation and stability. High counts in comparison to other products or the USP limits, or counts that seem to trend upward with time may raise a flag.

In the past three years, the author's discussions with peers who track such information indicate that NDA reviewer deficiency observations have included the following:

- Particulate matter data not submitted (collect during product development for *all* products?)
- Data suggests a relationship between particulate matter and degradants (explain, know what particles are)
- Explanation of OOS results lacking
- Why a high rate of light obscuration test failures
- Explanation of the application of in-house limits

Components of compliance in regard to particulate matter can be defined on the same basis as for other testing or monitoring that specifically ties in to cGMP or regulatory guidelines.

Airborne Particulate Matter Monitoring

The general conduct of airborne particulate matter and environmental monitoring should include OOS action, frequency of tests, and test plans. (A philosophical point apparently pursued here is that particle counts are believed to relate in some (variable) fashion to the potential for microbial contamination and that skin cells may be "rafts" on which microbes float and thrive.)

- Are HEPA filters challenged at least semi-annually with an oil aerosol?
- Averaging of counts to pass is forbidden.
- Is testing performed during manufacturing operations?
- Is there evidence of "testing until a room passes"?
- Is corrective action and investigation undertaken for OOS results?
- Is there rigorous testing of areas where product is vulnerable?
- Is a process in place for trending of data?
- Has validation of remote counting systems—especially sample transport—been performed?
- Is computer-aided data reduction utilized for optimal trending and pattern detection?
- Is continuous or intermittent monitoring used?
- Is interval of intermittent monitoring appropriate to detection of "spike" counts?
- Is there missing data?
- How are out-of-limits counts handled?
- Is handling and definition of outliers appropriate?
- Is the scheme for retesting appropriate?

Garbing and Personnel Protection

Garbing and personnel protection measures are questioned both in aseptic processes and those for terminally sterilized product. Questions might include the following:

- Are garbing procedures specified in an SOP?
- Do personnel receive training prior to work in clean areas?
- Investigator conduct of procedural observation.
- Is testing of individuals for comprehension carried out after training?
- Are personnel shown to be qualified to work in clean areas?
- Is particle generation by personnel monitored?
- Are there adequate records of training and monitoring?
- Does periodic retesting of workers at appropriate intervals occur?

Medical Devices

Recent device manufacturer inspections have touched to varying extents on the following areas of interest:

- What are the procedures for complaint follow-up?
- Are patient issues related to complaints addressed?
- What are the results of inspection of product on-site by auditors (e.g., with magnifying glasses)?
- Were samples prepared for FDA lab testing?
- What is the relationship of particulate matter levels to housekeeping GMPs?
- Is the packaging "clean" (e.g., blister packs)?
- Does the firm discriminate between "contaminants" and "unavoidable" particulate matter (e.g., package fragments)?

Product Complaints

The handling of product complaints for both drugs and devices may lead to the following lines of questioning:

- What are the medical issues (patient implications)?

- Is there a system for handling complaints in place?
- How is the customer addressed?
- Does the firm confirm or deny a complaint—"thread of logic"? (includes analytical follow-up)
- What is the company decision to reinspect product?
  - Retain samples?
  - Field samples?
- Is the process for complaint resolution specified in an SOP?
- Is there a process for historical review and trending of complaints?

Compliance with USP <788> for Solutions

- How do SOPs compare to the USP method?
  - Standards (dating, storage)
  - Understanding of microscopic test
  - Analyst training and qualification
  - Sampling plans and limits
- Is data recorded and reviewed?
- What level of light obscuration failures is acceptable?
- Handling OOS results—is there a specific SOP?
- Walk-though analyst questions (what are you doing?)
- Are analysts tested against standards?
- Is there a validation package for counters?
- Is there a calibration procedure for light obscuration counters—USP PCRS?
- How is "as-found" data for counter calibration handled?
- Are data review and retest procedures in place?

Visual Inspection

- Is the firm's interpretation of "essentially free" of visible particulate matter logical?
- Is the USP <1> "evidence of contamination" rejection criterion used?
- Is each unit visually inspected during manufacture?
- Is the visual inspection procedure described in an SOP or plant quality assurance specification?
- Is analyst training conducted on an annual basis and documented?
- What is the test for visual acuity (20/30 or better)? How?
- Are analysts screened for color blindness?
- Following training, is there procedural observation (by the supervisor)?
- Is there testing of inspectors using known reject samples?
- Is there inspection booth calibration performed?
- Do test procedures exist to define assay sensitivity (what size particle detected)?
- What are the in-house limits and rationale?

Overall Response to In-House Particulate Matter Issues

- Is corrective action taken per an SOP (reaction to problems)?
- Is there a resolution of particulate matter issues?
- Is problem particulate matter identified (composition)?
- Does routine testing show product below USP limits?
- Is the chemical composition of particulate matter innocuous?
- What amounts are present?
- Is the source when particle issues arise determined?
- Is there a documented mechanism for control and *elimination*?
- If particulate material increases with time, is the "why" determined and appropriate action taken (e.g., notification of the FDA with product implications)?

Liquid Filtration

- Are filter validation and selection processes defined in an SOP?
- Are there technically sound integrity test SOPs?
- Are filling head filters qualified as well as line filters?
- Do change-outs mirror industry practice? Procedures?
- Is there product sampling before and after change-out?

Facility Review

A facilities review will often be a component of a preinspection tour. A trained inspector will rapidly form an impression of the overall atmosphere of compliance (or lack thereof) based on a simple observation of key factors:

- Product contact area cleanliness
- Residues
- Nooks and crannies
- Employee behavioral and attitude factors
- Garbing
- Actions (general)
- Touch contamination potential
- Attitudes
- Training (knowledge)
- Number and location of inactive areas

The following items are popular requests, particularly in audits of aseptic processing:

- A blueprint of the facility physical layout
- Historical limits review summaries
- Process evaluation reports
- A description of room classifications versus use
- Environmental test records
- Statistical treatment summaries (trending?)

Record Keeping

Record keeping (completeness and accuracy) and tracking of process-related data are almost invariably an inspection focus in aseptic operations with regard to the following:

- What is the recording medium?
- What is the record style and format? Are electronic signatures used?
- Does record organization allow timely retrieval?
- Are corrective actions appropriate in timing and extent of documentation?
- Is there an environment of management awareness?
- Do periodic process evaluation reports exist?
- Have responsible management personnel reviewed and signed-off reports?
- Do documentation systems exist for demonstrating that problems/errors have been followed up, investigated, and appropriate corrective action taken?
- Are routine written process evaluations performed at a frequency that seems reasonable?
- Do the reports reflect both "good" and "bad" data?
- Do the firm's reactions seem reasonable when "bad" data (OOS) occur?

INSPECTIONAL OBSERVATIONS

The following excerpts represent inspectional observations that were taken from a number of sources. In some cases, the wording has been changed slightly in favor of a more readily understandable form or to protect the innocent. While it would be very time-consuming and counterproductive to work backward from inspectional observations to prepare for an audit or to track compliance, the nature of inspectors' questions does draw attention to specific shortcomings that should be avoided. They also provide, to some extent, an insight into the types of queries that may have to be addressed and the philosophy of inspections. Despite the fact that one can learn a great deal from previous observations, the reader must remember the old proverb: "Experience is a dear teacher; a fool will learn from no other." Systematic preparation and ongoing compliance is the only assurance of favorable audit outcomes.

Some of the observations provided here do not address particulate matter directly, but rather call on principles of laboratory practice and quality control that apply to particulate matter assays as well as other types of testing. Further, wording has been added to clarify a number of the items that were not readily understandable when separated from the context of the inspection in which they were made. Applicable references include various inspectional guidelines, FDA *Draft Guide on Repackaging of Solid Oral Dosage Form Drug Products*, FDA (Revised) *Guideline for Submitting Documentation of Sterilization Process Validation in Applications for Human and Veterinary Drug Products*, and the FDA Aseptic Fill Conference Transcript of October 12, 1993.

Manufacturing Operations

- Particle counts taken at various locations within a single aseptic sampling environment are averaged. The mean of particle counts taken at various locations within the single sampling environment is used to meet the Class 100 limit.
- A mean environmental particle count of $11{,}261/\text{ft}^3$ was obtained in solution filling. The class limit is 10,000. Documented corrective action indicates that the filters were changed and that a retest passed on X/XX/XX. A review of the available documentation for the filter change in the HVAC system in the room showed that the filters were not actually changed until X/XX/XX.
- There are no established limits for the pressure drop across the HEPA filters located in the CEAs. There are no set limits for the comparison and interpretation of the HEPA filter pressure drop data currently being generated so that filter wearout can be determined.
- During the filling of Nafcillin, lot #XXXX, the operators:
  - Repeatedly touched the equipment parts inside the primary barrier without proper sanitization of their gloved hands or changing their gloves.
  - Repeatedly touched items in the Class 10,000 area and then entered the primary barrier without proper sanitization of their gloved hands or changing their gloves.
  - Leaned directly over open empty final product vats on the filling line with gown or sleeve that had previously come in direct contact with materials in the Class 10,000 area.
  - Repeatedly entered and exited the filling room by pushing the door open with their gloved hands and subsequently entering the primary barrier.
- During this inspection, our observation revealed ashed labels on top of three of the HEPA filters inside the depyrogenation oven. This condition increased particulate matter contamination that could have compromised the Class 100 conditions inside the oven and is listed on this FDA 483 as item XX. This particulate matter may have been introduced to the filling room when the oven door was opened to the aseptic side as noted in this FDA 483 as item XX. Also, this particulate matter sits on the surface of empty stainless steel trays, which could raise the amount of nonviable particulate matter in the aseptic fill room.
- Record review revealed that the firm had not challenged the integrity of two HEPA filters on the depyrogenation oven. Additionally, these filters had not been tested in the firm's periodic HEPA filter testing program. This was noted on the FDA 483 as item #XX. Test results show only two of the HEPA filters were tested.
- The firm has no record for validation of the particulate matter testing for the depyrogenation oven. There are no written procedures providing directions for how to take particulate matter samples in the depyrogenation oven. Also, the sampling of the particulate matter at the south sampling depyrogenation oven is inadequate. The tube opening and the airflow face the downstream direction.
- At the beginning of this inspection, the firm did not have SOPs to describe the cleaning and sanitation of the depyrogenation oven.

- We also observed in the vial staging room numerous gaps in the false ceiling. Dust accumulations were visible above this false ceiling. The nonsterile side of the depyrogenation oven is directly under this ceiling. This is noted as item 120 on the FDA 483.
- A lot of drug that failed to meet release criteria, including the finding of high levels of visible particles in the product containers, was released.
- Despite the receipt of multiple customer complaints regarding the presence of black particles in the lyophilized powder cake of product XX, no failure investigation was undertaken.
- Air pressure differentials during production of XXXXX, lot #XXXX, and pressure differentials between rooms of different classification serviced by air handlers 10 and 11 were out of specification.
- There is no documentation as stipulated in specification XXXX that shows corrective action taken when the particle count mean exceeded the Class 10,000 "environmental" action limits. There is no documentation to show when the 85 percent efficiency filters were removed and replaced in the room HVAC system.
- Investigations for complaints of particles, determined to be a zinc-based compound used as a stabilizer in the blow-molded container material, are incomplete. The investigations did not include an analysis of file samples to determine if the entire lot was affected.
- Particle analysis SOPs did not define what constitutes a trend in particle out of alert limits results. An evaluation of batch records performed for complaint XXXX revealed that 5 consecutive batches, including the complaint lot, exceeded the 10 μm solution particulate alert limit. The complaint evaluation simply stated "no trend detected."
- There appears to be inadequate control of employees' plant uniforms, in that employees are allowed to wear uniforms outside the plant environment (i.e., at home or outside the building).
- The mean of particle counts taken at various locations within the single sampling environment is used in order to meet the 10,000 particles/ft^3 action limit. Averaging counts in this context is inappropriate. There are no procedures for initiation and subsequent documentation of corrective actions when single particle sampling sites within a single sampling environment of device production areas are in excess of the environmental 10,000 particles/ft^3 action limits.
- Specification #XXXX, dated XX/XX/XX does not provide for maximum allowable limits of airborne particulate matter, with procedural follow-up, in device and solution production areas.
- Specification #XXXX, issue date X/XX/XX, entitled, Particle Evaluation of Devices and Solutions, specifies only "sample one sterile unit per week per device category." This level of sampling is inadequate to determine particle levels given the variability of counts.
- Sample counts are averaged within the sample area and at sample points to achieve passing results.
- Dirt and/or lint is on the grill covering the intake prefilter of the particle analysis laboratory laminar flow HEPA–filtered hood.
- There are no maximum allowable limits or alert limits established for particles in finished medical device products.
- Up to 15 percent of a HEPA filter surface can be sealed with silicone RTV compound before specifications stipulate replacement for laminar flow hoods. This level of patching is inconsistent with industry practice.
- In dioctyl phthalate (DOP) challenge testing of HEPA filters, DOP is injected into the airstream as much as 45 ft upstream of the tested filter, levels directly behind the filter are not determined.
- Specification XXXX allows up to 100,000 particles/ft^3 for airborne particulate matter in manufacturing and assembly areas for device products in which product fluid path surfaces are exposed to the environment.
- After replacement of a HEPA filter, a particle counter test is used to detect leaks if an

immediate DOP challenge cannot be performed. Newly installed HEPA filters can be utilized for up to two weeks before being subjected to a DOP challenge test.

- Environmental particle alert limits are established at the uppermost limits of the specified room classification. Shutdown limits are established at the next highest classification.
- Class 100 aseptic processing areas are defined as those sites within an aseptic processing room that are protected by laminar flow HEPA–filtered air from overhead filters. These Class 100 areas have partial plastic curtains or a narrow higher velocity air curtain around their perimeters. The remaining areas of the aseptic processing rooms (all areas not under laminar flow HEPA–filtered air) are maintained as Class 10,000 areas. The air curtains or the plastic curtains are the only barrier separating the Class 100 area from the surrounding Class 10,000 area. This level of protection is not adequate for an aseptic process.
- Aseptic processing personnel are not required to wear goggles, and most are usually seen with a part of their forehead, cheeks, and eyebrows exposed.
- Open vials on the feed table/unscrambler are exposed to personnel. Operators continue to install parts and make adjustments to the equipment (star wheels, vial conveyor guides, etc.) while working within the Class 100 area, while the stoppers and vials were open and in place on the equipment.
- In the aseptic filling line, we observed the following conditions on X/XX/XX, during the setup for XXXX, lot XXXX:
  - There are paper identification tags taped to equipment, utensils, waste cans, windows, and so on.
  - There is an approximately 18 in. × 60 in. work table along the wall. Items on this table include
    - three trays with covers,
    - two loose-leaf ring binders,
    - a roll of 1-1/2 in. plastic tape with a paper core, and
    - a box of aluminum foil wrap.
  - There is an approximately 2 ft × 3 ft desk in the southwest corner. Items on this desk include
    - a balance used for net weights,
    - a tape dispenser with a roll of clear plastic tape,
    - an equipment log page,
    - a pen,
    - paper documents, and
    - a brown paper bag.
  - There are two metal chairs with padded plastic seats and backs that have stitched seams in the room; cushions of both are in need of repair.
- During the inspection, we observed a number of questionable practices by employees in the aseptic processing areas. The following are examples of employee practices that could contribute to increased levels of airborne particles or could expose products or components to contamination.
  - Aseptic filling operator opened door to filler cabinet and leaned inside to perform an adjustment, first with his hands then with a forceps. His entire upper body was inside the cabinet. The line had not been cleared of vials at the time.
  - One operator removed the protective wrap from the bottom of the sterilized stopper bowl outside the Class 100 area. The bowl was then turned right-side-up as it was moved into the Class 100 area. The remainder of the sterile wrap was then removed from the top of the bowl, and the bowl was installed. The operator then opened a tray of stoppers, but it appeared that part of the tray was in the clean area and part extended out into the Class 10,000 area. The stoppers were dumped into the bowl.
- The design of the aseptic processing rooms does not minimize the generation and accumulation of airborne particles and does not ensure that aseptic conditions are maintained throughout the room. The rooms are furnished with desks and work tables that provide large areas of horizontal surfaces for the accumulation of airborne particles.

These desks and tables are usually cluttered with an assortment of equipment, utensils, and difficult-to-sterilize items, such as paper documents, rolls of tape, and cardboard items. Unused equipment was also observed in some rooms during aseptic filling operations.

- Approx. 3 in. × 5 in. paper identification tags are taped to almost everything in the room, including equipment, utensils, waste containers, and the inside of the window. Tape residue/pieces from attaching previous tags remain on the windows.
- The high particulate counts found at Test Point 1 in the Class 100 area on line XXXX during the period of XXXX to XXXX were never thoroughly investigated. The cause of high counts was never determined, and no permanent corrective action was taken.
- Magnehelic gauge alarm settings were based on high room pressures, not pressure differentials between areas.
- Damper adjustments made to air handling unit 4 in response to the Alert/Action Response Form were not documented in the HEPA LAF Critical Systems Change Log.
- To date, neither nonviable particles nor microbial air quality are routinely monitored in the feeder bowl of the filling machines. Microbial air quality is monitored near the point of fill; however, there are no data available to show airflow within and between the cabinets.
- To date, nonviable particles are not monitored near the point of fill under normal operating conditions. Particles counts are taken at the face of the supply HEPA filter, which is about 10 ft from the fill nozzle. Current procedures allow 100 particles (0.5 μm or greater) at the HEPA filter face.
- On X/X/XX, we noted that the faces of employees in the fill room [Class 100 area] were not completely covered. Employee garments do not include eye coverings. The eye area, including eyebrows, of the employees is exposed. XXXX was being aseptically filled during this time.
- The firm has no written guidelines, procedures, or specifications for controlling environmental factors in their "clean room" manufacturing environment, including particulate levels, surface microbes, employee contamination, temperature, humidity, air velocity, and air exchange rates. There is no documentation to show that any appropriate and effective monitoring has been conducted for the above conditions, except for HEPA filter air velocity checks performed by firm's outside contractor.
- There is no documentation of monitoring of pressure differentials between the production cleanroom and flanking anterooms or between anterooms and outside plant area.
- There are no written procedures, guidelines, or specifications for "positive pressure" to be maintained in production cleanroom.
- There is no documentation that management routinely monitors and evaluates compliance with Personnel Practices Procedure XX or posted Employee Conduct Guidelines outside the cleanroom.
- There are no approved guidelines/specifications for gowning apparel worn by employees in the production cleanroom, including approved materials and suppliers, frequency of cleaning, control, and reuse.
- The "Monitoring Environment" procedure, number XXXX, requires that a team leader, process engineer, or production manager be notified immediately each time an out-of-tolerance environmental condition is found. The log of daily measurements of positive differential air pressure between the production area and adjacent noncontrol areas shows frequent out-of tolerance readings. The monthly environmental report sent to management summarizing these data for the last X months all state "yes" for "should exhibit positive pressure." There is no record that management was ever notified of these out-of-tolerance positive pressure readings.
- Other deficiencies include no SOPs to describe the DOP challenge of the HEPA filters in the aseptic filling room.
- The validation of the fill room found that the velocities at the HEPA filter face did not meet specifications in the Class 100 area. The velocities were later corrected, but airflow patterns and laminarity tests were not repeated after the filters were adjusted.

- Air velocities within the Class 100 area of the fill rooms are not taken at work height, that is, the height at which the vials are filled and stoppered. The velocities, which are checked semiannually, are taken six inches from the face of the HEPA filters.
- Written procedures that describe the monitoring of pressure differentials are not complete. The procedures do not describe the instruments used and do not include instructions for the calculations needed to determine the differentials between adjacent rooms. In addition, the calculations are not recorded, and the daily readings are not reviewed by anyone other than the person recording the readings.
- Plastic curtains hang from the HEPA filters along the perimeter of the fill line. However, the report of validation does not describe the curtains in use at the time of the validation. The drawings of the airflow patterns in this report do not include the location of the curtains. In addition, the curtains between the filling and stoppering stations are located on only one side of the conveyor and end about 2.5 ft above open, filled vials.

Personnel Protection

Many of the FDA's inspection efforts in areas related to particulate matter pertain to personnel protection and gowning. Gowning-related observation involve

- personnel gowning training and qualification,
- the retraining of operators in aseptic techniques, and
- personnel monitoring during aseptic filling (dynamic monitoring).

Personnel were observed entering production areas of various levels of cleanliness without observing the procedures listed in SOP XXXX "Personnel Procedures." It was also noted that gowning qualification had not been completed for all operators.

Laboratory Practice

- The firm cannot provide adequate justification for the selective reporting of results in those cases where assays for one or more units do not meet compendial limits. Results of analyses have been selected without justification in the laboratory notebooks and selectively reported so as to give a misleading impression on the overall conformance of the sample groups represented by the samples tested. For example:
  - We observed failure to use appropriate garments to perform the particulate matter test. Particulate-free garments and nonpowdered gloves are not worn to perform the test. No garments or gloves are used to perform the test. (Author note: Gloves and garments are not required.)
  - The HIAC automatic particulate counter instrument is not located under a HEPA filter hood when performing the test. (Author note: This is not required.)
  - The HEPA filter that is in the lab has not been certified since 199X and no DOP filter test has been performed since 199X.
  - Laboratory personnel routinely reject counts from the HIAC particle counters that, for some reason, appear to the analyst to be in error. Reasons for the rejection of counts are not documented in SOP XXXX, and supervisory review of rejected count data or documentation of this data is not required.
  - In a calibration of HIAC particle counter #XXXX performed on X/X/XX, the threshold channel settings were found to be significantly different from those of the previous calibration. No investigation was made to ascertain the effect of this change on data collected, and no report was filed.
- There is no action specified in SOP XXXX if count data collected by either the microscopic or light obscuration particle assays are in excess of the USP limits or in excess of the firm's internal limits. These alert limits are posted on the side of the counter but are not specified in any SOP.
- The pooling of a minimum of 10 containers for the particulate reading was not done. For SVPs, only one container per reading is tested instead of the minimum required.

- Five water samples of 5 mL each for the particulate control (system suitability) blank test were not taken. Only 3 consecutive water samples were taken instead of the required 5 samples.
- Invalidated, voided, or disregarded laboratory analyses for particle count data were not retained or explained.
- Original documentation associated with "voided" laboratory analyses was destroyed (e.g., sample and standard weight and calculation sheets).
- There was no written procedure addressing "voids" of analytical data (e.g., responsibility and criteria for voiding data). There were inconsistencies when "voids" of analyses or rejection of occurred.
- Stability particulate matter test reports were incomplete, unapproved, and improperly filed. This issue was related to a recent departure of a supervisor who had left unsigned paperwork in a desk.
- There was an incomplete investigation of an internal particle test failure that was later resolved (referee tested) by an R&D laboratory. This failure also was not documented in the Annual Review Evaluation.
- There is no specific written limitation on the time of testing for stability samples. The stability policy of the firm stated that "variations of frequency and time of testing limits" were the responsibility of senior management.
- Stability tests were not completed and test intervals were not met in a timely manner. (The firm was extremely "backlogged" in the area of stability testing.)
- A failure to collect initial stability samples (at time zero) resulted in the need to use retention samples for stability testing.
- A physical inspection at a specified stability test interval was not performed.
- Stability samples were not maintained in a controlled/monitored room until particle testing was performed. A temporary employee was removing samples from a controlled room and placing them in an uncontrolled area pending testing.
- Stability samples that "missed" testing intervals were not refrigerated to contain aging, thus ensuring accurate representation for that testing period. This observation was disputed by a supervisor stating that reduced temperatures may cause a negative effect due to condensate buildup.
- HIAC particle counters were not identified with a unique identification number.
- There was no written procedure addressing particle count variability and the ability to detect an unstable counter system.
- An inappropriate procedure was used to determine reproducibility and precision of the particle counting system.
- An outlier test was utilized in particle testing that did not specify the use of a statistical evaluation for discounting observations.
- There were no written instructions involving the rejection of counts from a particle counter due to air bubbles.
- There was no documentation for the calibration of particle counters.
- Test results for particle analysis testing that meet the USP values have been selected from a series of analyses for inclusion in the NDA such that the relative standard deviations of the analyses for a given sample group, as reported in the NDA, are within a narrower range of values than is actually the case. No SOP governing the rejection of data is available.
- A series of stability samples was assayed at the 30-day test point. The results from 2 samples were significantly in excess of limits, while 8 other results from this series were far below limits. The samples with the OOS results were reassayed using the microscopic method. Results of this retest were within limits, and this was reported in the NDA.
- Results from the analysis of units of one lot have been discarded in favor of a series of analyses of additional samples. The discarded results were declared void with no reason given. The assay results for other lots that have been discarded were flagged with "possible bad sample prep" beside the high results.

- The sample was reanalyzed, and a lower series of results was obtained. The lower results were reported as the assay results. No additional explanation is provided in the laboratory notebook.
- The analyst's finished product notebook was not reviewed for completeness of testing procedures and compliance with established SOPs prior to the issuance of the final report on the testing. SOPs followed to perform the testing are not identified.
- Partial counts of particulate matter are performed using the USP microscopic test, but the SOP does not explain how the decision of when to perform a partial count is made.
- Dilutions used in plotting the standard curve for calibrating the particle counter are not shown in the analyst notebook.
- Calibration of the HIAC solution particle counter failed to include a determination of accuracy after 6-month interval between calibrations.
- The HIAC light obscuration particle counting process per SOP XXXX has not been adequately validated; the transfer of data from counters to a computer database is routinely used.
- No training records are available to reflect the training of analysts who performed the USP <788> test on solutions for batch release.
- Specification XXXX does not instruct analysts how to conduct a particle count analysis by the membrane/microscopic test, even though this test is regularly performed.
- In the 6-month period between X/X/XX and X/X/XX, 32 batches failed HIAC particulate matter testing, necessitating microscopic retesting. No investigation action reports were filed addressing the reason(s) for the high rate of HIAC failures.
- Laboratory records lacked a description of the sample received for testing, including an identification of the source (where the sample was obtained), quantity received, identification of the container, or a distinctive number. Also, the batch record does not contain the amount collected for analysis.
- SOP "Repeat Analysis for Confirmation of Unexpected Results" does not establish a time limit for the investigation and uses resampling for rejecting analytical results. It also uses resampling to assume a sampling or preparation error.

Training

- Analysts are only required to read the firm's SOPs and sign a document to that effect.
- Firm records show that analysts cover more than 20 SOPs the same day.
- Firm records show that analysts received this training once they are hired but not thereafter.
- Training is not provided on a continuing basis and with sufficient frequency.

Visual Inspection

- The visual inspection procedure used contains no specification of detail of the following items:
  - Analyst training
  - Lighting intensity
  - Time of inspection
  - Particle detection capability
  - Limits for batch acceptability
- The firm had in use no less than three technically distinct methods for visual inspection, none of which were adequately described in procedural documents (SOPs). Inspection 1 was performed attendant to production with 100 percent inspection of the units produced. This inspection utilized black and white background light boxes, and the units were inspected for not more than 10 sec in each. Inspection 2 was performed for the purpose of determining whether an acceptable quality level had been achieved. This inspection allowed 30 sec for inspection in black and white booths different from those used for the line inspection. Inspection 3 was carried out attendant to stability studies using fiber optic illumination and allowing the analysts to observe the tested units "for however long

they feel necessary." When asked what the purpose of the stability inspection was, an analyst answered that he was looking for very small particles, such as those that might settle to the bottom of a vial and be detected as "smoke" or "swirls," haze in the solution, a lack of transparency, and color. Results of the three types of analyses were not compared or reconciled.

- The firm used both manual visual inspection and an automatic inspection machine. No SOP was in place for using the machine, and inspectors were allowed to observe units for variable periods of time. The procedure using the automatic machine had not been compared to the historical results obtained before use of the machine was implemented. The company has no rationale for what sizes or types of particles are rejected using the present inspection process. No limits are in place for acceptable quality levels. This company had not validated their automatic inspection machine.
- The firm's quality assurance inspection for visible particulate matter permits the release of product without investigation regardless of the defect rate in 100 percent inspection; the visual inspection has no action limits.
- The follow-up to customer complaints regarding visible particles in the product does not include an action plan to reduce levels of visible particles.
- The visible particulate matter in the product includes both protein agglomerates and contaminants such as fibers; no investigation has been conducted to determine the number of contaminant particles in the units or to discriminate between particle types.
- No limits are specified for batch suitability based on visual inspection results.
- The HIAC particle counters used are calibrated by the vendor, with no certification that the suitability tests specified in the USP, including count accuracy, are performed.
- Particle count standards used in the particle lab for calibrating the HIAC instruments are not controlled as to use, expiration dating, and conditions of storage. The USP particle count reference standards used are not controlled.

Filter Testing

- The firm did not conduct any testing nor have data or test reports from . . . to support acceptable filter integrity or bubble point testing limits for aseptic filters.
- The proposed filter validation studies to be conducted by the filter vendor have not been completed and are not anticipated to be completed for several months.
- The retrospective validation study of . . . filtration review has not been completed.
- There is no requirement for a review of filter integrity and . . . testing results by Quality Control Personnel prior to filling.
- Master and batch records fail to define and document filter flushing and . . . integrity test operations and results.
- Filter integrity entries in the batch records are not original test results but are transcriptions from the filter log. An FDA audit found discrepancies between the batch records and the filter integrity lot.

SUMMARY AND CONCLUSIONS

It should be kept in mind in any discussion of regulatory activity that the ultimate goal of the manufacturer and auditor is patient protection through generation of safe, pure, and efficacious product. For the manufacturer, this is an ongoing, unremitting effort. The only way to ensure a favorable outcome to an investigation is through thorough preparation and continuing compliance activities. The approach of "I'll worry about compliance tomorrow" is not practical; many firms have discovered the folly of attempting to correct deficiencies while an investigation is going on.

A number of conclusions can be reached based on a consideration of the U.S. regulatory perspective on particulate matter in drugs and devices. Particulate matter in itself appears rarely as a 483 item but may prove a very useful means by which the investigator can uncover non-GMP conditions, perhaps those that lead to the elevated particle burden, to document or initiate enforcement activity on the basis of. FDA investigators and management are now making considerable efforts to become more knowledgeable and scientific in their approach to regulating the drug and device industry. They are generating new guidelines more frequently, and the complexity of compliance has consequently increased.

Auditors today are routinely asking more in-depth questions, and the detail of investigations is far greater than was the case just a few short years ago. Where in the past only a few auditors could probe as deeply as Dr. Tetzlaff, the results of concerted efforts in training are becoming evident in the activities of field inspectors.

The FDA is working with regulatory groups in other countries to harmonize regulations. This activity will probably increase the complexities of compliance in the future. Incorporation of best demonstrated practices into inspector expectations occurs more rapidly with each passing year.

The philosophy of the FDA appears to be much like that of the manufacturer. Lowest levels of contaminant particles are achieved by using clean product components that are kept clean throughout the assembly of the final product. Reliance on the removal of contaminants, by any means, will ultimately result in a higher product particle burden. This philosophy must be followed in process design. Elimination of particulate matter through product protection must be achieved in all areas where the product is not protected by the final container. All personnel working in manufacturing areas critical to the particle burden of the product must be effectively trained with regard to particle sources and particulate matter control and elimination; lowest levels of particulate matter will be achieved only if each individual involved in production accepts a responsible role.

The FDA is extremely concerned regarding the potential dangers of aseptically filled product. It is impossible to keep larger process areas as clean as smaller process areas regardless of the costs for control measures that a manufacturer is willing to accept. Thus, the FDA expects the manufacturer to consider minienvironments, microenvironments, and isolators for their application across a wide range of clean production conditions (including critical devices). Process design to eliminate particle sources is believed critical to contaminant control.

REFERENCES

Avallone, H. L. 1980. Inspectional guidelines for sterile bulk antibiotics. *J. Parent. Drug Assoc.* 34:447–451.

Avallone, H. L. 1982. Clean room design, control and characterization. *Pharm. Eng.* (Sept./Oct.): 33–46.

Avallone, H. L. 1983a. GMP inspections: A field investigator's view. *Pharmaceutical Technology* (March): 48–53.

Avallone, H. L. 1983b. Microbiological practices for non-sterile manufacturing of topicals. *Pharm. Eng.* (Nov./Dec.): 29–31.

Avallone, H. L. 1984a. Regulatory issues arising from recent FDA inspections. *Pharm. Manuf.* 16–19.

Avallone, H. L. 1984b. Specific systems. *Pharm Eng.* (Sept./Oct.): 33–39.

Avallone, H. L. 1985. Control aspects of aseptically produced products. *J. Parent. Sci. Tech.* 39: 75–79.

Avallone, H. L. 1988a. Regulatory issues of parenteral equipment and systems. *J. Parent. Sci. Tech.* 42:89–93.

Avallone, H. L. 1988b. Regulatory aspects of sterile powders for injection. *Pharm. Eng.* (Mar./ Apr.): 50–57.

Avallone, H. L. 1989a. Current regulatory issues regarding parenteral inspections. *J. Parent. Sci. Tech.* 43:121–126.

Avallone, H. L. 1989b. Aseptic processing of non-preserved parenterals. *J. Parent. Sci. Tech.* 43: 113–118.

Avallone, H. L. 1990. Current regulatory issues regarding sterile products. *J. Parent. Sci. Tech.* 44: 228–232.

Avis, K. E., and J. W. Levchuk. 1995. Parenteral preparations. Chapter 87 in *Remington's Pharmaceutical Sciences,* 19th ed., edited by A. Gennaro. Easton, Penn., USA: Mack Publishing Co., pp. 1545–1578.

Barber, T. A. 1993. *Pharmaceutical particulate matter: Analysis and control.* Buffalo Grove, Ill., USA: Interpharm Press, Inc.

Fry, E. M. 1982a. The parenteral drug industry: Recent findings. *J. Parent. Sci. Tech.* 36:55–58.

Fry, E. M. 1982b. What we see that makes us nervous. *Pharm. Eng.* (May/June): 10–11.

Fry, E. M. 1984. General principles of process validation. *Pharm. Eng.* (May/June): 33–36.

Fry, E. M. 1985a. Process validation: The FDA's viewpoint. *D & CI* (July): 46–51.

Fry, E. M. 1985b. An FDA update on GMP's for aseptic processing. *J. Parent. Sci. Tech.* 39:154–157.

Levchuk, J. W. 1991. Training for GMPs. *J. Parent. Sci. Tech.* 45:270–275.

Michaels, D. L. 1982. Wither regulations of drugs? *Pharm. Eng.* (Mar./Apr.) : 10–12.

Michaels, D. L. 1984. A biased view of the future of drug regulations. *Pharm. Eng.* (Mar./Apr.): 15–19.

Michaels, D. L. 1985. Compliance issues in 1985. *J. Parent. Sci. Tech.* 39:158–160.

Munson, T. E., and R. L. Sorenson. 1991. Environmental monitoring: Regulatory issues. In *Sterile pharmaceutical manufacturing,* edited by M. J. Groves, W. P. Olson, and M. H. Anisfeld. Buffalo Grove, Ill., USA: Interpharm Press, Inc.

Nilsen, C. L. 1997. Acing your FDA laboratory inspection. *Pharm. and Cosmetic Qual.* (Nov.–Dec.): 16–17.

Sorenson, R. L. 1984. Federal Standard 209B and the need for a bioclean classification. *Pharm. Manuf.* 21:14–17.

Sorenson, R. L. 1985. FS209B: A part of total environment control. *Pharm. Eng.* (March–April): 21–23.

Sorenson, R. L. 1986. Update on the aseptic processing guidelines. Paper presented at ISPE/FDA Conference, 18–19 March in Crystal City, Va., USA.

Tetzlaff, R. F. 1983. Aseptic process validation. *Particulate & Microbial Control* (Sept./Oct.): 24–38.

Tetzlaff, R. F. 1984. Regulatory aspects of aseptic processing. *Pharm. Tech.* (Nov.): 38–44.

Tetzlaff, R. F. 1987. FDA regulatory inspections. In *Aseptic pharmaceutical manufacturing: Technology for the 1990s,* edited by W. P. Olson and M. J. Groves. Buffalo Grove, Ill., USA: Interpharm Press, Inc.

Tetzlaff, R. F. 1988. FDA inspections of aseptic processes. Paper presented at the 13th International GMP Conference, 22–25 February in Athens, Ga., USA.

Tetzlaff, R. F. 1993. Investigational trends: Environmental monitoring. Proceedings of Joint PDA/IES Conference on Environmental Contamination Control.

Turco, S. J., and N. M. Davis. 1971. Detrimental effects of particulate matter on the pulmonary circulation. *J. Amer. Med. Assoc.* 217:81–82.

COMPENDIAL DOCUMENTS

The Rules Governing Medicinal Products in the European Community, Volume IV. Good Manufacturing Practice for Medicinal Products. Brussels: ECSC-EEC-EAEC.

USP. 1995. *The United States Pharmacopeia,* 23rd ed. Easton, Penn., USA: Mack Printing Co.

FDA DOCUMENTS

Aseptic Fill Conference Transcript. October 12, 1993. 126 pages.

Current Good Manufacturing Practice in Manufacturing, Processing, Packing, and Holding of Drugs, Amendment of Certain Requirements for Finished Pharmaceuticals; January 20, 1995. 5 pages.

Draft Guideline Covering Stability Testing Requirements for New Dosage Forms and for Variations and Changes to New Drug Substances and New Drug Products. December 15, 1994. 8 pages.

Draft Guideline on Impurities in New Drug Substances (Revised). September 1994. 5 pages.

Electronic Signatures; Electronic Validation. January 1994; August 31, 1994. 18 pages.

Guideline on Sterile Drug Products Produced by Aseptic Processing. June 1987. 43 pages.

Guide to Inspection of Bulk Pharmaceutical Chemicals. September 1991. 29 pages.

Guide to Inspection of Dosage Form Drug Manufacturers. July, 1995. 19 pages.

Guide to Inspections of High Purity Water Systems. July 1993. 13 pages.

Guide to Inspections of Pharmaceutical Quality Control Laboratories. July 1993. 15 pages.

Guide to Inspections of Sterile Drug Substance Manufacturers. July 1994. 27 pages.

Human Drug GMP Notes, Vol. 1, No. 1–4; 1993. 24 pages.

Human Drug GMP Notes, Vol. 2, No. 1–4; 1994. 40 pages.

Investigations Operations Manual. September 1994. 541 pages.

U.S. v. Barr Laboratories, Inc. February 1993. Court decision, 79 pages.

21 CFR Part 211.113.

VII

VISUAL INSPECTION OF INJECTABLES AND DEVICES

Visual inspection is the most problematical analysis routinely performed on injectable products and medical devices due to its subjectivity. The precision (reproducibility) of any physical test for particulate matter will be dependent on the extent to which variables impact the test and the extent to which each variable can be controlled. Manual visual inspection is critically affected by a number of variables that are difficult to control and results in a high variability of data. There is some level of agreement between the U.S. Pharmacopeia (USP), manufacturers, and the Food and Drug Adminstration (FDA) that each unit of an injectable product or device should be inspected by a trained individual. The standards against which product is judged related to the occurrence of visible particulate matter are subject to individual interpretation. In the United States, "evidence" of visible particulate matter gives grounds for rejecting a unit. While limits for subvisible particles are uniform, thanks to the USP, one can expect to obtain different answers regarding visible particle limits from almost everyone queried.

The development of present-day inspection theory was begun by Julius Knapp, then at Schering Corporation. In a real sense, Mr. Knapp can be identified as the one person responsible for defining the theory of current visual (and machine) inspection practice. The author is indebted to Mr. Knapp for his persistent friendship over the years, his timely criticisms, and the use of materials for this chapter. The discussion in this chapter relies heavily on a simplified presentation of this investigator's findings. Those developing visual inspection methodologies are well advised to become familiar with the Knapp studies cited in the references.

In life, subjective assessments and judgments will be found to vary to the greatest extent between persons. The visual inspection of injectable products for particulate matter is, on this basis, the most variable and problematical test that is performed. There are no chemical standards available, and few physical ones; the "detector" used in the assay is the human inspector. For the manufacturer, the most important concept to grasp is that the visual inspection must be appropriately matched to the process and the product; no inspection or test method will add a quality attribute that is lacking; testing serves instead to verify that the process is operating within its control limits. In this regard, visual inspections are extremely important.

The visual inspection of injectables is closely linked to the perceptions of the end user of the product, the regulatory agency, and compendial groups. Let us consider for a minute the requirements of the USP. Solutions are to be rejected if they show "evidence" of contamination with visible particles, and injectables should be "essentially free" of materials detectable visually. Some opinions say that such compendial dispensations, without numerical limits or specified test methods are meaningless, but are they? The freedom of a product from particles that can be readily observed by an end user, given that relatively low light intensity, short inspection times, and untrained inspectors are involved, would seem to be highly desirable. In fact, units containing particles visible under the circumstances in intravenous (IV) therapy with less than optimal lighting by an untrained observer should not be considered appropriate for use. Particles of this size (let us use 200 μm as an example) not only are visually conspicuous, but represent sizes outside those predicted by the USP subvisible particle limits and are not to be expected with current Good Manufacturing Practice (cGMP) conditions that involve extensive container-closure cleaning and filtration. Furthermore, readily detectable particles may sometimes be tied to product instability or degradation.

Visible particles in injectables have been an historical concern. Interestingly, in 1978, task group #3 of the Parenteral Drug Association (PDA) published guidelines for visual inspections of injectable products for the manufacturer (Hamlin 1978). The information contained in this document is still valid today. The principles spelled out in this publication serve not only as a basis for manufacturers who must develop visual inspection schemes but also as a starting point for FDA development of regulatory expectations. In this latter regard, the manufacturer is very likely to be required to explain the occurrence of visible particles in numbers and/or sizes that lead to customer complaints. Also to be avoided are situations in which FDA field investigators detect visible particles in product taken for evaluation or lab testing.

There is a large amount of published information on this subject in the literature; the majority of the information that is quantitative and/or empirically proven for use was authored or coauthored by Julius Knapp. In historical and recent papers, Knapp deals in significant detail both with inspection methodology and probabilistic theory of particle detection as it applies to the inspection process. The author of this book, unlike Mr. Knapp, can claim only to be a generalist in the subject of visual inspection; his knowledge of the statistical principles of the inspection of large numbers of small-volume injection (SVI) units that provide idealized targets for visual inspection and statistical analysis is not sufficient in depth to allow either critical or constructive comment on the Knapp publications.

This shortcoming will hopefully not limit the value of the information in this chapter for many readers; often, the questions that arise in the subject area are based on a misunderstanding of basic principles. In this regard, the author's greatest concern is with the variables that critically impact the reproducibility and design of manual assays rather than statistical manipulations. Statistics, at least for the development of quality tests such as visual inspection, are best left to quality statisticians, who can readily and ably assist those responsible for performing the visual inspection assay.

Although the author's approach to the subject of visual inspection differs from that of Knapp in some respects, there is one principle that is invariant between the two considerations. Although the final quality level of the marketed product will be affected significantly by reject-accept procedures, the underlying and most important consideration is the level of quality built into the product. The best visual inspection strategy remains probabilistic in regard to defect removal; the only real assurance is provided by a process that generates extremely low numbers of defects.

VISIBLE PARTICULATE MATTER AS A QUALITY ATTRIBUTE

As discussed in previous chapters, particulate matter may be defined as discrete (visible or subvisible) material present in parenteral solutions (or devices). The USP definition hinges on random sources and heterogeneous compositions for particles of both visible and subvisible sizes as follows:

> Particulate matter consists of mobile, randomly sourced, extraneous substances, other than gas bubbles, that cannot be quantitated by chemical analysis due to the small amount of material that it represents and to its heterogeneous chemical composition.

There are two general contexts in which a visual inspection of injectable products may be performed. Determination of the presence or absence of particulate matter of visible size as defined by a specific visual inspection procedure (Standard Operating Procedure, SOP) per the USP requirement is the more common case. This is evaluation of product by attribute, i.e., particulate matter is either present or absent (a binary system). Determination of the presence or absence of particulate matter in combination with quantitation (and description in some cases) of the number of particles present in a unit may also be carried out. This analysis results in the assessment of particulate matter as a variable, i.e., in terms of quantities (or types or sizes) of particles present. A visual inspection is most often performed in the first context as a critical visual assessment of defect rate (percentage of units with visible particulate matter) as a judgment of batch suitability.

The reader should immediately hold suspect any "authority" who specifies that there is a certain specific size for visible particles. The size at which a particle becomes visible is dependent on many variables, and no one size may be summarily defined. Sensitivity, reproducibility, and precision of a visual inspection procedure for particulate matter are critically dependent on analyst training and visual acuity, lighting, and time during which the visual inspection is conducted, as well as other factors that will be discussed later in this chapter. A single minimum size for visible particles cannot be specified due to the fact that particle detection is critically dependent on multiple variables. One (obvious) key variable is the optical transparency of the solution container system being inspected. For example, a particle of ideal contrast such as a white particle viewed against a black background may be visible at a size of 75 μm in a glass container. Because of the decreased transparency of plastic containers, a particle of the same type may not be visible unless it is at least 200 μm or greater in size.

The four most widely used world compendia are the U.S. Pharmacopeia (USP 1995), the British Pharmacopoeia (BP 1993), the European Pharmacopoeia (EP 1996), and the Japanese Pharmacopoeia (JP 1992). None of these deal with visual inspection and the detection of visible-sized particles in any suitably definitive fashion, although the JP and BP do specify a visual inspection method:

> **USP:** Good pharmaceutical practice requires that each final container of injection be subjected individually to a physical inspection, whenever the nature of the container permits, and that every container whose contents show evidence of contamination with visible foreign matter be rejected; (reconstituted dry powder dosages are essentially free of particulate matter).
>
> **BP, EP:** Injectable preparations that are solutions, when examined under suitable conditions of visibility, are clear and practically free of particles. [See next page.]
>
> **JP:** When the outer surface of the container is cleaned, injectable solutions or solvents for drugs to be dissolved before use must be clear and free from foreign insoluble matter that is readily noticeable when inspected with unaided eyes at a luminous intensity of about 1000 luxes (93 foot-candles), directly under an incandescent electric bulb. As for injections contained in plastic containers, the inspection is performed with unaided eyes at a luminous intensity of 8,000 to 10,000 luxes (740 to 930 foot-candles) with an incandescent electric bulb placed at appropriate distances above and below the container.

In any inspection, the inspection timing, the amount of magnification, the visual acuity of the inspectors, the type of illumination and background that are used, as well as other factors, can have a significant effect on the detection of particulate matter. It is impossible to specify an absolute compendial requirement (i.e., absence of any visible particulate) due to the highly variable (probabilistic) nature of the inspection process for particles of visible size. This is a basic problem that the USP, BP, and JP have ignored for many years. While it is not possible to specify that product will have no visible particulate matter, the GMPs and the attention of manufacturers in the various countries subject to these compendia regarding the issue of large (visible-sized) particles in product has resulted in product that has an extremely low particle burden for particles of visible size.

The current EP contains the following description of philosophy and test methodology:

> Particulate contamination of injections and parenteral infusions consists of extraneous, mobile undissolved particles, other than gas bubbles, unintentionally present in the solutions. The types of preparation for which compliance with this test is required are stated in the relevant monograph.

(Author's note: This may occur in the future, but is not currently done.)

> The test is intended to provide a simple procedure for assessing the quality of parenteral solutions. The test is not intended for use by a manufacturer for batch release purposes. While a manufacturer will need to be confident that a product will meet the pharmacopeial specifications if and when tested, assurance of the quality of the product (with respect to visible particle matter) will be obtained either by 100 percent inspection and rejection of unsatisfactory items prior to release or by other appropriate means.

> **Apparatus**

> The apparatus (see Figure 7.1) consists of a viewing station comprising:

> — a matte black panel of appropriate size held in a vertical position.

> — a non-glare white panel of appropriate size half in a vertical position next to the blank panel.

> — an adjustable lampholder fitted with a suitable shaded, white-light source and with a suitable light diffuser (a viewing illuminator containing two 13 W fluorescent tubes, each 525 mm in length, is suitable).

Figure 7.1. European Pharmacopoeia proposal for inspection booth.

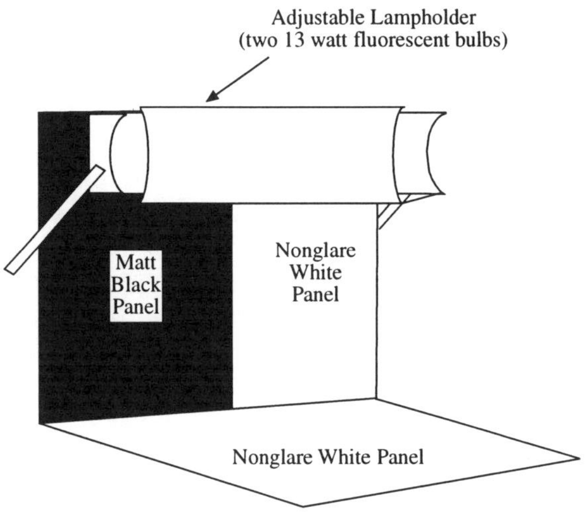

> The intensity of illumination at the viewing point is maintained between 2,000 lux and 3,750 lux, although higher values are preferable for coloured glass and plastic containers.

> **Method**

> Remove any adherent labels from a container and wash and dry the outside. Gently swirl or invert the container, ensuring that air bubbles are not introduced, and observe for about 5 sec in front of the white panel. Repeat the procedure in front of the blank panel. Record the presence of any particles. Repeat the procedure for a further 19 containers. The preparation fails the test if one or more particles are found in more than one container.

The USP contains the following general test specification in USP <1> Injections:

> **Foreign Matter in Injections**

> Every care should be exercised in (their) preparation . . . to prevent contamination with . . . foreign materials . . . Each final container (is to be) subjected individually to a physical inspection,

> whenever the nature of the container permits, and that every container whose contents show evidence of contamination with visible foreign material be rejected.

In USP <1151> Ophthalmic Preparations, the "essentially free" appears again:

> Ophthalmic solutions are . . . essentially free from foreign particles.

USP <771> Ophthalmic Ointments also specifies:

> The . . . test is designed to limit to a level considered to be unobjectionable the number and size of discrete metal particles . . . the requirements are met if the total number of metal particles that are 50 μm or larger in the total extruded contents of all 10 sampled tubes does not exceed 50; and if not more than 1 tube is found to contain more than 8 such particles. (There is also a repeat test provision on 20 tubes.)

USP <789> Particles in Ophthalmic Solutions (a proposed new chapter under consideration) also specifies requirements for subvisible particles:

> The proposed procedure involves passing the contents of a sample of containers (25) through a membrane filter and microscopically examining the entire membrane under 100× magnification. The proposed limits are 50 particles per mL equal to or larger than 10 μm, 5/mL for particles ≥ 25 μm, and 2/mL for particles ≥ 50 μm.

The EP provides that ophthalmic solutions are *practically free from particles,* while both ophthalmic ointments and suspensions must meet the following subvisible limits:

- Not more than 20 particles/10 μg of active ingredient ≥ 25 μm
- Not more than 2 particles/10 μg of active ingredient ≥ 50 μm
- No particles/10 μg of active ingredient ≥ 90 μm

(These limits are based on microscopic testing.)

What is "unacceptable" based on compendial language would appear to be particulate matter that is readily visible, which indicates solution degradation or physical instability, or the presence of large numbers of particles of subvisible size that are seen because of their collective light scattering effect. Containers of injectable solutions containing these types of materials will be judged "unacceptable" due to the inference of solution degradation or impurity.

The implication of the particle burden of units examined must ultimately be judged in the context of the acceptable small quantity of visible particulate matter that may be present in units produced by parenteral manufacture under GMP conditions. A most important aspect of the visual inspection procedure is the detection of any particulate matter that is related to solution degradation.

GENERAL CONSIDERATIONS IN VISUAL INSPECTION

The conduct of visual inspection may be considered in two categories:

1. Human visual inspection with the aided or unaided eye
2. Machine inspection by machines that have a validated capability to mimic the particle detection results of trained human inspectors.

These two types of inspection are considered separately in the following sections of this chapter.

There are many technical publications associated with visual inspection methodology. Some of the most useful references are listed at the end of this chapter. Again, Julius Knapp must be considered the "father" of modern (probabilistic) visual inspection theory. The original papers by Knapp and Kushner (1980a and b, 1980, 1987) and Knapp et al. (1983) are the most comprehensive references on this subject and deal with both manual and automated (machine) inspection and the relationship of these two methods. Historically, there has been a significant level of interest in particulate matter of visible size in injectable products (Kramer 1970; Sandell 1953, 1974). The level of visible particles in product is believed by many to provide an indication of process control; a significant increase in visible particles has been suggested to be an indicator of a poorly controlled manufacturing process.

Despite the availability of automatic inspection machines, human visual inspection is still (as of this writing in 1999) the most widely used technique in terms of the number of user companies. The training of analysts performing inspection must be sufficient to ensure maximum uniformity of inspection results between analysts and to make the analysis as objective as possible. Of all the types of particulate matter analysis that may be applied to injectable products, human (manual) visual inspection has long been accepted to have the potential for the most highly variable result (see Archambault and Dodd 1966).

There is no such thing as accuracy in any type of particulate matter assay (with the possible exception of microscopy). All of the assays that we perform, including visual inspection, are operational definitions; the result is dependent on the assay performed and is comparable only with other results from that same assay. All that we can strive for in the use of a particular particle counting or sizing assay is precision or repeatability. The goal is to obtain the same result each time a specific test is repeated on the same group of samples (Figure 7.2). This is particularly hard to achieve with visual inspection due to the many variables involved and the human factor.

Particles occur in injectables and devices in decreasing numbers as larger sizes are considered (Figure 7.3). Many more particles will be found in the 1–3 μm size range than in the 750–1,000 μm size ranges in any considered volume. The decline in number may look something like that in the figure, but generally there will be many more smaller particles than the figure illustrates and fewer larger ones. Log scales for both size and number are thus most often used for this type of data, causing the curved line of the figure to become a straight one. The Poisson or log-normal type of distribution that generally describes this number-size distribution is well known in the study of particle populations. More about this later.

If we test a number of human inspectors of varying capability (students, seniors, or professional inspectors), we would obtain a result like that depicted in Figures 7.4 and 7.5. Particles that are difficult to see under the conditions of inspection will be detected at a highly variable rate, with some inspectors seeing such particles at quite a small size, and other individuals not seeing them until the particles are of a much larger size (e.g., 500 μm). Small particles of indistinct morphology will give an especially high variation in detection rate (Figure 7.4). Particles that have optical or size properties making them more easily seen will be detected with much

Figure 7.2. Accuracy and precision.

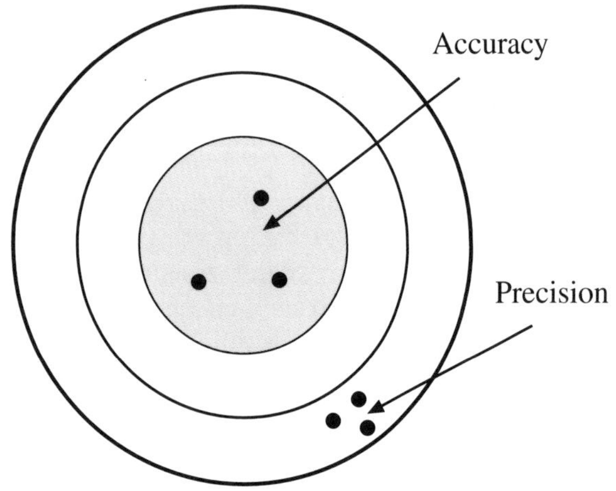

Figure 7.3. The Poisson distribution of particles in injectables.

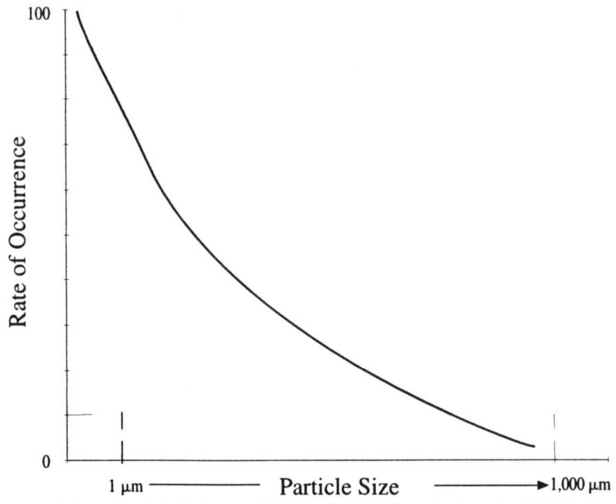

Figure 7.4. Inspector response for "difficult-to-see" particles.

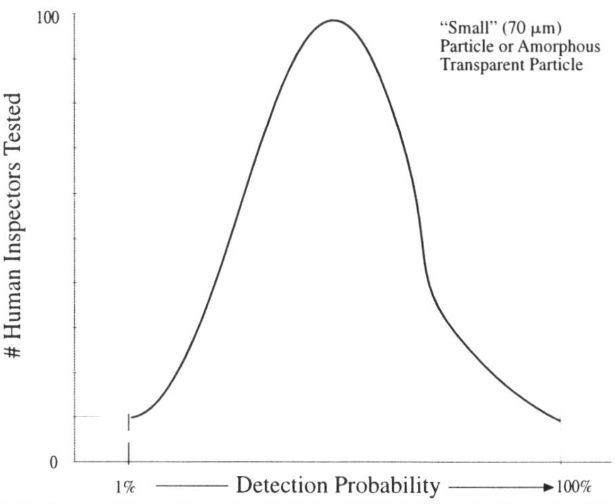

greater reproducibility (Figure 7.5). The downscale half of the curve shown in Figure 7.5 is much steeper than the corresponding portion of Figure 7.4; this "half" curve is, in fact, the response curve of the inspection method used.

Repeatability of Manual Inspection

With regard to the utility of visual inspection as a measurement of a unique quality attribute of an injectable product, reproducibility of the test

Figure 7.5. Inspector response for "easy-to-see" particles.

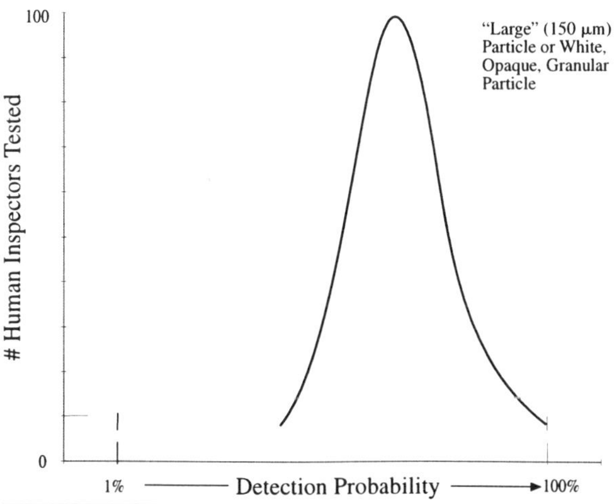

result is thus the most important characteristic (Boucher and Sloot 1965). Prior to the investigations of Knapp and Knapp et al., no focused consideration had been made of the physical parameters of the inspection process that critically inspected the precision of the result. Knapp and Kushner (1982) first investigated and quantified the effects of inspection variables, then proposed uniform detection strategies. Inspections, however conducted, must be designed to provide uniform detection of particles within the size range of concern. This applies to inspections by either man or machine. Once the inspection is adequately structured and controlled, detection of the target particle becomes uniform and we are in business.

Within the past 10 years, the author has been involved in a number of discussions with individuals and focus groups on the subject of visual inspection. A question that arises again and again has related to problematical variation in inspection results between labs or between a supplier and vendor. This invariably results from one or more of the following three errors in approach:

1. Failure to specify the inspection method adequately
2. Conducting inspections of different levels of criticality at different sites or in different labs
3. Lack of a definition of the target particle of minimum size to be detected.

This third item is of great importance. While there are numerous references dealing with minimum detectable sizes of ideal particles, such as latex spheres, that can be seen under specific conditions of inspection, there is a lack of information regarding

- compendial considerations,
- process capability,
- regulatory expectations,
- customer (pharmacist, physician, IV therapy nurse specialists) perception,
- best demonstrated practice, and
- cGMP in relation to visible particles.

All of these factors impact the minimal size of a particle to be detected. Specification of our expectation for the inspection result in terms of the sizes of particles to be detected defines how the inspection is to be conducted; these two factors are closely and inseparably joined in the design of the process.

An important observation with regard to larger databases and statistical data treatments available involves assay sensitivity and detection limits. The statistical treatments of visual inspection data sometimes focus on numerical relationships without a consideration of what is being detected. If the inspection parameters are altered, statistical considerations may change dramatically. Validation of a new machine method against an existing manual technique may heavily involve a procedure of simply adjusting the machine rejection rate (sensitivity) to equal the human inspection output.

If the same set of samples is examined multiple times under identical inspection conditions, one can define the rejection rate for an individual inspector or a group of inspectors. This information is critical to definition of the inspection process. In general, the performance of a manual inspection is not highly reproducible. This was clearly illustrated when a 1949 FDA court action against a U.S. manufacturer was dismissed because the FDA expert, while on the witness stand, could not distinguish his rejects in a group that included randomly selected materials. This scenario underscores the fact that there is variability in the capability of individual inspectors, and that the performance of each inspector can change significantly over time. For instance, rejection probability may vary from 20 percent to 70 percent for 100 μm particles and from 50 percent to 85 percent for particles of 150 μm size in a specific inspection. Knapp and Kushner (1982) found that for 23 inspectors testing the same group of vials, the reject rate varied from 13.7 percent to 44.9 percent.

In addition to failing to detect some particles reproducibly, human inspection systems also invariably have a definable false reject rate. The false reject containers may contain particles at or just below the lower size limit of detection for the test being performed, and thus have a very low detection probability. Several other factors can lead to the rejection of units that do not contain any detectable particles. Inspectors occasionally mistake an air bubble for a particle. If several samples are examined at the same time, human error may result in the removal of a good sample instead of a defective one (Hayashi 1980). In some drug formulations, immiscible components of the formulation will confuse the visual inspection. A critical issue in the inspection of reconstituted dry powder vials regards the differential solubility (different dissolution rate) of large and small particles of drug: large particles that dissolve more slowly may be perceived as extraneous particles.

There have been a number of attempts to assess the objectivity of the visual methods used for the inspection of ampoules. Early evidence of the lack of objectivity was presented by Sandell (1953), who used 6 inspectors to assess a group of 100 five mL ampoules. Each ampoule was marked and examined repeatedly by each of the inspectors, for a total of 20 inspections. The rejection rates varied between 2 percent and 23 percent (mean value of 11.6 percent), illustrating the uncertainty of the method. Of the 100 ampoules, 51 were rejected on at least 1 occasion, but the worst ampoule was only rejected 14 times out of 20; a probability of rejection of 14/20 or 70 percent.

The results of this test are shown in Table 7.1 and clearly indicate that the manual inspection yielded unreliable results. In addition, Sandell suggested that if the real probability of rejection for units of a batch fell below 50 percent, a single inspection would be ineffectual, so final inspection for particulate matter could be neglected. This suggestion has some merit, but it seems reasonable that visual inspection would be necessary to remove those individual units with a higher than 50 percent probability of rejection.

The data in Table 7.1 illustrate a basic principle of visual inspection. It is extremely important that those conducting visual inspections understand that the process is probabilistic, not deterministic, in nature. With a deterministic process,

Table 7.1. Rejection Rates for 100 Ampoules Inspected 20 Times by 6 Inspectors (Sandell 1953)

Number of ampoules	1	1	2	4	2	1	2	4	3	6	4	7	14
Number of times rejected	14	12	11	10	9	8	7	6	5	4	3	2	1

multiple inspection of the same set of containers would result in the same containers always being rejected. The rejection probability would be one of two values, 0 for good units and 1 for defective units. In contrast, with a probabilistic process, each container will have some associated variable rejection probability that may be any value between 0 and 1. An interesting observation here is that as the size of the target particle increases, the detection becomes more certain and reproducible; by making the inspection target large enough, we can make detection deterministic rather than probabilistic (i.e., a sure thing).

Knapp and Kushner (1980a), as will be discussed in greater detail later, carried out the first carefully designed experiments to determine whether human inspection was deterministic or probabilistic. In these studies, 1,000 vials were examined by 5 inspectors 10 times each, for a total of 50 inspections. The summary of these results, Table 7.2, demonstrates the range of rejection probabilities. Only 2 vials were rejected on 100 percent of the inspections; approximately 10 percent of the containers were rejected 10 percent of the time.

Knapp et al. (1983) also defined a relationship between rejection probability and particle size. Vials with the smallest particles fell into the lowest rejection probability groups; those with the largest particles had the highest rejection probability. In summary, larger particles resulted in a higher detection probability. The findings of Knapp and his coworkers are consistent with the function of the human visual process (Blackwell 1946, 1959). Accordingly, any description of visual inspection methods must include a consideration of the capability of the inspector, the size of the particle, the background illumination level, and the contrast of the particle against its background.

Table 7.2. Results of Knapp and Kushner (1980a) Experiments

Rejection Probability	Number of Vials
0.0	805
0.1	98
0.2	33
0.3	17
0.4	11
0.5	10
0.6	8
0.7	6
0.8	5
0.9	5
1.0	2

Variables in Human Visual Inspection

In Chapters 8 and 9, the author has included fairly detailed discussions of the principles of light-particle interactions. The same light-particle interaction principles pertain to visual inspection. Human visual inspection systems are almost invariably designed to function in a wide angle, light-scattering mode of detection, with the eye of the inspector collecting light deviated by the particle over a solid angle on an observation axis at 90° from the axis of illumination (Figure 7.6).

In the most common arrangement, the sample unit is illuminated vertically from above by a light source that is shielded from the inspector's eyes. The vial, ampoule, or bottle is viewed on an axis of 90° to the vertical. Other systems have been experimented with, orienting the unit between a light source and the observer; this arrangement is generally unsuitable for human visual inspection due to the high background illumination level in comparison to the diminution of light intensity caused by the particle and the relatively low level of light scattered.

Given that we are attempting to detect light scattered at a right angle by the particle, a number of optical effects become important. Obviously, the more intense the illumination, the higher the level of scatter. This effect is eventually subject to diminishing returns as light intensity increases, due to the increasing brightness of the background and the interior of the inspection booth. Illumination in the range of 2,200–4,500 lux or about 200–400 foot-candles is usually adequate.

Two effects in light-particle interaction will significantly affect visualization: reflection and scattering. Reflection, which will be either diffuse or specular (coherent), is most important. Particles that are smooth surfaced produce specular reflectance and are visualized as tiny mirrors. A particle with a finely roughened surface will appear white due to the combination of diffuse reflectance and interference of scattered light rays. The combination of these effects causes particles to appear white or light-

Figure 7.6. The manual visual inspection process.

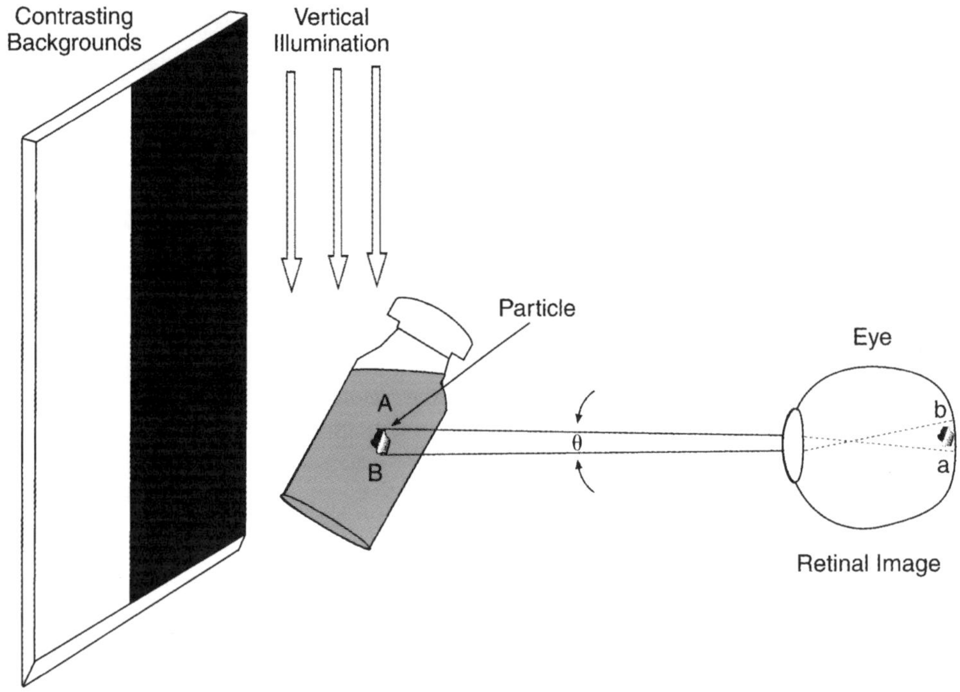

colored. The effects of color or light absorption decrease rapidly with particle size. Contrast of the background with the particle is of obvious importance. With commonly used human inspection techniques, most particles are detected against black backgrounds. Intensely colored particles may be seen as dark or black against white backgrounds.

Other factors also enter into the inspection process. Motion of the particle is extremely important, as the eye will be almost magically drawn to a moving object. The solution within the inspection volume (i.e., the volume inspected in a given time increment) and any particles in that volume should be kept in motion during inspection. Longer periods of inspection generally result in increased detection sensitivity, but variability in detection may also increase as well due to the increased probability of a favorable reflection being caught by the eye. Thus, pacing or structured timing of inspections is often applied as a uniforming factor. Magnifying lenses may be used to enhance human inspection, but the author's personal belief is that they are unnecessary. Some of the confusion about the size of visible particles results from studies done using magnifications of 1.5–2× to detect particles in the 50–60 μm size range. If magnification is varied, a wide range of magnification and detection sensitivities will result.

It is not uncommon to hear 50 μm quoted as the minimum particle size detectable in the visual inspection of injectable products. This figure is probably unduly optimistic. Particles far too small to be visually detected singly can be detected in high numbers due to their collective light-scattering effect (such as immiscible liquids) and will be visible to the human eye. For example, silicone oils employed to lubricate the container and/or closure may be seen visually as haze. Semisolids such as partially solubilized lyophilized cake or silicone-cake agglomerates in dry powder drugs are also in this category. Air bubbles are especially prone to cause false failures in visual inspection. Subvisible 5 μm crystals of precipitated drug material may be seen when present in large numbers due to specular reflection. Haziness or "tornado" swirls in the parenteral fluid are visible examples of these groups of small particles.

As we have discussed, all human visual inspection incorporates multiple variables that impact the precision and accuracy of the result. Particle detection in a visual inspection by a human analyst is critically dependent on a number of factors, including the following:

- Visual acuity
- Particle motion
- Volume of container inspected
- Type and intensity of lighting
- Background-particle contrast
- Time of inspection
- Particle movement
- The total background illumination
- Use of magnification

The human inspector determines the quality and success of the manual inspection process. Since the inspection process is subjective in nature, key limitations of the process lie with the vision, attitude, and training of the individual inspector. As a minimum standard, personnel assigned as inspectors should have good vision, corrected, if necessary, to acceptable standards. Inspectors must not be colorblind. Visual acuity should be tested at least on an annual basis. Good attitude and concentration cannot be overemphasized. One of the major limitations of human inspection for particulate matter is reduced efficiency of the inspector due to a lack of concentration. This negative factor may occur due to incidental or chronic distractions in the inspection area or due to "off-the-job" elements of concern. Unfortunately, the lab manager must often deal with inspector lack of focus on a subjective basis; help is available in industrial psychology texts and other reference areas; this human element in the inspection process makes an interesting subject for a literature search.

Inspector fatigue, which depends on such factors as inspection booth design and type of seating (chair) is also very important. Other psychological factors also come into play when humans are considered. If, for instance, the total number of reject units in a batch is ≥ 25 percent, the sensitivity of the inspectors is heightened, and units otherwise not judged defective may be rejected; if the total number of rejects is too low, boredom ensues, and the detection rate may decrease.

Lighting Intensity

There is a considerable variation in the lighting levels prescribed in the different compendia. Along with the visual acuity of the inspector, lighting, consequent illumination of particles to be detected, and contrast are probably the most critical factors in particle visibility. The earliest work of Knapp in the United States used a 225 foot-candle (2400 lux) illumination level produced by fluorescent bulbs. The JP prescribes 1,000 lux (93 foot-candles) produced by incandescent bulbs, and the EP calls out 2,000 lux (186 foot-candles). Knapp's 225 foot-candle inspection light level was selected from a table of task light specifications published by the Illumination Engineering Institute of North America. This level of illumination is called out in the specification for inspection of low contrast, very small size objects over prolonged periods and presumably constitutes a near-ideal balance of critical detection capability and eye fatigue. In Knapp's experiments, the light was produced by a two 1-1/2 in. diameter 20 W Cool White fluorescent lamps. Although the fluorescent lamps provided even illumination at a fixed inspection distance, illumination changed level with the line voltage. Further, the illumination was characterized by a 120 Hz "flicker." Lamp output intensity was checked monthly to ensure output was within acceptable limits.

While incandescent bulbs are used in many currently applied inspection booths with adequate results, they are probably less generally desirable for the application than if a fluorescent point source is used, which contains a prosaic of hot spots and dark areas. The fluorescent bulbs produce a "whiter" light spectral blend and far more even illumination.

The best-suited fluorescent lamps at present are about 16 mm in diameter and 500 mm long and produce approximately 30 percent more light for the same 40 W of power usage as the ones first used by Knapp (250–300 foot-candles at a reasonable inspection distance). Another significant improvement is the generation of flicker-free light if high frequency electronic ballasts are used. The light emission by this system is also subject to less intensity variation as the bulb ages or as line voltage fluctuates.

Visual Acuity and Particle Detection

Human vision allows an inspector with above-average visual acuity to see a particle as small as 30–50 μm in size, but only if lighting intensity, particle color, and background contrast are optimal. If we assume that a particle is of marginal visible size due to lighting intensity, color, or contrast, it will only be seen in a fraction of the inspections performed based on these factors alone. In a simplified consideration, if a particle is only visible at 50 percent efficiency on these grounds, and if the inspector can scan only 25 percent of the total inspection volumes appropriate for this size in a unit during the allotted amount of time, the empirical probability of detection approximates only 10 percent. Larger particles are more readily detected, not only because they are more easily seen but also due to the fact that they more quickly attract the eye when in motion. Thus, larger inspection volumes, less critical focus, and shorter inspection times may be appropriate if only large visible particles are to be reproducibly detected.

The study of the function of the human eye in visual inspection is a complex subject and is summarized well by Knapp (1987). It will be dealt with only briefly here. The most universal and probably the simplest description of visual acuity is that derived from the standard chart used in eye examinations. This chart is of almost universal distribution in the examining rooms of opticians, ophthalmologists, and others who deal with the assessment of eye function. The chart consists of a series of lines of letters (usually eight) that are used to rate a patient's eyesight in comparison to an "average" value. Measurement is based on a distance of 20 ft, and the vision of each eye is assessed separately. The first line, having the largest letters, is assigned a value of 20/120. A person with "normal" vision supposedly can read this line at 120 ft. If a patient can read only this line without correction at 20 ft, their vision is rated as 20/120. An inspector should have visual acuity (corrected or uncorrected) of 20/20 or better.

The near point of the human eye is another critical descriptor. This point, also called the close vision point, is the closest distance at which the eye (corrected or uncorrected) can achieve focus. For a normal-seeing individual, this is about 10 in. (20–25 cm). Far-sighted persons will have a greater near point distance than near-sighted persons. The near point represents the distance at which particles may be seen most critically. Thus, an inspector should (and automatically will) place a container to be inspected in a zone of the inspection box, into which the visual point falls. Some manufacturers add a near point requirement based on Figure 7.7 to inspector vision criteria. This is perhaps a better criterion for visual acuity at the short distance at which visual inspections are conducted.

Two other eye-related factors, motion of the particle and volume to be inspected, enter into our consideration. The eye will require some time to focus on different fields in a container to be inspected. The average fixation time of the human eye is approximately 0.3 sec. If the particle remains in the three-dimensional field, which can be examined critically, for less than this time, it will not be detected. To be seen, an object (particle) also has to have adequate control or difference in grayscale with regard to background. Contrast and the particle size, which can be seen, will differ dramatically with illumination intensity and background type; hence the importance of adequate illumination intensity background.

Initial particle detection is heavily dependent on particle motion. Because of the viscosity of the product, particle motion rapidly decreases after agitation. Therefore, the movement of solution (i.e., spinning of unit prior to inspection) should be adjusted to maximize the effect of motion. Also, in this regard, particle detection depends on the particle moving at a different rate than the solution after spinning is stopped. In very viscous solutions, this differential movement may not occur. The time that a particle remains in motion after manual agitation—the "time after spin"—or the time that a particle remains in motion after spinning is stopped with an automatic inspection or presentation machine is of overwhelming importance. All units inspected must be inverted, rotated, or otherwise agitated for the same time in identical fashion so that the effect of particle motion is the same for all inspections.

Inspection Volume

Another parameter to be considered is the inspection volume. A critical inspection volume is the volume of solution that an analyst can bring into sufficiently sharp focus to detect a

Figure 7.7. Near vision card (approximately 78 percent of actual size). (Courtesy of Bernell Optical, South Bend, Indiana)

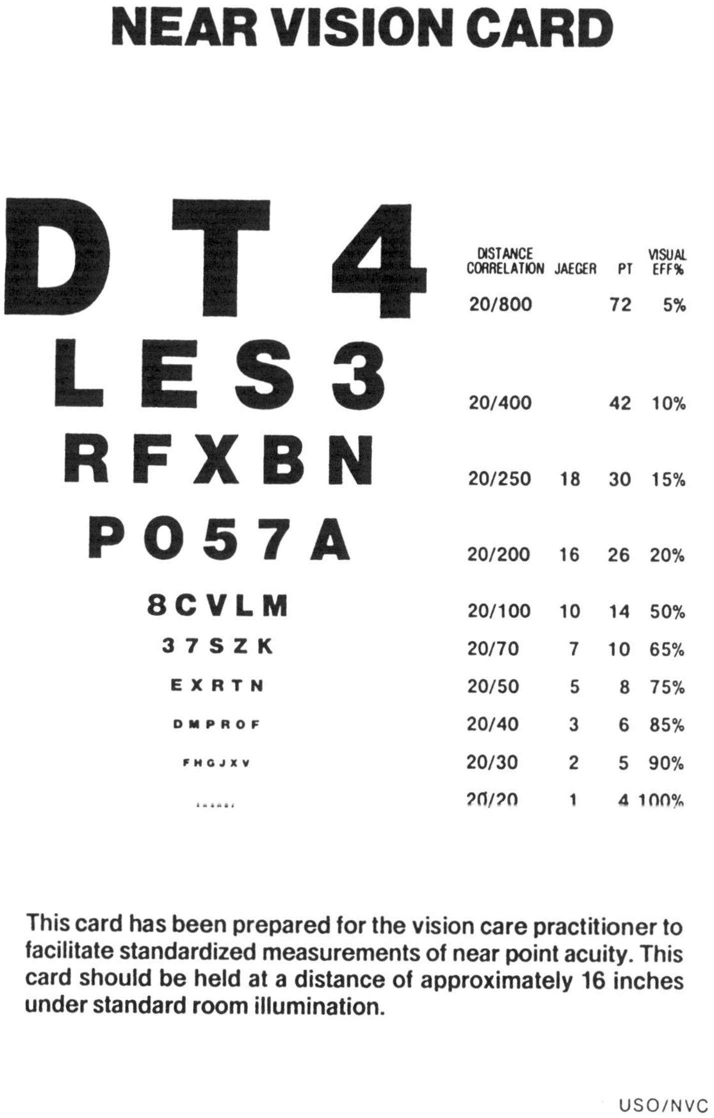

NEAR VISION CARD

	DISTANCE CORRELATION	JAEGER	PT	VISUAL EFF%
D T 4	20/800		72	5%
L E S 3	20/400		42	10%
R F X B N	20/250	18	30	15%
P O 5 7 A	20/200	16	26	20%
8 C V L M	20/100	10	14	50%
3 7 S Z K	20/70	7	10	65%
E X R T N	20/50	5	8	75%
D M P R O F	20/40	3	6	85%
F H G J X V	20/30	2	5	90%
[illegible]	20/20	1	4	100%

This card has been prepared for the vision care practitioner to facilitate standardized measurements of near point acuity. This card should be held at a distance of approximately 16 inches under standard room illumination.

USO/NVC

particle at the limit of human visual capability (e.g., 50 μm). Under ideal conditions, the human analyst's most critical inspection volume is theoretically that of a short cylinder with a diameter and height (depth) of about 4 mm that is perpendicular to the axis of view. For most critical detection, the movement of the solution volume during inspection must be such that particles will remain in motion long enough for a human analyst to scan the entire solution volume in these small cylindrical increments with a time of about 0.5 sec per detection volume. The conditions of many inspections do not approach this ideal and result in a smaller number of larger inspection volumes being visualized by an inspector in the allotted time. As inspection time decreases, the size of the particle that can be detected increases based on the larger inspection volumes, depending on the size of the particle under consideration.

The probability that a particle will be seen thus relates both to its size and the number of appropriately sized inspection volumes an analyst can critically scan in the period of time allowed for inspection. In a large-volume injection (LVI) container (e.g., 500 mL), a large number of inspection volumes exist; so more time must be allowed. If the procedure for agitation of a unit prior to inspection is properly specified, the experienced analyst will know where in the unit volume particles are most likely to be seen. In a swirled ampoule, for instance, most particles may be seen in the lower third of the solution volume when the unit is held upright.

Inspection Time

Variability may also be decreased by pacing or timing the inspection (Sokol and Kirsch 1964). Given sufficient time, particles with a low probability of detection can be detected with low reproducibility; the visual acuity of the analyst becomes the critical factor. Thus, the use of pacing in visual inspections is highly desirable, since it decreases variability. The time allowed may be adjusted so that the particles detected are those that are large enough to attract the eye of the inspector and are detected in the allotted period of time. Schemes in which the analyst can spend more time looking at a unit (if a given unit appears for some reason to deserve more scrutiny) should be avoided at all costs. In some cases, the unit receiving the additional time may contain no particles; in other cases, the increased time results in the detection of particles barely within the visible size range, contributing to a highly variable inspection result.

What Is "Visible Size?"

On the basis of the foregoing discussion, it becomes apparent that the size of a particle detected may vary widely depending on inspection conditions and the human inspector. In fact, no single size can be specified for the detection of nonideal, contaminant particles. Each inspection gives an envelope or range of detectable sizes, with particle morphology (fiber vs. sphere, color, reflectivity) and external parameters, such as light intensity, time, and volume of the container, entering the equation.

Under specific inspection conditions, the human inspector may see a reflective metallic particle or fragment of microporous filter material at the 50–70 μm size due to reflected light from the metal or diffuse reflectance from the highly textured filter material. Conversely, a much larger transparent or smooth-surfaced particle more closely approximating the refractive index of the solution may escape detection at a much larger size. Paper or cotton fibers are generally detected based on their length and the fact that they scatter significant amounts of light.

Several attempts have been made to quantify the size and concentration of particles that can be detected by the unaided eye (beginning with Brewer and Dunning 1947). Five mL ampoules containing 10–500 particles per mL of particle sizes between 5 μm and 50 μm (using polystyrene beads) were inspected by 17 inspectors in a standard booth. Based on a multiple linear analysis model that calculated the probability of rejecting an ampoule as a function of particle size and concentration, the sizes of particles detected at various concentration levels at 50 percent and 100 percent probability of rejection rates were predicted. The authors concluded that a 50 percent probability of rejection rate should be achieved with 20 μm particles in sample solutions in order for potential inspectors to be qualified for in-line inspection. However, it is interesting to note that a minimum particle size of 50 μm was required for all inspectors to reject all solutions containing this size of particle. It must be remembered that container size and type (e.g., glass ampoule vs. plastic LVI container) as well as specular reflection have a significant effect on detectable size.

These authors were also were surprised to find that inspectors were rejecting ampoules that contained multiple particles of glass 1–2 μm in size, using a 100 W lamp and examining against a black and white background. The authors had previously come to the conclusion from a literature survey and by consultation with ophthalmologists that a person with 20/20 vision under the inspection conditions should be able to detect particles of about 50 μm. That inspectors could detect very small glass particles is understandable based on the highly reflective nature of these particles and the high numbers that are occasionally found. Saylor (1966) concluded that clear solutions in glass vials effect some magnification and, under specific conditions, it was possible for some inspectors to detect particles as small as 50 μm in size. (The author questions the wide currency of belief in the lens effect, due to the placement of the particle within the lens rather than in its field of view).

Later work of Knapp et al. (1982, 1983) provided useful information concerning detection limits for human visual inspection. The ampoules that were inspected in this study were characterized beforehand with regard to particle content using a holographic technique. A 17 sec paced inspection of a single 2 mL vial with a 3× magnifying lens, a diffuse light source, and a white/black background was used in performing multiple inspection of the ampoules. The light intensity on the samples was approximately 225 foot-candles, and validated inspectors were used. In these studies, an approximate 70 percent rejection probability was obtained for a spherical particle with a diameter of 65 μm. Equivalent rejection probability using the same inspection conditions with the unaided eye would theoretically result in a 70 percent detection probability for a 100 μm spherical particle.

The visual inspection process for vials and ampoules has also been extensively studied by manufacturers. In some of the results discussed by Borchert et al. (1985, 1986, 1987) at Upjohn, a set of 1,000 ten mL ampoules containing particles as shown in Table 7.3 was used. The particles were fluorescent-dyed polystyrene divinylbenzene spheres. All inspectors examined the entire group without magnification in an inspection booth with standard lighting and background conditions. Paced inspection was used, with a "clip" of 10 ampoules being examined every 38 sec. The analysts chosen for the studies included both quality assurance and production inspectors. The results of one such study with 14 inspectors is shown in Table 7.4. Based on this data, the 70 percent rejection probability would occur for a spherical particle with a diameter between 100 and 165 μm. Despite the differences in inspection rates, magnification, and other conditions, these results were believed to be generally comparable with the earlier findings of Knapp. These experiments and those of Knapp et al. (1981a and b, 1983) suggest that 100 μm rather than 50 μm is a reasonable lower size limit for reproducible detection in a manual inspection.

Table 7.3. Ampoule Particulate Test Set (Borchert et al. 1987)

Number of Ampoules	Number of Particles/Ampoule	Size of Particles (μm)
50	1	165
75	1	100
875	—	—

Table 7.4. Average Results for Inspection of 10 mL Ampoule Material from Table 7.3 (Borchert et al. 1987)

Category	Mean Rejection Probability (%)
Good	1.1
One 100 μm particle	59.9
One 165 μm particle	82.0

The Probabilistic Detection Principle (Knapp et al.)

As stated earlier, Julius Knapp can be considered the original proponent of modern visual inspection theory. In the references, the author has cited a number of works that Knapp has authored or coauthored. The importance of this work should be briefly summarized. Prior to Knapp's first publications in the early 1980s, visual inspection was considered a simple attribute analysis yielding a binary, deterministic, result (i.e., a unit was either accepted or rejected based on whether or not a particle was seen in it). The most significant fact that emerged in the comprehensive studies of Knapp and coworkers (Knapp 1987; Knapp and Kushner 1980 a and b, 1983; and Knapp et al. 1981a and b) was that the manual visual inspection of injectable products is a probabilistic process. This means that in addition to recognizing absolutely good and absolutely bad quality, there must also be a range of variation between these quality extremes. When the rule-of-thumb industry quality categories, "good-good," "bad-bad," and "gray" were defined by these investigators, segments of the continuous quality variation curve were defined and a foundation for accurate comparisons and evaluations was established. The determination based on these definitions describes the inspection security achieved and hence the effectiveness of the accept/reject decision in discriminating acceptable from unacceptable quality units.

With the knowledge that the accept/reject decision of a unit by visual inspection is probabilistic, it became clear that comparing reject rates without reference to individual container rejection probabilities was an inadequate procedure for evaluating or comparing visual inspection results. Accepting the probabilistic nature of the accept/reject decision meant that experiments could be designed to measure both the inspection security and the discrimination achieved to any degree of precision required. For GMP validations, 95 percent confidence limits are judged adequate and were employed by Knapp and his coworkers.

In all of Knapp's work, the use of a carefully specified human inspection is emphasized as a standard. The reality remains, even in this day of high technology, automated inspection machines, that the standard of performance in visual inspection remains the skilled and validated human inspector. The performance of the skilled inspector must continually be used in the validation procedures that will be described as the basic performance yardstick.

Since the inspector's performance in terms of reproducibility and sensitivity of particle detection cannot be divorced from the inspection conditions provided, an accurate definition of the conditions prevailing in the inspection booth are also essential to any review of the results obtained. In much of Knapp's work, a dual 20 W fluorescent fixture with an open white enameled luminaire yielding 14° crosswise light shielding was employed. A pacer was used to establish the inspection period. A selectable white/black background was used to optimize contrast. The test containers were positioned 8 in. below the center of the lamps in a region yielding a constant illuminance of 225 foot-candles. A 2-lamp ballast for a fluorescent lighting fixture was found to be a design prerequisite, which results in a flicker rate of 240/sec that is above the visual perception threshold. Using a single lamp or two individual ballasts results in a flicker rate of 120/sec that is visually perceptible and increases inspector fatigue. Since the manual inspection procedure is affected by inspector fatigue, booth and chair design, incoming material and inspected material container placement, and lens height (if used), adjustability must be carefully reviewed before implementing a method.

The pharmaceutical industry, in general, has relied over the years on the establishment of "standard" conditions for manual inspection that leveraged the Knapp studies. An example, for 2 mL and 5 mL ampoules, is a 20 sec inspection at an intensity of ≥ 225 foot-candles with black and white background. In one "modernized" inspection booth reported by Knapp (personal communication), the backgrounds switch automatically between black and white, saving the inspector critical time in sample placement. In any consideration, the conditions of inspection will define the particle that is reproducibility detected and must be tied into the size of the particle that the manufacturer desires to detect. (The latter decision should be well based in view of the possible need to explain it to a regulatory inspector).

Lastly, this author emphasized that those designing an inspection must consider peripherals. Analyst seating arrangements (i.e., chair) should be the best possible and adjustable in height, so that the inspector will be least subject to fatigue and can look horizontally (not up or down) at the unit being inspected. The chair must be placed in such a position in front of the booth that the analyst can inspect in a normal upright sitting position, using the support of the chair back; under no circumstances should the inspector lean forward into the booth, resting elbows on the tabletop. This latter position increases both muscle and eye fatigue. Human factors must be carefully considered. If, for instance, the analyst is allowed to rest elbows inside the inspection "box," taller inspectors will show a higher reject rate due to inspections at a higher light intensity.

Knapp's most recent presentation of this subject (Knapp and Abramson 1996) deals both with probability theory and inspection method validation and presents a number of evolutionary refinements regarding the application of probability theory to visual inspection. This publication serves as a useful summary of the state of knowledge to this date. Based on this presentation and earlier data (Knapp 1987), spherical particles for unassisted visual inspections may be separated into three categories: subvisible, proportionately visible, and visible. The first category, subvisible, is for spheroids ≤ 50 μm. Knapp's "reject zone" particulate matter (i.e., particles routinely detectable in inspection) commenced at 100 μm. The reader is cautioned that in many of Knapp's studies, magnifying lenses up to 3× were used. This assisted inspection, and the detection size of particles for this type of test is significantly lower than with the unaided eye. Using a 3 diopter lens makes a significant difference in detection sensitivity.

The existence of rejection probability zones based on the size of the particle in a container was the original basic premise introduced by Knapp and his coworkers. This key principle is commonly used for assessing the effectiveness of visual inspection systems. The range of the rejection probability, *p*, was divided into three regions:

$0.0 \leq p < 0.3$: accept zone

$0.3 \leq p < 0.7$: gray zone

$0.7 \leq p \leq 1.0$: reject zone

The region of low rejection probability, into which the great majority of units from a controlled manufacturing process will fall, is the "accept zone." Higher rejection probability, the "gray zone," is critically affected by changes in the inspection process and includes a wide range of rejection probabilities. The region of most certain rejection probability was termed the "reject zone." This group of samples will be rejected at high probability by an effective visual inspection process.

These three quality categories now have wide industry usage. An "accept zone" for two inspections may be defined as one with all containers that have less than 1 chance in 10 of being rejected in two sequential inspections (i.e., a rejection probability of 0.10 or lower). A "reject zone" is defined as one with all containers that have at least 1 chance in 2 of being rejected in two sequential trials (i.e., a rejection probability of 0.50 or greater). The "gray zone" is defined as one where the containers could be accepted but that can be sacrificed to ensure the security of the "reject zone" rejections. The "gray zone," therefore, exists in the rejection probability region greater than 10 percent and less than 50 percent in two sequential inspections.

An interesting and simple (classical) empirical demonstration of the probabilistic nature of inspection can be made by simply having 2 analysts inspect a group of 100 units of product. If the inspection is not appropriately designed to effect reproducible detection of units in the reject zone, a result such as that shown in Figure 7.8 will result. Each inspector rejects 5 percent of the units, but only 1 percent are rejected by both. Thus, the uncertainty of detection is unacceptably high, and the inspection result is unacceptable in terms of reproducibility.

Figure 7.8. Venn diagram showing nonreproducible inspection results.

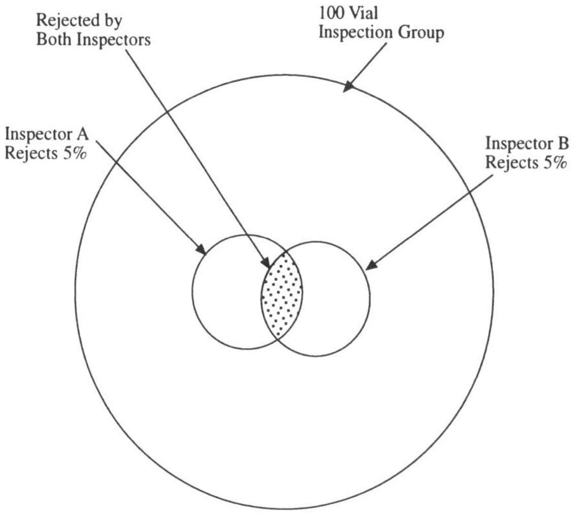

To facilitate closer scrutiny of manual inspection and to characterize the process empirically, Knapp's early studies recorded the accept/reject record of each vial in test groups through multiple, sequential inspections. To generate these records under a set of standard conditions, the accept/reject status of each vial was recorded multiple times. In a typical experiment, as many as 6 experienced inspectors inspected 250–1000 vial test groups as many as 12 times. The test results for all inspectors were then merged. The total inspections yielded a score for each vial from 0 to 100 percent rejection. The mean rejection score stated as a decimal fraction was used as an estimate of the experimental rejection probability for the vial. The error of this estimate is sufficiently low so that it could be treated as a true probability. A summary of the data generated by Knapp for this typical experiment is shown in Figure 7.9.

This analysis, which was repeated many times, characterizes the particulate contamination inspection as a probabilistic inspection. Simply to state that a particle has been observed is an inadequate summary of essential data. The multiple inspection technique step that made possible the data summarized in Figure 7.9 was

Figure 7.9. Histogram of rejection probability for a test group of 1,000 units.

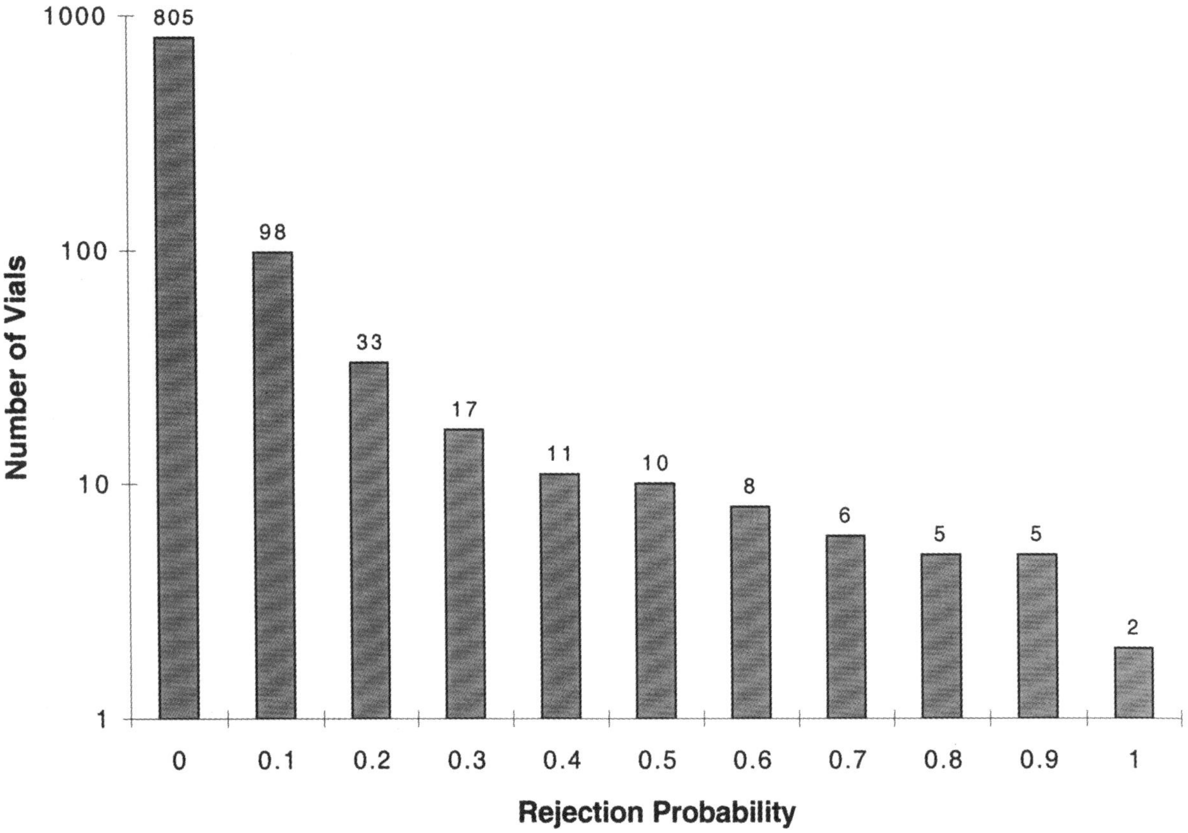

the identification of each container in the test group. Since each container was individually identified, its accept/reject history in multiple manual inspections became possible. This accept/reject history permitted the segregation of the containers into rejection probability groups, as shown in Figure 7.10. The senior inspectors participating, given unlimited time, were able to establish a size distribution from very small to large particles, paralleling the detection probability progression from 0 to 1.0. This summary histogram is the foundation on which Knapp's evaluation of visual inspection systems rests.

In Figure 7.10, the relative rejection probabilities of a single and two sequential inspections are shown. The downward inflection of the rejection probability curve for two sequential inspections shows that serious impairment to the security of the inspection in the high rejection rate zone might be the price paid for obtaining discrimination improvement at the low rejection probability (large particle) end of the quality spectrum. This effect is much reduced with real data because of the preponderance of good material produced in the well-controlled parenteral manufacturing environment.

A further extension of the logic on which Knapp based his probabilistic detection theory is that the average number of vials rejected in a single inspection in each probability group is the product of the rejection probability for that group and the number of containers within the group. These data may be summarized by a dashed line histogram of the type shown in Figure 7.11. The average rejection rate for a single inspection is then the sum of the contributions from each segment of the distribution.

A summary of Knapp's investigations can be obtained based on a simple numerical consideration. Since each zone is now numerically defined, the total population in each zone can be easily determined. The "reject zone" population is symbolized as RZN (reject zone number); for this distribution, RZN is 18 containers. The

Figure 7.10. Rejection probability theory (Knapp 1996).

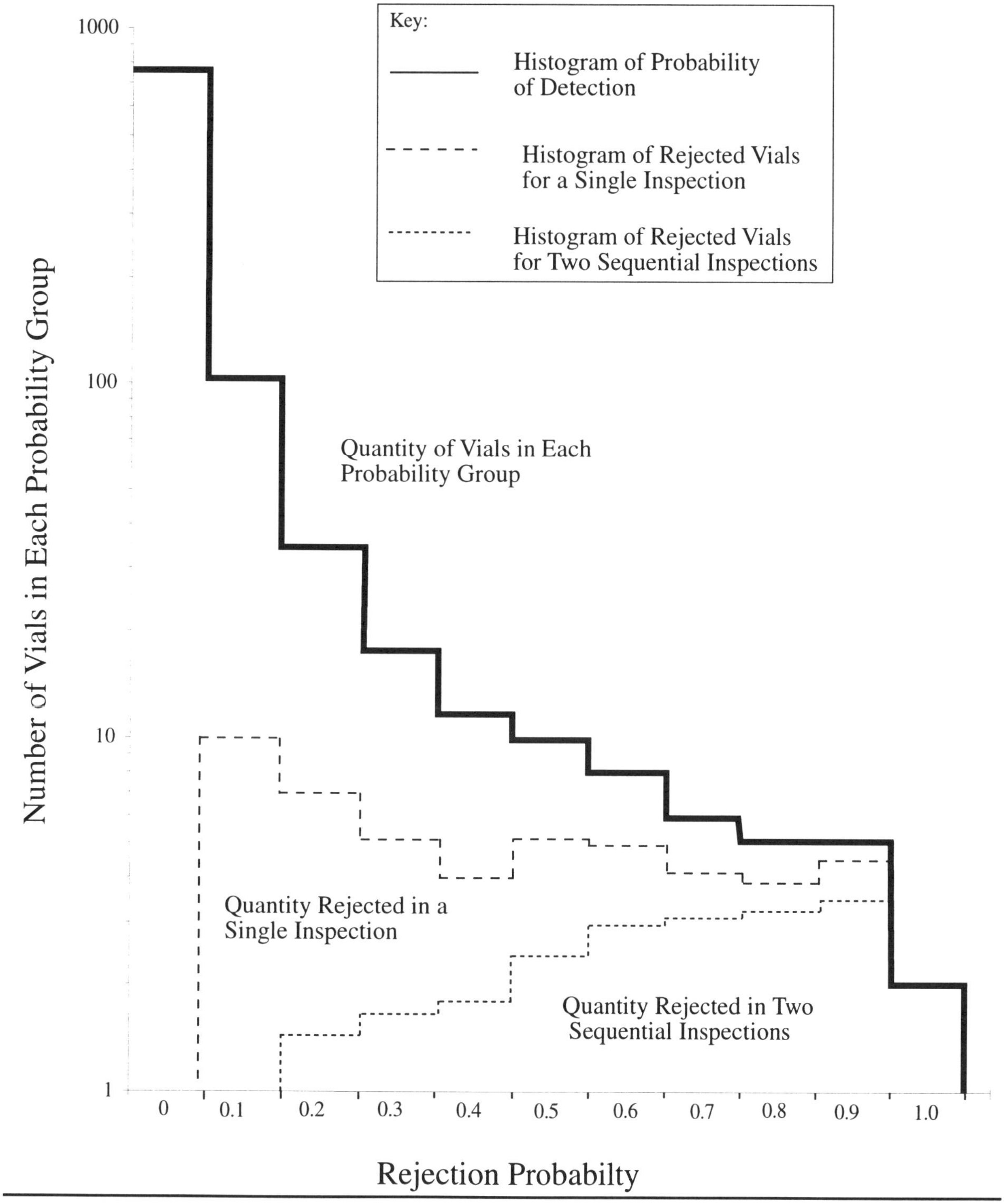

Figure 7.11. Single inspective cumulative rejection probability (Knapp 1996).

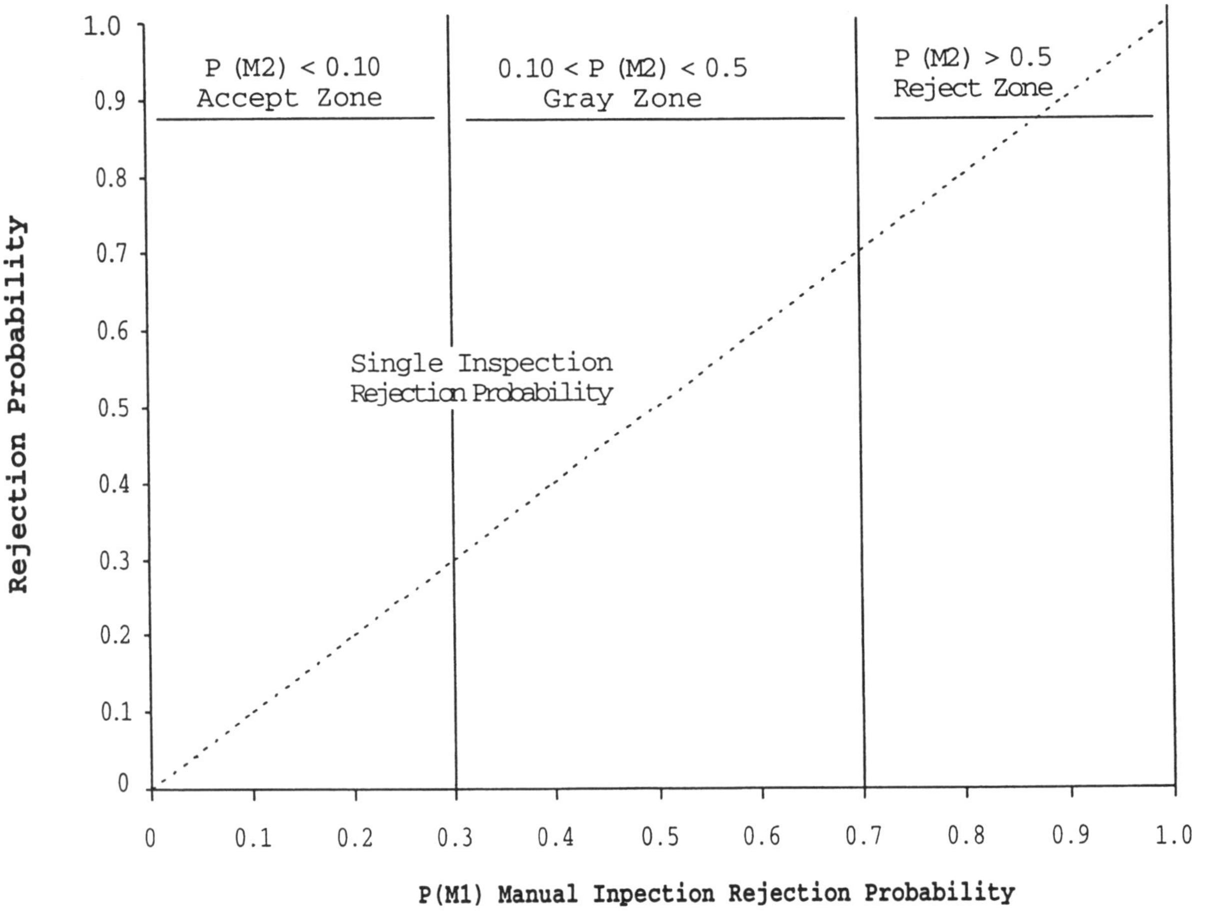

average reject rate for the "reject zone" can be computed as the sum of the contributions from each rejection probability segment in the zone. Shown here is the reject rate for one and two sequential inspections. This reject rate is symbolized as RZR followed by number in parentheses to indicate that it corresponds to the first or second manual inspection; i.e., RZR(1) = 14.7, RZR(2) = 12.2. With the population and the average reject rate for one and two inspections established, the relative security of inspection could be evaluated numerically.

With both the "reject zone" population and the average reject rate established for each inspection, a relative efficiency factor can be calculated for each sequential inspection. This is the ratio of the number of containers rejected in the zone to the total population within it. This ratio is the average rejection probability for all containers in the "reject zone" and is symbolized as RZE(Method 1 or M1)

$$\mathrm{RZE(M1)} = \frac{\mathrm{RZR(M1)}}{\mathrm{RZN}} = \frac{14.7}{18} = 81.7\%$$

The corresponding efficiency for two sequential inspections is defined as

$$\mathrm{RZE(M2)} = \frac{\mathrm{RZR(M2)}}{\mathrm{RZN}} = \frac{12.2}{18} = 67.7\%$$

Knapp stated the opinion that matching or exceeding this objective measure within an acceptable confidence interval of the inspection security achieved by manual inspection for the containers identified as occurring in the "reject zone" should be the only requirement for validating any alternate inspection technique or process (see Table 7.5).

Table 7.5 summarizes the concepts and definitions presented so far and lists additional information for real-world applications of this methodology. For the "reject zone," the single inspection rejection probability bounds are from 0.70 to 1.0, the number of containers and the security parameter RZE (reject zone efficiency) is the ratio of RZR to RZN. The ideal RZE is 1.00; this means that every container in the "reject zone" is rejected with a probability of 1.00 in every inspection. In actual practice, RZEs range from 0.80 to 0.95 for a single inspection.

From Knapp's analyses, it is clear that the final delivered parenteral product quality is not determined or controlled by particulate matter inspections. Implicit here is the finding that product quality will have a significant influence on the statistical data resulting from any test. The principles of cGMP place responsibility on the user to establish equivalence between a proposed and an existing technique or process. The methodology Knapp described provides for a numerical evaluation of existing manual inspection techniques. Importantly, to satisfy cGMP validation requirements, the RZE established for a standard manual method must be matched or exceeded by an alternative method. Through a shift in emphasis from overall reject rates to the average reject rate within defined rejection probability of "accept," "gray," and "reject zones," Knapp made a statistically definable, reproducible analysis available to the user. With this analysis, the security achieved in the particulate inspection of parenteral products can be evaluated with any degree of precision required, and the problem posed by the result when particulate inspection replicability was evaluated with the Venn diagram of Figure 7.8 can be addressed.

One additional requirement discussed by Knapp in the 1996 reference is the choice of inspection strategies. Strategy refers to the matrix in which the inspection is performed—a single pass test with rejected units being scrapped, or a multiple pass test with either the defect or good units from prior inspection(s) being reinspected. Today, many large manufacturers apply multiple inspection methods to reach internally defined quality levels in this regard. The most common approaches are as follows:

- "Accept in 2"—units are rejected only after being rejected in a total batch inspection and in a second "cull inspection" of the original rejects. In this process, units found to be free of particles on a first inspection are released from the inspection as good; the defect unit group is reinspected a second time (once termed a "cull" inspection), and those units found to be free of particles also released to stock.
- "Reject in 2 (or 3)"—units are accepted only after being found free of visible particles in two or more complete inspections of the entire batch. In the most common application of this inspection sequence, the batch is

Table 7.5. Basic Definitions and Relationships Required for Objective Validation of Probabilistic Inspection Systems (Knapp 1996)

	Zone		
	Accept	**Gray**	**Reject**
Zone Rejection Probability	$0.0 \le p \le 0.3$	$0.3 < p < 0.7$	$0.7 \le p \le 1.0$
Population	AZN	GZN	RZN
	AGZN = AZN + GZN		
Average Rejected in a Single Inspection	AGZN – R	RZN – R	
Average Zone Rejection Probability	$\text{RAG}^{**} = \frac{\text{AGZN} - \text{R}}{\text{AGZN}}$		$\text{RZE}^{**} = \frac{\text{RZN} - \text{R}}{\text{RZN}}$

inspected a first time, with all defective units being scrapped. The procedure is then repeated on the remaining units, with defects again being scrapped.

The first procedure obviously will result in some units that might be wrongfully destroyed being saved, it leads to speculation regarding the fate of units falsely accepted in the first pass total batch inspection. The second procedure is a more common strategy when high-speed machine (automated) inspection is performed; combinations of machine and human inspection may be performed to attain the desired confidence level with regard to reject removal.

Importantly, in different inspection schemes, different controlling probabilities come into play. In Knapp's original work, the controlling probability was simply the probability of rejection in either a single pass or "accept in 2" manual inspection. If a "reject in 2" approach is used, the controlling probability becomes that of accepting units that are actually in the reject zone.

Validation of method equivalency will be required if a new machine or manual method replaces an existing manual technique; should the method of inspection be changed, validation of equivalency provides the necessary link between new and historical data that ensures maintenance of product quality levels. In this regard, the databases of large companies and statistical data treatments defined by Knapp are of significant value.

In regard to validation, the RZNI tables provided in the Knapp and Abramson (1996) publication are very useful. While an appropriately detailed consideration of the use of these tables in a specific validation is best undertaken with a quality statistician, a summary of their application is useful. These tables relate the number of repeat inspections of a test group of reject zone containers to an estimate of the controlling probability of the inspection(s) performed at a 0.05 level of probability on a test group of 56 reject zone units. The tables provide a numerical listing of the equal number of inspections needed by each method in validation to be confident at a 95 percent level of equivalent RZE. As an example of the simplest use of the tables, Table RZNI-1.0 considers single inspection validations (Method 1 Vs. Alternate 1 or M1/A1). To achieve our benchmark (0.85 RZE), we see from the table that 1,104 total inspections will be required for validation at the specified levels of equivalency. If there are 56 containers in the group of RZE units, the number of repeat inspections of the group will be 1104/56 or 19.9; this is rounded up to 20. If 2 manual methods are considered, then 5 inspections of the test group by each of 4 inspectors with each of the 2 methods will be required.

As a final caution to those performing the statistical manipulations necessary for validation based on historical databases, it must be kept in mind that these data were collected with specific human inspection method or specific machine settings; care must be taken that data from different methods are not lumped together. If inspection parameters are altered, statistical considerations may change significantly.

Detection of Particles in Translucent Containers

The data reviewed by Knapp in the majority of his publications were based on the examination of clear glass ampoules or vials. Translucent containers offer a less ideal subject for the detection of particulate matter. This effect was examined briefly in experiments on syringes, which are summarized in Table 7.6. In addition to normal lighting, light at 2,000 lumens was delivered through 2 lateral, rectangular-orifice fiber optics. Using the glass syringe inspection as a comparison standard, it is fully matched by the data shown in line 3 taken with 8 TPX syringes having a newly manufactured surface. In line 2 are the inspection results for TPX syringes with a slightly worn (crazed) surface. The presence of crazed surfaces reduced visibility of 40 μm spheres from excellent (always seen) to good (seen more than half the inspection time). By comparison, the 4 lower lines for polypropylene syringes show detection for 40 μm and 80 μm spheres to be equal in probability to 10 μm spheres in clear transparent containers. The translucent polypropylene syringe walls significantly curtailed the effectiveness of the visual inspection. The JP offers a specific inspection method for translucent containers with a 8,000–10,000 lux illumination level.

Methods of Manual Inspection

Manual (human) inspection of injectable solutions for particles of visible size is currently performed by many smaller manufacturers on

Table 7.6. Transparent/Translucent Container Effect on Particulate Detection in 2 mL Syringes

		Detection Probability			
Material	**Clarity**	**80 μm**	**80 μm**	**40 μm**	**10 μm**
Glass	TP	E	E	G	F
TPX	TP	E	G	G	F
TPX	TP	E	E	G	F
PP JOOd	TL	F	F	B	Z
PP 6513	TL	F	F	B	Z
PP DN6005	TL	F	F	B	Z
PP DN6006	TL	F	F	B	Z
PP PB 397-1	TL	F	F	B	Z

Legend: TP = transparent

TL = translucent

E = excellent—particles visible during entire inspection

G = good—seen more than half of inspection time

F = fair—seen less than half of inspection time but more than 10% of inspection time

B = bad—seen less than 10% of inspection time

Z = zero—never seen

PP = polypropylene

SVI products and is the method of choice for LVI solutions. For reasons apparent from the previous discussion and to be discussed in the following section, containers ≥ 100 mL are generally not good subjects for a machine inspection, and the adequacy of the automated inspection of large volume units of injectable solutions generally decreases as the solution volume increases. Additionally, many of the LVIs on the market are marketed in plastic containers produced by blow molding or by sealing vinyl or composite-laminated or coextruded films. The flexible nature of these materials, which hampers uniform handling for setting particles in motion, and their variable transparency makes them generally unsuitable for machine inspection.

Smaller volume containers are currently not as frequently inspected by manual means as was the case in the past, and automated or "machine" inspection is widely applied. When manual inspection of ampoules was common practice, "clips" that held 10 units were often used, and an inspection time of 30–50 sec was commonly employed for a group of 10 units. Despite efforts by manufacturers to standardize visual inspection methodologies, almost as large a number of manual techniques is used today as was the case in the past. Typically, these techniques have in common the use of separate black and white backgrounds, illumination normal (at 90°) to the viewing axis so that particles are visualized based on the light that they scatter or reflect, and some means of pacing (timing) so that each unit is inspected for the same interval. Most companies using manual inspection have documented criteria for inspector training, and some have requirements for inspector visual acuity.

Visual examination in diffuse white light is most common. This method enables larger particles (i.e., ≥ 100 μm) or populations of smaller ones to be detected. Garvan and Gunner (1963) used a light-scattering apparatus for their inspections that introduced the intense light beam through the bottom of the bottle with the unit to be viewed in a horizontal position; particles as small as 10 μm in diameter and of moderately high refractive index can be seen as twinkling spots of light by this method and may appear to change color as the angle of viewing changes.

Figure 7.12 shows a laboratory inspector using one type of "light box" in current use. Historically, many different types of devices have been designed and used (see Figures 7.13 and 7.14) for visual inspection in pharmacies. The devices shown in the latter two figures are primarily of historical interest; although they provide for critical inspection, they do not allow easy manipulation of the unit being inspected and do not enable the analyst to work quickly enough to inspect large numbers of units. The latter is almost invariably a key consideration in manual inspection.

Methods for visual inspection may vary widely between companies. White and black backgrounds with lighting from above are the standard environment for visually inspecting product containers. The white background aids in the detection of dark-colored particles. Light-colored or reflective particles will appear in better contrast against the black background. A standard inspection booth (or light box) typically has a front opening for inspector access. A vertical background in the back of the booth is half black and half white.

Light almost invariably is projected vertically downward with a baffle to protect the observer's eyes from direct illumination. Inspection cabinets should have black side walls. The sides of the booth serve to prevent visual distractions. A magnifying lens at 2.5–3× magnification may be set at eye level to aid the inspector in viewing the container in front of the white/black background. With the lens, acuity of vision is increased to improve the level of discrimination. The author is not a proponent of magnification as a general principle. It is arguable that the level of discrimination may become too great, that is, containers are rejected that would not have been rejected had no magnification been used.

Lighting may be fluorescent, incandescent, spot, and/or polarized. The most common source of light is fluorescent. The range of light intensity as specified by various sources ranges between 100 and 350 foot-candles, although the author considers a range of 200–400 foot-candles (approximately 2,200–4,500 lux) to be more suitable, depending on the type and size of the light box. The intensity can be achieved either with 150 W, inside-frosted incandescent light bulbs, or with three 15 W or two 20 W fluorescent bulbs (with the container held about

Figure 7.12. Manual visual inspection using a simple black and white background light box.

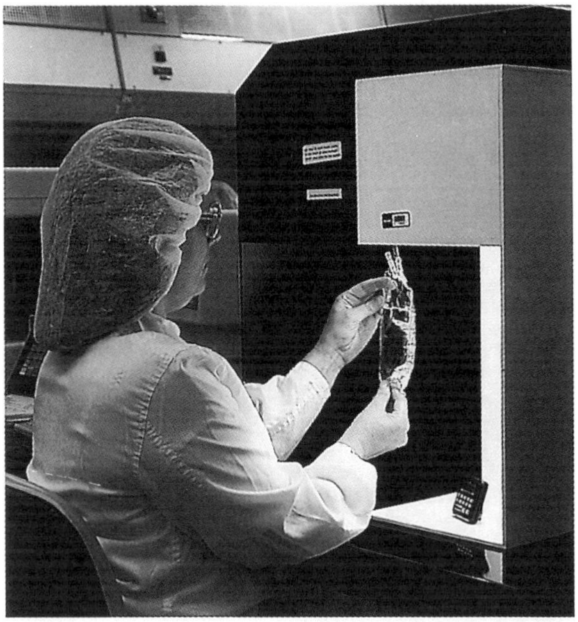

Figure 7.13. Godding visual inspection device (1945).

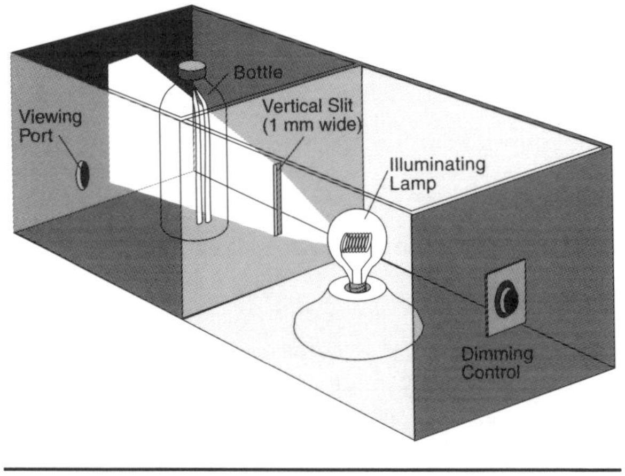

Figure 7.14. Mastenbroek device (1959).

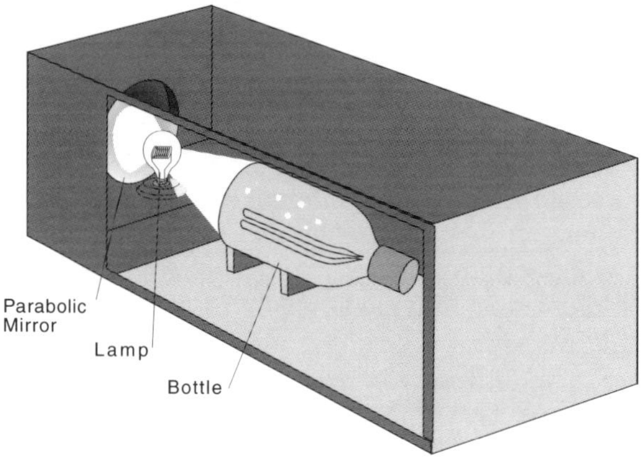

10 in. below the light source) or with other combinations of bulbs that give the same power output. A note here is that these are approximate figures, and that inspection boxes should be clearly specified as to illumination intensity and calibrated to provide the desired light intensity before use. As light intensity begins to weaken, due to age or usage, the lamps should be replaced. Good practice demands that inspection lamps be monitored periodically. Fluorescent lamps are considered by most users to provide a better light source because the light is more diffuse than that from incandescent lamps. Certain types of products (e.g., colored solutions) or certain types of containers (e.g., amber) require increased light intensity over that normally used.

A generalized inspection procedure for manual inspection of product in glass containers might be as follows:

1. Any labels must be removed, and the container thoroughly cleaned. A dampened, nonlinting wipe may be used to remove external particles. (If the inspection is conducted prior to labeling, this step is omitted.)
2. Holding the container by the neck, the contents are set in circular motion by rotating the wrist. Note: too vigorous swirling will draw air bubbles into the vortex of the solution; this must be avoided. Air bubbles will rise to the surface of the liquid slowly; differentiating bubbles from particles by this means is tenuous and should be avoided.
3. The container is inspected while being held at an angle of 45–60° from the vertical about 10 in. below the light source in front of white and black backgrounds. Note: the eyes of the inspector must be shielded from direct light, and the solution volume viewed at an angle to that of the incident light, such that particles are detected on the basis of scattered light.
4. If no particles are observed at this stage, the container may be inverted slowly and inspected for any heavy particles that may not have been suspended by swirling.
5. Any container seen to contain visible particles at any time during the inspection process is rejected.
6. Representative times of inspections are 5–10 sec for small-volume vials and ampoules, 10–20 sec for 50–100 mL samples, and 30 sec to 1 min for 250 mL to 1 L samples.

One other type of device for manual inspection should be mentioned here. It is manufactured by the P. W. Allen Co. of London, England. The Allen ampoule viewer allows inspection of single ampoules in polarized light at a magnification of approximately 2×. The incorporation of magnification provides enhanced detection. Single manual inspection of ampoules is, of course, impractical for inspecting large numbers of units. These devices, especially the liquid viewer, receive fairly wide use in the United Kingdom; their application in lieu of a conventional inspection using indirect white light will obviously require validation by anyone contemplating a change in their visual inspection method.

This device allows an analyst to view the unit being inspected between two (crossed) polarizer sheets and provides enhanced detection of some types of particles (such as paper fibers) in glass LVI units. It is thus a potential enhancement for conducting inspections for the purpose of classifying particles. One potential disadvantage of this unit is that it may not enhance detection of isotropic particles above that attained in indirect unpolarized light. In fact, it may, under some circumstances, give lower detection sensitivity for this type of particle than conventional inspection. The extruded plastic sheeting or tubing of which many LVI containers are made may be anisotropic and produce extinction patterns in polarized light that confuse the analyses.

Although not documented, the concern has been raised by some that the polarized light method may be unsuitable for the purpose of examining compounded solutions such as total parenteral nutrition (TPN) admixtures. The basis of this concern is that freshly compounded units may contain fine crystalline precipitates (e.g., some forms of calcium phosphate) that are of diminished visibility to the pharmacist using polarized light due to either their isotropic nature or the decreased illumination intensity of the inspection. In conventional inspection, particles of this type may be visualized against the black background of an inspection box based on their collective reflective light-scattering effect, even if they individually are below visible size (70–90 μm). This is a potential issue to be considered, since the very fine precipitates present in the freshly compounded units may grow with time to constitute a patient issue.

Applications of Visual Inspection

The manual visual inspection process can be employed in a number of roles. Most human visual inspection processes are off-line inspections, in which the inspection procedure occurs at the completion of the manufacturing, filling, and sealing process. In-line inspection of container components can also be done, especially if the production process can be suitably applied to achieve results without increasing the risk of microbial and particulate contamination. Obviously, the removal of defective containers, such as those showing cracks or the presence of particles, prior to the filling of the product assures product quality and minimizes loss of expensive drug products. Some firms inspect stoppers for adherent particles prior to use. While stoppers are a historically notorious source of visible contaminant particles that can enter the final product, such inspections are very difficult to conduct effectively. A far better means of eliminating particles from this source is an efficient wash cycle validated to exhaustively remove particles of visible size from the closures before use.

The context (i.e., intended purpose for which the data will be used) in which visual inspection is being performed is extremely important. The previous discussion directly relates to the inspection of product as manufactured (i.e., collection of data for batch release). In this regard, the manufacturer must make a determination of whether the product has a suitably low particle burden, based on process capability, process control limits, and pharmacopeial language. The number of extraneous particles present in units as manufactured will not increase over the shelf life of the product, since these particles are denied access once the container is sealed. This initial inspection thus provides no assessment of the longer-term suitability of the batch with regard to particulate matter content.

Different types of visual inspection tests must be considered acceptable and appropriate if they are well defined as to purpose and conduct and are performed by well-trained individuals. The principles of the "accept in 2" and the "reject in 2 (3)" inspection procedures were discussed earlier. While these inspection strategies are of fairly wide use and yield data amenable to statistical analysis, they are not by any means the only inspection formats employed. The most straightforward assay of finished product is a single manual inspection of known detection capability and defined reproducibility of results. Although inferior to multiple inspection schemes based on the work of Knapp et al. (1981a and b), this procedure is still widely used. In practice, this consists of a single critical inspection, with containers found to contain particulate matter discarded. In this type of inspection strategy, 85 percent or higher probability of rejection may be achieved based on careful inspection design for particles of the target size. The author reiterates here that inspection design includes not only physical factors, such as the lighting and timing of the inspection, but also the whole umbrella of inspector qualification items, beginning with appropriate training. The goal of the single inspection is the assessment that the batch is "essentially free" of visible particles through the rejection of defect units, and a comparison of the defect rate detected to historical data from the manufacturing process involved.

Some (relatively few) firms practice what is termed a "cull" inspection at one level of sensitivity and a more critical inspection based on a sample of the remaining units (e.g., 100 units, 200 units) against an acceptable quality level (AQL) standard for the purpose of defining acceptability. Philosophical difficulties with this process are that two different particle sizes are theoretically of interest, and that particles detected in the second inspection (and presumably occurring in some fraction of all of the untested units of the batch) do not elicit action unless the AQL is exceeded.

Another (uncommon) variation of this rationale is the initial application of an initial (manufacturing) inspection at a 100 percent level, which is designed to detect particles of a size sufficiently large to indicate a departure from GMP conditions or particles at a size larger than predicted by the USP <788> subvisible particle limits. This 100 percent inspection is followed by a more critical (quality assurance) inspection of smaller sample groups (e.g., 100 units) for particles of smaller size that are expected to be present in lower numbers than would be predicted by the <788> limits extrapolation.

What about inspections for purposes other than batch release? An example might be visual inspection attendant to a stability study performed on marketed product. Importantly, the purpose of this inspection may be to determine if particles indicative of physical instability of the product or product-container interactions are present in a unit. These particles may take the form of a fine precipitate that swirls up from the bottom of a vial, then vanishes. Amorphous particles, to which light obscuration counters may be relatively insensitive, might be present, visible as gelatinous appearing, stringy colorless material. Solution haze may indicate light scattering by subvisible-sized material, due to the application of silicone oil to the closure. The detection of these types of particles may, in some cases, call for different methods of visual inspection than those applied on-line for the detection of visible extraneous particles.

If different visual inspection methods are applied to a single product at different points in its life cycle, the justification for the different inspection procedures as well as the differing limits must be a matter of record in the form of SOPs. In conducting multiple visual inspections, the manufacturer should be aware that a number of 483 observations have been written based on the application of different type of inspections without appropriate written rationales or the specification of limits and action levels. Limits and requirements that differ from those of the on-line inspection should be justified in the appropriate authorizing document format. Such specialized analyses must be clearly differentiated from others that might be conducted at lower levels of sensitivity or criticality.

A clear philosophical distinction should be made by the manufacturer (before implementing inspections) regarding the impact on batch suitability that detection of particles in different types of inspections will have. For example, batch suitability determinations cannot be made based on the analysis of a small number of units typically evaluated after release during a stability study. The stability assay may, however, indicate the need for further testing. Careful consideration should be given in the design of any specialized analysis as to how the occurrence of isolated extraneous particles will be reported, since this more critical inspection could detect extraneous particles at a higher sensitivity than this on-line inspection and thus have implications for a released batch. Careful management of methods, data reporting, and documentation is obviously in order.

Yet another potential complexity is the application of a referee inspection following the detection of defects in a stability study inspection. This will give a batch a third level of scrutiny, which should be carefully evaluated as to size detection, limits, training of inspectors, and other factors. The purpose of this analysis would presumably be to assess the implications of drug- or container-related particulate matter detected for released materials or for the suitability of a specific formulation. The usefulness of a another visual inspection at this point is questionable. This is, in fact, the point at which the use of instrumental or microscopic test methods should be considered. While these methods are destructive and somewhat more time-consuming to apply than visual inspection, they provide results that may be compared to defined compendial limits for the assessment of sample acceptability.

The conduct of one of the most important types of inspection—the inspection conducted in a pharmacy by a pharmacist or technician to determine the suitability of an admixture—is often overlooked. Pharmacy inspections are most often conducted for the critical purpose of determining if admixture incompatibilities have resulted in precipitation; although these inspections are often of insufficient sensitivity to detect single particles < 150 μm size, they still may effectively serve their intended purpose of ensuring that high numbers of particles in subvisible sizes are not present in solutions for infusion. The presence of precipitates in admixtures is one of the very few situations in which particles can present a threat to patients; an admixture may contain overwhelming numbers of particles in the size range that causes the massive occlusion of organ capillary beds (see Chapter 1).

AUTOMATED (MACHINE) INSPECTION

Many manufacturers have progressed to using automated inspection machines. These machines vary in the extent of automation they provide, throughput, cost, and complexity of operation. As with other expensive, technically evolved instruments or production machinery,

Figure 7.15. Seidenader tableteop inspection machine V 90 T.

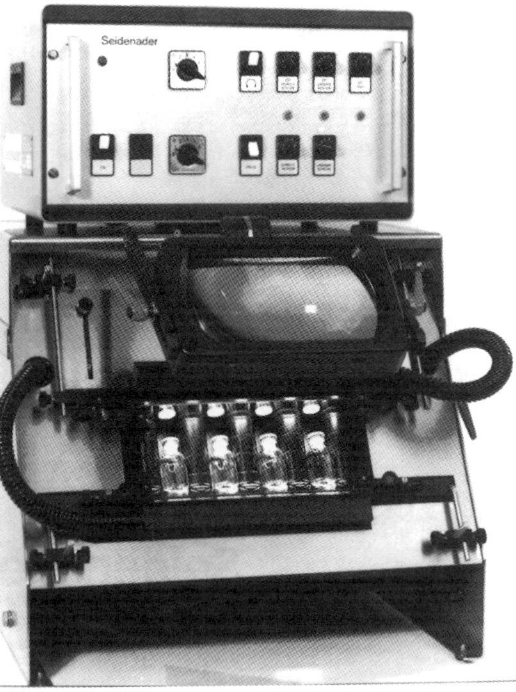

the automated inspection machine cannot be dealt with as a black box. The automation and the consequent savings in man-hours are achieved at a cost of ongoing validation and the operational modifications that must invariably be made to integrate it into production operations. As discussed earlier, appropriate inspection procedures should specify the adjustment of the machine's operating parameters as required to achieve a security of inspection that is at least equivalent to that resulting from a previously established manual inspection procedure. Ongoing monitoring of machine performance is necessary using standard test sets, vials, or ampoules containing standard test particles of known size. Importantly, a structured human inspection remains the only comparison standard for the machine inspection, and the benchmark against the machine must be validated.

The previous discussion has served to establish that manual inspection is a process that is complex but of significant utility if its application is appropriately structured. Automated inspection is no less complex in application. Although the automatic process would appear intuitively to be more reproducible than human inspection, detection efficiency and sensitivity (both of which are functions of machine design) remain current issues (see Faesen 1978; Klein and Reuter 1978; Akers 1985; Knapp and Abramson 1996). The following discussion is neither technically detailed nor comprehensive; the author's intent is simply to discuss principles of operation in sufficient detail to familiarize the reader with principles of machine inspection. The use of a specific vendor's machines as examples of principles of operation is not intended to specify a recommendation for a given vendor's product; operating characteristics of these machines must be carefully matched to the pharmaceutical manufacturer's product, and inspection philosophy and suitability may vary extensively based on specific user requirements.

A number of machines using different detection methods are commercially available. Presentation or tabletop machines running at a rate of up to 1,800 ampoules or vials/hour (Figure 7.15) span the gap between unassisted human inspection and large, fully automated machines that require significant capital outlay and validation. The presentation machines, such as the tabletop model shown Figure 7.15 facilitate uniform test unit orientation to the analyst and particle suspension and provide uniform lighting conditions; however, considerations of analyst qualification and physical parameters of the inspection remain similar to those for the manual method.

Although larger, fully automatic machines are based on different electronics and principles of particle detection, there are common mechanical features between models and vendors.

The papers by Knapp (1987) and Knapp and Abramson (1996) provide useful descriptions of available automatic and semiautomatic inspection machines, vendor listings, and the principles of operation for automated machines. The units inspected are typically spun at a high rate of speed, and the movement of the container is stopped immediately before the sample is examined. This has the effect of causing particles to rise from the bottom of the container to the top of the solution, then spiral downward after container rotation has stopped. This spin-stop procedure results in particles being sensed as moving objects and significantly enhances detection. On most systems, the spin speed can be varied. The interval between the deceleration of the container and the observation time is typically short in duration. This is necessary in order to detect heavy particles, such as glass fragments, that will settle quickly. Conversely, if the interval between braking and detection is too short, bubbles will not have sufficient time to escape the solution and will not be distinguished from particulate contamination, thus resulting in false rejects.

The validation of machine inspection against a manual inspection is complex and any detailed consideration is beyond the scope of this chapter. One ongoing concern is the confidence levels of the two methods in terms of the range of probability of detection at a specific particle size. For large particles, the spread of detection values will be slight, with both methods detecting at near 100 percent. At smaller sizes, the range will broaden; for example, for a 50 μm particle, the spread of probabilities may range from 40 percent to 90 percent for a given inspection.

The Eisai AIM series of inspection machines are widely used both in the United States and other countries (see Knapp and Abramson 1996). The functional layout of an Eisai machine is shown in Figure 7.16. Figure 7.17 illustrates how the proprietary static division particle detection head functions. Immediately before reaching the light beam, each container is spun at high speed and stopped with proper timing so that only the liquid in the ampoule or vial enters the light path. If there is any particulate matter rotating with the liquid, the light

Figure 7.16. Functional layout of Eisai Automatic Inspection Machine (AIM) for ampoules (Eisai, Ltd.).

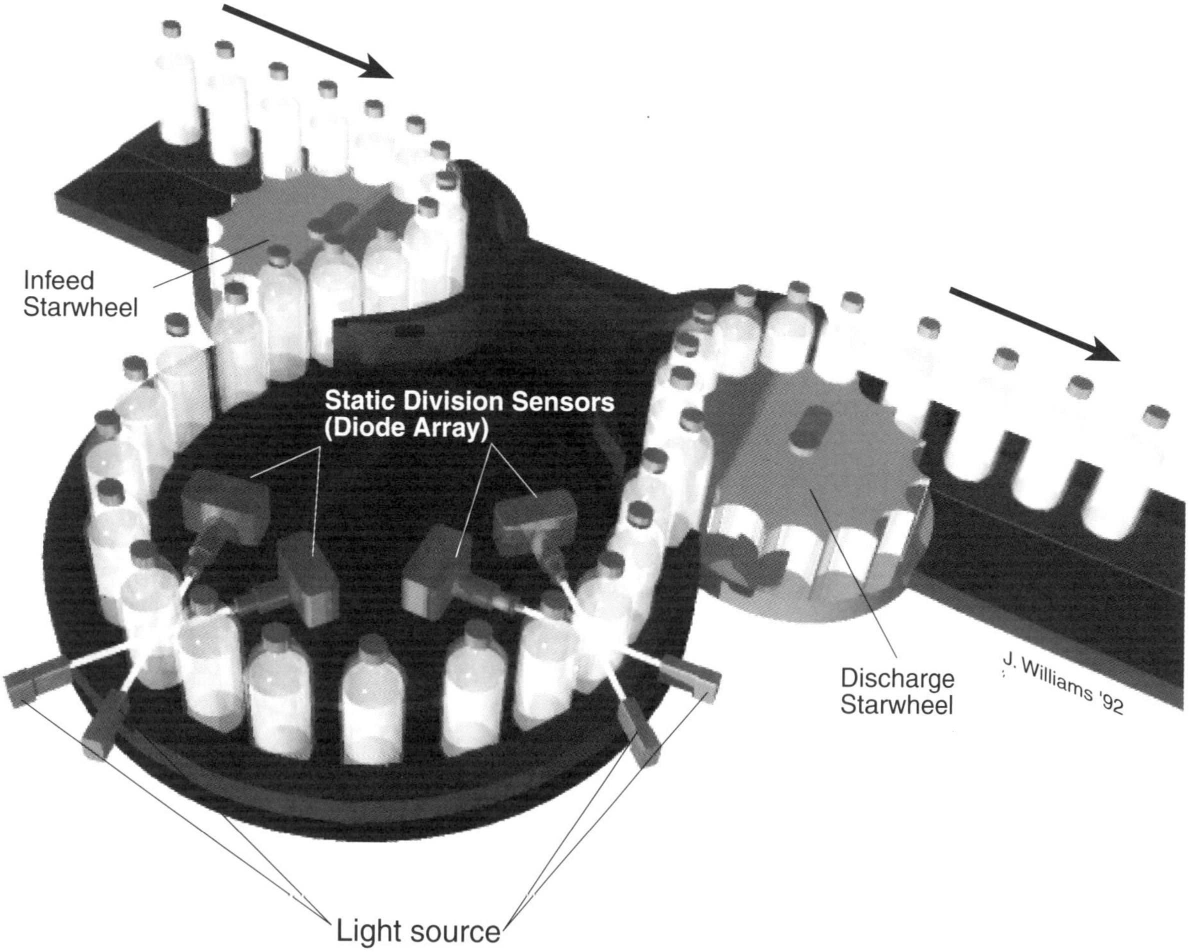

Figure 7.17. Detection system of Eisai AIM (Eisai, Ltd.).

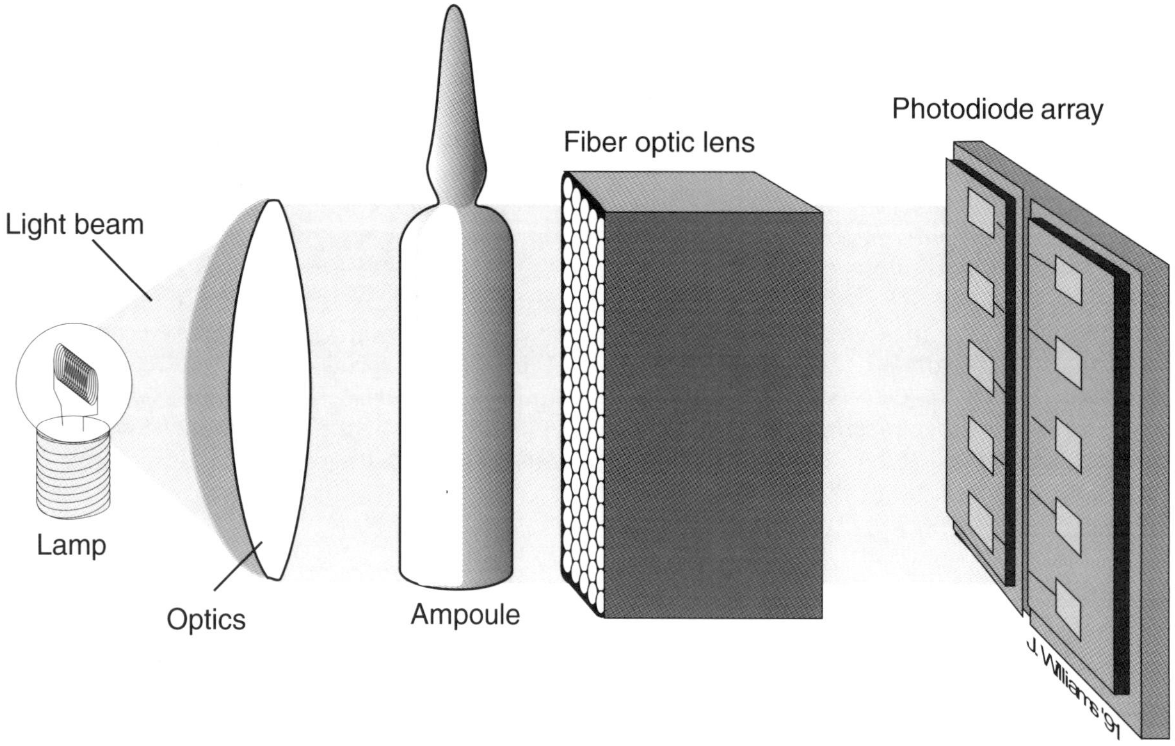

transmitted through the liquid is blocked, and a shadow is cast by the moving particulate matter. Most Eisai machines employ a double check system that inspects each vial twice to ensure a maximum detection rate. Performance of the test is not affected by the presence of flaws or stains on the surface of the container or the color of the particulate matter. According to the vendor, colored particulate matter is detected as accurately as noncolored particulate matter. The machine provides solution level detection simultaneously with particulate inspection, so that empty, overfilled, or underfilled units may be rejected.

The detector consists of fiber optic light guides partitioned into blocks to cover the area cast by moving shadows. The light guides are connected to a photodiode array that converts the shadow to an electrical signal. The photodiode array senses a moving shadow and evaluates its size in accordance with a preset detection sensitivity, which may be based on particle size or on a manual inspection result. Only those ampoules/vials containing particulate matter outside of predetermined sensitivity levels are rejected. The photodiode array design results in an increased sensitivity over designs using a single diode; the shadow cast by a particle most frequently falls on only one diode of the array, so the signal-to-noise ratio is significantly increased over that which would result if a single large photodiode were used.

In addition to high throughput (5,000 to 10,000 units/h), two other operational factors typically enter into a manufacturer's decision to purchase an inspection machine. First, the potential sensitivity of some machines is greater than that of a human analyst. In experiments conducted by Eisai (Table 7.7), a machine set for maximum sensitivity proved capable of near 100 percent detection of particles of sizes that were not detected with good efficiency by trained inspectors. Although the size of the insoluble foreign material detected was not specified by the vendor, the relative detection efficiency of multiple manual inspections and a single machine inspection was notably in favor of the machine. This high level of detection by the machine was based on two inspections; under some circumstances, double

Table 7.7. Detection Ratio of Foreign Matter in 2 mL Ampoule (Eisai, Ltd.)

Insoluble Foreign Matter*	Inspection by the Naked Eye	Automatic Inspection Machine
Small size glass	46%	98.8%
Middle size glass	54	100.0
Large size glass	65	100.0
Small size dust**	42	99.1
Middle size dust	60	100.0
Large size dust	89	100.0

* Small = 20 μm to 40 μm; middle size = 40 μm to 60 μm; and large size = > 60 μm

** Nonglass particulate

inspection by two human inspectors will significantly increase the level of detection in manual inspections. Second, when percent detection for particles of a given size is plotted against particle size threshold, machines (depending on the model in question) often show a very steep detection rate curve (Figure 7.18). It is interesting that even with these highly efficient machines, the detection process remains probabilistic.

Significant differences between automatic systems are found in the manner in which particles are detected. In the case of the Eisai machine, as discussed above, a light beam passes through the container and illuminates a linear array of photodiodes. A particle will cast a shadow on one of more of the detectors, resulting in a signal that is compared to a threshold voltage. If the signal is greater than the threshold voltage, the sample is rejected. Thus, with the Eisai machine, the threshold voltage is the sensitivity parameter. Besides this method of detection, image analysis technology and light scattering have also been utilized.

The other general mode of detection used in automated inspection machines involves high resolution imaging. With a device of this type marketed at one time in the United States by Takeda (Figure 7.19), a video image of the solution unit, illuminated from below, is recorded with a TV camera. This image, which is projected onto a master frame, is digitized in memory. The master is then compared to successive frames of digitized video data from the unit inspected. Two inspections and 24 frames/inspection are used. If any of the digitized images are not identical to the master, an error signal results. Rejection sensitivity is determined by defining the number of error signals that correspond to a defect. Video detection systems have the potential advantage of allowing particles to be visualized and, in some cases, identified on-line as a basis for troubleshooting; the usefulness of this capability in practice is unclear.

The Seidenader automated inspection machines (see Figure 7.20) also employ high resolution imaging and image processing for particle detection. These machines enjoy popularity in the pharmaceutical industry, both in the United States and Europe; this product currently benefits from active development and

Figure 7.18. Particulate detection curve for AIM (Eisai, Ltd.).

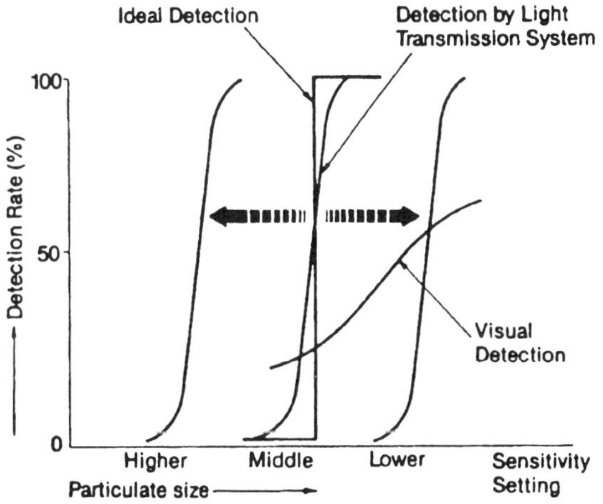

Figure 7.19. Operation of Takeda AK Series Inspection Machine (MTC Corporation).

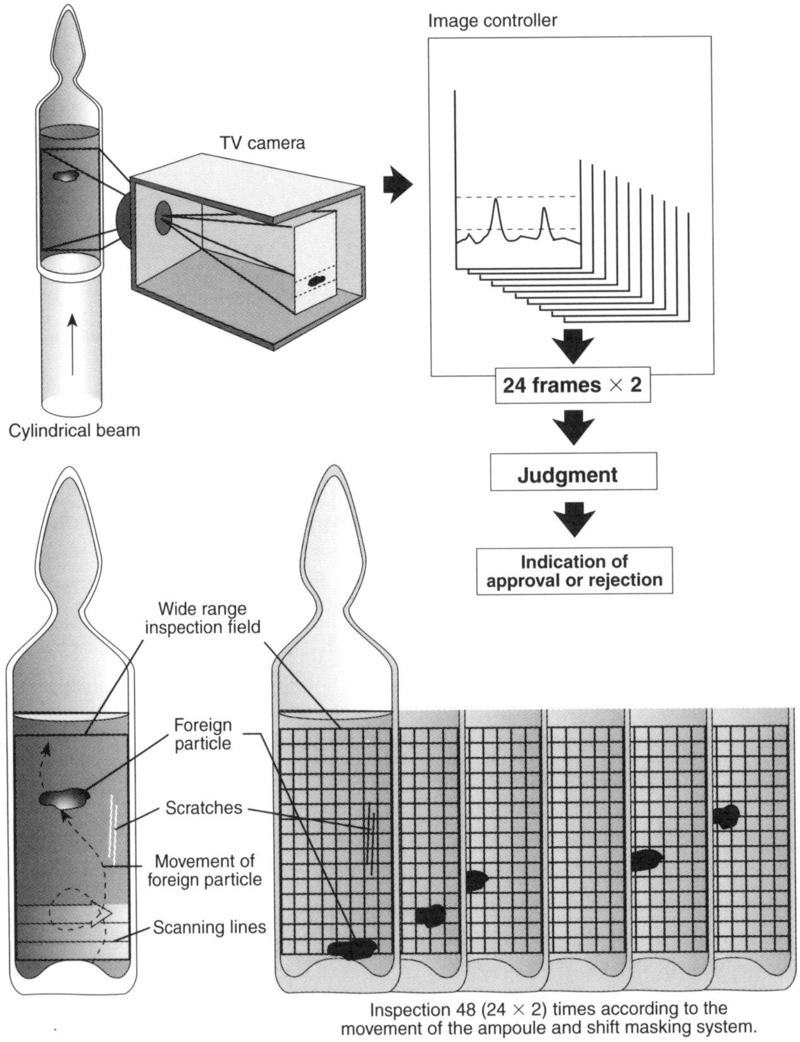

marketing programs. For particle inspection, a separate window for rapidly settling, large, heavy particles is available, as well as the more conventional window for the entire container volume; in addition to inspecting the product unit for particles, the ampoule tip, vial neck, sidewall, and container bottom may also be inspected. Seidenader machines are also available for inspecting the cake in lyophilized product. Lighting and camera positions are optimized for each type of inspection. For particles, as shown in the figure, a "spin-stop" appropriate to the size and volume of the unit are applied; two images are generated by high resolution cameras and subjected to image analysis, which defines particles based on motion between the two frames.

As with human inspection, machines must be evaluated in terms of their critical key characteristics, including size detection limit, reproducibility of particle detection, and false reject rate. In general, the detection limit of machines is as low or lower than the human process. For any particular automated system, the detection limit depends primarily on the setting of the sensitivity parameter. If the response of a human inspection is well characterized, a machine inspection may be adjusted to the same sensitivity based on the use of spherical polystyrene latex particles of appropriate size. For example,

Figure 7.20. Seidenader automated inspection machine particle detection: 1. container rotation; 2. container stop; 3. image "A" generated by camera 1; 4. image "B" generated by camera 2; 5. comparison of images A and B in image processor.

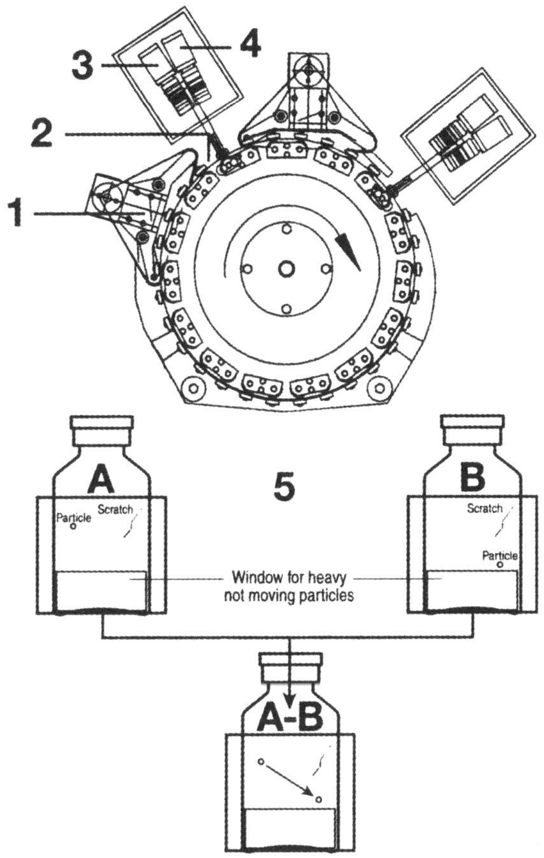

if a 100 μm particle is the smallest particle that falls in the "reject zone" of the normal manual inspection, the machine sensitivity may simply be adjusted to give an equivalent reject rate for particles of this size.

It is interesting to note that manufacturers of visual inspection machines discuss precision and sensitivity of inspection primarily with regard to ampoule or small volume vial product. Both the human eye and automated detection systems are limited by the size of the three-dimensional solution volume that they can critically inspect within a given time and considerations of geometric optics. Ampoule inspection is, in effect, a relatively easy procedure, since swirling of the ampoule will tend to ensure motion of a particle within the focal volume of the instrument or of the human eye. The thin column of solution in an ampoule requires a shallow depth of field for inspection, and the small solution volume results in minimal scattering of both the illuminating and imaging radiation (light). The fact that many of the particles found in ampoules are dark also tends to enhance levels of detection with automated machines (Knapp, personal communication, 1997).

Eisai has overcome some of the difficulties associated with the inspection of large volume units through extensive modification of their standard machine that is used for small volume product. The large volume container machine uses paired detectors for each detection station in place of the single detectors of the ampoule inspection device. This design has the effect of significantly increasing the critical view volume of the machine and reportedly results in a high detection efficiency for particles of ≥ 40 μm in size in the critical lower half of the volume of the large containers.

"ESSENTIALLY FREE"—THE CONCEPT AND THE COMPENDIUM

The USP requires that injectables be "essentially free" of visible particulate matter. We also know, based on earlier considerations of the pharmacopeial language, that any units showing evidence of visible contamination should be rejected. The premise of contamination as monotypic or single-sourced particles has also been discussed. The key points of this discussion of subvisible particles also pertains to visible ones. According to Groves (1993), one of the most cited authorities in this area, evidence of contamination may be provided by occurrence of particles at sizes and/or frequencies above those predicted by a log-log data plot based on the USP limits.

The quality concepts of process control, validation, and capability are interwoven with the visual inspection process and the "essentially free" concept. A general definition of quality is that it represents the degree of conformance of manufactured product to the standards for that product. Single-sourced particles, as well as those of the same chemical composition, are generally considered both by manufacturers and regulators to be "contamination" when present in numbers that approach the USP limits. As an example of this occurrence, we can refer to the hypothetical failure of a bottle washer. In this case, particularly when LVI are considered, each unit may contain one to several visible particles. This is unacceptable (single source and high number of rejects) and may result in scrapping the batch. Similarly, suppose that minor components of a stopper elastomer (e.g., stearates) are leached or extracted and deposited on the solution contact surface of a stopper. This situation (significant numbers of particles of the same composition present due to a component defect) is unacceptable.

The allowable particle burden of units tested must ultimately be judged with respect to the acceptable small quantity of visible particulate matter that may be present in unit produced by parenteral manufacture under cGMP conditions. The most important aspect of the visual inspection procedure is the detection of any particulate matter that is related to solution degradation or any particulate matter present in sufficient quantity or of sufficient size to constitute a non-GMP condition.

It is important to remember that the decision regarding essentially free is intended to be made on the basis of a single unit according to the USP <1> definition. A doctor, pharmacist, or IV therapy nurse makes this decision. A single extraneous particle may make a specific unit unacceptable, but the same is not true by implication of the batch from which the unit came, which contains a much larger number of units. Since some small number of units produced under current GLP may be expected to contain visible process-related particles, small numbers of units with a single visible particle that are discarded may actually be acceptable. Despite rigorous cleaning procedures either at the vendor's plant or in-house, glass vials and stoppers occasionally will bear or contain a single visible particle following filling, stoppering or lyophilization, and these particles may become free in the solution. Such single visible particles of random isolated occurrence are analogous to the "allowable" particle burden under USP <788>: A unit bearing such an isolated single visible particle must be rejected. On the basis of the low level of occurrence of such visible particles in the batch of manufactured materials from which the unit came, however, the batch may still be considered essentially, substantially, and practically "free" of visible particulate material.

A USP proposal for visual inspection in 1987 suggested that 2 percent be an acceptable level of occurrence of visible particulate matter. The current EP proposal would allow a 5 percent occurrence of visible particles. Obsolete Federal Standard No. 142a allowed a visible particulate defect in aqueous solutions of an AQL level of 1.0 percent. World Health Organization (WHO) General Monograph Pharm/92.206 Rev. 2 and Annex 1 and 2 describes a method for scoring the visible particulate matter in injectables, with acceptance or rejection of the batch depending on the results of the scoring by human inspectors.

As discussed earlier, the inspection concept of some manufacturers involves a single 100 percent "cull" inspection in which all defect units are removed. This serves as one (empirical) basis for saying the remaining batch (which will be released) is "essentially free" of particulate matter. Obviously, if we want this "cull" inspection to be reproducible in the context of the Knapp investigations, multiple inspections are required; otherwise, it will mean only that we have discarded an unacceptably high percentage of good units and have left may undetected "bad" ones in the released material. The key here is always to match the desired detection probability at a specific particle size to the inspection process and vice versa. Following a reproducible initial inspection with discard of rejects, partial sampling plans (e.g. pull 50 units, accept on 1 defect or 2 percent, etc.) may be used on the remainder of the batch to specify "essentially free." Sufficient units must be inspected in this inspection to ensure that the original defect rate has been lowered to an "acceptable" level.

The decision regarding "essentially free" must consider both extraneous and product-related particles. Most of the discussion above has been directed at routine products and manufacturing environments (i.e., product inspected immediately after manufacture, before release, at which time extraneous particles or container-related particulate material dominate). In this regard, stoppers (closures) and containers have historically been a source of visible particles. In the July/August 1983 issue of the *Journal of Parenteral Science and Technology,* Whyte reported that the most cost-effective way of increasing the cleanliness of products in a pharmaceutical clean area was to improve the quality of the containers and closures. This assessment probably remains generally correct for many manufacturing situations today. However, at present, the increasing numbers of novel drugs and dosage forms on the market or in development dictate a higher level of concern with particles relating to the physical instability of complex products. Visible particles perceived as "swirls," or fine dispersions of reflective material consisting of individual particles below the classical visible sizes, or faintly visible "strands" or "strings" of transparent amorphous material, may result from product-container interactions or reactions producing vanishingly low levels of insoluble degradants. These materials may require specialized inspection strategies. Detection of these types of materials visually is extremely difficult, and attempts to do so result in highly variable results. A microscope is a far better tool for this purpose than visual inspection.

Can some types of products be considered to be "essentially free" of particles even if they contain dispersions of subvisible-sized amorphous particles due to the predictable degradation of some trace quantity of the active material? The USP <788> requirement does not pertain to products for which the monograph states that a final filter must be used. Some biopharmaceuticals, monoclonals, and blood or plasma fractions may have an inherent particle burden that is higher than simple aqueous drug or electrolyte solutions due to the complex nature of the process and product. The best approach is to document a consideration of the nature of the product thoroughly and build from that base with regard to visual inspection procedures.

The manufacturer may, based on the unique nature of this type of product, be forced to consider an inspection process incorporating detection variability based on particle type as the only effective means of inspection. Discriminatory procedures that allow the unavoidable inherent materials to be excluded from consideration while extraneous contaminants are detected are extremely difficult to design and execute; these procedures may involve sophisticated lighting and sample presentation or a two-stage inspection.

Defining "Essentially Free"

With the concepts up to this point in the chapter in mind, the author has struggled for a number of years (as have others) with the problems of reproducible visual inspections and "essentially," "substantially," or "practically" free. The text below summarizes some of the current thinking of the author and others on the subject; the considerations are those of the author and are not to be construed as recommendations for the manufacturer. The approach taken involves a discussion of possible rationales for defining essentially free and tests that might be used to demonstrate the condition. Although this section provides a basis for considering how to establish a visual inspection process and limits based on the compendial language, it does not provide actual protocols to be followed. The author also reminds the reader that all roads lead back to process control. No particle enumeration assay provides any level of control of the particle burden; what may be provided in the best case is a measurement on which the degree of process control in place can be assessed.

The best general approach to establishing limits for visible particles that will serve to define the "essentially free" condition would seem to be one that considers process factors, the product under scrutiny, and the existing compendial limits for subvisible particles. An important consideration in the latter regard is that subvisible particle limits provide our only size-number measure for allowable versus not allowable, and that numbers of larger (visible) particles may be expected to relate in some fashion (perhaps through a simple log-log plot) to subvisible particles. Key items to be reviewed by the manufacturer in an assessment of the "essentially free" condition thus include the following:

- Definitions of allowable particulate matter
- Process capability and degree of process control
- Compendial specifications for particulate matter
- Best demonstrated practices in regard to particulate matter control

These items are interrelated to the extent that they cannot be considered separately, since all relate to the assessment of the numbers and sizes of visible particles that will be allowed within the "essentially free" definition.

Evaluation of Process Capability and Process Control

A validated, FDA–approved process that complies with cGMP and applies best available technology will generate units that may contain visible particles. The great majority of the units produced will not. Capability in this sense describes the extent to which the process in question uniformly generates product that complies with internal and compendial limits. As a simple measure of capability, monitoring the process output with microscopic testing will provide an objective assessment of the particle burden of units from the process and will allow a plot of particle number versus size to be established. Setting a simple Gaussian confidence limit on this distribution may be used to define what is the largest acceptable particle. A visible inspection process may then be designed to detect particles of the specified size with a > 85 percent probability. By this procedure, the defect particle size may be established; statistical manipulation will allow acceptable numbers of such particles in a batch to the determined.

The principle applied here is to leverage the higher counts and more size-critical microscopic or light obscuration test results. In addition, the more critical microscopic test will allow types and sizes of particles present in the population of product units generated by the process to be characterized. Using the microscopic and light obscuration data, process control values and the degree of process control may be established.

Process control indicators include the number of particles present, size distributions, the types of particles present, and variability of any of these parameters between units. Appropriate low variability of numbers of particles present and the presence of similar types and size distributions between units also defines process control. Statistical assessment is an arbitrary consideration here; quality management statisticians will invariably have the tools to make the necessary judgments involving the prediction of numbers of particles of visible size that will occur based on the empirical determinations provided by microscopic testing. At this point, visible inspection methods may be designed to "pass" units containing no particles larger than the size predicted by the control limits and to detect particles of a larger size.

Compendial Allowances for Particulate Matter

Determination of the expected occurrence of visible particles (numbers and sizes) in product may also be made based on the USP subvisible particle limits as well as microscopic particle count data generated by microscopic testing. A sample calculation is shown below using the USP limits numbers for the microscopic test.

If c is the particle number per mL at some diameter d specified in some limit sets, then occurrence of larger particles may be predicted by:

$$K = -\left(\frac{\log \frac{c_1}{c_2}}{\log \frac{d_1}{d_2}} \right)$$

where

$a = c(d)^k$

Then

$$\text{limit, particles/mL} = a(d)^{-k}$$

$$\text{limit, mL/particle} = \frac{1}{a(d)^{-k}}$$

Thus, for d and c of specified sizes*:

d	c
10	12
25	2

$K = 1.955$
$a = 1.083$

Using these estimates, the resulting limits for larger particles are given in Table 7.8.

The justification for using the USP <788> subvisible particulate matter limits in this fashion is as follows: These limits, when plotted on a log-log grid, will indicate that as particles larger than 25 μm are considered, these will be found to exist as fractions of a particle per mL. For example, when the current USP microscopic limits are considered, (not more than 12 particles/mL ≥ 10 μm and not more than 2 particles/mL ≥ 25 μm), 1 particle of 50 μm size may be found per 2 mL; if we consider a 10 mL volume, a single particle of 100 μm may occur. This procedure of relating number and size to volume may be extended to the whole unit volume. If a

*Some may recognize the USP <788> microscopic limits per mL for LVI here.

Table 7.8. Limits for Particulate Matter of "Visible" Size

d = Diameter (μm)	Limit per 1 mL (c)	Limit: mL/particle (1/c)
50	0.515	1.939
100	0.133	7.521
1,000	0.001	678.760

1 L unit is considered, then a 1,000 μm particle might be predicted to be present.

It should be observed that this calculation generates a line plot that describes the distribution of particles and assumes a slope that is defined by only 2 points. This line will always be a "best fit." With process control data, if the ≥ 25 μm data remains unchanged and the ≥ 10 μm mean count increases, then smaller numbers of large particles will be predicted. If the ≥ 25 μm count increases and the smaller size count is unchanged higher numbers of larger particles are predicted. Stated another way, we can say that the line pivots around the data point that is unchanged. Thus, while this calculation is useful as an approximation, care must be exercised in its use if process data departs from the USP <788> distribution and count ratio.

Definition of Allowable Particulate Matter

The definition involves a choice of "allowable" size for "essentially free" product. Importantly, in recognition both of the compendial intent and the variability of the visual assay, this is a limits test, not a test to determine numbers of units with particles of a certain size. The output of an appropriately designed visual inspection assay (including limits) should define whether the numbers of particles detected are within the process control envelope. The particle size that we choose as a lower limit for a "defect" should relate to the process directly and to the USP limits distribution in some definable fashion per the discussions above. In the illustration above, the USP <788> LVI microscopic limits suggest that a single particle of 1 mm (1 000 μm) in size might occur in each 1 L container. Again, few manufacturers will be willing to accept this level of occurrence of large particle defects, although the figure is useful for the purpose of assessing theoretical particle size distribution shape.

Use of the manufacturer's actual historical counts or control limits at the ≥ 10 μm and ≥ 25 μm sizes seen in product for our definition of the allowable rate of occurrence of visible particles may result in a more realistic yardstick. For an LVI, the numbers plugged into the equation above might be 1 particle/mL ≥ 10 μm in size and 0.05 particles/mL ≥ 25 μm. For example, assume that our process evaluation and revised calculation allows 1 particle per 1 L at the 50 μm size and one particle 150 μm in size per 12 one L units. Thus, an inspection based on reproducible detection of a 150 μm particle would be expected to show in the range of 8 percent occurrence. This is a closer approach to reality, and might be considered acceptable by some for LVI units, given the large total surface area of the container and the terminal sterilization of most product of this type.

This "limit" may, of course, be adjusted downward based on the theoretical occurrence of large particles (e.g., ≥ 50 μm, ≥ 100 μm, ≥ 175 μm) at a rate lower than that predicted by the subvisible count or by using actual microscopic counts of particles at large sizes determined in microscopic testing of product to provide data for modeling distributions. Given the latter type of data for multiple size ranges, a statistician can predict the rate occurrence of larger particles based on Poisson statistics. This empirically determined distribution will most often result in a significant decrease of the size/number inspection limit predicted by ≥ 10 μm and ≥ 25 μm counts or by compendial limits.

With SVI, the scenario is somewhat more complex than for LVI. Higher surface-area-to-solution-volume ratios pertain, and these have historically been pointed out as relating to higher particle burdens per unit volume for ampoules and vial product. This judgment has stood the test of time, and the additional factors of manipulation and difficulties in clearing may also pertain. Accordingly, USP limits per <788> are higher for smaller volume product. The prediction of occurrence of sizes and numbers of larger particles in SVI, as with LVI, will ideally be based on historical data as well as compendial limits.

Best Demonstrated Practice

Best demonstrated practice (and user perception) is somewhat nebulous and, hence, may not be the best tools to work with in our deliberations. Particle sizes and numbers selected based on a manufacturer's process capability or compendial limits generally will be reasonably similar to those determined based on an evaluation of data from the product from other manufacturers due to common practices in applying cGMP, filtration, container and closure cleaning, and other process factors. Competitor data are sometimes hard to come by and constitute only a "snapshot" in time. A useful exception to this situation is the case in which manufacturers work together through industry associations such as the Parenteral Drug Association (PDA) as members of task forces or working groups. The PDA is currently addressing the "essentially free" criterion of the USP through this approach, and the sharing of data by manufacturers will presumably provide a very useful input to the process. Without a broad-based knowledge of the competitor process and product, it is dangerous to base decisions on a relatively small number of samples. A manufacturer is often better advised to learn more about their own product than to attempt to use sporadic data gathered from competitor product in such decision making.

There is also a very real issue with using customer perception as an input in our determination of appropriate inspection criteria. Customer inspections typically incorporate far less critical and uniform conditions than do those of the manufacturers and may reflect extreme variability of detection. Thus, while customer perception is a concern, a manufacturer's visual inspection process must be far more critical and uniform than those applied by customers. If the manufacturer's inspection detects particles only as they reach the size at which customers will uniformly detect them in marketed product, this manufacturer will face the same level of regulatory risk that running an uncontrolled process would entail.

DESIGN OF INSPECTION METHODS

The design of inspection methods may be based on the use of standards or on the inspection result from product as discussed above. Standard particles with general properties similar to those of the particles to be detected are selected and inserted into test units of the type to be inspected. Visual inspection is then designed (using multiple inspectors and inspections) to give a > 85 percent probability of detection for the standard. A more rigorous empirical procedure (and one requiring a high level of technical skill) is to adjust inspection characteristics followed by microscopic evaluation of defect units to determine what was actually detected. Using an inverted microscope may allow this to be done without entering units; the filtration procedure with small volume units is prone to add visible-sized particles and must be performed very cleanly if this approach is chosen.

Referring back to Figure 7.18, which shows the detection characteristic curve for an Eisai automated inspection machine, one of the obvious and basic operating principles of visual inspection is illustrated. The detection sensitivity (sensitivity setting) of the machine may be varied so that particles of different sizes are detected as defects. While there are no measurement units on the x-axis of the graph shown, a favorable impression of the detection response to a particle of specific size is elicited. Ideal selectivity or discrimination would result in a vertical slope of the response curve; the response for the machine on the figure approaches the ideal.

The key to repeatable inspection results with either human inspectors or machine vision is a high selectivity with regard to the judgment between units that contain and do not contain a particle. Selectivity may be defined as the level of resolution of the accept-reject decision. The point here that is of value in our discussion of "essentially free" is that inspections may be designed so that size selectivity is optimized; when selectivity is optimized, the presence or absence of a particle at a certain size can be most reproducibly judged. The selectivity of the human inspection process will depend on the variables involved in the detection process discussed earlier. Given constant operating conditions in terms of lighting, time of inspection, training, and particle optical properties, a population of inspectors will respond more uniformly to particles of larger size. As increasingly larger particles are considered, the response curve becomes steeper.

Thus, in our pursuit of the determination of "essentially free," there are two basic steps to be taken. First, on the basis of process capability, compendial instructions, and cGMP, we should define what sizes of particles will be allowed specific rates of occurrence in the units output from the manufacturing process. Secondly, the visual inspection must be designed to provide the highest possible selectivity of detection for particles at the designated sizes.

The sensitivity of the inspection with regard to reproducibility of detection of particles of specific size may be adjusted based on the critical parameters of the inspection discussed earlier, including, notably, lighting and time of inspection. This is simple in principle but more difficult in execution. With human inspectors, particles differing in optical properties will have different detection probabilities at the same size. Particle shape (e.g., fibers vs. spheres) will also enter into the consideration. The variations in size, shape, and morphology between particles that will occur in inspected units will have the effect decreasing the slope of our selectivity curve and making detection less reproducible. Those investigating visual inspection have historically faced these same difficulties.

Two methods of calibrating the inspection response to particles of selected sizes have historically been followed. The first involves selecting a standard particle, such as a latex or polyacrylamide sphere that is of a size for which inspection selectivity is to be maximized. Training and selection of inspection conditions then simply focuses on reproducible detection of particles of this size and larger and minimizing the probability of detection of others at some smaller size. Smaller particles of the same type may be used as a negative control, representing a smaller size for which detection should be minimized. The ultimate objective is to design an inspection that gives the most significant difference in detection probability between standards of minimal size difference.

A second approach involves selecting units from the process to be monitored bearing particles of size, shape, or origin deemed to be unacceptable. The latter judgment must be based on a knowledge of the process as well as careful characterization and selection of the particles represented in the defect units that will be used for training. Once this selection of the training samples is made, multiple inspections are carried out to maximize the resolution or discrimination between particles in the test group and those representing the "allowable" particle population. Here, as with the use of standard particles, it is obviously desirable to have a second test group of containers with particles in a smaller size range so that relative detection probabilities can be assessed. Due to the existence of multiple variables that come into play, the best result that can be attained is the optimization of discrimination between particles representing the two test groups. Such calibration of the visual inspection response of human analysts will always be difficult; nonetheless, it offers the only method for making uniform the inspector response for particles in specific size ranges.

An important principle that impacts the calibration of human inspection systems is that of the probabilistic detection of particles based on size as the primary determinant of detection rate (Knapp et al. 1983). Given that lighting, time of inspection, particle type, and visual acuity remain constant, the detection probability for particles of different sizes will increase in a generally linear progression as the size of the target particles increases. Conversely, as we vary selected parameters of the inspection (the author hastens to remind the reader that it is not considered technically appropriate to vary inspection sensitivity), detection probability will also vary, with the particle size detected 85–100 percent of the time (i.e., the "reject zone" particle), shifting to larger sizes as lighting and time of inspection decrease and to smaller sizes as the intensity of illumination and interval of inspection increase. For example, consider a mixed low-numbered population (1 particle at each size per unit) with standard particles with 35, 50, 70, 100, and 200 μm sizes represented. At a given inspection time and lighting intensity, 35 μm particles will never be seen; the detection of the 50 μm and 70 μm particles will be at a 50 percent probability; 100 μm specimens will be at 70 percent; and the 200 μm particles will always be seen.

Appropriate decreases of the two varied parameters will shift the sub-200 μm particles toward the zones of lower probability of detection and/or invisibility and leave the 200 μm particles at a relatively high level of detectability. By further "tweaking" the inspection conditions, we will find a set of conditions under which the latter target particles elicit a maximal detection rate; smaller size particles are detected at a significantly lower rate. Repeated inspections of a defect unit group initially selected under higher illumination and longer inspection times will allow the discrimination of the larger particles to be achieved on a statistically significant basis (i.e., will maximize the selectivity of the process). This simple principle may be the basis of an inspection designed to define the number of visible particles above a size selected as the lower limit of the acceptable range.

FDA EXPECTATIONS FOR VISUAL INSPECTION

Based on inspectional observations, presentations, and industry experience with questions asked by inspectors, the FDA has a number of expectations regarding the manufacturer's inspection of newly manufactured product. Based on the author's perception, these expectations may be summarized as below.

Visual detection or examination methods are variable in nature and are carried out on unopened solution containers. These methods are used to ascertain the presence of particles and may allow estimation of their size, but they allow only general quantitation. Fibers or other particles of large size will rarely be present in large enough numbers to be detectable by particle counters, and so visual examination will always be required as a part of routine testing procedures. One hundred percent visual inspection is required by the USP. Two types of inspection generally are expected to be performed by manufacturers:

1. Nondestructive (lot-based) inspections (whenever the container permits, including containers with solid contents)
2. Destructive (sample-based) inspections (when container does not permit, such as many ophthalmic containers and constituted solutions from solid contents)

Critical factors in human visual 100 percent in-line or off-line inspection include considerations of the following, not necessarily in the given order of importance:

- Apparatus and technique
- White and black backgrounds
- Appropriate light intensity
- Effective light shielding and glare minimization
- Magnification
- Time
- Nondistracting environment (minimize noise, visiting, commotion, bright lighting, etc.)
- Inspector training (documentation)
- Particle morphology

Three general inspectional principles appear to direct the focus of audits of visual inspection processes:

1. A well-planned and maintained inspection apparatus and environment should be used to ensure constant inspectional conditions.
2. Procedures for training and conduct of the inspection process and the expected criticality and precision of the inspection should be available.
3. Operators are the most critical inherent variable due to threshold of visibility (which can vary from about 35 to 50 μm in the belief of some investigators), fatigue, boredom, and attention span; emotional state at the moment, general attitude, and so on are also important. Control of the inherent operator variability and detectability differences among particles is critical and involves the following process inputs:
    - Eye exams and vision requirements
    - Training
    - Procedural observation (i.e., supervisor observes inspector performing inspections)
    - Output assessment against known reference (known defect units or "doped" standards")
    - Output audits (review of inspector results and variability)

Historical audit items of interest as reflected in 483 observations indicate that one critical component that will be looked for by an FDA auditor is the firm's audit of the inspectional output. This may involve independent reinspection by internal auditors (i.e., inspection of random samples of nonrejected units from identified segments of each lot). This consideration should include review of the number of reject units based on historical records versus present results. No visible foreign matter should appear in sampled, nonrejected units. Exceeding an action level in the auditor's reinspection should result in 100 percent reinspection by the firm of that portion of the lot identifiable with the subject operator, with auditing of at least double the usual sample size and requalification of the operator and inspection of the inspection apparatus and conditions if necessary.

There are, predictably, also expectations regarding alert and action limits. Alert and action levels in place for the product should indicate whenever the manufacturing process is no longer under control in regard to particulate matter. Patient safety concern must be placed above consideration of economics (i.e. cost of discard). Investigation and corrective action in response to exceeding action levels should include the following:

- Evaluate if lot should be reinspected.
- Some potential manufacturing conditions to scrutinize and correct as needed are as follows:
  - In-line particulate filter investigation (e.g., location, integrity testing)
  - Porosity (5 μm is usually acceptable for particle exclusion)
  - Cleaning, handling, and packaging of containers and stoppers
  - Cleaning and assembly of filling apparatus and all fluid pathways
  - Corrosion, adjustment, or other contribution of filling apparatus
  - Generation of particles in the vicinity of the filling operation—point contamination

A word is also in order regarding the application of different types of visual inspection for a single product, as was discussed earlier. A number of 483 observations have been written regarding the application of different inspection procedures by different functional groups within a company (e.g., R&D, Stability Operations, and Manufacturing). Observations have also been made based on the fact that inspection procedures were not clearly defined and that limits or standards were not in place. There are often reasons for different types of visual inspection at different points in the product life cycle in development, manufacturing, and shelf life.

On a practical note, the expectation of FDA auditors appears to involve an expectation that the output of any visual inspection process must be evaluated periodically. An assay with many variables will show more tendency to drift. The premise to be periodically reevaluated here is that the visual inspection performed effects significant reduction in the original defect occurrence in the manufactured batch. If this does not occur, i.e., if a reinspection of a batch or subsample yields the same defect level (or higher) by the same method, something is amiss.

In this simplistic consideration, a pivotal philosophical concern is a review of the result of the inspection in rejection of defective units. The FDA may refer to this as auditing the result of the inspection output; the manufacturer may term it reinspection or a "cull" inspection, or define it in terms of an "accept in 2" approach. One means of accomplishing this is a reinspection of a batch that yielded a specific defect rate, such as 40 units out of 1,000 with particles of a size giving 85 percent probability of detection. If this batch yields the same defect rate on reinspection after the removal of rejects during the first inspection, the inspection is probably ineffectual. If the first inspection yields 40×0.85 or 34 rejects, the second should theoretically yield 6 rejects/966 units.

In the event that a first 100 percent inspection indicates that an upward trend of defect number may be taking place (e.g., a result of 9 percent rejects against a historical figure of 2–3 percent), not only should the cause be investigated, but the inspection result should be scrutinized to ensure that the elevated reject rate did not affect the result of the inspection. In the data of the example, the 9 percent result may actually have been low, due to the inspectors being used, seeing only a much lower number of rejects, and failing to detect reproducibly at the higher level.

The author will comment in closing regarding the regulatory interpretation of the "essentially free" compendial language. Auditors well know that no visual inspection is conducted at the 100 percent detection level for particles of a size that maybe picked up later by a user of product in the field. Thus, the subjectivity of compendial wording allows the FDA reasonable latitude in making judgments with regard to batch suitability based on the detection of small numbers of particles in the field. Some particles detected in the field may also have been persistently adherent to the stopper or in the neck area of a unit and thus were not free in solution for detection during the inspection, despite the best efforts of the analyst to dislodge them.

VISIBLE PARTICLES AS ATTRIBUTES VERSUS VARIABLES

A further word is necessary regarding the two major categories of visual inspection. If the inspection system gives a binary (yes/no) result, the inspector is said to be based on attributes. Only one of two results, i.e., the presence or absence of particulate matter, is possible. This is the common application of visual inspection and has been the subject of discussion in this chapter thus far. While the defect rate may be based to some extent on the size of particle detectable, the basic judgment is defective versus acceptable.

On occasion, however, it may be necessary to investigate the occurrence of visual particulate matter with respect to its source and/or chemical composition. Generally, this is undertaken if higher-than-normal numbers of particles are observed in visual inspection. Typically, if higher numbers of particles are seen, the average size of particles detected will also increase. More detailed visual inspections may be performed to gather more information on particle morphology as a time-effective prelude to more complex analytical procedures.

If the inspection result has a number of possible results (e.g., measurement of the size of a particle), the inspection may be said to be a variable-based one. Visual inspections may be performed on this basis, with the desired result being not only the number of defective units detected but also a judgment of the size of particles. In the USP, for instance, an inspection of ophthalmic ointments is directed at metal fragments ≥ 50 μm in size with an allowable number. In such a visual inspection of product to determine variable properties of particles seen, the size and morphology of particles are typically recorded. Morphology is described in terms of descriptors that analysts are trained to use in standardized fashion.

- *Particle:* Fragments of material that have a generally regular, equant, or cuboidal morphology with no evidence of roundness, flattening, or elongation.
- *Fiber:* Particulate matter that appears elongated by inspection to the extent that its length (greatest dimension) is in excess of 10× its width.
- *Flake* (or *Disk*): Particulate matter characterized by a platelike or flattened visual appearance.
- *Crystal:* Particle having an angular, regular shape.
- *Spherical:* A sphere-shaped particle or one that approximates this morphology.
- *Agglomerate* (also called *floccule*): A loose association of smaller particles that may be dissociated by agitation.
- *Precipitate:* Collective term describing a particle population consisting of a large number of particles of closely similar morphology (e.g., crystals, flakes) resulting from chemical precipitation or solution degradation.
- *Color:* Analyst visual impression of color, also including general terms: "light colored," "white," "black," or "dark colored."
- *Large:* Particulate matter that, based on analyst experience and/or judgment exceeds the size expected to be present in a sample unit. Also, particulate matter that is readily observed in a short inspection or immediately upon beginning inspection.
- *Small:* Difficult to visualize particles, fibers, flakes, or other particulate matter that may require long periods of inspection to detect.
- *Glitter* (also known as "swirl"): Multiple (numerous) reflective or shiny particles, often of very small size.
- *Numerous:* Particulate matter present in numbers too high to enumerate accurately.
- *Reflective:* Particulate matter that appears to redirect or reflect the incident light depending on particle orientation.
- *Transparent:* Particulate matter that appears to provide little or no hindrance to the passage of light.
- *Opaque:* Nontransparent; blocking the passage of light.
- *Swirl:* A collective term most often applied to populations of very fine particles that exist as sediment in a vial or ampoule and are visible as the container is rotated.

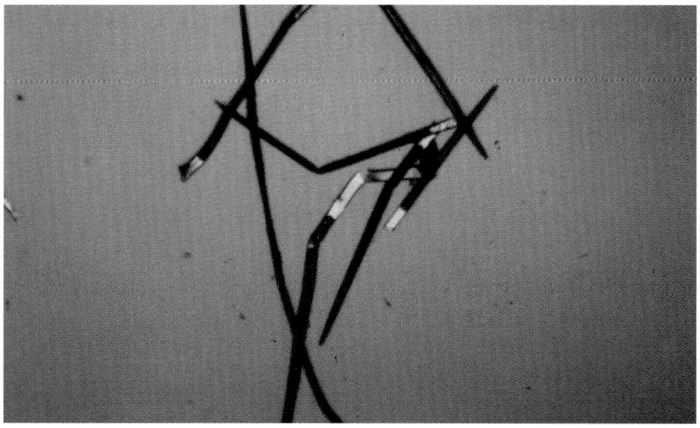

Exhibit 1. Insect Parts

Exhibit 2. Rust (Iron Oxide)

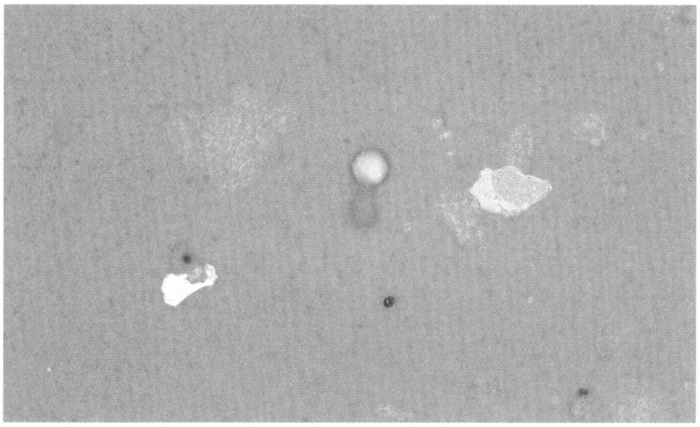

Exhibit 3. Cosmetic Residue

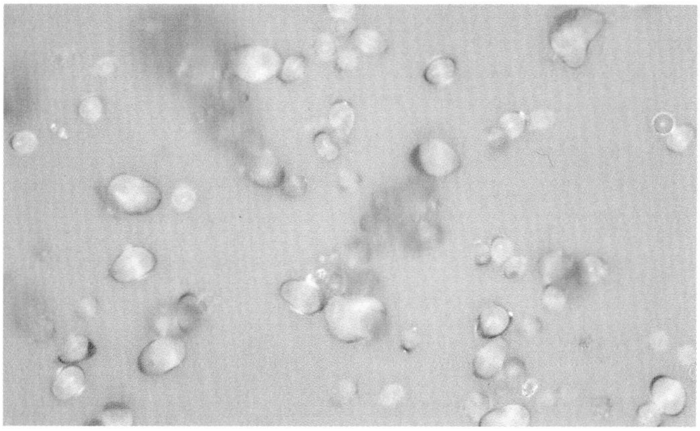

Exhibit 4. Starch

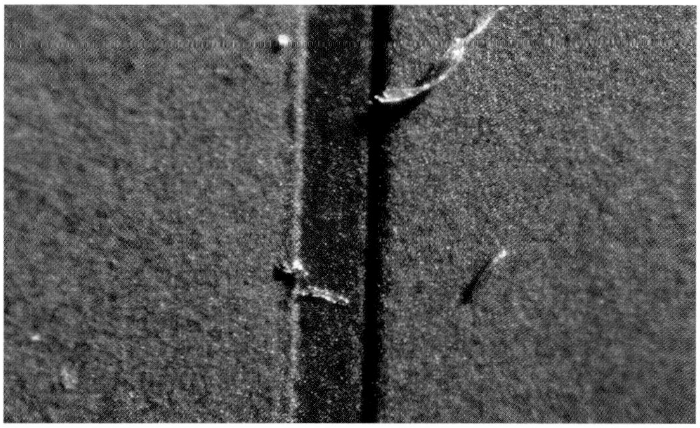

Exhibit 5. Paper Fragments

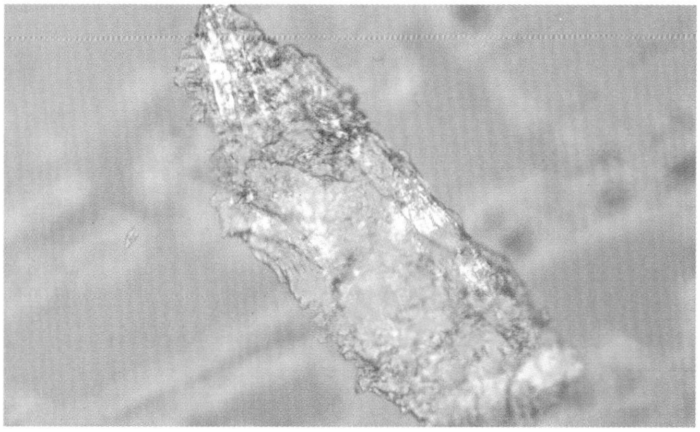

Exhibit 6. Teflon® Flakes

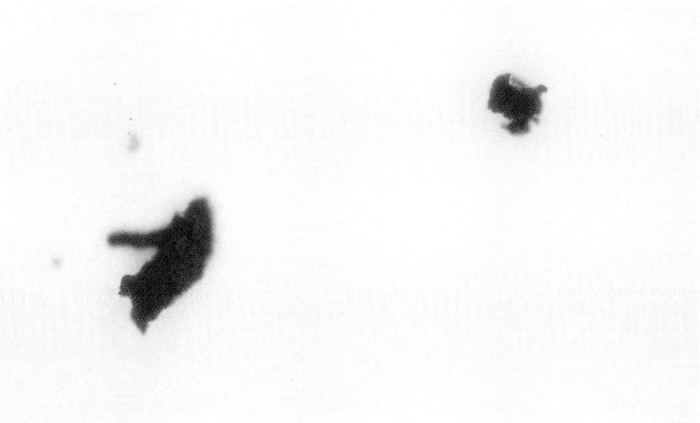

Exhibit 7. Stopper Fragments

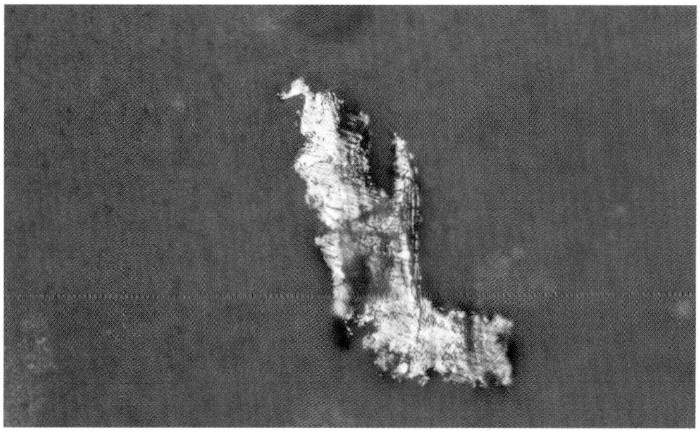

Exhibit 8. Stainless Steel

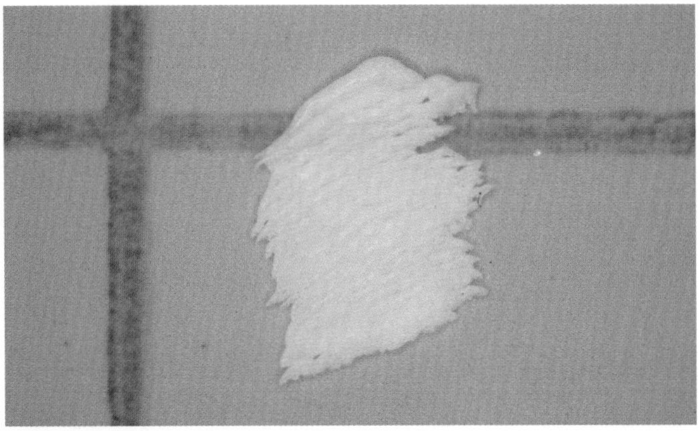

Exhibit 9. Polyethylene

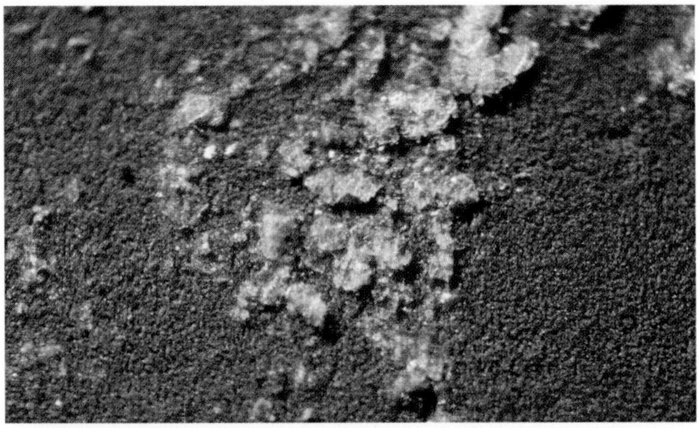

Exhibit 10. Calcium Carbonate

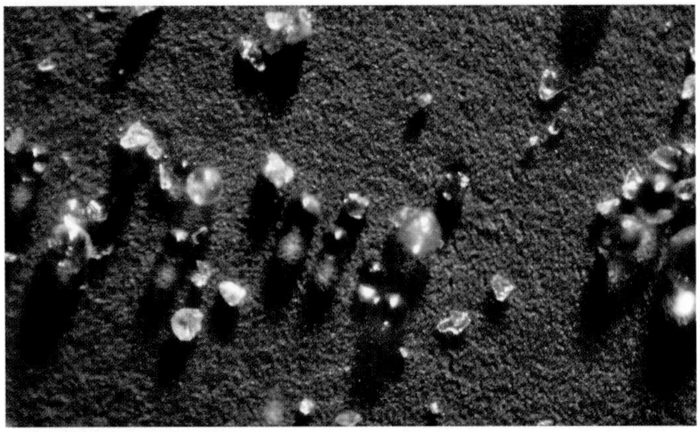

Exhibit 11. Magnesium Phosphate

Exhibit 12. Corn Starch

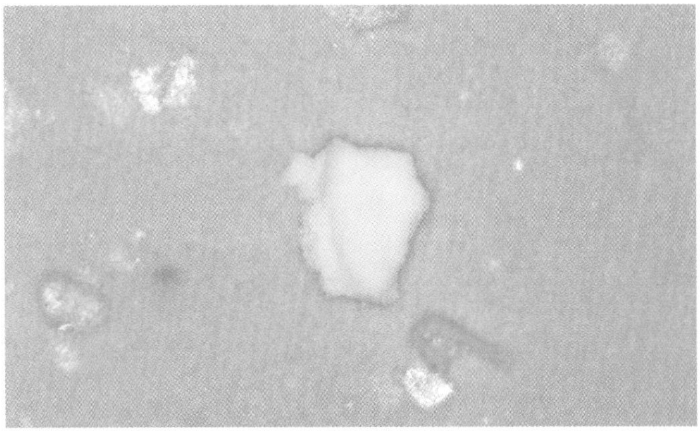

Exhibit 13. Filter Membrane Fragments

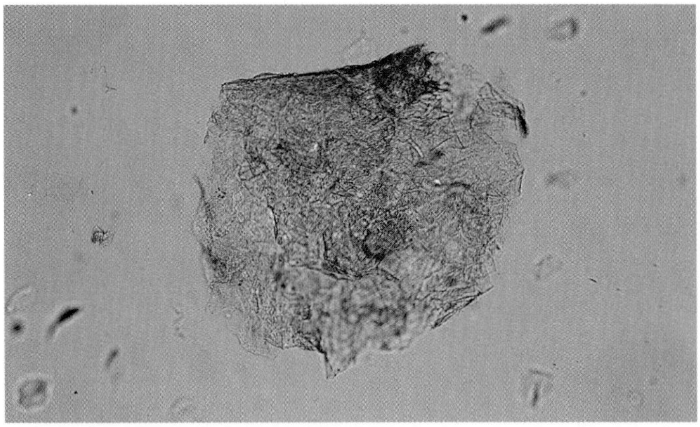

Exhibit 14. Dandruff

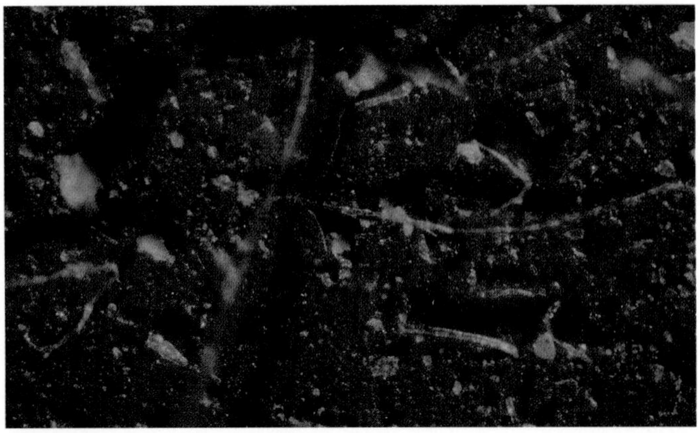

Exhibit 15. Drug Residue

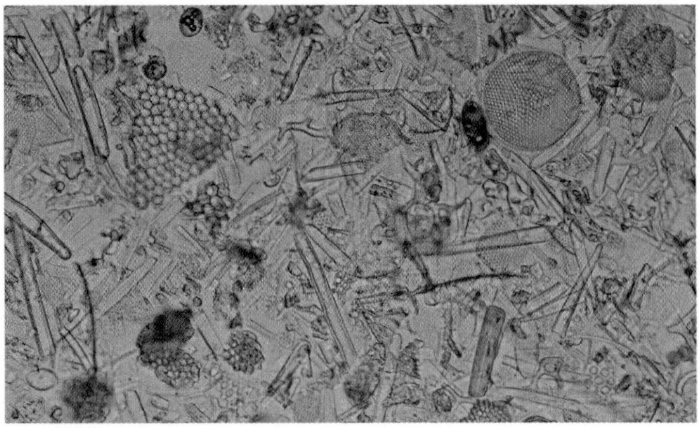

Exhibit 16. Diatoms

Exhibit 17. Fungal Hyphae

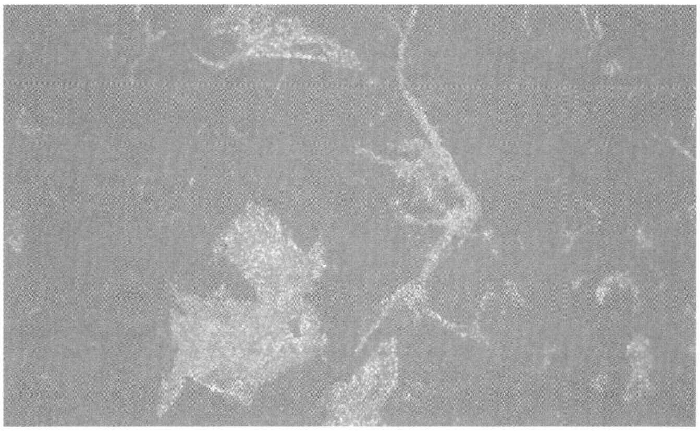

Exhibit 18. Amorphous Material

Exhibit 19. Hair (Human, Caucasian)

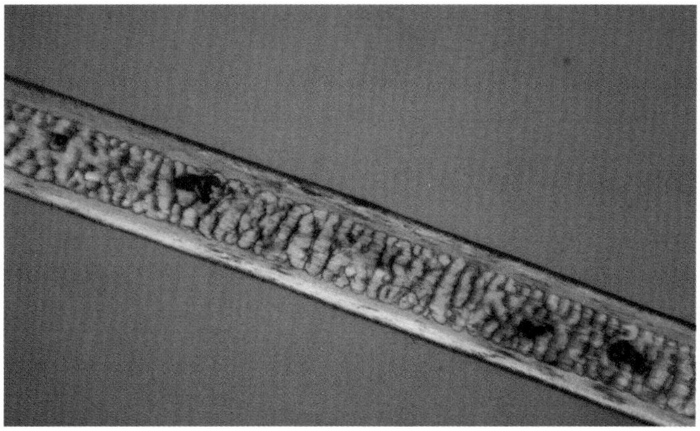

Exhibit 20. Rat Hair

Exhibit 21. Paper (Coarse, Hardwood)

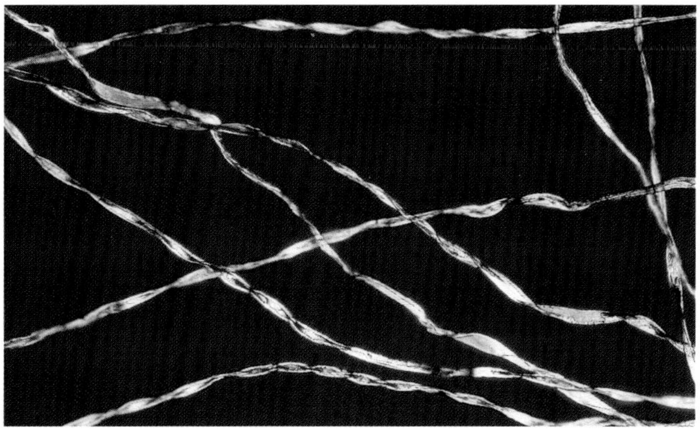

Exhibit 22. Cotton

Exhibit 23. Paper and Cotton Mixed

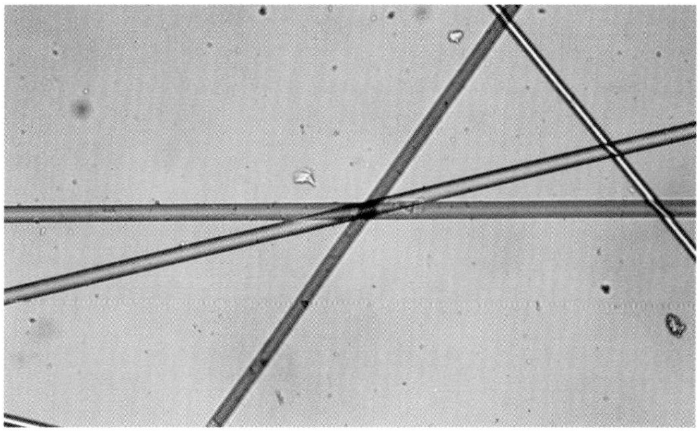

Exhibit 24. Glass Fibers

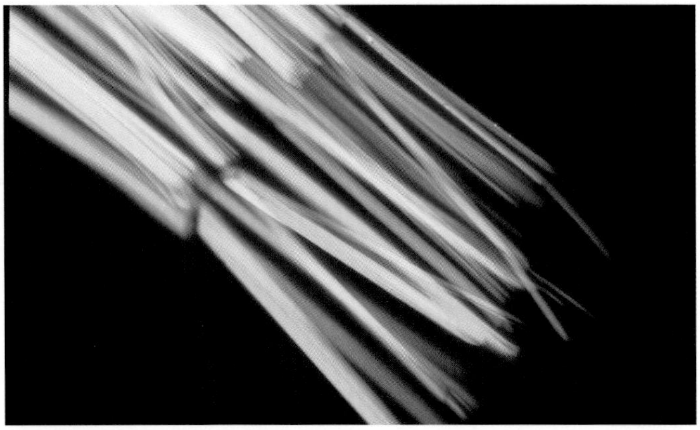

Exhibit 25. Asbestos

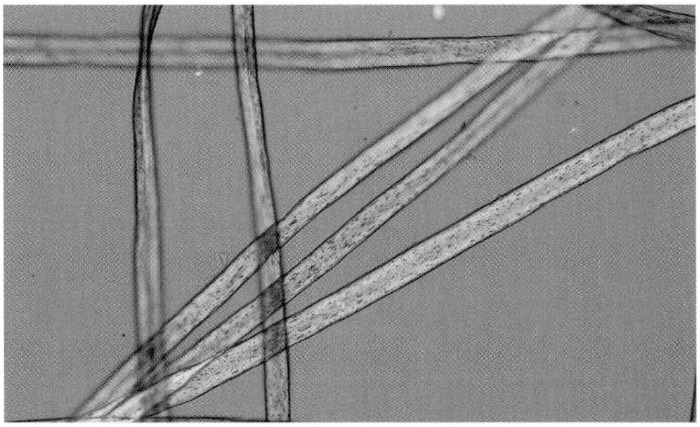

Exhibit 26. Acrylic Fiber (Orlon®)

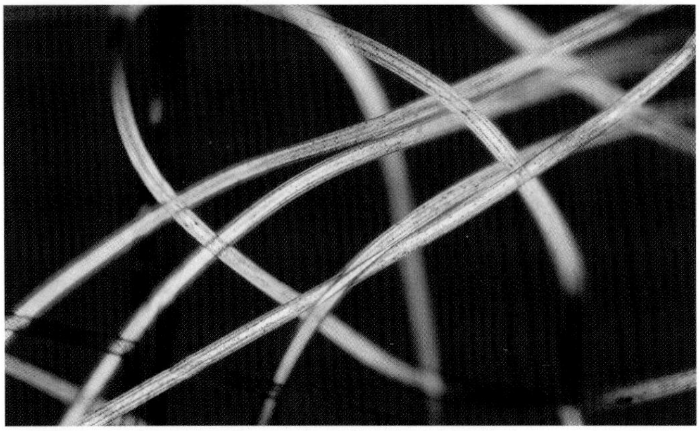

Exhibit 27. Polyester (Dacron®)

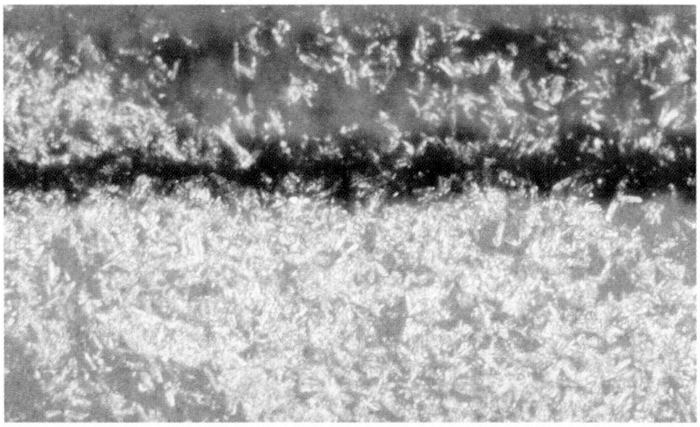

Exhibit 28. Bulk Powder Drug Residue

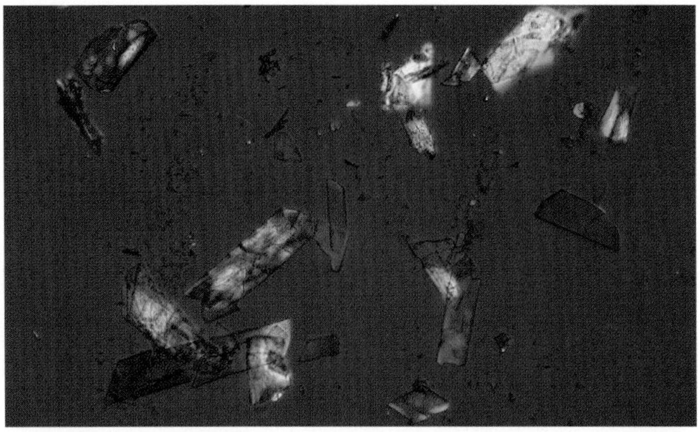

Exhibit 29. Talc

Exhibit 30. Calcium Oxalate

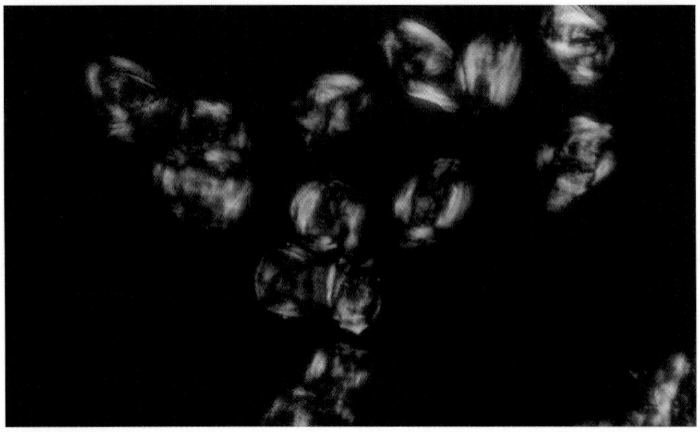

Exhibit 31. Skin Cells

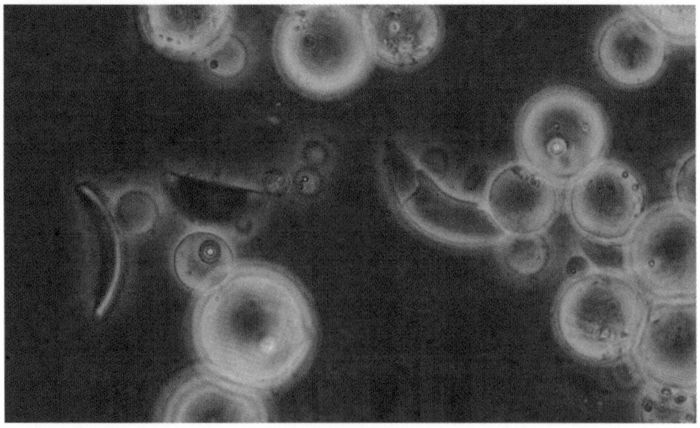

Exhibit 32. Glass Balloons

Particle size in visual inspections is most reproducibly made if the analyst is supplied with a visual size reference chart for immediate reference. One means of providing this is to use standard size comparison charts such as the TAPPI dirt chart (Technical Association of the Pulp and Paper Industry, Atlanta, Ga.). This chart is black on white and may be photographically reproduced to give a white-on-black scale. The vendor should be contacted regarding the copyright applications with regard to such modified use. A similar scale may be prepared in-house using calibrated photographic reductions or computer graphics.

VISUAL INSPECTION OF MEDICAL DEVICES

Medical device visual inspections are more frequently carried out on an attribute basis than are inspections of injectable products. Either components of a set, finished assemblies, or packaged devices may be inspected. Visual inspections are typically made without magnification, under good lighting conditions, at a distance of 12–18 in. and by analysts with 20/20 or corrected vision. Inspection may be conducted on each batch of items, and samples are selected randomly to be representative of the sampled group. One of the key factors that allows devices to be inspected by attribute is the fact that particles may be observed while "standing still" (i.e., they will not be moving or settling as is the case when units of an injectable are tested).

Consider the case of a simple solution administration set in a blister pack that is enclosed in the master pack of a procedure tray. When a particle is identified on or in the finished set assembly, it may be classified as loose or embedded and as a fiber or a nonfiber. After classification, the particle's length and diameter may be measured or estimated. Knowing the finished assembly's blood fluid path (if one exists) and the particle's classification and dimensions, a matrix such as that in Table 7.9 can be used to determine if the sample with particulate matter is to be rejected. If the sample is rejected, then the number of rejected samples needed to reject the group of parts being sampled comes into consideration. At times, specified components, e.g., a filter, that are an inherent part of a design of a finished assembly, may be a source of self-generating particulate matter. Particulate matter that can be traced to such sources may be allowed based on its source.

An AQL is often applied to devices. When a continuous series of lots is considered, the AQL is the quality level that, for the purposes of sampling inspection, is the limit of a satisfactory process average. A sampling plan and an AQL are chosen in accordance with the risk assumed. Use of a value of AQL for a certain defect or group of defects indicates that the sampling plan will accept the great majority of the lots or batches provided the process average level of percent defective (or defects per hundred units) in these lots or batches is no greater than the designated value of the AQL; thus, the AQL is a designated value of percent defective (or defects per hundred units) for which lots will be accepted most of the time by the sampling procedure being used. The sampling plans provided are generally arranged that the probability of acceptance at the designated AQL value depends on the sample size, being generally higher for large samples than for small ones, for a given AQL. The AQL alone does not identify the chances of accepting or rejecting individual lots

Table 7.9. Finished Device Free Particle Inspection Criteria

	Blood Path				Nonblood Path			
	Fiber		Nonfiber		Fiber		Nonfiber	
AQL	.65	.35	1.5	1.0	1.5	1.0	2.5	2.0
Particle Size (Major Dimension)	Length > 1.5 mm but < 5 mm	Length ≥ 5 mm	> 1 mm but < 2.5 mm	> 2.5 mm	Length 0.5 mm but < 2.5 mm	Length > 5 mm	> 1 mm but < 5 mm	> 5 mm
Observations to Reject	2	1	2	1	2	1	3	2

or batches; it more directly relates to what might be expected from a series of lots or batches. Inspection matrix examples for packaging embedded particles and particles free in the solution contact path of a device are shown in Tables 7.9 and 7.10.

SUMMARY

Visual inspection of injectable products and devices is a complex undertaking. Manual and automated automatic analyses here significant potential for error. With manual procedures, extensive training as well as validation of inspectors and the inspection methodology are necessary if a reproducible result is to be obtained. The primary consideration must be the definition of the reject probability for particles of a given size with whatever system or inspector method used. Objective evaluation of automated systems can only be performed on the basis of a thoroughly validated manual process. Automatic systems are highly effective for the inspection of small volume liquid vials and ampoules, i.e., < 20 mL in volume, but are less readily adapted to units > 100 mL in volume.

The importance of the assessment that a visual inspection (manual or machine) as a probabilistic process cannot be overemphasized. The most critical implication of this finding is that standards requiring "no visible particles" cannot be implemented or enforced. Total freedom of a product from any defect can only be achieved if the defect is an attribute, that is, it is the means of inspection employed either present or absent with absolute certainty. A deterministic inspection procedure is essential to either the total elimination of a defect or the allowance of a specific level. A probabilistic process such as visual inspection cannot achieve either of these ends. Another key consideration (particularly with manual inspection) is the occurrence of false rejection, or the rejection of good product. An important observation with regard to larger databases and statistical data treatments is assay sensitivity and detection limits. The statistical considerations ideally must not focus on numerical relationships alone without a consideration of what is being detected.

Table 7.10. Finished Device Embedded Particle Inspection Criteria*

Application	Blood Path*	Nonblood Path	Exterior
AQL	1.5	2.5	5.0
Particle Size (Major Dimension)	> 7.5 mm	> 10 mm	> 25 mm
Observations to Reject	1	2	4

*Loose particulate observed on the exterior of the individual device package, yet within the kit master package, will be inspected to an AQL of 5.0.

To be a devil's advocate, those who say it is not possible to manufacture injectables with absolute freedom from particles are wrong, strictly speaking. The problem is that the small percentage of units not destroyed in testing or dedicated to stability studies would be available only to the very rich or famous at a very high cost. Ultimately, only 1 bottle from 1,000 might survive testing and be available for sale. Since manufacturing controls rather than testing determine quality, what about the practicality of this approach? It may be calculated that the increased costs of manufacturing required to achieve absolute freedom from particles would have the effect of increasing healthcare costs without benefit and denying reasonably priced health care to the masses.

The USP, FDA, and manufacturers share the same basic concerns over visible particulate matter and product quality. The USP has for some years taken the ethical and defensible position that injectable solutions should be "essentially free" from visible particles; for the manufacturer, there are multiple, practical considerations in defining what *essentially free* "means." The USP has established this specification for particles in injectable solutions by administrative decree, rather than on the basis of scientific data. The upper size limit for subvisible particles in SVI and LVI has been set at 25 μm (visible particles are generally accepted as being ≥ 50 um in size).

The 50 μm size for visible particles has no greater technical or scientific support than does the term "essentially free." The particle size distribution established for subvisible particles at the ≥ 10 μm and ≥ 25 μm sizes provides the manufacturer with one basis for a size and number definition of allowable numbers of visible particles. Extrapolation of the subvisible units set in combination with considerations of process control limits provide a rational basis for allowable numbers and sizes of visible particles and allows inspection methods to be designed and validated.

In a general overview, the approach of the FDA to visual inspection processes would seem to incorporate two simplistic and effective principles: (1) a consideration of critical factors, such as specification of the inspection method, and analyst training, which will indicate whether a uniform, controlled inspection is being performed, and (2) a determination of whether an assessment of reproducibility of the assay result is periodically performed.

In all of these deliberations, it is important for the pharmaceutical industry and regulators to recognize that there are no particle-free parenteral solutions. There is, furthermore, no evidence of any patient issue related to infusion of a small number of inert particles within the current or previous USP limits. If 600 particles per container ≥ 25 μm size are acceptable per SVI limits, there must also reasonably be acceptance of smaller number of particles at larger sizes. Particles of approximately the visible size range must also be allowed at low numbers.

In the United States and the United Kingdom, an objective of GMP for the past 30 years has been the manufacture of injectable solutions with the lowest possible particulate burden based on best demonstrated practice and current technology. As technology has improved in its ability to exclude and detect particles, it has become apparent that it is not possible to produce "particle-free" solutions. This goal is, however, being approached asymptomatically, and increases in the quality of injectables (and devices) may be expected.

REFERENCES

Akers, M. J. 1985. *Parenteral quality control: Sterility, pyrogen. particulate, and package integrity testing.* New York: Marcel Dekker, Inc., pp. 143–197.

Archambault, G. F., and A. W. Dodds. 1966. Macroscopic light-testing procedure for large volume parenterals. In *Safety of Large Volume Parenteral Solutions.* Rockville, Md., USA: Food and Drug Administration, pp. 15–17.

Blackwell, H. R. 1946. Contrast thresholds of the human eye. *J. Opt. Soc. Am.* 36:624–643.

Blackwell, H. R. 1959. Development and use of a quantitative method for specification of interior illumination levels on the basis of performance data. *Illum. Eng.* LVI:317–353.

Borchert, S. J., R. Gaines, and J. R. Kraska. 1985. Validation of the Eisai visual inspection system. In *Proceedings of the Europe Conference on Visible and Subvisible Particles in Parenteral Products,* pp. 224–260.

Borchert, S. J., R. J. Maxwell, R. L. Davison, and D. S. Aldrich. 1986. Standard particulate sets for visual inspection systems: Their preparation, evaluation, and applications. *J. Paren. Sci. Tech.* 40:6.

Borchert, S. J., A. Childers, L. Fox, D. Myer, and A. Reynhout. 1987. Preparation of standards and metrology. In *Proceedings of the PDA Conference on Liquid Borne Particle Detection and Metrology,* pp. 122–166.

Boucher, C. L., and H. A. Sloot 1965. An improved method for visual inspection of injections. *Pharm. Weekblad Ned.* 100:253–262.

Brewer, J. H., and J. H. F. Dunning. 1947. An in vitro and in vivo study of glass particles in ampoules. *J. Amer. Pharm. Assoc.* 36:289–299.

BP. 1980. *British Pharmacopoeia,* Vol II:578 and A120. Cambridge: Cambridge University Press.

BP. 1988. *British Pharmacopoeia,* Vol II:756. London: Her Majesty's Stationery Office.

Faesen. 1978. Reproducibility of visual inspection of parenterals. *J. Parent. Drug Assoc.* 32:75.

Garvan, J. M. and Gunner, B. W. 1963. Intravenous fluids: a solution containing such particles must not be used. *Med. J. Aust.* 2:140–145.

Godding, E. W. 1945. Foreign matter in solutions of injection. *Pharm. Jour.* 154:124–132.

Graham, W. D., L. G. Chatten, M. Pernarowski, C. E. Cox, and J. M. Airth. 1959. A collaborative study on the detection of particles in impuled solutions. *Drug Standards* 27:61–66.

Groves, M. J. 1993. *Particulate matter: Sources and resources for healthcare manufacturers.* Buffalo Grove, Ill., USA: Interpharm Press, Inc.

Hamlin, W. E. 1978. General guidelines for the visual inspection of parenteral products in final containers and in-line inspection of container components. *J. Paren. Drug Assoc.* 32:63–66.

Hayashi, T. 1980. Studies on the particulate matter in parenteral solutions: Part I. Occurrence and size distribution of particulate matter in parenteral solutions. *Yakuzaigaku* 40:62–67.

JP. 1996. *The Pharmacopoeia of Japan,* 13th ed. Tokyo: Takeji Nippo, Ltd., pp. 20–21.

Klein, H. J., and E. W. Reuter. 1978. Automatic electronic inspection device for the detection of particles in ampoules. *Pharm. Ind.* 40:1357 1366.

Knapp, J. 1987. Process control by non-destructive testing. In *Proceedings of the PDA Conference on*

Liquid Borne Particle Inspection and Metrology, pp. 521–562.

Knapp, J. Z., and H. K. Kushner. 1980a. Implementation and automation of a particle detection system for parenteral products, *J. Paren. Drug Assoc.* 34:369.

Knapp, J. Z., and H. K. Kushner. 1980b. Generalized methodology for evaluation of parenteral inspection procedures. *J. Paren. Drug Assoc.* 34:14–61.

Knapp, J. Z., and H. K. Kushner. 1982. Particulate inspection of parenteral products: From biophysics to automation. *J. Paren. Sci. Tech.* 36:121–127.

Knapp, J. Z., and L. R. Abramson. 1996. Evaluation and validation of nondestructive particle inspection methods and systems. In *Liquid- and surface-borne particle measurement handbook,* edited by J. Z. Knapp, T. A. Barber, and A. Lieberman. New York: Marcel Dekker, pp. 295–450.

Knapp, J. Z., H. K. Kushner, and L. R. Abramson. 1981a. Automated particulate detection for ampuls with the use of the probabilistic particulate detection model. *J. Paren. Sci. Tech.* 35:21–35.

Knapp, J. Z., H. K. Kushner, and L. R. Abramson. 1981b. Particulate inspection of parenteral products: An assessment. *J. Paren. Sci. Tech.* 35:176–185.

Knapp, J. Z., J. C. Zeiss, B. J. Thompson, J. S. Crane, and P. Dunn. 1983. Inventory and measurement of particulates in sealed sterile containers. *J. Paren. Sci. Tech.* 37:170–179.

Kramer, W. 1970. Inspection for particulate matter essential to I.V. additive program. *Drug Intel. & Clin. Pharm.* 4:311–313.

Martyn, G. W. 1970. Utilization of TONDO unscrambler and Strunck units in ampoule inspection. *Bull. Paren. Drug Assoc.* 24:231–244.

Mastenbroek, G. G. A. 1959. The problem of particles in infusion solutions. Ph.D. dissertation, University of Amsterdam.

McGinn, A. B. 1973. Panel discussion: Mechanical inspection of ampuls II—Ampule inspection using the Rota machine. *Bull. Paren. Drug Assoc.* 27:247.

Rothrock, C. H., R. Gaines, and T. Greer. 1983. Evaluating different inspection parameters. *J. Paren. Sci. Tech.* 37:64.

Sandell, E. 1953. Inspection control during ampoule filling. *Farmacevtisk Revy* 52:859–870.

Sandell, E., and B. Ashlund. 1974. Visual inspection of ampoules. *Acta Pharm. Sued.* 11:504–508.

Saylor, H. M. 1966. Particulate matter, II—Visual inspection. *Bull. Paren. Drug Assoc.* 20:31–44.

Schipper, D., and R. Gaines. 1978. Comparison of optical electronic inspection and manual visual inspection. *J. Paren. Drug Assoc.* 32:118.

Sokol, M., and N. Kirsch. 1964. The pacing of mechanized pharmaceutical production operations. *Bull. Paren. Drug Assoc.* 18:13–19.

USP. 1995. *The United States Pharmacopeia,* 23rd revision, <788>. Easton, Penn., USA: Mack Printing Company, p. 1586.

Visual particulate matter in large and small volume injections. 1985. *Pharm. Forum* (July–August): 567.

Whyte, W. 1983. A multicentered investigation of cleanroom requirements for terminally sterilized pharmaceuticals. *J. Paren. Sci. Tech.* 37:184–197.

VIII

LIGHT EXTINCTION PARTICLE COUNTING OF LIQUIDS

There are a number of methods for enumerating and sizing particles in liquids based on the use of electronic instruments. The instruments measure light scattering, light extinction, resistance modulation, or a combination of light scattering and light extinction. In writing this chapter, both because of the general wide use of light extinction methods and their almost exclusive use in the medical products industry, the author chose to deal only with this type of particle counter. The application of light extinction counters is complex; a specific critical concern involves error sources that may cause the count data collected to be of decreased value or, to various extents, misleading. The subject of light extinction particle counting involves more than enough information for one chapter.

A point that should be considered at the beginning is the terminology that is applied to this form of particle enumeration. The term *light extinction* implies to those practiced in the art that the particle measurement being made is based on a combination of light-scattering effects as well as those of geometric optics (for particles equal to or greater than about 1.5 μm in size). Historically, these instruments were referred to as "light blockage" devices, and the functional analogy of a shadow cast by a particle in the instrument sensor had wide currency. The reference to "shadow" or "blockage" effects that occur due to light-particle interactions are but a poor description of the actual complex physical basis of the instrument's operation. The U.S. Pharmacopeia (USP) chose another descriptive term, "light obscuration" (LO), to designate this type of instrument. The term *obscuration* is another way of saying "blockage." Apologetically, even though the term *obscuration* is in some sense technically incorrect, the author has used it throughout most of this chapter in deference to the USP choice and to avoid possible confusion by users who might be involved with regulatory auditors.

Just as is the case with chemical or other physical assays, we are concerned with the accuracy and precision of the data gathered by LO counting. As discussed earlier, there is no real accuracy parameter with particle counting methods; all methods simply provide mechanistic, operational determinations that depend on the type of instrument used. Thus, the terms *particle count accuracy* or *validation of count accuracy* are misleading, even if only one specific instrument is considered. With LO instruments, the size assigned a particle in the sensor view volume will depend on its optical properties just as much as its size. In fact, while we purport to measure the size of particles in the sensor of the counter, we are actually measuring the interactive properties of a particle with light of the wavelength(s) used. Thus, if polystyrene latex (PSL) spheres are used as calibrants, and our contaminants are also PSL spheres, the data will be most meaningful. With real particles of varying refractive index, size, and opacity, a variable response is achieved, and counts are more properly assessed in terms of magnitude rather than number.

In this author's opinion, recent papers by Knapp and Abramson (1996a, b) may have had somewhat misleading titles in that the words *validation* and *count accuracy* both appeared. What these authors were in fact writing about, as we will see later, was the determination of coincidence count error based on a novel particle size–particle concentration model. While their theory provides for a sound understanding of coincidence effects (i.e., counting multiple particles as a single particle) than was previously available, it provides no meaningful data or inferences regarding count accuracy, especially at the low count levels of optically heterogeneous particles found in contemporary injectables or devices made under current Good Manufacturing Practice (cGMP). In fact, the users of LO counters are more critically concerned with increases of counts at the limits sizes (i.e., ≥ 5 μm, ≥ 10 μm, and ≥ 25 μm) due to immiscibles or other artifacts rather than count loss.

Readers are asked to consider the contents of this chapter and to develop their own assessment of the appropriate role of LO counting in the testing of injectables and devices. Many will come to the conclusion that in routine use, the instrument is best applied in the detection of significant (statistically and mechanically) trends in levels of counts that signal a change in product characteristics. The numbers of particles present in injectables (e.g., 1–3 particles/mL ≥ 10 μm in large-volume injections [LVIs]) are generally very low in relation to limits; because of the difficulty of making statistical determinations with these small numbers, comparison of the actual numbers of counts obtained from different codes, lots, and batches is often tenuous.

THE USP <788> METHOD

The USP <788> requirement for particulate matter in small-volume injections (SVIs) was implemented on January 1, 1986, in USP/NF XXII. The requirement was based on an LO assay, a pooled 10-unit sample, and one set of limits for all types of SVI product to which the test applied (see also USP 1990). The requirement covered a wide variety of product types, including 100 mL liquid drug codes in plastic or glass containers, small volume dry powder vials, and ampoules as small as 0.5 mL. Since this 1986 introduction, LO particle counting of injectable products and devices has remained an evolving technology, in terms of the pharmacopeial description of the method, user application, and the instrumentation itself. Even the descriptive term for the test has changed since 1986, when the term *light blockage* was widely used. With increasing realization of the actual complexities of particle sensing by this instrumentation, LO counting came to be applied. The USP 23rd version of <788> retains the latter term, which is of limited use and not fully descriptive of the method of particle sensing involved.

The LO methodology, whatever its shortcomings, constitutes a time-effective means of performing particulate matter testing against pharmacopeial limits. While USP <788> is currently applicable to a range of SVI and LVI product, many injectables contain particles of unusual morphology, immiscible fluids, and materials with surfactant activity. Thus, importantly, inherent errors of the test method often negatively impact the analysis of some types SVI. Some injection products cannot be accurately tested by the methodology due to their nature (i.e., viscosity, content of suspended materials). The most dangerous error in applying this method is a failure to recognize erroneous counts. Hence, the author has included a discussion of factors that may result in "bad" data, as well as principles of operation and use of the LO instruments.

The number of particles seen and counted by LO counters is deceptive with regard to the actual chemical levels of materials present; herein lies the advantage of chemical assays for the concentration of materials and the rationale behind USP <788> specifying that particles are materials "not amenable to analysis by chemical means." As an example for unit density material:

1 part per billion (ppb) =

2×10^3 10 μm particles/L,

2×10^6 1.0 μm particles/L, and

2×10^9 0.1 μm particles/L.

The primary advantages of electronic LO particle counters are their automated characteristics and the rapidity with which they allow the performance of particulate matter analysis on product to ensure that USP, British Pharmacopoeia (BP), or other limits are met. There also are disadvantages to their use. Electronic particle counters cannot discriminate between various types of particles; data collected are subject to artifacts and require review by a trained analyst. The instruments are relatively complex in operation, and they require a time-consuming calibration at regular intervals. The data collected are subject to a number of serious artifacts. Comparisons of the count data obtained by instrumental and microscopic methods of counting particles in parenteral products are discussed by DeLuca (1977), DeLuca et al. (1988), Schroeder and DeLuca (1980), and Montanari et al. (1986).

LIGHT OBSCURATION SENSORS

Principles of Operation

The original reference for LO counting was presented by Carver in 1969. The sensors of the first and current instruments operate based on the interaction of an intense collimated and focused beam of light with a particle suspended in a liquid medium. In current designs, the light is supplied by a laser diode source, and particles are constrained to pass through a narrow view volume between light source and detector, which is characteristically a photodiode. The variation in light intensity is converted by the photodiode to a voltage signal. Thus, particle size is not measured directly, but rather the voltage variation is related to the size of ideal standard particles used in calibration. Both focusing and receiving optics may be used. (The reader is referred to Chapter 10 for a more specific discussion of the interaction of light and matter.)

The sensor is thus an electromechanical device of sorts; the voltage response of the sensor (pulse amplitude) and the rate at which it can count (particles/mL) depend on straightforward electrical and mechanical functions. For particles that exceed about 2 μm in size, the interaction of a solution-borne particle with light in the sensor of an LO counter is characterized by complex geometric optical effects—diffraction, refraction, absorption, and reflection (Figure 8.1). For particles smaller than 2 μm, scattering effects become important. The total combined attenuation of incident light results from a combination of these four light-particle interactions. The complexity of this interaction has been

Figure 8.1. Interaction of > 2 μm particles with light (geometric optics).

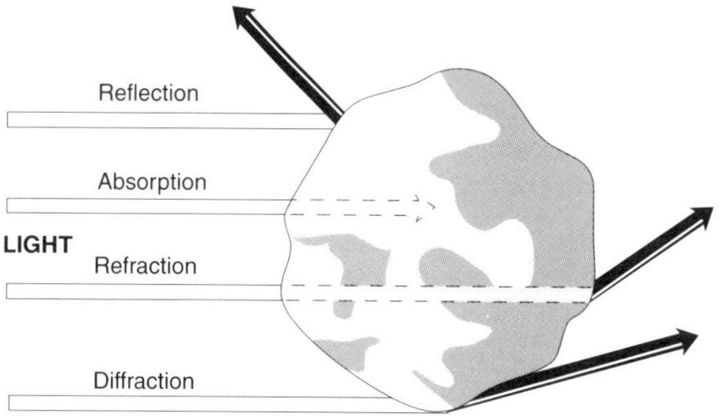

recognized by the evolution of terminology over the years since Carver's invention. The interaction of light with the particle is further modulated by the nature of the fluid in which the particle is suspended.

For transparent particles, refraction effects predominate. Scattering phenomena are of concern primarily for particles that approximate the wavelength of the illuminating radiation (e.g., 1.5 μm or smaller); they result in redirection of light incident on a particle in forward or rearward direction, as described in Chapter 9, which deals with counting of airborne particles. Obscuration efficiency for strongly absorbing nontransparent particles of this size (i.e., carbon) is primarily the sum of the scattering and absorption efficiencies.

Above the 5 μm minimum size of current concern in pharmaceutical particle counting, geometric optics rather than scattering effects predominate; for transparent or semitransparent particles, refraction effects result in light being deviated from the central optical axis and from the beam illuminating the detector. The variation of light scattering with particle transparency, refractive index, and size has serious implications for the sizing accuracy of LO counting when particles of different transparencies, sizes, colors, and surface textures are present in a sample.

If the light incident on one particle is redirected along some linear path in the direction of its source, the interaction may be termed *reflection*. Diffraction is the bending or spreading of light rays in front of an opaque object that it has passed. Diffraction effects significantly influence the extent to which an opaque particle "blocks" light incident on it and affects the degree of light extinction produced by a particle as opacity increases. Forward-scattering effects from this source are difficult to predict mathematically, but the net effect is decreased attenuation of light between a given particle and the photodiode of an LO sensor. Refractive index becomes a minor concern with opaque particles. Nontransparent particles (e.g., carbon or stopper elastomer) must be expected to result in more light extinction than transparent ones of the same size. The intense light in the sensor may cause particles that are opaque to ambient light become translucent. Metal particles are both opaque and reflective; carbon particles are opaque and absorptive. Some particles and solutions can absorb at the wavelength of the laser source and fluoresce at other wavelengths to which the detector is sensitive.

The data obtained from LO counters thus result from complex interactions between a small particle moving at high velocity and an intense light beam in the narrow confines of the particle counter sensor (Figure 8.2). In the sensor, the sample fluid travels through a narrow rectangular passage, typically of 90–800 μm in width and of 15–50 μm in height in front of an illuminated window. The length of the light path through the cell varies from 400 to 1,000 μm. The light from an incandescent lamp or laser is formed into a collimated beam at the window and directed through the liquid sample stream onto a photodiode. As long as the number of particles in suspension does not exceed a specified concentration, typically in the range of 7,000–10,000 particles/mL, most of the particles will pass through the sensor view volume individually. Whenever a particle traverses the light beam, the intensity of light reaching the photodiode is reduced, and an amplified voltage pulse is produced by the sensor. In a simplified practical consideration, the amplitude of the pulse is proportional to the projected area of the particle in a plane normal to the light beam, and the particle size is registered as the diameter of a circle having an equivalent projected area. The sample solution may be passed through the sensor cell by applying either pressure or vacuum; pressure sampling has obvious advantages for viscous liquids (e.g., dextrose solutions).

What about electronic and physical considerations in LO counting? Due to the extremely small cross-section of the flow channel in the view volume, a flow rate of 10–25 mL/min will equate to a particle velocity of up to 60 m/sec in the view volume. As a consequence, high-speed counting electronics are required to prevent pulse pileup (e.g., "dead time") or coincidence counting whereby two or more particles in the view volume during a single count event may be recorded as a single pulse. This is electronic coincidence, and the key variable to consider is the speed with which the pulse-counting electronics can reset after a count is recorded and be ready for the next count. The electronics of current generation counters are so fast that electronic coincidence is not a primary consideration. Optical or physical coincidence, however, is far more likely to occur. In this case, multiple particles are simultaneously present in the view volume or sensing zone of a sensor and are sensed as one particle. This results in the collection of an artificial count at a size generally corresponding to the cumulative cross-sectional area of the particles being considered.

Figure 8.2. Light obscuration particle sensor.

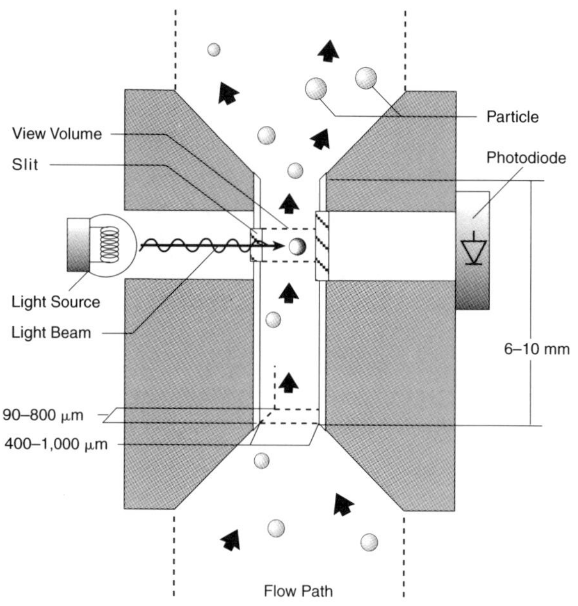

The analog signal received from the sensor is screened in the counter and converted into a digital form that can be read and stored in memory by a dedicated microprocessor. Pulses are sorted and counted within channels determined by preset voltage amplitude thresholds in order to measure particle size distributions. In the differential mode of counting, the pulse produced by the particle is presented as a count only in that channel with a range in mV that includes the pulse produced by the particle; in the total or cumulative mode, the pulse from the particle is presented as a count in all channels, with a threshold setting in mV below that of the pulse produced by the particle. Some vendors of contemporary LO sensors claim capability for counts of up to 24,000 particles/mL at a flow rate of 25 mL/min with less than 10 percent coincidence.

A (generalized) theoretical relationship between the size of the particle and the amplitude of the voltage pulse in LO counting is given as

$$E_o = \frac{\lambda}{A} E_b$$

where E_o is the pulse amplitude from photodetector; λ is the maximum projected area of the particle; A is the area of the window; and E_b is the base voltage from the photodetector.

Based on the earlier discussion, the interaction of a particle with light in the view volume of an LO sensor is not as simple as this relationship suggests (Figure 8.2). This equation is generally appropriate, but it does not adequately take into consideration refractive index, light absorption factors, and light-scattering effects—all of which can impact the result obtained in LO counting.

Before proceeding, a brief mention should be made on the wide availability today of LO counting systems suitable for use in counting particulate matter in pharmaceuticals and medical products. The following discussion of laser and white light sensors is based largely on HIAC (HIAC Royco, a division of Pacific Scientific, Silver Spring, Maryland) instrumentation, but the principles of operation of sensors manufactured by other vendors are sufficiently similar that the same considerations apply. Other vendors of counters with light and/or laser sensors include Climet, Malvern, Met-one, Particle Measuring Systems, and Particle Sizing Systems. Rion instruments are made in Japan and, based on the limited reports the author has received, are appropriate for pharmaceutical applications. Because the author lacks experience with Rion instruments and does not know if maintenance support is available, personal recommendations cannot be made. Counters manufactured by PAMAS in Germany are also marketed in the United States and, based on the author's limited knowledge, appear generally suitable for counting particles in pharmaceuticals; these instruments are notable for the high concentration limits claimed (up to 120,000 particles/mL).

Sensor Construction and Function

There are a number of ways to make a sensor for LO counting; any simplistic description belies the actual physical difficulty of the process. In the following discussion, the author has relied heavily on information furnished by vendor technical representatives: Dan Berdovich and Alvin Lieberman of Particle Measuring Systems, Dave Nicoli of Particle Sizing Systems, and Charles Montague and Peter Rossi of HIAC. The prerequisites are a small, linear flow channel ("capillary") and an intense, zone-restricted area of intense and uniform illumination. As an example, four flat pieces of transparent quartz may be "sandwiched" using a transparent adhesive shown in Figure 8.3. Pieces A and B serve as the support for the capillary and are separated by pieces C and D, which define the length and width of the flow channel. The collimated laser beam itself defines view volume height (Figure 8.4).

Figure 8.4 shows the interaction of light and the particles passing through the LO sensor view volume. Depending on the nature of the particle, the various optical interactions (refraction, reflection, diffraction, absorption) play a greater or lesser role in the production of the signal output of the sensor. In the absence of a particle in the view volume, the output of the detector is the "baseline" d.c. (direct current) level, which corresponds to full illumination of the detector by the "ribbon" of incident light, of typical thickness (height) 30–50 μm and width 90–800 μm (depending on sensor design). Arrival of a particle in the view volume gives rise to a small, negative-going pulse superimposed on the baseline. In this context, the sensor behaves like a turbidity meter, in which diminution of the light beam is caused by suspended particles. The measurement is thus inherently of decreased sensitivity, since a small change in signal is being compared to the intensity of the entire undeviated beam. For particles smaller than about 1.5 μm, light-scattering counters as described in Chapter 9 provide the high sensitivity required for detection and sizing. These instruments differ significantly in function from obscuration counters; in the scattering mode of sensing, the near-forward or wide angle scatter is compared to a "dark"

Figure 8.3. Capillary construction: (a) Vertical view showing sample tube diameter superimposed on "sandwich" formed by two quartz blocks (A, B) and two spacers (C, D); (b) Lateral view showing vertical capillary path and view volume.

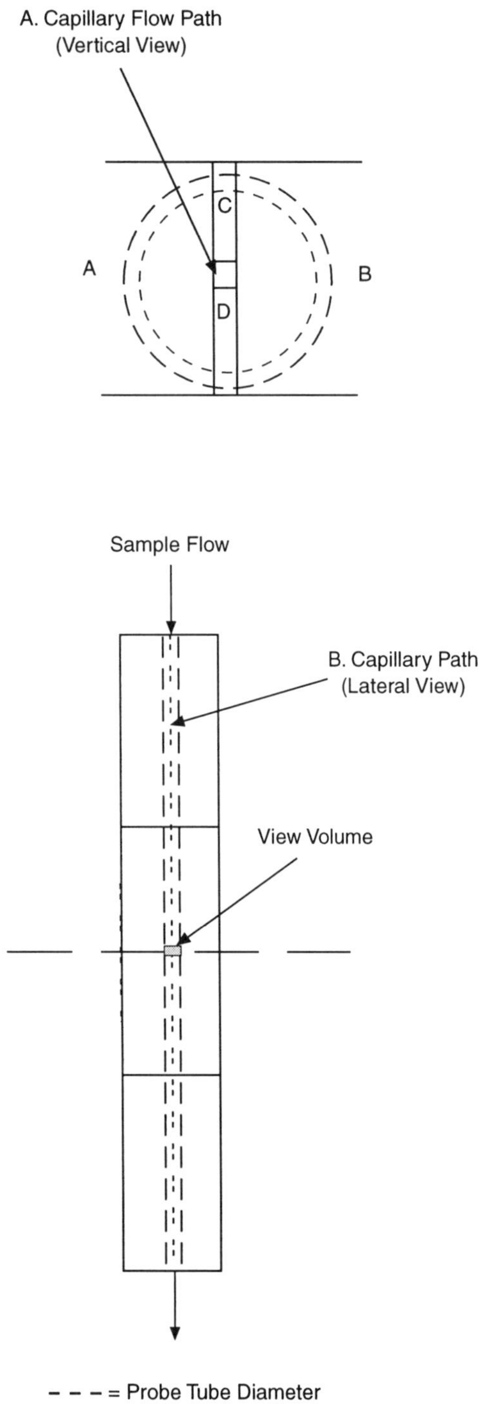

Figure 8.4. Sensor view volume or sensing zone.

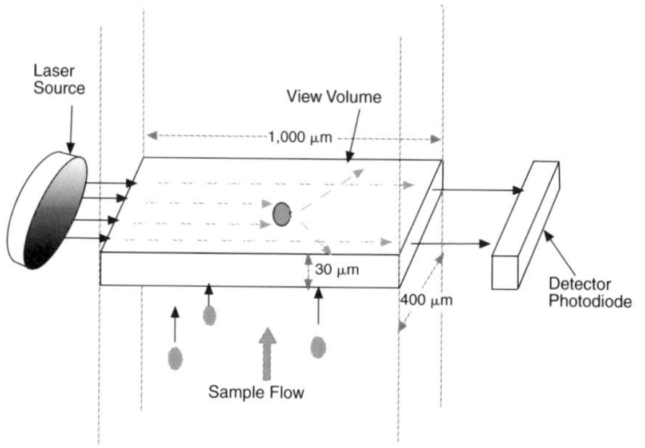

baseline, allowing a much more sensitive measurement of sub-2 μm particles.

For particles significantly larger than the wavelength of the incident light (typically 780–830 nm for a laser diode source), geometric optical effects shown in Figure 8.1 prevail. The predominant light-particle interaction that describes the behavior of an LO sensor for transparent and semitransparent particles is optical refraction. In considering refraction effects, each transparent particle that enters the active view volume acts like a tiny lens, able to deflect much of the light that is incident on it away from the detector that intercepts the illuminating light beam, thereby causing a decrease in the detected intensity. The function of a particle in acting as a lens is shown in Figure 8.5.

For the sake of simplicity, the particle being measured is most often assumed to be spherical in shape. We arbitrarily assume that the refractive index of the particle, n_2, is 1.50, a typical value for glass. The value of n_2 for the PSL beads that are typically used to calibrate all LO instruments is 1.59 at 590 nm (and very similar at 780 nm). The refractive index of the surrounding fluid, generally referred to as n_1, is assumed to be 1.33, which is the value for water. Assuming that the diameter of the particle is much larger than the wavelength of the incident light (780 nm), simple geometrical optics and Snell's law for refraction can be used to calculate the angle of bending of a light ray, from fluid to particle and vice versa. Simply stated, the greater the refractive index difference between the particle and the suspending

Figure 8.5. Light obscuration due to refractive index difference of particle and medium.

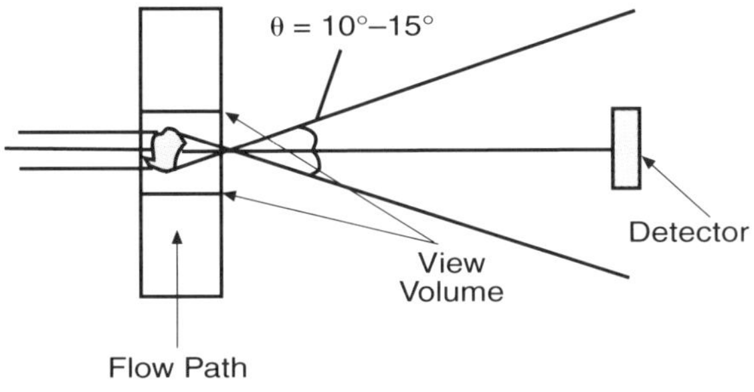

medium ($n_2 - n_1$), the larger the angle at which light rays passing through the particle are deviated from the beam axis, and the more significant the resulting decrease of the illumination intensity falling on the detector.

This discussion represents, of course, a serious oversimplification. Particles do not have ideal transparency. Even latex spheres will be seen to be reflective when suspended in water and viewed in a light microscope. Opaque particles do not transmit light and do not cause refraction. Translucent particles will absorb and refract light to some degree. Particles counted are unlikely to be spherical, and even if completely transparent irregular will give a wide range of pulse responses depending on the aspect viewed by the sensor.

At some point, as the refractive index of counted particles becomes very close to that of the surrounding fluid, the particle is increasingly ineffective in its ability to refract incident light rays away from the detector. The average angle of refraction of the light rays becomes so small that they do not diverge outside the area of acceptance of the detector, allowing some of the deviated light to reach the detector. Therefore, the particle is no longer effective in "obscuring" light. The effect on the measurement is clear: The pulse height in the case of particles close to the refractive index of the suspending medium V is smaller than would ideally be the case, causing the particle to be undersized, relative to a "standard" PSL particle or other material that gives a more pronounced refraction effect. This signal diminution occurs notably in the analysis of pharmaceutical solutions when amorphous, highly hydrated particles are contained in the solution analyzed; these particles are not only very thin but often have a refractive index very close to that of the aqueous suspending medium.

In those rare cases where the refractive index of the particles is very close to that of the fluid, one can correct for the resulting undersizing of the particles for either of two methods. First, one can isolate particles of the material of interest for each of several known sizes (determined, for example, by optical microscopy) and use these as "standards" for creating of a new calibration curve, log V versus log diameter, using the dispersing fluid to which the particles are (unfortunately) nearly refractive index matched. A second, less time-consuming, approach involves finding a standard material with refractive index matching or approximating that of the problem particle. Four or five different points on this calibration curve, spread out over the size range of interest, should suffice. This new calibration curve will substitute for the one that is typically obtained using PSL beads dispersed in water.

Another technique for recalibrating an LO instrument is to calculate a new, approximate calibration curve from the existing, "normal" one, using a simple model to account for the effect of inadequate refraction of the particles. This typically involves the use of a scale factor, which is multiplied by the pulse height V due to the reduced "light-blocking" power of the particles under investigation. This approach has been used to size particles that were nearly index matched to water.

View Volume and Sensor Response

The view volume of a sensor is most often defined laterally by the opaque walls of the flow passage and vertically by either beam height or the height of the window (physical opening) through which light passes. This height, in sensors used for pharmaceutical counting applications, is typically in the range of 15–50 μm. The width of the beam may range from 90 to 800 μm, and length of the cell along the optical axis may be as great as 1 mm. The response of a sensor with this view volume geometry is complex, with pulse amplitude being proportional to the cross-sectional area of the particle in the beam ($E = d^2$) until the diameter exceeds beam height (Figure 8.6). This steep slope of the area-dependent response can be described by a diameter squared (a power law function). As the diameter of the particle exceeds the cell height, the

Figure 8.6. Relationship of particle diameter (D) and view volume height (H).

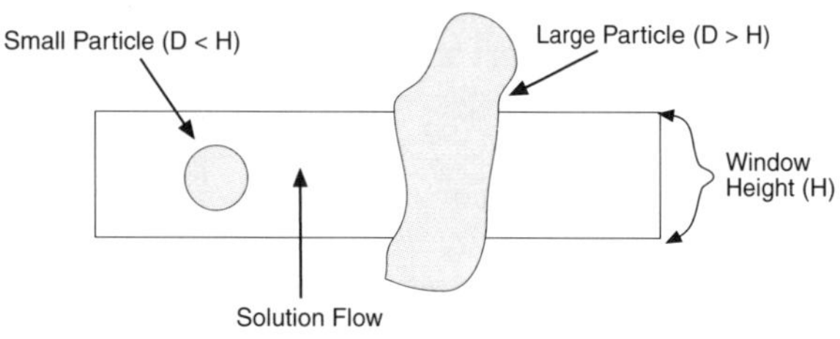

response becomes proportional to the width of the particle in the light beam ($E = d$).

Particle light obscuration thus relates to pulse height as the square of the diameter of the particle up to the sensing zone height; beyond that, it increases directly as the diameter increases. As an example, for a 30 μm sensing zone (view volume) height, size versus response might be as follows:

- 10 μm → 100 mV
- 20 μm → 400 mV
- 40 μm → 800 mV

This relationship is shown in Figure 8.7.

While the flow regime in the view volume is not laminar (i.e., linear), some alignment of elongate particles is effected, so that their narrow axis obscures the beam, and the maximum dimensions of fibers are rarely measured. As the depth of the view volume decreases, the range of pulse amplitudes for particles of irregular shapes that have a dimension exceeding the height of the view volume increases. In this case, the sensor essentially makes a linear measurement of the particle perpendicular to the axis of the particle that is aligned with the direction of sample flow. Importantly, a shallow view volume results in decreased potential for physical coincidence (which decreases with the illuminated volume in the sensing zone).

The dynamic range of a sensor is the range of particle sizes over which pulse amplitude can be shown to have a predictable relationship to particle size. A pulse height of 2× the noise level may be used for the low endpoint, and the upper size limit may be determined based on the narrowest dimension of the view volume. In recent years, significant gains have been made in the reduction of noise levels in LO sensors. Based on design factors, sensor concentration limits and dynamic range are determined by the manufacturer through design considerations and involve both electronic and physical parameters of the sensor.

Figure 8.7. Sensor response curve based on particle diameter relative to view volume height.

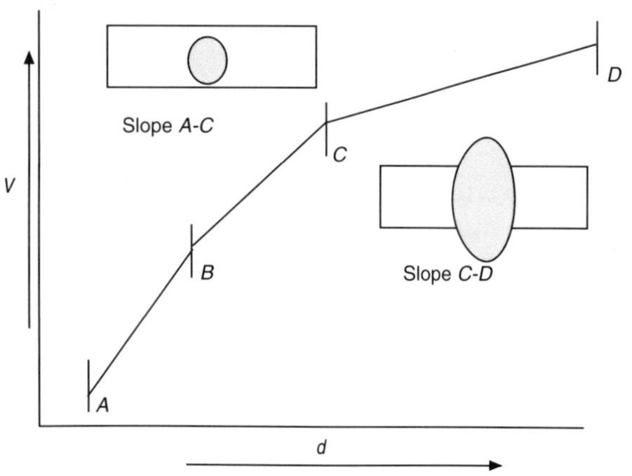

Laser Versus White Light Obscuration Sensors

The current generation of LO sensors are universally of the laser diode type. This type of sensor has almost completely replaced the older model white light sensors in the pharmaceutical industry based on customer preference and attrition, the latter due to the fact that white light sensors with incandescent or quartz-halogen sources are no longer available. The function of the two types of sensors is discussed in readily understandable fashion in the papers by Holger Sommer (currently of TEAM Consulting) (Sommer 1990a, b; Sommer 1991).

Based on the preceding discussion, it must reasonably be concluded that specific kinds of particles will be detected with differing sensitivity by the two types of sensors. Before adopting laser sensors in our laboratory, we performed the testing of suspensions by instruments equipped with the two designs of sensors. Representative data are presented in Table 8.1. Test suspensions used included NIST (National Institute of Standards and Technology) glass spheres, AC fine test dust, and carbon particles; parenteral materials ranged from sterile water to dry powder antibiotics. The result suggested that laser-based sensors are readily applicable to counting extraneous particles in parenteral products or in other medical products and did not give results dramatically different from white light sensors, which is important in considering historical versus current data. It should be noted that the ampicillin drug in Table 8.1 contained very thin (≤ 1 μm) crystalline particles, which were presumably more critically detected by the laser diode.

The laser diode sensors embody a number of advantages. Since monochromatic light is produced by laser diodes, all optical elements can be optimized for the laser diode wavelength. This theoretically results in better resolution, since the measuring volume is more accurately defined as to height and is more uniformly illuminated; particles of the same size passing through different parts of the view volume would be expected to give pulses of similar magnitude. (In all current generation LO sensors, the height of the view volume or sensing zone is defined by the thickness of the ribbon of illuminating radiation in the solution path, rather than by an aperture.) The dimension of the radiating area of the laser diode (laser cavity) is an order of magnitude smaller than the tungsten filament of an incandescent light bulb, resulting in a smaller measuring volume height. This improves both the concentration limit and the dynamic range of the sensors.

Long source lifetime is probably the most desirable feature of the laser diode. White light LO sensors generally suffer from short bulb life (approximately 250 h maximum, in the author's experience—this figure may be pessimistic, based on previous responsibility years ago to change the bulbs in sensors of several LO counters when they burned out). Heat has a detrimental effect on bulb life because the bulb is enclosed and typically not well cooled. The sensitivity of the filament to mechanical vibrations also affects the performance of white light LO sensors and limits their applications in environments with mechanical vibration. The laser

Table 8.1. Comparative Counts (Laser Versus White Light Sensors)

Sample	Sensor	2	5	10	20	25
Carbon Particles						
	Laser Diode	1,206	468	97	8	3
	White Light	1,345	472	101	8	3
Glass Beads						
	Laser Diode	1,839	709	78	64	47
	White Light	1,670	653	55	41	28
ACFTD*						
	Laser Diode	813	192	28	4	2
	White Light	810	203	26	3	1
Normal Saline						
	Laser Diode	475	120	20	3	1
	White Light	369	100	22	4	0
5% Dextrose						
	Laser Diode	359	13	2	0	0
	White Light	276	28	3	0	0
Na Ampicillin (Reconstituted)	Laser Diode	2,021	846	123	41	16
	White Light	616	189	39	6	2

*AC Fine Test Dust (no longer available, but a similar polydisperse quartz reference standard is available from NIST).

diode is insensitive to vibrations within reasonable limits.

Despite the considerable advantages of laser diode sensors, the user should be aware that these sensors may detect and count some types of particles differently than white light sensors. Their calibration also differs significantly from that of white light models.

Figure 8.8 shows one optical arrangement of a laser diode sensor. The laser diode is mounted on a rigid, thermally conductive metal support to keep it near ambient temperature, thus prolonging its lifetime. The typical diode source life at 25°C is optimistically estimated by the vendor to be 1,000,000 h or longer at the nominal 3 mV output of the laser. While this figure may not be achieved, the source life is far greater than that of an incandescent bulb. The diode in this model sensor is a gallium aluminum arsenide type, with an active laser cavity of about 20 μm × 10 μm in cross-section. The illumination from the laser diode is uncollimated. It emerges in a flattened cone with a half angle in the range of 10° by 40°. This source is imaged by high numerical aperture optics (collimating and focusing lenses) and focused into the center of the capillary view volume. To provide uniform illumination of the entire depth of the view volume, a negative cylindrical lens is placed between the cell and the collimation optics. This optical arrangement provides a uniform ribbon of light in the flow passage, providing for excellent particle size resolution.

The effect of view volume height on sensor response was discussed above. This first-order effect is the major factor in determining the sensor response curve shape, but more subtle optical effects are also reflected in the shape of the curve. An actual response curve (as shown in Figure 8.9) is more complex than the simple one diagrammed earlier.

The reason this curve represents an increase in complexity over the simplified one shown in Figure 8.6 is found in the principles of light scattering and extinction phenomena. Extinction, as we have seen, is the loss of light intensity, due to the removal of a portion of the light from the illuminating beam in the sensor by a particle. It is a complex combination of physical phenomena (absorption, refraction, reflection, etc.) as discussed earlier. In the size range between 1 μm and 2 μm, the slope of the curve is approximately 6. This indicates that the extinction signal is predominated by Mie scattering effects. (Again, see Chapter 9 on counting of airborne particles for an explanation of scattering effects.) Generally, for particles in this size range, the uniform increase of scattering and removal of light from the sensor beam results in a uniform increase of pulse height with increases in particle size. For particles greater than about 2 μm in size, the effect of scattering in removing light from the beam becomes less. Particles in this size range tend to scatter a small amount of light in a forward direction, toward the detector; thus

Figure 8.8. Laser diode sensor construction.

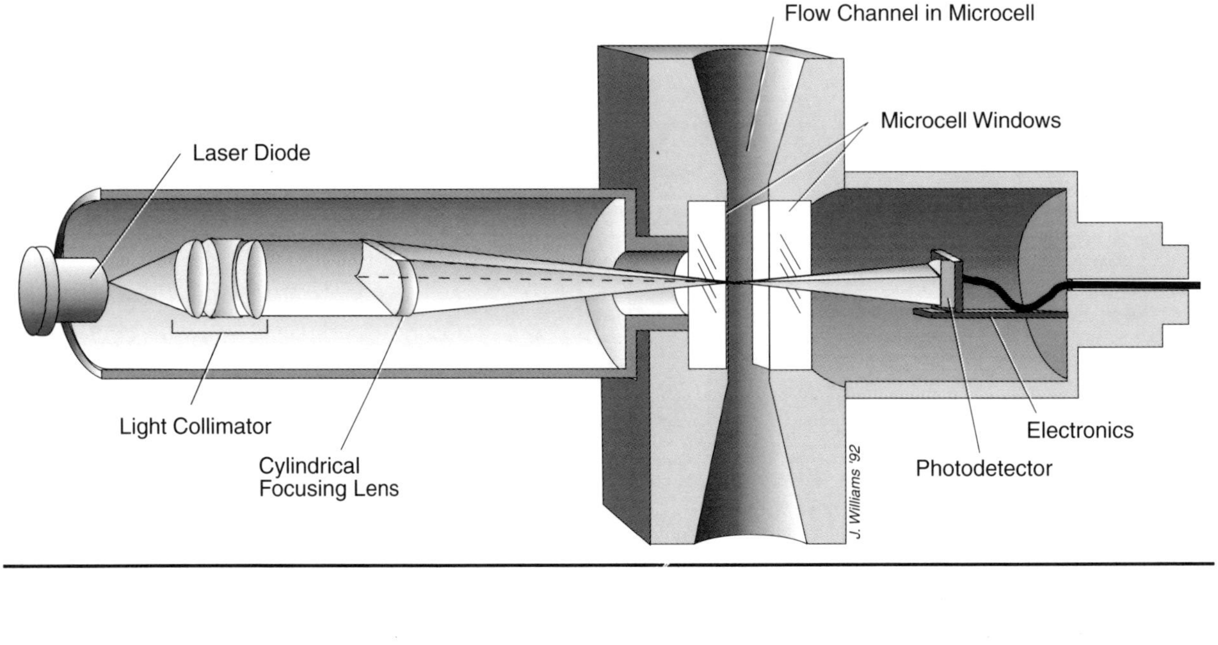

Figure 8.9. Representative calibration curve for laser diode sensor (HIAC/Royco model HR/LD 1-150).

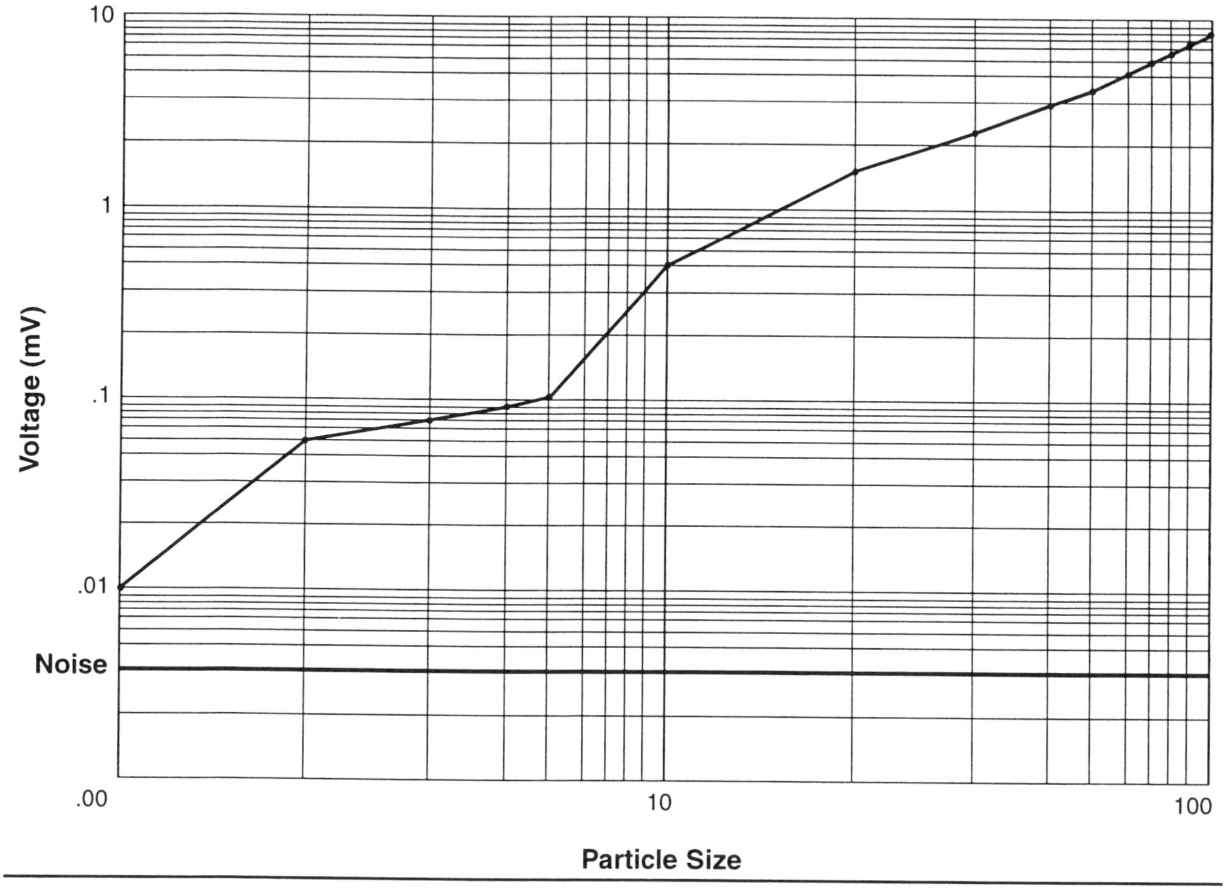

much of the scattered light is not removed from the beam. Furthermore, in the 2–5 μm range, particles begin to interact with light according to the principles of geometric optics as described earlier.

Thus, within the 2 μm to 5 μm range, some smaller particles may generate higher extinction signals than larger ones (due to a removal of light from the beam that is disproportionate to their size). This effect is not noted with white light sensors, due to the compensating effect of the various wavelengths of height in the illuminating beam. With these sensors, the mixture of wavelengths of light in the beam from the source tends to block out the nonlinear regions of the curve that may be seen with some laser sensor in the 2–5 μm region of the curve.

In the region of the curve from 5 μm to 20 μm, the calibration curve follows a square law dependency, much like a white light sensor. The height of the measuring volume of most laser diode extinction sensors (approximately 20 μm) is smaller than the largest particle diameter allowed passage by the microcell width dimension. As shown earlier in Figure 8.6, a transition from square law dependency to linear dependency of pulse height on particle diameter occurs for particles with a diameter comparable to the height of the measurement volume. Thus, particles larger than 20 μm elicit a linear response. Between 20 μm and 50 μm, a slope of 1 is observed, indicating that in this particle size range the measuring volume height is defining the maximum particle dimension that can be measured. In the vertical direction, the complexity of the curve may be further increased if different rates of gain (electronic signal amplification) are applied to particles of different sizes. According to one design, one (higher) rate of gain is applied for pulses from 2 μm to about 13 μm, and a second gain rate with lower slope is applied for larger particles.

Depending on the application, the selection of calibration procedures for the laser diode sensor becomes extremely important. The foregoing discussion explained that the calibration response of the laser sensors was significantly more complex than that of the earlier white light models. As is obvious from the curve in Figure 8.9, a larger number of size standards will be required to characterize the responses of the laser sensor if particles below 5 μm or greater than 30 μm are to be counted.

The apparent greater "low-end" sensitivity of the laser sensors may also be problematical when historical data are considered. Where white light sensors were often limited in sensitivity at 2 μm, their practical low-end sensitivity limit, laser sensors (by virtue of a stronger signal-to-noise ratio at smaller particle sizes) may provide higher count sensitivity at the 2 μm size. Realistically, laser sensors thus must be assumed to have some potential for acquiring higher counts at lower sizes. It will be up to the individual user to determine whether or not this occurs.

Because of the nonlinear response curve of laser sensors, it is not possible to derive a simple best-fit calibration line, even if only the ≥ 10 μm and ≥ 25 μm particle sizes will be counted. A curve can be empirically fit using a segmented line constructed by simply drawing straight line segments between supporting calibration points. A better fit can be obtained using the software routines introduced as part of the automation package in the HIAC Model 8000 counters, which are based on a cubic spine equation. The latter is simply a mathematical means of fitting a curve to points not on a straight line. The greatest problem with either method of calibration is that between any two calibration points, the curve only provides an approximation of the true response. If a laser sensor is to be used to size particles in multiple channels below 20 μm (i.e., for population analysis), the use of more standards should be considered. The most precise sizing can be achieved only by using standard particles at the individual threshold settings. This complex response of laser sensors can be a significant consideration if the user intends to use the USP nominal 15 μm monosize particle count standard (PCRS). It may prove impossible to split the 15 μm peak if a 15 μm calibration point is not generated with standard spheres and included in the calibration curve. Although the potential exists for the resolution of laser sensors to be better than that of white light sensors, this higher level of performance cannot be taken for granted.

In our own experience, the only types of particles for which laser diode sensors give significantly higher counts are transparent flake-like particles in some drug solutions and amorphous particles in protein solutions or biopharmaceuticals. These monotypic materials of high transparency tend to be counted with greater efficiency by laser counters. The amorphous particles are globular protein or drug agglomerates that consist largely of water. These materials would be expected to have an "average" refractive index very close to that of an aqueous suspending medium. This effect may prove to be an issue with the present <788> requirement's use of LO counters to resolve the microscopic count problems arising with dextrose solutions that contain a similar amorphous material.

Refractive Index Dependency of the LO Measurement

Considerations of how the optical properties of particles relate to light obscuration have dealt primarily with refractive index (n) effects (Knollenberg and Gallant 1990). A difference of > 1.25 percent in refractive index between a given particle and the carrier fluid is generally accepted as being necessary and sufficient for counting. This simplistic approach obviously ignores critical interactions of the particle and incident light involving diffraction and particle transparency (or opacity) and particle size; the difference in refractive index required would be expected to show at least some measure of particle size dependence. Refractive index effect experiments conducted using KIMAX glass particles in a butanol-toluene carrier (Table 8.2) suggest that a refractive index difference between particles and the suspending fluid of > 1.25 percent is sufficient for counting, but suggest that the difference necessary for accurate sizing may be greater than this figure. (Glycerol or other materials can also be used to vary the refractive index of aqueous solutions.)

The data in Table 8.2 suggest that the counting of 10 μm and smaller particles (> 2 μm, > 5 μm) is enhanced by refractive index differences above 1.25 percent. Common parenteral solutions can vary in refractive index over a range of 1.33 to 1.45, which constitutes a

Table 8.2. Dependence of LO Counts on Refractive Index Difference—Glass in Butanol-Toluene

	Approximate Solution vs. Particulate Matter Refractive Index Difference (%)		
Particle Size (μm)	**1.25**	**2.50**	**5.0**
> 2	1,280	2,401	3,479
> 5	194	280	1,682
> 10	20	23	121
> 20	16	14	23
> 25	3	2	6

variation of approximately 10 percent. This range can reasonably be expected to introduce some variability into the LO result for particulate matter with a refractive index of 1.3 to 1.5. The use of light obscuration to perform assays on solutions that might contain transparent materials, such as glass flakes, fragments of stopper coating, filtration aids, or siliceous prefilter material is thus subject to considerable variability. Our results using equal concentrations of NIST glass spheres suspended in distilled water and concentrated dextrose solutions showed lower courts for the particles in the 70 percent dextrose solution than in water. Microscopic counts for the two solutions varied < 10 percent at the ≥ 10 μm and ≥ 25 μm sizes (Table 8.3).

As a practical note, interstices in sensors (e.g., seals, mating surfaces, or threads) and in sample transport lines between the sample container and the sensor may require prolonged washout of calibrant solutions or solutions of different refractive indices from the one being currently run to avoid mixing artifact (i.e., "schlieren") or carryover of calibrant particles that may be counted as contaminant particles.

Importantly, our consideration of the complexity of the interaction between a particle and incident light in an LO sensor to this point begins to suggest that no clearly defined relationship between particle size and light obscuration can exist, particularly when particles of miscellaneous shape, composition, and optical properties are being counted. As previously discussed, particle physical characteristics significantly influence how they are sized and counted. The concept of offsetting (or complex) variables is important here. In routine counting, so many variables come into play in each individual count event that single considerations, even that of refractive index differences, are of decreased importance. In this context, the magnitude of counts obtained should be the focus rather than the individual numerical values.

Coincidence Counting

The simultaneous presence of two or more particles in the sensor view volume of an LO particle counter during a count event results in

Table 8.3. Refractive Index Dependence of Counts—NIST Glass Spheres in Water and Concentrated Dextrose Solution

		(Counts/mL)				
Test Suspension		**> 2 μm**	**> 5 μm**	**> 10 μm**	**> 20 μm**	**> 25 μm**
Filtered Water	HIAC	1107	748	253	15	1
	MICRO	—	—	283	—	2
70% Dextrose Injection						
	HIAC	801	424	142	4	1
	MICRO	—	—	269	—	2

Refractive Index Values:

0.45 μm Filtered Distilled H_2O—1.33

70% Dextrose Injection—1.42

NIST Standard Glass Spheres—1.55

physical coincidence counting. Raasch and Umhauer (1984), Willeke and Liu (1975), Lieberman (1975), and Knapp and Abramson (1996) provide useful summary discussions of the mechanism of this mode of coincidence counting and its effect on data. The error effect from electronic coincidence is due to signal superimposition. It can be shown, using an oscilloscope or multichannel peak height analyzer, that larger particle counts are registered with a concurrent decrease in numbers of smaller particles when either of these conditions occurs. As a result, measured particle distributions are systematically shifted to indicate larger particles as small particle concentration increases. As numbers of particles in tested solutions increase, coincidence effects become an increasingly important issue; problems can occur due to high numbers both of particles of a size for which thresholds are set and particles of a size below the lowest threshold setting. One issue of understandable significant concern is that subcountable-sized (i.e., submicrometer) microdroplets of immiscibles or crystallites of undissolved drug powder may cause counts at sizes within the countable size range.

In LO counters, coincidence effects are related to particle concentration, sensor view volume size, sample flow rate, and sensor signal-to-noise ratio. The first two of these factors may be characterized as contributing to physical coincidence and the latter two as affecting the occurrence of electronic coincidence. With regard to the latter, the ratio of signal produced by a particle of a given size to the inherent sensor electronic noise is the critical determinant of the lower size limit for a sensor. Noise is proportional to the square root of the sensor signal bandwidth. The bandwidth of the pulses that will be accepted by a counter is limited to approximately 20 kHz to reduce noise effect; this limitation of signal bandwidth and the time required for current generation in the sensor photodiode dictate that pulses produced by the sensor will have a finite rise time on the order of 10 μsec to 20 μsec. Sensor flow rate thus is typically adjusted so that particles cannot exit the view volume before the maximum pulse height is reached; if not, they will be undersized. It is possible to specifically calibrate for higher flow rates to minimize this effect. Double counting of single, large particles is typically prevented by auxiliary comparator circuits that require that the pulse level drop to below the voltage threshold settings before another count can be registered in a given channel.

Occurrence of simultaneous multiple particle passage through the sensor of an LO counter is critically dependent on view volume size. Flow rates of 10–50 mL/min result in a particle residence time in the sensor of 100 μsec or less, and an average pulse duration of 50. Smaller view volumes allow larger numbers of particles to be counted singly, since the probability of two particles being present in the smaller view volume is reduced. The smaller view volume sensors, however, require that a lower flow rate be used to ensure if pulses are to have an opportunity to rise to maximum height before the particle exits the sensor.

The mathematical probability of coincidence has historically been calculated based on Poisson statistics or by the linear volume model of Raasch and Umhauer (1984). For the former method, the probabilistic coincidence effect due to two or more particles occurring in a single sensor volume at a concentration of 1 particle/10 cell volumes is approximately 5 percent; for the latter method, it is about 10 percent. While these simplistic models allow an approximation of the effects of coincidence at different particle concentrations to be obtained, they fail to address variables related to the height of the view volume adequately in relation to particle size and the probability that a second particle can begin to enter the sensor zone or view volume before a first exits. The triggered Poisson model of Knapp and Abramson (1996a, b) results in a far more realistic calculation that considers these latter effects. This model shows excellent agreement with empirical results and defines a concentration limit for most sensors significantly lower than that based on a Poisson calculation. The effect of particle size increases as the particle size to view volume height ratio increases.

The paper by Knapp and Abramson (1996a) is of great interest to those interested in the LO counting of solutions that may contain numbers of particles approaching the USP limits (e.g., a 1 mL ampoule with an excess of 5,000 counts ≥ 10 μm per unit). While counts at this level are rarely encountered, the user should understand the mechanics of count loss at smaller sizes and count increase at larger sizes. The classical model for particle counting described in the literature is the geometric Poisson model of Jaenicke (1972) and extended by Lieberman (1975). Knapp's publications have shown that calculations based on this model do not agree with experimental data. According to Knapp and Abramson, the single particle counting error estimate for USP 23 SVI limits, using Jaenicke's geometric model to evaluate a commercial laser-sourced sensor, is 9.32 percent; the single particle count error estimate for this detector using the experimentally validated particle-triggered Poisson model is 19 percent. The estimated count error increases for particles larger than 10 μm.

The novel basis of the triggered Poisson model is the relationship of detector height and particle size to coincidence. In every earlier Poisson model of the counting process, the ratio of particles to detector volumes in the test solution is used to estimate the effect of the concentration-dependent coincidence error. However, the physical or geometric detector volume is affected by the size of the measured particle. This interaction, shown in Figure 8.10, illustrates the fact that the effective height of the detector, the dimension in the flow direction, varies with the diameter of the particle being measured. In the figure, a second particle has begun to enter the sensor beam before the one preceding it has completely exited, and a coincidence pulse results.

Stated another way, the effective beam height will be an additive dimension based on the geometric beam height and the size of the first of the two particles being considered. The first particle must have time to completely exit the view volume window before the second enters, or the net effect will be a coincidence count just as if both particles were simultaneously present within a single geometric view volume as defined by the geometric window height, width, and length. Since the effective detector height varies, the effective detector volume and thus the particle concentration effect also varies. This variable detector volume is defined by the authors as the effective detector volume. An important conclusion that can be drawn from this figure is that concentration-limited particle counting accuracy decreases as particle size increases.

One additional factor must be considered when calculating the effective detector volume. There is an interaction between the particle size chosen as an analysis threshold and the effective concentration. The basic particle size–geometric window height interaction is shown in Figure 8.11a. When the diameter of the particle size selected as the analysis threshold is 10 percent or less of the size of particle being counted, the effect is negligible. For accurate results with larger ratios, as Figure 8.11b shows, the effective reduction in the area of the particle being measured must be considered. The effect of the analysis threshold is a reduction in the measured particle area by twice the threshold particle area.

The definition of the variation of the effective detector volume with the particle size being measured is an important step forward. This understanding must support accurate functional measurements of the geometric volume and the beam height.

Figure 8.10. Particle size–view volume height interactions: Effective view volume.

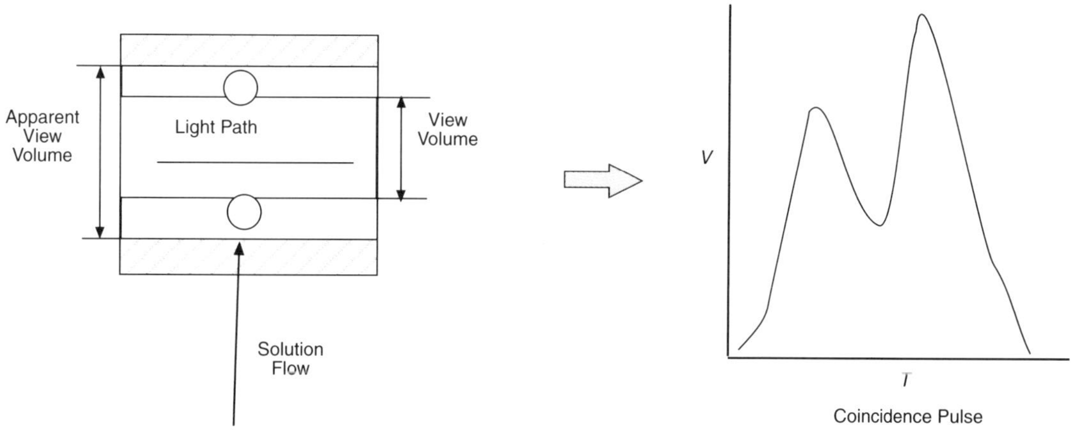

Figure 8.11. Particle size–view volume height interactions: Threshold effects (after Knapp and Abramson 1996b).

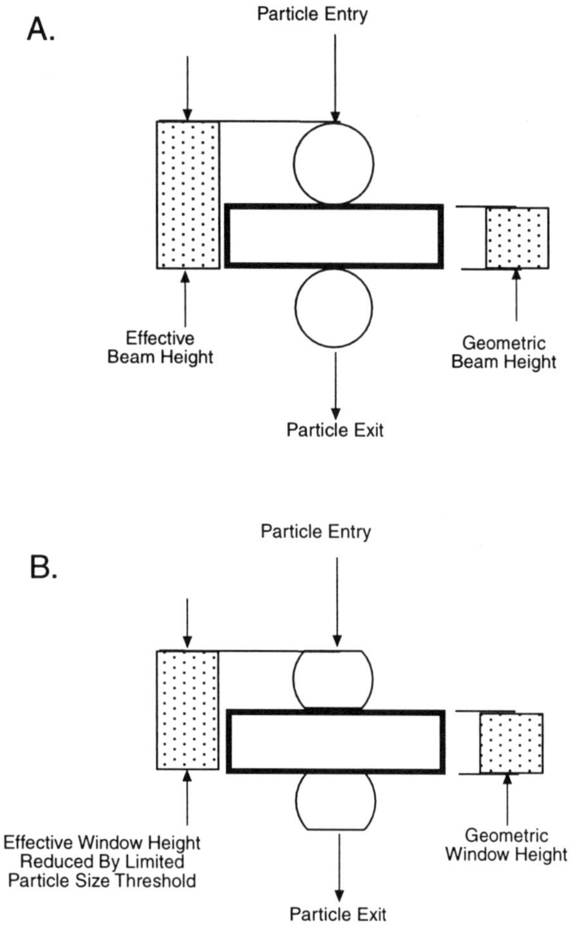

Since the 1996 publication of the Knapp and Abramson paper, some particle counter manufacturers have apparently revised or reviewed their methods for quantitating the coincidence effect, with a consequent decrease in concentration effects for some sensors. The basic method in empirical determination of the true concentration limit is to perform sequential, accurate dilution of a concentrated test particle suspension. At some particle number concentration, the counts obtained will begin to decrease appropriate to the dilution performed. When the count of particles per unit volume at these lower concentrations is known, the true particle number at higher concentrations and the count loss may be determined.

What about count increases due to coincidence? Since LO counters assign a particle size related to the total area of light obscured by particles in the sensor, a simplified calculation of the minimum number of particles of any countable size required to count as larger particles may be estimated. For example, six 2 μm spheres would have approximately the same projected area as one 5 μm sphere in the sensing zone, and four 5 μm spheres would be approximately equivalent to one 10 μm sphere.

An example of coincidence effect is provided by the sample data in Table 8.4. The test system consisted of 7 μm latex spheres (with a coefficient of variation [CV] of < 1 percent) that were thoroughly disaggregated before counting. The test particle concentration was approximately 10,000 particles/mL, and the published sensor concentration limit for the sensor used was stated by the vendor to be 10,000 particles/mL for ≤ 10 percent coincidence.

This result shows reasonable agreement with the value calculated using Poisson statistics; the combined LO area of two 7 μm particles is sufficient to give one count at > 10 μm. While the majority of counts from the 7 μm spheres in Table 8.4 were collected in the correct channels (> 2 μm, > 5 μm), an appreciable number of counts occurred in the next higher size channel (> 10 μm). The data indicate that for this monosize latex test system containing numbers of particles roughly equal to the specified sensor concentration limit, a significant coincidence count existed at ≥ 10 μm. It is unlikely that this concentration of particles between ≥ 5 μm and ≥ 10 μm will be encountered in parenteral solutions, but it must be emphasized that similar artifacts can occur to a lesser extent with particle concentrations at lower levels.

It is instructive to conduct a simple experiment that will allow the user to become familiar with the count distributions that indicate coincidence. If a sufficient range of thresholds is set, as in Table 8.4, and particle concentration in a

Table 8.4. Coincidence Effect with 7 μm PSL Spheres

(Counts/mL)				
> 2 μm	**> 5 μm**	**> 10 μm**	**> 20 μm**	**> 25 μm**
10,488	10,301	487	2	0

test suspension is increased incrementally by the addition of more test particles (e.g., PSL spheres), the following conditions will be observed on inspection of the count data:

- Below concentration limits, particles are typically counted singly and pulse-per-particle conditions exist.
- As sensor concentration limits are approached, a higher proportion of count events occur with multiple particles in the view volume so that apparent counts of larger particles are slightly increased, and counts for particles of a given size are lost.
- With further increases in particle concentration, more count events reflect the presence of multiple particles in the view volume, and particle size distribution data become seriously affected.
- As the limits of electronic coincidence are approached, and the majority of the counts recorded represent multiple particles and counts over smaller size (e.g., 2 μm, 5 μm) channels decrease and often become similar between adjacent higher channels. Importantly, if this condition occurs, sequential dilution will result either in a decrease of counts disproportional to the dilution or in an increase in counts in lower channels where count loss was occurring.

"VALIDATION" AND "COUNT ACCURACY"

The Knapp and Abramson (1996) publication comprises a suitably detailed consideration of coincidence or concentration effects. These are, at best, a minor practical consideration with LO counting per USP <788> due to the cleanliness of solutions tested and the low allowable level of counts generally recorded. A particle counter vendor's assessment of coincidence effect is, in most cases, generally acceptable based on the Knapp and Abramson method or other approximation due to the low numbers of particles being counted.

The title of the Knapp and Abramson paper, containing the phrase *validation of count accuracy,* may prove somewhat misleading to some users of these instruments. As discussed at the beginning of this chapter, it is important that the analyst grasp the concept that there is no such parameter as accuracy in counting of real-world particles of heterogeneous optical properties. Obscuration counters do not, in fact, size and count particles; they size and count voltage pulses, which are referenced to standard particles. There are, in this case, different operational determinations based on the type of instrument used. Precision with a given method may be enhanced or ensured, but accuracy applies only to the instrument under consideration. Given the high number of variables in the LO assay, the use of the term *accuracy* seems somewhat inappropriate. Further, validation of an instrument's function consists of verifying that key operating parameters do not change during intervals of operation and that specified instrument functions are dependable over a range of test conditions. Examples are provided by sample volume, flow rate, calibration thresholds, and resolution. These must be evaluated and verified regardless of whether or not they are software driven.

Let the user beware of another potential snare. There is a major point that should be made with regard to definitions of method validation, software validation, and instrument validation with respect to LO particle counting (see chapter 11). The method itself (i.e., the USP <788> methodology) is considered to be validated; however, modifications to the method will need to be validated for theoretical practical equivalency. Software validation must be dealt with in some detail. More importantly, it may become necessary to validate the function of a specific instrument to show it capable to perform the USP <788> test. The instrument must be shown to perform as intended.

RESOLUTION EFFECTS

During the writing of the original USP XXI <788> requirement for LO counting in 1985, resolution was correctly chosen as a critical parameter of instrument performance that might have a significant effect on count data. The term is borrowed from microscopy, in which resolution is the capability of the instrument to separate two closely opposed image points. In LO counting, resolution describes the ability of a sensor to separate the pulses of particles of similar size. Resolution is measured as the artificial increase in the width of a voltage pulse distribution from monosize calibrant particles due to sensor counter effects. As an example, if the true distribution of sizes in a "monosize" 10 μm standard is ±0.1 μm and the spread of voltage pulses produced is equivalent to the mean size ±0.5 μm, the resolution is 8 percent.

The term *resolution* is intended, then, to describe the range of pulse voltages that will be produced by the sensor for particles of a single size. Poor resolution (i.e., broadening of the pulse response distribution) will result in decreased precision of count data. Nonuniform illumination in the sensor is the primary cause of variances in sensor response for particles of the same size. This variance is typically several percent of the mean diameter of particles in the ≥ 5 μm to ≥ 20 μm range. Under normal sampling conditions, the particles are randomly distributed throughout the view volume, and a statistical averaging process applies to the pulses generated.

The factors that detrimentally affect resolution in white light sensors also affect the laser diode type:

- Nonuniform light intensity across the measuring volume due to source nonuniformity
- Finite depth of field of the illuminating optics
- Dark areas at the periphery of the view volume
- Nonuniform response of the photodetector surface

The first three factors result in uneven illumination in the sensor view volume. This nonuniform optical performance causes particles traveling at identical velocities through different portions of the measuring cell to generate different signal pulse heights. With a white light HIAC HA-type sensor, resolution of 4–5 percent is routinely attainable. In a sense, the multiple wavelengths in the white light source and the poor source collimation result in an "averaging" of illumination intensity in the view volume. Because of the more precise collimation of the laser beam in the view volume, the laser diode output pulse amplitude will be more strongly dependent on the first three factors. Poorer resolution may theoretically result than with the white light sensor, unless the sensor and source elements are critically aligned to take advantage of the unit's higher performance potential. Unlike the situation with white light sensors, resolution for laser diode sensors may vary significantly over the dynamic range. The best resolution is typically found in the portion of the laser diode sensor curve that follows a linear relationship (≥ 10 μm in Figure 8.9). In this portion of the curve, there is less signal change for change in particle size. The calculated resolution of the laser diode sensors with a segmented line calibration curve will typically be less than if a spline fit is used.

Thus, despite the higher resolution that might be predicted for the laser diode sensor (based on the monochromatic nature of light employed), some production laser diode sensors may actually have significantly lower resolution than white light models of similar dynamic range. This assumedly results from the existence of a non-Gaussian beam intensity profile in the view volume. This would presumably result from diffraction effects that occur when a beam of monochromatic light passes though a lens system. The end result is that in addition to the central distribution of the focused light, peripheral rings of decreased intensity occur (diffraction rings or fringes).

In addition to specifying widened pulse distribution for monosize standards, low resolution implies that particles below a specific size threshold may be counted as being above the threshold (Figure 8.11). Theoretically, lower resolution at a given size could result in higher counts at that size, due to the log normal or Poisson distribution of the particles in most injectable materials.

As shown in Figure 8.11, particles are present at sizes below about ≥ 1.5 μm in numbers higher than would be predicted, based on direct extrapolation of the size-number curve for particles above this size. Further, the distribution of particles below this approximate ≥ 1.5 μm cutoff is distinct from those of larger size, so that the size distribution curve is discontinuous.

Sources of Erroneous Count Data

It is absolutely essential that the user be aware of the potential for collecting erroneous data with the LO method. Errors in counts can arise based on a significant number of factors, including particle size and shape, particle overconcentration, immiscible fluids, and air or gas bubbles. The author's choice of most important error sources are discussed below. The discussion is not comprehensive, in that there are probably significant

Figure 8.12. Resolution effects on count data.

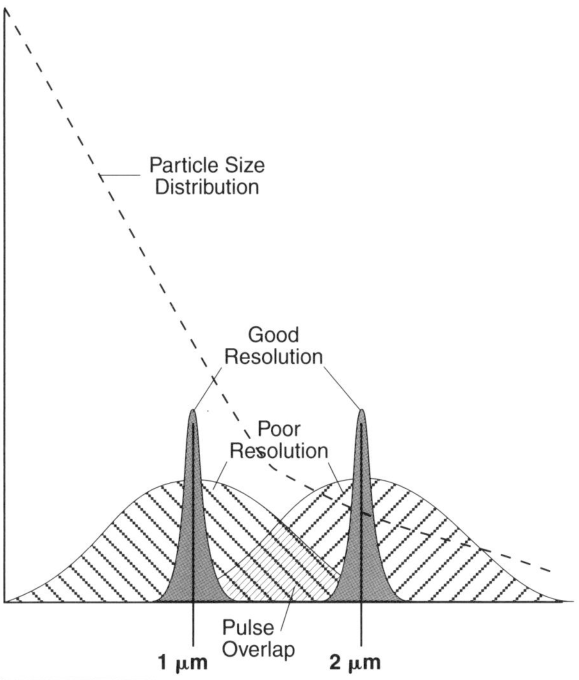

error sources known to other LO users that do not appear here.

The results of the instrumental assay cannot be expected to agree in any predictable fashion with those of a microscopic test (Rebagay et al. 1977; DeLuca 1977). The papers by DeLuca et al. (1987) and Schroeder and DeLuca (1980) are also useful in this regard. Various sources of artifacts inherent in the method have been previously reported (Barber 1987; Chrai et al. 1987; DeLuca et al. 1987; Grundelman and Goldsmith 1987; Barber and Williams 1990). Product types analyzed must be well understood with respect to inherent sources of artifacts, including the production of gas bubbles, refractive index effects, and droplets of dispersed immiscible liquids. The user should also be aware that although LO sensors with a similar dynamic range have a generally similar response to particles found in parenterals, there may be specific sensitivity differences related to particle refractive index or type of sensor illumination used (white light, diode laser, gas laser) that should be investigated before an instrument is purchased or applied to a broad range of product types.

Particle types significantly undercounted or not detected by the LO test may include

- Particles of amorphous or gelatinous morphology
- Transparent, flakelike particles
- Large contaminant particles, such as paper fibers

In addition to product-related sources of error, mechanical artifacts are also related to the way in which light extinction counters operate. It appears that some sources of artifacts inherent in the method can never be totally eliminated.

Particle Size and Shape Bias in Light Obscuration Counting

Spherical standard particles are the only particles that are counted and sized in a generally predictable fashion by LO counters. Spherical particles rarely occur in injectable solutions. With respect to any type of particle other than a sphere, light obscuration may assign sizes that are lower than those obtained by a microscopic assay (West 1990; Schroeder and DeLuca 1980; Delly 1980; Blanchard et al. 1976). Further, since randomly determined aspects of irregular particles are viewed by a LO counter as they pass through the sensor view volume, a broadened distribution of counts will be recorded for a population of nonspherical particles of the same size and shape. This randomness of particle sizing by LO counters is an unavoidable source of variability.

The undersizing of irregularly shaped particles by LO instruments becomes apparent with test samples such as quartz grains that are generally cuboid (equant); it is also more pronounced with fibers and flakes. The method error is most serious with fibers. In an extreme case, the pulse height produced by a long fiber that completely spans the sensor window will be proportional to the fiber width times some dimension of the sensor window rather than the actual particle length. The undersizing of fibrous particles in tested parenteral solutions may be a serious error, given the indication that fibers may provide proof of non-GMP manufacturing processes.

Air or Gas Bubbles

Air bubbles can be counted as solid particles. Such bubble counts can result from cavitation, high (unstable) concentrations of dissolved gases as present in aerated water, and chemical reactions (e.g., bicarbonate). Cavitation in this context refers to a condition whereby the negative pressure of a syringe sampler drops below the vapor pressure of the fluid being passed through the sensor. Air or other gases present in dissolved form in a solution may be vaporized by the negative pressure differential in a sampler to form bubbles in the sensor. Solutions may simply entrain bubbles due to agitation. This problem sometimes occurs with LVI solutions in flexible containers that have an air headspace volume. An example of this is provided by the per mL count data obtained before and after degassing a sample of 5 percent dextrose from a flexible container in a vacuum of 15 in. Hg (Table 8.5) for 1 h.

One importance of the bubble issue is the fact that the USP <788> requirement for SVI calls for the inversion of test units before counting; this agitation may serve to disperse a considerable amount of air in the form of bubbles. While vacuum and ultrasound are effective in removing air bubbles from solutions, the time required for this additional step increases the time required for the LO assay and requires rigid adherence to a testing protocol to ensure that all solutions tested are treated similarly. The difficulty of removing air bubbles is increased relative to size; bubbles of 10 μm or smaller are difficult to eliminate. Pressure sampling may alleviate the difficulties encountered with air bubbles to some extent, but it is generally more cumbersome and time consuming than vacuum sampling.

Sodium bicarbonate may be added to various parenteral or physiologic solutions to establish a buffer system, according to the following reactions:

1. $NaHCO_3 + H_2O \leftrightarrow Na^+ + HCO_3^- + H_2O \leftrightarrow Na^+ + H_2CO_3 + OH^-$
2. $H_2CO_3 \leftrightarrow H_2O + CO_2(g)$

The important thing about this reaction in our present consideration is that each time a H^+ ion is taken up to maintain the pH of the system, a molecule of CO_2 gas may be generated. This system is poised with respect not only to pH but also with regard to physical shock or pressure. If the pressure on the solution system containing the bicarbonate is decreased, the equilibrium of equation (2) moves toward the right, releasing gaseous CO_2. If pressure is increased, CO_2 gas is less likely to be released. If the solution container is subjected to physical shock (i.e., shaking, a blow, or rapid acceleration), bubbles of CO_2 may also be evolved. This is due to cavitation effects whereby cavities or bubbles may be formed in a solution as the local vapor pressure of the solution is exceeded.

Passage of a bicarbonate-containing solution through a LO particle counter thus favors the release of CO_2 bubbles, both through application of negative pressure and the rapid flow of the solution through the narrow orifice of the instrument's sensor cell. Bubbles may also be formed by the fusion of molecules of CO_2 already present, due to the effect of vacuum. Carbonate reactions are a classic and well-known artifact in instrumental particle counting of solutions that contain bicarbonate buffer. They may be avoided to some extent by letting the solution sit undisturbed for some time at atmospheric pressure, by sonication, or, as mentioned above, by use of a pressure sampler that does not apply a vacuum as do conventional LO samplers.

In the case of solutions containing bicarbonate, flexible bags or other nonrigid solution containers may be distended by the pressure of the CO_2 gas when received for testing, bubbles may also be observed to form in bottles when the stopper or closure is removed. Either of these observations should indicate to the user that a potential problem exists. After the pressure is released by opening and allowing the CO_2 in the solution to equilibrate with atmospheric pressure, normal

Table 8.5. LO Counting Due to Presence of Air Bubbles Before Degassing Solution

	(Counts/mL)				
	≥ 2 μm	≥ 5 μm	≥ 10 μm	≥ 20 μm	≥ 25 μm
Before Degassing	1,026	418	42	4	3
After Degassing	428	118	3	0	0

counts may be obtained; times required for equilibration may be prolonged. If the solutions are tested immediately on opening the units, high counts due to bubbles are generally recorded. The particle burden of the affected solutions may be verified by microscopic particle counting. The counts obtained due the presence of bubbles do not represent particles and thus cannot be considered with regard to the USP limits.

Elevated instrumental counts may also be obtained when some drug solutions not containing bicarbonate are tested. There is a pronounced tendency of solutions of some drugs to form bubbles or foam on agitation based on the nature of the drugs themselves. As was the case with the bicarbonate-buffered solutions discussed above, smaller subvisible bubbles, in the ≥ 10 μm and ≥ 25 μm size ranges, are sensed as particles by LO particle counter sensors causing artificially elevated counts.

The primary consideration regarding the mechanism of foaming is the chemical nature of the drug itself. The issue here is generally persistence of bubbles rather than bubble generation per se. An example in the author's experience is provided by drug compounds that are sodium salts of carboxylic acids and ionize to form molecules that have both hydrophobic and hydrophilic properties, due to the presence of both charged and neutral regions. Such compounds are termed amphiphiles, a term indicating that this type of molecule has some degree of affinity for both polar and apolar solvents. In the case of most drugs of this type, the balance of properties favors water solubility. These materials, however, by virtue of their amphiphilic properties, significantly reduce the surface tension of the water in the solution.

The mechanism of bubble generation in the such drug solutions is straightforward. On agitation, air bubbles are dispersed in the liquid. The drug molecules function as surfactants, orienting at the interface between the air bubbles and the liquid, decreasing the interfacial free energy of the system. In a functional perspective, the molecules form a film or "skin" on the air bubbles that tends to prevent their fusion and elimination from the liquid. The overall effect is the production of a stabilized air-in-water emulsion (foam), with bubbles of various sizes dispersed in the solution. Because of the reduced surface tension of the water in the system, the bubbles tend to remain dispersed and persist for some time after agitation.

The persistence of a bubble within a solution volume will be inversely related to size; smaller bubbles take far longer to leave the solution than larger ones. The rate at which bubbles are cleared from solution is strongly size dependent according to the relationship:

$$\alpha = \frac{1}{d^3 \Delta\rho g}$$

where α is the relative time required for clearance, δ is the bubble diameter, $\Delta\rho$ is the density difference between the bubble and the suspending medium, and g is the force of gravity.

This, of course, pertains only to systems where bubbles are free to move without restraining influences, such as adherence to container surfaces. Bubbles above about 5 μm tend to be cleared from the solution by moving vertically; smaller bubbles may be more likely to be removed by dissolution. Thus, while the foam formed is most obvious in the headspace of an agitated unit, bubbles of both visible and subvisible size persist for significant intervals following agitation under the solution surface within the solution volume of an agitated unit. Due to the decreased surface tension of the water in solutions with surfactant properties, bubbles of a wide range of sizes are readily sheared off an air mass that moves through the solution volume. Bubbles of subvisible (i.e., 10 μm to 50 μm) size may become adherent to the inner surface of plastic containers and require significant energy (e.g., via sonication) for removal.

The persistent adherence of subvisible-sized bubbles to some plastics results partially from the properties of the polymer surface. Polyethylene, for instance, is a hydrophobic, highly ordered polymer, so air bubbles produced by agitation tend to adhere to it on contact, because of their greater affinity for the polyethylene than for the aqueous drug solution. This adherence is enhanced even further with bubbles in solutions with surfactant activity due to the hydrophobic properties of the molecules in the "skin" at the bubble surface. In consideration of the bubble-related artifacts that may result with LO particle counting of these types of solutions in plastic, such solutions may be tested after simply being poured from the container into ultracleaned beakers. The solutions in the beakers can be sonicated as a further precaution to ensure the removal of air bubbles.

In a final note with regard to bubble artifacts, a word of warning is also in order regarding sampling fluid transport systems with LO counters. Fluid in these lines tend to contain dissolved gases under pressure that may form bubbles when the pressure is released (i.e., when a sampling point valve is opened). The sampling system should be set up so that the valve is downstream of the sensor for dynamic sampling rather than upstream (Table 8.6). Otherwise, bubbles will be released upstream of the sensor, resulting in artificial count data. (Be sure the sensor can withstand the line pressure.)

Variability Due to Sampling Effects

In particle counting of parenteral solutions, as in any endeavor involving statistical inference, larger samples more accurately represent the population from which they are taken (Wheeler and Chambers 1986). Hence, sampling small aliquots of a solution unit gives a more variable result and a less accurate estimate of the particle population present than will a larger sample. The sampling mechanism of LO counters compounds this problem, in that the sample aliquot is taken from a small region of the solution volume. Thus, unless continuous agitation is used, the settling of particles ≥ 20 μm in size may remove them from the area of the sample probe. Thus, a LO counter probe placed at a given depth in a container may collect a sample that gives an inaccurate estimate of the particle population diversity. Even if continuous stirring is used, particles tend not to be suspended evenly in the solution volume; smaller particles are suspended or dispersed to a greater extent than larger particles.

The bottle sampling mechanisms most frequently used with LO counters invariably result in anisokinetic sampling (see Chapter 9). Larger particles are not effectively drawn into the sample probe due to their higher mass and the relatively low sample flow rate. This effect is significant whether the particles fall past the probe after manual inversion or are propelled past it in circular orbits by stirring. By definition, rotational stirring at a rate sufficient to cause even dispersion of particles cannot be used, since the vortex formed will entrain air. Thus, probe location within the solution volume can influence count data, due to the nonhomogeneous, size-dependent distribution of particles in the container (stratification or zonation effect).

To study this effect, suspensions of NIST standard glass spheres were used (NIST Standard Material #1003) in filtered water in a 1 L beaker. During the sampling process, the position of the sample probe tip was placed at different points in the solution volume. Significant variability was noted in the counts obtained (Table 8.7).

An interesting observation to be made at this point is that microscopic methodology is far less limited with regard to sample volume than is LO testing. The potential here exists for reducing the variability of results obtained simply through an increase in the volume of the aliquot sampled (i.e., the entire volume of the tested unit).

Issues Relating to Nonaqueous Vehicles, Color, and Viscosity

Generally, caution should be exercised when analyzing any product that is not a pure solution with a clarity and viscosity approximating that of water. This obviously includes preparations containing suspended materials or emulsified particles that may be sensed by the counter and contribute to background counts or counts in the size channels of compendial interest. Some SVI products are formulated with organic vehicles to obtain drug solubility. In many instances, these vehicles are cosolvent systems incorporating

Table 8.6. Effect of Sample Point Valve Location on Count Data

	(Counts/mL)				
	≥ 2 μm	≥ 5 μm	≥ 10 μm	≥ 20 μm	≥ 25 μm
Valve upstream of sensor	1,250	805	138	21	14
Valve downstream of sensor	128	56	3	0	0

Table 8.7. Particle Stratification Effect: NBS Glass Spheres in Water in a 1 L Beaker (HR-120 HA Sensor)

	Counts/5 mL				
Probe Tip Location	**≥ 2 μm**	**≥ 5 μm**	**≥ 10 μm**	**≥ 20 μm**	**≥ 25 μm**
1″ from top center	5,807	5,911	2,661	186	47
2″ from top center	5,809	6,509	3,842	417	84
1″ from bottom center	8,461	6,554	2,265	200	56
1/4″ from bottom sidewall	7,036	3,886	1,129	952	616
1″ from top sidewall	6,576	3,301	709	72	25
2″ from top sidewall	7,191	3,974	1,064	88	28
1″ from bottom sidewall	7,021	3,783	1,088	146	41
Top center (Vortex)	7,075	3,633	757	53	26
Average	7,547	4,694	1,689	264	115
C.V. (%)	10.64	29.37	66.39	190.11	176.15

polyethylene glycol, propylene glycol, ethanol, and/or surfactants. These products are not normally amenable to LO counting, as contact with water in the sampler/sensor may result in drug precipitation. Products for which LO counting may not be appropriate may be generally grouped into three categories:

1. Products that are not pure solutions
   - Emulsions (parenteral) or admixtures containing emulsions
   - Colloidal preparations
   - Liposomal preparations
   - Products containing macromolecular materials (e.g., dextrans, starches, povidone iodine, etc.)
   - Products containing micellular materials, such as benzalkonium chloride, or other surface active agents
2. Products that can cause physical sensor effects
   - Colored solutions, such as iron dextran and povidone iodine
   - Solutions of high viscosity (e.g., concentrated dextrose, starch)
   - Solutions that differ in refractive index from water by more than ±10 percent
   - Solutions containing pressure-active components, such as bicarbonate, or solutions that can incorporate persistent air bubbles
   - Formulations containing organic cosolvent systems, such as polyethylene glycol, propylene glycol, ethanol, glycerol, or benzyl alcohol
3. Products that cannot be entered without introducing particulate matter, such as product contained in glass ampoules or liquid-powder devices

Intensely colored products that strongly absorb light are generally not appropriately tested by LO counting. The technique is dependent on light transmission, and while there is a certain latitude within a sensor to increase the light intensity in order to maintain a constant output to the photodetector, this correction is limited in range and can seriously shorten the lamp life of white light sensors. A product such as iron dextran would be an example of this type of solution.

As discussed earlier, sensors are calibrated at specific flow rates, and counts must be made at ±10 percent of that flow rate to ensure sensor performance and calibration accuracy. Some product solutions have a viscosity that is significantly higher than water. This can result in a reduction in the flow rate through the sampling device tubing and sensor orifice. Samples that function by using negative pressure to aspirate sample are more sensitive to this than positive displacement or pressure samplers. Concentrated dextrose (50–70 percent) and dextran solutions are examples.

Interferences from Subcountable-Sized Particles

Interference from particles below the size that can be counted in LO counters is not well understood. It is important to note that large numbers of evenly distributed subcountable particles present in the test suspension can result in the sensor view volume containing similar numbers of particles for each count event. In this situation, counts may be closely similar in the lower channels of the instrument and show a significant decrease for the first channel in which the combined pulse is not sufficient to cause a count (Table 8.8).

Individual particles of submicrometer size do not cause sufficient light obscuration to trigger a pulse in LO counters. Particles of this size will result in LO pulses of the same magnitude as sensor noise. The observed effect is unlikely to be due to coincidence, since sixteen 0.5 μm particles would be required to give an LO pulse equivalent to a 2 μm particle, and 400 would be required to equal one 10 μm particle. Two other mechanisms for counting this type of particle may be suggested: (a) the random superimposition of pulses from particles of this size on noise pulses to give pulses of countable size; and (b) the scattering interaction of submicrometer particles and white light resulting in a diminished or varying intensity of light that the LO counter senses as particles. Although counts in the > 10 μm size range due purely to coincidence from this source appear unlikely, the probability of an additive error effect through mixed coincidence of submicrometer particles with larger particles of 5–10 μm in size will be increased in the presence of high concentrations of submicrometer particles.

Table 8.8. Count Artifacts Due to Particles of Subcountable Size (0.5 μm PSL Spheres)*

≥ 2 μm	≥ 5 μm	≥ 10 μm	≥ 20 μm	≥ 25 μm
23,308	1,429	137	5	0

*Data are abstracted from the *Proceedings of the 1990 PDA/IES Conference on Particle Detection, Metrology, and Control.*

Artifacts due to submicrometer particles are potentially important due to the common occurrence of small quantities of silicone in some rubber-stoppered SVI products. Dispersed particles of this material are often of submicrometer size, and USP failure counts could, in theory, be obtained when allowable numbers of particles at the lowest compendial size (> 10 μm) are actually present. The effect due to submicrometer particles is typically indicated, as is coincidence counting, by inordinately high counts in the lowest size channel on an LO instrument. This effect may also be important with submicrometer crystals of undissolved drug powders. In the author's experience, some models of sensors are more seriously affected by submicrometer particles than others. Submicrometer particles dictate that the LO assay will not be usable for parenteral products such as emulsions; colloidal, micellar, or liposomal suspensions; or microencapsulated drug material.

Immiscible liquids in drug formulations, such as a liquid silicone lubricant from elastomeric closures, can cause serious artifacts. Many of the droplets of dispersed immiscibles are of subcountable size. Agitation or storage of solutions in containers with silicone-coated closures can result in a transient, unstable dispersion of immiscibles in the solution. The presence of extremely low chemical concentrations of silicone oil dispersed as microdroplets can result in significant count artifacts (Chrai et al. 1987. As an example, only 0.1 mg/mL of silicone per mL is sufficient to create particulate counts at the USP SVI limits under some conditions of dispersion. In one experiment we have conducted, 0.5 ppm of silicone oil was dispersed as ≥ 0.4 μm to ≥ 2.0 μm microdroplets in an aqueous system using a bench top ultrasonic homogenizer. The resulting LO counts per mL are shown in Table 8.9. In these data, a counting error is suggested by closely similar counts in the 2 μm and 5 μm channels. This is caused by the coincident, simultaneous passage through the sensor of multiple microdroplets of ≤ 2 μm in diameter, creating a false response.

In this experiment, analysis of a diluted sample of the test material with the Coulter counter (with a much higher concentration limit) indicated that an average of less than 10 particles > 10 μm and less than 40 particles > 5 μm per mL were present in the test system. The basis of these LO artifacts remains unclear. The conditions of flow convergence and increasing sample velocity (resulting from the 10-fold or greater

Table 8.9. LO Counts from 0.4 μm to 2.0 μm Silicone Microdroplets (Counts/mL)*

	≥ 2 μm	≥ 5 μm	≥ 10 μm	≥ 20 μm	≥ 25 μm
LO	10,036	9,670	148	18	1
Coulter	763	39	7		

*Data are abstracted from the *Proceedings of the 1990 PDA/IES Conference on Particle Detection, Metrology, and Control.*

reduction in the cross-section of the solution path between a sensor probe tube and view volume) may result in the fusion of dispersed immiscible microdroplets. Questions regarding the coalescence of immiscible liquid microdroplets during the sampling process and a consequent increase in counts at larger sizes cannot be conclusively addressed with the data at hand. Immiscibles may, under some circumstances, also create a coating on sensor windows. Artifacts from immiscible particles should be suspected whenever counts in the smallest size channel of an LO instrument (≥ 2 μm and ≥ 5 μm) are present in excess of the SVI limit. This is particularly true if these counts are near the sensor concentration limit.

The numbers of particles detected appear to be more critically related to the extent of dispersion of the immiscible material than to its concentration. To evaluate the role of silicone in producing the high instrumental counts (10 μm) noted for samples of SVI drug product that we recently tested, stoppers from units of these codes were tested for the release of silicone in the absence of drug. First, stoppers from 30 units from each code were removed and the drug powder discarded. The vials were then half filled with normal saline diluent and subjected to mechanical agitation at 60 cycles/min at 35°C while lying on their sides on a platform shaker. Groups of 10 samples were removed after 1, 5, and 10 min agitation, and the solution was pooled for instrumental counting. The volume of test solution remaining after counting was tested for silicone by extraction and infrared spectroscopy. This test was positive for the presence of silicone in each of the three solutions. The silicone-related counts obtained in this test are shown in Table 8.10. These data suggest that (1) counts obtained from this source are dependent on the extent of agitation the container undergoes; (2) higher (10 μm) counts may result with less agitation; and (3) different microdroplet size distributions may result depending on the mode of agitation prior to counting.

With regard to the counting of silicone microdroplets and silicone-related materials in SVI dosage forms, it must be kept in mind that both liquid and powder materials may evidence artifacts from this source. Silicone oil itself may be removed from the inner surfaces of closures by a liquid (in some cases) more readily than by a powder, so that liquid drugs show higher levels of the material than do powders. Silicone emulsions applied as stopper components may contain reactive impurities that can combine with the drug to form amorphous materials that are also problematical.

Intermittent Instrument Problems

Count errors contributing to data variability may arise sporadically due to external influences, such as vibration (as from the impeller in a laminar flow hood), sounds, radio frequency energy or electrical fields from nearby equipment, instrumental electronic problems, or temporary sensor blockage by large particles. The result in each case are "nonsense" counts that will be detected by a trained analyst. The erroneous data sets are usually noted due to a close similarity of counts (approximating exact agreement) or an unusual relationship of counts between channels (Table 8.11).

In terms of troubleshooting, sporadically occurring counts of this type are often due to electronic problems and may be triggered by the starting pulse of the syringe or syringe motor that delivers the sample. If the problem is caused by vibration, (gentle) tapping of the sample beaker or the bench top near the instrument may cause reoccurrence. Sensor blockage by solution residue typically causes a nonintermittent problem.

Table 8.10. Generation of Silicone Microdroplets by Mechanical Agitation (Light Obscuration Counts/mL)*

Product Code	Minutes Agitation	≥ 2 μm	≥ 5 μm	≥ 10 μm	≥ 20 μm	≥ 25 μm
	1	10,161	4,372	183	2	0
3	5	6,715	2,703	2	0	0
	10	6,944	1,540	3	0	0
	1	2,960	2,743	1,809	17	1
5	5	3,025	3,352	673	6	2
	10	5,674	2,137	2	0	0
	1	2,250	2,150	1,846	163	39
8	5	2,222	2,120	1,777	93	16
	10	4,961	2,723	6	0	0

*Data are abstracted from the *Proceedings of the 1990 PDA/IES Conference on Particle Detection, Metrology, and Control.*

Table 8.11. Count Artifacts Due to Electronic Problem ("cross-talk" between instrument comparators) (Counts/mL)*

≥ 2 μm	≥ 5 μm	≥ 10 μm	≥ 20 μm	≥ 25 μm
8,362	8,360	8,333	8,341	8,317

*Data are abstracted from the *Proceedings of the 1990 PDA/IES Conference on Particle Detection, Metrology, and Control.*

LIGHT OBSCURATION COUNT DATA

Dry Powder Dosage Forms

Reconstituted dry powder dosage forms produced by either lyophilization or crystallization have been known for many years to have a higher particle burden than injectable solutions that are terminally filtered. The relatively high particle burden of dry powder SVI antibiotics has been commented on by a number of investigators (Signoretti et al. 1987; Jamet 1988; Parkins and Taylor 1987; Montanari et al. 1986; Masuda and Backerman 1973; Backhouse et al. 1987; Pavanetto et al. 1989). Over the past 10 years, the author's lab has been involved in the use of LO methodology to perform assays on bulk dry powder antibiotics and powder vials from domestic and foreign vendors (see Barber and Williams 1990). Evaluations have indicated both raw material powders and some SVI finished product containing significant numbers of particles not detected by LO counting. In the case of a number of products and raw material lots evaluated, serious discrepancies were noted between results obtained using HIAC counters and a microscopic test. On the basis of previous experience with the test, one reason for the difference in counts was believed to be the variable detection efficiency of the LO counter for different types of particles. Another mechanism responsible for different counts obtained with LO and microscopic methods appeared to be the presence of immiscible microdroplets and air in some of the solutions tested.

The materials tested represented products manufactured in Europe, Japan, and the United States. The test articles consisted of both lyophilized and crystallized drug (penicillins, cephalosporins, and proprietary antibiotics). Samples of bulk antibiotic raw material powders produced in the United States, Japan, the United Kingdom, Italy, Austria, and Holland were also tested.

Prior to testing, SVI product units were reconstituted with specified diluents per the <788> test method, and 5- or 10-unit pools were prepared in blanked beakers of appropriate size. Powders were visually inspected for completeness of dissolution prior to testing. All diluents used for reconstitution were filtered through 0.22 μm filters to preclude the possibility of contamination from this source. Pooled samples were degassed by ultrasound prior to counting. Sampling for LO counting was conducted by taking 4 consecutive 5 mL aliquots and averaging the counts for aliquots 2, 3, and 4.

Twenty-five mL aliquots for microscopic testing were withdrawn from 10-unit sample pools before and after instrumental testing, and the two aliquots were filtered through a single 0.8 μm porosity-gridded membrane filter to collect particulate matter for counting. After screening to verify an even distribution of particulate matter on the analysis membranes, particles were analyzed by counting 10 fields at 100× and extrapolating this particle count to the total area of the membrane. Particles were sized microscopically using the USP 23 method, with a combination of lateral low-angle reflected illumination and episcopic bright field illumination. This type of illumination has the advantage of revealing some particle types visible (such as thin, transparent, or amorphous particles) that are not readily visible using reflected illumination. Based on a comparison of microscopic and LO count data obtained in this testing, additional experiments were also performed to determine the effects of immiscibles and air bubbles on count data.

Inspection of data obtained from dry powder vials allowed particle counts to be separated into one of three categories: (1) cases of good agreement between LO and microscopic counts; (2) instances in which microscopic counts were significantly higher than LO results; and (3) the occurrence of LO counts that were high in relation to microscopic results. The first case is illustrated by the data in Table 8.12. All 5 codes for which data are presented in this table were domestically produced antibiotic powders. The close count similarity at the 10 μm size is representative of the best agreement noted between the 2 methods. For these 5 product codes, the 10 μm and 25 μm counts were extremely low in relation to the USP 23 limits for SVI of not more than 6,000 particles/container 10 μm in size, and not more than 600 particles/container 25 μm in size.

It was determined that microscopic and instrumental counts disagreed if the instrumental counts at 10 μm were more than an order of magnitude higher than microscopic counts, as shown in Table 8.13. In the case of some sample groups, the per container instrumental count at the 10 μm size indicated tens of thousands of particles per container present, while microscopic counts generally showed less than 1,000 particles per container at this size. The counts in the first 3 channels of the LO instrument were often similar for these products and did not show the expected decrease between size ranges. As discussed earlier, this result occurs most frequently with SVI dosage forms when immiscibles, such as silicone stopper lubricant, are present as suspended microdroplets in the tested solutions; an effect resembling coincidence counting occurs. One report suggests that as high as 50 percent of the 10 μm counts obtained from dry powder vials of antibiotic may be due to silicone (Signoretti et al. 1987)

Amorphous Material

Complex count data can also result over time as a result of the formation of amorphous, semiliquid, highly hydrated material in protein- or amino acid–containing solutions such as monoclonal antibodies or amino acid mixtures for infusion. By the USP definition, fragments of amorphous material do not constitute particles in the conventional sense. This type of material is typically detected only by microscopy, and specific dispensations are provided for dealing with its occurrence using the LO test method. In

Table 8.12. Comparative LO and Microscopic Counts (Particles/Container)*

		Light Obscuration					Microscopic	
Code	Vol./mL	≥ 2 μm	≥ 5 μm	≥ 10 μm	≥ 20 μm	≥ 25 μm	≥ 10 μm	≥ 25 μm
7	10	3,010	326	86	36	10	95	7
9	10	4,322	460	114	20	10	106	19
11	10	1,140	262	82	10	10	89	4
12	30	4,098	282	66	42	6	105	13
16	50	2,710	210	110	50	0	95	3

*Data are abstracted from the *Proceedings of the 1990 PDA/IES Conference on Particle Detection, Metrology, and Control.*

Table 8.13. Comparative LO and Microscopic Counts (Particles/Container)

		Light Obscuration					Microscopic	
Code	Vol./mL	≥ 2 μm	≥ 5 μm	≥ 10 μm	≥ 20 μm	≥ 25 μm	≥ 10 μm	≥ 25 μm
3	10	60,642	56,180	18,880	540	60	1,078	160
5	10	56,980	54,240	22,094	266	40	479	32
8	15	68,914	54,186	26,790	105	30	720	56

fact, most biopharmaceuticals are in reality micromolecule suspensions rather than solutions. In the example shown in Table 8.14, solutions were noted to be colored at 3 months, but it was not until 9 months later that LO counting detected the amorphous material, which was by that time detectable by visual inspection. By this time, the solutions were deeply colored. Color sometimes indicates intermediates in the degradation process that forms an amorphous material and may be present before the material forms.

This type of count disagreement also occurred in studies of drug powders (see Table 8.15). In this case, microscopic results were orders of magnitude higher than the instrumental counts. The materials responsible were typically transparent, flakelike particles or the amorphous (globular) material of high transparency.

Sample Pooling

High microscopic counts from pooled samples raises another interesting issue regarding the performance of the present USP <788> method. The particle burden of individual dry powder vials may show an order of magnitude variation between units. This is illustrated by the data in Table 8.16 excerpted from the publication of Pavanetto et al. (1989). The pooling method of the <788> requirement effectively suppresses any measure of between-unit variability and theoretically could allow units with higher-than-limit counts to be released. Thus, although the limits are stated on a per-container basis, the particle burden of individual containers is not determined.

Data Variability Versus Sample Volume

Yet another important difficulty with the USP <788> LO test is in regard to the variable statistical strength of the test result obtained from a pooled sample. If a random distribution of particulate matter in the pool is assumed, some variability of counts will result based simply on the proportional relationship of sample volume to pool volume. Thus, the lower the sample volume–to–pool volume ratio, the less well the sample represents the population, and the

Table 8.14. Formation of Color and Amorphous Material in Protein Solution

	(20 mL units; Particle Counts/mL)						
Interval	≥ 2 μm	≥ 5 μm	≥ 10 μm	≥ 20 μm	≥ 25 μm	Color	Visual
1 month	286	37	6	1	0	—	—
3 months	361	40	10	2	1	+	—
6 months	580	78	26	3	1	++	—
9 months	1,021	285	41	8	3	++	—
1 year	4,081	865	297	84	21	++	+
USP Limits			300		30		

Table 8.15. Comparative LO and Microscopic Counts (Particles/Container)*

		Light Obscuration					Microscopic	
Code	Vol./mL	≥ 2 µm	≥ 5 µm	≥ 10 µm	≥ 20 µm	≥ 25 µm	≥ 10 µm	≥ 25 µm
18	30	8,886	4,230	480	120	24	TNTC*	TNTC
19	35	1,519	630	35	0	0	1,809	271
20	50	2,050	750	250	50	0	TNTC	406
21	50	15,820	7,350	310	150	0	1,056	784
22	50	27,250	2,270	250	100	0	1,289	472

*Too numerous to count

Table 8.16. Cumulative Particle Number Per Container at Different Size Levels

#	Product	Level	≥ 2 µm	≥ 3.5 µm	≥ 5 µm	≥ 10 µm	≥ 20 µm	≥ 25 µm
6	Dextrose 20%[a]	Min	40	30	20	0	0	0
		Max	3,260	2,040	1,340	370	40	30
		M[c]	728	409	260	73	12	5
2P	Ceftazidinium[b]	Min	6,820	1,000	540	60	0	0
		Max	23,400	4,080	3,420	2,850	2,820	2,800
		M[c]	12,952	2,611	1,625	997	852	834
3P	Ethacrynate Sodium[b]	Min	13,220	9,800	1,400	460	20	20
		Max	82,060	37,140	20,680	4,640	700	286
		M[c]	35,301	16,023	8,226	1,705	209	71

[a]Liquid products

[b]Powder products

[c]Mean contamination value of the batch

higher the statistical variability of the test result. In Figure 8.13, this effect is graphically approximated based on the assumption of a binomial, log normal, or Poisson distribution of count events and random particle distribution in a pool, using the proportion of the pool sampled as the independent variable. The dependent variable is the coefficient of variation (CV) where N is the total number of particles at a given size in the solution pool volume. The CV of the count data obtained varies with the number of counts available. If 100 particles > 25 µm are available in a 100 mL solution pool and a 10 mL sample is drawn, a 30 percent CV results based on sampling statistics alone. This calculation assumes a random distribution of particles in the pool, which often does not occur due to stratification effects.

User Interpretation of Complex Light Obscuration Count Data

An important point that must be mentioned again is that addressing particulate matter issues and interpreting LO count data may require a detailed knowledge of the production process and the container-closure system as well as the drug product involved. Some drug formulations may be particularly sensitive to the presence of divalent cations. As an example in this regard, calcium, which may be included in the drug formulation and/or leached from the glass of a container, may cause the drug material to precipitate or may result in the formation of calcium phosphates or sulfates of low solubility. Such problems may be exceedingly complex. A number of companies have brought in a consultant to solve

Figure 8.13. CV of particle count per unit volume given the sample volume relative to container volume.

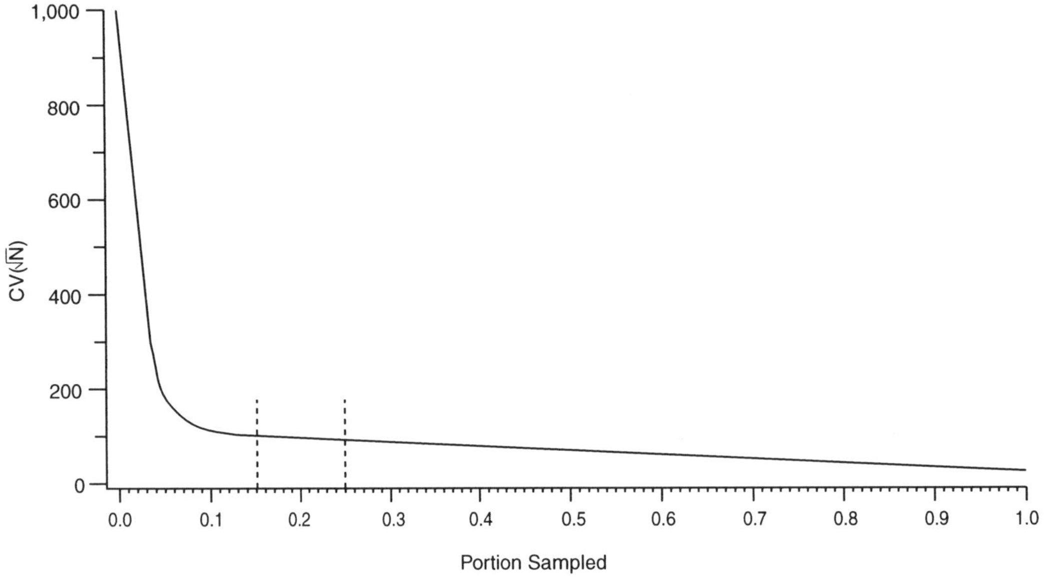

N = Total Number of Particles of a Certain Size

a difficult particulate matter problem, only to have the consultant ask seemingly endless questions about the process itself. When the consultant offers a solution to the problem based on knowledge of the process rather than the nature of the particles themselves, those who hired the consultant may feel a little sheepish. Most experienced consultants adept at sourcing or defining the origin of particle problems will use such a "holistic" approach.

Immiscibles and surfactant type materials will probably continue to plague the user of LO counters well into the next century. Often, the count artifact is highly variable between batches and individual units; it is also time dependent. In stability studies performed on drug products, the manufacturer is concerned with counts at the various test intervals of a study over time (e.g., initial, 30 days, 90 days, 6 months, 1 year, etc.). Simple degradation of a drug most often results in an upward trend in counts obtained with time. If surfactant materials, silicone from stoppers, or complex reaction products are involved, the count trends observed with time may not be simple or straightforward in interpretation. As a hypothetical example, suppose the counts in Table 8.17 are obtained from a liquid drug product in 50 mL glass vials.

One of the mechanisms that would obviously be suspect in the case illustrated is the presence of silicone oil (perhaps even surfactant) on the stoppers of the units tested. The surfactant may be a material added to the silicone oil to disperse it (i.e., to form an emulsion) or to be extracted from the stopper (e.g., a stearate). At the zero time initial test interval, the 10 μm counts are low, since the contaminating material has not been released. With release over time at 30 days, the increasing numbers of silicone droplets result in counts at 10 μm. At 90 days, the release of small (2 μm, 5 μm) droplets has proceeded to such an extent that coincidence counting (indicated by similar counts in the ≥ 2 μm, ≥ 5 μm, and ≥ 10 μm channels) is occurring. At 6 months, the ≥ 10 μm counts have greatly decreased due to fusion of the silicone droplets in the 2-phase system or their adsorption onto the glass container surface. The fusion of the silicone droplets results in the release of free surfactant and, as a consequence, the formation of micelles and high ≥ 5 μm counts again at the 1-year interval. Sounds complex and imaginative doesn't it? This scenario actually happened to a manufacturer and required both serious analytical work and an expert consultant to figure it out.

Table 8.17. Variability of LO Counts in SVI

Time	Counts/Container at Various Sizes				
	> 2 μm	> 5 μm	> 10 μm	> 20 μm	25 μm
0	7,257	86	19	3	0
30 days	8,600	2,320	288	18	5
90 days	8,990	7,820	2,480	206	35
6 months	6,620	8,65	78	14	6
1 year	7,840	3,064	900	28	16

Imagine the difficulty of trying to interpret the trend of ≥ 10 μm counts if the ≥ 2 μm and ≥ 5 μm counts weren't available (Figure 8.14). Problems of this type may often be elucidated by testing individual units and counting at smaller sizes (i.e., ≥ 2 μm, ≥ 5 μm) as well as the compendial sizes.

The following questions may be useful in the investigation of such unusual counts:

- Do replicate runs of the same sample stay closely similar (< 10 percent variation) or "jump around"?
- Does an "overconcentration" or "baseline fail" indicator come up with high counts?
- Was the previously run solution thoroughly rinsed out so that mixing (schlieren) effects are not an issue?
- Is coincidence counting suspect? This may be indicated by "flat" or similar counts in low channels or by a single channel count significantly above the sensor concentration limit. (Try serial dilution—if coincidence is occurring, count decreases with serial dilution will not occur in the expected incremental fashion.)

Figure 8.14. A complex count trend.

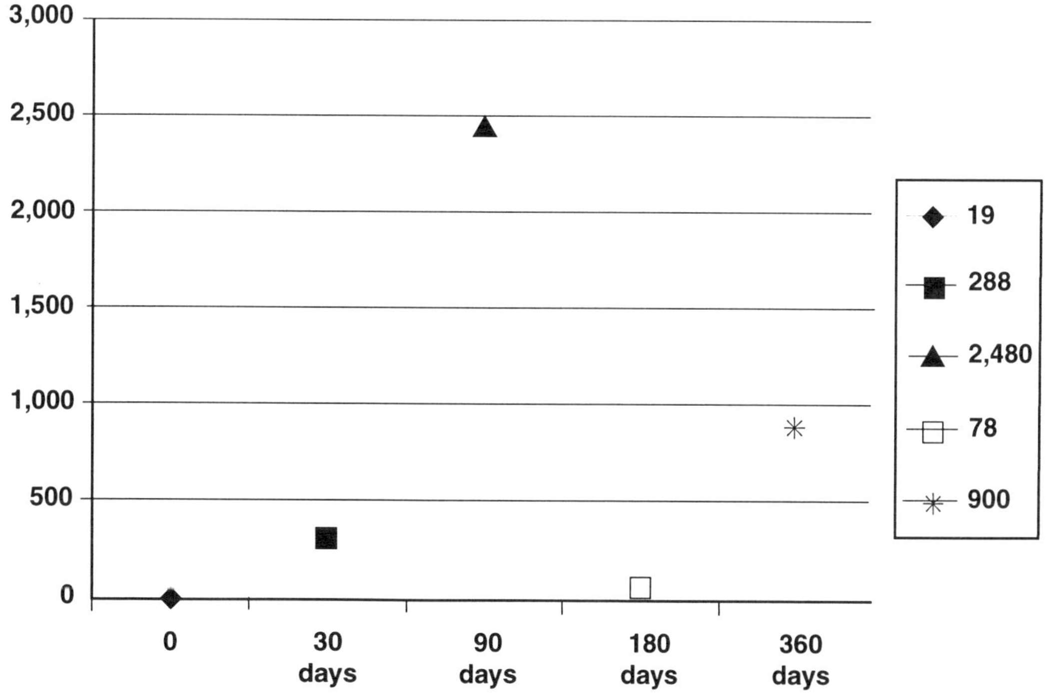

- Is the solution itself unusually colored, a cosolvent system, or a surfactant?
- Was the diluent clean and compatible with the solution (count the diluent alone)?
- Were micro air bubbles responsible? (Let the solution sit undisturbed for 1 h and sonicate for 5 min to check.)
- Does the solution contain a bicarbonate buffer system? (If so, try letting it or use a pressure sampler.)
- Does the solution contain immiscibles (e.g., silicone)?

Some dosage forms may consist of an apolar drug dispersed (i.e., dissolved) in a cosolvent system of miscible polar and apolar liquids (e.g., an alcohol and water). The dilution of such drugs with water or aqueous systems can destabilize them to form drug micelles, so that higher dilutions give higher counts and dilution with organic solutions such as dextrose give lower counts than dilution with saline (Table 8.18).

Changes in Calibration Points—Threshold Shift

Many companies calibrate their LO counters on a 6-month schedule. What if the new calibration shows a significant threshold shift? As an example, Table 8.19 shows a review of the data regarding hypothetical 6/17/97 and 12/16/96 calibrations of a HIAC instrument and associated component. It shows the differences in the calibration (threshold) values between calibrations.

Based on these data, the higher threshold settings for the 6/16/97 calibration lead to two conclusions: (1) A small amount of instrumental drift occurred in the interval between calibrations, and (2) the drift was toward higher threshold settings. This means that counts collected during the 6-month interval between calibrations subsequent to the shift have incorporated artificially elevated data at the ≥ 10 μm and ≥ 25 μm particle sizes. Stated another way, one can say that as a result of the shift, counts were higher than those that would have resulted if the settings obtained in the 12/17/96 calibration had

Table 8.18. Example of Destabilization of Drug in Cosolvent System

		Counts/mL				
High Drug Concentration (1:5)		**> 2 μm**	**> 5 μm**	**> 10 μm**	**> 20 μm**	**25 μm**
	Saline	2,602	862	159	26	8
	Dextrose	538	140	25	6	1
Low Drug Concentration (1:20)	*Saline	18,402	17,821	2,602	580	74
	Dextrose	6,840	1,121	203	34	7

*Flat counts in first two channels (≥ 2 μm and ≥ 5 μm) show overconcentration and coincidence.

Table 8.19. Threshold Settings for 6/16/97 and 12/17/96 Calibrations

6/16/97 Calibration		12/17/96 Calibration		
Standard (mm)	**Threshold**	**Standard (mm)**	**Threshold**	**Approximate % Difference**
2.013	150B	2.013	134B	11%
3.004	201B	3.004	183B	9.8%
4.991	294B	5.010	278B	N/A
9.975	496B	9.685	443B	N/A
15.03	744B	15.03	682B	9.09%
25.09	1450	25.09	131C	10.6%

Note: The letters B and C refer to the mV level.

remained unchanged. As an example, after the change, an 8 μm particle might have given a pulse equal in size to that of a 10 μm particle based on the old calibration curve. This represents an unfavorable bias (i.e., against the tested product); larger numbers of particles than the true count would have been recorded due to the increased numbers of particles at smaller sizes in the particle size distribution on which the USP limits are based and that actually occurs in product.

To acquire additional supporting information, a statistical analysis of the historical data to determine the mean, SD, and count distribution may be performed. This will allow one to determine the effect on average count data that the drift would have caused. While the change in calibration thresholds was greater than ±5 percent, the actual change in counts at the ≥ 10 μm and ≥ 25 μm would be expected to be inconsequential with regard to the data collected from product tested if historical counts are far below the USP limits. This consideration is based both on evaluation of the worst-case bias resulting from the shift and the extremely low level of historical count data gathered between the calibrations in comparison to the LVI limits. With this type of problem, it should be kept in mind that allowable drift in threshold settings does not translate directly into shifts of particle sizes counted at specific thresholds.

ERROR SOURCES CHECKLIST

In the event that unexpected counts are collected, the possibility that a product tested is in some way different from others of the same code historically counted must, of course, be thoroughly investigated. In terms of both Good Laboratory Practice (GLP) compliance and the Barr Laboratories court decision, it would seem a good idea to record all unusual count data that elicit investigation. The tested solution or other product may be vindicated in some cases only after a comprehensive investigation of instrument function and possible methodologic sources of artifacts. The checklists below can be used as an initial basis for such an investigation; they are separated into instrument-related and sample-related categories.

Instrument-related errors

- Sensor opening partially blocked
- Leaks (sensor or sampler feed lines and entrained air)
- Dirty sample tube (dried residue?)
- Electronic noise
- Mechanical shock or vibration
- Sensor dry out (may cause a dirty sensor)
- Flow rate changes due to sampler problems (compressed air line pressure variation, syringe wear)
- Dirty sensor cell in view volume area
- Mixing of different refractive index solutions in sensor (residue, "dead legs")
- O-ring, tubing, or seal degradation
- Line current fluctuations
- Electronic instability in the instrument
- Cavitation (below vapor pressure of liquid)
- Radio frequency (RF) energy interference

Sample-related error sources

- Nonuniform mixing of dry powder vial product
- Undissolved drug from powder vials (what does package insert say about time required for dissolution?)
- Bubbles (or dissolved gases) in lab water supply
- Erratic stirring (constant at ≅ 120 rpm?)
- Shedding of glass flakes in beaker
- Failure to blank pooling vessels
- Different shape/size of pooling vessels
- Residue carryover in beaker
- A serial oxidation of product
- Manipulation (gloves?)
- Closure removal
- Paper labels?
- Nonuniform placement of probe tube (validate uniform position)
- Different size stirring bars

- Particle stratification in beaker or solution unit
- Sampler flow rate changes due to differing sample viscosity

Laboratory Technique

Sources of variability may also be introduced by the environment and with the techniques employed. The outcome of any particulate matter assay for injectables is critically dependent on the cleanliness of any materials that contact the sample and of the general environment. On this basis, standards and technical publications that deal with the particulate matter testing of solutions or fluids specify background particle control measures (Pavanetto et al. 1989; DeLuca et al. 1980; National Fluid Powder Association 1972; Millipore Corporation 1991). The 1980 paper by DeLuca and associates is an extremely valuable reference in this regard. These procedures specify allowable background particulate matter levels and the means of denying particles access to the test materials. When determining counts for individual samples, the test should be conducted in a general laboratory environment equipped with high efficiency particulate air (HEPA) filters and a filtered laminar flow hood. The hood is not necessary for performing the test, but it provides a clean enclosure for the cleaning and storage of test equipment and samples. Low-particle filtered water should be used as a reference to verify the cleanliness of the operating environment. It may also be run between samples and should be available in the HEPA filtered hood.

Cleaning of Glassware

In the <788> test, the most critical factor with regard to background particulate matter sources is the vessels (glass or plastic) that will be used for sample pooling. General use laboratory glassware should never be used in the particle counting assay. Surface scratches contribute particles, and dedicated glassware that is stored clean under HEPA protection is far easier to blank. The cleaning of glassware and other equipment used in the test may be conducted as described by DeLuca et al. (1980), whereby cleaning is carried out by first placing the glassware in an ultrasonic bath containing warm detergent solution for 1 min. Both the outside and inside of the glassware are then rinsed with hot water and then with distilled water and transferred to the hood. In the hood, the glassware undergoes a final rinse with ultraclean water. The cleaned glassware is placed in the hood upstream of all other operations and allowed to dry.

The general principle involved in cleaning is simply the exhaustive removal of adherent particles from any glassware, vessels, or apparatus to be used in the test. In lieu of the ultrasonic method, cleaning may be accomplished by soaking in a detergent solution, rinsing in distilled water, and a final rinse with a jet of filtered distilled water (0.45 μm or lower retention-rated filter, see Figure 8.15). The final rinse should result in the water "sheeting" the entire inner surface of the object being washed. A pressure vessel with a filter on the dispenser nozzle (FilterJet Solvent Dispenser, Millipore Corp.) may also be used as a means of final rinsing.

Figure 8.15. Glassware final rinse.

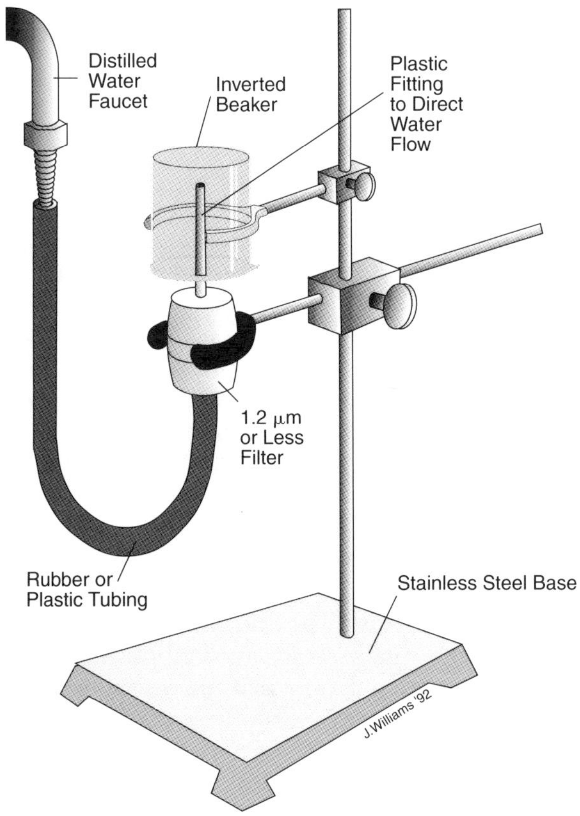

In this case, after cleaning in warm detergent solution and then rinsing in flowing warm tap water, the final rinse is performed with filtered water from a pressure vessel in a HEPA–filtered hood. Use a back-and-forth motion to direct the spray while working from top to bottom. If an item of glassware is needed in a short period of time, a rinse may be performed with filtered isopropyl alcohol (IPA), which will dry quickly. The rinsed articles are allowed to dry in the hood upstream of other operations.

In using a cartridge or other type of ≥ 0.45 μm or ≥ 0.22 μm filter to "clean up" distilled water for use in the <788> test, the first thing the technician should keep in mind is that the first volumes of water flowing through a new filter cannot be expected to be free of particles larger than the filter retention rating. Some bits of the filter media will be shed at first, and particles will also be released from the filter housing and from fittings and tubing downstream of the filter. The latter are best kept at a minimum. Importantly, whenever a filter is changed or before the first use of filtered water for the day, it is advisable to clean out the filter and lines by passing several liters of water through them to drain. Most stopcock-type valves are capable of releasing significant numbers of particles.

Another piece of equipment required for glassware sample preparation is a laminar airflow hood. Such laminar flow enclosures are available from a number of vendors of cleanroom supplies. The minimum useful size is a width of 48 in. and a depth of 24 in. The laminar airflow of 90 ± 10 ft/min will provide adequate velocity to prevent particles from the general environment from entering. A shield of clear acrylic plastic sheeting that may be lowered over the opening when the hood is not in use is helpful in keeping the work area clean. The hood should contain and provide ready access to pressure vessels, filtered water, glassware, and other items to be used in the test.

Ideally, the analyst will wear gloves while performing any procedure in which there is exposure of the solution being tested (such as sample pooling). Touching the mouth of sample vials or of a cleaned vessel with an ungloved finger can deposit thousands of skin cells that will provide sporadically high levels of background count artifact. The best gloves for this purpose are nonpowdered, vinyl gloves that are essentially particle free as purchased. Any type of glove used should be rinsed under filtered distilled water after donning to remove the surface adherent particles present due to the manufacture of the glove or the cleaning procedure. Dependent on the method of sampling used, HEPA protection and the wearing of a head cover and polyester smock may be desirable. The wearing of gloves, a hair cover, and a smock constitutes best practice. The smocks should have snaps at both the cuffs and the (high) collar.

The burden is also placed on those performing the test to develop a rational and technically valid approach that results in a blank as good or cleaner than that specified in the <788> test. For SVI, the blank will generally entail use of a cleaned vessel that will be used to collect the 10-unit solution pool with a 50 mL volume of solution equal to what the vessel will hold in the test (e.g., from ten 10 mL vials, a 50 mL volume of filtered distilled water used to rinse the glassware before the test is collected) (see Chapter 4). Then, a test volume for the blank should be drawn that is equal to the sample volume. If two 5 mL aliquots will be counted for record, then two 5 mL aliquots should be used for the blank. If the glassware has been cleaned properly and the water is of appropriate cleanliness, then the combined blank volume should have no particles 10 μm in size.

If the blank criteria are met, the test system is satisfactory for proceeding with sample determinations. If the criteria not met, either the environment is not suitable or the filtered water and/or glassware have not been properly prepared. Repeat the preparatory steps until the criteria are met, before proceeding with sample analysis. Many labs consider the USP allowable blank to be too high and substitute a standard of no particles 10 μm in size. They believe that if any 10 μm particles are allowed, an unpredictable blank contribution to counts for the record will be present.

Another significant contribution to blank counts can result from the manipulations associated with opening the container. This is unfortunate but unavoidable, since particles introduced from this source are inevitable. This contribution can be assessed by filling blank vials, ampoules, and the like with filtered water and then counting the number of particles added by the opening procedure. Ideally, the blank vials should be filled under production conditions, and the solution should be removed from containers tested in a fashion that imitates the use of the product. As an example, dry powder vials would be reconstituted and emptied using a cleaned and blanked syringe. The caution here is that each component used in the test must be blanked for background particle burden.

Sampling of Injectable Products

Although less sensitive to contamination artifacts than microscopic testing, LO testing of injections per <788> must be carefully conducted to prevent extraneous particles from inflicting a bias on the count data. In addition to the cleaning of glassware, another critical step involves the point at which containers are entered and the contents are pooled in beakers for testing. If the product to be sampled is a small volume, dry powder vial, the removal of the stopper for reconstitution or fluid removal will result in some particles being shed from the stopper. If ampoules are not opened carefully, large numbers of glass particles can contribute to the analytical results. Some laboratories have developed a technique for entering stoppered containers by using pencil point needles that generate extremely low levels of particles.

Considerations to be observed when testing different types of injectable product containers are discussed below by product type. Before sampling any liquid product for LO testing, it is standard practice in many labs to examine each unit visually against white and black backgrounds for the presence of particulate matter. This simple inspection is not intended to detect particles of visible size but rather to ensure that the solution appears normal with regard to clarity and color. This is a sound practice.

Ampoules

The outer surface of each ampoule should be cleaned with warm detergent solution and rinsed before transferring to the hood. After scoring the glass, chips are rinsed from the outer surface of the ampoule, and the solution then withdrawn with a plastic syringe and needle that have been thoroughly cleaned and rinsed with ultraclean water. If a syringe is used, a suitable blank must be incorporated. A better procedure involves the use of an oxy-propane pencil point torch to flame a hole in the ampoule from which the solution is withdrawn.

Solutions Packaged in Vials and Bottles

The outer surface of each vial is cleaned with a warm detergent solution and rinsed before transferring to the hood. The tear-off seal and inner seal from the test vial are then removed, and the outer surface is cleaned with a jet of IPA. The rubber closure and the glass-closure junction are rinsed with distilled water from a FilterJet apparatus and allowed to dry. The test solution is then inverted 20× to resuspend the particles, then the closure is removed with the aid of a forceps. The counter probe may then be inserted in the bottle, or the contents poured into a beaker (pooled).

Dry Powders Packaged in Vials

The outer surface is cleaned with detergent as directed above. The tear-off seal and the inner seal are removed from the test vial, the outer surface of each vial is cleaned with a lint-free cloth moistened with IPA, and then the vial is transferred to the hood and rinsed with ultraclean water. The closure and the glass-closure junction are rinsed with ultraclean water and allowed to dry. The closure is then removed with a forceps and placed upside down within the hood. The powder is dissolved with ultraclean water or other appropriate solvents (e.g., particle-free 5 percent dextrose or 0.9 percent sodium chloride solutions) at the appropriate concentration recommended by the manufacturer. The closure is replaced, and the vial is inverted 20× or until all the powder dissolves. The closure is then removed, and the solution is poured into the collection vessel.

Powder in Vials with Diluent in a Separate Ampoule

Both containers are first cleaned with detergent solution and rinsed with water. The tear-off seal and inner seal are removed from the test vial, and the outer surface of the vial and ampoule of diluent are wiped with a lint-free cloth moistened with IPA. The containers are then transferred to the hood and rinsed with ultraclean water. The ampoule is scored lightly with a file or diamond scribe at the breaking point, then rinsed and dried. After breaking the ampoule, the diluent is withdrawn with a blanked plastic syringe and needle. A blanked needle filter is then put in place on the syringe. After opening the vial (as described earlier), the diluent is injected into the vial. The closure is then replaced, and the vial agitated until all of the powder dissolves. The samples are then pooled as required for testing.

Prefilled Syringes

The containers are cleaned with detergent as for ampoules and vials. The outer surface of the syringe and syringe parts are then wiped with a lint-free cloth moistened with IPA and transferred into the hood. The outer surface of the test articles are then rinsed with ultraclean water and allowed to dry. The syringe parts are then assembled according to the manufacturer's instructions. The syringe is inverted three times to resuspend the particles. The needle sheath is then removed, and the solution is aspirated into the collection vessel.

USING PARTICLE SIZE STANDARDS

Particle size standards made of polymeric materials play an important role in the testing of parenteral products. Particle standards suitable for calibration purposes in pharmaceutical particle analysis are available from several vendors, including Duke Scientific (Palo Alto, Calif.), Coulter Electronics (Hialeah, Fla.), and Seradyn Inc. (Indianapolis, Ind.). A discussion of the use of standard reference materials for counter calibration procedures is provided by Barber et al. (1990). A great deal of information regarding the manufacture, handling, and use of polymer particle-size standards is available from vendors (see also Duke 1988; Duke and Layendecker 1989; Duke et al. 1989; Bangs 1988).

Standard polymer particles are typically furnished by vendors as aqueous suspensions, stabilized by surface-active agents. Room temperature storage is recommended. Refrigerated storage is also acceptable if one guards against freezing. The bottles should be tightly capped to prevent water loss and a consequent change in the percent of solids or the drying of particles. Dried particles will be irreversibly agglomerated. Dilutions of particles should be made with deionized water; electrolytes can also cause the agglomeration of particles (Wilkison et al. 1987).

Standard latex particles larger than approximately ≥ 1.0 μm will settle after standing. Agitation using a table shaker, wrist-action shaker or rollers, or manual agitation will effect resuspension. Depending on particle diameter and settling time, it may take several hours of agitation to redisperse settled, agglomerated particles fully. Mild ultrasonic agitation may also be used, but ultrasonic energy should be used carefully. If too intense, it can overheat and irreversibly flocculate the particles. The dispersion of particles can be checked by examining them under a light microscope.

For particle counters used to determine the levels of particulate matter in parenteral products according to the compendial requirements, service engineers or users generally adjust the responses of the instruments using certified, monodisperse PSL dispersions with diameters approximating 10 μm, 15 μm, and 25 μm (for USP counting) or 5 μm and 10 μm (for BP counting); additional sizes may be required by the vendor for specific counters. Newer instruments allow the user to perform the calibration procedures based on internal software. Unfortunately, some users are still reluctant to carry out the calibration procedures themselves, despite the relative simplicity of the procedure and the need to check any particle counting instrument at frequent intervals.

The use of polymeric microspheres for the calibration of automatic particle counters will always be questionable, because these materials represent ideal particles unrelated in shape and refractive index to contaminant particles that will be counted or recorded. These standard particles, nonetheless, represent the only uniform approach that is widely available to users, and they have distinct advantages over other reference materials, such as the AC fine test dust. The user constantly must be aware of the aspects of using polymeric microspheres that may result in erroneous calibration data. The two most serious issues are the tendency of the spheres to form agglomerates (doublets, triplets, or larger clumps) under some conditions and to adhere tenaciously to parts of the test system, including the sensor probe, windows, and glasses.

In 1991, the USP began marketing a reference standard (PCRS) intended to allow a count accuracy check for LO particle counters without resorting to the difficult comparison of the microscopic test. The results of testing this standard are specified in the May–June 1990 issue of *Pharmacopeial Forum*. The experience with this standard to date includes some difficulties due to the leakage of diluent from sample bottles, but user experience in general indicates the standard to be useful and practical.

The PCRS is composed of nominal 15 μm particles. This size is sufficiently above the ≥ 10 μm threshold so that, in theory, all of the particles will be counted at the 10 μm threshold, and half of the particles will be counted at the ≥ 15 μm threshold. The current PCRS consists of a vial that contains 25 mL of a dilute suspension of monodispersed PSL spheres with a mean diameter of 15 ± 0.18 μm. The diluent is particle-free water with added dispersants and preservatives. A sample blank is provided in an identical container and consists of the diluent without particles.

One inherent difficulty with the PCRS would seem to be that monosize standard spheres with a narrow distribution make poor count standards. The threshold setting for a specific particle size that is determined from a calibration "curve" may differ from the threshold that will split the pulse distribution of a monosize standard of the same size. Thus, with the PCRS, a fairly slight deviation in the slope of an instrument calibration curve with regard to the actual thresholds may result in a failure to meet the count ratio criteria. Further, minor instrumental "drift" that has no practical effect regarding the outcome of a compendial test may also result in a failure to split a monosize standard pulse distribution at a specific size.

A more appropriate control for particle sizing and counting for LO counters would seem to be a polydisperse material with a distribution of sizes over the size range of interest (e.g., ≥ 2 μm to ≥ 50 μm). Such a material with a distribution approximating log normality would be insensitive to minor drift during day-to-day operations, but it would indicate significant changes in the particle size response curve. Such a product would meet the need for a sample to check the calibration and precision of parenteral product particle counters on a daily or weekly basis. Duke Scientific is presently marketing a standard of this type.

Another alternative to use of the PCRS is to prepare standards in one's own laboratory. This operation has historically been avoided by counter users, but can be used to obtain a standard of good precision and stability. One vendor of standard spheres (Duke Scientific) provides certified specifications for the number of spheres in each bottle of standard with a ±5 percent SD. Given that the number of spheres in a bottle of standard is known, and the material evenly suspended, careful dilution procedures will allow a standard with a precision equal to or exceeding that of the USP standard to be prepared in the laboratory for in-use validation and calibration of LO counters.

The materials needed are as follows:

- Ultraclean sample cuvettes (e.g., sample cuvettes from Coulter Electronics)
- 0.1 mm filtered water from a cartridge filter or pressure vessel
- 0.1 mm filtered alcohol
- Analytical balance with 1 mg or better accuracy
- Low intensity sonic bath
- HEPA hood or bench
- Clean wipes
- Fixed volume micropipettes for 100 mL delivery
- Appropriate standard materials of known concentration

All containers used should be thoroughly cleaned with the filtered alcohol and water and allowed to dry. Sphere suspensions are mixed by a combination of gentle rolling on a flat surface, slow inversion 20–30 times, and 20–30 sec sonication at low energy (500 W). Once the standard is mixed, 100 mL of standard suspension is pipetted from the bottle (after the top is removed) and added to the tared sample container. Inversion of the stock suspension must be continued until the moment the standard is pipetted, since spheres of 5 mm size or greater will settle very rapidly. The sample tubes plus standard are weighed, and 25 mL of diluent is added, followed by a final weighing. After sonication, the number of particles in the diluted standard may then be determined based on the dilution factor applied to the original suspension. The precision of this method, in practice, is typically good, with multiple diluted samples from a single vial of standard giving better then ±5 percent reproducibility. Several vendors offer such standard solutions (Berdovich and Associates, Micro Measurement Laboratories, Wheeling Ill.)

CURRENT GENERATION LIGHT OBSCURATION COUNTERS

Over the past five years, technical advances in LO particle counting and an increased user knowledge base related to these instruments have made possible great strides in the application of this mode of particle counting in pharmaceutical labs. Improvements in instruments systems include the following:

- Compact, user-friendly system design
- Integrated sensor design (counter and sensor in one unit)
- Microcomputer data processing with wide range of capabilities
- Storage of calibration data for interim checks of instrument calibration
- Interactive software function (Windows)
- Real time multichannel data presentation
- Reporting and data presentation options

Benefits that accrue to the use of these instruments include not only simple, relatively trouble-free operation, but the ability to perform the USP <788> calibration and resolution determination in minutes instead of the hours originally required.

The user of these instruments for performance of the USP <788> test is occasionally faced with the choice of a new instrument for replacing existing obsolete systems or for acquiring additional capacity for analysis. The author cannot recommend one system over another due to the fact that circumstances of use and the area in the world in which the instrument will be used impact the choice heavily. However, some criteria on which the choice may be based can be provided. There are two key criteria: (1) the extent to which the instrument "fits" the mode of operation of the lab in which it will be used and (2) the availability of service support. With regard to the first criterion, many laboratories are called on to perform relatively few analyses on a daily or weekly basis. An example might be a small company with one or two novel products that are produced on a scale of one or two batches per week. With this sample load, the time savings produced by counters with automatic (computer-based) functions may not pay off, and validation (see Chapter 11) might not be realized. The company might opt for a simple instrument amenable to manual operation and data transcription. The automated functions of some current designs may, on the other hand, be extremely attractive to corporate level labs that perform hundreds of analyses per day on stability studies or for formulation or preformulation samples.

The second criterion, the availability of service, will be critical in any regimen of use. Service is needed on a routine basis, and some companies require the vendor to perform calibration and system suitability determinations per USP 23 <788>. There is also the possibility of instrument failure in service, which can generate an emergency with regard to batch release and getting product to the patient. For international facilities, the availability of local service is critical; competent local service is generally much preferred to sending an instrument back to the vendor at a site in another country thousands of miles remote from the plant. When no local service is available for any of the instruments available, some manufacturers have opted to purchase two instruments so that a "spare" will be available for backup while one instrument is being repaired.

The initial cost of an LO counting system is generally not, nor should it be, a significant factor in the decision to purchase. Reliability in service is extremely important. An important item in deliberations on an instrument for purchase is obtaining a customer list of pharmaceutical uses from the vendor. Other users of systems under consideration can provide valuable insight with regard to what may be expected. While this discussion is directed primarily at the selection of systems to be applied for performing the USP test, the user should keep in mind that this test is not the only one that may be applied for the purposes of "contamination control" in pharmaceutical solutions and medical devices. If other tests are to be performed, the necessary versatility must be within reach of the counter selected. In some cases, the presence of contaminant particles may be based on a population analysis in which counting and sizing of particles across a broad size range is necessary, and 16- or 32-size channels may be required. Such requirements may not be met by instruments that are routinely capable of the 2-channel test specified in <788>, and a more versatile instrument may have to be considered.

A number of vendors produce systems that may be suitable for application in a specific use situation. The most widely used systems in the United States are those of HIAC Royco and Particle Measuring Systems. HIAC Royco has a product line offering systems that may be used in stand-alone fashion with minimal microprocessor-assisted operation (8103™ System) and computer-based models (9703 System). The widely used 8103™ System is shown in Figure 8.16. This system has a good record of reliable service in the field and is simple in operation.

Particle Measuring Systems offers the Automatic Particle Sampling System (APSS 200™) with a laser diode sensor (Figure 8.17). The system has a syringe-operated solution sampler and a laser-diode sensor that contains all of the electronics. This system utilizes a solid state diode laser light source.

Figure 8.16. HIAC/Royco's scientific model 8103™ LO counting system.

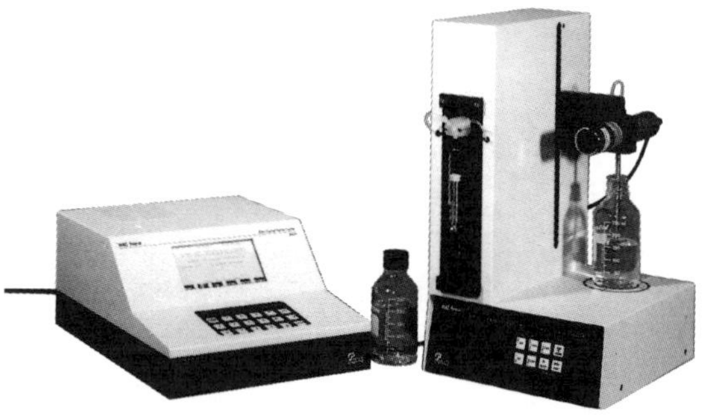

Figure 8.17. Particle Measuring Systems's APSS 200™ with Liquilaz™ sensor.

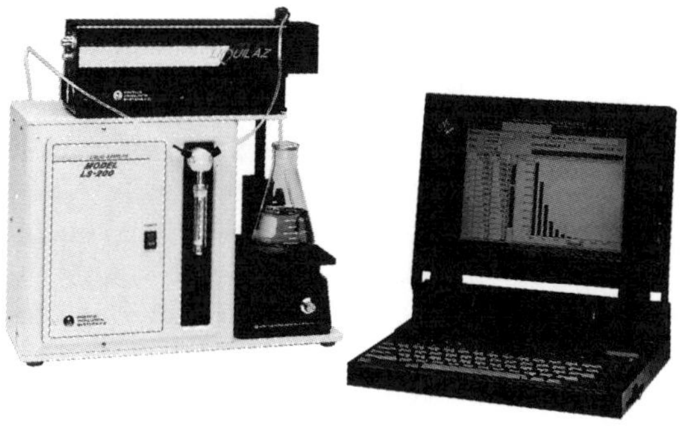

The development and application of LO counting remains an evolving technology. For example, Particle Sizing Systems (Santa Barbara, Calif.) has developed a hybrid sensor that combines light scattering and LO signals to decrease the lower limit of the dynamic range to 0.5 μm. While of limited interest in counting at compendial sizes, this advance is significant for the user desiring to apply the rapid assessment of light extinction counting at below the 2.0 μm practical limit of conventional counters. The sensor produces a continuous response curve for particles ranging from 0.5 μm to 400 μm in size. This combination of scattering and obscuration signals in the sensor rather than by postsoftware signal processing decreases the possibility of sensor artifacts in the crossover range of the two signals. The sensor is combined with the Accusizer™ system, which provides the capability for automated sample dilution to the optimal concentration range of the sensor.

Both the APSS 200™ and the 8103™ systems offer significant refinements in data handling. (See Chapter 11 for a discussion of validation procedures.) The data systems are microcomputers, and use screens are manipulated using the Windows™ operating system. Both incorporate software control of particle size thresholds. The user simply provides calibration solutions with particles of the proper size and known variance, and the calibration curve for operation of the instrument is automatically determined. Both of these instruments also offer calibration-standardization routines tailored to the USP <788> requirement.

SUMMARY

LO counters provide a time-effective means of counting extraneous particles in parenteral solutions. The LO assay is used worldwide in the enumeration of particles in parenteral solutions. The advantages for using LO particle counters generally outweigh the disadvantages. The assay is easy to perform, and when applied in a controlled environment to a product that is well characterized with regard to the expected particle burden, it is an effective tool for routine analyses. The method is the official USP test for particulate matter. It has proven generally suitable in this role. When used by appropriately skilled personnel so that the accuracy of data is maximized, the speed of the LO test becomes a significant advantage. Since microscopic and LO tests measure different physical parameters of particles, no precise agreement can be expected between counts obtained by the two methods.

Based on findings reported in the literature (see, for example, Chrai et al. 1987), the LO test cannot be relied on as a stand-alone method for the enumeration of particulate matter in parenteral products. Both large contaminant particles and drug-related insoluble particles of certain morphologies are undersized or not detected by LO counters. Dispersed immiscible fluids may result in overcounting. Thus, only if a product and process are well characterized with regard to these materials will a manufacturer be able to place confidence in the LO test result. On the other hand, if these materials are known to occur in a given product or group of products, the manufacturer is best advised to perform some level of microscopic testing to verify the LO counts.

LO counts obtained from SVI solutions containing dispersed silicone oils provide another subject for discussion. This material constitutes a single-source chemical entity that is beyond the intended application of a particulate matter test. It can only be quantified accurately by chemical tests, such as infrared spectroscopy or inductively coupled plasma spectroscopy. While high counts due to silicone suggest that excessive amounts of this material may be present, this implication cannot be verified without a chemical test. The actual counts resulting from a given amount of silicone oil appear to be dependent on the extent of dispersion rather than the level of silicone present.

In addition to count artifacts, which may result when specific product types are tested, errors may be encountered based on the principles of LO counting and specific components of laboratory technique. The instruments will invariably be subject to considerations regarding sampling statistics, particle transport, immiscible fluids, undersizing of large particles, and differences in calibration technique (laboratory technique, particularly cleanliness of the test area, and sampling). The existence of predominant particle types in SVI dry powder drugs that cannot be detected by the compendial test method raises a critical issue with regard to this assay. In effect, the LO test ignores the presence of some particle types, while providing artificially high counts based on dispersed microdroplets of silicone. The implication of this fact is clear for both the regulatory agency and for manufacturers: Interpretation of a count result must be tempered with the knowledge that an error potential with regard to poorly visualized particle types exists. In the light of the stated objectives of the FDA in regard to the application of compendial particle testing (Jamet 1988), this shortcoming appears worthy of further investigation.

Despite their drawbacks and detractors (as the author is considered by many), LO counters provide a useful, practical, and time-effective means of sizing and enumerating particles in medical devices and injectable products. This discussion of the known defects of the method should not be taken in a negative context. We know that, in fact, all particulate matter tests produce data that are only useful to the extent of our understanding of the method used. If the user completely understands the nature of the product tested and defects in the test method, there is little chance that data accepted will contain significant errors or artifacts.

REFERENCES

Backhouse, C., P. Ball, S. Booth, M. Kelshaw, S. Potter, and C. McCollum. 1987. Particulate contaminants of intravenous medications and infusions. *J. Pharm Pharmacol.* 39:241–245.

Bangs, L. D. 1988. *Uniform latex particles.* Indianapolis, Ind., USA: Seradyn, Inc. (Particle Technology Division), p. 60.

Barber, T. A. 1987. Limitations of light blockage particle counting in the analysis of parenteral solutions. In *Proceedings of the PDA Conference on Liquid Borne Particle Inspection and Metrology,* pp. 317–375.

Barber, T. A., and J. Williams. 1990. Analysis of dry powder antibiotics by light obscuration counting. In *Proceedings of the PDA International Conference on Particle Detection, Metrology, and Control,* pp. 502–537.

Barber, T. A., M. D. Lannis, J. Williams, and J. Ryan. 1990. Application of improved standardization methods and instrumentation in the USP particulate test for SVI. *J. Paren. Sci. Technol.* 44:185–203.

Blanchard, J., C. Thompson, and J. Schwartz. 1976. Comparison of methods for detection of particulate matter in large volume parenteral products. *Am. J. Hosp. Pharm.* 33:144–150.

Carver, L. D. 1969. Light blockage of particles as a measurement tool. *Ann. N.Y. Acad. Sci.* 158 (3): 710–721.

Chrai, S., R. Clayton, L. Mestrandrea, T. Myers, R. Raskin, M. Sokol, and C. Willis. 1987. Limitations in the use of HIAC for product particle counting. *J. Paren. Sci. Technol.* 41 (6):209–214.

Delly, J. G. 1980. *Parenteral solutions: Problem of sizing particulate using USP methodology.* Project Report MA 8861. Walter C. McCrone Associates, Inc. p. 36.

DeLuca, P. P. 1977. Need for improved microscopic methods and understanding of correlations between microscopic and automatic methods. *Bull. Paren. Drug Assoc.* 31:173–178.

DeLuca, P. P., S. Boddapatti, and S. Im. 1980. Guideline for the identification of particles in parenterals. *FDA By-Lines,* No. 3 (July).

DeLuca, P. P., B. Conti, and J. Z. Knapp. 1987. An overview of technical issues in particle detection. In *Proceedings of the PDA Conference on Liquid Borne Particle Inspection and Metrology,* pp. 376–380.

Duke, S. D. 1988. Particle retention testing of 0.05 to 0.5 micrometer membrane filters. In *Proceedings of the International Technical Conference on Filtration and Separation,* pp. 525–532.

Duke, S. D., and E. B. Layendecker. 1989. Improved array method for size calibration of monodisperse spherical particles by optical microscope. *Part. Sci. and Technol.* 7:209.

Duke, S. D., R. E. Brown, and E. Layendecker. 1989. Calibration of spherical particles by light scattering using photon correlation spectroscopy. *Part. Sci. and Technol.* 7:223.

Grundelman, G. P., and S. H. Goldsmith. 1987. User experience with liquid borne particle counters. In *Proceedings of the PDA Conference on Liquid Borne Particle Inspection and Metrology,* pp. 495–506.

Jamet, X. 1988. Influence of packaging material and of appropriate measures during manufacturing on the particulate matter content of small volume parenterals. *Pharm. Ind.* 50:108–110.

Jaenicke, R. 1972. The optical particle counter: Cross sensitivity and coincidence. *Aerosol Sci.* 30: 95–111.

Knapp, J. Z., and L. Abramson. 1990. A systems analysis of light extinction particle detection systems. In *Proceedings of the PDA International Conference on Particle Detection, Metrology, and Control,* pp. 283–352.

Knapp, J., and L. Abramson. 1996a. A new coincidence model for single particle counters, Part III: Realization of single particle counting accuracy. *J. Paren. Sci. Technol.* 50:99–122.

Knapp, J. Z., and L. R. Abramson. 1996b. Validation of counting accuracy in single-particle counters: Application of a new coincidence model. In *Liquid and Surface-Borne Particle Measurement Handbook.* New York: Marcel Dekker, chapter 10.

Knollenberg, R. G., and R. C. Gallant, 1990. Refractive index effects on particle size measurement by optical extinction. In *Proceedings of the PDA International Conference on Particle Detection, Metrology, and Control,* pp. 154–182.

Lieberman, A. 1975. Flow rate and concentration effects in automatic particle counters. *Proceedings of the National Conference on Fluid Power.*

Masuda, J., and J. Backerman. 1973. Particulate matter in commercial antibiotic injectable products. *Am. J. Hosp. Pharm.* 30:72–76.

Millipore Corporation. 1991. *Detection and analysis of particulate contamination.* Technical Bulletin AD030. Bedford, Mass., USA: Millipore Corporation.

Montague, C. 1997. Personal communication.

Montanari, L., F. Pavanetto, B. Conti, and R. Ponci. 1986. Determination of the degree of particulate contamination in small volume intravenous injections. *Pharma* 2:456–458.

National Fluid Powder Association. 1972. *Procedure for qualifying and controlling cleaning methods for hydraulic power fluid sample containers.* Document 893–20. Milwaukee: National Fluid Powder Association.

Parkins, D., and A. Taylor, 1987. Particulate matter content of 11 cephalosporin injections: Conformance with USP limits. *Am. J. Hosp. Pharm.* 44:1111–1118.

Pavenetto, F., B. Conti, I. Genta, R. Ponci, L. Montanari, and S. Vianello. 1989. Particulate matter contamination in small volume parenterals. *Int. J. Pharm.* 51:55–61.

Raasch, J., and H. Umhauer. 1984. Errors in determination of particle sizing distribution caused by coincidence in optical particle counters. *Part. Character.* 1:53.

Rebagay, T., H. G. Schroeder, and P. P. DeLuca. 1977. Particulate matter monitoring II. Correlation of microscopic and automatic counting methods. *Bull. Paren. Drug Assoc.* 31:150–155.

Schroeder, H. G., and P. P. DeLuca. 1980. Theoretical aspects of particulate matter monitoring by microscopic and instrumental methods. *J. Parent. Drug Assoc.* 34:183.

Signoretti, C. E., I. Montanari, and M. Grande. 1987. Particulate contamination in small volume parenteral solutions. *Boll. Chim. Farm.* 126:399–402.

Sommer, H. T. 1990a. Optical sizing of single particles (HIAC Laser). In *Proceedings of the 2nd International Congress on Optical Particle Sizing,* pp. 612–618.

Sommer, H. T. 1990b. Performance of monochromatic and white light extinction particle counters. In *Proceedings of the PDA International Conference on Particle Detection, Metrology, and Control.*

Sommer, H. T. 1991. Performance of optical particle counters: Comparison of theory and instrument. In *Proceedings of the IES 36th Annual Meeting,* HIAC Royco-Pacific Scientific. Silver Springs, Mass., USA: HIAC Corp.

USP. 1990. *The United States Pharmacopeia,* 23rd Revision, Chapter <788>. Rockville, Md., USA: United States Pharmacopeial convention, Inc., pp. 1596–1598.

West, J. 1990. Practical consideration in evaluating a liquid borne particle counter. In *Proceedings of the PDA International Conference on Particle Detection, Metrology, and Control,* pp. 372–394.

Wheeler, D. J., and D. S. Chambers. 1986. *Understanding statistical process control.* Knoxville, Tenn., USA: Statistical Process Controls, Inc.

Whitby, K. J., and B. Y. H. Liu. 1967. Generation of countable pulses by high concentration of subcountable sized particles in the sensing volume of optical counters. *J. Colloid Interface Sci.* 25:537.

Wilkison, M. C., J. Hearn, F. H. Karpowicz, and M. Charley. 1987. The stability of latex particulate in aqueous suspensions. *Part. Sci. and Technol.* 5:65–82.

Willeke, K., and B. Y. H. Liu. 1985. *Single particle optical counter: Principle and application.* Particle Technology Laboratory Publication No. 264. Minneapolis, Minn., USA: University of Minnesota, Dept. of Mech. Engineering, p. 37.

IX

AIRBORNE PARTICLE COUNTING AND ENVIRONMENTAL MONITORING

The extraneous particle burden of final product consists of particles from multiple sources. These particles come from the ambient air in the manufacturing facility, containers, filling machines, filters, process piping, and people. If our ultimate concern is limiting total product particle burden, it is necessary first to quantitate the different components of the total particle burden.

The term *environmental* is typically used to describe particles as randomly dispersed airborne (i.e., aerosolized) contaminants in the ambient air of a process room or production facility. These particles range from submicrometer fragments of glass, plastic, metal, or biological material to large, visible fibers of paper or cotton. This particulate population is notably distinct from process- or point-generated particles that may occur at high levels in local areas due to human or machine operations. The control of environmental particles is based on several key procedures:

- Continuously supplying clean air to the process area by means of high efficiency particulate air (HEPA) filters
- Removing contaminated air through return plenums
- Personnel protection
- Material control
- Directional pressurized airflow
- Monitoring airborne particle numbers with optical particle counters (OPCs)

Particle counting is not, of course, a control measure, but it enables the manufacturer to assess the effectiveness of cleanliness procedures and to direct efforts in this regard.

Figure 9.1 is a Venn diagram of sorts. The process encompasses all sources of particle matter to which the product is exposed. There are four sources of particulate matter represented:

1. People (the most prolific source)
2. Process points and workstations
3. Materials
4. Ambient air

Figure 9.1. Interaction of particle populations in CEAs.

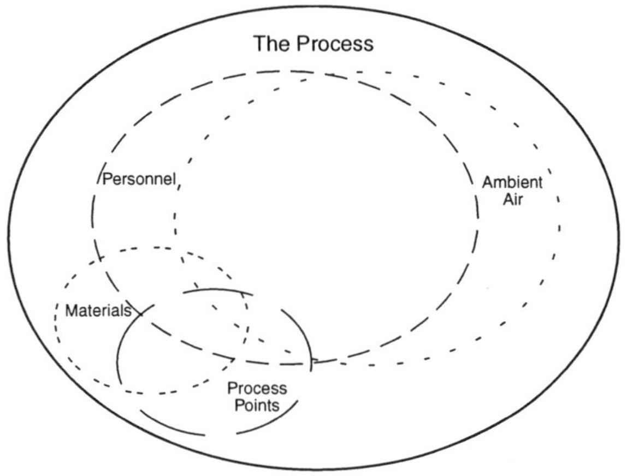

The process includes the cumulative particle burden from low-level miscellaneous sources as well as those falling into the above four categories. Note the interaction between people and the environment: personnel generate most of the particles counted in air samples at ≥ 0.5 μm. Process point particle buildup shows a strong interaction between people and materials; many of the particles generated here are large and never become airborne but can make their way into product. The process encompasses all sources occurring in the cleanroom. This includes those particles sourced to people, machinery, carts, re-entrained materials, and particles from sources such as machines, moving wear surfaces, and particles entering from adjacent areas.

The environment that the product sees is not just environmental air; it includes everything surrounding the product (i.e., air, surface, people, process machinery) (Ensor et al. 1987). An unavoidable problem with considering the individual components of the total population of particles to which a product is exposed is found in the fact that the populations merge, overlap, or interact as in Figure 9.1. In the dynamic equilibrium that exists in a controlled environment area (CEA), airborne particles collect on surfaces, deposited particles become airborne, personnel contribute particles to the airborne and surface populations, and particles from process machinery and materials become universally distributed.

Liquid-borne particle counting, which was discussed in Chapter 8, most often is applied as an assessment of the particle burden in the final product; the airborne particle burden is an intermediate or contributory component, and airborne particle count data must be considered in this light. Despite these considerations, the monitoring of aerosol particles in pharmaceutical manufacturing environments has been and remains critical to maintaining Good Manufacturing Practice (GMP) principles and the assurance of product quality (Sorenson 1986; FDA 1987a; Federal Standard 209D and E). The primary reason for the concern with monitoring aerosol particles is simply that these materials can adhere to surfaces that should be clean or enter product via the container or closure. Although such airborne particles are generally nonviable organic or inorganic materials, they may transport bacteria and conceivably could be chemically reactive. The GMP regulations enforced by the Food and Drug Administration (FDA) deal specifically with monitoring in controlled environments, over filling lines, and of compressed air lines. The GMPs, in some cases, specify maximum particle concentration limits and provide procedural recommendations for monitoring. Nonviable particle monitoring to ensure that air quality is maintained is critical in aseptic fill processes.

Airborne particles range in size from 0.001 μm to several hundred micrometers. Suspended particles settle onto containers, closures, and exposed surfaces at a rate that depends on the size and density of the particle. For example, according to Stokes Law, a spherical particle of 50 μm size with a unit density would take 1 min or less to settle 10 ft, while a particle in the 1 μm range could take 15 or 20 h to move the same distance, and may never settle out due to Brownian motion, turbulence, and/or directional airflow. Before any methods of environmental control of airborne particles can be applied successfully, a decision must be made as to how critical particulate matter is to the process or operation. The allowable size and number of airborne particles at a point within an area depends on the sensitivity of the process to be performed at that particular point. At the same time, the quantity of the particles of a given size that might originate at process points within the area of concern must be considered (Lieberman 1987a and b).

Particles from sources both internal and external to the process are important. Despite filtration and other safeguards, gross atmospheric contamination and particles from surrounding areas tend to find their way into the protected environment. External contamination, such as combustion particles, may be brought in through the air-conditioning system that supplies the process room with makeup air. In addition to the air-conditioning system, contamination from adjacent areas will enter through doors and pass-throughs. The external contamination from adjacent areas to which a process is exposed is critically affected by the type of filtration used, overpressure gradients, and personnel activity.

People and and their activities are the greatest source of contamination due to the fact that they continually produce and shed particles (both viable and nonviable). The shedding of skin cells generates particles in the 1–10 μm range, and exhaled air contains large quantities of particles ranging in size up to 100 μm. In addition to people, moving contact between surfaces creates contamination. Writing with a pencil on a piece of paper generates an aerosol cloud of fine carbon particles and paper fiber fragments. The movement of a metal part on a bearing surface sometimes generates particulate matter in the form of a fine metallic dust. Particles of dry powder drugs are notorious as an all-pervasive contaminant in powder fill facilities.

In a "clean" or "controlled" working environment, air movement results from personnel activity, operation of process machinery, and the air movement generated by a heating, ventilation, and air-conditioning (HVAC) system. All of these cause the air to move locally with random velocity and direction in the work space. Particles caught in random currents within a given work space may easily move from one area to another unless they are controlled by some means. This movement of contamination via random air currents from one part of the clean space to another or to adjacent areas is termed "cross-contamination" and is a significant contributor to the general process contamination level. In this regard, unidirectional airflow is used to eliminate cross-contamination and "sweep" particles away to return plenums. Over time, contamination from all sources will build up in the process area on surfaces and in the air and reach a steady state condition, with particles being added to the environmental air as quickly as they are removed. Depending on the type of manufacturing environment, the steady state count may reach as high as 300,000 particles ≥ 0.5 μm/ft^3.

There are important distinctions to be made between "large" and "small" particles and between those particles that comprise the natural population of environmental air in a facility and those that represent "contamination." The term *large particles* in airborne particle monitoring almost invariably refers to those ≥ 5 μm in size. This is different from the terminology used with injectables, whereby large particles are commonly defined as those of visible size or those approaching the visible size range. The majority of this chapter deals with the counting and monitoring of particles in the range of 0.5 μm to 5 μm in the air of process areas. In the business of pharmaceutical manufacturing, the 0.5 μm and greater size are almost invariably believed by manufacturers to be those of importance in defining facility and process cleanliness. In aseptic processing areas of current designs, manufacturers may face the dilemma of having such low counts that they are unable to separate fluctuations in particle levels from random statistical fluctuations.

A number of methods are applicable for monitoring larger particles. In this case, the technologist is not only faced with issues relating to the small number statistics and transporting particles of larger sizes. While some sampling situations in the United Kingdom once required collecting counts of ≥ 25 μm particles, the data on these particles collected in a conventional cubic feet per minute (CFM) sample flow may be very difficult to interpret. Thus, specialized sampling methods such as witness plates or particle fallout meters may be employed.

The "normal" population in a process area will depend on both the human activities and mechanical operations in place and will be defined based on historical data. The FDA will expect all counts collected in an aseptic area (or filling areas for sterilized product) to be within established limits. This would be < 100 ≥ 0.5 μm particles/ft^3 for an aseptic process. Counts that approach this limit are dimly viewed, since statistical variation in counts may be expected to result in some counts that are higher than a historical average. Thus, counts that approach limits values can be viewed as due to a "contaminant" component.

Despite such necessary philosophical considerations regarding the complex nature of the total particle burden, the manufacturer is still forced to separate and consider individual and principal components of the whole. The monitoring of airborne particles represents one means of separation.

The principal subjects of this chapter are the detection and enumeration of airborne particles and considerations in airborne particle monitoring. It is presented generally, with a consideration in some detail of how particles are transported and counted. Numerous standards have been generated to deal with particulate control in CEAs and clean rooms. The two most widely used are Federal Standard (FS) 209E and British Standard (BS) 5295 (Part 1, now incorporated into the EEC GMPs). The implementation of ISO (International Organization for Standardization) 14644 promises to add both a greater degree of harmony and complexity to the situation. These standards specify allowable particle levels at specific sizes for various classifications. Environmental standards were discussed in some detail in Chapter 5 and will not be covered here (also see De Valle 1997).

MECHANISMS OF AIRBORNE PARTICLE DETECTION

As mentioned in the previous chapter, light extinction methods are relatively insensitive to particles less than about 1.5 μm in size. There are two reasons for this: (1) Particles below this size do not cause the deviation of significant amounts of light from the illuminating beam of the instrument. (2) The sensitivity of the light extinction instrument is limited by the basic detection mechanism, which is based on a diminution of intensity of the total amount of light falling on a photodetector. Hence, smaller particles (0.1–1.0 μm) that must be counted and sized in air and gases, as specified in various environmental standards, require a more sensitive type of counter. Light-scattering particle counting instruments fill this need. In the vernacular of environmental monitoring, these instruments are called "optical particle counters" or OPCs; the author is using this term throughout the chapter.

The basic mechanism of light-scattering particle detection is inherently more sensitive than light extinction. The signal recorded by the photodetector in this mode of counting is the small amount of light scattered by a particle referenced against the "dark" signal of the photodetector, which is prevented from receiving light by elaborate systems of baffles and light-absorbing dark (black) surfaces. In this system, the electronic noise that dictates the lower size sensitivity threshold is the dark current of the photodetection and amplification circuits (Bemer et al. 1990; Buettner 1990; Caldow 1990). This noise signal is very small with current solid-state electronics. To better understand the function of these counters, it is necessary to know something of the nature of light-scattering phenomena.

The terms and definitions given below pertain to scattering mechanisms and light-particle interaction. For the purpose of this discussion, we will assume light to have a wavelength of 500–800 nm depending on the source. The majority of sources used today are either diode lasers or red helium-neon (HeNe) lasers. The latter are advantageous for more sensitive counters because of the very high beam photon density that they can achieve. Since smaller particles, in general, scatter less light, and the amount of light scattered depends on the intensity of light incident on them, a more intense beam of illuminating radiation results in more scatter and a higher signal-to-noise ratio. The latter limits the low end sensitivity of any particle counter.

The following definitions are critical to the discussion of the mechanisms of light scattering (see Willeke and Baron [1993] as a general reference):

- *Scattering:* The collective interaction between light and aerosol particles that results in the deflection of light from its original path by processes of diffraction, absorption, reflection, and refraction (note the term *collective*). While the latter two mechanisms were discussed as predominant geometric optical effects, their role in producing light scatter from particles < 2 μm is less important. Emphasis in light-scattering particle detection is placed on collection of the signal from the deviated light; in extinction detection, it is based on source beam diminution.
- *Scattering angle:* The angle formed by the optical axis of the illuminating light beam and the linear axis of scattered light detection, with the particle at the vertex of the angle. By convention, a scattering angle of 0° would describe a case where the detector looks directly into the light source (as in extinction).

- *Rayleigh scattering:* Light scattering by particles whose size is much smaller than the wavelength of the illuminating light beam (e.g., air molecules illuminated by visible light).
- *Mie scattering:* Light scattering by particles whose size approximates the dimension of the wavelength of the illuminating beam (e.g., smoke particles illuminated by visible light).
- *Size parameter:* Defined as the ratio of the particle circumference to the wavelength of the illuminating light. Expressed as $a = \pi d/\lambda$, where d is the particle diameter and λ is the wavelength of light.
- *Opacity:* The fraction (usually expressed as a percentage) of the intensity of a light beam that is removed from that beam by absorption and/or scattering while passing through an attenuating medium.
- *Absorption:* The interaction between light and aerosol particles that results in the retention of part of the light within the particles by conversion into other forms of energy (e.g., heat, fluorescence).
- *Attenuation:* A measure of diminution of intensity based on opacity.
- *Extinction:* A multifactored measurement of diminution of intensity based on scattering and geometric optics (see Chapter 8).
- *Extinction efficiency:* The sum of scattering efficiency and absorption efficiency.
- *Transmittance:* The ratio between transmitted to incident light or the fraction of a light beam that, in passing through a medium, is *not* removed from the beam. Transmittance is equal to 1 minus the opacity (see Table 9.1).
- *Optical density:* The logarithm (base 10) of the reciprocal of the transmittance (see Table 9.1).
- *Particle index of refraction (or refractive index):* An optical property that is the ratio of the speed of light in a vacuum to the speed of light within the particle. The refractive index is a complex number (i.e., it consists of a real and an imaginary part). For example, the index of refraction of coal dust is expressed as 2 – 0.66*i*). The real part describes the scattering properties, and the imaginary part describes the absorption characteristics of the particle material. The index of refraction of a given material varies with the wavelength of light. The indexes of refraction of various selected materials are shown in Table 9.2. Particles with a negligible imaginary component are dielectric or nonconductive particles and exhibit essentially no light absorption. Particles with a complex index of refraction (i.e., with a real and an imaginary part) are conductive and exhibit both scattering and absorption. Importantly, there are no substances that as small particles are pure absorbers of visible light.

Table 9.1. Relationship of Optical Transmittance Parameters (Liu 1996)

Opacity	Transmittance	Optical Density
0	1.0	0
10	0.9	0.046
20	0.8	0.097
30	0.7	0.155
40	0.6	0.222
50	0.5	0.301
60	0.4	0.398
70	0.3	0.523
80	0.2	0.699
90	0.1	1.000
100	0	infinity

TYPES OF SCATTERING

The scattering by very small, spherical particles, those with a size parameter smaller than about 0.2 (i.e., whose circumference is less than about 2/10 the wavelength of the illuminating light) is characterized by a relatively simple equation, developed by Rayleigh:

$$I_{rs} = I_o \frac{9\pi^2 V^2}{2\Pi^2 \lambda^4} \left(\frac{m^2 - 1}{m^2 + 2} \right)^2 \left(1 + \cos^2 \theta \right) \qquad \text{(eq. 1)}$$

where I_{rs} is the scattered light intensity, I_o is the intensity of the illuminating light beam

Table 9.2. Refractive Indices of Various Materials (Liu 1996)

		Refractive Index	
Substance	**Wavelength of Light (nm)**	**Real**	**Imaginary**
Aluminum Oxide	589.3	1.77	10^{-6}
Carbon (amorphous)	436	1.90	0.68
Carbon (amorphous)	623	2.00	0.66
Quartz	589.3	1.544	negligible
Sulfuric acid	589.3	1.437	negligible
Benzene	589.3	1.501	negligible
Water	589.3	1.333	negligible

(unpolarized), V is the particle volume, R is the distance between the particle and the observer, λ is the wavelength of the illuminating light, m is the particle index of refraction, and θ is the scattering angle. Note: For the peak of the mixed wavelength solar spectrum with a peak intensity at about 550 nm, an a of 0.2 corresponds to a particle diameter d of 350 mm.

Although the illuminating light is usually unpolarized, the Rayleigh scattered light becomes polarized into two components; one of these has its light vibrations perpendicular to the plane of observation, the other is parallel (the plane of observation is defined by an angle normal to the axis of illumination). The angular distribution of a Rayleigh scatterer is shown in Figure 9.2, which shows that forward- and backscatter intensities are the same.

Two additional aspects are noteworthy about Rayleigh scattering as characterized by Equation 1. First, the scattered intensity varies as a function of the particle volume squared (i.e., as the sixth power of particle diameter). As a consequence, a cloud composed of particles 200 nm in diameter will scatter 64 times as much as a cloud with the same number of 100 nm diameter particles. This property explains why scattering by individual particles as small as a few nanometers in diameter can be detected in the presence of innumerable air molecules whose exceedingly small diameter (2–3 nm) makes them extremely weak scatterers. Second, the intensity of light scattered by small particles (i.e., in the Rayleigh regime) varies inversely as the wavelength of light to the fourth power. Thus, blue light (shorter wavelength light) is scattered much more intensely than red light (longer wavelength light). This accounts for the blue color of the sky when illuminated by a nearly white light (all visible wavelengths) source, the sun.

Figure 9.2. Angular light-scattering pattern for a small Rayleigh scatterer.

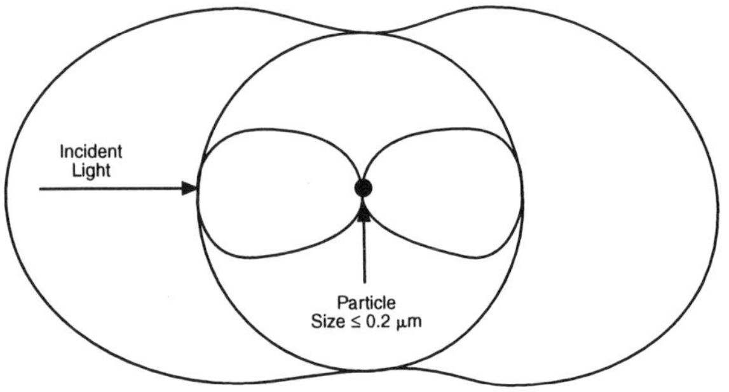

Rayleigh scattering is of importance in the sizing and enumeration of particles < 200 nm in size. For scattering of light by particles whose size exceeds about 200 nm, the relatively simple Rayleigh model no longer is applicable. A comprehensive theoretical treatment (wherein the Rayleigh model is included as a limiting case) for all sized particles was developed by Mie (see Umhauer and Bottlinger 1990). This theory, however, is not nearly as simple as the Rayleigh model and is subject to application by computer. The solutions of these equations and the tabulation of scattering coefficients for a variety of particle sizes, refractive indexes, and shape parameters has had to await the availability of digital computers.

The equation from which the Mie scattering intensity of a particle (or of an ensemble of particles) can be calculated is

$$I_{ms} = I_o \frac{\lambda^2}{4\pi^2 R^2}\left(\frac{i_1 + i_2}{2}\right) \qquad \text{(eq. 2)}$$

where I_{ms} is the scattered light intensity (Mie) and i_1 and i_2 are the scattering coefficients for the two polarization components of the scattered light. The other parameters are as defined in Equation 1. The scattering coefficients i_1 and i_2 are (even for spherical particles) complicated functions of the particle size parameter, the particle complex refractive index, and the scattering angle. These coefficients have been computed and tabulated mainly for spherical particles and for representative indexes of refraction.

The complexity of the scattering behavior for particles exceeding about 0.1 μm in size arises from the wave interference patterns that become significant when particles and waves exhibit comparable dimensions. The most salient characteristic of Mie scattering is that the angular scattering pattern (i.e., the scattering intensity as a function of the scattering angle) is nonsymmetrical between the forward and backward directions. In fact and without exception, forward-scattering intensity exceeds back-scatter intensity, and the ratio of these two intensities increases with particle size. This behavior is illustrated by Figures 9.3 and 9.4. Figure 9.3 shows the angular scattering pattern for a small Mie scatterer (a = 1 or about 200 nm diameter). This pattern may be compared with the symmetrical pattern for the Rayleigh scatterer depicted in Figure 9.2. Figure 9.4 shows the pattern for a larger Mie scatterer (0.8–1 μm). There is a gradual transition from the Rayleigh regime to the Mie regime that is evident in Figures 9.2, 9.3, and 9.4.

A photodetector placed in some forward or wide angle approximation to the scattering event, shown in Figure 9.4, sees light scattered in a number of forward-angled lobes or intensity maxima. These occur at various angles from the optical axis dependent on the size of the particle with regard to its refractive index and the wavelength of the incident light. This lobed pattern is responsible for the "cross-over" response observed with photodetectors that collect light at narrow forward angles. In specific cases, a lobe of scattered light intensity may fall directly on the detector of a particle counter, giving a strong scatter signal. The scattered light from a relatively larger particle may not be as favorably placed with regard to the detector and thus may give a weaker signal.

Figure 9.3. Angular light-scattering pattern for a small Mie scatterer.

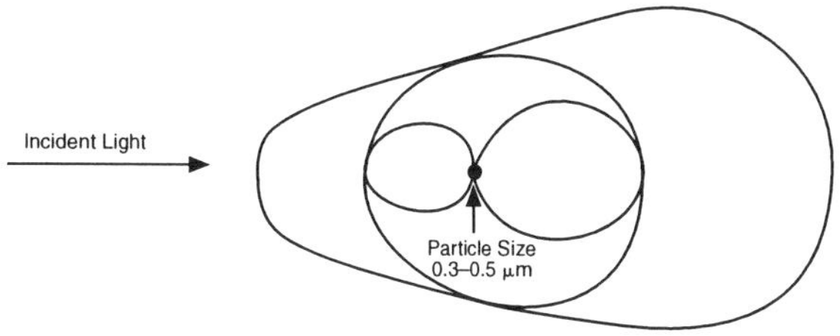

Figure 9.4. Larger Mie scatterer with lobed scattering envelope.

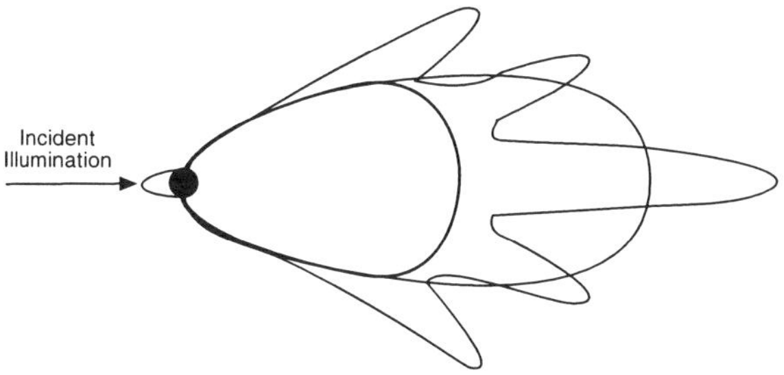

The predominance of forward scattering for particles whose size parameter approaches that of the illuminating radiation is caused by diffraction. This phenomenon, in turn, results in the so-called "extinction paradox," which states that a particle whose size is large compared with the wavelength of light removes from the beam twice the amount of light intercepted by its geometrical cross-sectional area (i.e., the extinction efficiency is equal to 2). For particles of sizes that are comparable to the wavelength of light, this extinction efficiency first rises as a function of particle size, then oscillates as size increases, finally settling to the value of 2 for very large particles. This relationship is shown in Figure 9.5. This parameter is thus equivalent to the scattering efficiency for nonabsorbing particles (e.g., carbon black, coal). The extinction efficiency is the sum of the scattering and absorption efficiencies, absorption being an energy degradation process (i.e., internal conversion to heat).

An important observation here is that the variation in scattering envelope shape that

Figure 9.5. Particle extinction efficiencies of spherical particles in air as a function of particle diameter for m = 1.33, m = 1.5, and m = 2 – 1i (which approximates the value for coal) (after Liu 1996).

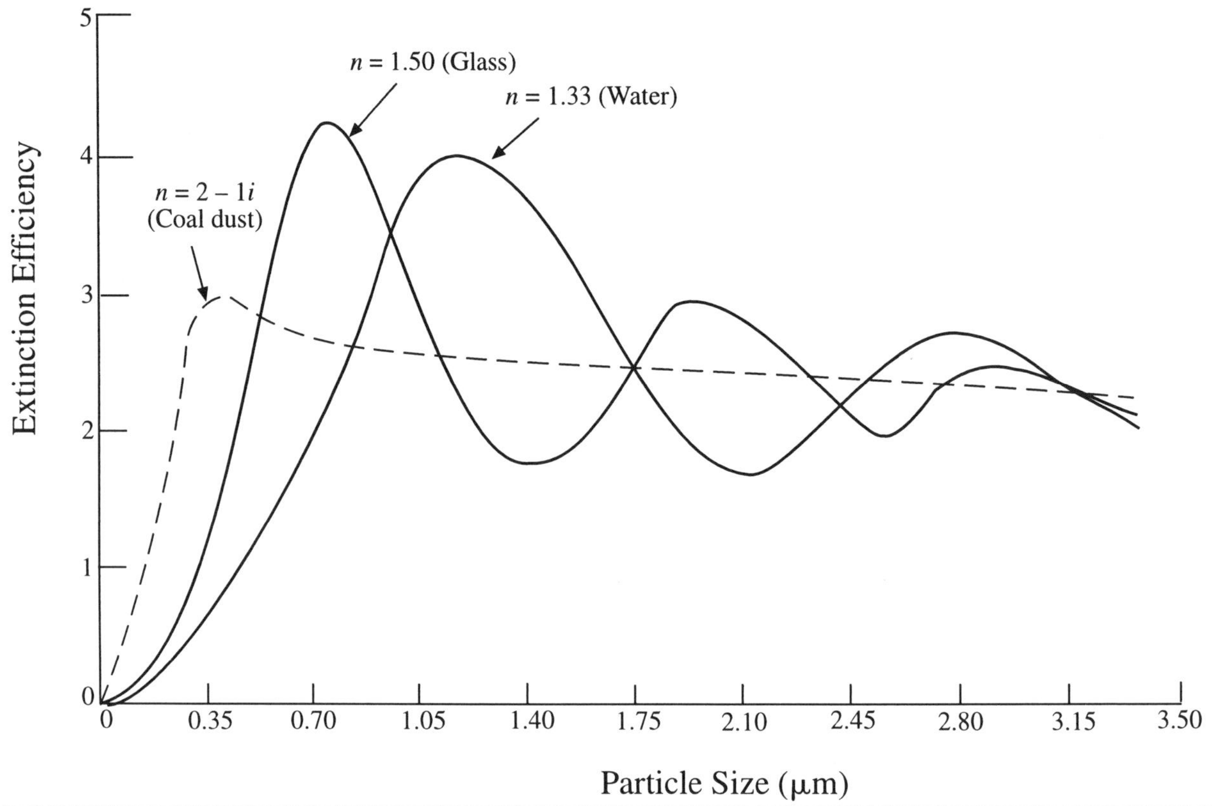

occurs due to refractive index effects will cause different amounts of scatter by particles of the same size, thus leading to sizing errors. The lobed scattering intensity pattern for larger Mie scatterers as discussed earlier will vary in angular distribution due to refractive index effects, and the size assigned will be highly dependent on detector geometry. The overall effect is probably unimportant in most situations, due to the heterogeneous nature of the particle population. White light OPCs, because of the polychromatic, less well-collimated nature of the source radiation, generally are less affected than laser counters.

Lower Detection Limits of Light-Scattering Instruments

The lower detection limit of an OPC is determined by noise (Szymanski and Wagner 1990). In most cases, this limit is reached when the detector noise becomes comparable to the scattering signal from the particles. Well-designed, high sensitivity instruments are also limited by Rayleigh scattering from air molecules within the scattering volume. Rayleigh scattering results in nearly steady state noise levels that vary slowly as a function of air density, limiting the minimum detectable signal from aerosol particles.

For single particle counters, the lower detection limit (i.e., the smallest detectable particle size) is determined by the so-called shot noise, a random statistical, broadband fluctuation of detector output associated with either dark current– or signal current–induced photoemissions. This noise signal follows the relationship

$$i_n = \left(2eGI\Delta f\right)^{1/2} \qquad \text{(eq. 3)}$$

where i_n is the root mean square (r.m.s.) value of the fluctuating noise current, e is the unit electronic charge (1.6 × 10^{-19} C), G is the detector gain or amplification factor, i is the direct current signal or dark current, and Δf is the bandwidth of the detection circuit (Hz). This equation applies to both photomultiplier type tubes and semiconductor diode detectors.

OPTICAL PARTICLE COUNTERS

OPCs are the most commonly used cleanroom air monitors. They use the light-particle interactions discussed above to determine the size and concentration of particles suspended in air. Although OPCs cannot distinguish between viable and nonviable particles, they provide real-time, in situ monitoring of air quality (see Fitch 1991; Lieberman 1988; Sommer and Harrison 1991; Montague and Harrison 1989). While other types of counts may be used for smaller or larger particles, the universally applied basic instrument is the light-scattering OPC, with a sensitivity of ≥ 0.3 μm or ≥ 0.5 μm.

When sampling air in a cleanroom or clean area, the definition of the classification level is usually based on the measured concentration of particles in threshold size ranges from 0.1 μm to 0.5 μm, as defined in FS-209E[1]. However, it is sometimes necessary to define the population of particles ≥ 5 μm. Many cleanroom air measurements are made solely with an OPC with inlet sample flow rates ranging from 2.8 to 28.3 L/min (or 0.1 to 1.0 cubic feet per minute [CFM]. Many operators have stated that OPCs cannot detect particles larger than 1–2 μm in diameter. This is not generally correct. Once the particle has been drawn into the instrument, it will be counted and sized. The problem is instead one of poor sample collection by the inlet line to the counter. Most counters are fitted with an inlet nozzle approximately 6–8 mm in internal diameter, located a few centimeters away from the counter case. Operators usually add a flexible extension tube with a blunt end to the inlet and may orient the tube randomly in the air parcel being sampled. For the higher flow rate (i.e., 1 CFM) counter, the normal inlet sampling efficiency for 5 μm particles may be less than 90 percent efficiency and can be much less for lower flow rate instruments.

The primary descriptors of instrument function are sensitivity and resolution; flow rate is also critical because of the more favorable sampling characteristics of large samples. Sensitivity is the smallest particle that can reproducibly be separated from noise on the basis of its pulse size. Accuracy is not included with resolution and sensitivity in this discussion. Because of differing operational principles, each model and type of OPC provides in its count data a unique descriptor of the particle population counted; thus, there is no such thing as accuracy! A comparison of count data from counters made by different vendors and even from two counters of the same model may become a complex issue. The resolution of a counter is defined as the smallest difference in particle sizes that can be reproducibly detected. The sizing function of the instrument is inextricably linked with the count data that will result, just as is the case with liquid-borne light extinction counters.

The generalized function of a light-scattering OPC is shown in Figure 9.6. Particles pass through the light beam (sensing or view volume) one at a time. The intensity of scattered light is measured by a photodetector. Photodetector pulses are analyzed by a pulse height analyzer (PHA) that sends count data to appropriate size channels. Calibration permits conversion from pulse height to particle size and from number of counts to concentration. Relating the pulse height data to particle size information yields a response curve for the instrument and allows particles to be counted at given size thresholds.

Airborne particle counters must give accurate particle size and concentration information. They should also be reliable, requiring only annual or semiannual maintenance. Low maintenance and stable calibration of an OPC can be accomplished by using a solid-state laser-diode light source equipped with a feedback circuit that maintains a constant light intensity. OPCs used for monitoring should be easy to install and should operate automatically. They should also be small and relatively inexpensive so that they can be placed in all critical production locations.

In order to meet the needs of the pharmaceutical industry, newer OPC models have been designed to provide the local display of particle concentrations, easy-to-clean surfaces, high measurement stability as well as light source and flow rate status indicators. At this time, the most sensitive particle counting instruments are capable of counting and sizing particles in air ≥ 0.05 μm in size at a flow rate of 2.8 L/min (0.1 CFM). Instruments of this sensitivity are rarely applied in monitoring pharmaceutical

1. The author's distinction between cleanrooms and clean areas or CEAs is simple: areas classified as Class 100 or cleaner are cleanrooms.

Figure 9.6. Basic function of an optical particle counter (Dp = pulse distribution).

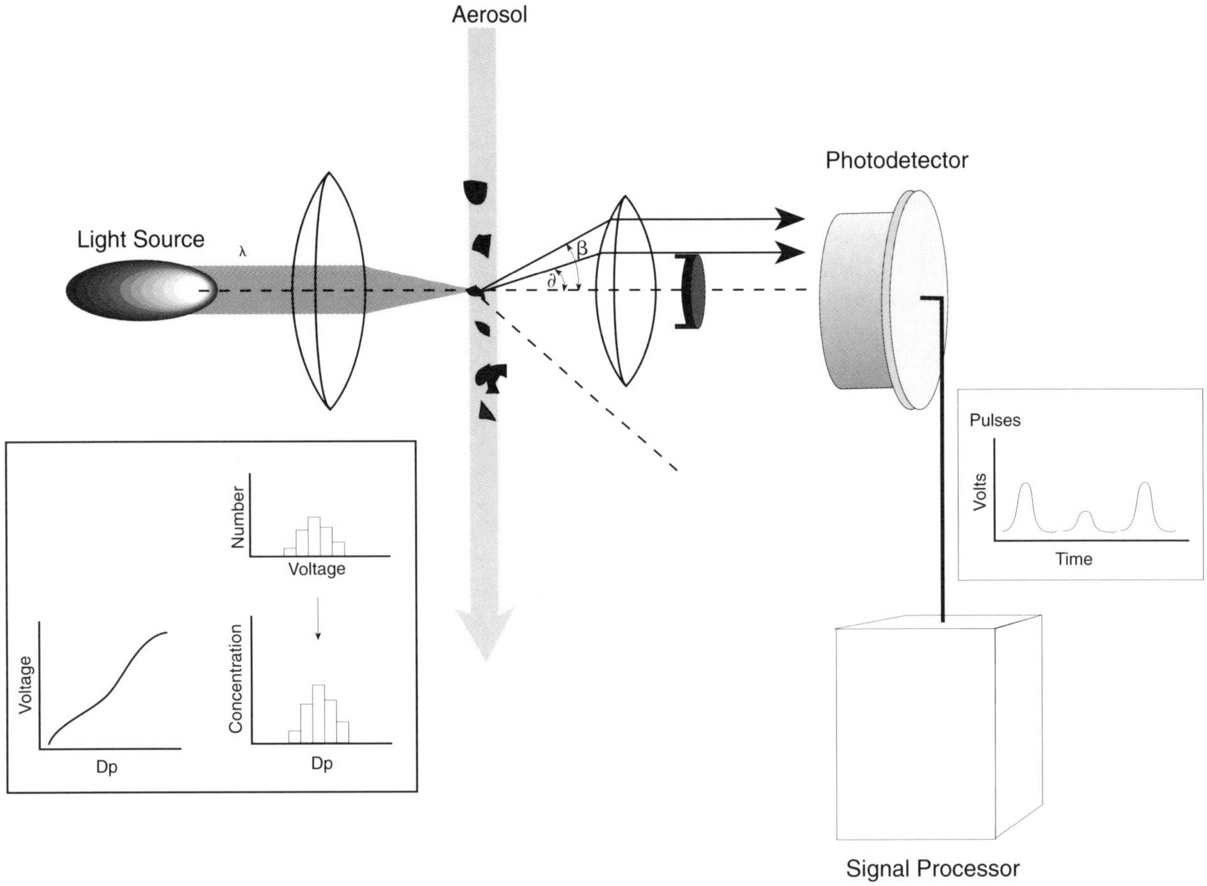

manufacturing environments. A number of particle counting instruments are available for measuring particles ≥ 0.3 μm in size at sample flow rates of 28.3 L/min (1.0 CFM). These instruments are well suited to verifying and monitoring areas of Class 100 and higher. There are trade-offs in counting applications. Low sampling flow rates require longer sampling intervals in order to reach a given level of statistical confidence in data. Low count rates also mean that the counting rate will be closer to the noise level of the instrument—an undesirable condition. A key premise to be kept in mind during the selection process is the following: A current-generation counter with higher sensitivity (e.g., 0.3 μm) should be suspect of a potential for giving higher counts at the ≥ 0.5 μm size used to define CEA classifications than one stated by the manufacturer to have a ≥ 0.5 μm low-end sensitivity.

The type of OPC used for monitoring is an extremely important consideration. The light-scattering OPCs widely used for size distribution measurements of aerosol particles incorporate a wide variety of sensor designs. The most critical difference between single particle OPCs in use today involve source illumination (laser or white light) and the geometry of scattered light collection (forward, near forward, or wide angle). While quartz-halogen bulbs were typically used with older white light instruments, present-day laser counters may use either "hard-seal" HeNe lasers or a laser diode. Currently available models are almost invariably based on laser illumination.

Based on our earlier discussion, the differing wavelengths of light involved in source illumination obviously will involve differences in scattered light intensity for a given particle. Other factors being equal, sensitivity (i.e., the smallest particle that may be sensed and counted) will be dependent on the intensity of the illuminating radiation and its wavelength range. A hard-seal laser used in the active cavity mode to give a near monochromatic light with extremely high photon density generally gives the best sensitivity available (0.05 μm). This is based on the fact that the more intense illuminating radiation simply results in a higher level of scattered light intensity for particles of a given size.

Counters are also classified based on the angle at which light is gathered by the collection optics. This angle is defined with the optical axis of illumination (a line drawn to the center of the detector from the source). Near forward–scattering instruments are defined as those collecting light scattered by particles at 5°–35° angles; wide angle scatter collection is defined as the collection of scattered light above 35°. The collection of light scattered at 10° or less is defined as the forward-scattering collection mode. Off-axis collection is typically defined as the collection of 90° scattered light. Forward-scattering collection geometry in OPC tends to produce response curves that vary polytonically with particle size; wide angle collection tends to result in a monotonic response.

The response of an OPC to aerosol particles is thus a complex relationship between the functional properties of the instrument and physical and optical properties of the particles in question. (See Hovenac [1990] and Liu and Szymanski [1987] for general discussions.) Considerations that limit the sensitivity and resolution of OPC measurement and directly affect rate include the following:

- Instrument properties:
  - optical design,
  - type and wavelength of illumination source (λ),
  - gain of the electronics, and
  - flow rate.
- Particle properties:
  - size,
  - refractive index,
  - complex refractive index,
  - shape, and
  - orientation of nonspherical particles relative to the incident beam.
- Multivalued responses: result of optical interference effects for nonabsorbing, spherical particles at visible wavelengths. The nonmonotonic response usually occurs close to the 1 μm size and can lead to substantial sizing errors. Stronger, more pronounced responses are exhibited in forward-scattering geometries.
- Limited resolution: Instruments with wide angle geometries tend to suffer from "flat spots" in response curves for smaller particles with sizes comparable to the wavelength of illumination. For particles smaller than the wavelength, the resolution is usually good due to the fact that scattered light is a very strong function of size and is independent of particle shape.
- Uncertainty in the refractive index limits counting accuracy and efficiency and may lead to large sizing errors for submicrometer particles, thus making the lower limit of detection uncertain.
- The size of the OPC sensing volume reduces resolution due to variability in scattering angles for particles entering different sensing volume areas.
- Nonuniform illumination of the sensing volume due to uneven intensity in the illuminating beam.
- Noise in photodetectors and circuits (electronic) as well as optical noise from light scattered by subcountable-size particles and air molecules.
- Uncertainty in sample flow rate influences the accuracy of concentration measurements.
- Coincidence losses (physical) occur in situations when two or more particles are in the sensing volume at the same time (see Bader et al. 1972) as with liquid-borne particle counters.
- Statistical limitations on count accuracy for low particle concentrations.
- Nonrepresentative sampling and/or particle losses in transport from the area sampled to the OPC sensor.

The elimination of discontinuities and resonances in the counter response curve generally involves a "trade-off" between sensitivity and response linearity. Generally, as collection angles exceed 35° (i.e., wide angle detection), response becomes more linear, and resonances are eliminated; at the same time, however, sensitivity decreases. This occurs because the strongest scattering of particles above about 0.2 μm in size is in the forward direction; lateral scattering is much weaker. With particles approaching the wavelength of the illumination, not only is scattering heavily concentrated in the forward direction, but the pronounced maxima and minima (lobes) with varying regular distribution cause the detected intensity to vary based on particle optical properties.

Because of the latter phenomena, and the variation of scattered light intensity dependent on the angle of observation, the position of a detector to quantitate scattered light may be critical. Because of the variable displacement of intensity maxima (Figure 9.4), particles of smaller size may be seen to scatter more light than larger particles. The intensity of scattered light detected by an OPC for a particle of given size is thus strongly dependent on the angular direction at which the light is received by the detector. Figure 9.7 shows the type of variation in OPC response to particle size that is dependent on whether collection of forward scatter (curves a and b) or wide angle (curve c) is used.

This discussion has dealt with scattering phenomena as is appropriate considering the sizing and counting of particle ≤ 2 μm in size on which environmental monitoring is based; in counting larger particles, geometric optical effects become important, as considered in Chapter 8. The response curve of an OPC is typically nonmonotonic for particles that have different refractive indices at sizes ≥ 1 μm. At sizes < 1 μm, particles of different sizes with different refractive indices can give the same pulse height due to resonances. The shape and orientation of nonspherical particles also affect the response. If particles have dimensions less than the wavelength of illumination, the measured size may be considered equal to the volume equivalent diameter. For particles with dimensions above the wavelength of illumination, the measured particle size approximates the projected area equivalent diameter for instruments with wide angle collection optics.

In summary, counters calibrated differently or of different design will yield different counts for the same particle populations. While users should be generally familiar with how these

Figure 9.7. Scattering analysis of 0.3 μm particles with different OPC collection geometry.

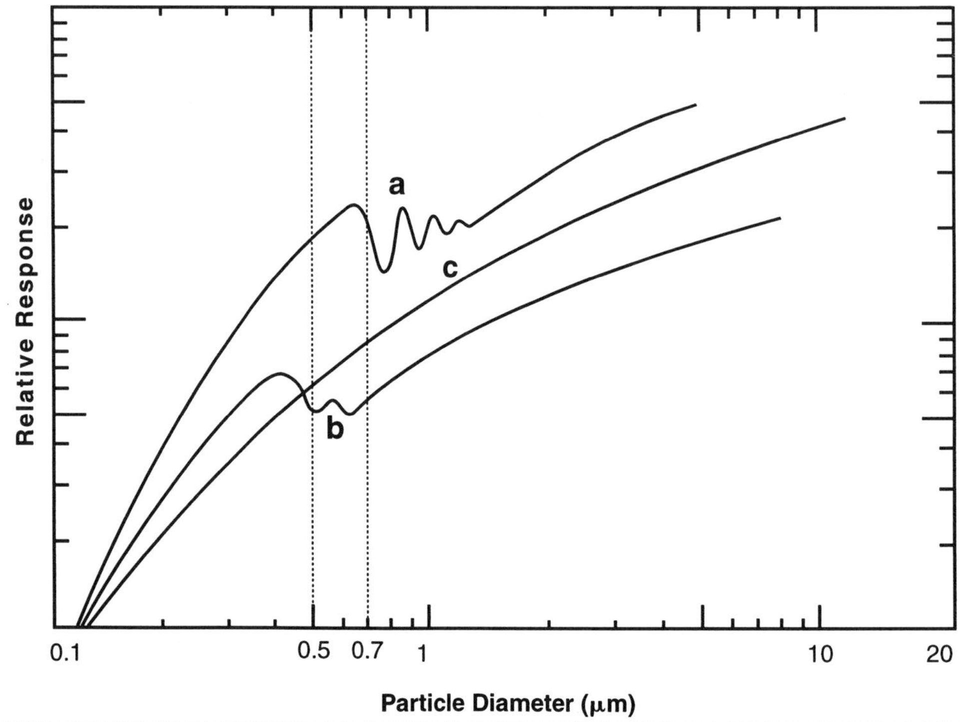

instruments work, they will not generally be willing to spend the time to delve deeply into this area. For more information regarding the function of OPCs of different types, the papers by Knollenberg (1970, 1976, and 1989) are highly recommended, as are those by Sommer (1989) and Liu and Szymanski (1987).

Light Sources

Light source types have a significant effect on OPC performance. The properties of light sources used in OPCs are summarized in Table 9.3.

Function of Collector Optics

The schematic in Figure 9.8 represents a historical white light sensor design. This is Climet's ellipsoidal mirror collection system used on the 208 series instrument and also used on the current model 7300. When introduced, this design, with its large, solid angle collection geometry represented a "quantum leap." Although it is a dated design, this geometry gives good sensitivity (0.3 μm at 1 CFM) and count accuracy comparable to some laser sensors. The extremely wide angle collection optics and the use of white light result in a near monotonic response for the smallest particles counted. This design has stood the test of time and is considered by many to be the only white light sensor currently worthy of consideration.

A forward-scattering configuration of a more current design is shown in Figure 9.9 (see Keady et al. 1990). This design was used in TSI's highly successful 3753 and 3755 counters. Other sensors with a similar designs are the CI-3100, the Met One 202B, and HIAC/Royco models 2250 and 5250. In this type of design, a beam stop blacks out the unscattered illumination and light scattering in the forward direction as an angle of 4°–10° is collected and focused on the photodetector. An instrument design format illustrating 90 collection by Climet Instruments is shown in Figure 9.10. Dual photomultiplier detectors and lens systems are used to collect wide angle scattered light centered on an axis oriented at 90° to the illuminating beam.

The design shown in Figure 9.11 is a complex active cavity laser variation of the off-axis design used in the PMS Micro LPC-HS counter. (See Schuster and Knollenberg [1997] for an early discussion of this principle.) This sophisticated design uses ellipsoidal collection mirrors with an active cavity laser sensor and wide angle collection to achieve extreme sensitivity to small (0.05 μm) particles. At this high level of sensitivity, the scattering contribution of air molecules and its contribution to noise becomes important. Two mechanisms are employed to minimize the effect of scatter by air molecules. The first is a closely collimated rectangular airflow; the second is a mosaic array of detectors (Figure 9.12), each of which sees a fraction of the view volume, thus increasing the ratio of light scattered by particles to that scattered by air. Light scattered by a particle will affect two detectors on opposite sides of the detector; random noise scatter affects only one so that the count event is not confirmed.

Calibration

Calibration is determining the instrument's response to particles of known size, refractive index, and, when possible, concentration. PSL solid particles (refractive index = 1.55) are most frequently used as a standard for calibrating OPCs. Some sizes are traceable to the National Institute of Standards and Technology (NIST). They are very uniform in size and close to perfect spheres in shape. They are available over a

Table 9.3. OPC Light Source Characteristics

Source	Life (hours)	Sensitivity	General
Incandescent (quartz halogen)	1,000	0.3 μm	Cost-effective, needs internal compensation for multiple wave lengths to reduce width of response curve.
Laser Diode	10,000	0.2 μm	Compact, cost-effective, efficient, less intense light than gas laser; temperature sensitive output and source life
Gas Laser	10,000	0.05 μm	Extreme sensitivity, may give multivalued response due to monochromaticity; expensive active or passive cavity configurations

Figure 9.8. Schematic diagram of extremely wide light-scattering optics (15–150°) used in Climet CI-7300 with incandescent light source.

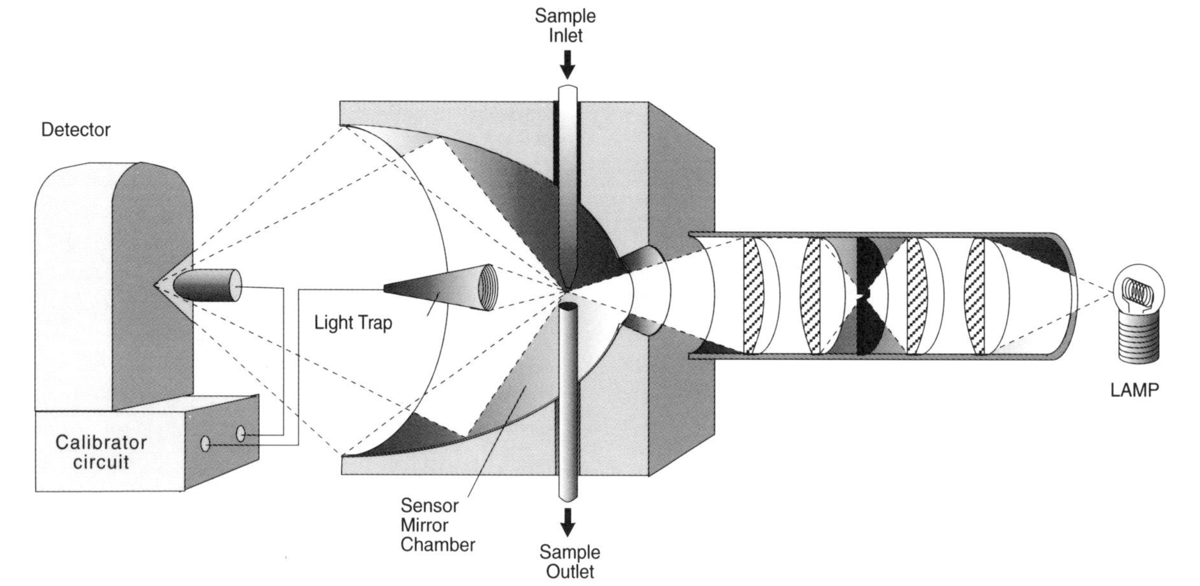

Figure 9.9. Forward light-scattering optics of TSI LPC 3753.

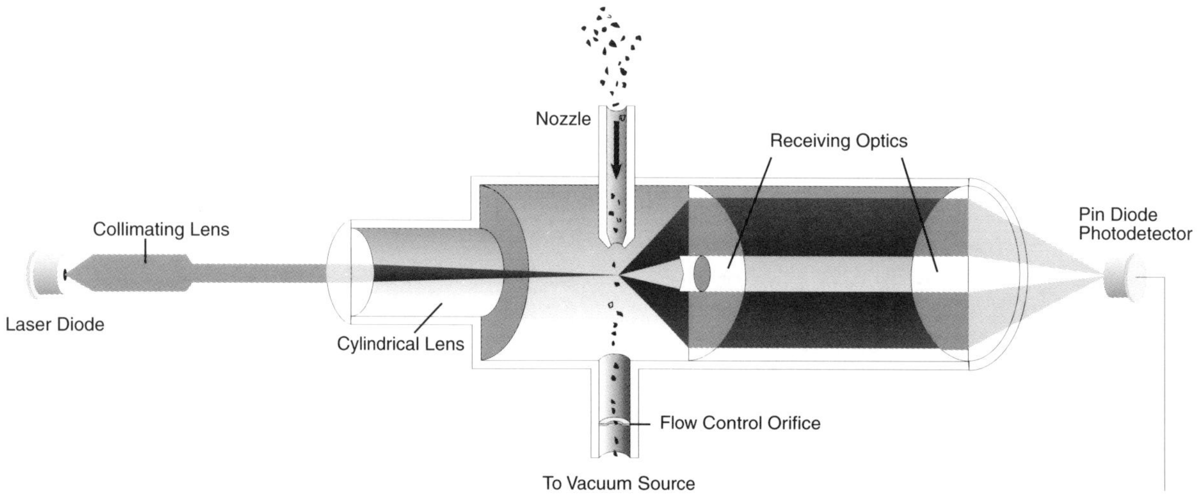

wide range of sizes, typically from about 0.03 µm to 3 µm, packaged in aqueous solutions, and easy to aerosolize. Other sources of monodispersed calibration particles are differential mobility analyzers (DMAs) that can be used to generate liquid or solid, nonspherical, particles in the size range below about 1 µm and vibrating orifice aerosol generators (VOAGs) used above about 2 µm. The papers by Berglund and Liu (1973), Bemer et al. (1990) and Liu et al. (1985) provide useful discussion on calibration theory and principle.

A key concern for the pharmaceutical manufacturer in upgrading OPC technology is the higher count levels that may be obtained with some current generation counters versus their older counterparts. The situation may arise whereby a new counter, applied in a Class 100 area, indicates counts approaching limits, whereas counts with an older model were well below 100 particles ≥ 0.5 µm/ft^3. The purchaser is well advised to test new units thoroughly prior to purchase and become familiar with operational principles and calibration techniques

Figure 9.10. Wide axis collection using twin collection optics and detectors (Climet Instruments).

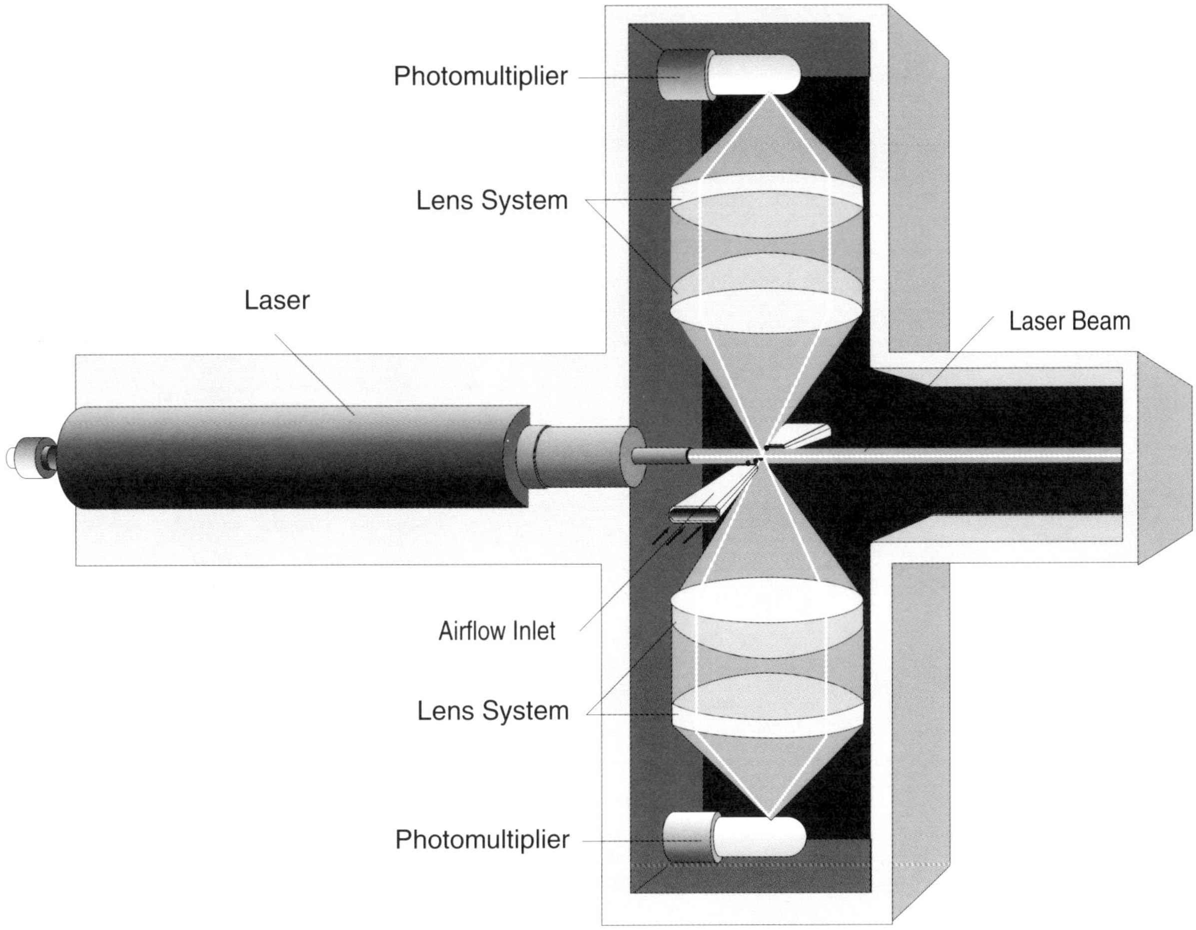

(Cooper and Grotzinger 1989). Comparison calibration procedures may also be used (ASTM F649).

When U.S. Federal Standard 209B was prepared a number of years ago, most OPCs were capable of measuring ≥ 0.5 μm particles at a sample flow rate of 1 CFM or 0.3 μm particles at 0.01–0.25 CFM. Tungsten filament illumination was used almost exclusively. No universal particle size calibration procedures were available. Each vendor used a slightly different procedure. In 1980, the American Society for Testing and Materials (ASTM) released two accepted procedures for calibrating OPCs. A primary method (ASTM F328) for counting and sizing accuracy and a secondary method (ASTM F649) for aiding interinstrument correlation were provided. ASTM F328 specifies the generation of monodisperse aerosols of latex spheres and defines the pulse height distribution produced by the counter for each size range of particles. In this way, an equivalent optical diameter base is established for the counter output. ASTM F649 specifies operation of a test counter with a standard counter, while both are measuring the same ambient air particle concentrations. The gain levels of the test counter are adjusted so that concentration data from the two counters agree reasonably well.

ASTM F328 also includes a method so that particle concentration could be compared with that developed by an independent referee method. This comparison procedure requires appreciable personnel skill and access to some specialized equipment for successful operation. For this reason, it is seldom carried out except by some manufacturers to verify the operation of an in-house standard instrument. Data from other instruments may then be compared with those from the standard. A complete revision

Figure 9.11. PMS active cavity laser sensor design with a diode array detector.

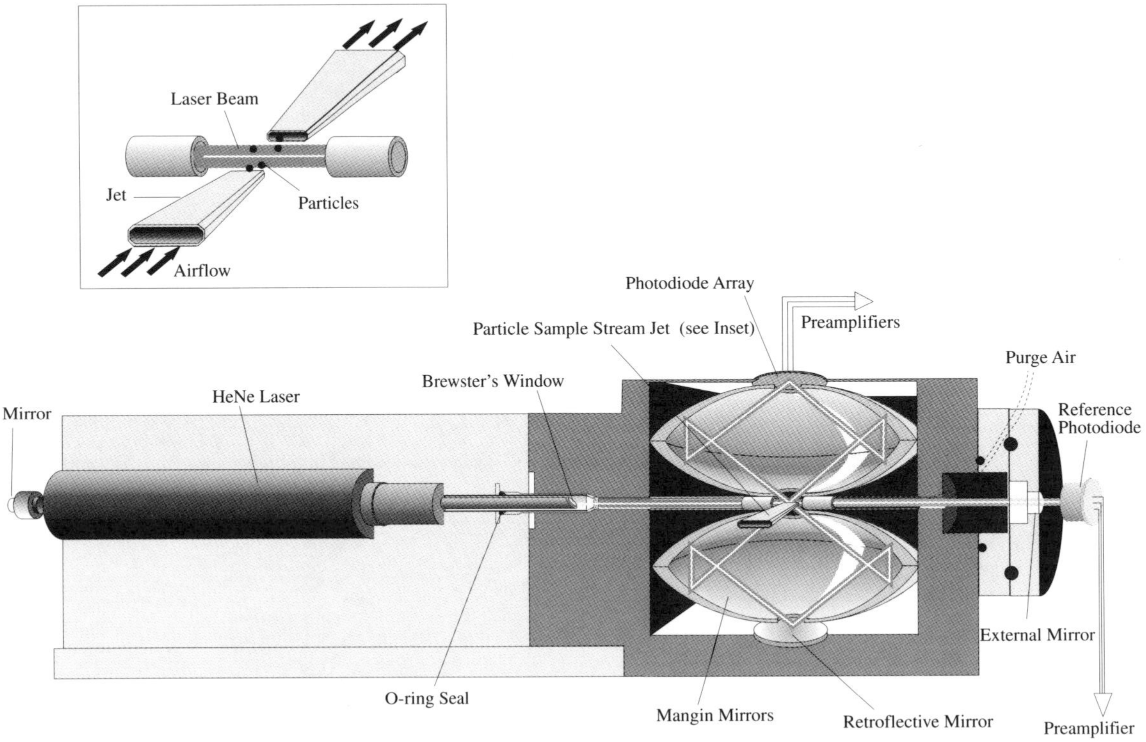

Figure 9.12. PMS diode array detection system.

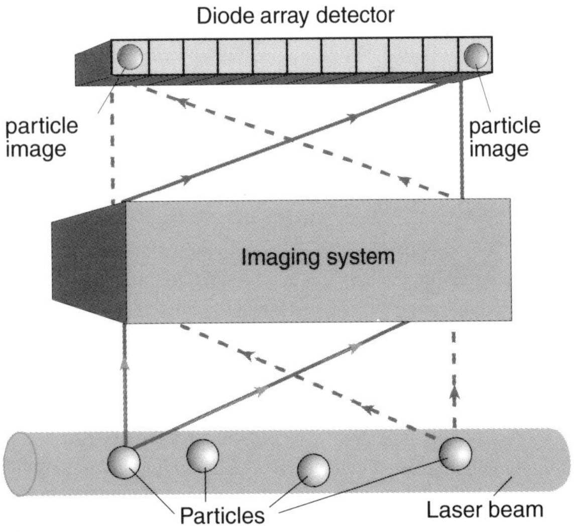

and expansion of this standard will be released in 1999.

The use of standard calibration procedures has continued, and work has been carried out to show the effects of some counter operating parameters on the performance of these instruments. In particular, the need for good sizing resolution has been shown to be important in producing more accurate concentration data at the lower limits of counter sensitivity. The means of defining counting efficiency more accurately and the requirement for an adequate signal-to-noise ratio at that lower limit have also been studied. As a result of this work, counter-to-counter correlations in ambient air of 10 percent or better are possible for the same instrument designs (Liu and Szymanski 1987). When different optical systems are used, then application of ASTM F649 is required in order to achieve acceptable intra-instrument correlations.

Vendor use of different calibration procedures, coupled with the fact that different counters have different sensitivities to particles of different sizes, shapes, and refractive indexes, can translate to a variable counter response to a given aerosol. This variability is illustrated by the data in Figure 9.13, which were obtained in the testing of four different models of 1 CFM counters from different vendors. These counts were collected in a Class 100,000 area for the purpose of statistical strength; variability of this magnitude might be more important in a Class 100 area.

Secondary (comparison) calibration of OPCs is often useful in a situation where consistency between counters, and the comparison of counters to a historical database is more important than the count levels generated by a specific counter. When the OPC is calibrated and used with only monodisperse, smooth spherical particles, then the median pulse height may be used as a reference level for sizing and reasonable correlation in counting is obtained between different OPCs. However, ambient aerosol is composed of irregular particles with variations in optical properties. Thus, the combination of effects of particle composition, particle size distribution, and instrument resolution may result in appreciable differences in the number reported by two OPCs that are calibrated in terms of sizing to an accuracy of 5 percent or better.

Figure 9.13. Variability in counts from 4 different models of counters over a 10-hour period (see also Wen and Kasper 1986)

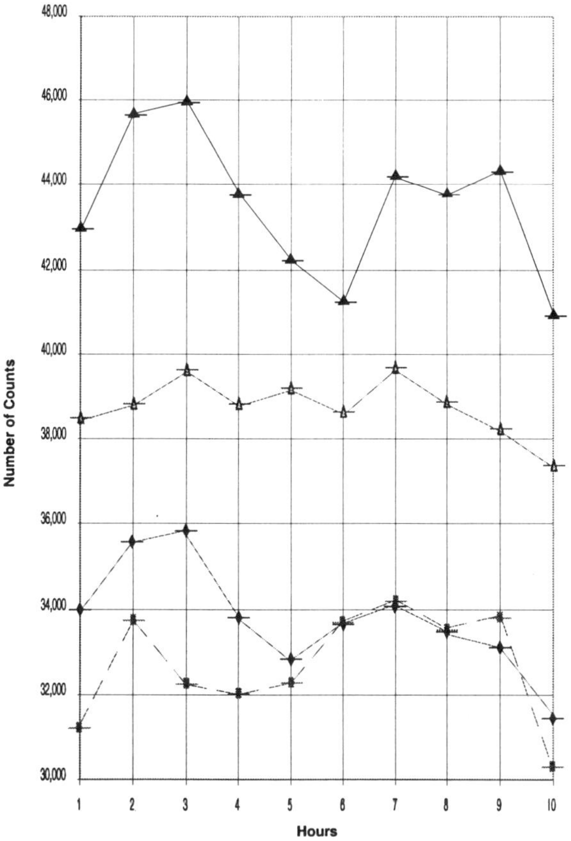

Of special note here is the new generation of small, easily portable counters that have come onto the market in recent years. These devices range in suitability for application from toys (random number generators) at the low end of the quality scale to instruments that are the equal of their larger siblings in terms of sensitivity and resolution at the upper end. The buyer must beware in this regard. Only in-use testing will allow the user to ascertain whether or not a specific model is suitable. Those choosing one of these models for purchase should ensure that count data are consistent with those from older models (perhaps through "match" calibration) and that flow rates are appropriate (e.g., 1 CFM) for optimal data collection.

PARTICLE TRANSPORT AND SAMPLE ACQUISITION

An air sample collected by an OPC should be adequately representative of all particles in the environment being sampled and be of sufficient size to contain reasonable numbers of particles for analysis. These simple sample requirements are complicated by the fact that particles are not uniformly distributed in space; particles are not uniformly entrained in gas-stream lines, and they are subject to settling and aerodynamic forces (Table 9.4).

With regard to sample transport, environmental sampling requirements can be divided into three procedural categories. (1) It is necessary to procure a sample from a static volume of air. (2) It is necessary to sample from an open duct that transports air at ambient pressures and at reasonably constant velocities. (3) It is necessary to examine air flowing from the face of a HEPA filter. If one is attempting to obtain samples only for particles 1 or 2 μm in diameter or smaller, problems with sampling efficiency, sample losses, and so on are generally negligible. For particles larger than 5 μm, however, losses on collector probe surfaces or on the walls of sample tubing, as well as losses from differential particle inertia, may be significant.

Table 9.4. Settling Rates of Airborne Particles

Diameter of Particles (μm)	Velocity of Settling	
	ft/min	in./h
0.1	0.00016	0.115
0.2	0.00036	0.259
0.4	0.0013	0.936
0.6	0.002	1.44
0.8	0.005	3.60
1.0	0.007	5.04
2.0	0.024	17.0
4.0	0.095	68.7
6.0	0.21	
8.0	0.38	
10.0	0.59	
20.0	2.4	
40.0	9.5	
60.0	21.	
80.0	37.9	
100.0	59.2	
200.0	352.0	
400.0	498.0	

Note: This table is based on "Size and Characteristics of Airborne Solids," by W. G. Frank, published in the *Smithsonian Meteorological Tables*. Rates are for spherical particles having a density of 1.0 and settling in air at a temperature of 70°F.

The first situation above relates to sample collection from areas with relatively low airflow rates, such as a room with minimal air circulation. If we consider the sample inlet as a sample point, then aerosol particles must be drawn into the inlet tube from a volume that can be considered spherical, with the tube inlet at the center of the sphere. The motion of 0.5 μm particles in still air has no discernible direction; it is essentially Brownian in nature. Common practice in still air sampling is to use a vertical probe whose inlet is directed upward. The inlet orifice diameter is approximately 1 cm, and the inlet tube edge should be as sharp as is practical. Under these conditions, particles as large as 25 μm in diameter can be sampled at rates of 2.5 to 30.0 L/min (i.e., 0.1 CFM to 1.0 CFM), and sampling errors should not exceed 10 percent.

Sampling of air flowing from ducts or from the face of a HEPA filter is in some ways easier than that for static air volumes. One simply ensures that the axis of the sampling probe is facing the flow, parallel to the flow lines within the duct or from the HEPA filter and that the velocity of the air into the sample probe is approximately equal to the duct flow velocity. In this situation, sampling is said to be isokinetic, and the sample probe collection efficiency approaches 100 percent. This raises the issue of situations in which the flow volume into the sample probe is anisokinetic, i.e. either greater than or less than that of the flow rate of the sampled air mass. FS-209E contains a considerable discussion of isokinetic and anisokinetic sampling (see also Hinds [1982], Sehmel [1970], and Willeke [1993]). Briefly, the principle of interest here is that sampling velocity and flow rate of the air being sampled must be matched if the representative nature of particle counts is to be attained. The effects of isokinetic and anisokinetic sampling probe velocities are shown in Figure 9.14. Importantly, the flow is assumed to be isoaxial, that is, the probe faces directly into the airstream being sampled.

Generally, if the sample probe velocity is subisokinetic, some of the smaller particles approaching the sample probe opening will be entrained in the streamlines passing around the probe and avoid the nozzle; undercounting of small particles and elevated counts of larger particles (3–5 μm) will result. If velocity through the probe is greater than that of the air approaching the probe, small particles will be drawn into the nozzle opening from streamlines beyond the probe opening diameter. This results in count errors whereby the proportional presence of large particles is decreased. In the opposite case, if the probe velocity is lower than ambient, the relative numbers of large particles are increased. The detrimental effects of nonisokinetic sampling pertain primarily to particle counts at larger sizes (≥ 3 μm). Probe velocity can vary significantly (20–40 percent) from that of the ambient airflow without causing appreciable errors for particles of 0.5 μm and less in size.

Obviously, a requirement for isokinetic sampling might lead to very complex manipulations if the probe velocity had to be matched to the velocity of the air being sampled in all situations. A simplified, widely used sampling technique usually gives suitably accurate results. The

Figure 9.14. Flow lines at probe nozzle for various relative sample and probe velocities.

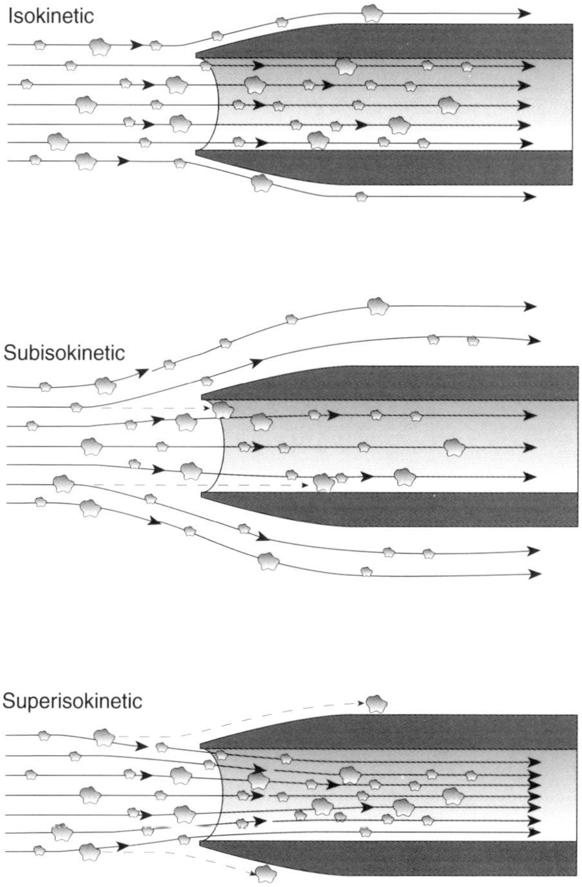

velocity of unidirectional airflow that is most often sampled is the 90 ft/min ± 20 percent that is delivered by HEPA filters. All manufacturers of OPCs market "isokinetic" probes that serve to match up the 1 CFM airflow of a counter with the velocity of HEPA airflow. These are typically conical probes of 25 to 30 mm inlet diameter that serve to collect a volume of air from the HEPA airflow that is equivalent to the counter sample flow rate and smoothly accelerate particles of all sizes into the sample tube. It is interesting to consider that when OPCs are used to sample ambient air with a probe of the same inner diameter as the sample tube, probe flow patterns are always superisokinetic.

The situation shown in Figure 9.15 is pertinent with regard to sampling of turbulent air (Agarawal and Liu 1980). In the figure, S_t is the Stokes number. The Stokes number is the

Figure 9.15. Sampling efficiency curves for various combinations of S_t and V_s/V_t.

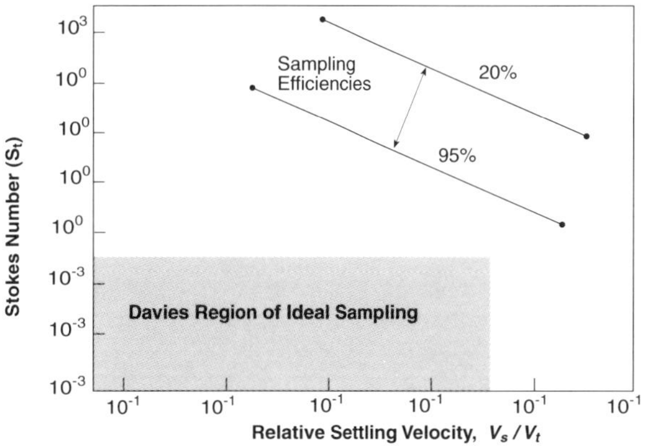

product of the particle relaxation time (time for a particle to respond to a change in air velocity and/or direction) and the particle velocity divided by the diameter of the sample tube inlet:

$$S_t = tV/R \qquad \text{(eq. 4)}$$

where t is the particle relaxation time, V is the free stream gas velocity, and R is the radius of the probe.

If the relative settling velocity of a particle (V_s/V_t, where V_s is the particle settling velocity and V_t is the inlet sample tube velocity) is plotted against the Stokes number, a family of sampling efficiency curves can be developed for particles with specific characteristics (Stokes numbers). Generally, as the probe radius increases and particle relaxation time and velocity decrease, sampling efficiency increases such that representative sampling can be carried out over a wide range of V_s/V_t ratios. As described by Davies (1964), a zone of "perfect" sampling exists for V_s/V_t ratios of 10^{-5} to 0.004 and S_t numbers of 10^{-4} to 0.032.

Transport of Particles in Tubing

The use of tubing to transport samples to OPCs from remote sample sites within a cleanroom has historically raised concerns regarding particle loss and/or segregation (Bergin 1987; Fissan and Schwientek 1987) and may be a concern with regulatory auditors as well. The Reynolds number is a dimensionless descriptor of flow characteristics in tubing, piping, or flow channels. It is calculated for airflow in tubing as

$$R = \frac{VDp}{u} \qquad \text{(eq. 5)}$$

where V is the tubing airflow velocity, D is the diameter, p is the density of air, and u is the viscosity of air. Representative values of R for different flow rates and tubing diameters are shown in Table 9.5. Generally, Reynolds numbers in excess of 2,000 are deemed sufficient to produce turbulent flow characteristics in the flow channel; turbulent flow minimizes the possibility of particle loss from the sample air volume.

FS-209E indicates that both the probe and transit tube should be configured so that the

Table 9.5. Kinetics of Particle Transport in Tubing (Courtesy of Climet Instruments Inc.)

Flow Rate (CFM)	Tube ID (in.)	Tube Length (ft)	Velocity (ft/sec)	Residence Time* (sec)	Reynolds Number
0.1	0.125	25	19.56	1.28	1,274.88
0.25	0.125	25	48.89	0.51	3,187.20
1	0.125	25	195.57	0.13	12,748.78
0.1	0.25	25	4.89	5.11	637.44
0.25	0.25	25	12.22	2.05	1,593.60
1	0.25	25	48.89	0.51	6,374.39
0.1	0.375	25	2.17	11.50	424.96
0.25	0.375	25	5.43	4.60	1,062.40
1	0.375	25	21.73	1.15	4,249.59
0.1	0.125	50	19.56	2.56	1,274.88
0.25	0.125	50	48.89	1.02	3,187.20
1	0.125	50	195.57	0.26	12,748.78
0.1	0.25	50	4.89	10.23	637.44
0.25	0.25	50	12.22	4.09	1,593.60
1	0.25	50	48.89	1.02	6,374.39
0.1	0.375	50	2.17	23.01	424.96
0.25	0.375	50	5.43	9.20	1,062.40
1	0.375	50	21.73	2.30	4,249.59
0.1	0.125	100	19.66	5.11	1,274.88
0.25	0.125	100	48.89	2.05	3,187.20
1	0.125	100	195.57	0.51	12,748.78
0.1	0.25	100	4.89	20.45	637.44
0.25	0.25	100	12.22	8.18	1,593.60
1	0.25	100	48.89	2.05	6,374.39
0.1	0.375	100	2.17	46.02	424.96
0.25	0.375	100	5.43	18.41	1,062.40
1	0.375	100	21.75	4.60	4,249.59

*FS-209D specifies a residence time of 5 sec or less; this requirement was deleted in FS-209E.

Reynolds number is between 5,000 and 25,000. For particles in the range of 0.1 µm to 1 µm and for a flow rate of 28.3 L/min (1.0 CFM), a sampling tube of up to 30 m long may be used per the standard; for particles in the range of 2 µm to 10 µm, the tube should be no longer than 3 m. Under these conditions, losses of small particles by diffusion and of large particles by sedimentation and impaction are predicted to be no more than 5 percent during transit through the tube. This transport efficiency is adequate for most situations that will be encountered in airborne particle monitoring in pharmaceutical production.

Note in Table 9.5 that the Reynolds number decreases by a factor of nearly 3 as the inside tubing diameter increases by a factor of about 2.5; at the same time, the residence time per unit length of tube increases by a factor of almost 6 (inversely proportional to gas velocity). To ensure good collection and transport, sample probes should be designed for isokinetic sampling, and the sample tubing should have minimum radii of curvature in transporting the particles from the point of collection to the counter. Additionally, the sample line should be as short as possible so as to minimize transit time, wall loss, and pressure drop within the tubing.

Two factors (static charge attraction and inertial impaction) are responsible for most particle loss in transport tubing (Liu et al. 1985). Losses in 150 ft of 3/8 in. flexible tubing at an airflow rate of 1.0 CFM have been empirically shown to be 4.2 percent at 0.5 µm and much greater (25 percent) in the 3.0 µm to 5.0 µm range (Zweers 1985). Tubing losses generally remain constant throughout the working life of the tubing, however, and do not vary nearly as widely as do OPC electronic sensor outputs over time. Particle losses at 5.0 µm in tubing are large, but the measurement error is not significant in the light of the poor statistical validity of the few 5.0 µm particles present in the usual 1 min/ft^3 sample. Particle problems are almost invariably indicated by the generation of 0.5 µm or smaller particles.

Flow rate variation may occur because of differing pressure drops in different lengths of tubing (Table 9.6). By minimizing variation in tubing length, the facility planner can effectively minimize differences in pressure drop and sample volume. This effect is generally not significant. For instance, between a 5 ft length of 1/4 in. tubing and a 50 ft length of 5/16 in. tubing, the pressure differential is only about 0.44 psi. The resultant airflow will be on the order of ±0.02 CFM or 2 percent at 1.0 CFM. Again, such variances in airflow are generally not significant when related to normal variances between individual electronic OPC sensors that are often 10 percent or greater.

Another parameter of interest to the facility planner is the particle residence time in tubing (Table 9.5). The sample tube volume can be quickly calculated for any tube/diameter combination. For example, calculations at 1 CFM airflow on a 100 foot length of 3/8 in. diameter tubing show the tube to be completely cleared in less than 5 sec. Manifold systems should, therefore, include a short delay time (5 sec as a minimum in the preceding example) to purge the tube

Table 9.6. Flow Parameters: Effects of Sample Line Diameter (flow rate = 1.0 CFM, or 28.3 L/min)

Inside Diameter	Reynolds Number	Pressure Loss (psi/m)	Gas Velocity (m/sec)
4 mm	9,150	0.980	40.35
5 mm	7,360	0.340	25.90
6 mm	6,130	0.150	18.00
1/4 in.	5,780	0.110	16.00
7 mm	5,250	0.070	13.20
5/16 in.	4,585	0.040	10.10
9 mm	4,070	0.020	8.00
3/8 in.	3,365	0.016	7.20
10 mm	3,670	0.013	6.50

of the prior sample. Commonly, 2–3 volumes are purged to ensure a representative sample.

The designer of a remote sampling system will also need to consider the physical aspects of the transport tubing to include materials, installation attitude, port placement, bend radius, and cost. Table 9.7 lists commonly used materials in ascending order of relative particle loss (i.e., stainless steel, minimum losses; aluminum, maximum losses). One cost-effective, efficient, and easy-to-install material is polyester; conductive carbon-impregnated ethylene vinyl acetate (EVA) tubing lined with polyethylene is also available. Suppliers of this type of flexible tubing include Dupont Chemical (Wilmington, Del.) and Thermoplastic Scientifics, Inc. (Warren, N.J.) under the tradename Bev-A-Line. Stainless steel, at about 10 times the cost per foot of polymer tubing (not including installation) will reduce losses at 0.5 μm by only a fraction of a percent. However, it may be the material of choice in cleanroom situations utilizing steam or chemical sterilization techniques. Flexible tubing, of course, allows the planner more latitude to reposition the tubes to meet changing manufacturing needs. The user of polyvinyl chloride (PVC) tubing should keep in mind that heating tubing can result in the escape of plasticizer vapor, which will impact particle count data at the 0.5 μm size and may result in the inner surface of the tubing becoming "sticky."

Table 9.7. Transport Tubing Material Selection (Lowest to Highest Particle Loss)

Lowest	Stainless steel
↓	Conductive plastic
	Polyester
	Polyurethane
	Polyethylene
	Copper
	Glass
	Nylon
	Tygon (PVC)
Highest	Aluminum

MONITORING METHODS

When the U.S. GMPs were first released in 1976, there was some ambiguity in the statements describing sampling methodology both for verification and monitoring-type activities as specified in FS-209B. Sample size or sampling frequency as a function of airflow capacity was not defined. Sample acquisition procedures were not considered. Sample probe inlet and transport line system particle losses were not mentioned, even though aerosol physics publications had pointed out sampling efficiency and line loss considerations. Calibration was based on individual instrument manufacturer procedures, resulting in uncontrolled variability in data when a variety of instruments were used in the same area. Measurement of particle content in compressed gases was required, but there were no instruments capable of measuring at line pressure. Satisfactory devices nor acceptable procedures were available for reducing line pressure to the point where instruments of the time could operate.

Today, many of these problems have been resolved. FS-209 was revised in 1992 as Version E. Cleanroom sample size, location, and number for verification of the area classification level are now defined more clearly in this document. ISO 14644 also contains clear specifications as to sample number and count criteria as well as metrology and monitoring information. Clean area sample acquisition procedures are also defined in this standard. A monitoring protocol is required for subsequent operation. Calibration and operation methods for particle counters are defined in ASTM procedures, and the quality of size calibration materials is excellent. Relative standard deviations (RSDs) in the single digit area are attainable, and traceability to NIST standards is available. Operating pressure reduction devices have been developed that will allow counting and sizing of particles in pressurized gas lines.

There are three general methods of airborne particle monitoring applied in cleanrooms or CEAs practiced today:

1. Mobile OPCs mounted on "carts" that are rolled from point to point by a technician
2. Dedicated remote facility monitoring systems (FMS) combined with individual counters at critical points of use
3. Totally centralized FMS with remote sensors or manifold type sampling devices interfaced to computers

Methods (2) and (3) are the most commonly used in critical applications such as aseptic processing areas.

Stand-alone OPCs comprise individual portable units that are used to acquire periodic samples at different locations. This is usually a labor-intensive process for obtaining, recording, and analyzing the data. This procedure can also introduce contamination into an area during production because an operator must be present to obtain air samples and also because the counters contain contaminating internal pumps, fans, and printers. The FMS can provide real-time data regarding high count levels, so that corrective action can be taken before the room or process must be shut down, and costly product is lost. A centralized FMS is particularly cost-effective when large numbers of locations must be monitored. Whether or not an FMS should be considered for purchase depends on the risk (cost) of product exposure and the type and level of monitoring required.

The advantages of a remote FMS generally include the following:

- Improved product yield
- Lower reprocessing costs
- Reduced manpower costs for data collection/monitoring
- Reduced manpower required for data interpretation/analysis/recording

For electronics and aerospace facilities, for example, an automated FMS can produce the desired product improvements and savings and is easily justified. For pharmaceutical facilities where other (manual) techniques must be used in conjunction with particle monitoring (i.e., microbiological, viable), justification for an FMS is often based almost solely on the reduced manpower costs. In some cases, this has not been a sufficient justification. In order to justify the cost of an FMS, a consideration must be made of the various forms of monitoring possible.

With regard to an FMS, remote monitoring involves several possible approaches (Figure 9.16). One approach often taken is to use one particle counter that is connected to several sampling locations with transport tubing (i.e., a sampling manifold) (Figure 9.16a). Samples are drawn to the particle counter, and a controller determines the sequence and duration of the sampling locations. Tubing not in use may be kept clean by "flushing" with air at an elevated

Figure 9.16a. Remote particle counting system: Aerosol manifold. (Courtesy of Particle Measuring Systems, Inc.)

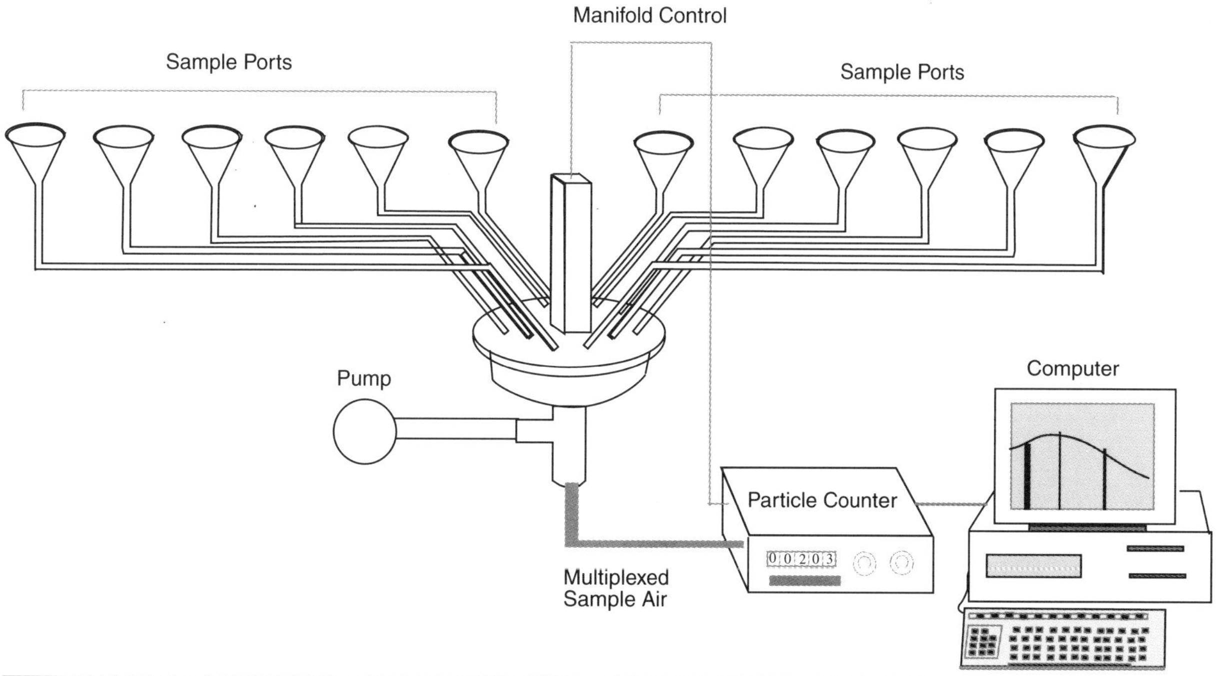

flow rate. Because sampling is not continuous, high count events may go unnoticed in some locations. Another problem with this type of monitoring is that particles must be brought to the detector via tubing, and transport losses as discussed earlier may occur. This method is generally termed *remote sampling* or *manifold sampling.*

Another approach, called *multipoint monitoring* or *point-of-use monitoring,* is to place individual sensors at every critical production location (Figure 9.16b). Sampling at every location is continuous, and data are sent to a control computer for display, analysis, and storage. This type of monitoring system gives comprehensive knowledge of the entire production environment. Although pharmaceutical manufacturers may develop their own daily sampling procedure, continuous multipoint monitoring offers several distinct advantages over the other methods. Because cleanroom personnel are a significant source of contamination, automated systems are advantageous over manual sampling procedures. Since immediate alarms from all locations can limit the size of quarantined lots and reduce waste, continuous multipoint systems are preferred over sequential manifold systems.

In an infrequently applied FMS approach to the problem of detecting count variation and high counts, two particle counters are used (Figure 9.16c). One measures an aggregate sample from a number of sample tubes, and the other measures the particle concentration from a specific tube. The claimed advantage of this approach is that the mixed sample can be used to detect a problem; the specific location can then be found by testing specific sample tubes. The obvious detractor here is the fact that if many sites are being sampled, an excursion in one area may not affect the aggregate count and may not be elevated enough to trigger a search.

The selection of particle counting equipment for facility monitoring applications obviously requires careful consideration of a number of instrument and area dependent functions. Key parameters are number of sample points (to be discussed later in this chapter), sensitivity, sample flow rate, background noise counts, and system response time. Obviously, the counter must have adequate sensitivity at the lowest particle size to be counted. The second parameter in collection efficiency is the flow rate. An instrument with a flow rate of 1 CFM will sample air (and hence collect particle count data) at 10 times the rate of an instrument with a flow rate of 0.1 CFM. When the effects of sensitivity and flow rate are combined, the instruments offered will be found to have a wide range of data

Figure 9.16b. Remote particle counting system: Multiplexed sensors. (Courtesy of Particle Measuring Systems, Inc.)

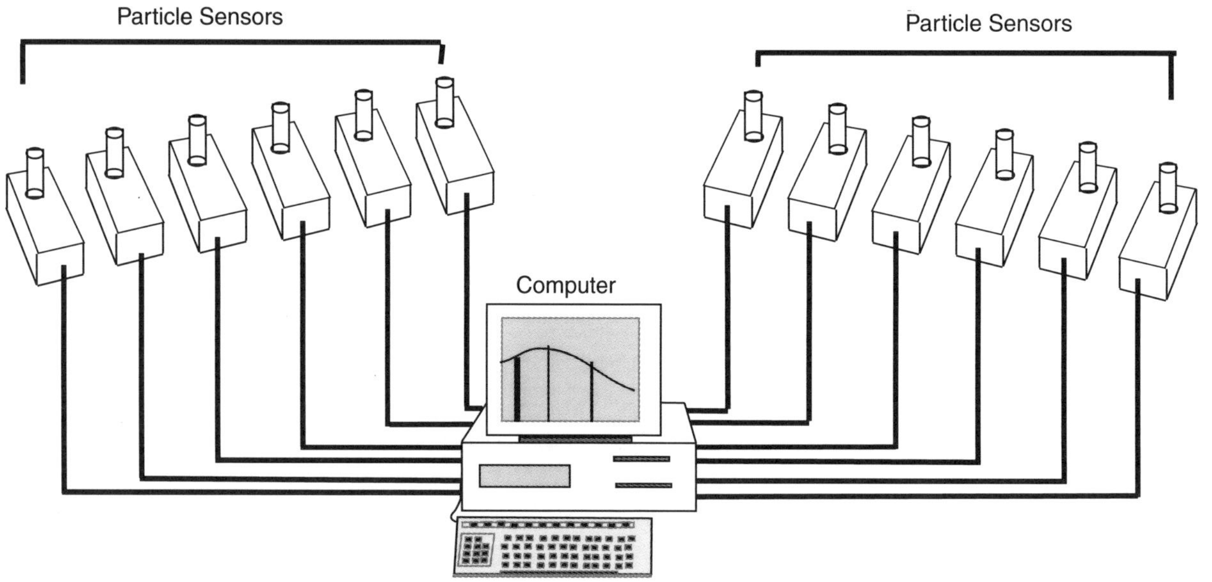

Figure 9.16c. Remote particle counting system: Continous/multiplexed aerosol manifold. (Courtesy of Particle Measuring Systems, Inc.)

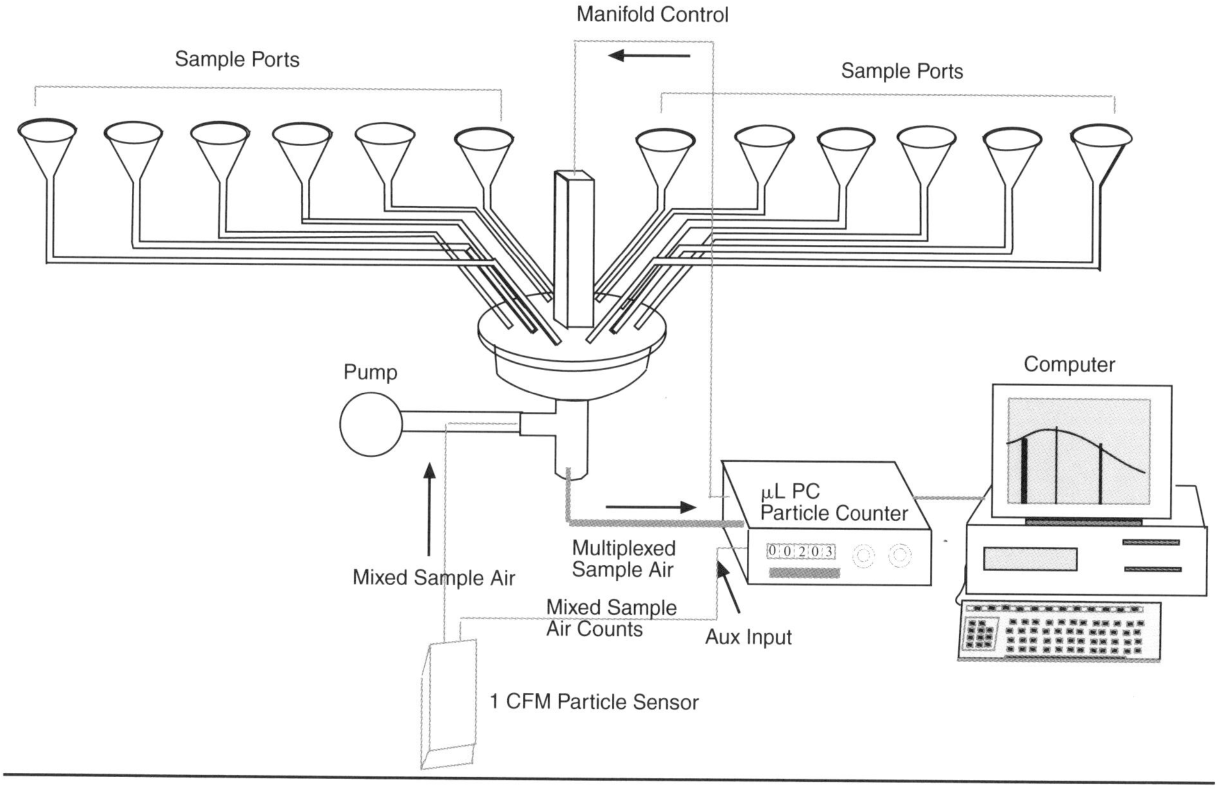

collection efficiencies. A counter with a sensitivity of 0.1 μm and a flow rate of 1 CFM has a data collection efficiency of about 350 times that of one with a sensitivity of 0.5 μm and a flow rate of 0.1 CFM, based on the ratio of occurrence of particles at different sizes (Table 9.8).

To make a decision about what sensitivity and flow rate to select, we must go one step further. FS-209E has a table that contains the minimum sampling volumes required when sampling at various sensitivities. Part of that table is reproduced in Table 9.9. The units of the numbers inside the table indicate cubic feet of air in the required sample volume to contain 20 particles at the class limits.

FS-209E requires that class levels be defined within an upper confidence limit of 95 percent. For measurements where particle counts have values less than approximately 20, the number of standard errors that must be added to the mean value of the measurements becomes excessive. For values of 20 or higher, less than 2 standard errors are added to the mean. Since standard error is a direct function of the square root of the mean, the precision of the data is poor for small counts. The standard sample interval used by most FMS software is 1 min, although sample intervals ranging from 10 sec to 10 min may be

Table 9.8. Collection Efficiency and Count Number Versus Particle Size

Measured Particle Size (μm)	0.1	0.2	0.3	0.5
Ratio	35	7.5	3	1

Table 9.9. Required Sample Volumes

	Measured Particle Size (μm)			
Class	**0.1**	**0.2**	**0.3**	**0.5**
1	0.6	3.0	7.0	20.0
10	0.1	0.3	0.7	2.0
100	NA	0.1	0.1	0.2

chosen. To provide an example, suppose a Class 100 area is being monitored and one decides to collect 20 particles in 1 min to obtain both adequate statistical significance and good response time. From Table 9.9, a sensitivity of 0.5 μm and a flow rate of 1 CFM yields a sampling time of 0.2 min, which is less than 1 min, but meets the requirements. Note that selecting a sensor with a flow rate of 0.1 CFM and a sensitivity of 0.5 μm would require 20 min to gather 20 particles in a Class 10 area.

REMOTE SYSTEMS ANALYSIS—GENERAL

The following considerations are involved in the decision to purchase a remote system:

- Does the system work in combination with existing particle counters?
- Can it use the features of existing counters or sensors (e.g., temperature, humidity), as well as those being added to the system?
- How are the data used and archived?
- Capital equipment needed.
- Installation costs.
- Training requirements.
- Maintenance/calibration expense.
- Operational factors.
- Manpower/personnel cost savings.
- Cost of handling data (routine and alarm).
- Cost of the lack of timely availability of data (time to obtain needed information).
- Cost of maintaining data (additional Standard Operating Procedures [SOPs] needed, file storage, servers).
- Cost of maintaining system security.

Cost Considerations: Manual Versus Automated Monitoring

The cost of a mobile cart–mounted particle counter is difficult to measure. The following components figure into the total cost:

- The hourly technician time to perform the tests at a given point.
- Each point monitored.
- The frequency at which each point is monitored.
- Points requiring different length sample times or special equipment arrangements, such as environmental probe readings (temperature, humidity, etc.).
- The time spent to prepare the equipment for the testing.
- The time spent to move the equipment to and between test points.
- The time spent setting up the equipment at each point, then taking it down.
- The man-hours lost while waiting for sample acquisition at each point.
- The time spent decontaminating the equipment for use inside an aseptic area.
- The time spent for the technician to gown and move the cart into the cleanroom.
- The time spent for the technician to degown once the room data are obtained.
- The time required for someone to reformat the data from paper tape into data files.
- The time spent to transcribe the data (without errors), or to type it into a spreadsheet (also without errors).
- For those facilities just starting out, the time needed to determine where to monitor, where the sample points should be, and how frequently to take samples at each one.
- The time required to reduce and collate data to draw conclusions.
- The cost of product scrapped due to failure to promptly detect out-of-limits conditions.

Validation of Remote Sampling or Counting

The following constitutes a partial list of considerations that may be involved in the validation of a remote monitoring system:

- What equipment and/or methods are currently being used to monitor the areas for contamination and environmental conditions?

- Who will receive the routine monitoring data/information/reports?
- How does he or she use the data collected (daily/weekly reports, etc.)?
- Why were these methods and/or techniques chosen?
- What class level (FS-209E, etc.) of cleanliness is needed?
- Is there a listing of sample points monitored and why (Grotzinger and Cooper 1992)?
- What are the limits of contamination and the guidelines for establishing them?
- What are the environmental conditions for each location?
- What is the rationale for the monitoring plan as it relates to the following?
  - Locations
  - Frequency
  - Ease of access
  - Statistics
  - Trend analysis
  - Interference with operations
  - Data reports to supervisors
  - Alarm response/acknowledgment
  - Action items resulting from alarms—follow-up
  - Independent particle generator studies (i.e., impact on routine)
  - Monitoring
- What effect will the above considerations have on the choice of particle counting hardware and/or the FMS?
- What type of work is currently being done in the room or area?
- How many locations must be monitored? Why?
- Is there a location qualification form for each monitored location?
- What individuals review the form and approve the monitoring method and frequency?
- Who will be reviewing the data?
- How will personnel determine out-of-limits count levels?
- What provisions are maintained for alarm response and follow-up?
- What determines when the room/area is again usable after out of limits?
- What particle size should be used for routine monitoring versus certification validation?
- Once the readings are taken, what form are the data put in?
- What does the user do with the data once obtained?
  - Routine acceptable readings
  - Alarm or out-of-limit readings
- What does the user do when an out-of-limit reading is obtained?
  - Action required
  - How does he or she know what to do?
- How are these limits set?
- What personnel are/were involved in establishing them?
- When an alarm reading is obtained, how many additional readings are to be taken to verify the condition indicated?
  - For how long?
  - What specifications must be met? Are they voluntary or compendial?
- What calibration methods are used?
- How are count data transferred from counter to computer verified (data loss, misidentification or corruption of data)?
- Who performs qualification of installation, operation, and performance? How is it done?
- How is the remote count for a sample point validated as being the actual count at that point?
- How is sample transport validated (no particle loss)?

The last two items are extremely important to the FDA. Field auditors may be expected to question sample transport on any system involving tubing (Munson and Sorenson 1990).

One means of validating transport and data transfer that has been employed is the use of a stand-alone counter next to a remote sensor or sample port and the simultaneous collection of of samples for the comparison of data. The two counts by the two counters will not agree exactly but can be statistically shown not to be significantly different with a high level of confidence, thus showing equivalency. Similarly, installation, performance, and operational qualification must be addressed.

SELECTION OF SAMPLING PLANS

In the previous sections, the author has discussed the effect of instrumental variables and sample transport on count data. To summarize, we expect OPCs of different types to count differently based on differences in optical properties of the particles counted as well as their differing modes of operation. Sample transport resulting in representative samples is also critical. Once these issues have been dealt with, the third component in our deliberations becomes the choice of appropriate sample plans. Locations of sample sites, numbers of samples to be taken, and the timing and duration of sampling all must be considered under the general heading of sample plans and/or sampling theory. References here are critical, probably more so than with any often subject we have discussed. In this regard, the author favors the various discussions by two authors: Bzik (1985, 1986, 1988) and Cooper (1990). Other papers coauthored by Cooper (Cooper and Grotzinger [1989], Cooper and Millholland [1991], and Cooper et al. [1990]) are also useful and have historically served as references for those devising sampling plans for historically pharmaceutical or medical device manufacturing areas. The discussions presented by Bzik (see especially the 1988 reference) is, in the author's humble opinion, the most readily understandable of sampling statistics available for particle counting. The author is indebted to Mr. Bzik for furnishing the information that is the basis for the discussion in this section (Bzik 1994).

The first and foremost consideration is deciding where to sample. One of the most glaring differences between the philosophy of FS-209E and ISO 14644-1 and that of the pharmaceutical manufacturer is the "uniform" placement of sample sites specified in these standards. The intent of these standards is to determine whether or not uniform air cleanliness exists across the area in question. This is as it should be for umbrella standards, but "uniform" placement does not meet our specialized interests. The pharmaceutical quality management individual in charge of monitoring will want to know if elevated counts occur in any critical area of the process, which might indicate that protection of the product was compromised. In this regard, samples are taken at locations where high risk to product might occur (e.g., in unstoppered vials on a feedway), where human activity might cause particle release, or at entrances to the aseptic site where particles might enter from a less clean area. Thus, sample sites are most often carefully matched to points in the process area that are sensitive in one way or another. Baseline levels of counts in these different areas may differ extensively.

The whole ethic of monitoring involves comparison of data: data gathered with different times during a fill run, data gathered with different levels of activity, counts during line intervention versus those collected when the line is running smoothly, and counts collected for specific sites during different process runs. Obviously, before data are compared, we must ensure that technical variables are controlled to the extent that the sample (the aerosol) will be the only significant one remaining. From the preceding discussion, we know two different OPCs must be matched to give the same results, even with standard PSL calibration particles. If this is not done, a native aerosol with its particles of different refractive indices and shapes may produce widely differing responses from different optical systems. Samples must be large enough so that variability, either as a result of short sample duration or of low flow rate, is limited. Unless all of these factors are addressed, the variability between sample air volumes will result in undesirably wide differences between measurements. Statistical variability based on the number of counts recorded is always an unavoidable issue.

Let's assume that all mechanical or instrumental variables have been eliminated to the maximum extent possible, and that sample sites have been placed in locations that are critical to defining process protection. Areas such as those in immediate proximity to particle generating devices (e.g., a powder filler, a turntable, or perhaps a feeder bowl or stoppering machine) that will reflect counts due to process function have been thoroughly tested for cleanliness in either the as-built or at-rest states (see ISO 14644-1 or FS-209E) without the particle-generating mechanism present. At this point, our statistical considerations come into play.

The statistical strength of a particle number estimate is for airborne particles commonly defined on the basis of a Poisson distribution, also called a spatially random distribution or a distribution of rare events. One property of this distribution is the equality between the variance and the mean. The standard deviation of the particle count thus is set equal to the square root of the mean; that is, a count of 100 particles has a relative standard deviation of 10 percent; a count of 1,000 particles has a relative standard deviation of 3 percent; a count of 10,000 particles has a relative standard deviation of 1 percent. As with a normal distribution, values will fall within 1 standard deviation of the mean value approximately 68 percent of the time and within 2 standard deviations of the mean approximately 95 percent of the time, dependent on collecting a sufficient number of counts. The concentration of the aerosol to be sampled must, of course, always be limited below the point where optical coincidence will cause interference with correct counting and sizing (see Whitby and Liu [1967]). For most counters used for a ≥ 0.5 µm particle measurement, the concentration limit is in the range of 106 particles/ft^3. Exceeding concentration limits is thus not normally an issue; in aseptic process areas where ≥ 0.5 µm counts must be less than 100 per ft^3, the issue becomes the high variability of the data based on statistical concerns alone. Was our count of 70 particles/ft^3 ≥ 0.5 µm 70, 78, or 62? How do we apply alert limits on this sliding scale?

Particle enumeration at ≥ 0.5 µm per FS-209E is a commonly accepted practice in the United States for many applications. As discussed earlier, this standard is of limited usefulness for defining the cleanliness in a pharmaceutical cleanroom. Isokinetic sampling may or may not be used, depending on the nature of the airflow, as also discussed earlier in this chapter. Neither the microscopic-membrane technique, which is useful for particles 5 µm and larger, nor condensation nucleus counters, which are useful for counting particles 0.01 µm and larger, are suitable for counting particles in the ≥ 0.5 µm and larger range in pharmaceutical cleanrooms. If the velocity or flow direction of the air being sampled vary such that isokinetic sampling cannot be performed, the user generally assumes that the airflow velocity is varying in a random fashion and that the only nonrandom factor remaining that will affect particle motion is the effect of gravity. In this situation, the sample probe may simply be a tube with a diameter equal to the inner diameter of the sample line to the OPC that is pointed upward. (Recall our earlier discussion of probe shape and inlet velocities.)

The pharmaceutical application of cleanroom classification is generally fairly uniform between companies. Aseptic production areas are invariably defined as a Class 100 areas per FS-209E, where concentrations of particles ≥ 0.5 µm are not greater than 100/ft^3 of air. Generally, these areas are operated with average airborne particle concentrations far lower than the Class 100 limit. For some applications, clean areas or clean zones may be enclosed within areas that are less clean than the Class 100 level. These areas may be Class 1,000 or Class 10,000. Air classification levels are established by particle counting in accordance with valid sampling techniques and measurement plans per a local SOP. Terminally sterilized LVI product is often filled in a Class 10,000 area (see Chapter 3).

FS-209E specifies that for unidirectional airflow clean areas, the required number of sample points is the lesser of (1) the area of the entrance plane perpendicular to the airflow (in ft^2) divided by 25 or (b) the area of the entrance plane (in ft^2) divided by the square root of the airborne particulate cleanliness classification. For nonunidirectional airflow rooms where the mixing of air is more thorough, the number of sample locations is equal to the square feet of floor area of the clean zone divided by the square root of the airborne particulate cleanliness class designation. The sample size and number of samples are specified to permit definition of the cleanliness classification level within an upper confidence limit of 95%. Table 9.10 shows the minimum number of sample points required to verify classification of cleanrooms of various sizes at several classification levels. The required number of sample measurement points can be very large for cleanroom verification; the pharmaceutical industry has historically objected to this specification of sample points on the grounds that clean (aseptic) areas or enclosures for sterility testing may have too few

Table 9.10. Federal Standard 209E Sampling Point Number Requirements

	Minimum Number of Sampling Points for Cleanroom Classes			
Area (ft²)	**1, 10, 100**	**1,000**	**10,000**	**100,000**
100	4	3	2	2
300	8	6	2	2
400	15	13	4	2
1,000	40	32	10	3
2,000	80	63	20	6
4,000	160	126	40	13
10,000	400	316	100	32

sample points, and higher classification areas may have excessively high numbers of sample points. Witness the case of a Class 10,000 room from the table with an area of 10,000 ft^2: 100 sample points would be required, at a cost of a large number of man-hours in an area used for noncritical processes.

The number and location of sample points for monitoring as opposed to verification is not specified in FS-209E. Monitoring is simply testing at periodic intervals to demonstrate continuing compliance. The standard requires that a meaningful monitoring protocol be established and implemented during cleanroom operation, and the monitoring SOPs of most pharmaceutical manufacturers already comply with this provision. Monitoring produces information to define overall cleanliness trends in the cleanroom or CEA (once "representative" sample locations have been selected). The same type of information can be obtained by either manual sampling or simultaneous sampling at several locations as described earlier by remote counters (FMS), with information transmitted to a microcomputer.

The synopsis of this consideration is that monitoring plans are critical, having much greater impact with regard to daily operations than verification. The obvious consideration here is that verification test results have clear relationships to the in-use condition. Insofar as the regulatory agency and the manufacturer are concerned, the most important factor in airborne counting is the demonstration of appropriate cleanliness in use, with no test data exceeding specified limits.

An appropriate protocol for the validation of monitoring plans might consist of the following sequence (Bzik 1994):

1. Identify the process areas where count data are to be collected and the associated allowable count levels, based on at-rest testing and process or regulatory requirements. Decide if data test points are to be recorded continuously, when the process is operating, or when alert limits are exceeded. An FMS system for continuous data recording and reduction should be considered as outlined earlier. Importantly ISO 14644-1(2) allows less frequent recertification if continuous monitoring is carried out.

2. The specific sample locations and rationale for their selection must be defined based on airflow patterns, the nature of the work carried out in an area, process flow, and personnel and material movements. In unidirectional flow areas, sample points should be placed in close proximity to critical locations and approximately 1 ft above the work surface. Before identifying sample locations in critical areas such as filling lines, the area should be tested during placebo fills with personnel present so that the appropriate sample site locations and numbers can be empirically chosen and count levels to be expected are known. Larger areas will require higher numbers and sample points.

3. Based on this information, sample number, volume, frequency, and sequence of counting

at different locations by the OPC are then selected. Sample volumes can be based on those in FS-209E or ISO 14644-1. In aseptic fill areas, sampling plans that ensure that Class 100 cleanliness is maintained are essential. Obviously, sample frequency and sample acquisition sequence will depend on the operations in the area being tested; these may vary between sites, and a combination of manual testing and an FMS may be used.

4. Identify the particle counting system best suited for the monitoring plan. A suitable computer for the FMS system must be purchased from the vendor or from a commercial supplier. Appropriate software, generated in-house or purchased, is essential. With an FMS system, sequential or simultaneous monitoring may be chosen. Simultaneous sampling gives more complete coverage for an aseptic process but generates reams of data that may be reviewed during a regulatory audit.

5. For manual systems, it is essential that a performance qualification (PQ) be carried out to evaluate the generation of the system and area variability. With an FMS system, counts obtained with a stand-alone counter must be compared to counts from a remote sensor or sample tube at the same location. The protocol for such should take account of the level of variability between identical counters testing the same aerosol. One manufacturer conducting this exercise employed two stand-alone counters at the remote sample point and compared their counts to each other as well as to the remote system.

NUMERICAL EVALUATION OF COUNT DATA

(The author is indebted to Thomas Bzik [1988, 1994] for the following discussion of count uncertainties, which has been only slightly modified by the author).

The methods by which sets of particulate data are collected and treated affect the kinds of inferences and decisions that can subsequently be drawn from these data. Two data resolution categories, time and particle size, must be considered. Particle size resolution in data is influenced by two factors: (1) Has one recorded all of the resolution data that should be obtained from the measuring instrument? Particulate data from an OPC's multiple channels should not be aggregated before the data are stored; if this is done, discrete information on particle size distribution will be lost. Counts from each channel should be kept separate for later scrutiny. (2) This factor, which is generally more critical, centers on the issue of which particle size range to investigate. If one picks an overall range that is too broad, the difference in data based on specific size channels may be lost. Hence, some manufacturers today have abandoned counting all particles ≥ 0.5 μm and count at a number of sizes. If the range selected is too narrow, the size resolution within that range will be excellent, but the data may not cover enough sizes to be useful. The sampling strategy should thus be selected on the basis of what information is relevant.

A Poisson distribution is commonly used to describe observed steady state particle contamination variability. The Poisson distribution is useful in estimating the probability of x occurrences of an event in a specified sample volume or time interval if the following assumptions can be made: (1) the time interval or volume may be subdivided such that the probability of an occurrence in any given subinterval is small; (2) the occurrence of an interval in a subinterval does not affect the probability that one occurrence will be observed in any other subinterval; (3) the probability of a single occurrence within any subinterval is constant; and (4) the probability of more than one occurrence within a subinterval is zero.

Suppose confidence intervals are to be constructed for the true unknown average arrival rate when a process is operating in steady state. Confidence intervals for Poisson-distributed arrivals can be located such as that shown in Table 9.11. The estimated standard deviation (SD) of the count (N) is given by its square root. Suppose, then, that a single measurement is taken for some stated volume, and a count of 20 is observed. From Table 9.11, a 95 percent confidence interval for the unknown average arrival rate would be given by (12.216, 30.889). This would signify that one can be 95 percent certain, on the basis of observing a count of 20, that the true unknown average arrival range lies between 12.216 and 30.889. If this interval is unacceptably wide, the only available solution is to collect additional data by counting for longer times or counting smaller particles.

Table 9.11. Ninety-Five Percent Confidence Limits for Average Arrival Rates of Poisson-Distributed Data

Count	Low	High
1	0.025	5.572
10	4.795	18.391
20	12.216	30.889
30	20.240	42.827
40	28.576	54.469
50	37.110	65.919
60	44.913	76.106
70	54.568	88.441
80	63.435	99.567
90	72.370	110.626
100	81.363	121.627

Often a normal approximation is made to Poisson-distributed data. Typically, this is computed under the assumption that the upper and lower confidence limits could be represented by the 0.025 and 0.975 percentiles from a normal distribution, with a mean equal to the observed count and with a population SD equal to the square root of the observed count. Expressed in terms of the observed count, the normal approximation results in the following confidence interval: count – 1.96(sq. rt. [count]), count + 1.96(sq. rt. [count]). This is a very rough approximation for low levels of microcontamination.

Another common variation of the above means of calculating a confidence interval is as follows: Replace 1.96 with 2 in the normal approximation above and then round the lower confidence limit down to the next integer and the upper confidence limit up to the next integer. Continuing the previous example with an observed count of 20, this method yields a confidence interval of (11, 29) rather than the correct interval of (12.216, 30.889) from Table 9.11. This variation of the normal approximation still results in an unacceptably large error in approximating confidence intervals for count data originating from an ultraclean process, especially for small numbers of observed counts.

The close approximation to confidence limits based on a Poisson distribution that is necessary for SPC (Table 9.11) may be based on Bzik's forumlas:

$$LCL = (TC+1) - \left(\frac{1.96 \times \sqrt{TC}}{n}\right) \text{ and}$$

$$MCL = (TC+1) + \left(\frac{1.96 \times \sqrt{TC}}{n}\right)$$

where LCL is the lower confidence limit, MCL is the upper confidence limit, TC is the total count, and *n* is the number of observations represented in the total count (i.e., number of samples taken at one sample point)

In dealing with microcontamination data from a clean process, it is important to be aware of the assumptions made by the quantitative method of analysis about the statistical nature of the data. Many statistical process control (SPC) charting or analysis methods, for example, are not intended for use with count data when the observed counts are relatively small; a Gaussian distribution is also commonly used as a basis for SPC. All such assumptions must be checked; if not checked, decisions may be made on the basis of information from an incorrect or inappropriate tool.

A common error in data management is to fail to realize that when count data are rescaled, the uncertainty of these data is rescaled as well. Suppose, for example, that 0.1 ft^3 is sampled in a OPC on a process that is in steady state. The observed particle count is then rescaled by a factor of 10 to put it into reference to cubic feet. Suppose, then, that a count of 1 is observed. How well has the average particle arrival rate per cubic foot been established? Using Table 9.11, based on a count of 1, a 95 percent confidence interval for the average arrival rate per 0.1 ft^3 is given by (0.025, 5.572). The estimated count for per cubic foot is 10 × (0.025, 5,572) = (0.25, 55.72). Count data reliability must be assessed before rescaling and must then be rescaled itself by the same multiplier or divisor. If the scaled count of 10 had been used erroneously with Table 9.11 in the previous example, the uncertainty in the average arrival rate per cubic foot would have been perceived to be only (4.795, 18.391) instead of the correct (0.25, 55.72).

Measured microcontamination levels have been categorized into four states: steady, drift, stepped, and burst. These states, which describe the mode of arrival of particles over time, are applicable to microcontamination measurements taken from a cleanroom, a gas pipeline, or a cylinder.

The following sets of 10 observed counts illustrate the importance of time resolution (Bzik 1986):

- 17, 14, 12, 19, 18, 16, 20, 14, 15, 16 (steady state behavior)
- 2, 5, 10, 9, 15, 25, 22, 35, 31 (time drift behavior)
- 3, 4, 28, 26, 25, 29, 35, 4, 5, 2 (stepped behavior)
- 3, 56, 4, 2, 5, 44, 7, 3, 5, 32 (burst behavior)

If these sets of 10 counts in Bzik's example had been aggregated, the total count for each set would have been the same—161, yielding an average of 16.1 in all four cases. Four distinctly different, time-related behaviors would thus have been masked. Thus, continuous monitoring of 1 CFM samples in aseptic processing areas is emphasized.

With regard to time resolution, an OPC will record counts at some preset time interval for a fixed number of channels that corresponds to subdivisions of some preset particle-size range of interest. As mentioned earlier, the time resolution at which the data are collected is crucial to a complete analysis of microcontamination data. Central to this issue is the fact that within the recorded time interval, particle counts are aggregated. Thus, if recording intervals are long, occurrences such as bursts will often be buried by the data aggregation process. Other states can be obscured as well; for example, a time drift may appear as a step jump, and even step jumps may become disguised. Thus, the requirement for time resolution and obtaining a statistically strong number of counts are opposing concerns.

Being in steady state implies that the particle counts observed over time can be considered to be random arrivals from an unchanging, underlying distribution. Thus, in any given size range of particles under investigation, the average arrival rate for a process in steady state remains constant over the time period in question. Being in steady state does not, however, imply that the process will yield the same sizes and quantities of particles at equally spaced time intervals because there is no physical mechanism at work that uniformly produces and disperses different-size particles over a volume.

In the step jump, the level of particles produced by the process changes abruptly and then remains the same. This would result from a change to the process that remains in effect for some period of time, as would occur following damage to a HEPA filter. The burst state is a temporary phenomenon, often appearing as a spike in the microcontamination data. The opening of a valve or a sudden pressure fluctuation, for example, could cause a burst to be observed.

Figure 9.17 illustrates Bzik's four time-count states in graphical form with minor deviation, per the author's terminology. In the case of the alternate terminology, the original Bzik term is parenthetically given. The author has added 9.17e (unstable behavior), which is essentially two spike subsequences because of unpleasant real-world experiences. Per Bzik (1988), it should be noted that a relatively constant (steady state) average arrival rate per time interval was used to generate Figure 9.17a. When particles are randomly distributed over time, however, as in Figure 9.17b, one will not observe

Figure 9.17a. Count profiles from OPC monitoring: Steady state. (redrawn from Bzik 1986)

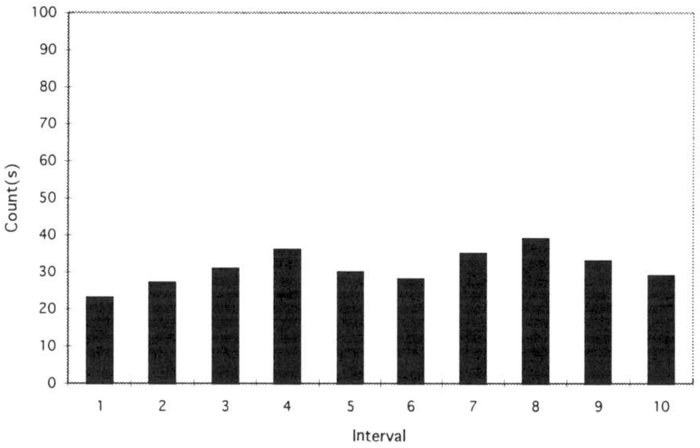

Figure 9.17b. Count profiles from OPC monitoring: Trend (time drift state). (redrawn from Bzik 1986)

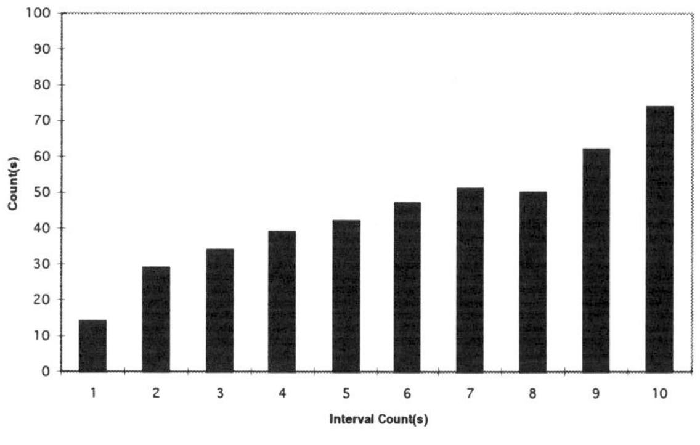

Figure 9.17c. Count profiles from OPC monitoring: Step jump state. (redrawn from Bzik 1986)

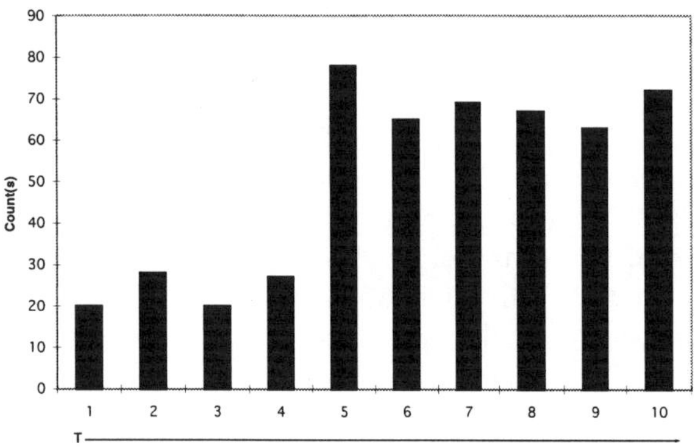

Figure 9.17d. Count profiles from OPC monitoring: Spike (burst state). (redrawn from Bzik 1986)

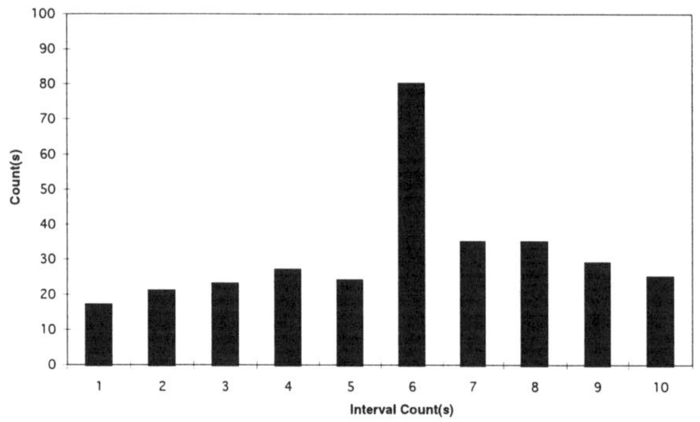

Figure 9.17e. Count profiles from OPC monitoring: Unstable condition. (redrawn from Bzik 1986)

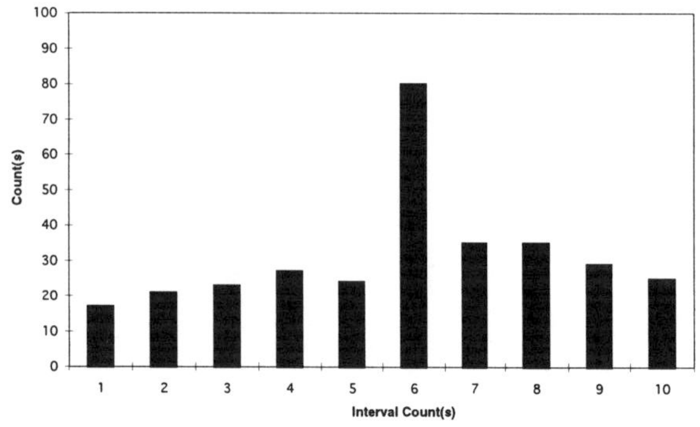

the same number of particles in each interval. Note also that Bzik's four time states represent only a rough categorization scheme; real data collected over time may reflect some or all of these behaviors.

One final observation should be made regarding Bzik's presentation (see especially the 1988 reference) of statistical principles of airborne particle count data collection. As mentioned above, one of the most serious drawbacks of the sampling plans specified in both FS-209E and ISO 14644 is the larger numbers of sample points uniformly spaced over the area to be tested that may be required. Manufacturers of medical devices and pharmaceuticals tend to focus sampling with regard to critical areas for a "clean" process rather than considering the total number of sample points; in higher classification areas that enclose less critical operations, minimal numbers of sample points below those required by the standards may be used. All samples values are typically required to be within the class limits, with no averaging permitted. Such sampling schemes, with lower numbers of sample points and single observations at each point may, in fact, be acceptable, in that they represent a bias in favor of a higher estimated count and thus against the manufacturer. Point-to-point count variability is a key consideration in this regard. Thus, as the author has advised the reader earlier on several occasions, a professional statistician should be consulted before sampling plans are implemented or revised, which also includes the use of additional sampling sites.

In summary, particulate matter data are discrete count data, a factor that strongly influences the types of statistical distributions that can reasonably represent or approximate such data. When relatively low levels of microcontamination are in question, the discrete nature of the data must be taken into account. Continuous approximating distributions, such as the normal distribution, are at their worst for low levels of microcontamination and may thus be inappropriate in this context. Differing time behaviors will also call for different statistical approaches. In general, however, relatively large quantities of data will be needed for the quantification of low levels of microcontamination.

Some additional summary considerations are the following: Data capture is, in general, an automated function on an OPC. Ideally, this capture takes place electronically, not by hand copy or printer tapes. Excessive labor and, historically, longer recording intervals have been favored for two reasons: (1) Aggregation of particle counts can average out much of the viability ("noise") inherent in the sampling and measurement of microcontamination; with longer time intervals, the relative reliability of the average particle count (for some stated unit volume and particle size range) increases. This is often made necessary by the extremely "noisy" nature of particulate data and by the processes from which these data arise. (2) If longer time intervals are used, fewer data must be processed, stored, and analyzed. While responsible, this process may be objected to by a regulatory agency as having the potential to mask jumps or spikes.

If one has a process that truly operates in a steady state, i.e., a controlled process, no price will be paid for the use of longer recording intervals. In many instances, however, processes change over time, in which case valuable information will be lost through inadequate time resolution. How, then, does one resolve the trade-off between the reliability of the long term data signal and the data resolution that one must obtain to detect and locate time-related phenomena such as burst states? It is best to use the shortest time interval that is useful in the detection of any important time-related phenomena. Using a computer, the data can also be averaged over time. A number of analyses can then be performed on a set of data: one based on short time intervals and another that aggregates some number of shorter time intervals into longer intervals to obtain more reliable long-term data signals. Short time count intervals can be averaged as part of a data analysis, but long recording intervals cannot be broken down further.

A great potential for introducing error results from the need to re-create microcontamination databases from outprints. Once count data are stored in some logical structure, a statistics package or a high-level language with statistical routines can be used to analyze, graph, and regroup data as necessary. A good package, such as the Statistical Analysis System (SAS Institute, Inc., Gary, N.C.), will save much of the cost and trouble associated with data analysis.

As described above, it is possible to establish a remote FMS network of sample acquisition locations within a cleanroom. Particle data can be obtained by using either a number of OPC sensors or by using a single sensor counter and multiplexing air sample lines to it. The statistical considerations discussed also apply here. For operations with multiple sensors, simultaneous measurement can be carried out at each sample point, but a cost problem may be present. For operations with multiple sample lines, sequential measurements are carried out at each location, and large particles may lost in sample lines, but no intersensor correlation of the count is required, and costs are appreciably lower.

Data can be recorded and processed by the controlling computer at a central location to describe particle concentrations and/or ambient temperature, relative humidity, air pressure,and so on at each point on a routine schedule or as specific operations are carried out. Trend information can be compared with the option of reporting or acting upon warning or alarm levels. The computer can also display data from a number of similar areas or can show point-to-point comparisons. This type of information is particularly useful when evaluating different process components in similar operations and may be requested in audits. At the same time, the computer can be programmed so that the monitoring data can be used with or without product information for input to either a standard or a specialized statistical quality control program.

SAMPLING OF COMPRESSED GASES

Isokinetic sampling is highly desirable if compressed air or other gases are to be sampled from a line. The historical methodology used, with an in-line pressure reducer, is less than ideal (Figure 9.18). In this situation, the sampling effectiveness (E) is defined as the percentage of the true count for a particle of a given size that is represented by the count. For this exercise, counter efficiency is assumed to be 100 percent. Count loss in various parts of the sample acquisition path may be a much more serious error here than is the case for simple tubing transport systems in Figure 9.18.

In Figure 9.18, the pressure reducer is of simple design, serving only to reduce the line pressure to ambient, so that the pump of the counter can collect an unpressurized sample. This general design incorporates a number of undesirable features, including tubing bends, a butterfly valve, and large diameter supply tubing that allows an excessive volume of compressed gas to enter the chamber, with resultant

Figure 9.18. Sampling of compressed gas line with pressure reducer.

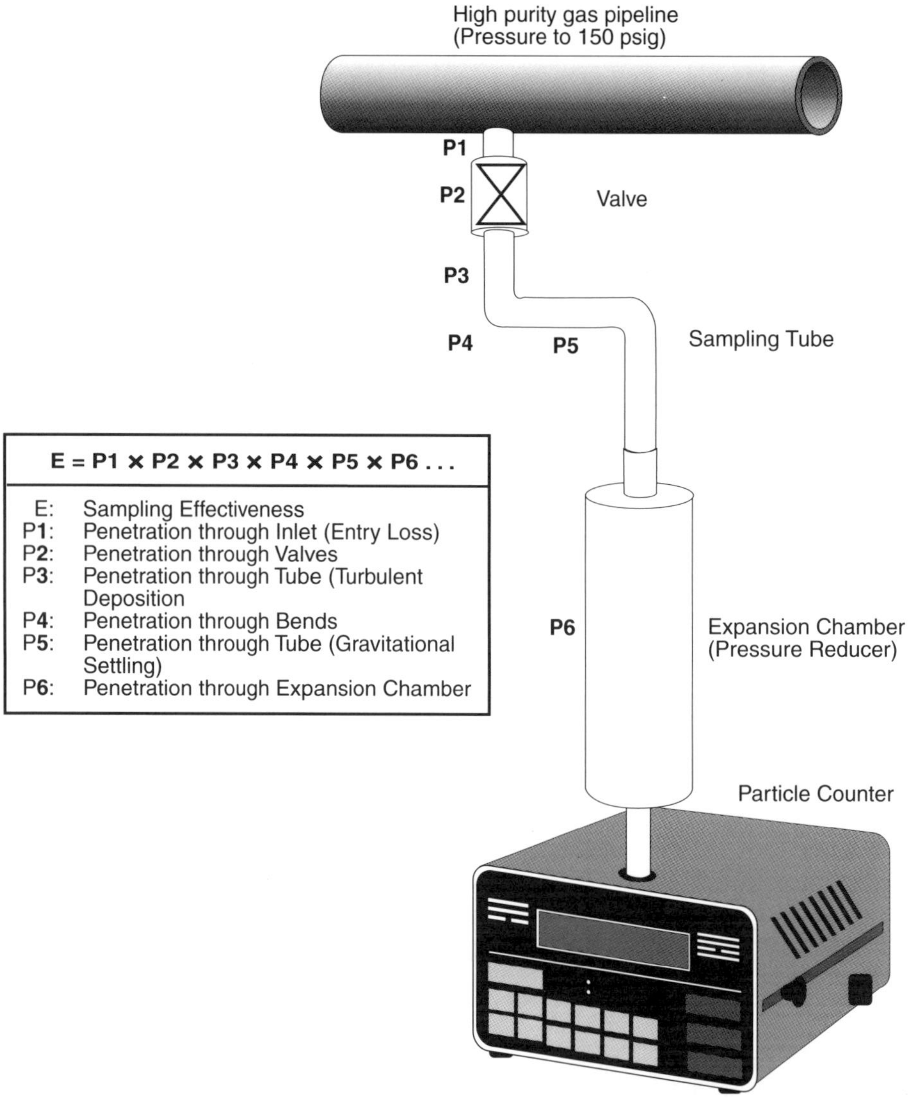

turbulent impaction and loss of particles. Valves in-line between the gas stream to be sampled can both generate and trap particles and are thus problematical. Obviously, with this type of sampling, it is essential that the sample tubing, valve, and expansion chamber are purged before counts are recorded. (Adequate purging is indicated by low variability between successive counts.) There is a potential for particle generation or entrapment counts, which may overwhelm any beneficial effects of anisokinetic sampling. Care must also be taken that ambient air is not drawn into the exhaust ports of the pressure reducer at low sample flow rates.

More refined pressure reducers incorporate a sonic flow restriction (critical orifice) at the orifice plate or a capillary orifice (Figure 9.19). The orifice-type pressure reducer makes use of a thin orifice plate to reduce the pressure while the capillary-type reducer makes use of a capillary tube. The evaluation by Rubow et al. (1990) showed that the orifice-type reducer gives both lower background counts and count loss than the capillary-type reducer, due to the much smaller surface area that the high purity gas comes into contact with during expansion through the devices. Particles are sampled by a sample extraction tube located at the axis of the chamber, with

Figure 9.19. Design concept of pressure reducers for isokinetic sampling.

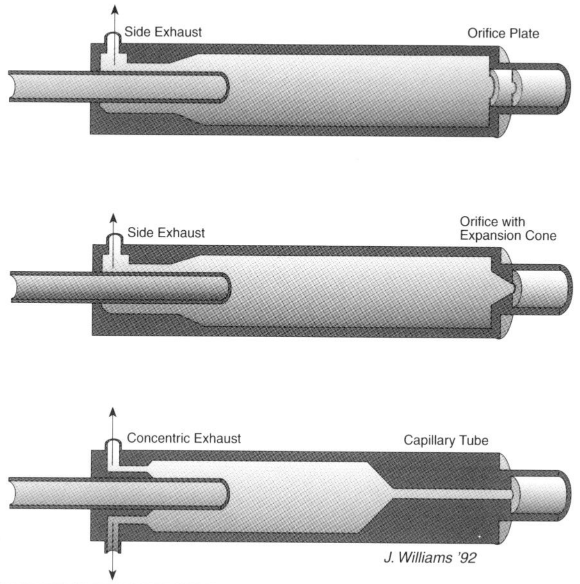

the excess gas flow exiting through a 90° bend or through a lateral buffer or holes. The pressure reducer typically reduces the absolute pressure from above 100 psi in the high purity gas line to an ambient pressure of 14.7 psi, allowing the amounts of gas or air in excess of that required by the particle counter to escape. The pressure ratio exceeds the critical ratio, and a sonic flow restriction is formed at the orifice.

The efficiency of penetration through all of these types of reducers is generally good for particles up to 10 µm in size. Since the 0.5 µm particle size is of primary concern, these reducers are entirely adequate for most pharmaceutical compressed gas sampling. One caution in the use of pressure reducers for compressed air or gas sampling regards adiabatic cooling and the condensation of moisture from the air to form water droplets. This condition should be suspected when the flow of air into the pressure reducer at full line pressure of 80 to 105 psi results in counts that are 10-fold or more higher than those observed when the pressure in the reducer is decreased by one-half or one-third. Care must also be taken to keep reducers clean, given the low count limits generally placed on compressed gases.

Both the design of the pressure reducer and the sampling point tubing geometry are critical to isokinetic sampling and the enumeration of particles in their correct representative proportion of the population. Figure 9.20 illustrates some of the sampling point considerations regarding isokinetic sampling of compressed gas lines. If carefully designed, right-angled sample collection tubes may prove suitable; if the sample is caused to exit at too high a velocity, however, significant centrifugal impaction may occur at the 90° bend. In regard to compressed air or gas lines, isokinetic sampling is a more elusive goal than in free air volumes in CEAs. The air in the line may not, in fact, be moving at a constant rate, but rather may also be subject to velocity changes due to rate of gas use. Similarly, actual pressure in the line will also be subject to variation due to the gas is being withdrawn at points of use. In some extreme cases, the only movement of the air or gas from the line may be at the inlet to the sampling point. Thus, in reality, large particles are very unlikely to be entrained in the sample stream, and small particles will be adequately withdrawn by a number of sample tube designs.

Many pharmaceutical users consider the flow kinetics and particle transport mechanics in the process gas line to be less practically important than particle sizes and numbers that exit the line through a typical point-of-use fitting, be it hose barb, quick-attach fittings, or other coupling. In this consideration, it is important that the sample exiting the point-of-use fitting be representative of gas exiting this sample point rather than particle numbers present within the line. In this context, it is more important to ensure that the sample measured from the transport line be subject to isokinetic sampling conditions free from error and artifacts than that within a line.

In consideration of a final application, it may be necessary to sample the low pressure gas volume used to apply an inert gas sparge to a mix tank of a solution to be filled, to purge containers of air before filling, or to provide a nitrogen "overlay" in the headspace of various dosage forms before filling. Since these applications typically involve process gas at low pressure, considerations are much the same as for sampling CEA or cleanroom air volumes at ambient pressure. In this case, a simple T or right angle connection to the vessel may be applied.

Figure 9.20. Sampling point geometries for isokinetic sampling.

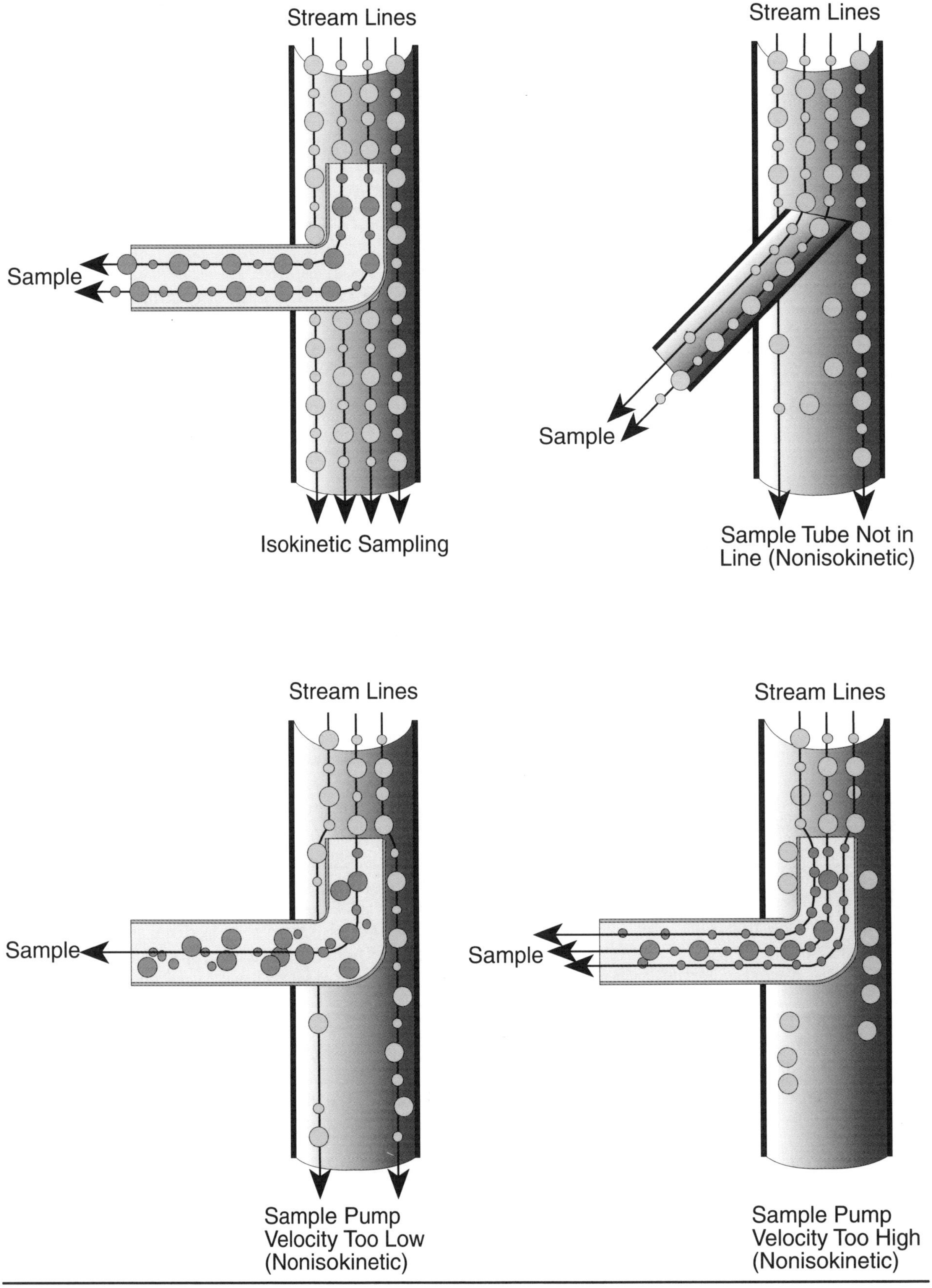

Vapors as a Source of Artifactual Counts

In some manufacturing environments, notably for large-volume injections (LVIs) or devices, the presence of persistent vapors from cleaning, steam, or peroxide sterilization may cause erroneous counts when OPCs are applied. In these cases, vaporized liquid particles, whether from blow molding, extrusion, or washing or filling procedures, become dispersed in the ambient air. These droplets appear to the OPC as counts indistinguishable from skin cells, glass fragments, aerosolized drug powders, or submicrometer-sized contaminants from outside air. These artifact populations represent condensed vapors and may be eliminated from consideration if revolatilized before passage through the OPC.

Elimination of these "particles" may be effected simply by revolatilizing, vaporizing, or returning them to a gaseous state prior to passage of the sampled air volume through the OPC. First, identification of the interfering moiety must be established. The simplest case is water vapor that persists or exists due to a washing or filling procedure (e.g., water vapor in a high-speed LVI filling room). We know that water vaporizes at about 100°C. Thus heating the droplet bearing air to this temperature will eliminate the vapor bearing the water microdroplet counts. A short helix (e.g., 30 in.) of stainless steel tubing with the same I.U. as the OPC sample line tubing coiled around an electrical block heater will be effective in this regard. Do not neglect a validation process for the technique, including a thermistor to measure temperature applied and characterization of the effect of the heating on the identified artifact source. The method of count elimination should also be verified to have no effect on the "in-use" or "as-built" counts in the CEA of interest due to solid particles. These counts are presumably due to "real" particles. Microscopic testing, as allowed by FS-209E but not ISO 14644 might also be applied to resolve this artifact.

LARGE PARTICLE MONITORING

The rate at which large particles (10–100 μm) settle onto surfaces in a process area may also be measured using electronic particle counters. Because the numbers of particles in this size range are far lower than total numbers of airborne particles ≥ 0.5 μm, the statistical strength of the data collected will be decreased. Results obtained in different portions of a process area or for samples taken at different times of a day may also be expected to show higher variation (Borden et al. 1989). One such device is the Model 11A fallout sensor marketed previously by HIAC/Royco. This device is shown in Figure 9.21. It consists of a laser particle detector mounted below a stagnation column. The stagnation column (sampling tube), which is simply a cylindrical tube 2 in. high and 1 in. in diameter, sorts settling particles from the particles drifting with airflow. The settling particles are then counted with a laser light-scattering detector. The column and bottom cover create a dead air space around the detector, which is located in the sensor body. The entire sensor fits on a mounting base with dimensions of only 5 in. × 3 in.

Figure 9.21 also shows the optical configuration of the sensor mounted under the stagnation column. It is similar to particle detectors used in vacuum process equipment for the microelectronics industry, but it is detuned to reduce its sensitivity to particles smaller than about 10 μm in diameter. It consists of a laser diode source and lenses that project a 30 mW laser beam at a 780 nm wavelength back and forth between two mirrors. The beam repeatedly transverses the sensing zone by multiple reflection between flat mirrors. This creates a "net" of light under the stagnation column. Particles settling through the beam scatter light onto twin photocells mounted above the mirrors. This creates electrical signals that are amplified in a preamplifier and then sent to a control unit up to 100 ft away for processing.

The rate at which large particles (10–100 μm) settle onto surfaces in a process area may also be measured using other types of particle counters described for the purpose (i.e., fallout counters or witness plates, as discussed in Chapter 11 that deals with sample collection). Because the numbers of particles in this size range are far lower than the total number of airborne particles ≥ 0.5 μm, the statistical strength of the data collected will be decreased. Results obtained in different portions of a process area or for samples taken at different times of a day may also be expected to show higher variation (Borden et al. 1989).

PERSONNEL MONITORING

Personnel monitoring is discussed in a later chapter and will not be dealt with in any detail here. It

Figure 9.21. Electronic real-time fallout monitor (HIAC/Royco).

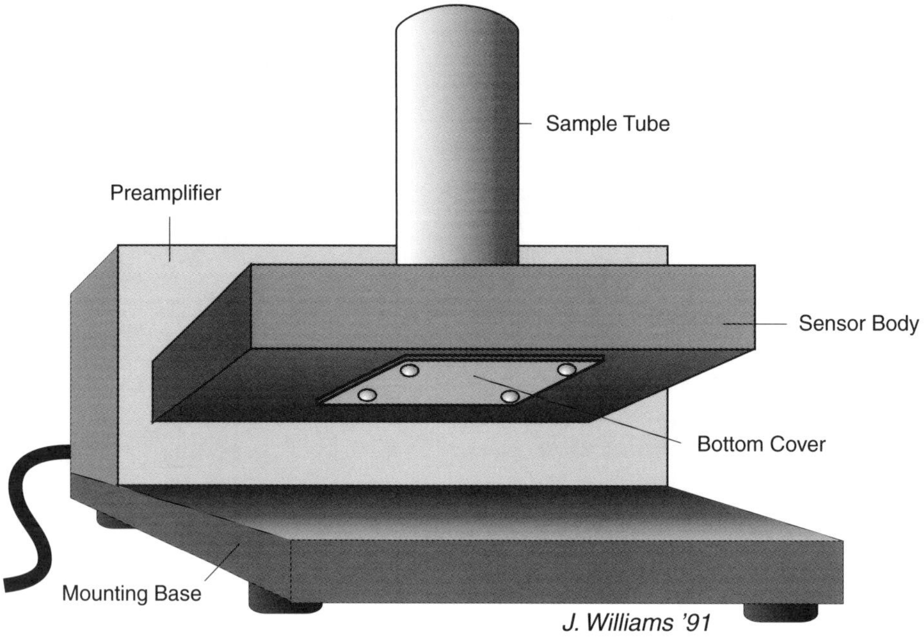

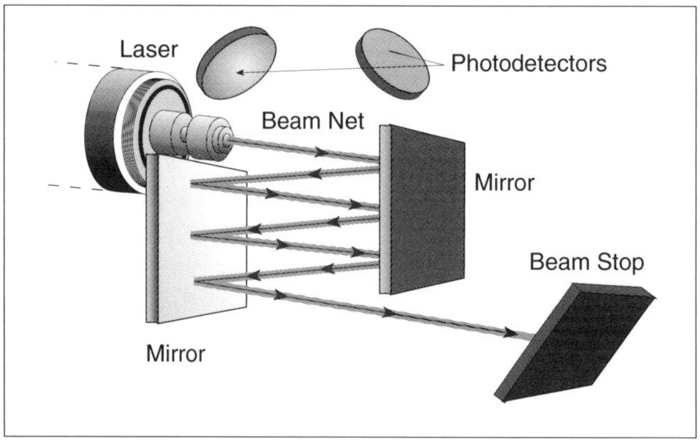

is overwhelmingly important in Class 100 or cleaner areas and worthy of mention. Personnel are the largest single source of particles in a cleanroom. Most of the particles shed by people are skin cells, which have been identified as a source of bacterial contamination, and fibers. There is an extremely high level of sensitivity to human contamination in the pharmaceutical industry and on the part of regulatory agencies. The viable and nonviable particles that are shed from people has been described as "human dust" that is continually being shed by personnel at a rate of 10^5–10^7 cells/day. The skin cells that make up most of this "dust" transport microbes that are present in significant numbers on the human body. No monitoring program for aseptic fill areas is complete unless there are provisions for personnel testing (Dixon 1990). This monitoring provides information both on the effectiveness of gowning procedures and employee aseptic practices.

Monitoring should be performed monthly or at more frequent intervals on workers in an aseptic fill area; as with the cleanroom itself, action and alert limits should be established. Each area of overgarments worn should have different limits based on proximity to the critical area during filling manufacturing. Gloves must have the lowest limits, while boots have the highest limits. Although the ultimate goal of personnel protection is to have zero counts, action and alert limits must be set at realistic levels; the goal is simply to keep contamination at a minimum and under control. If the results for an individual exceed limits or if a trend is indicated, the person may be counseled or required to take a review course in gowning and aseptic practices. Importantly, personnel testing data can sometimes be correlated with the failure rate of media fills performed to validate aseptic fill processes.

SUMMARY

The monitoring of the numbers of airborne particles at sizes ≥ 0.5 μm affords the pharmaceutical or medical device manufacturer an effective means of determining whether the environmental air in production facilities meets the requirements of various standards and compendia. Monitoring is critical in defining appropriate conditions for aseptic filling of drug products. The concern is that the number of viable particles (i.e., microorganisms) may track total particle numbers of ≥ 0.5 μm particles in some general fashion. Product particle burden will reflect the combined effects of the particle number to which the product is exposed, product vulnerability, and time of exposure; no specific numerical relationship has ever been shown to exist between viable and total counts (Whyte 1983). Furthermore, two other significant uncertainties exist with regard to the interpretation of monitoring data. Particles larger than 0.5 μm (e.g., ≥ 10 μm, ≥ 25 μm, ≥ 50 μm) have no predetermined numerical relationships to total counts ≥ 0.5 μm. The numbers of particles at compendial sizes or of visible size may or may not track high or low instrumental counts. A second particular concern is that large process-related particles that cause visible defects in product almost never become airborne and, if they do, are poorly sampled by OPCs.

The premise that HEPA filter control of the environmental air particulate burden and monitoring of airborne particle counts ≥ 0.5 μm are key elements in the manufacture of pharmaceutical products remains true. It must be kept in mind that neither of these measures has any significant effect on large particles that occur randomly in the process environment and are not effectively controlled by HEPA filtration, process- or human-generated particles that occur in intimate proximity to the product, or particles adherent to containers as received.

The forces that adhere particles to surfaces are a significant issue when particles up to 50 μm are considered. The velocity of HEPA airflow is insufficient to remove particles once they have adhered to a container or device. Large particles are not effectively entrained by HEPA airflow and are difficult to remove from surfaces.

Based on differences in sensor design, calibration methodology, and a differing response based on particle size, shape, and refractive index, different OPCs from a single vendor or from different vendors can be expected to show a wide count disparity. The selection of counter type for a given application or for upgrading monitoring technology must be based on a consideration of the historical database, the role in which the counter will be employed, and vendor reputation and support. Present generation counts can be considered only relative because of air quality, even for Class 100 environments. It is incumbent on the user to establish valid sampling plans based on the uncertainty of OPC measurements and appropriate statistical data reduction based on the level and variability of counts collected and the classification and use of the area tested.

A variety of particle collection methods are available for the remote sampling of pharmaceutical CEAs. The choice of an appropriate system must be based on the area classification, the degree of particulate control required, the historical database, product sensitivity, and vendor support. Appropriate calibration technology is essential if minimal variation is to be obtained between two or more OPCs.

Remote monitoring of airborne particulate matter with remote sensors that monitor continuously or by means of tubing sample transport is widely used in the pharmaceutical industry. The remote system has many advantages over manual sampling of preselected sites, which was prevalent at the beginning of this decade. The implementation of a remote system is a nontrivial undertaking and requires a considerable expenditure of funds as well as careful planning and validation. Generally, four approaches are used in the monitoring of particle concentrations in pharmaceutical CEAs. Whatever remote system is selected for use, validation is a significant undertaking.

Statistical methods are valuable as a structured means for summarizing data, comparing data to standards or historical data, and tests of hypothesis. Even results of statistical tests have to be interpreted with caution. Consulting with a statistician before tests are run is essential. In the absence of statistical tests, data are likely to be given subjective evaluations that conform to the preconceptions of the investigators rather than to information presented by the data. The future regulatory climate as well as the implementation of international standards suggests that increased emphasis will be placed on monitoring of smaller particulate sizes and with more sample points.

REFERENCES

Agarwal, J. K., and Liu, B. Y. H. 1980. A criterion for accurate aerosol sampling in still air. *AIHA Journal* 41 (3):191–197.

ASTM F328. 1980. *Standard practice for determining counting and sizing accuracy of an airborne particle counter using near-monodisperse spherical particulate standards.* Philadelphia: American Society for Testing and Materials.

ASTM F649. 1979. *Standard practice for secondary calibration of airborne particle counter using comparison procedures.* Philadelphia: American Society for Testing and Materials.

Bader, H., H. R. Gordon, and O. B. Brown. 1972. Theory of coincidence counts and simple practical methods of coincidence count correction for optical and resistive pulse particle counters. *Rev. Sci. Instr.* 43:1407–1412.

Bemer, D., J. F. Favries, and A. Renoux. 1990. Calculation of the theoretical response of an optical counter and its practical usefulness. *J. Aerosol Sci.* 21 (5):689–700.

Berglund, R. N., and B. Y. H. Liu. 1973. Generation of monodisperse aerosol standards. *Environ. Sci. Technol.* 7:147–153.

Bergin, M. H. 1987. Evaluation of aerosol particle penetration through PFA tubing and antistatic PFA tubing. *Microcontamination* (Feb.): 38–43.

Borden, P., J. Munson, D. Bartelson, and M. McClellan. 1989. Real time monitoring of large particle fallout for aerospace applications. *Proceedings of the Annual Meeting of the Institute of Environmental Science,* pp. 394–396.

BS 5295. 1989. *Environmental cleanliness in enclosed spaces, Part 1: Specifications for cleanrooms and clean air devices.* London: British Standards Institute.

Buettner, H. 1990. Measurement of the size of fine nonspherical particles with a light-scattering particle counter. *Aerosol Sci. Tech.* 12 (2):413–421.

Bzik, T. J. 1985. Statistical management and analysis of particle count data in ultraclean environments. *Proceedings of the Microcontamination Conference and Exposition,* pp. 93–118.

Bzik, T. J. 1986. Statistical management and analysis of particle count data in ultraclean environments, Part I. *Microcontamination* (May): 89–90.

Bzik, T. J. 1988. Statistical analysis methods for ultraclean environments: Course methods for ultraclean environments. Course 107, Institute of Environmental Sciences, 2 May, King of Prussia, Penn.

Bzik, T. J. 1994. Personal communication.

Caldow, R., and J. Blesner. 1989. A procedure to verify the lower counting limit of optical particle counters. *J. Paren. Sci. Technol.* 43:174–179.

21 CFR. 1996. Code of Federal Regulations, Section 210.1. Washington, D.C.: U.S. Government Printing Office.

Cooper, D. 1990. Particle statistics for contamination control: An introduction. *Proceedings of the PDA International Conference on Particle Detection, Metrology, and Control,* pp. 183–208.

Cooper, D. W., and S. J. Grotzinger. 1989. Comparing particle counters: Cost vs. reproducibility. *J. Environmental Sci.* 35 (5):32–34.

Cooper, D. W., and D. C. Milholland. 1990. Sequential sampling for Federal Standard 209 for cleanrooms. *J. Inst. Env. Sci.* 33 (5):28–32.

Cooper, D. W., R. J. Miller, and J. J. Wu. 1991. Comparing three condensation nucleus counters and optical particle counters in the measurement of small particles. *Microcontamination* 9 (4):19–26, 86.

Davies, C. N. 1964. The aspiration of heavy airborne particles into a paint sink. *Royal Society London Proc.* (series A) 279:413–428.

Del Valle, M. A. 1997. Effects of the 1997 revision of the European Community GMPs on HVAC design for fill/finish facilities. *Pharm. Eng.* (Nov.–Dec.): 34–41.

Dixon, A. M. 1990. Human contamination issues in cleanroom manufacturing. *Proceedings of the PDA International Conference on Particle Detection, Metrology, and Control,* pp. 65–81.

Ensor, D. S., R. P. Donovan, and B. R. Locke. 1987. Particle size distributions in clean rooms. *J. Environ. Sci.* 30 (6):44–49.

FDA. 1987a. *Guideline on sterile drug products produced by aseptic processing.* Rockville, Md., USA: Food and Drug Administration.

FDA. 1987b. *Guideline on general principles of process validation.* Rockville, Md., USA: Food and Drug Administration.

Fissan, H., and Schwientek, G. 1987. Sampling and transport of aerosols. *TSI Journ. of Particle Instrumentation* 2 (2):13–15.

Fitch, H. D. 1991. Measurement and instrumentation in contamination control. *Clean Rooms* (Feb.): 64–68.

FS-209D. 1989. *Cleanroom and work station requirements: Controlled environments.* Washington, D.C.: General Services Administration.

FS-209E. 1992. *Cleanroom and work station requirements: Controlled environments.* Washington, D.C.: General Services Administration.

Grotzinger, S. J., and D. W. Cooper. 1992. Selecting a cost-effective number of samples to use at preselected locations. *J. of the IES* (Jan.–Feb.): 41–49.

Hinds, W. C. 1982. *Aerosol technology.* New York: John Wiley.

Hovenac, E. A. 1990. Scattering from nonspherical particles. *Proceedings of the 2nd International Congress on Optical Particle Sizing*, pp. 108–117.

Keady, P. B., P. A. Nelson, and J. Blesner. 1990. State-of-the-art in sensor technology for multipoint monitoring. *ICCCS* 90:32–35.

Knollenberg, R. G. 1970. The optical array: An alternative to scattering or extinction for airborne particle size determination. *J. Appl. Meteorology* 9:86–103.

Knollenberg, R. G. 1976. Open cavity laser "active" scattering particle spectrometry from 0.05 to 5 microns. In *Fine particles: Aerosol generation, measurement, sampling and analysis,* edited by B. Y. H. Liu. New York: Academic Press.

Knollenberg, R. G. 1989. The measurement of latex particle sizes using scattering ratios in the Rayleigh scattering size range. *J. Aerosol Sci.* 20 (3):331–345.

Lieberman, A. W. 1979a. Free air monitoring of nonviable aerosol particles, Part I. *Pharm. Technol.* 21:71–77.

Lieberman, A. W. 1979b. Free air monitoring of nonviable aerosol particles, Part II. *Pharm. Technol.* 2161–66.

Lieberman, A. 1988. Requirements, instrument capabilities and potentials for particle monitoring in pharmaceutical manufacturing areas. *Pharm. Eng.* (Jul–Aug.): 65–78.

Liu, B. Y. H. 1996. Aerosol monitoring short course. Minneapolis: University of Minnesota, Department of Mechanical Engineering, Particle Technology Laboratory.

Liu, B. Y. H., and W. W. Szymanski. 1987. Counting efficiency, lower detection limit and noise level of optical particle counters. *Proceedings of the Meeting of the Institute of Environmental Science*, pp. 417–421.

Liu, B. Y. H., W. W. Szymanski, K. H. Ahn. 1985. On aerosol size distribution measurements by laser and white light optical particle counters. *J. Environ. Sci.* 28:19–24.

Montague, W., and H. Sommer. 1989. Performance parameters of optical aerosol particle counters. *Filt. News* (Nov./Dec.): 26–30.

Munson, T. E., and R. L. Sorenson. 1990. Environmental monitoring: Regulatory issues. In *Sterile pharmaceutical manufacturing: Applications for the 1990s,* edited by M. J. Groves, W. P. Olson, and M. H. Anisfeld. Buffalo Grove, Ill., USA: Interpharm Press, Inc., pp. 163–184.

Rubow, K. L., J. Lee, D. Y. H. Pui, and B. Y. H. Liu. 1990. Performance evaluation and comparative study of pressure reducers for aerosol sampling from high purity gasses. *Proceedings of 10th International Symposium on Contamination Control*, pp. 168–173.

Schuster, B. G., and R. Knollenberg. 1972. Detection and sizing of small particles in an open cavity gas laser. *Applied Optics* 11:1515–1520.

Sehmel, G. A. 1970. Particle sampling bias introduced by anisokinetic sampling and disposition within the sampling line. *Am. Ind. Hyg. Assoc. J.* 31 (6): 758.

Sommer, H. T. 1989. Resolution, sensitivity, counting efficiency and coincidence limit of optical aerosol particle counters. *Proceedings of the Fine Particle Society Conference*, pp. 126–131.

Sommer, H. T., and C. F. Harrison. 1991. Aerosol size concentration measurements from scattered light signals. *Proceedings of the Conference Sensor '91*, pp. 56–59.

Sorenson, R. L. 1986. Contamination control in pharmaceutical manufacturing. *Proceedings of the IES*, pp. 565–567.

Szymanski, W. W., and P. E. Wagner. 1990. Absolute aerosol number concentration measurement by simultaneous observation of extinction and scattered light. *J. Aerosol Sci.* 21 (3):441–451.

Umhauer, H., and M. Bottlinger. 1990. The effect of particle shape and structure on the results of single particle light scattering analysis. *Proceedings of the 2nd International Congress on Optical Particle Sizing*, pp. 425–434.

Wen, H. Y., and G. Kasper. 1986. Counting efficiencies of six commercial particle counters. *J. Aerosol Sci.* 187 (6):947–961.

Whitby, K. T., and B. Y. H. Liu. 1967. Generation of countable pulses by high concentrations of subcountable sized particles in the sensing volume of optical counters. *J. Colloid Interface Sci.* 25:537–546.

Whyte, W. 1983. A multicentered investigation of cleanroom requirements for terminally sterilized pharmaceuticals. *J. Paren. Sci. Technol.* 37 (4): 184–197.

Willeke, K., and P. A. Baron, eds. 1993. *Aerosol measurement: Principles, techniques and applications.* New York: Van Nostrand Reinhold.

Zweers, J. R. 1985. Personal communication.

X

LIGHT MICROSCOPY

This chapter provides a summary discussion of the basic principles of identifying, characterizing, and enumerating particles using visible light microscopy. For many years, prior to the invention and development of the various modern modes of microchemical analysis, the particle analyst was limited to using stereomicroscopes and compound light microscopes for the analysis of particulate matter. Methods for particle identification, sizing, and counting using light microscopy were highly developed during the first half of this century, and the methods developed remain applicable today.

Of all the methods of particle analysis, with regard to number, size, or composition, light microscopy is exceeded only by visual inspection in the extent that subjective judgments and analyst competence affect the outcome. While this chapter stresses the intuitive rather than the quantitative use of the microscope, the reader is cautioned that the microscope and its use cannot be approached like many other analytical instruments that are "black boxes" with regard to function and perform their analyses based on pressing a button. For the beginner in this endeavor and, indeed, those with some level of basic experience, a significant amount of reading is necessary in order to understand the principles of operation of the instrument. The other piece of the puzzle that must be mastered is the use of the instrument in counting, sizing, and identifying particles. This involves comprehending how particles appear under different instrumental conditions and/or the acquisition of the necessary skills to size and enumerate particles with the level of precision necessary to acquire the required data. This skill is gained only through practice in the use of the instrument on known and unknown samples; such practice requires the expenditure of time, which will be repaid manyfold when the capability for rapid or intuitive identifications is acquired.

This chapter is intended to give the beginning analyst basic information on how the microscope works and how to apply it in the analysis of pharmaceutical particulate matter. A brief description of the simple applications of the polarizing light microscope and microscopic tests related to the identification of particles are provided. Some of the critical terminology that relates to the use of microscopes is also defined. Much of the more detailed information that an analyst should be exposed to is available from the following references: Rebegay et al. (1977a, 1977b), Schroeder and DeLuca (1980), McCrone and Delly (1973a, 1973b, 1973c, 1973d), McCrone and Stewart (1974), McCrone et al. (1979), and Crutcher (1975, 1978a, 1978b, 1979, 1986). Of particular value are the publications by McCrone and Delly, McCrone and Stewart, McCrone et al., and Crutcher (1986). Aldrich and Smith (1995) provides an interesting insight into the coupling of microscopy with other forms of analysis. In addition to literature references, a list of standards and recommended practices is provided. These applications and techniques are pertinent to both counting and identifying particles using light microscopy.

A specific mention should be made here of operational concerns related to the application of microscopy. The use of the microscope for particle identification in pharmaceutical companies is often relegated to a supportive role, whereby the microscope supports higher technology instrumentation, such as mass spectrometry or X-ray spectrometry. This is unfortunate in several senses: It is more fitting that the instrumental forms of analysis be used to support microscopy. As a result of this perception, the laboratory space provided for microscopy is sometimes poorly suited or at best marginal for its application. The user is cautioned that the results obtained will be directly related in value to the emphasis given not only to the training of analysts but also to the nature of the laboratory space and peripherals provided for the microscopist. Delly (1996) provides detail on how to equip and utilize a microscopy lab and is as valuable a reference as any of those dealing with the use of the microscope.

The presence of elevated numbers of particulate contaminants in product is typically unforeseen, and their occurrence is often associated with a serious production or product problem. To identify particles and solve problems in a cost-effective manner, the analyst needs as much information as possible, quickly, from a small sample. The technique used must be able to characterize the sample unequivocally. No interferences can be ruled out; a single contaminant source cannot be assumed; in fact, mixed populations occur more frequently than monotypic ones. If assumptions are made too soon, the analyst may prove your assumptions right and fail to solve the problem or, worse yet, reach the wrong conclusion (Overington 1981).

A word with regard to the organization of this chapter and related materials: In the text, principles of microscope operation are illustrated with line drawings; in the appendix following the chapter text, some case history information is given for the color plates illustrating common or illustrative particles that appear in a separate section in this book. Keyed to the color plates, the case histories will allow the analyst to better understand the actual principles of identification based on how a particle looks under different conditions of observation.

The light microscope is ideal for this type of analysis. The key to the effective application of the light microscope in particle analysis is the availability of an experienced analyst. The mental database of the human analyst allows quick identification of particles previously seen. An experienced analyst can size and count particles with a speed and precision approaching that of an automated microscopic image analysis system (Barber et al. 1989). The problem is that experienced microscopists with comprehensive skills applicable to particle analysis are not easily found. Without well-directed training, years of experience may be required to acquire competence in particle analysis with the light microscope. In the majority of pharmaceutical particle analysis labs today, reliance is typically placed on expensive, complex microanalytical instrumentation to provide analytical answers, rather than human experience. This situation occurs primarily because of the limited number of skilled microscopists available.

There is a simple solution to the problem posed by the short supply of experienced microscopists. This involves training laboratory technicians to utilize light microscopy in a relatively narrow range of analytical endeavors in which the instrument is most useful. In a laboratory that is constrained with regard to capital for the acquisition of expensive instruments, the lab manager is well advised to acquire a light microscope and develop structured training of individuals who will be responsible for its subsequent application. Although extensive training and years of experience may be required to acquire a broad-based expertise in light microscopy, the basic skills required for the identification of a very high percentage of all particles encountered in pharmaceutical products or manufacturing environments may be acquired in a relatively short time. Similarly, near-automatic skills for rapidly and accurately sizing particles may be acquired quickly. The actual time required for training will vary depending on how frequently the analyst is required to use the microscope. In the author's experience, analysts can become adept at particle enumeration in about two months with daily practice. Largely because the microscope is used to identify particles on a less frequent basis, acquisition of the ability to use the microscope in an analytical role will take longer (4–6 months).

The author would be remiss here if a mention of an extremely useful training aid was not made. The original hard copy of *The Particle Atlas,* authored by McCrone and Delly in the 1970s, is out of print due to the original master materials' being destroyed in a fire. Copies of the the *Atlas* are highly prized among particle analysts and jealously guarded (and difficult to obtain on loan, even from a friend). The *Atlas* has, however, been made available in CD-ROM format for IBM-compatible computers. In this form, it provides access to the entire *Atlas* on a single disk. Both pictures and text are available and may be searched electronically in a small fraction of the time necessary for a search in the actual book. The electronic search is, in fact, comprehensive as well as rapid. A number of students of the microscopist's art with whom the author has talked use the electronic volume of the *Atlas* as a valuable training aid; this broad information base regarding microscopic particle identification can be kept close by the microscope so that its color plates and descriptive information may be quickly compared with known or unknown specimen materials.

Acquiring the skills to make effective use of light microscopes quickly is dependent on teaching the use and application of the instrument as a technology rather than as a science. Simply stated, the best progress will be made by teaching an analyst only the basic skills needed to operate a light microscope and to identify particles by sight or by simple manipulations of polarized light. A basic feature of this approach is the use of comparison standards that allow the analyst to see what a particle looks like and file this information in his or her mental database. The human observer has an almost incredible ability to discriminate one object from another. The difficulty is that unless the observer's brain has been trained to recognize the relevant features in the field of view, they may pass without an observation being made. This training in particle recognition is the difference between a microscopist who is able to use the instrument in identification and a person who only occasionally uses a microscope.

THE POWER OF VISUAL OBSERVATION

The visual detection of contaminant particles is probably the most common cause of samples requiring an analysis of contaminants. Humans are incredibly sensitive detectors of deviations from the norm. The eye is surprisingly sensitive and is the most acute sensory organ of the human. The visual detection of contaminants is probably the most common cause of why pharmaceutical samples require contaminant analysis. Although the normal human eye can only resolve particles of about 150 μm, it can detect collections of particles below 10 μm if the lighting is optimal. The eye can also detect slight changes in color or reflectivity. The eye's sensitivity to color results in the ability to detect films on smooth surfaces with a thickness of about 1 μm to about 0.1 μm by virtue of the interference color. The most common example is an oil film on water. Human vision is a very complex activity involving the eye as the sensor and the visual cortex of the brain as the decoder. The observer is only aware of the decoded signal.

The view we see is a creation of the brain (Overington 1981). The image processing that the brain performs is very sensitive to suggestion. If a defect is expected, it may be found as a new problem, when, in fact, it is a common condition that has simply not been recognized before.

Most people are unaware of the sensitivity of the eye or the way light can "play tricks" on the eye. As a result, reports on the appearance of a contaminant must be considered carefully by the analyst. A dark film on a rough surface of a microporous filter membrane may be a colorless liquid. If it is observed with the light reflecting from the surface of the filter, the oil spot is dark. If it is observed with transmitted light, the oil spot is bright because the oil has made the filter more transparent. When viewed with reflected light, more light passes through the oil spot and is not reflected, making it look darker.

If it were not for the visual cortex's ability to simplify, generalize, and derive three-dimensional data from a two-dimensional sensory array, the world would be a very confusing visual experience. As it is, the incoming signal is screened and reduced to a recognizable array of generally familiar objects, colors, textures, and sizes by coordination of the brain and eye. If the observer expects to see something in the field of view not immediately apparent, other parts of the brain can override the visual cortex and force it to reprocess or add to the signal incoming from the eye (Overington 1981). This effect results in the multiple images we receive from optical illusions.

The principle underlying particle analysis using the light microscope is the comparison of the beam of light changed by a particle to the unmodified beam. The human eye can detect a 5 percent change in intensity for objects that are not adjacent in a field of view. The brightness sensitivity of the eye requires an intensity of 1.5 percent difference between the background and the particle or feature. The color or frequency sensitivity of the eye is about 6 nm of wavelength with a 1 percent saturation over a region of about 1° of arc. The human eye can distinguish about 300 different colors (hues) at about 5.5 different saturations and 300 different brightness levels (Crutcher 1986). A combination of these factors results in a total of more than 90,000 individually distinguishable combinations of frequency and brightness before even considering size, shape, texture, or the large variety of other parameters available.

No other form of data collection is so heavily dependent on analyst subjective judgment, data manipulation, and interpretive decision making as microscopy or the interpretation of photomicrographs. The judgments and assessments are made by the eye-brain link of the trained human observer.

Image interpretation is reflective not only of what we see (objective) but what we perceive (subjective). Perceptions will be grounded in objective scientific judgments to a greater or lesser extent dependent on the expertise of the microscopist (observer), but human subjectivity will always be a component. This is as it should be. The human visual-mental capability in this regard is several evolutionary stages ahead of the fastest Pentium II®–based image analysis systems. This final visual assessment of a microscopic image must not be taken out of the human analyst's hands.

An important consideration here is the nature of an analog image captured on photographic recording media versus a digital image file. Whereas the digital data output from many instruments, such as an infrared spectrophotometer or mass spectrometer, may provide more information than the corresponding hard copy printout, such is not the case with a digital image file. Digital image files are limited in resolution, such that the finest detail of the image field cannot be observed. Photographic (analog) images contain a great deal more detail because of the inherently higher resolution of the photographic emulsion. Thus, the most critical final morphological interpretation must often be made from the micrograph itself, with the aid of appropriate lighting, not from the instrument viewing screen or from a digital image.

THE LIGHT MICROSCOPE

There are two basic types of light microscopes: a stereoscopic microscope and a compound microscope. Both types are essential to the analysis of contaminants. The stereoscopic microscope is used for preliminary observations of relationships and sample preparation, while the compound microscope is used to identify the materials present and to refine the preliminary observations. Stereomicroscopes are also called dissecting microscopes or macroscopes.

Stereomicroscopes

The stereoscopic microscope has two optical paths, one for each eye. These paths join at the back focal plane of the objective lens system in some instruments. In some older designs, such as the one shown in Figure 10.1, two complete objective ocular systems are present. If a single objective lens system is used, the light path must be sufficiently large to allow for the angular divergence of the ocular paths. These paths diverge at an angle of about 14–16° in most models. The visual effect is analogous to human binocular vision and provides a three-dimensional view of a sample. This angle effectively limits the resolution of these systems to a useful magnification range of 6× to 150× (a numerical aperture [NA] of about 0.14). With this instrument, the image is not reversed, simplifying the manipulation of the sample during sample preparation.

Stereomicroscopes are available with fixed or zoom (variable) magnification and transmitted, reflected, or coaxial illumination. Historically, the fixed magnification microscope had more fully corrected optics and provided a better quality image; this is no longer true. The high-quality variable power stereomicroscopes currently produced by Wild, Nikon, Olympus, and some other manufacturers produce

Figure 10.1. Ray paths through a binocular stereomicroscope.

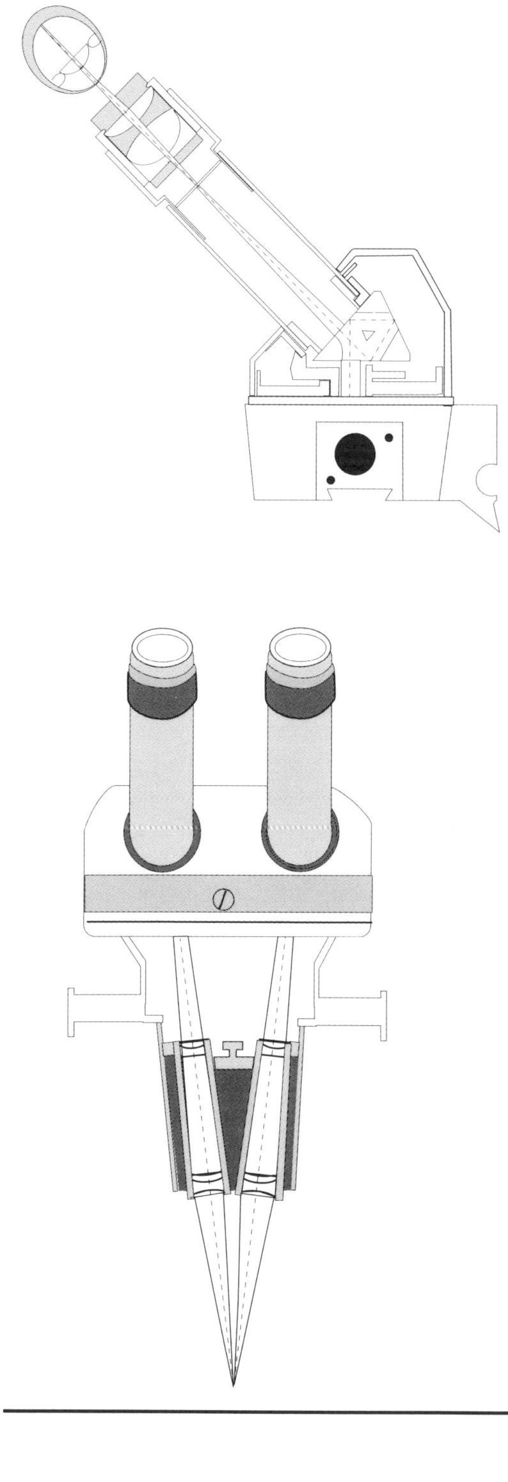

excellent images and allow magnification change in steps or continuously. The zoom stereomicroscope has the advantage of easily changing the magnification to fit the requirements of the sample. Zoom stereomicroscopes are more expensive than fixed magnification instruments but are generally well worth the cost. The major parts of the stereoscopic microscope are the body, the oculars, the stand, and the single objective. Illumination can be provided by inexpensive gooseneck lamps or sophisticated coaxial fiber-optic systems.

A brief philosophical discussion is in order regarding the process of particle identification and the uses of various instrumentation. (In other sections of this book, a stepwise approach to particle identification has been discussed.) The two components of an optimally effective analytical path that are omitted or abused are obtaining adequate background on the sample and circumstances of its occurrence and stereomicroscopic observations. While stereomicroscopy is often used for the isolation of particles for instrumental analysis, it is often underutilized as a means of collecting information. The instrument provides, because of the noninverted image and magnification bridging the range between gross and compound microscopic observations, a means of tying the sample submitter's description of particles to higher magnification microscopy and instrumental findings. The image from a high-quality stereomicroscope may be readily matched to critical human vision with regard to color, surface textures and contours, and dimensional perspectives. Stated another way, this image is usefully magnified but intuitively interpretable.

The Compound Microscope

Microscopes that do not provide for stereovision are commonly referred to as "compound" microscopes. The compound microscope is based on the magnifying power of complex combinations of single lenses arranged so that their magnification is multiplied or compounded. The basic principle of operation, however, is the formation of magnified images by single biconvex lenses (Figure 10.2). The magnification, *M*, of an image of an object produced by the lens is given by the following relationship:

$$M = \frac{\text{image size}}{\text{object size}} = \frac{\text{image distance}}{\text{object distance}}$$

The light microscope is an optical instrument designed to produce a beam of light with any of a wide variety of specific characteristics,

Figure 10.2. Formation of a projected image by a biconvex lens.

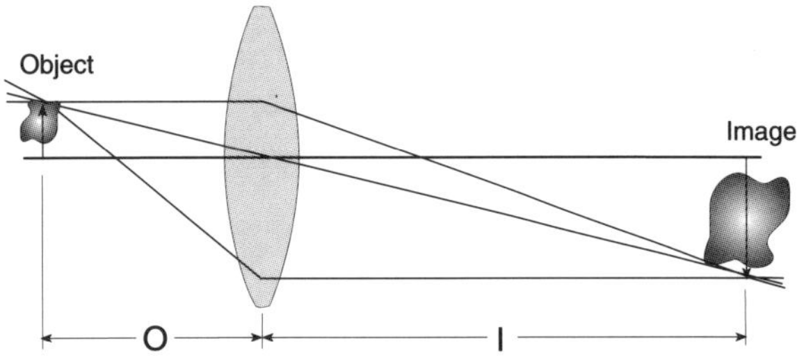

Figure 10.3. Image formation in the compound microscope.

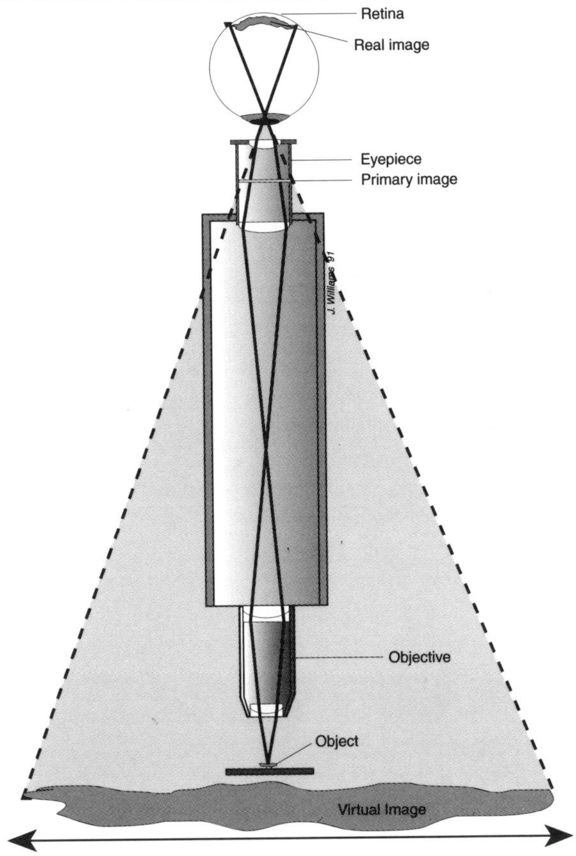

including beam frequency, direction, aperture, intensity, polarization, and modulation or interference effects (Needham 1958). This beam is focused onto the stage of the microscope directly at the preparation being studied. After passing through the preparation, the light beam is collected at a specific angle and aperture with respect to the particle being analyzed and the incident beam. The collected light is then processed to amplify specific optical effects. This image is then projected onto an optical measuring grating and through a final lens system into the sensor. The sensor may be a human observer, a photographic camera, or a photodetector.

In a compound magnifying system, magnification takes place in two or more stages. A second lens may be placed so that it produces a further magnified image of the real image produced by the double convex lens shown in Figure 10.3. The total magnification is the product of the magnification of the first lens and the second lens. This is the basic principle of a compound magnifying system. The analyst looks at the primary image with a lens that produces an enlarged secondary image or virtual image. The brain sees this virtual image, rather than the real image formed on the retina; there is no real image at the point where the virtual image appears to be (Figure 10.3).

Light rays emerging from the eyepiece converge at a point called the eyepoint. In practical use, the lens of the eye is placed exactly at this spot. The distance from the eyepoint to the virtual image, or final image, within the microscope system is about 250 mm, which is standard, close-viewing distance. The first lens in a microscope is called the objective, since it is near the object being magnified. This lens projects a magnified image of the specimen to a fixed position, called the primary or intermediate image plane. The primary image is located about 1 cm from the top of the body tube of the microscope. The distance from the objective upper focal plane to the primary image is the optical tube length (see Figure 10.4).

Lens Aberrations

A lens aberration is a failure of a lens to produce an exact point-to-point correspondence between an object and its image. Spherical aberration is shown in Figure 10.5. Light in ray paths near the center of the lens focus at different points on the optical axis compared to those in paths near the periphery. This can be reduced using only the central zones of the lens, placing spherical surfaces in the lens system that compensate for the defect, or through the use of glasses with

Figure 10.4. Optical parts and path of light through the compound microscope.

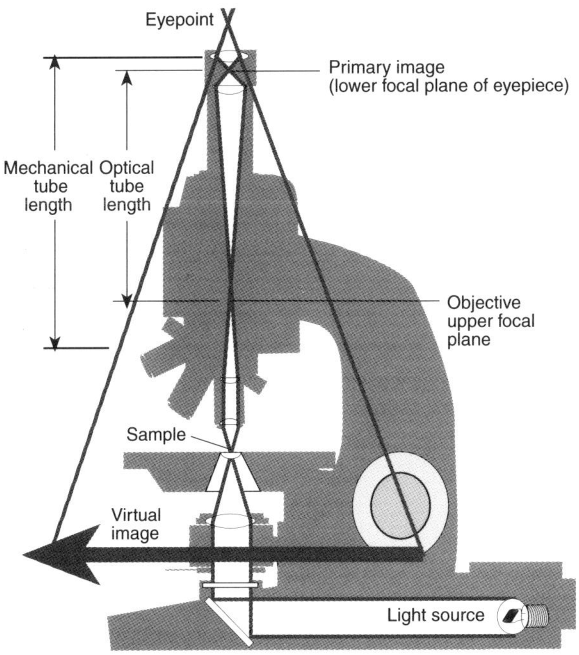

Figure 10.5. Spherical aberration.

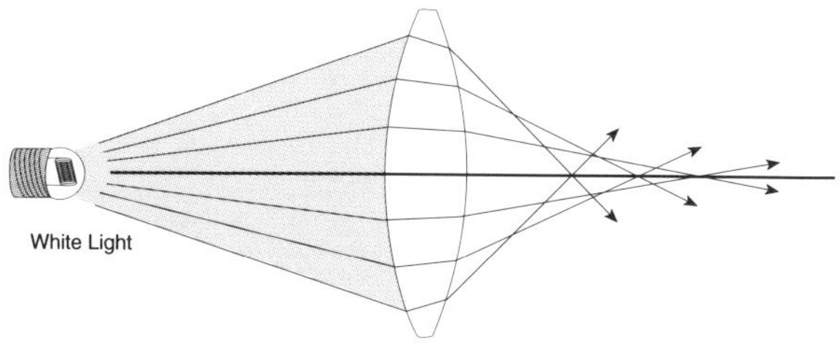

Figure 10.6. Chromatic aberration.

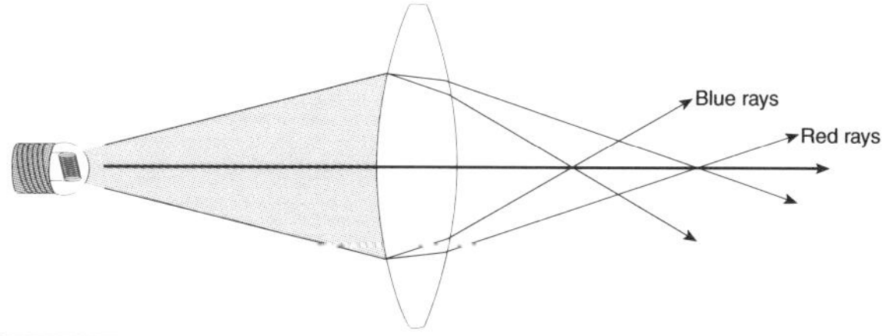

different dispersion (e.g., crown glass and flint glass) and curvature in a doublet or triplet array.

Chromatic aberration (Figure 10.6) is caused by refractive index variation of the lens material with wavelength. Thus, a lens accepting white light from an object will form a blue image closest to the lens and a red one further away. Achromatic lenses, employed to minimize this effect, are combinations of two or more lens elements of materials having different dispersive powers and refractive indices.

Objectives are divided into types according to how they are corrected for the various aberrations. There are achromats, semiachromats or fluorites, and apochromats. In achromatic lenses, chromatic aberrations are generally corrected for two colors, and spherical aberration is corrected for one color. Semiapochromatic lenses incorporate correction for both defects in two colors. Apochromats are corrected for chromatic aberrations in three colors and spherical aberration in two.

Images from objectives exhibit some curvature of field due to the different lens thicknesses through which the on-and-off axis image-forming rays must pass (Figure 10.7). When the image is sharp in the center of the field, it may be less sharp at the periphery (plane of focus P in the figure); if the image is made sharp at the periphery, it may be blurred at the center of the field of view (plane of focus C in the figure). In addition to the several types of objectives classified according to chromatic and spherical corrections, there are also plano or flatfield objectives. These flatten the image, correcting for field curvature through the use of additional lenses; these lenses are usually within the objective itself, but may also be in the body tube.

Figure 10.7. Curvature of field.

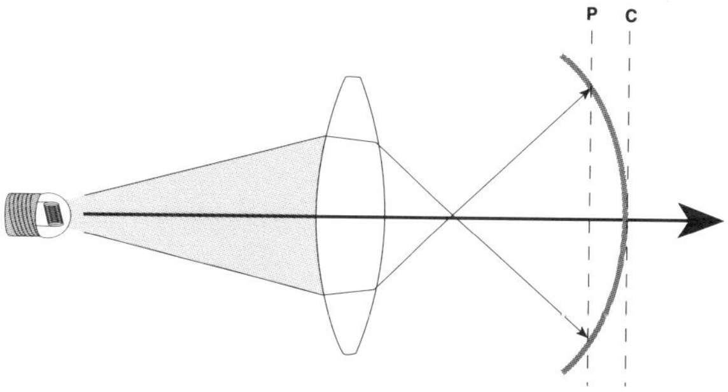

Plano or flatfield objectives are important in photomicrography, where it is essential to have the image in focus across the entire field.

Objective Lenses

Objectives are the most important optical component of the compound microscope. They form the intermediate or primary image of the object that is subsequently magnified by the ocular. The magnification of a single lens or lens system depends on its focal length (Table 10.1).

Magnification is achieved through the curvature of the lens—the higher the curvature, the shorter the focal length. Lenses of different focal length, therefore, have different angular acceptance of diffracted light waves from the specimen (Figure 10.8). Lenses of shorter focal length (higher magnification) have the greatest angular apertures; that is, the largest acceptance angle of image-forming rays.

NA is related to angular aperture by

$$NA = n \sin \frac{AA}{2}$$

where n is the refractive index of the space between the coverslip and the objective front lens and AA is the angular aperture of the lens, that is, the angle through which it accepts illuminating radiation. The highest theoretical NA for a dry objective, where the refractive index of air is 1.00, is 1.00. In actual practice, an NA of 0.95 is the highest available in a dry objective. NA is related to resolving power: The higher the NA, the greater the resolving power; therefore, a high NA is desirable. To increase the NA, increase the AA and/or increase the refractive index of the space between the specimen and the objective. Given a practical maximum focal length and AA, the best method of increasing the NA is by increasing the refractive index of the medium in the space between the coverslip and the objective. Objectives intended for this use are called immersion objectives. Oil immersion objectives have a practical maximum NA of 1.4.

Working Distance of Objectives

The distance between the front lens or lens mount of an objective and the top surface of the cover glass on a specimen slide or surface of a membrane is the working distance for the objective. Working distance governs the allowable movement of the objective in obtaining critical focus of the specimen image. Generally speaking, working distance decreases as the focal length of the objective decreases and magnification increases. For oil-immersion objectives, the working distance is measured in fractions of a millimeter.

Ocular Lenses (Eyepieces)

The ocular, or eyepiece, provides the second stage of magnification. It forms a magnified image from the primary image formed by the objective. The usual magnifications available in oculars are from 5× to 30×. The most useful oculars are in the 10–20× range. A rule of thumb for oculars is that the maximum useful magnification for a microscope is 1,000 times the NA of the objective. Magnification in excess of this value gives no additional resolving power and

Table 10.1. Representative Performance of Microscope Objective Lenses

Focal Length (mm)	Magnification	Angular Aperture (degrees)	Numerical Aperture	Working Distance (mm)	Depth of Field (μm)	Field of View (mm*)	Resolving Power (μm)
30	5	10	0.10	25	16	5	3.0
15	10	30	0.25	7	8	2	1.22
8	20	60	0.50	1.3	2	1	0.61
4	40	80	0.65	0.7	1	0.8	0.47
4	40	116	0.85	0.5	1	0.6	0.36
1.6	90	120	1.28	0.2	0.4	0.35	0.20

* Measured with a 10× eyepiece.

Figure 10.8. Angular and numerical aperture of objectives.

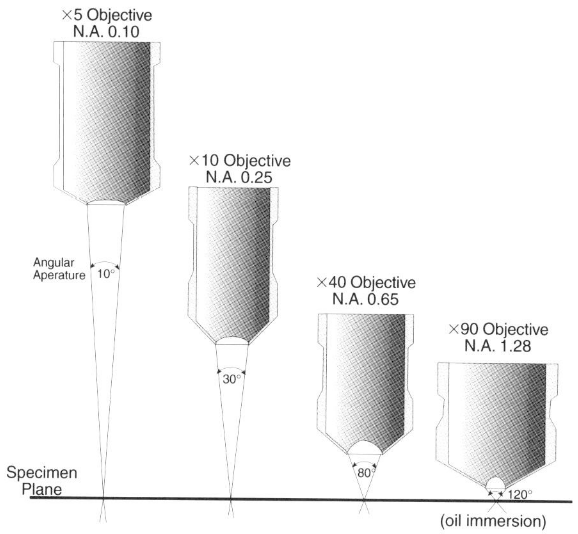

thus results in useless magnification. Only higher-power oculars can give full use of the resolving power of high NA objectives. A 10×, or even a 15×, ocular may provide insufficient magnification for the eye to see detail resolved by high-quality objectives.

Wide-field eyepieces cover large areas and are useful for scanning membrane filters and particle counting. High-eyepoint eyepieces are made for those who must keep their glasses on when viewing through the microscope. Eyeglass viewing is necessary when the user has eye defects that cannot be corrected with the microscope, such as astigmatism. Analysts who are simply nearsighted or farsighted may use a microscope without glasses.

Condenser Lenses

The third major optical component of a compound microscope is the condenser lens assembly or substage condenser. The condenser provides a cone of light that illuminates the specimen. Light from the condenser converges on the specimen, and the light diverges in passing through the specimen to form an inverted cone, the included angle of which fills the objective lens. The angular size of the illuminating cone is controlled by a variable diaphragm located beneath or within the condenser. This diaphragm is called the aperture diaphragm or substage diaphragm. The correct focus of the condenser and the proper opening of the aperture diaphragm are of extreme importance in the microscopical viewing of a sample.

One of the most important points to consider in choosing a condenser is the NA. The AA of the illuminating cone of light from the condenser must be matched to the AA of the objective for best image resolution. If a condenser is used dry with an oil immersion NA 1.40 objective, the NA realized will not be greater than 1.0. To obtain the benefits of objectives with NAs greater than 1.00, a condenser with an NA greater than 1.0 must be used, and oil must be placed between the condenser and the bottom of the slide.

Resolving Power

In geometric optics, it is assumed that light travels in straight lines, but this is not always true. The bending of light rays is the basis of the diffraction effect that limits microscopical resolution. A beam passing through a pinhole aperture creates a bright spot larger than the pinhole, with alternating bright and dark rings appearing on either side of the central bright spot. The intensity of successive rings decreases as a function of the distance from the central spot. For this reason, the image of a point of light produced by a lens is not a point but rather a larger spot of light surrounded by dark and bright rings. The diameter, D, of the first dark ring may be defined as

$$D = \frac{1.22\lambda}{\sin\theta}$$

where λ is the wavelength of the light and θ is one-half the lens AA. For a minimal diffraction disk size to be maintained with light of a given wavelength, the lens aperture must be as large as possible. Shorter wavelength light by definition produces a smaller diameter disk pattern and, theoretically, better resolution.

Diffraction effects determine image resolution. If two points are to be distinguished in an image, their diffraction disks must not overlap more than one-half of their diameter. The ability to distinguish image points is called resolving power and is expressed as one-half the diameter of the diffraction disk. The limit of resolution for two discrete object points, a distance R apart, is defined as

$$R = \frac{0.61\lambda}{\text{NA}}$$

where λ is the wavelength of the light and NA is the numerical aperture of the objective. Thus, with a wavelength of approximately 550 nm for white light and assuming an NA of 1.2, two points separated by about 200 nm (0.2 μm) can be resolved. Better resolving power can be achieved only with light of shorter wavelength. Short wavelength ultraviolet (UV) (200 nm) lowers the resolution limit to about 0.1 μm. Diffraction is the single most critical factor limiting resolution; the practical resolution of detail observed in a microscopic image also depends on the illuminating conditions, specimen contrast, condition of the optics, and acuity of the human eye.

Illumination

Transmitted, nonpolarized light is of limited usefulness for the analysis of particles. There are two general procedures for illuminating particles from above. One makes use of auxiliary illuminators; this is generally called lateral or reflected illumination. The second makes use of illumination from the objective lens of the microscope. This procedure is commonly called vertical, episcopic, or incident illumination. Bright field incident illumination is directed downward onto the specimen through the objective lens using a half-silvered mirror; the dark-field incident uses mirrors or lenses at the periphery of the objective lens to project light on the sample at angles of 20–45° (Figure 10.9).

Particles on a membrane filter, or dispersed on a slide, can best be examined by vertical lighting. Auxiliary light sources may be added to any microscope. For particle analysis, external illuminators may be placed at the side, with the light beams directed downward from above the microscope stage onto the top of the particles. The source must deliver sufficiently strong illumination and should be used at as high an angle to the horizontal as possible. The current U.S. Pharmacopeia (USP) <788> method for counting particles collected on filters prescribes low-angle reflected illumination used in combination with bright field episcopic lighting.

Dark-field transmitted light technique produces a black background image against objects that are brilliantly illuminated. This is accomplished by equipping the light microscope with a specific type of condenser that directs the light path from the source of illumination through the specimen but outside the angle of acceptance of the objective (Figure 10.10). Thus, if the specimen is completely transparent and homogeneous, the light directed through the condenser does not enter the objective, and the entire field of view is dark. If, however, the transparent medium contains objects that differ from it in refractive index, there will be a scattering of light by reflection and refraction. The scattered light will enter the objective, and thus the object will

Figure 10.9. Vertical illuminators.

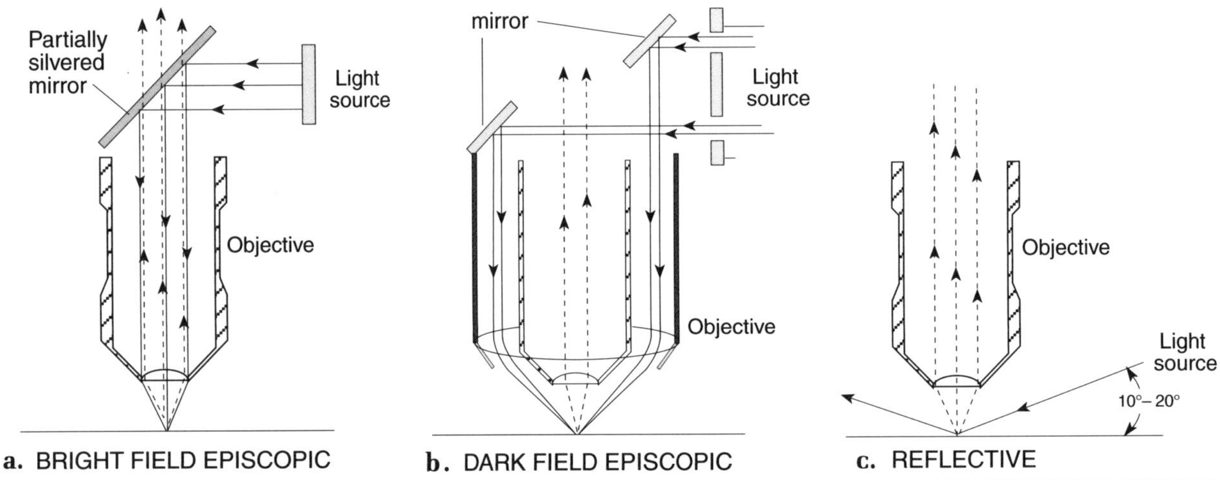

Figure 10.10. The path of light through a dark-field condenser.

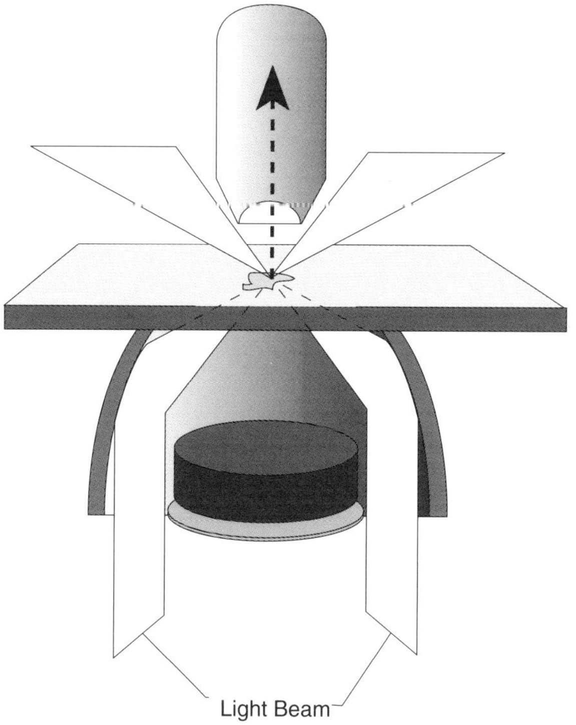

appear bright in the otherwise dark microscopic field. Dark-field microscopy may be useful for examining particles in a transparent film or on a slide.

MICROSCOPIC VISUAL DESCRIPTORS (MORPHOLOGY)

Morphology is defined as the science of form and structure (Kremp 1965; Whalley 1972). Many microscopic particle identifications may be made on the basis of simple observations of morphologic descriptors, such as size, shape, color, reflectivity, texture, or transparency. The outward form or shape of any object is the result of the internal and external forces that have acted on the object. The structural unit of a material is modified by distortions such as voids or misalignments whose frequency is a function of the unit volume. As linear dimensions increase, the number of defects increases as a cube function (Crutcher 1986). Smaller particles, as a result, are more likely to exhibit a form that indicates the structural units of its compositional material. These structural units may be a biological cell or a characteristic crystal symmetry. The key to characterizing a particle is to recognize distinctive features.

Particle Size

Size is a complex parameter (Herdan 1961). Size is well defined for a sphere, but for any other shapes, it is ambiguous unless the method of establishing the size of the particle is specifically described (Herdan 1961; Whalley 1972). An ellipse, for example, may be described by its long axis, short axis, or an arithmetic or geometric average axis. The diameter of a sphere with an equivalent volume, the diameter of a sphere with an equal surface area, aerodynamic diameter, or some other index may also indicate the "size" of the particle. However, each would result in a different number. For accurate and precise sizing of particles, a methodology should be chosen that allows the analyst to size particles based on comparison to a circular rather than a linear scale, making a direct smaller-than–bigger-than comparison rather than an incremental judgment.

The measurement of particle size varies in complexity depending on particle shape. The size of a sphere is defined by its diameter. The size of a cube may be expressed by the length of an edge or a diagonal; the surface area, volume, and weight (if density is known) of a cube or sphere can thus be calculated based on a measured dimension. If the particles of interest are irregular in shape, the task becomes more complex. An irregularly shaped particle has a number of different dimensions that might be measured as "diameters." Figure 10.11 shows four statistical diameters commonly used in determining particle size.

Feret's diameter is the distance between parallel lines tangent to the particle profile and perpendicular to the ocular scale. The present USP measurement of particle size is, in effect, the maximum Feret's diameter, with no regard given to the orientation of the tangents. The maximum horizontal intercept is the longest horizontal diameter from edge to edge of the particle, parallel to the ocular reference line. Martin's diameter is one of the most frequently used for particle sizing. It is measured as the dimension (parallel to the ocular scale) that divides a randomly oriented particle into two equal, projected areas. Projected area diameters are found by comparing

Figure 10.11. Particle size measurements.

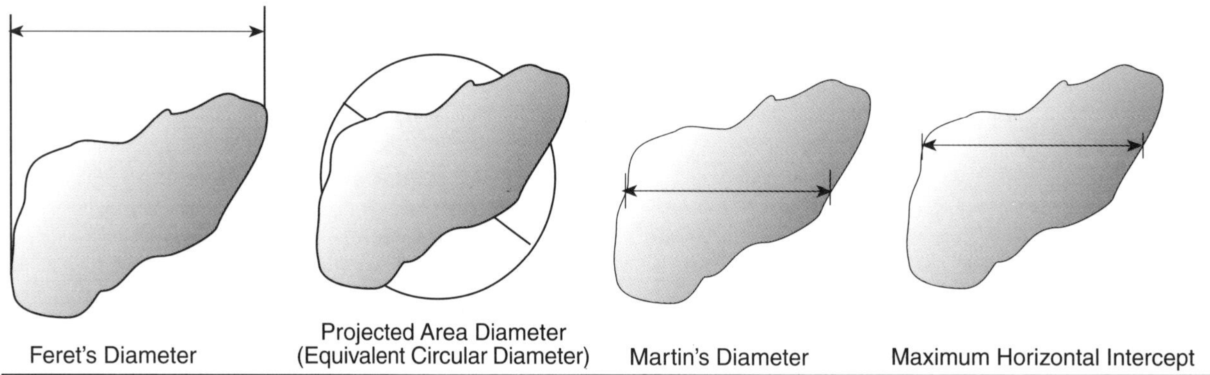

the projected area of the particle with the areas of reference circles on an ocular graticule.

Shape

The outward form or shape of a particle is the result of the internal and external forces that have acted upon the object, or its origin, and often allows rapid identification of an unknown. The key to characterizing a particle's shape is to know distinctive features and morphologic descriptors; the description of particle shape is a technology of words. There are some differences in the terminology used by different authors, but there is sufficient commonality so that the multiple terms for a single morphology are readily understood. Some examples are shown in Figure 10.12.

Another important physical effect of the small size of particles is the increased importance of surface active forces over physical factors dependent on volume or mass. Surface tension and surface chemistry become dominant over gravity and bulk chemistry. This has a profound effect on morphology as well as some analytical approaches. A most common phenomenon is the formation of particle associations. Loose, highly hydrated associations are termed "flocs" or "floccules." Associations of solid particles held together by weak physical forces are termed "agglomerates," and chemical bonding of solid particles results in cementation to form an "aggregate" (Figure 10.13).

Particle Color

The color of a particle is the difference in the spectral composition of light coming from the particle with respect to that from the illumination source or background (Bouma 1971; Nassau 1980). Color is affected by the angle of incidence for the illuminating beam of light, the angle of observation, the wavelength-dependent absorption coefficient for the object, the optical properties of the medium around the particle, and the crystallographic orientation (see also Chapters 8 and 9 for light-particle interactions). The perceived color is also dependent on the background intensity, the wavelength dependent sensitivity of the eye, and the intensity of the light from the particle to the observer. The color of a particle in transmitted light is the result of one or more features of its basic atomic or molecular structure. These include electron bond energies, energy states, lattice defects, and chemical composition. Another source of color effects related to these properties or the adjacent materials is the absorption of color.

According to Crutcher (1986), absorption color (seen with transmitted light) can be a very sensitive criterion, but it can be equally misleading if misused. This author discusses the pleochroic properties of different compounds. As an example, silicon carbide, a common industrial abrasive, has six forms, with a color variation from colorless to blue to yellow to brown to black. These colors correspond to slight changes in electron vibration frequencies with different crystal phases in pure silicon carbide, not to the presence of impurities.

Figure 10.12. Morphologic descriptive terms used in cognitive recognition.

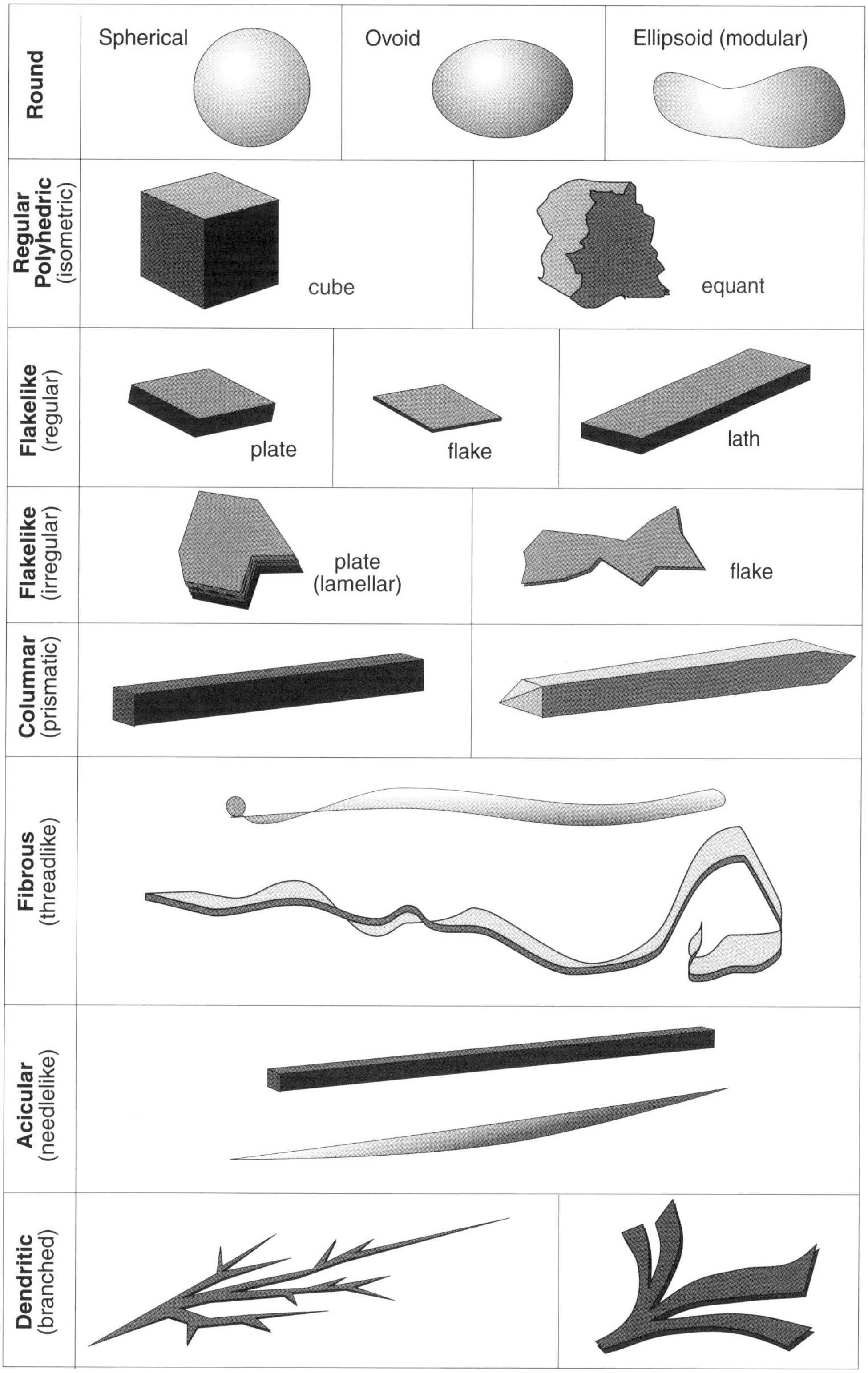

Figure 10.13. Agglomerated and aggregated particles.

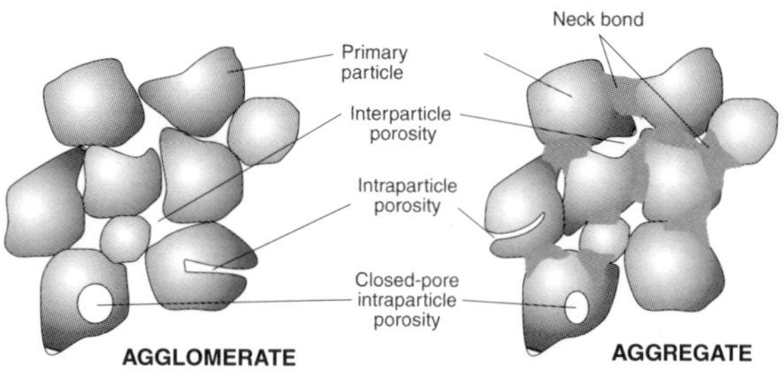

Potassium bromide is another case in point. If it is radiated with X rays, it turns from colorless to pale blue. This change in color is the result of single electrons occupying sites normally occupied by displaced bromine ions. The resultant low energy electron is a strong adsorber in the red range of the spectrum. Based on molecular arrangement, the same component may show different absorption colors. The transmitted color of a particle is the color seen when light is passed through the particle. This color has a hue dependent on the colors transmitted and a saturation dependent on the efficiency of the absorption of all other colors per unit path length and the thickness of that particle.

For some particles, color can be an indicator of the source of a material. Iron (III) oxide (Fe_2O_3) is a good example (Crutcher 1978a, 1979). This material has two basic sources, oxidation through burning and oxidation through corrosion (hydration). Iron (III) oxide from burning has made the transition from iron to iron (II) oxide (FeO) to Fe_3O_4 to Fe_2O_3. This transition is accompanied by a change from a very high absorption coefficient to a moderate absorption coefficient, with very high absorption in the blue-green. The result is a deep red-brown color. Iron corrosion products vary in color from green-yellow to orange to red and black. The colors in these iron corrosion products are characteristic of their progression through the whole serial oxidation reaction.

Diffraction colors are from the monochromatic spectrum, first documented by Newton. With the microscope, these colors are identified by their characteristic hue and by their change as the particle is moved in the field of view or the angle of the illuminating beam is changed. The diffraction color seen is dependent on the geometry of the fine structure and its orientation with respect to the illuminating beam and the objective. Dispersion colors are the complementary color pairs generated by dispersion differences in the object and the media in which it is mounted.

If the illumination used for microscopy is reflected rather than transmitted, interference colors and reflected colors will be seen. Interference colors, for example, are represented as the colors of oil films on water. The same color series is used in microscopy to characterize thin films and the crystallographic properties of materials viewed between crossed polarizing filters. The color seen reflecting from a particle may or may not be similar to the transmitted color. The reflected color may also be the complement of the transmitted color. Many particles do not transmit light effectively and may appear to be opaque and bright in transmitted light but are distinctly colored in reflected light.

Refractive Index

Refractive index, also a morphologic descriptor, is due to the fact that light travels more rapidly in a vacuum or in air than in solid or liquid media. The ratio of the speed of light in a vacuum to its speed in a transparent solid or liquid medium is termed the refractive index (n). That is,

$$n = \frac{V_v}{V_m}$$

where V_v is the speed of light in a vacuum and V_m is the velocity in another medium. This property is responsible for the deviation of light rays by glass lenses; it is also responsible for a number of phenomena observed when light microscopes are used in particle identification. Glass, for example, with a refractive index of 1.52, allows light to travel at a velocity of about 64 percent of its speed in a vacuum.

Transparent, colorless particles will be visible in water only if the n of the particle differs to some extent from that of the liquid. Glass fragments may be seen in water due to the relatively large difference in n between the two materials (1.5 vs. 1.33). A series of refractive index liquids (1.3 to 2.10) are available that have closely specified refractive indices (I_R), which may be used as suspending media to determine the I_R of unknown particles.

Reflectivity

The reflectivity of a material is a function of its refractive index and its microstructure (texture). Thus, very finely divided materials may reflect light very efficiently if they are not strong absorbers of light. This gives rise to diffuse reflection. Diffuse reflection is reflection in many directions with an incident beam from one direction. Rough surfaces (fragments of filter membrane) or heterogeneous materials consisting of dispersed systems (such as pharmaceutical emulsions) are examples; the diffuse reflection of white light results in the white appearance of the particles.

As the refractive index of a material increases, so does the amount of light it reflects. This can be seen in the following equation:

$$R = \frac{(n_1 - n_2)^2}{(n_1 + n_2)^2}$$

where R is the percentage of the incident intensity reflected, n_1 is the the refractive index the incident medium, and n_2 is the the refractive index of the reflecting medium. This equation describes what is known as specular reflection. Specular reflection is a highly oriented, directional reflection like that from a flat mirror. The angle of a specular reflection will equal the angle of incidence of the reflected light. Cleavage surfaces, crystal faces, and smooth glassy surfaces reflect light specularly.

Luster is also a term commonly used to describe the way a material reflects light (Crutcher 1986). Luster is the result of the efficiency of the reflection. For smooth particles when the difference between the refractive index of the particle and the mounting medium is large, the luster is "adamantine"; a diamond is an example. When the refractive indices are closer, it is "vitreous" or glassy.

Crystals and Crystal Morphology

Crystalline materials are characterized by a highly ordered structure based on a characteristic, unique internal molecular and atomic arrangement. This ordered structure results in a diagnostic external morphology that may be observed with a compound microscope using either reflected or transmitted white (unpolarized) light. This mode of microscopy results in observation of the crystal habit or external morphology. Polarized light microscopy, described in a later section, is extensively used for the study and identification of crystals. A very useful general text describing crystal morphology is Glusker et al. (1994).

The shape of a crystalline solid results from its molecular, crystalline structure. The first and most characteristic expression is in its growth habit (Flint 1971; Winchell and Winchell 1964). This is a reflection of the molecular arrangement in a compound and can be sufficient to establish the identify of the compound. A second descriptor is the arrangement of the cleavage planes in the material (Kerry 1959). Cleavage planes reflect the relative bonding energies through the lattice and are generally less characteristic of a specific compound.

Crystal morphology also depends on the generation mechanism. To exhibit a characteristic growth habit, the crystal must grow free of physical constraints. Examples of such conditions include the growth of crystals from the vapor phase, solution by precipitation, or evaporation of airborne droplets of a solution. Typically, the slower the growth rate, the larger and more perfect the crystal's structure.

Cleavage is the result of the force or stress applied to the material. Crushing or grinding operations typically produce cleavage fragments in those materials that exhibit cleavage. In some materials, such as the micas, the bonding energies are so different along different paths in the crystal that only cleavage fragments occur. In other materials, such as the feldspars, cleavage may or may not be present.

The habits (external morphology) of crystals are shown in Figure 10.14. Terms used to describe crystals (Crutcher 1986) include the following:

- *Crystal form:* A set of symmetrically equivalent faces in a single crystal. A crystal may display several crystal forms.
- *Crystal structure:* The pattern of the regular arrangement of atoms and ions in space. It usually refers to the basic (or average) structure of solids (and the ordered positions of liquids) without reference to minor localized deviations.
- *Polycrystalline:* A substance composed of two or more single crystals.

Figure 10.14. The six crystal forms.

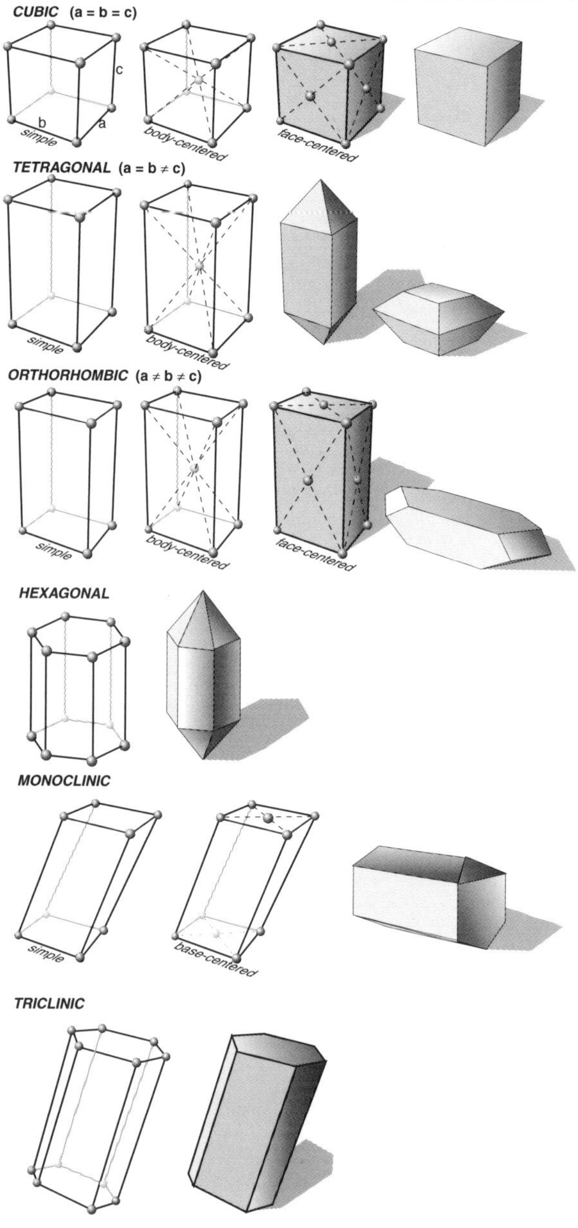

- *Polymorphs:* Two crystals with identical chemical composition but different crystal structures.
- *Polytypes:* Polymorphism with essentially identical structural components, like layers, which are arranged in various patterns, such as the stacking of layers.
- *Single crystal:* A crystal containing an uninterrupted crystal structure in a single orientation.
- *Twinned crystal:* A crystal composed of two or more single crystals, where adjacent crystals share a plane that is an integral part of both orientations of the crystal structure.
- *Crystal habit:* The actual outward shape or form assumed by a crystal or group of crystals as a result of the growth of dominant crystal forms (faces).
- *Aciculate:* An extremely slender crystal with small, cross-sectional dimensions (a special case of prismatic form). Aciculate crystals may be blunt ended or pointed. The term *needlelike* refers to an acicular crystal with pointed termination at one or both ends.
- *Blade:* A bladelike crystal with one longer dimension and two unequal, much shorter dimensions.
- *Equant:* A single crystal or grain with three approximately equal space dimensions.
- *Filiform:* Threadlike (mineral fibers).
- *Lamellar:* A very thin, platey crystal.
- *Platey:* A crystal with one short dimension and two longer, approximately equal dimensions. Chlorite micas and talc usually crystallize into platey shapes.
- *Prismatic:* A single crystal with one elongated dimension and two shorter, approximately equal dimensions.
- *Cleavage:* The tendency of a crystal to break in definite directions that are related to the crystal structure and are always parallel to possible crystal faces.
- *Crystal aggregate:* A cohesive mass of individual crystals or grains.
- *Columnar:* The arrangement of a group of approximately parallel, prismatic, acicular, or bladed crystals.
- *Fibrous:* An elongated and flexible form of matter.
- *Massive:* Homogeneous structure without stratification, flowbanding, foliation, or schistosity. Also crystals or crystalline grains that are tightly packed and scarcely distinguishable.
- *Radiating:* An arrangement of prismatic, acicular, or bladed crystals that appear to be diverging from a common center.

- *Reticulated:* The pattern of a crisscross network of acicular, prismatic, or bladed crystals.
- *Spherolite:* A spherical arrangement of long chain molecules or small acicular crystals about a nucleation point.

Fibers

Fibers, a widely occurring contaminant in pharmaceutical products, are characterized by their aspect ratio (length/height), diameter, internal structure, surface structure, termination characteristics, and cross-sectional shape. Fibers can be placed into five groups: plant, hair, insect, synthetic, and mineral fibers (Crutcher 1986; McCrone and Delly 1973b). Plant fiber is a category that includes all of the natural cellulose fibers. These fibers represent many different functional parts of plants: the woody pulp of trees (mostly paper pulp); grass (straw, cane, bamboo); leaf hairs (mullen); leaf fiber (sisal); seed fiber (cotton, cottonwood seed); and blast fibers (flax, hemp, jute). Plant fibers vary from single celled to multicellular in structure, but they all consist of the dried cells of functioning plant parts. As a result, their morphologies are all related. They generally have a hollow core called the lumen. They are all composed of cellulose fibrils and may have associated structures that indicate their function in the plant.

To analyze fibers, the sample material is first examined at low magnification (50 to 60×). Selected groups of fibers from the textile material are then mounted and examined at a higher magnification (250 to 500×). The longitudinal appearance of individual fibers is noted and compared with that of known fiber comparison standards. If a more careful examination is required, the cross-sectional appearance may also be determined. Examination of the fibers by microscope can provide positive identification of the principal natural fibers. The appearance of man-made fibers, however, may vary depending on variations in the manufacturing processes of different companies.

Most synthetic fibers are plastics of various types. Glass fiber is the most notable exception, though carbon and metal fiber are increasingly common today. They all lack the complex and detailed structure of biological fibers. Their form is the result of mass production practices. Mineral fibers are individual crystals or bundles of similar individual crystals and, as such, have a structure that is dictated by the laws of crystallography.

Hair

Strictly defined, hair is a fibrous body-surface covering peculiar to mammals. Given the diversity of animal species, hairs of a wide variety of morphotypes occur. "Fur" animals typically have two types of hair: coarser guard hairs and a much finer type referred to as fleece hair. Despite the wide variety of types of hair, only a very narrow range of hairs is of a interest to the pharmaceutical particle analyst; yet all hairs have generally negative implications if present in pharmaceutical products. As an example of the importance of identifying hairs, consider a case where a hair is noted by a customer in the blister pack of a medical device. What kind of hair is it? If it is a mouse or rat hair, its presence has serious negative implications for the manufacturing process. If it is a cat or dog hair, the situation is not as dark, due to the fact that pet hairs ride on the street clothes of workers. A human hair simply dictates that better garb or garbing procedures should be implemented. The discrimination between these hair types is simple, and the author can remember a situation when the simple discrimination between human hair and animal hair became very important in a regulatory context.

"Mammalian" hair may be divided into either two or three structural zones or layers depending on the author involved. There is an outer layer, or cortex, and an inner core or canal called the medulla. Some authors discriminate an outer layer as a cuticle or shell composed of a shingled array of plates or scales that are more or less prominent based on the species of origin (see Exhibits 19 and 20 in the section of color plates in this book as examples). In addition to the scale pattern, the identification of different kinds of hairs will be based on diameter, a round or flattened cross-section, curl, length, and color. The pigment granules that give hair color are located in the cortex; if they are dark and dense, the color of the hair will be dark or black. Light-colored granules or smaller numbers of granules will result in a blond, red, or brown hair.

The structure of the medulla or central core of a hair is often diagnostic for the species of origin. The medulla may be continuous, segmented, or broken, and in some large hairs of some species so small as to be invisible without careful observation. The medulla pattern is obvious and diagnostic for rodent hair, consisting of a regular and linear series of blocks or segments shaped like the letters T and H. Human hairs show a much less regular medulla pattern. The morphologic descriptors of human hairs vary to a significant extent between different races and subraces. Forensic scientists have made extensive studies of hair, and many important criminal cases have been solved by tracing hair found at a crime scene to the culprit.

The termination of a hair provides some evidence as to its source and a great deal of information regarding its history. The cortex of a hair is composed of elongated protein fibers cemented together by a polyamide adhesive. This cementing material is subject to dissolution by bases, so exposure to shampoo or soap will have the effect of swelling or gradually disrupting the hair structure. This is evident on a mature human scalp hair as "split ends." Freshly cut hairs have sharp, angular terminations; after cutting, the cut becomes rounded and less angular.

The refractive indices of most hairs are in the range of 1.54 to 1.55. Human hair ranges from as little as 50 μm to greater than 150 μm in diameter.

The analyst desiring to obtain expertise in the identification of hair is counseled to simply practice with hairs of different types until a mental database based on sight identification is developed. This can be done in a surprisingly short period of time. The distinct morphology of mammalian hair sets it apart from other animal (nonvegetative) fibers. Such materials as insect-generated fibers (silk, spider web) have a structure more akin to synthetic fibers. The hairs of insects are very different from mammalian hairs, being, for the most part, hollow pointed or tapering transparent cylinders. The characteristics of plant hairs, as discussed above, are at a smaller scale, i.e., striations, pits, cross-markings, and highly variable extinction patterns.

Analysis of Fibers

Identification of fibers can be performed on the basis of distinctive morphology, extinction patterns or sign of elongation, or a combination of these. Paper, the most common pharmaceutical contaminant fiber, shows no extinction in crossed polars. Cotton, the second most common, is of smaller diameter, does not show extinction, and often displays a regularly twisted ribbon morphology. A number of morphological distinctions on the basis of fiber morphology in unpolarized light are made in Figure 10.15. The key is intended to serve only as an example of how such a structural approach may be used to expedite an analysis. Some microscopists among the author's aquaintances who work with pharmaceutical products have found it useful to devise their own keys that are customized with regard to the fiber descriptors used, the types of fibers included, and the points of dichotomy (e.g., twisted versus not twisted, etc.).

Biologicals

Biological particles may, in a general overview, be discriminated from those of physical or chemical origins by virtue of the relationship of their morphology to their functional role or mechanism of function in the living organism. The plant fibers discussed earlier are a good example. Paper fibers generally reflect the structural-functional elements of plants from which they come, such as channels for conducting fluids, pores for respiration, pits providing channels for cross-flow with other fibers, and sites of adhesion between the fiber cells. Further, the general form is exceedingly complex in comparison to mineral or synthetic fibers. The morphology of biological materials is typically more varied and complex than inanimate objects, but some crossover occurs. Starch grains, which are in effect a chemical compound produced by living cells, show similarity in size, structure, and shape but incorporate minute irregularities, thus making each grain an individual.

Human skin cells may be generally characterized as transparent ovals 10 μm to 20 μm in size, but on more detailed examination show nuclear remnants, pigment, and, in polarized light, a complex extinction pattern due to the structure of their biomolecules of which they are built.

Simple unicellular organisms, such as bacteria and yeasts, are of generally regular structure but display size differences and characteristics of a life process such as budding or fission. Diatom shells are composed almost entirely of SiO_2 but display a highly complex and ordered substructure that cannot reasonably be confused with a particle generated by physical forces or

Figure 10.15. Fiber morphology key (courtesy of Dr. Philip Austin).

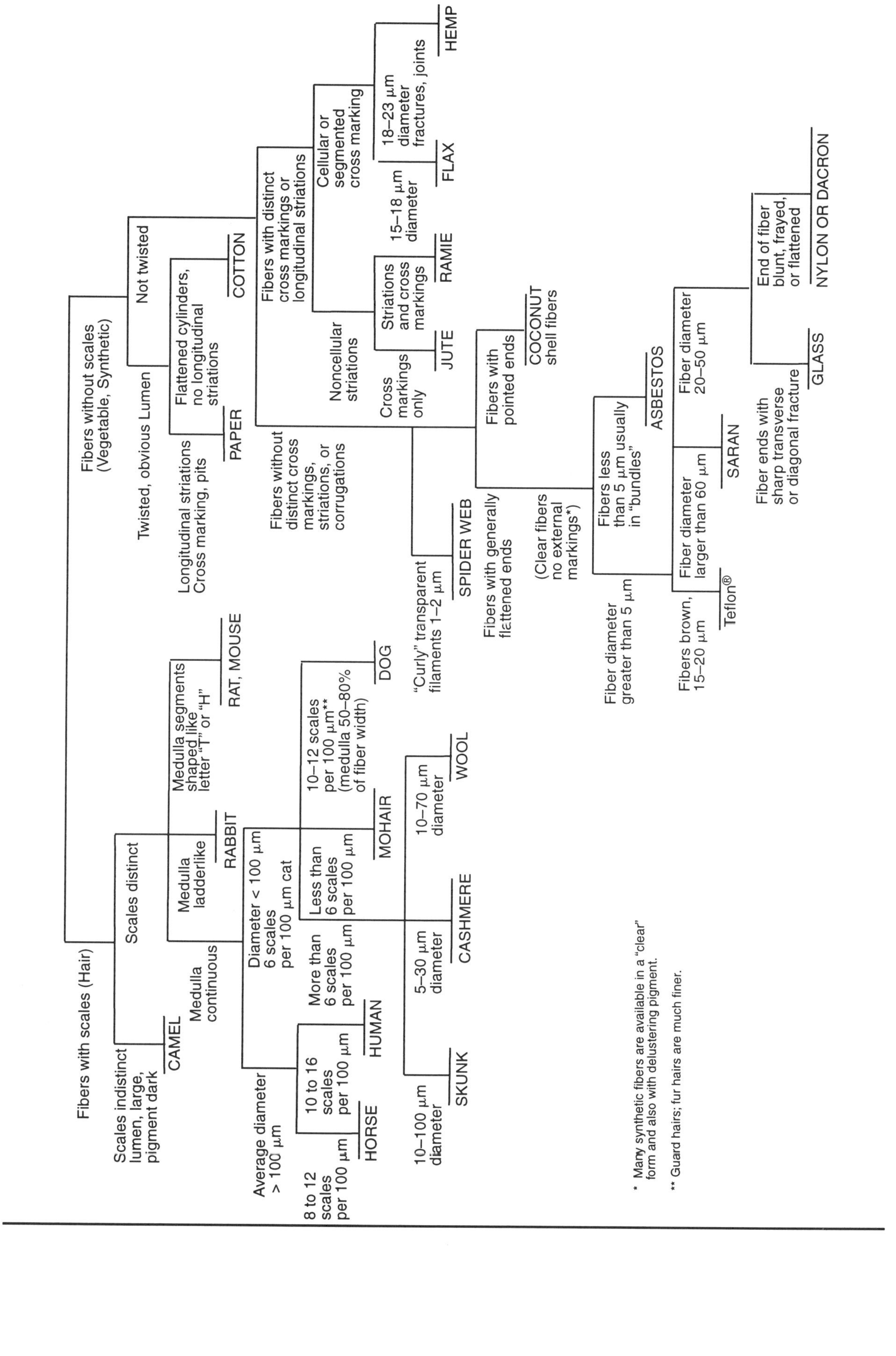

chemical reactions. Many biological morphotypes are distinguishable by a substructure that is generally ordered but random in orientation, e.g., a hyphal mass of mold. Finally, some materials of biological origin have a morphology that can be associated with the gross visual appearance of the organism from which they came. Insect parts, such as the segmented appendages of a beetle's leg or the scales of a moth ring, fall into this category.

Biological reaction products form a separate subcategory of morphology. As an example, the humoral components of mammalian blood (including human) react to form various types of solid or semisolid aggregates. The best known example is a blood thrombosis or clot, which may result in a customer observation with a medical device during use. Here the particle analyst is able to call on the technology available to the histologist and the pathologist. Selective stains with color-generating reactions or based on selective uptake of dyes by a certain biochemical structure will allow the analyst to discriminate between and identify materials that have no distinct morphology.

Other Morphologically Distinct Particles

Having discussed in some detail the structural characteristics of crystalline and biological particulate materials, it is also necessary to mention a class of particles that falls into neither of these two categories on the basis of morphology. Generally, these particles have no morphologic descriptors relating to the materials of which they are composed but rather reflect in form their means of generation or source of origin. These generally fall into two classes: (1) those formed by combustion or thermal processes and (2) others generated by the operation of machines or man-made devices. Crutcher (1986) uses the term *morphologic sourcing* to describe the characterization of particles the form of which leads the analyst back to the source.

Let's first consider the generation of particles by thermal processes. Sources of these particles can vary over a range from oil or coal-fired furnaces to molding or heat-sealing operations in medical device manufacture. The simplest case is likely a coal-fired furnace. The coal contains not only carbon but also levels of sulfur, silicon, calcium, aluminum, iron, and other elements. Consider silicon as a case in point. If the temperature of the combustion process exceeds that necessary to liquefy the material, silicon can be converted to a glassy or liquid state. The concomitant release of superheated combustion gases provides a mechanism for the production of bubbles or balloons of the glassy material that, on cooling, form hollow, glassy spheres akin to minute Christmas tree balls. These are termed cenospheres (Exhibit 32). Liquid fuels such as those used in diesel engines generate similar spherical particles of carbonaceous material that may be hollow or solid. This latter material is the "soot" we see issuing from large trucks on the highway as they accelerate.

Similar particles may be generated by the processes used to manufacture glass containers for injectable dosage forms. In this case, glass cenospheres form part of the particle population found in some bottles and vials prior to washing. The high-temperature glass molding process may also result in spherical particles with a high iron or carbon content, with these particles having their origin in iron oxides on the vessels used to hold the molten glass on conveyor belts or in carbonized compounds that build up at process points.

Other thermal or melting processes at lower temperatures also generate spherical particles. Examples are blow-fill-seal molding processes or the heat sealing or extrusion of plastics. In these cases, tiny, sometimes submicrometer, spheres of molten polymers result. The commonality of spherical shape is due to the molten materials' following the physical dictate to assume a shape with a minimal surface-area-to-volume ratio with the lowest free energy while they are in the molten state.

A second grouping of particles in this general class is also based generally on mechanisms of volatilization. As an example, one can consider a high-speed filling, stoppering, or vial handling machine. Moving parts of such machines are in contact at high linear or rotational speeds and must be lubricated. Submicrometer- or micrometer-sized particles consisting of materials worn from the moving surfaces in conjunction with the volatilized lubricant may be ejected from the wear surfaces and aerosolized by directional airflow. These aerosol populations may be responsible for high counts collected using optical particle counters in clean process areas during process operations and may be collected and identified using filtrative collection techniques.

POLARIZED LIGHT MICROSCOPY

Polarization

The polarizing light microscope is an invaluable aid in identifying particles based on the principle of the polarization of light. Ordinary unpolarized light consists of a bundle of rays having a common propagation direction (linear axis) but different vibration directions (axis perpendicular to linear axis). The rays of polarized light have a single vibration plane. Polarized light may be produced from ordinary light by reflection, double refraction with a suitable crystal, or absorption.

The completeness of the polarization of reflected light depends on the optical properties of the reflecting surface and on the angle of incidence (and hence, the angle of reflection). Transparent substances at a specific characteristic angle of incidence may be nearly 100 percent effective as polarizers. For each transparent substance, the particular angle of reflection giving maximum polarization is called the Brewster angle. The angle for maximum polarization, *i*, is about 57° for most optical glasses with a refractive index of approximately 1.5.

Most crystalline materials show different properties of light absorption and transmission on different axes. Such crystals are called pleochroic. A pleochroic compound showing very strong absorption in one direction and very weak absorption perpendicular to that direction will transmit polarized white light with the plane of vibration corresponding to the weak absorption direction. Early polarizing filters were composed of such crystals. Polarization by absorption depends on all visible wavelengths being absorbed, but strong polarization of light in the visible range does not necessarily equate to polarization in the UV and infrared (invisible) ranges. The function of one type of polarizer in producing polarized light is shown in Figure 10.16.

Figure 10.16. Polarization of light by a Nicol prism.

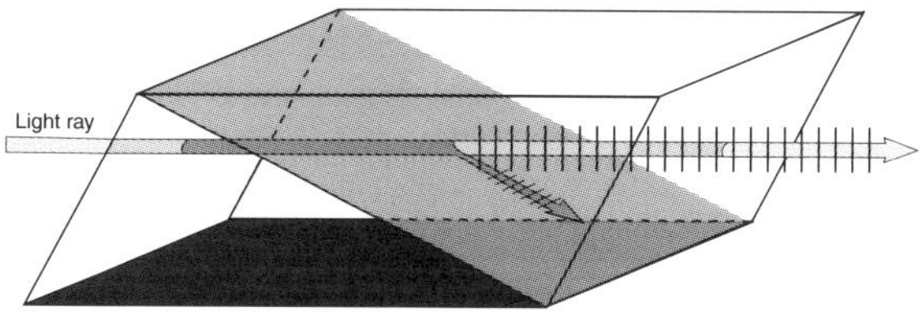

Interference

In addition to polarization, interference between light waves traveling in mutually parallel ray paths is a second key principle in using the polarizing microscope. Two light rays from a coherent source arriving at a point in phase agreement will reinforce each other. This is known as constructive interference (Figure 10.17). If they are completely out of phase, they show destructive interference and cancel each other (Figure 10.18). Two light rays from a single source can interfere destructively by various mechanisms. One is reflection of a light ray by a thin, transparent film or by transparent layers of a crystalline material. One ray is reflected from the top surface and one from the bottom surface of the film or crystal. The distance traveled by the first ray in excess of the second ray is twice the thickness of the film or greater if the incidence is at an angle (Figure 10.19). With white light, the difference in distance traveled may cause the light in the ray paths of different length to form colors when they recombine (interference colors).

Figure 10.17. Constructive interference between two in-phase light beams.

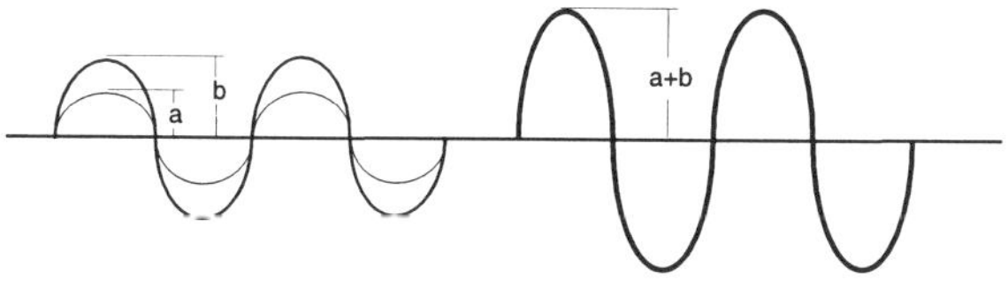

Construction of the Polarizing Light Microscope

Knowing the principles of the polarization of light and interference between light of various wavelengths in mutually parallel ray paths, it is relatively easy to understand the use of the polarizing microscope in particle identification. The simplified design of a polarizing

Figure 10.18. Destructive interference.

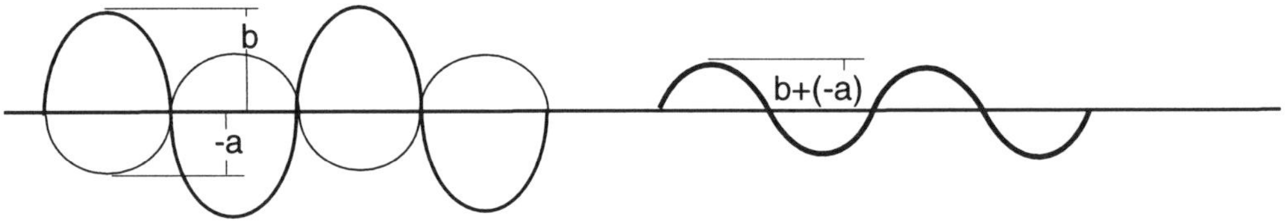

Figure 10.19. Thin film interference.

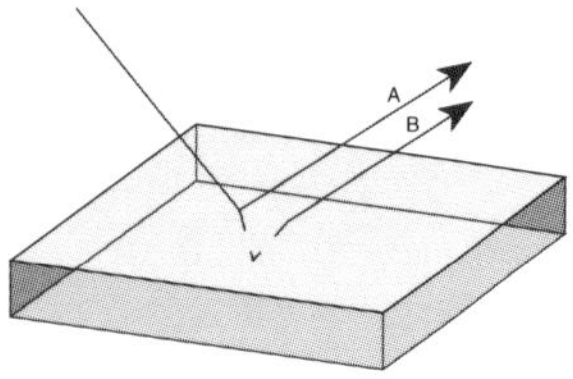

microscope involves a polarizing crystal plate inserted between the sample and the light source (polarizer). This serves to polarize the incident light that passes through the specimen (Figure 10.20). A second polar analyzer is inserted above the specimen so that light that has passed through the specimen may also be polarized. The analyzer is a linear polarizing filter that should be easily removed and rotatable through a full 180°. This is the second polarizing filter in a polarized light microscope and is used in conjunction with the polarizer below the substage condenser.

Only light that has been polarized (it vibrates in planes parallel to the axis of polarization of the analyzer) will pass into the ocular. If the vibration direction for the two polarizing plates is perpendicular (crossed polars), the field will appear dark.

An analytical light microscope equipped with polarizing filters is shown schematically in Figure 10.21. The compensator filters are special colorless sheets of a crystalline material that are used as reference standards for evaluating the structure and orientation of properties of anisotropic particles. A quarter-wave and full-wave plate compensator are most often used.

The back focal plane of the objective is where the Fourier transform of the field image is formed as well as the location of the focused image of the substage iris. Special optical stops are placed in this plane for some types of dispersion staining. This plane is also where the interference figure of a crystalline particle is imaged. The bertrand flip-in lens on most polarizing microscopes can be used to see the back focal plane of the objective without having to remove the eyepiece of the microscope.

The stage of an analytical microscope should be axis centered or centerable and rotatable. A vernier mechanical stage is also desirable. The rotatable stage is required to establish optical properties in anisotropic particles. The verniered stage is used to standardize the transverse pattern and to ensure that the same field is not analyzed twice. The substage condenser and stage should move in unison while focusing the microscope. If the unit does not move during the focusing, the scope will not be adequate for the convenient mounting of any photodetection instrumentation.

The substage condenser must be capable of supplying a wider angled cone of light than the objective can collect and must be free of strain. The wider cone of light is needed for dark-field types of illumination. Strain in this lens system introduces aberrant effects because the linear polarizer is located beneath it.

The count of morphological and optical detail collected is limited by the geometric and physical optical of the objective and the positional geometry of the illumination particle objective system. On most microscopes, the polarizer is built into the bottom of the substage condenser. It is often more convenient to have this filter in a position where it is easily rotated or removed from the light path.

The field iris is an aperture-controlling device. One of its uses is to limit the amount of the field illuminated. Normally, it is adjusted to the edge of the field of view. For special tests, such

Figure 10.20. Principle of the polarizing light microscope.

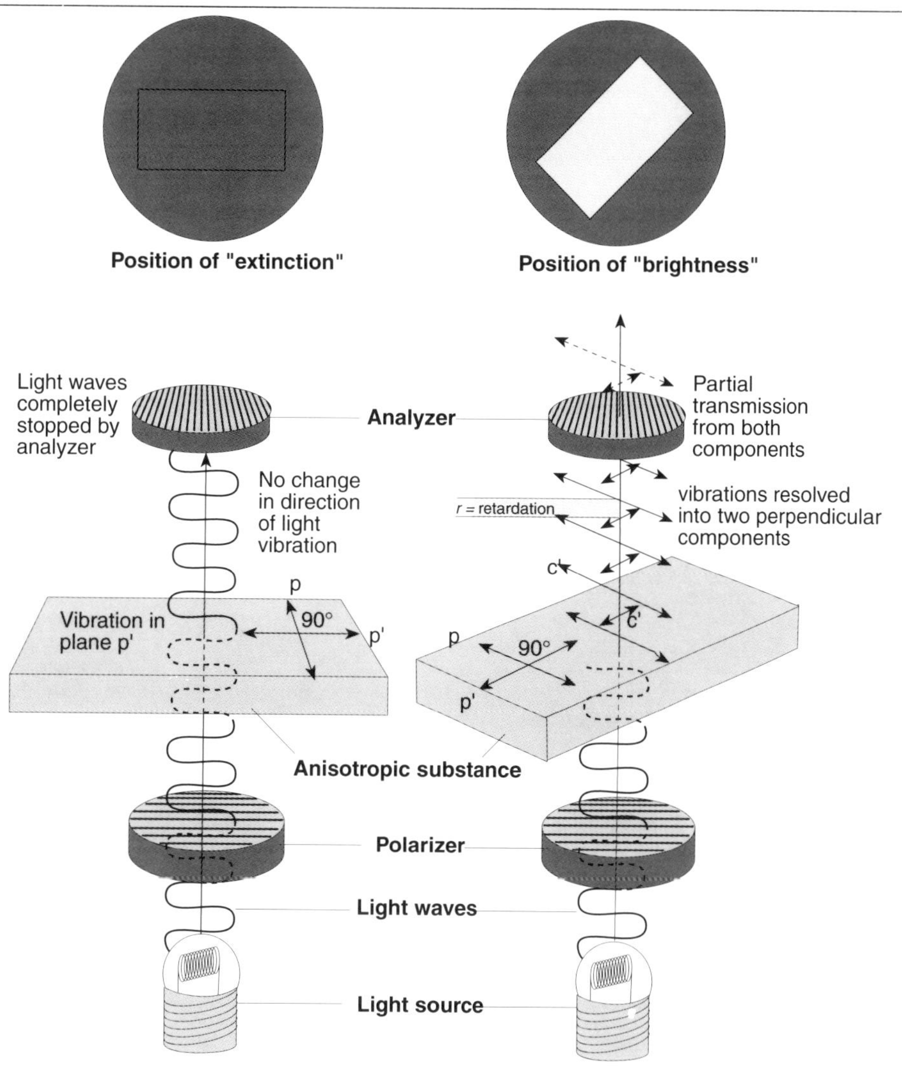

as an examination of an interference figure, it may be used to isolate a specific particle. Above the iris are the substage condenser filters. (Microscope construction will vary with manufacturer.) The filters normally included at various times are neutral density filters, a blue color compensating filter, and/or rotating linear polarizing filter called the "polarizer."

Above this is the substage condenser filter pack. The configuration of this filter pack may be as complex as a set of slide-in filters built into the microscope stand or as simple as a set of filter stacks positioned above the field iris by the operator.

The first component in the system is the illuminator. This is not simply a source of light, but it must be source of a reasonably constant and characteristic distribution of the visible wavelengths. It should have a radiation temperature above 3,000 Kelvin as a minimum and be relatively intense. A 30–100 W quartz halogen illuminator would meet these requirements. Other types of illuminators may be useful, such as a UV source, for specific applications. The illuminator includes a lens system called the lamp condenser. The bulb and lamp condenser must be in proper adjustment to optimize the illumination of the particle.

Figure 10.21. Analytical compound microscope.

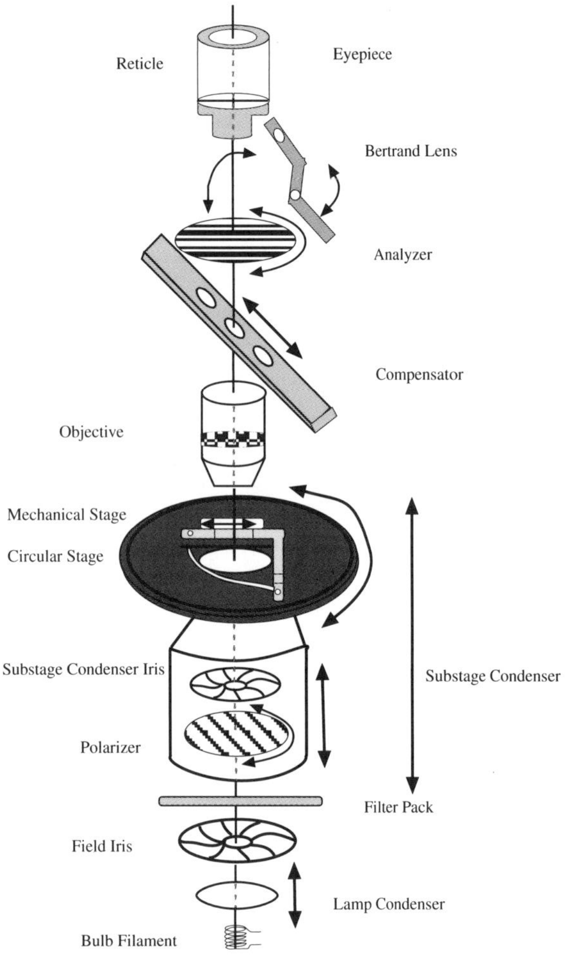

Figure 10.22. Köhler illumination.

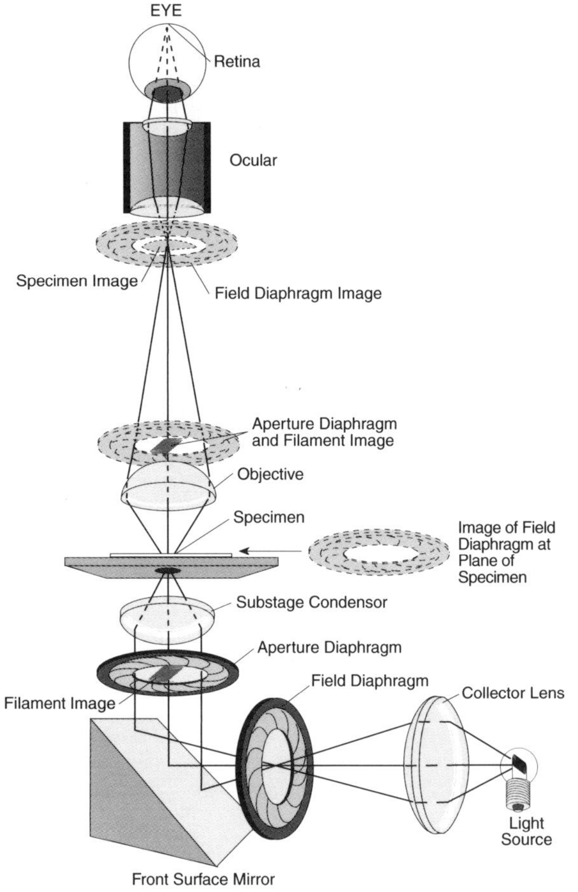

Illumination in Polarized Light Microscopy

A very important consideration at this point is that the quality of the image seen with a polarizing microscope and its interpretive value will depend on the quality of illumination. The alignment of illumination of a polarizing microscope is critical to obtaining uniform, reproducible images. The most commonly used form of illumination is Köhler illumination, so termed for its originator (Figure 10.22). First, the field diaphragm is focused in the plane of the specimen using the substage condenser focus. The substage condenser iris of a polarizing microscope is located in the substage condenser in a specific position such that its image can also be focused and visualized. When the microscope is in focus and the substage condenser is properly tuned, the iris is focused at the back focal plane of the objective. This can be tested by removing the eyepiece and looking down the body tube of the microscope. The substage condenser iris should be seen in focus. This configuration is essential to the evaluation of the particles in the field and is a fundamental requirement for Köhler, Köhler-like, or good diffuse illumination systems. The substage iris location is used for many special analytical techniques, such as phase contrast, dark-field illumination, and dispersion staining. Each of these techniques requires a special modification to the light beam at the substage iris position.

Particle Identification with Polarized Light Microscopy

If transparent particles are placed on a slide between crossed polars, some crystals will appear colored, others will appear bright, and some will be invisible against the black background. The crystals that appear bright or colored are anisotropic and must have at least two principal refractive indices. Those that are invisible are isotropic and are either glassy, cubic crystals or unoriented polymers. If the orientation of the crystals that appear bright or colored between crossed polars is changed by rotating the stage, they will seem to disappear (become black) four times during the complete rotation of the stage. These positions, 90° apart, are called extinction positions and show the vibration axes of each crystal.

When polarized light enters an anisotropic crystal (a crystal with different indices of refraction on different axes), the light is resolved into components vibrating in two perpendicular planes (Figure 10.23). This splitting of plane polarized light into two components is called double refraction. Since the components follow two principal vibration directions having different refractive indices, they move through the crystal at different speeds; they emerge with one retarded by a definite amount that depends on the difference in the two refractive indices, $n_2 - n_1$

Figure 10.23. Slow and fast crystal vibration axes.

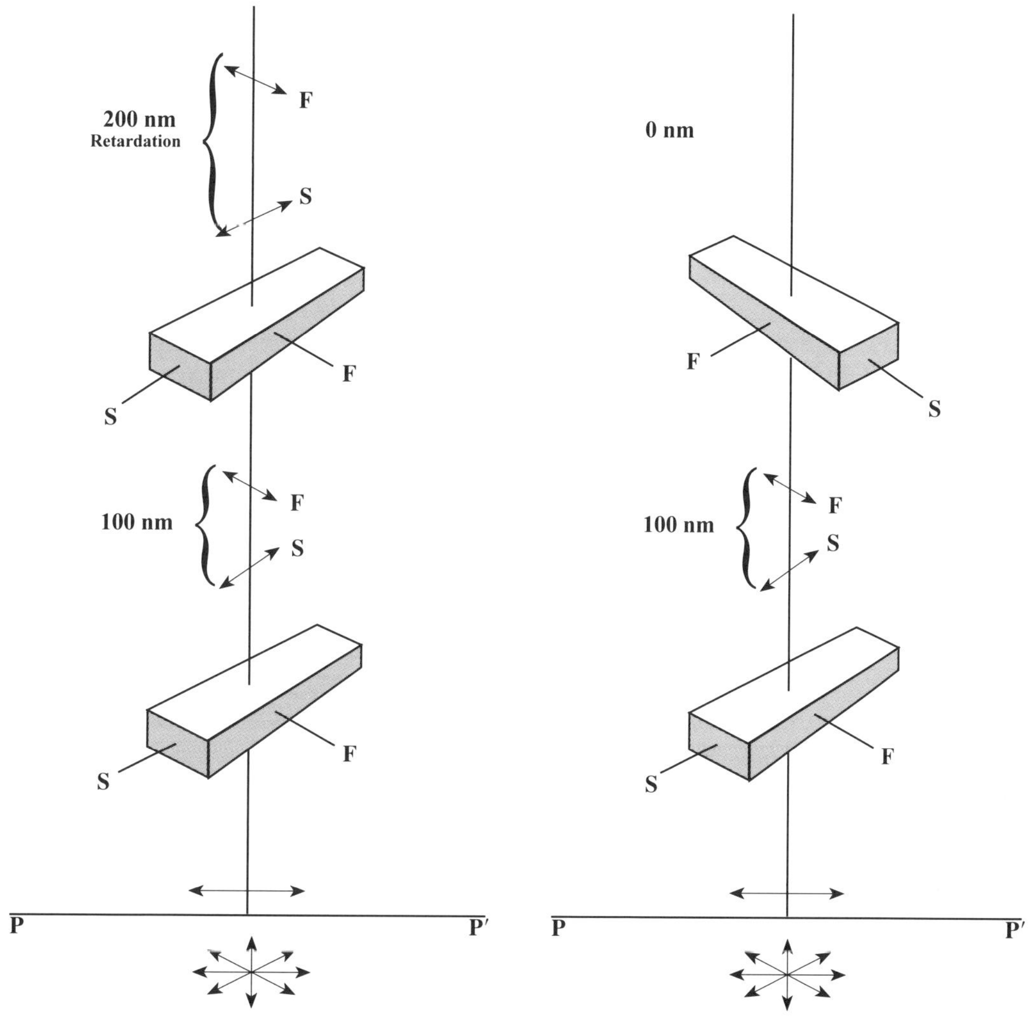

(birefringence), and the thickness. The actual offset of the wave fronts is called the retardation.

If the polars are crossed and the crystal is oriented so that one of its principal refractive indices is parallel to the vibration direction of the polarizer, the second vector component becomes zero. All light emerging from the crystal has the same vibration direction as the polarizer; it will be absorbed by the analyzer plate. The analyzer plate has a vibration direction perpendicular to that of the polarizer. Both the crystal and the background will appear dark when the crystal is in the extinction position.

If a birefringent (anisotropic) crystal is oriented so that a principal axis of refraction is not parallel to the vibration direction of the polarizer, the emerging components of illumination will recombine in the vibration plane of the analyzer. Interference will destroy some wavelengths of light and reinforce others. Since one component is retarded, interference on recombination of the two components by the analyzer may cause the image to appear colored. The colors seen will depend on the retardation, which, in turn, depends on both the thickness and the differences in refractive indices (birefringence). If the crystal varies in thickness, several colors may be observed. The colors will be brightest when the crystal is between extinction positions.

Both interference colors and patterns of extinction are characteristics of different transparent materials. Calcium carbonate, calcium oxalate, skin cells, penicillin powder, starch, and other materials of interest to the particle analyst have their own interference colors and patterns. It is possible to use the polarizing microscope to perform detailed studies regarding the properties of crystalline materials. The methodology used for this purpose is covered in some detail by Winchell and Winchell (1964), McCormack et al. (1976), and McCrone et al. (1973a, 1973b, 1973c, 1973d). The pharmaceutical particle analyst does not commonly wish to perform studies on the crystalline properties of an unknown; only identification is desired. For this purpose, a relatively unsophisticated use of the polarizing microscope is sufficient.

The light microscope was originally developed to examine the magnified image of a small object or feature. This is still its principal use, but the understanding of the physical properties of the image has evolved to the point where the morphology of the particle can be secondary to its optical properties. These optical properties are the basis of particle identification by polarized light microscopy and provide information on the molecular or crystalline structure and elemental composition of the material as well as often providing historical information regarding its generation and exposure to the environment since generation or emission. In the case of submicrometer particles, the visual morphology has no significance; only optical properties can be measured. See Leigh-Dugmore (1961), McCrone et. al. (1969), and Saylor (1967) for further information on the uses of polarized light microscopy.

An example of this type of use is illustrated by McCrone's *The Particle Atlas*. Some of the recent volumes of this series are abundantly illustrated with color plates of many types of particles as seen under a polarizing microscope. Almost without exception, these micrographs were taken with "slightly uncrossed polars." The use of slightly uncrossed polars allows enough unpolarized light to pass through the specimen so that both anisotropic and isotropic particles are visible, yet the extent of polarization present allows extinction patterns and interference colors to be seen. Thus, an analyst can readily compare the morphology of an unknown with a picture. More importantly, the morphology of the unknown is easily compared with a particle of known identity.

The use of comparison standards and slightly uncrossed polars is a key principle in the identification of particles encountered in a pharmaceutical environment. Comparison standards to which an unknown may be compared are readily available and are based on a little inductive reasoning. The situation is even further simplified by the fact that many pharmaceutical particles are identifiable on sight (e.g., skin fragments, starch, talc, glass fragments, and hair).

Because of the historical development of this form of analysis, a considerable reference base exists in the literature that can be used to gain a better understanding of the principle of polarized light microscopy in particle identification. With regard to cognitive identification (identification of particles by sight), McCrone's *Particle Atlas* is invaluable. The *Handbook of Chemical Microscopy* (Mason 1983) provides information regarding a number of useful chemical methods. The paper published by DeLuca and Bodapatti as an FDA guideline is of historical interest, as are those of Godfray (1979), Oles (1978), and Trasen (1968a and 1968b). Many of the "spot tests" described by Feigl and Anger (1972) are useful for the microscopic identification of particles.

Definitions

There are a number of commonly used terms that apply to the application of the polarizing microscope. The analyst should be familiar with these.

- *Crystallinity:* Any crystal belonging to the hexagonal, tetragonal, orthorhombic, monoclinic, or triclinic families will be birefringent. Polymers that exhibit a preferential alignment of molecules will also be birefringent.
- *Interference figure:* A pattern that forms at the back focal plane of the objective. This pattern contains the Fourier transform of all the optical information that will be visible in the final image, as well as data on the crystal class and orientation of the particle. To use the interference figure analytically, a single particle must be isolated from others in the field.
- *Optic axis:* An imaginary line through the material to which all optical properties are the same. Viewed down this axis with a narrow parallel light beam, anisotropic material would not appear to be anisotropic.
- *Optic sign:* The optic sign of a crystal relates the refractive indices of the crystal to part of the crystal's molecular structure. Stated simply, the crystal is optically positioned if its low refractive index characterizes a plane of symmetry; for biaxial crystals, the two lowest refractive indices are much closer together than either is to the highest refractive index.
- *Pleochroism:* A property of a particle that is exhibited when that particle changes color when viewed in different orientations under a linear polarizing filter.
- *Polarized light:* This term refers to linear polarized light. Light is polarized if all of its field vectors are aligned parallel to one another.
- *Polarizing filter:* A polarizing filter only allows light polarized parallel to one line to pass through. Polarizing light microscopes contain a filter below the stage called the polarizer and one above the stage called the analyzer.
- *Retardation color:* An interference color effect seen in anisotropic materials viewed between crossed polars produced by a combination of the thickness of the particle and the difference between the two refractive indices measurable in that specific orientation of the material. The change of colors with increases in retardation follows a modified Newtonian interference color scale known as a Michel-Lévy chart sequence.
- *Sign of elongation:* Materials that have an obvious long axis can be easily tested to determine if their high refractive index is oriented more closely to that axis. If it is, the material has a positive sign of elongation. If not, the sign of elongation is negative.
- *Uniaxial crystal:* A crystal that has only one optic axis is uniaxial. Hexagonal, tetragonal, and trigonal crystal systems are uniaxial.
- *Anisotropy:* An anisotropic material is bright when viewed between cross polars. It indicates that the material has more than one refractive index.
- *Biabsorption:* The difference between the maximum and minimum absorption exhibited by a material in any possible orientation. This is a fundamental property of crystalline materials.
- *Biaxial crystal:* A crystal that has two optic axes is biaxial. The orthorhombic, monoclinic, and triclinic crystal systems are biaxial.
- *Birefringence:* The difference between the maximum and minimum refractive indices exhibited by a material in any possible orientation. This is also a fundamental property of crystalline materials. These values can be used to identify the compound from tables of optical properties. The birefringence of a mineral is not always constant for all wavelengths. In fact, slight wavelength-dependent variations in birefringence are common. These differences produce dispersion of the birefringence and, in some materials, anomalous retardation colors that can be very useful in the identification of these compounds. The mineral melilite is a

common example of a combustion residue showing anomalous colors under polarized light. The term *birefringence* is synonymous with *anisotropy.*

- *Cross-polarized light:* The use of a linear polarized filter below the subject and another above the subject with the angle between the polarization direction for the two filters at 90° to one another. The result is that no light can pass through both filters except where modified by anisotropic material.
- *Extinction position:* As anisotropic material is rotated between crossed polarizing filters, there are four positions in which the linear polarized beam of illumination is transmitted with a single phase velocity. This occurs when the vibration direction in the beam of illuminating light matches one of the preferred vibrational directions of the material. In that position, all of the incident energy is transmitted along that path, and there can be no interference. When the material is in one of these positions, it becomes black against the black background and is said to be in an extinction position. If the material exhibits a characteristic morphology, the relationship between that morphology and the vibration direction of the illuminating beam of light producing extinction can be a very useful characteristic not parallel to or bisecting a characteristic feature. Oblique extinction can be more characteristic than other types of extinction, in that a specific angle between a characteristic morphology and the vibration plane of the illuminating beam can be identified.

MORPHOLOGY AND PARTICLE IDENTIFICATION

Anisotropic Substances

Any anisotropic substance will appear bright while the background and isotropic materials, such as glass, become black between crossed polars (linear polarizing filters with their polarization direction at 90° to one another). Anisotropy is a property that is immediately apparent using cross-polarized light and indicates the presence of a number of additional properties that can be used to characterize the anisotropic particle. Anisotropy indicates a directionally dependent difference in the material's ability to respond to the incoming light. A stretched piece of plastic, glass under stress (Durelli and Riley 1965), crystals (Bloss 1961; Hallimond 1970; Shubnilkov 1960), aligned polymers (Thetford and Simmens 1969), and so on all have the property that light polarized in one direction will move with a phase velocity greater (lower refractive index) than light polarized at right angles to that direction (higher refractive index). The phase velocity difference introduces a relative retardation in one wave front with respect to the other. This retardation results in an interference effect when these beams are recombined as they pass through a linearly polarizing filter.

Other types of extinction patterns are important in the characterization of particles, including the "Maltese cross" pattern seen in spherulites such as starch and around pores in wood fiber and fanning extinction seen in many plant fibers. Some plant structures grow from a center by sending single, acicular crystals in all directions from that center. The result is that in any position, four zones of crystals are in an extinction position. This produces a Maltese cross pattern with the nucleation center at the center of the spherical or radial structure. Fanning extinction is a term that describes the sweep of a black extinction line across a particle as it is rotated, as compared to a fixed pattern or full extinction position. This is due to the variety of fibril overlapping in plant fibers. Cotton has no extinction position at all. The presence of twinning, zoning, or interbedding in a crystal are most obvious near the extinction position of one of the phases.

Identification by Refractive Index

Objects are only visible if they are surrounded by a medium of dissimilar refractive index. That condition is satisfied for nearly everything in our day-to-day experience. Furthermore, everything is surrounded by a medium of lower refractive index—air (1.00) or water (1.33). Under the microscope, this is not the case. The medium selected to mount a sample may be lower or higher in refractive index than that of the sample. It important to know which condition is satisfied.

Small particles mounted in a liquid behave like small lenses. They will either focus a beam of light to a point or spread the light depending on where the liquid has a higher or lower refractive index. This is known as the lens effect and is very useful in the characterization of small particles (McCrone et al. 1969). A more generally applicable method used for larger particles and for very irregularly shaped particles is the Becke line method.

The refractive index of a particle can be determined using a variety of observable effects, including absorption, relief, Becke line motion, lens effect, dispersion, and relative phase shift. The most obvious of these is relief. Relief is seen as a sharp, dark border at the edge of the particle. If the border is a wide black band, the sample is said to have a high relief, which indicates the particle's refractive index is much higher or much lower than the medium. Low relief, a very thin black border, indicates a smaller difference in refractive index.

Absorption differences are the result of the sample or particle absorbing more light than the mounting media. This effect is seen in the middle of the particle and is independent of relief. Absorption is a measure of the imaginary refractive index, while all of the other effects are the result of the real part of the refractive index.

The polarizing microscope may also be used to determine particle refractive indices. This is performed using the polarizer and a range of refractive index oils that bracket the refractive index of the particle. The test performed is called the Becke line test. The Becke line is a bright ring of light (halo) at the periphery of a transparent particle that moves with respect to the particle edge as the objective focal point is moved up and down through the particle (Figure 10.24). The ring will always move into the higher refractive index medium as the focal point is raised and into the lower refractive index medium when the objective is lowered.

The formation of the Becke line is due to optical effects that concentrate light at an interface between two materials of different refractive index. When the microscope is critically focused on the crystal, no Becke line is visible. Raising the focus shifts the image in focus to the region above the crystal, where light is concentrated within the boundary of the higher index medium. Lowering the focus to a point below the best focus for the crystal causes the Becke line to cross the interface to the lower index medium. Using the Becke line test, the analyst can readily compare the refractive index of an unknown particle to that of a known liquid.

Figure 10.24. Becke line test: a. particle in focus, b. movement of line as focal plane raised (refractive index of particle higher than that of medium), and c. movement of line as focal plane raised (refractive index of medium or higher).

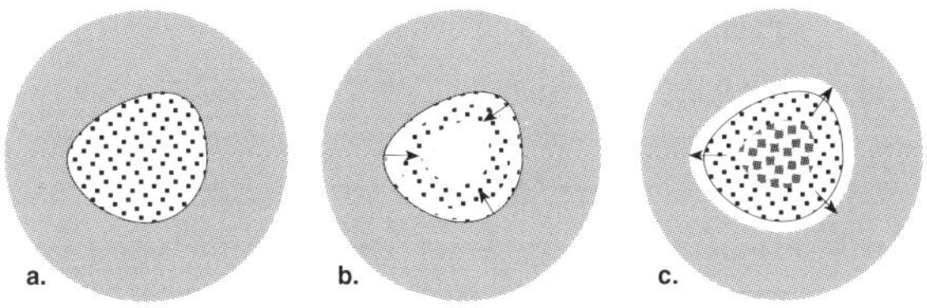

Refractive index media are available in a wide range of values that span the refractive index range of pharmaceutical contaminant particles.

When the refractive index of the particle matches the refractive index of the mounting medium for one wavelength but not the others, the Becke line breaks into two lines. One of the lines, a reddish one, generally moves into the particle, and the bluish one moves into the mounting media, or vice versa, depending on the relative refractive indices of the particle and the medium. This is the result of the relative dispersion of the particle in the mounting medium (see section on dispersion staining).

The most common method of measuring the refractive index of a particle or observing Becke lines is to immerse the particle in a liquid of well-known refractive index values and then to compare the particle to that liquid. The liquid around that particle is then changed, or another mount is made of identical particles in a different liquid until a match is found. Typically, the sequence below is followed.

A small drop of the refractive index standard liquid is placed on a clean slide, and a selected collection of fibers and particles is added to the liquid and dispersed using a probe. A coverslip is then added to cover the sample. The sample is placed on the stage of the polarizing microscope, and Köhler illumination is checked. The sample is examined using cross-polarized light. If the particle is birefringent, polarized light must be used to measure the refractive index.

For the Becke line method, the particle to be tested is placed in an extinction position, and the substage iris is closed down. One of the polarizing filters is then removed, and the direction being tested in the particle is noted. The distance between the objective and the slide is increased, and the movement of the Becke line is noted. The particle or the polarizing filter is rotated 90°, and the other refractive index direction in the particle is tested by repeating the prior sequence.

Dispersion Staining

Transparent media (and particles) have different refractive indices for light of different wavelengths. This effect is called dispersion. Dispersion staining is a simple particle identification technique based on the refractive index characteristics of a particle and of the liquid medium in which the particle is immersed. To produce dispersion staining colors, the particles and immersion liquids must have different dispersion curves that intersect in the visible light region. At λ_0, the wavelength at which both the particle and the liquid have the same refractive index, the particle-liquid preparation becomes homogeneous with regard to the refraction of light of that wavelength.

If a particle is mounted in a liquid that matches the particle's refractive index at some wavelength, but a significantly different dispersion color fringes the particle, (colored) Becke lines will result. At wavelengths shorter than the match, the liquid will have a higher index of refraction than the solid. At wavelengths longer than the match, the solid with have a higher index. As discussed above with regard to the Becke line, refractive index effects will be noted at the particle liquid boundary if the two do not match in refractive index. If a particle and refractive index liquid match for yellow light, this color will be undeviated at the boundary, and an axial aperture (stop) in the objective lens will portray the particles with a yellow boundary in a clear field. A central stop will allow only light that is deviated beyond a certain angle to pass (e.g., blue—the particles in question will appear blue) (Figure 10.25).

When the match is not in the visible wavelengths, the solid has a higher or lower index than the liquid for all visible wavelengths; thus, the Becke line is white. If the match is in the visible range of wavelengths, there are two Becke lines, one blue to greenish and one yellow to reddish, traveling in opposite directions. The color of these Becke lines indicates the extent to which the liquid and the solid match in their index of refraction and dispersion. The dispersion of the particle relative to that of the liquid can be seen as different colored Becke lines in the near matching liquid. The bluish Becke line will move into the liquid, while the reddish Becke line will move into the particle. By placing a dispersion staining stop at the back focal plane of the objective, the dispersion effects can be greatly enhanced. The refractive index listed for a compound is usually the sodium D line wavelength value, which corresponds to an orange color. When the dispersion colors show a blue-green on the side of an oblique dispersion stop and an orange or orange-red on the other side with standard low dispersion liquid, the refractive index is probably within 0.01 units of that of the liquid. Figure 10.26 illustrates the optical basis for the effects observed in dispersion staining.

Figure 10.25. Dispersion staining objective.

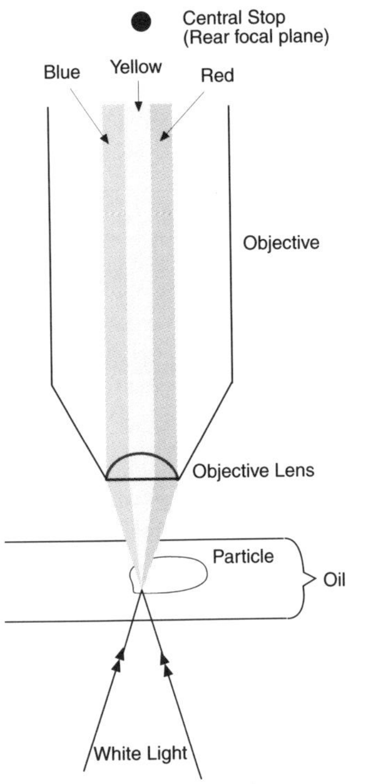

Obviously, the color dispersion characteristics for an anisotropic particle will be more complex than those of an isotropic material. This procedure is particularly valuable for identifying fibers and is widely used in forensic microscopy (Fong 1982). Simplified for this

Figure 10.26. Dispersion curves for a solid and liquid showing a match in refractive index at λ_0.

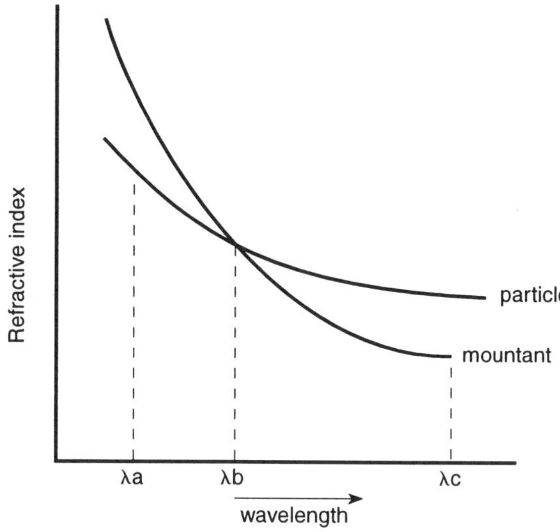

purpose, the technique allows the identification of a wide variety of fibers by means of the color observed in a single refractive index liquid. In this simplified application, a table is used to determine what color will be observed with a specific refractive index liquid-fiber combination.

The approach in all its applications consists of three steps: (1) selection of a significant index of refraction, (2) selection of a suitable liquid with the proper index of refraction and dispersion characteristics, and (3) mounting and examining the sample. The mounting solution is selected to have a dispersion curve that intersects that of the solid at the greatest angle. The higher the angle of intersection, the more precise the data collected.

When using dispersion staining with anisotropic substances, the polarizer's position with respect to the particle becomes significant. In order to see the predicted dispersion of color, the alignment of the particle in the plane-polarized field must be such that the desired refractive index is seen. Very small anisotropic mineral particles can be tested for birefringence using dispersion staining. Unlike interference colors, dispersion is not size dependent. In practice, particles at or just below the resolution limit of light microscopy can still be characterized by dispersion staining.

The application of this technique requires a knowledge of the optical properties for the compound in question. These properties can be taken from tables in the literature or from tests performed on the known compound. It is not necessary to know the name or identity of the compound if reference standards of the material are available. For this reason dispersion staining has proven very valuable in evaluating sources of unusual contaminants.

Microchemical and Microphysical Tests

Microchemical tests include molecular refraction, optical crystallography, and ultramicrochemical tests (see Chamot and Mason 1958; McCrone 1971). With regard to molecular refraction, Sir Isaac Newton first noted the relationship between the density of a material and its refractive index. Optical crystallography was covered in a previous section. The result of the analysis is the identification of a chemical compound rather than a characterization (i.e., a specific rather than a general description).

Microchemical tests, which involve a chemical reaction under the microscope, can also be conducted with the light microscope sensitive to picograms of a specific element (see Feigl and Anger 1968; Chamot and Mason 1958). Exposed to hydrochloric acid, iron oxide dissolves, leaving a yellow color; carbonates evolve gas. Nylon is soluble in formic acid; vinyl in cyclohexanone. One very simple ultramicrochemical test described by Crutcher (1986) involves the use of the electromotive series (Table 10.2). Any metal higher on the list will replace a metal lower on the list. This is a good test for copper compounds in solution. If an iron filing is added to a microdrop of a test solution containing copper, the copper will plate onto the filing or crystallize as free copper in the solution.

Another quick test if a large sample is available is a fusion test conducted in a capillary tube. A few particles are added to the end of a capillary tube, and the tube is fused in a gas flame. The presence of water or other volatiles can be seen as condensates on the tube. Organics will char, leaving a black residue on the glass. Inorganics will fuse with the glass, often exhibiting the characteristic color of any transition element present.

The most common physical tests involve simple manipulations, such as deforming the

Table 10.2. Metal Electromotive Series (Crutcher 1986)

1. Cesium	15. Iron
2. Rubidium	16. Cobalt
3. Potassium	17. Nickel
4. Sodium	18. Tin
5. Lithium	19. Lead
6. Barium	20. Antimony
7. Strontium	21. Bismuth
8. Calcium	22. Arsenic
9. Magnesium	23. Copper
10. Aluminum	24. Mercury
11. Chromium	25. Silver
12. Manganese	26. Platinum
13. Zinc	27. Gold
14. Cadmium	

particle with a tungsten needle or testing its magnetism. Such tests are easy, quick, and provide valuable information about the possible composition of the particle. Any black particle isolated from a small-volume injection (SVI) unit or a large-volume injection (LVI) unit with an elastomeric stopper will retain the elastomeric properties of its parent when pressed with a tungsten probe. Plastic particles show plastic rather than elastic deformation when tested the same way. Many "particles" seen on visual inspection are really agglomerates consisting of an organic matrixing material and include particles of metallic or inorganic origin or fine particles. With experience, the analyst devises solvent dissolution schemes that can be used to dissolve the matrix material and free the included particles for analysis by polarized light microscopy or other means. For example, given the wide use of silicone greases and oils, Freon®, or methylene chloride rinses are a logical first approach to dissolve large agglomerated particles seen on visual inspection of a product.

The polarized light methods described above work well for crystalline materials or for materials that have characteristic interference or extinction patterns, such as skin fragments. Polymeric materials typically have neither of these properties. For most polymers, however, the melting point is a key parameter that may be sufficiently unique to provide identification. In its simplest application, hot stage microscopy consists of determining the melting point of a particle in question and comparing it to a known comparison standard or to published data. This can be performed on a precise device, such as the Mettler hot stage, or using less sophisticated methods devised in one's own laboratory. One (crude) method involves placing the unknown particle(s) and a comparison standard particle sandwiched between two glass slides that are then heated with a soldering pencil. The idea is to begin with the soldering pencil in contact with the edge of the slides, then move it slowly toward the particles so that an even rise in temperature results. Similar methods using a hot wire suspended above the slide holding the particles or an electrical cauterizing needle have also been used.

Totally satisfactory hot stage microscopy can be done without the purchase of a commercial hot stage, which is highly useful but also expensive. The analytical alternative here is using an electrically coated hot stage device. The use and calibration of this device is described by McCrone (1981). The basis of the "device" is a 25 × 75 mm microscope slide that has a thin surface coating of chemically deposited tin oxide. The bond between the glass and the metallic oxide is persistent and electrically conducting; the passage of current through this thin layer generates heat that may be regulated with a variable transformer to give suitably precise control for melting point determination. Metal oxide–coated slides are available from McCrone Research Institute (Chicago).

In use, the lead wires for a Variac are attached to the slide with conducting cement. A calibration is then effected using chemical melting point standards to generate a plot of voltage versus temperature. The device is then ready for use in melting point determination, using either published melting point tables or comparison standards to identify unknowns. A resistance thermometer may also be used to obtain direct temperature readouts.

The identification of fibers can be verified by determining whether the textile material, or fibers from the material, still dissolve or completely disintegrate in selected liquids. A partial list of liquids that still dissolve or disintegrate each class of fibers is given in Table 10.3.

A brief note is helpful here regarding the source of the excerpted data in Table 10.3.

Table 10.3. Some Solubilities for the Analysis of Textile Fibers

Test procedure: Microscopic fibers may be placed in an excess of liquid on a slide or in a microscope hot stage; interpretation of the test is subjective, and more specific tests will often be based on the tentative identification obtained.

Effect of liquid: I = fiber insoluble; P = partial dissolution, swellings, or fragmentation; S = fiber soluble, complete dissolution.

Caution: These are hazardous liquids and should be handled with care. Chemical laboratory exhaust hoods, gloves, aprons, and goggles should be used for fiber solubility work.

Chemical Agent	Concentration (% by weight)	Temperature (°F)	Acetate	Acrylic	Modacrylic	Nylon	Nytril	Olefin	Polyester	Rayon	Saran	Spandex	Cotton & Flax	Silk	Wool
Acetic acid glacial	*	75	S	I	I	I	I	I	I	I	I	I	I	I	I
Acetic acid, glacial	*	200	S	I	I	P	I	I	I	I	I	I	I	I	I
Acetone	*	133	S	I	S	I	P	I	I	I	I	I	I	I	I
Acetone	65	75	S	I	I	I	I	I	I	I	I	I	I	I	I
Acetonitrile	*	75	S	I	I	I	I	I	I	I	I	I	I	I	I
Acetonitrile	*	180	S	I	S	I	S	—	I	I	I	I	I	I	I
Ammonium thiocyanate	70	200	I	S	I	I	I	I	I	I	I	I	I	I	I
Benzyl alcohol	*	200	S	I	I	I	I	—	I	I	I	I	I	I	I
Butyrolactone	*	75	S	I	S	I	S	I	I	I	I	I	I	I	I
Butyrolactone	*	200	S	S	S	I	S	I	I	I	I	I	I	I	I
Carbon tetrachloride	*	170	I	I	I	I	I	S	I	I	I	I	I	I	I
Chloroform	*	75	I	I	I	I	I	—	I	I	I	I	I	I	I
Cresol (meta)	*	75	S	I	I	S	I	I	I	I	I	I	I	I	I
Cresol (meta)	*	200	S	I	P	S	P	I	S	I	I	P	I	I	I
Cyclohexanone	*	75	S	I	P	I	I	I	I	I	S	I	I	I	I
Cyclohexanone	*	200	S	I	S	I	S	I	I	I	S	I	I	I	I
Dimethylacetamide	*	75	S	I	S	I	S	—	I	I	—	I	I	I	I
Dimethylformamide	*	75	S	P	S	I	S	I	I	I	P	I	I	I	I
Dimethylformamide	*	200	S	S	S	I	S	I	I	I	S	S	I	I	I
Dimethylsulfoxide	*	75	S	P	S	I	S	—	I	I	I	I	I	I	I

Table continued on next page.

Table continued from previous page.

Chemical Agent	Concentration (% by weight)	Temperature (°F)	Acetate	Acrylic	Modacrylic	Nylon	Nytril	Olefin	Polyester	Rayon	Saran	Spandex	Cotton & Flax	Silk	Wool
Ethylene carbonate	*	200	S	S	S	I	S	—	I	I	I	I	I	I	I
Formic acid	85	75	S	I	I	S	I	I	I	I	I	I	I	I	I
Hydrochloric acid	37–38	75	S	I	I	S	I	I	I	S	I	I	I	I	I
Monochlorobenzene	*	269	P	I	P	I	I	S	I	I	S	I	I	I	I
Nitric acid	70	75	S	S	I	S	I	I	I	I	I	P	I	S	S
Pyridine	*	200	S	I	S	—	I	—	I	I	S	I	I	I	I
Sodium hydroxide	40	at boil	P	I	I	I	—	I	S	I	I	P	I	S	S
Sodium hypochlorite solution (pH = 2)	3.5 or 5.0 (avail. chlorine)	75	I	I	I	I	I	I	I	I	I	I	I	S	S
Sulfuric acid	75	75	S	—	I	S	I	I	I	S	I	P	S	S	I
Sulfuric acid	40	75	I	I	I	S	I	I	I	I	I	I	I	I	I
Xylene (meta)	*	200	I	I	I	I	I	P	I	I	P	I	I	I	I

*neat

Tabular data excerpted from DuPont Technical Bulletin X-156, December 1961. Courtesy of Dr. Terry Allen.

During the 1960s and 1970s, the Textile Fibers Division of the DuPont Company (E.I. DuPont De Nemours & Co.) made available a significant number of technical bulletins relating to the chemical and physical properties of fibers. Those wishing to develop a differential solubility scheme for fibers are cautioned that (1) polarized light microscopy is generally more easily applied to simple fibers, and (2) a solubility key should be attempted only with the original DuPont publication available, or with other comprehensive sources.

When small clumps of fibers or individual fibers are used in the test, they should be selected carefully to ensure that each of the different classes of fibers in the textile material is tested in every liquid. Good illumination is required for observing the effect of the liquids on single-fiber specimens. The liquids must be used in the numerical sequence specified if each class of fibers is to be systematically removed. The order of removal of the fiber classes by the series of liquids is also shown in the table. If certain fibers are known to be absent from the textile material, the solvents for these fibers can be omitted from the tests.

The solubility tests provide a positive means for identifying man-made fibers by generic class. They may also be used to confirm the microscopic identification of natural fibers. Solubility tests can also be used to distinguish between different fibers within the same generic class.

There are also chemical and physical tests for nonfibrous particles. The chemical characterization of a particle using the light microscope is not like any other approach. It can be the identification of specific elements in a specific particle. It can be the specific identification of a chemical compound, thereby identifying the elements present, their relative ratios and their specific configurational relationships within the molecules. Finally, it can be the identification of the chemical groups constituting the material. The chemical characterization of materials generally involves a mixture of these approaches and a variety of physical tests such as magnetic properties.

Characterization and Sourcing of Mixed Particle Populations

Generally, the source of contaminants is more important than its identity. A property of most particle sources is that in addition to major particle types, a variety of other types of particles are emitted. Often, the presence of these other types of particles will permit a simple type of mass balance to help distribute the contents of a category with a number of different sources. Quartz is an example; in an environment that contains cement plants, glass manufacturing, or rock crushing, these particles can end up in pharmaceutical or medical products. Assemblage analysis (Crutcher 1978b, 1979) (also referred to as "collective," "population," or "ensemble" analysis) can be performed on such single-sourced materials containing multiple discrete particle types.

This type of population analysis uses sample heterogeneity as an essential source labeling device and can be applied to trace particle sources once the source has been characterized. *Assemblage analysis* is a term Crutcher borrowed from archaeology, where it refers to a technique used to monitor cultural development and change through time. Here it refers to a methodology used to identify sources of contamination through the recognition of different types of particulate assemblages characteristic of specific sources. A "source" can be a point of origin, a generation mechanism, a transport mechanism, or some combination of the three. The most important issue is almost invariably to determine how the particles entered the product and eliminate the source. There are four basic assumptions that underlie Crutcher's assemblage analysis procedure:

1. A source generates more than one type of particulate matter or aerosol. The emissions may vary over time, but their variability is less than that between sources.
2. The particulate matter generated by a source reflects the mechanism of production, the material being acted upon, and the environmental conditions at the time and place of generation.
3. Once generated, the particulate matter behaves in a predictable manner, subject to gravitation, filtration, and chemical or physical alteration.
4. Specific transport mechanisms deliver particles that have specific physical parameters that reflect the method of transport, the distance of transport, and obstacles or scrubbing devices designed to clean the mobile phase in the transport system.

The application of these principles in sourcing and characterizing heterogeneous groups of particles is generally the same for an environmental scientist concerned with air pollutants or the particle analyst concerned with contaminant deposition in the process areas of a manufacturing facility. Even in controlled areas such as an aseptic processing suite there is considerable potential for the generation of mixed particle populations to which the four principles above apply.

Consider a high-speed filling line for a drug powder. Beginning at the vial infeed, the movement of the containers on the accumulator turntable generates not only glass particles from contact of the containers but also wear particles from moving surfaces and from contact of vial surfaces with plastic or metal transport surfaces. The movement of stoppering punches and their actuators not only generates particles but may provide a means of dispersal. Contact of the vacuum pickup heads of the stoppering punches with stoppers may aerosolize fine particles of stopper elastomer, and, on each release of vacuum as a stopper is placed on a vial, these accumulated particles may be released. Feeder bowls become reservoirs for stopper debris. The aerosolization of the drug powder itself is a prolific and well-known particle source.

The most important philosophical consideration in the application of microscopy to this type of analysis is that not only microscopic technique but also deductive reasoning must be applied. The analyst is well advised to carefully consider circumstances of occurrence of the particles prior to beginning an analysis. One caution here is that the means of collecting particles for microscopic analysis is critical (Chapter 12). An excursion of particle counts above Class 100 in a clean area can result from a very low total number of particles; conventional means of collecting these particles for microscopic analysis, such as filtration, must be carried out for prolonged periods, and sampling sites must be judiciously selected if success is to be achieved.

In the pharmaceutical industry, assemblages of particles result from a multitude of sources external to the product. Personnel not properly garbed for the role that they are performing can deposit skin cells, fibers from street clothes, hairs, and cosmetic residues. Wear debris from machinery may consist of silicone oil or grease, stainless steel, Teflon®, and rubber. Line filter breakage will result in the presence of all of the contaminants present in the bulk drug or chemical raw material in the product. This can include coarse paper, glass, metal, rust, and polyethylene fragments. Each of these industries has its own typical assemblage or different parts of its processes.

MICROSCOPE CALIBRATION

Light microscopes are easily calibrated using a stage micrometer scale (Bovey 1962). Such stage scales commonly have major divisions 100 μm (0.1 mm) apart, subdivided into 10 μm (0.01 mm) or finer divisions. These are used as a standard against which the divisions of the ocular scale (graticule) are calibrated. Each objective must be separately calibrated based on the correspondence between the stage scale and the ocular scale (Lanier et al. 1978). The ocular scale is first focused, using the adjusting ocular of the microscope. Then, starting with the lowest power objective, the stage scale is focused, and the two scales are superimposed in parallel alignment (Figure 10.27). At some point, it will be possible to find a number of ocular divisions exactly equal to some whole number of divisions of the stage scale, expressed in micrometers (μm).

The calibration consists of calculating the number of micrometers (μm) per ocular scale

Figure 10.27. Comparison of stage scale with ocular scale.

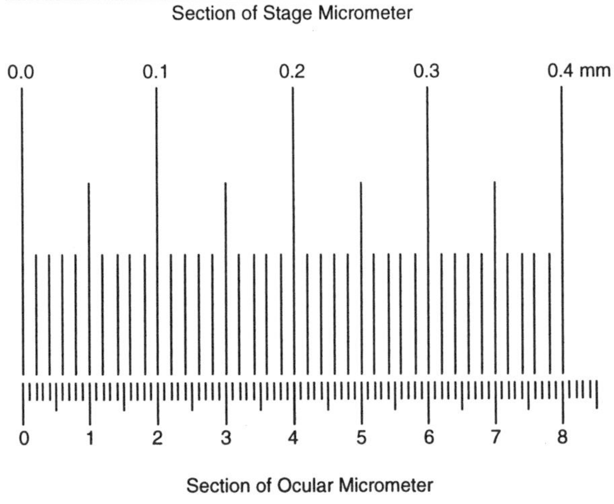

division. To make the comparison as accurate as possible, a large part of the length of each scale must be used. For instance, in Figure 10.27, if four large divisions of the stage scale equal 80 ocular scale divisions, then

80 ocular scale divisions = 400 μm

1 ocular scale division = 400/80 = 5 μm

Thus, when this ocular and scale are used with this specific objective and microscope tube length, each division of the ocular scale equals 5 μm, and the scale can be used to measure an object on the microscope stage. A particle, for example, observed with the calibrated objective measuring 10 divisions on the ocular scale, is 10×5 or 50 μm in diameter.

ISOLATION AND HANDLING OF PARTICLES

Teetsov (1977) provides an excellent discussion of how particles may be isolated for analytical procedures. The following brief information should also be useful. For a bulk material such as a drug powder, a representative sample may be removed with a microspatula from different regions of the dry powder, or the material may be dispersed in a liquid in which it is insoluble, and a drop of the resulting suspension transferred to a slide before settling occurs. When the particulate sample is dry, a drop of mounting medium is then placed over the particles and may be further dispersed by sliding a coverslip over the drop of mountant in a circular motion (as necessary).

The most suitable mounting medium is a viscous liquid with a high refractive index. The Aroclors®, a series of chlorinated diphenyls or polyphenyls manufactured by Monsanto, have fallen out of favor due to environmental concerns. Other mountants that may be substituted are glycerol, Permount®, Karo® syrup, Canada balsam, or silicone oil. Aroclors® may still be purchased, but their sale is controlled. Some very useful alternative polychlorobiphenyl-free mountants are those of the Meltmount® series from Cargille. They are available in refractive indices ranging from 1.539 to 1.704. They have the beneficial properties of low cost and low melting point (65°C) and contain no volatile solvents.

SUMMARY

The combination of a light microscope and an analytical microscopist is the single most powerful tool available to a modern laboratory. The light microscope can be used at a number of levels of sophistication. Only a brief introduction to the light microscope is necessary. A much more thorough description can be found in the references cited.

Light microscopy constitutes the simplest and most cost-effective means for enumerating, characterizing, and identifying particles. The methods required can be applied by an analyst having only basic training in microscopy; the majority of the particulate matter encountered in parenterals may be identified on sight or through simple manipulations involving the polarizing light microscope. Microscopic methods of particle enumeration require more time than does the application of automatic particle counters, but microscopic methods are not subject to false counts due to air bubbles or artifacts due to immiscible liquids. Improvements to the current USP <788> microscopic assay for LVIs have resulted in an improved test method that shows promise as a future compendial particulate matter referee test. In this era of high-tech analytical instrumentation, light microscopy remains a means by which a laboratory with limited funds and expertise can acquire the capability of analyzing and counting pharmaceutical particulate matter with a minimal outlay for equipment and personnel.

REFERENCES

Aldrich, D. S., and M. A. Smith. 1995. Pharmaceutical applications of infrared microspectroscopy. In *Practical guide to infrared microspectroscopy,* edited by H. Humecki. New York: Marcel Dekker, pp. 323–375.

Barber, T. A., M. D. Lannis, and J. G. Williams. 1989. Method evaluation: Automated microscopy as a compendial test for particulates in parenteral solutions. *J. Paren. Sci. Technol.* 43 (1):27–41.

Bloss, D. F. 1961. *An introduction to the methods of optical crystallography.* New York: Holt, Rinehart, and Winston.

Bouma, F. J. 1971. *Physical aspects of colour.* London: Macmillan.

Bovey, E. 1962. Graticules and fine scales: Their production and application in modern measuring systems. *J. Sci. Instrum.* 39:405–413.

Chamot, E. M., and C. W. Mason. 1958. *Handbook of chemical microscopy,* Vol. I and Vol. II, 3rd ed. New York: John Wiley and Sons.

Crutcher, E. R. 1975. *Optical and density study of airborne particulate in the Duwamish Basin.* Boeing Document D-18-18611-1. Everett, Wash., USA: The Boeing Company.

Crutcher, E. R. 1978a. *Analysis of airborne particulate collected on glass fiber filters.* Document WEF-8-03211-B. Washington, D.C.: Environmental Protection Agency, Region X.

Crutcher, E. R. 1978b. Assemblage analysis: Identification of contamination sources. *Proceedings of the 2nd International Spacecraft Contamination Conference.*

Crutcher, E. R. 1979. A standardized approach to airborne particulate analysis using quantitative light. *Proceedings of the Institute of Environmental Sciences,* pp. 78–121.

Crutcher, E. R. 1986. Analysis of particulate contaminants: Microscopical methods. *Proceedings of the Annual Technical Meeting of the Institute of Environmental Sciences,* pp. 36–42.

Delly, J. G. 1996. Microscopy: The setup and operation of the polarized-light microscopy lab for particle identification. In *Liquid and surface-borne particle measurement handbook,* edited by J. Z. Knapp, T. A., Barber, and A. Lieberman. New York: Marcel Dekker.

DeLuca, P. P., S. Boddapatti, and S. Im. 1980. Guidelines for the identification of particles in parenterals. *FDA By-Lines.* Rockville, Md., USA: Food and Drug Administration, Department of Health and Human Services.

DuPont. 1961. *Identification of fibers in textile materials.* Technical Information Bulletin X-156. Wilmington, Del., USA: DuPont, Textile Fibers Division.

Durelli, A. J., and W. F. Riley. 1965. *Introduction to photomechanics.* Englewood Cliffs, N.J., USA: Prentice-Hall, Inc.

Feigl, F., and V. Anger. 1972. *Spot tests in inorganic analysis,* 6th ed. New York: Elsevier.

Flint, E. 1971. *Essentials of crystallography.* Moscow, Russia: MIF Publishers.

Fong, W. 1982. Rapid microscopic identification of synthetic fibers in a single liquid mount. *J. of Foren. Sci. 27* (2):257–263.

Glusker, J. P., M. Lewis, and M. Rossi. 1994. *Crystal structure for chemists and biologists.* New York: VCM Publishers, Inc.

Godfray, M. F. 1979. Microscopy of contaminants of pharmaceuticals and their characterization. *Proc. Analyst. Div. Chem. Soc.* 16:160–161.

Hallimond, A. F. 1970. *The polarizing microscope.* York, UK: Vickers Ltd.

Herdan, G. 1961. *Small particle statistics,* 2nd ed. New York: Academic Press.

Kerry, P. F. 1959. *Optical mineralogy.* New York: McGraw-Hill.

Kremp, G. P. 1965. *Morphological encyclopedia of palynology.* Tucson, Ariz.: University of Arizona Press.

Lanier, J. M., G. S. Oxborrow, and L. T. Kononen. 1978. Calibration of microscopes for measuring particles found in parenteral solutions. *J. Paren. Drug Assoc.* 32:145–148.

Leigh-Dugmore, C. H. 1961. *Microscopy of rubber.* London: W. Heffer and Sons.

Mason, C. W. 1983. *Handbook of chemical microscopy,* vol. I. New York: Wiley Interscience Publishers.

McCormack, J., J. E. C. Harris, and H. J. Sullivan. 1976. Single particle characterization by optical microscopy and associated techniques. *Proc. Analst. Div. Chem. Soc.* 13:344–348.

McCrone, W. C. 1971. Ultramicrominiaturization of microchemical tests. *Microscope* 19:235–241.

McCrone, W. C. 1981. Calibration of the EC slide hot stage. *The Microscope* 39:43–61.

McCrone, W. C. 1987. *Polarized light microscopy.* Chicago: McCrone Research Institute.

McCrone, W. C., and J. G. Delly. 1973a. *Principles and techniques,* Vol. I: *The particle atlas,* 2nd ed. Ann Arbor, Mich., USA: Ann Arbor Science Publishers.

McCrone, W. C., and J. G. Delly. 1973b. *The light microscopy atlas,* Vol. 2: *The particle atlas,* 2nd ed. Ann Arbor, Mich., USA: Ann Arbor Science Publishers.

McCrone, W. C., and J. G. Delly. 1973c. *The electron microscopy atlas,* Vol. III: *The particle atlas,* 2nd ed. Ann Arbor, Mich., USA: Ann Arbor Science Publishers.

McCrone, W. C., and J. G. Delly. 1973d. *The particle analyst's handbook,* Vol. IV: *The particle atlas,* 2nd ed. Ann Arbor Mich., USA: Ann Arbor Science Publishers.

McCrone, W., and W. Hudson. 1969. The analytical use of density gradient separations. *J. Forensic Science* 14:370–384.

McCrone, W. C., J. G. Delly, and S. J. Palenik. 1979. *Light microscopy atlas and techniques,* Vol. V: *The particle atlas,* 2nd ed. Ann Arbor, Mich, USA: Ann Arbor Science Publishers.

Nassau, K. 1980. The causes of color. *Scientific American* (Oct.): 124–154.

Needham, G. H. 1958. *The practical use of the microscope including photomicroscopy.* Springfield, Ill., USA: Thomas Publishers.

Oles, P. J. 1978. Particle analysis and identification in the pharmaceutical industry. *Microscope* 26:41–48.

Overington, I. 1981. Image quality and observer performance. *SFIE* 310:108–117.

Rebegay, T., H. G. Schroeder, S. Im, and P. P. DeLuca. 1977a. Particulate matter monitoring I. Evaluation of some membrane filters and microscopic techniques. *Bull. Paren. Drug Assoc.* 31:57–69.

Rebegay, T., H. G. Schroeder, S. Im, and P. P. DeLuca. 1977b. Particulate matter monitoring II. Correlation of microscopic and automatic counting methods. *Bull. Paren. Drug Assoc.* 31:150–155.

Saylor, C. P. 1967. Accurate microscopical determination of optical properties on one small crystal. In *Advances in optical and electron microscopy,* vol. 1, edited by P. Bare and V. E. Cosslett. New York: Academic Press, pp 41–76.

Schroeder, H. G., and P. P. DeLuca. 1980. Theoretical aspects of particulate matter monitoring by microscopic and instrumental methods. *J. Paren. Drug. Assoc.* 34:185-191.

Teetsov, A. S. 1977. Techniques of small particle manipulation. *Microscope* 25:103–113.

The United States Pharmacopeia. 1990. 22nd revision. Rockville, Md., USA: United States Pharmacopeial Convention, <788>:1596.

Thetford, A., and S. C. Simmens. 1969. Birefringence phenomena in cylindrical fibers. *J. of Microscopy* 89 (Feb.):143–150.

Trasen, B. 1968a. Detection and reduction of particulate matter in pharmaceuticals. *Chem. Eng. Progress* 64 (2):64–68.

Trasen, B. 1968b. Reduction of particulate matter in pharmaceuticals. *Drug and Cosm. Ind.* 102 (6): 40–162.

Whalley, W. B. 1972. The description and measurement of sedimentary particles and the connect of form. *Subdimentary Petrology* 42 (4):961–965.

Winchell, A. N. 1931. *The microscopic characters of artificial inorganic solid substances or artificial minerals.* New York: John Wiley and Sons, Inc.

Winchell, A. N., and H. Winchell. 1964. *The microscopical characters of artificial inorganic solid substances: Optical properties,* 3rd ed. London: Academic Press.

SPECIFICATIONS AND STANDARDS

Air Force Technical Order 00-25-203. *Contamination control of aerospace facilities.*

ASTM F 24-65. *Measuring and counting particulate contamination on surfaces.* Philadelphia: American Society for Testing and Materials.

ASTM F 25-68. *Sizing and counting airborne particulate contamination in clean rooms and other dust-controlled areas designed for electronic and similar applications.* Philadelphia: American Society for Testing and Materials.

ASTM F 51-68. *Sizing and counting particulate contaminant in and on clean room garments.* Philadelphia: American Society for Testing and Materials.

ASTM F 154-76. *Identification of structures and contaminants seen on specular silicon surfaces.* Philadelphia: American Society for Testing and Materials.

ASTM F 311-78. *Processing aerospace liquid samples for particulate contamination analysis using membrane filters.* Philadelphia: American Society for Testing and Materials.

ASTM F 312-69. *Microscopical sizing and counting particles from aerospace fluids on membrane filters.* Philadelphia: American Society for Testing and Materials.

ASTM F 314-70. *Identification of metallic and fibrous contaminants in aerospace fluids.* Philadelphia: American Society for Testing and Materials.

ASTM F 315-70. *Identification of solder and solder flux contaminants in aerospace fluids.* Philadelphia: American Society for Testing and Materials.

ASTM F 318-78. *Sampling airborne particulate contamination in clean rooms for handling aerospace fluids.* Philadelphia: American Society for Testing and Materials.

Federal Standard 209E. 1991. *Cleanroom and workstation requirements, controlled environment.* Washington D.C.: General Services Administration.

MIL-STD-1246. *Product cleanliness levels and contamination control program.*

NAS 1638. *Cleanliness requirements of parts used in hydraulic systems.* Aerospace Industries Association of America, National Aerospace Committee.

SAE ARP-598. *The determination of particulate contamination in liquids by the particle count method.* Aerospace Recommended Practice. Warrendale, Penn., USA: Society of Automotive Engineers.

SAE ARP-743. *Procedure for the determination of particulate contamination of air in dust controlled spaces by the particle count method.* Aerospace Recommended Practice. Warrendale, Penn., USA: Society of Automotive Engineers.

APPENDIX: DESCRIPTIONS OF PHOTOMICROGRAPHS OF COMMONLY ENCOUNTERED PARTICULATE MATTER

Thomas A. Barber
Damian Neuberger
Baxter Healthcare Corp.

The photomicrographs reproduced in the section of color plates in this book were recorded on film using an Olympus Vanox microscope with a PM-10ADS camera system to illustrate commonly encountered particulate matter. Various optical methods were used with either transmitted or reflected light optics, including polarized transmitted light, phase contrast, bright-field (BF) episcopic, and dark-field (DF) episcopic microscopies. Polarized and plane-polarized light were also used with BF episcopic illumination to control reflections from particulate matter with highly reflective surfaces, e.g., metal flakes.

Kodak Ektachrome Professional Daylight EPR 135 daylight film with an 81A + 20B + 5M filter pack was used due to the unique color shift when imaging with partially uncrossed polars. Photomicrographs using various optical methods were also recorded on Tungsten EPY 135 film with an 81D + 81E film. Additional instrument settings included crossed polars or polars uncrossed by 8° to 14° and a 10–11 V transformer setting for the 100 W quartz halogen illuminator. Photomicrographs can be recorded on a wide variety of films, depending on the particular requirements for their use. However, there are a few requirements for obtaining suitable results. Purchase professional grade film, which should be stored in a refrigerator by the vendor to maintain a predetermined rate of color shift as the film ages. The selected film should be purchased in a quantity and of the same lot sufficient to take a series of test photos (to determine the proper filtration, voltage setting, and other camera and microscope settings) as well as all of the specimens to be recorded.

The plates depict materials resting either on a microporous filter membrane, in which case the background appears gray or off-white, or on a glass microslide (for polarized light). With polarized light, the background color will appear black, bright, or colored depending on the degree to which polars are crossed and whether a wave plate was used. The plates emphasize the identification of particles by recognition rather than obtaining precise morphologic descriptors.

Two other color tools that are helpful in identifying particles are the Michel-Lévy Birefringence Chart and the Dispersion Staining Colors chart[1]. The Michel-Lévy chart uses the formula $r = 1{,}000t \times B$, where r is the retardation in nm (color), t is the thickness in μm of the specimen, and B is the birefringence. Birefringence is the difference between the high and low refractive indices ($n_2 - n_1$) of anisotropic particulate matter. Given two elements of the equation, the third factor can be determined. For example, given the interference color (retardation) as observed with a polarized light microscope and the thickness of a fiber as measured in cross-section, the birefringence can be determined and the fiber identified. The Dispersion Staining Colors chart is used to identify particles in a liquid medium and is based on the difference between the refractive index dispersion of the particle and several liquid media of known refractive index (n_D). This technique also requires a microscope objective lens fitted with either an annular or central stop in the back focal plane. These lenses produce a colored particle boundary that allows for the systematic identification of transparent particles using existing data tables for numerous particles and fibers.

The following is descriptive information for each photomicrograph.

1. Insect Parts

Insect parts exhibit various morphologies, but if they are large enough, an organized structure will be visible. The most characteristic aspect of insects is their jointed appendages, such as the legs shown in this photomicrograph, that may be smooth or hairy. Antennae may be featherlike, filamentous, or bulbous. The main body components may be smooth, scaled, hairy, sculpted, perforated, and with a hard or soft exoskeleton or shell. Wings are thin films that are almost isotropic and may be smooth or covered with hairs, scales, or ridges. The color

1. McCrone, W. C., and J. G. Delly. 1973. *The particle atlas,* 2nd ed. Ann Arbor, Mich., USA: Ann Arbor Science Publishers, Inc.

may range from colorless to opaque, but it is usually yellow or brown. Birefringence may be moderate to low or nil. Nearly all of the refractive indices are 1.54–1.55, with a few lower than this. The insect parts most frequently encountered are chitinous body parts, fragments of wings and legs, isolated scales, and hairs. An interesting example of insect parts were hastisetae from dermestid beetle larvae identified from packaging material of warehouse stored materials. The initial identification as insect parts was based on the segmented nature of the particulate matter.

2. Rust (Iron Oxide)

Rust particles are usually aggregates of iron oxides and were obtained by filtration onto a Millipore® membrane and photographed with DF episcopic illumination. The particles (mostly yellow, orange, and various shades of bright red to brown red) are opaque aggregates of rounded to equant grains approximately 2 μm in size; a few metal gray to black iron particles may also be present as seen in this example. Fe_2O_3 particles are soluble in HCl, and their identification can be confirmed by the presence of Fe and O peaks in energy dispersive X-ray spectroscopy (EDXS) spectra.

3. Cosmetic Residue: Talc (Magnesium Aluminum Silicate) and Skin Cells

Epithelial tissues cover the human body and line the digestive tract. The cells are of several different kinds, but the most important are from the outer human skin. Skin tissue is stratified, with a basal layer of newly produced columnar cells. These pass successively outward, become flattened, and are cornified by the time they reach the surface. It is this cornified top layer that scales off as dandruff or as a single epithelial cell. Cells from the scalp usually flake off in sheets. Individual cells or aggregates of cells come from the fingers and other parts of the body or from weathered dandruff. They may be deeply folded on themselves or rolled up like scrolls. They are transparent, colorless, or light tan polygonal flakes. This photomicrograph of skin cells on a filter membrane taken with BF episcopic illumination appears as transparent, colorless, slightly reflective thin sheets. The associated talc appears as flat, highly reflective birefringent (white to pink interference colors) sharp-edged flakes.

4. Starch

This photomicrograph shows a mixture of potato and rice starch on a filter membrane and observed with BF episcopic illumination. These particles are transparent colorless, mostly subspherical to polyhedral grains ranging in size from 5 to 20 μm in size. The refractive index is approximately 1.53. Starch is composed of long, unbranched, helically coiled polysaccharide chains that may be the basis for the grain formation. Starch grains may show a concentric layering of starch around a centric to eccentric dark central point (hilum). The morphologic variation of starch is so extensive that its origin may be identified as to the species of plant and plant part. Starch identification may be confirmed by the iodine–potassium iodide (I_2KI) test, in which starch stains a blue to deep purple color.

5. Paper Fragments

Cells of softwoods (conifers) and hardwoods (dicots) have basic structural differences. The terms *hardwood* and *softwood* do not express the degree of density of these woods; balsa, a tropical dicot, is one of the softest woods, and slash pine is harder than some hardwoods. The wood is chemically treated (macerated) to dissolve the intercellular binding material (middle lamella), resulting in relatively intact cells called chemical fibers. The morphology of wood cells can range from long, tapered, smooth-walled fibers to short, wide cells with numerous pores and perforated ends. This can make identification difficult for some types of isolated fibers. The refractive indices of fibers range from 1.53 to 1.58; all fibers have a positive sign of elongation. Paper in injectables most often occurs as short, straight, or twisted fragments, like the ones in the photomicrograph.

6. Teflon® (DuPont) Flakes

These particles are colorless, transparent particles of fluorocarbon resin (PTFE). The particles are scrapings or filings and, as such, yield shapes that are generally elongated with frayed, fiberlike pieces protruding from the rough edges. Thicker fragments of Teflon® material are translucent in transmitted light, whereas in reflected light they appear a waxy white. Thin fragments on a filter membrane, such as the one shown in this photomicrograph, exhibit a pale blue to pale red color under episcopic illumination. The refractive indices are low, about 1.34–1.38, and the birefringence is moderately high. Characteristically, Teflon® is not affected by common reagents or solvents; it has no melting point but chars at very high temperatures.

7. Stopper Fragments (Black Chlorobutyl Rubber)

This sample is from an LVI stopper. Bits of such rubber are infrequently found as suspended particles in parenteral solutions. Occasionally, there is an adverse reaction between the components of the septum and the solution, leading to precipitation, shortening of shelf life, loss of potency, and so on. This sample is an opaque, amorphous elastomer; the black color is due to the presence of carbon black added to strengthen the elastomer matrix. Physically, the material is elastomeric and recovers when deformed with a probe or between a slide and a coverglass.

8. Stainless Steel

Metal fragments look like the parent material in reflected light. Stainless steel is a highly reflective silver color, brass is a pale gold, and aluminum exhibits a dull white to grayish white luster. The fragment surface often bears striations or wear marks from the process that produced them. This photomicrograph was recorded using plane polarized BF illumination of a thin, flat fragment that exhibits a shiny, grooved surface.

9. Polyethylene (cutting fragment)

This fragment of polyethylene material on a filter membrane was produced by the cutting of an airway tube for an LVI unit at the manufacturing plant. It is a white, opaque flake and exhibits an accordion-pleated pattern from the cutting process. The fragment resembles the morphology of the parent bulk material. Such plastics are typically of low melting point and sensitive to organic solvents. Fragments of this size are readily identified by micro Fourier transform infrared (FT-IR) spectroscopy.

10. Calcium Carbonate

This reflected light photomicrograph shows aggregates of a precipitated form of calcium carbonate. These crystal aggregates appear as colorless, transparent to white opaque, irregularly shaped particles. Precipitated $CaCO_3$ crystals are better formed than a ground sample of limestone and exhibit a six-sided crystal habit differing from a cube only in that the interfacial angles are not 90°. Refractive indices are 1.659 ω and 1.486 ε, with a concomitantly high birefringence that gives a blue-green dispersion stain color (see Dispersion Staining Colors chart) in Arochlor® 5442 with annular stop.

11. Magnesium Phosphate

This particulate sample on a filter membrane was precipitated from a total nutrient admixture (TNA). The transparent spherical crystals shown in this reflected light photomicrograph are characteristic of this crystalline material, the identification of which can be confirmed by the presence of Mg and P peaks in EDXS spectra.

12. Corn Starch

These particles are transparent colorless, mostly subspherical to polyhedral grains, ranging from 5 to 25 μm in size. The refractive index is approximately 1.53. Starch is composed of long, unbranched, helically coiled, polysaccharide chains that are laid down in concentric layers with a centric to eccentric circular- to triangular-shaped air bubble or hilum. The result of this type of development is that a dark cross pattern (✦) is seen between crossed polars, and the grains are gray to white in color. The morphologic variation of starch is so extensive that its origin may be identified as to the species of plant and plant part. Starch identification may be confirmed by the iodine–potassium iodide (I_2KI) test, in which starch stains a blue to deep purple color.

13. Filter Membrane Fragments

This white, opaque, irregularly shaped particle was shed from a microporous membrane filter; fragments are shed or released if filters rupture. This material appears snowy white in BF episcopic illumination due to light scattering by its complex fine structure.

14. Dandruff

Dandruff scales are composed of epithelial cells that have flaked off from the outer surface of the skin. Single cells may also be present, but one more frequently sees cornified, flattened sheets of epithelial cells. The scales may be folded, curled, flat, polygonal to irregular in shape, transparent, and colorless to light tan in color. The interference colors are first-order white. Under polarized light, the epithelial cells exhibit characteristic bands or striations of weakly birefringent collagen bundles within the cells. Extinction can be seen for portions of the collagen bundles. Microscopic preparations of dandruff usually contain a great variety of artifacts, especially paper and textile fibers, because oily skin causes loose particles to adhere. When dandruff is shed, a number of these other particles come with it.

15. Drug Residue

This is an assemblage of different particle types: synthetic fibers, glass, rust, and insoluble drug impurities. These materials are removed by filters prior to the filling process for a solution made with the drug.

16. Diatoms

Diatoms are mostly unicellular organisms (some colonial forms occur) of freshwater and marine habitats. The cell walls of diatoms are unique double shells called frustules, which are composed of polymerized, opaline silica ($SiO_2 \cdot nH_2O$); the two halves or valves fit together, one on top of the other, similar to a culture dish. The valves are colorless; transparent; usually isotropic; and may be circular, oval, crescentic, linear, triangular, or square. The silica shells of diatoms have accumulated over millions of years to form diatomaceous earth, a fine, powdery substance used as an abrasive, filtering, or insulating material. Scanning electron microscopy (SEM) has shown the shells to be delicately sculpted with large numbers of minute, intricately shaped depressions, pores, and/or passageways that connect the living protoplasm within the shell to the outside environment. These markings, which serve to classify species, have been used by microscopists to test the resolution capabilities of objective lenses on optical microscopes.

17. Fungal Hyphae

Many fungi or molds form a mass of cottony growth, known as a *mycelium*. The mycelium is composed of individual strands (hyphae) of cells that may be filamentous (multicellular) or coenocytic (unicellular). A specialized hyphal branch, a conidiophore, produces strands of conidia, usually by abstriction. The conidia are asexual spores that ultimately develop new hyphae. The filaments in this sample are about 5 μm wide, with nuclei, vacuoles, and droplets of storage material.

18. Amorphous Material (Drug Residue)

This tan, yellow, or light brown amorphous material may be filtered from dextrose solutions and SVI–reconstituted dry powder drugs. This material may appear as faintly visible, reflective surface discolorations on the filter membrane when viewed with BF episcopic illumination, as shown in this photomicrograph. This material is formed as a result of a complex reaction of reducing sugars with amino acids to form numerous compounds, among which are furfural derivatives and N-glycosides.

19. Hair (Human, Caucasian)

These fibers consist of an outer cuticle composed of layers of tightly overlapping scales surrounding a layer of tightly packed cells called the cortex. The central core or medulla is composed of loosely packed cells and will appear dark opaque if filled with air. The medulla may be continuous, especially in larger hairs, discontinuous, or absent. The cuticular scales in human hair do not protrude much from the shaft. The cross-section profile varies in shape from circular to oval, and hair types vary depending on the area of the body, age, sex, and race. Hairs may be transparent to almost opaque; refractive indices are about 1.554 (parallel) and 1.542 (perpendicular); birefringence is 0.012, and the sign of elongation is positive. Extinction is usually complete and parallel, but depending on the shaft morphology, it may not be uniformly complete. Human hair ranges from 50 to 150 μm in diameter.

20. Rat Hair

These hairs are transparent, scaly cylinders that taper at the end. The scales are large, with approximately 3–7 scales/100 μm, and form a sharp-toothed edge. The medulla consists of blocklike segments, each of which is constricted to form a block-shaped "T." The medulla may appear black due to entrapped air. Pigment granules, when present, are rather large and compact and are usually at one end of the medullary segment; the other end is clear. The diameter of rat hairs ranges from 10 to 70 μm. Extinction is parallel. The refractive indices are about 1.54 (crosswise) and 1.55 (lengthwise), with concomitant low birefringence yielding low, first-order interference colors; the sign of elongation is positive.

21. Paper (Coarse, Hardwood)

Dicot woods exhibit a greater variety of thick-walled, fiberlike cells than do coniferous woods. These include vessels, tracheids, and fibers. The vessel and tracheary cells have thick secondary walls, lack living protoplasts at maturity, and have pits in their walls that exhibit a black cross-shaped (✦) pattern between crossed polars. Interference colors are often first-order gray. In addition, vessels generally have perforations in the end walls, but perforations may also be found on the lateral walls. Since perforations are holes in the cell wall that connect one cell to the next, they do not exhibit polarization patterns as do the pits. Vessels are broad and flattened as a result of processing the wood into pulp, whereas tracheids are narrow and flat, with more highly tapered end walls. Fibers are long, narrow, tapered cells with a polygonal cross-section profile and few or no pits; they exhibit low, second-order interference colors. The refractive indices of fibers range from 1.53 to 1.58, have a positive sign of elongation, and exhibit complete extinction. This photomicrograph shows a bundle of fibers on the left, with a single tracheid along the left edge and horizontal striations that represent parenchyma cells. On the right is a large vessel with numerous pits (dark spots) in the cell wall (bright). Most isolated, dicot, fiberlike cells originate from paper or paper products.

22. Cotton

These fibers are transparent, colorless, and are flattened into regularly to irregularly twisted ribbons with a central lumen or canal. The cross-sectional profile ranges from narrowly elliptical to circular, the latter exhibiting a smaller cell lumen. Cotton fibers are not wood-type fibers; they are cotton seed trichomes (hairs), approximately 12–60 μm long (depending of the variety of cotton) and 10–40 μm in diameter. The cell walls of cotton fibers consist of a primary cell wall and a thick secondary wall of almost pure cellulose. The surface of the fiber wall occasionally has transverse markings, but the cell wall does not have pits. A characteristic feature of cotton fibers that distinguishes them from paper is the lack of complete extinction between crossed polars. The refractive indices are 1.578 (parallel) and 1.532 (perpendicular).

23. Paper and Cotton Mixed

This photomicrograph shows the most frequently encountered types of fibers—cotton and paper. The larger diameter paper fiber (diagonally oriented) exhibits extinction in the central region of the cell fragment; however, when this fiber was rotated 45° between the crossed polars, complete extinction (essentially a total darkening of the cell wall) was observed. In contrast, the narrower, twisted cotton fibers did not show any extinction when rotated between the crossed polars.

24. Glass Fibers

These fibers are transparent, colorless, isotropic cylinders, almost always smooth and uniform in diameter. The fiber ends usually show a clean transverse or diagonal break, but they may be chipped due to their brittle nature. The cross-sections of the fibers are perfectly round, 5–35 μm in diameter, and of varying length (5 μm to > 1.2 mm). The refractive index is approximately 1.55, although it varies widely depending on the composition of the glass, normally about 1.47 for borosilicate and greater than 1.52 for soft glasses.

25. Asbestos

Chrysotile asbestos is one of the more common forms of asbestos. It consists of bundles of straight, parallel fibrils with nonperiodic crimps; wavy bands of fibrils; and single, randomly twisted fibrils. The colorless, transparent, cylindrical fibrils are less than 1 μm in diameter and have smooth surfaces. The structure of the larger fibrils as bundles of smaller and smaller groups of fine fibrils is apparent as magnification is increased, especially at the broken ends of larger bundles as seen in this photomicrograph that also shows a crimp band. Chrysotile is monoclinic; its refractive indices are 1.529–1.559 (α), 1.530–1.564 (β), and 1.537–1.567 (γ parallel to fiber length); the sign of elongation is positive, and extinction is parallel. Chrysotile has a low refractive index, which is the best key to its identification.

26. Acrylic Fiber (Orlon®, Delustered)

These fibers are colorless, transparent, smooth cylinders with a flattened dog-bone or dumbbell cross-section profile. Other fibers may be trilobalor mushroom–shaped in profile. The cross-section may be deduced from the longitudinal view. The fiber surface appears to have a thin, narrow groove down the center of the fiber. As the focus is adjusted above and below the best focus, the central portion will correspondingly become dark then light. Extinction is parallel and complete, and the sign of elongation is negative; Orlon® is one of the few fibers that exhibits a negative sign of elongation. The refractive indices are 1.505–1.515 (lengthwise) and 1.507–1.517 (crosswise), resulting in a low birefringence. The interference colors for the 30 μm wide, heavily delustered fibers in this sample are low, first-order gray.

27. Polyester (Dacron®)

These fibers are transparent, colorless cylinders with a smooth surface, except where delustering pigment particles have broken through and are either exposed on the surface or have become detached to form an irregular surface texture. The refractive indices are 1.710 (lengthwise) and 1.535 (crosswise); the sign of elongation is positive, and extinction is complete. The birefringence is extremely high for a fiber, 0.175, resulting in fourth-order and fifth-order light green and rose interference colors. The fibers in this sample are 45–48 μm.

28. Bulk Powder Drug Residue

This photomicrograph shows a membrane filter bearing the free acid crystals precipitated from a drug formulated as a sodium salt. It is monoclinic and crystallizes as colorless, birefringent prisms. The crystals are very soft (Moh 1). The refractive indices are 1.502 (α), 1.640 (β), and 1.652 (γ); there is a negative sign of elongation, and birefringence is high (0.150).

29. Talc

This sample of talc in polarized light with crossed polars shows the flat, platey, sharp-edged shape of these crystals. Refractive indices are 1.539 (α), 1.589 (β), and 1.589 (γ), with a positive sign of elongation and a birefringence of 0.05. Interference colors are high-order reds and blues due to the thickness of these crystals.

30. Calcium Oxalate

Calcium oxalate is formed in TNA formulations through the degradation of ascorbic acid. Crystal habits in this sample consist primarily of bipyramidal forms. Calcium oxalate is soluble in acid but insoluble in water and acetic acid. It is monoclinic and crystallizes as colorless, birefringent tablets. Ground crystals are conchoidally fractured. The refractive indices are 1.490 (α), 1.555 (β), and 1.650 (γ); they exhibit a positive sign of elongation, and the birefringence is high at 0.160. The crystals are soft (Moh 2.5) and have a specific gravity of 2.23. The crystal in this photomicrograph is a bipyramid on edge showing higher-order colors due to the high birefringence and the increasing thickness toward the center of the crystal.

31. Skin Cells

The ovoid, 10–20 μm transparent cells show a distinct cross-hatched or interwoven extinction pattern when rotated between crossed polars; this pattern is due to bands or striations of birefringent collagen bundles around the periphery of the cells. The interference color is first-order white; refractive indices are about 1.54. The central dark void indicates the location of the nuclear material in the living cell.

32. Glass Balloons (Cenospheres)

These transparent, colorless, isotropic, hollow glass spheres range from 10 μm to > 800 μm in diameter. They are characteristic of similar hollow spheres seen in combustion particulate processes that produce both gas and melted siliceous material. Most appear opaque black except for a central light spot. This is due to the refractive index difference between the entrapped gas (air) and the medium (a red aqueous dye). This photomicrograph was taken using phase contrast optics, resulting in a bright halo around the periphery of the sphere with a darker central core. The surface shows a variety of flaws and defects including striae and pinhead glass droplets. The broken fragments shown in this photomicrograph are curved and sharp-edged and reveal the hollow structure of the spheres. Glass balloons have a refractive index of about 1.52.

XI

VALIDATION AND ITS APPLICATION TO PARTICLE COUNTING INSTRUMENT SYSTEMS

Pharmaceutical manufacture has historically incorporated rigorous safeguards that address the surety that product generated by a specific manufacturing process is safe, efficacious, and not misbranded or adulterated. What is the meaning of the term *validation* in this context? In a general overview, the manufacture of a drug or medical device is conducted within a well-defined process envelope that includes all of the associated inputs and operations.

The process envelope typically includes a number of subprocesses, some relating to the actual manufacturing operation (e.g., aseptic filling, container washing, or sterilization); others are assays or tests conducted to demonstrate that the product meets requirements. Particle counting assays are in this latter grouping. These individual components of the larger process become relatively complex as they are considered in greater detail. Instrument, machine, and human inputs and interactions are involved. Some are controlled or overseen by a computer using a combination of software, hardware, and firmware.

For the total production process and for each of the subcomponent operations and processes, there is some level of associated risk that unexpected, unwanted, detrimental events or errors in output will occur. The term *validation* in generic broadscale use is simply a procedure undertaken to minimize risk. In another equally simplistic consideration, validation at whatever level it is carried out must produce documented evidence of a high degree of assurance that a specific operation, device, instrument, and/or computer system or software will consistently produce a product meeting predetermined specification and quality attributes. An important philosophical concept here is that risk can be shown to be reduced to a very low level (e.g., a 10^{-6} sterility assurance level or SAL); its elimination cannot be demonstrated. Fortunately, however, errors or deviations from the expected or acceptable result can be demonstrated and eliminated during testing.

Given this general conceptual framework for validation, the term may be used in reference to number of process components, including the following:

- Instruments (and instrument function)
- Assays (tests)
- Methods

- Analytical systems
- Production processes
- Computer systems
- Computer software

The potential complexity of validation activities, considering the absolute necessity of generating safe, pure, and efficacious product, ongoing compliance, and the interrelation of these different systems/processes are significant. One of the most significant decision points we encounter is "How much validation is enough?" The validation of a process as directed specifically or generally by Good Manufacturing Practices (GMPs) simply constitutes ensuring that a process performs as intended and that checks are in place to ensure that the process continues to run within clearly defined limits (Trill 1994). The control of the process in effect controls the quality of the product.

Historically, compendial particle counting was often viewed by FDA inspectors as a "special" endeavor that yielded assay results of such high variability that only general scrutiny of the data and the assays was necessary. This historical perspective on the part of some industry and regulatory personnel that particle counting was an assay not subject to the classical tests for precision, ruggedness, and so on has changed. Particle counting has, in the past 2–3 years, begun to come under a much higher level of scrutiny in audits, with inspectors questioning standards, instrument validation, and "as-found" data that are reflected in a change in instrument threshold settings between calibrations. Hence, the definite need for the user and lab manager to consider the information in this chapter. A section on Food and Drug Administration (FDA) perspectives regarding validation of computer-based systems and software has been included and should be both interesting and useful.

Most pharmaceutical companies currently have groups that deal specifically with assay or software (also termed computer) validation. In planning a validation of a particle counting assay/instrument system, it is critical that representatives of these groups be included and involved heavily in the decision-making process. In order for these experts to do their valuable job, the technologist must make the commitment to spend whatever time necessary to ensure that the unique concepts peculiar to particle counting assays are understood. What the company should strive for is a systematic approach to validations that reflects common elements in all cases.

Based on this preliminary discussion, a scenario evolves in which we must consider the computer-driven analytical system in terms of the individual subunits of which it is composed. These subunits will be dealt with specifically and in different levels of detail in validation. A computerized particle counting system may be broken up into the components shown in Figure 11.1. (Obviously, if the particle counting system is not software driven, this element will be removed from the deliberations, but other components of the system will remain.) Note that compendial information is considered to be a component; along with this are training, Standard Operating Procedures (SOPs), standards, and other items.

With this introduction of the topic, the author begins with what he believes to be a logical discussion of the specific components of validation of instrument systems in general and particle counting systems in particular. The discussion does not go into detail for any specific step in the process (e.g., the requirements definition, the validation rationale or risk assessment, etc.), and, in fact, there is no need to; a number of the references provided give step-by-step instructions. The customary caveats regarding the content of the chapter apply here. While the attempt has been made to reflect the perceptions and policies of regulatory, industry, and compendial groups accurately, no guarantee is provided that designing a validation approach according to the information in this chapter will result in acceptance by representatives of any of these groups. The principles discussed, however, should be uniformly valuable to those approaching a validation project.

THE TERMINOLOGY

A good part of the difficulty in understanding the validation process is due to some level of misuse or at least overbroad use of the word *validation*. Historically, in the pharmaceutical and medical device industries, the function of critical systems has been ensured in some rigorous fashion prior to acceptance for use. Examples of such systems are provided by solution filters and filtration processes, environmental controls such as room pressurization, and vial-stopper systems

Figure 11.1. Elements of a particle counting system.

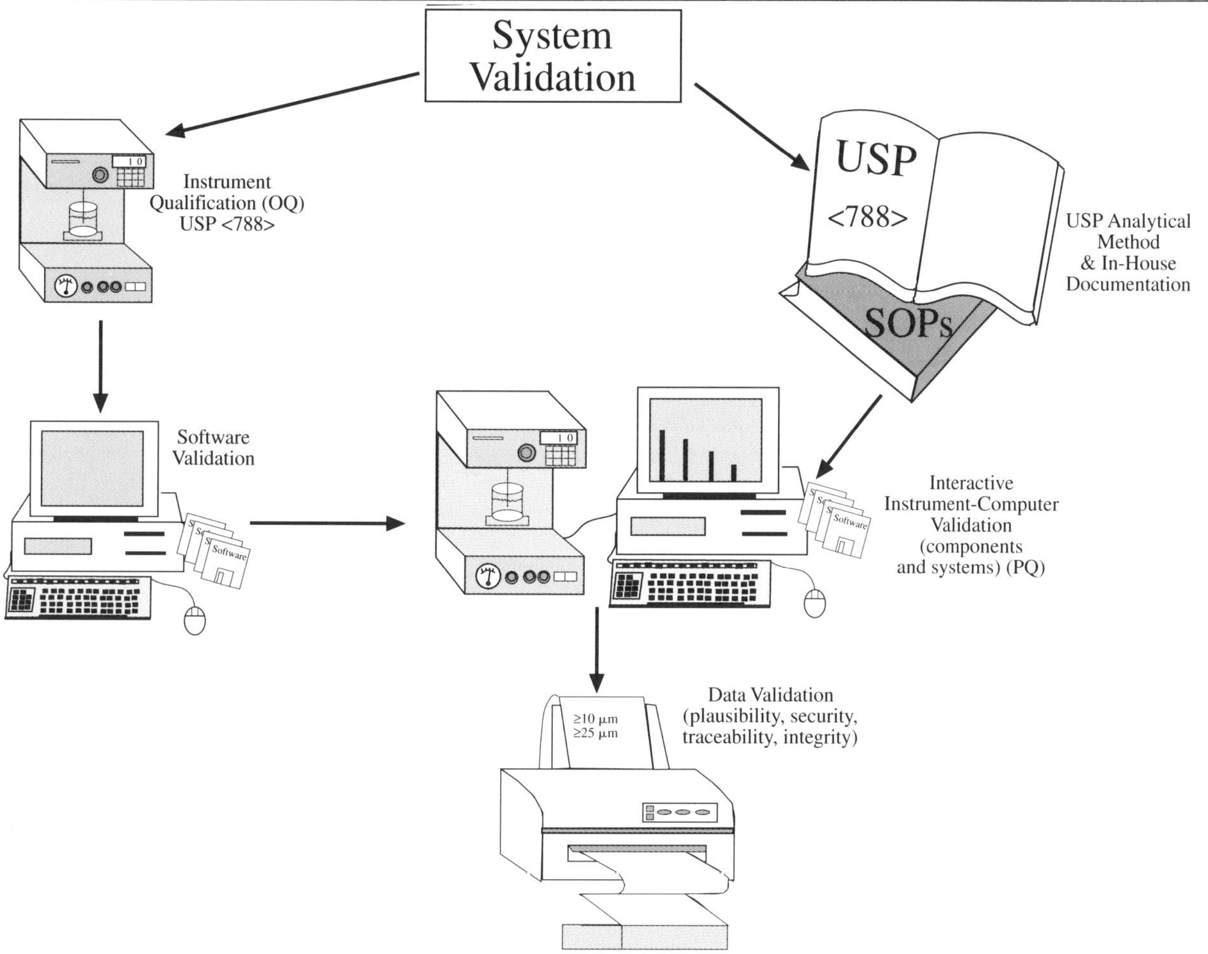

used for aseptically filled drugs or terminally sterilized large-volume injectables (LVIs).

Obtaining such assurance was variously termed *certification, qualification, verification,* or *assurance testing.* Today, all such activities are too often generically grouped under the umbrella of validation without appropriate differentiation. This is probably a generally acceptable practice, but a technical manager or quality management director must make the appropriate mental discrimination between the types of activities involved.

The validation of computer-controlled analytical systems, whether they be high performance liquid chromatography (HPLC) systems or particle counters, must include testing and calibration of the analytical instrument hardware as well as software. Thus, in this chapter, strategies for testing and instrument calibration are discussed as part of the overall validation process. While this principle is only implicit in much of the text, the reader should keep it firmly in mind.

There is some confusion regarding the meaning of the terms *testing, verification,* and *qualification* within the validation process, and definitions are helpful (Alford and Kline 1990a; PICSVF 1995). *Testing,* as defined in ISO/IEC Guide 25, consists generally of the determination that specific performance characteristics comply with specified procedures or descriptions. The same source says that *verification* is confirmation that specific requirements have been met; for an analytical instrument, this reduces to comparing test results to a specification. *Qualification* is system specific and comprises a test of the suitability of work (or data) output under the conditions specified for a certain application. For a particle counter, this might involve the ongoing assurance of the validity of particle count data obtained in compliance with U.S. Pharmacopeia (USP) <788>.

VALIDATION: THE COMPENDIAL INFORMATION

By a logical interpretation in terms of USP <788>, performance qualification (PQ) implies that the entire test method, including suitability testing, sample selection, blanks, analyst training, data reduction, and other specified components, be shown to function as intended to give an appropriate data output in evaluation of batch suitability. Along these lines, USP <1225> discusses in some detail the validation of test methods, particularly those considered suitable for use as compendial methods. Method validation is basically distinct from the validation of software or software-driven systems, and deals with items that have been the classical concerns of the analytical chemist—namely accuracy and precision, reproducibility, and ruggedness of a test method. Importantly, the general chapter states,

> Also, according to these regulations [21 CFR 211.194(a)(2)], users of analytical methods described in the USP and the NF are not required to validate accuracy and reliability of these methods, but merely verify their suitability under actual conditions of use.

By inference, based on this USP <1225> information, many manufacturers assume the method itself to be validated, and pursue software-interactive function validation as <788> instructs for automated determinations of resolution and calibration. USP <1225> goes further to define validation and to suggest test parameters that should be considered.

> Validation of an analytical method is the process by which it is established, by laboratory studies, that the performance characteristics of the method meet the requirements for the intended analytical applications. Performance characteristics are expressed in terms of analytical parameters.

A number of parameters to be considered in test validation are given and defined in USP <1225>:

- Accuracy
- Precision
- Specificity
- Limit of detection
- Limit of quantitation
- Linearity
- Range
- Ruggedness
- Robustness

These items are classical criteria of instrumental and "wet chemical" analytical methods and are used in characterizing methods of analysis in many applications other than those of the pharmaceutical industry. As stressed earlier in this book, particle counting assays are physical rather than chemical tests. Data from such tests, including those of <788>, are subject to complex sources of variability, including, with regard to the light obscuration (LO) assay, the optical properties of the particles counted and their suspending fluid. This test in fact represents an abstraction, in which particles are not directly sized but assigned a size based on their light "obscuring" properties. The microscopic assay of <788> sizes particles directly but suffers from a high level of variability due to sample manipulation and human variation in microscope use.

What is the impact of compendial information on validation of these assays? In a basic vein, without consideration of computers, <1225> instructs us to "verify the suitability of the methods under actual conditions of use." Such verification reasonably incorporates analyst training, user SOPs, and basic assessments of variability due to application factors such as different analysts (with different means of entering or agitating units) and between-instrument variability. For instance, the same standard suspension might be run on several LO counters, and analysts performing the microscopic assay might be required to count a series of standard filters to quantitate and control variability. Such measures represent tests of assay precision and are quite useful. They are not, per se, validation. They serve to give us a handle on the variability of the assay, and in simple terms tell us how close to the USP limits values our counts can be before we become concerned with product quality.

The situation with regard to assay validation becomes much more complex when computers are considered. The USP 23 version of <788> calls out "suitable validation" with regard to software-based calculations of LO counter resolution and the calibration curve, which, in the case of the resolution measurement, requires that the determination should be "equivalent" to the manual method. Thus, based on the specific instruction of the compendium and the general regulatory climate, some level of validation of computer-based LO particle counters is called for. A caution here repeated here is that validation of software-based counter functions should not be considered stand-alone verification of instrument suitability in-use. The validation of the software should be considered and undertaken as a single component of the assay/instrument suitability assessment.

THE LITERATURE

A point is in order regarding the information base on which validation activities must be based. The regulatory, compendial, and industry philosophies on validation are still evolving with each passing day, as publications, reports of industry task forces, and regulatory agency guidelines are issued. This large, rapidly expanding body of information effectively will remain beyond those of us who specialize in the analytical technologies relating to contamination control and particle analysis. The author's approach has been to rely on validation coordinators and specialists to provide the detailed interpretations and instruction on validation activity that ensure compliance with the current thinking and to contribute the technical information needed to ensure scientifically valid approaches.

The references are intended to be useful to the reader rather than comprehensive. In addition to the literature cited, there is also a bibliography of other guidance documents containing a cross-section of more current references, many pertaining to specific analytical systems. The most valuable texts with regard to the treatment of instrument systems are those of Huber (1995, 1999). Chamberlain (1993) is also very valuable in that it deals in some detail with the breakdown of individual steps in the major processes of validation, such as the validation plan, and provides examples of many of the forms needed to record test data. Huber's books contain the most useful chronological outline and reference list that the author has yet found under a single cover. Further, the basic discussions in these texts are directed specifically at the function of analytical instrumentation, and precise examples of how components of validation (e.g., execution of a test plan) are conducted. Huber is employed by Hewlett-Packard, a corporation with a significant dedication to validation as one component of product quality, and he is able to draw on that experience base. To obtain the most current information, the reader is recommended to the following sources and others listed in the bibliography:

- FDA policy guides, inspectional guidelines, and cGMP documents (also similar documents from the European Community, Japan, the United Kingdom, and Australia)
- Publications of the Pharmaceutical Manufacturing Association's (PMA) Computer System Validation Committee (CSVC)
- Institute of Electrical and Electronic Engineers (IEEE) standards
- International Organization for Standardization (ISO) documents such as ISO 9000-3: 1991—*Guidelines for the Application of ISO 9001 to the Development, Supply, and Maintenance of Software*
- Publications of the U.S. Environmental Protection Agency (EPA) regarding operations of automated laboratories
- Periodicals, e.g., *The Journal of Validation Technology*

The general GMP documents of national and international organizations will be found to contain the basic requirements for validation in both laboratory and manufacturing practice. Referenced in this category are the publications by Anisfeld (1990), Matsuda (1994), the *EEC Guide to Good Manufacturing Practices* (current issue), the UK PICSVF guidelines, and the Code of Federal Regulations (CFR) citations. Chapters 3 and 5 of this book contain references to and discussion of the GMP documents that impact device and drug manufacture.

The reader who is or will be involved with the validation process as applied to particle counting systems, the specialized nature of these systems not withstanding, is advised to become familiar with some of the general literature pertaining to validation in the manufacture of devices and injectables. Some references in this category include Agalloco (1987), Drug Information Association (1988), FIPS Publications (1976, 1983), Furman et al. (1994), ISO/IEC Guide 25 (1990), Koseisho (1988), and Kuzel (1985, 1987). The PMA CSVC document (1986) also contains a good deal of useful information. Although some of these documents are early ones, basic principles of validation remain unchanged, and the author's intent here is to make the reader familiar with the principles and pattern of validation activity. These considerations are very well presented in the earlier documents.

This mass of information on the subject of validation made this chapter somewhat difficult for a nonexpert to write; I attempted to include examples and considerations peculiar to particle assays and instruments and to deal only superficially with the literature on the subject. It is hoped that this approach achieves the desired end with regard to information transfer. Once again, the information presented in the following sections represents the author's perspective rather than any policy, expressed or implicit, of the FDA. The format includes a general discussion on the principles of validation followed by a discussion of specifics pertaining to particle counters. I used this approach in the belief that, knowing the principles, users may then proceed with their own "customized" application to particle counters based on both technical perceptions and company philosophy.

The requirement for validation in the United States is found in the Drug and Device GMPs. One of the many definitions of validation is found here: "Documented evidence that shows a high degree of confidence that systems are accurate and reliable for their intended function." Validation of software and computer-based systems is required to obtain such reasonable assurance that the software or system functions as intended and that data generated are accurate. An FDA 483 (inspectional observation) may be given on the basis that some deficiency in data generated or in system function exists that could result in a drug or device product being adulterated, mislabeled, or unsafe. However, the statement "adulterated, mislabeled, or unsafe" on a 483 may also be used simply to indicate a failure to meet GMP for other reasons.

For the purposes of this requirement, a computer is defined as an electronic device that assembles, stores, or correlates data or performs mathematical or logical calculations related to the reduction of data. A computer-based system consists of a combination of a computer and an analytical instrument or device to reduce data that are gathered. Computer-based systems may be used to conduct testing, manufacture test articles, collect data, analyze results of laboratory experimentation, perform calculations on original data, and/or in some way change the representation of data from the form in which it is originally collected to another format. The function of hardware and software components of a computer-based system as well as the interactive function of the instrument(s) associated with the computer must be validated.

The extent of the validation activity required, then, will depend on specific regulatory or compendial directives that apply to the system, the complexity of the computer-based function of the system, and the extent of documented applicable validation that the vendor(s) of the system or components has previously performed. (See the general 1986 reference by PMA's CSVC.) It is incumbent upon the manager responsible for a specific system to design the extent of validation appropriate and to determine the extent to which validation activities performed by a vendor may preclude the need for validation or components of validation performed in-house. For the purpose of defining validation activities, a critical and obvious discrimination is made between computer-based systems, such as an infrared spectrophotometer or laboratory robot, and simple microprocessor-controlled laboratory instruments, such as a digital stopwatch or thermometer in which the microprocessor function is an analog to digital conversion that is defined by the vendor at time of manufacture. The correct function of microprocessor-controlled devices is verified by calibration activities conducted according to the vendor's recommendations regarding the interval of and procedure for calibration.

The application of computer-based systems in the pharmaceutical analysis laboratory depends on their ability to meet criteria set forth by the FDA's cGMPs. The criteria include completing validation trials. The regulations require that the purpose of the system be defined and the functions of the system be validated in a scientific manner. The Code of Federal Regulations [21 CFR 211.160(b) and 211.168(a)] for automatic, mechanical, and electronic equipment say such equipment "shall be routinely calibrated, inspected, or checked according to a written program designed to assure proper performance." Written records of these calibration checks must be maintained. Although the FDA requires users to demonstrate that validation was and is being performed, they do not provide guidelines for completing system validation. They do not specify the methodology to follow when conducting system validation. The responsibility instead lies with the user of the system. This approach is understandable based on the wide variety of systems used in the pharmaceutical industry and the multitude of validation approaches required. The user is best equipped to design and conduct a validation. While the FDA does not provide instructions, there are common items that are expected to be dealt with from all systems; these apply to particle counters as well as to analytical systems of other types.

As the industry began to replace manual analytical procedures and databases with automated ones, the FDA began to inspect these systems and ask questions regarding their development, maintenance, and documentation. One of the results was the issuance by the FDA, in 1983, of the *Guide to Inspection of Computerized Drug Processes.* Since then, there has been a great deal of activity in both the industry and the FDA (Masters and Figarioic 1986).

Much of the activity regarding validation in industry was started by two committees of the PMA. These two working groups, the Computer Systems Validation Committee and the R&D Computer Systems Validation Committee, initially produced two working papers (1986) that had a significant impact on the thinking of the pharmaceutical and medical device industries. These groups have had a great deal of dialogue with the FDA on the subject of computer systems validation and, specifically, the interpretation of the "Blue Book." During this time, articles continued to appear in the literature, such as those in *Pharmaceutical Technology* (Kuzel 1985, 1987) and those published by the PMA. Although many of these articles are oriented toward manufacturing, the principles are obviously the same. Huber (1995) contains a large bibliography of relevant documents, and the IEEE standards (1985, 1989, 1990) contain relatively recent material on software validation and quality assurance. FDA documents provide insight into their philosophy on validation in general and with regard to instrument systems (Furman et al. 1994); the remainder of the references allow the reader to trace the philosophy of software assurance documentation and validation as it has developed in the pharmaceutical industry.

TO VALIDATE OR NOT TO VALIDATE

The decision not to validate means to the FDA, in essence, that the system evaluated is not required to be accurate and reliable, and that data generated or processed have no implications regarding the safety or efficacy of a product. The inspector may mentally question why a company that manufactures a drug or device product would have computer systems that do not have a relationship to the product they produce. The decision not to validate some programs, such as widely used word processors or spreadsheets, is understandable if one only considers the role of such software.

Risk assessment, as discussed earlier, is used at the starting point of the validation process as a means of determining the criticality of a specific validation effort in terms of effects of incorrect data, implications with regard to R&D, and the relative importance of validation of the system in question in comparison to other systems. The best course to follow in the planning stage for a validation is to consider validation incorporating the elements listed earlier as essential for any computer-based system that produces, handles, or stores data. The variant quality between systems will then be a question of the degree or level of validation to be pursued.

"Gap analysis" is a similar procedure to risk assessment. The "gap analysis" is documentation of specific areas in validations already in place where requirements have not been completely met or in which a validation component has not been completed. Any gap analysis will serve as a flag of some magnitude to an investigator. In any assessment of validations already in place, the requirements definition is frequently found to be incomplete or not current; revision of this document is of the utmost importance. The best way to deal with this is to break the existing documents into logical subsections that can be handled individually, then revise and refine these components and reassemble the document, and to identify and deal with gaps in an objective fashion.

For critical mainframe systems that reduce, store, or archive data relating to batch release or R&D operations and are based on custom applications software, each component of validation must be dealt with in some depth based on in-house testing and documentation as befits the critical and unique nature of such systems. At the other end of the queue of systems to be validated are items such as particle counters or spectrophotometers that have only limited (dedicated) system software or hardware that cannot be modified by the user. The validation plan for these systems will undoubtedly leverage in-use standards (also referred to as system suitability tests or SST), vendor-provided information, and much more simplistic test plans and requirements definitions. Nonetheless, an approach incorporating consideration of each validation component should be conducted. The key is to consider that each validation must be driven by and based on the key components of validation and to avoid categorizations such as GMP/non-GMP, critical/noncritical, and category 1/category 2. All computer-based systems, regardless of complexity, are validated to establish and document a high degree of confidence that the component functions as intended.

The following discussion of validation principles for users of particle counting systems is not intended to be comprehensive or to conform to the specific policies of any company or corporation. The author warns that any user who chooses not to validate particle counting of instruments must be aware that data being gathered might not be usable as documentation for a regulatory submission or other regulatory purpose or, ultimately, for batch release. In a small number of recent audits, FDA auditors have asked for validation "packages" related to particle counting systems that were not software based; the likelihood is extremely high that hardware or software-driven systems will be scrutinized in future inspections.

FDA Perspectives and Enforcement Activity

The regulatory concept of "validation" of critical processes as essential in the pharmaceutical industry is not new. The cGMP regulations were implemented in 1963. Since that time, the FDA has considered it to be a cGMP requirement that the equipment, facilities, processes, and procedures used in production and control be properly designed and appropriately tested to ensure that drug products have suitable identity, strength, quality, and purity. Obtaining this assurance is what validation is all about, irrespective of the term used to describe the process. Historically, long before "validation" became such a popular topic, the FDA expected companies to have adequate documentation to show that scientifically sound procedures were being used to ensure drug quality. This concern was evident long before validation language was first introduced in the 1976 proposed revision to the drug cGMP regulations, which were adopted in 1978.

In 1987, the FDA published its *Guideline on General Principles of Process Validation* (Blue Book) (FDA 1987) to describe practices that were acceptable to the FDA for process validation. Importantly, the guideline again noted that no single document could list all of the specific validation elements that would be applicable and that it was intentionally limited to principles for process validation. Therefore, by default, a range of validation strategies are acceptable as long as they are based on sound scientific principles and supported by adequate documentation.

The FDA regulatory definition of validation contained in this guideline focused on process validation. However, there are other important validation activities, not process specific (e.g., support and control operations), that may be encountered in pharmaceutical applications. Even though support and control operations are not directly associated with specific production process, they affect overall operations for the drug manufacturing facility. Support operations are frequently qualified (or validated) independently from specific production processes. For example, the validation of a laboratory test method is applicable to all products monitored with such method.

One FDA document that gives insight into FDA perception is the draft *Validation Documentation Inspection Guide* developed in 1993. While this document is available through the Parenteral Drug Association (PDA), the caveat that the document was never circulated and approved by the FDA is added. The overall focus presented in this document seems to be that "validated" is best used to describe a condition that is achieved when sufficient data, records, and information are available to demonstrate an overall state of control. In this 1993 FDA document, one category of a support and control operation that may be validated is defined as physical, chemical, and biological test methods.

A validated condition is stated to be one supported by documented evidence that shows that various aspects of manufacturing and control operations have remained within acceptable boundaries (e.g., within limits, tolerances, specifications, or other predetermined conditions).

The amount of testing and the degree of control necessary to attain a state of control depends on fundamental validation and documentation elements that must be satisfied before a "validated" condition can be established. The FDA accepts validation programs that are diverse, since each is custom designed to fit a company's particular applications. Because of this, the FDA appears to believe that each validation program needs to be evaluated on a case-by-case basis. The basic elements as listed in the 1993 guide, however, are generally similar between cases:

- Define goals.
- Perform testing.
- Document results.
- Verify accuracy.
- Compare against test criteria.
- List conclusions.
- Certify/approve results.
- Conduct periodic evaluations.

The similarity of this list of essential elements to others generated by standards organizations, manufacturer's associations, and individual companies will become obvious to the reader in the following discussions in this chapter. The primary objective of those setting out to acquire a working knowledge in this area is, in fact, to understand this commonality of approaches. Different groups have different terms for steps or elements in the validation process, but the actual testing, planning, and definition of critical data to be gathered are often closely similar.

THE PROCESS OF VALIDATION (Chamberlain 1991)

Chamberlain poses and addresses four crucial questions that are the basis of understanding the validation process:

1. *How do you establish "evidence" to the satisfaction of the company and the FDA?* The notion of establishing evidence is actually a simple one. First, document how you plan to establish evidence. That is, what kind of documentation, testing, and procedures you plan to implement (the validation protocol). Second, do what is outlined in the validation protocol.
2. *What is a "computer system"?* As discussed earlier, a computer system consists of instrumentation (e.g., chromatographic system, infrared spectrophotometer, particle counter) and the associated computer software and hardware that provide direction for the operation of the instrument and data transfer storage and reduction. This includes hardware and software components that are used to manufacture test articles, conduct testing, collect data, store data, analyze results of laboratory experimentation, and prepare calculated or original data to be used in final reports.
3. *How do you describe what a system purports to do? To what level of detail?*
   - Produce a User Requirements Specification.
   - Produce a system requirements document (general design).
   - Generate a detailed design document.
   - Implement the system.
   - Test the system.
   - Accept the system.
   - Evaluate the final product.
   - Perform ongoing maintenance of the validation.
4. *How do you show that a system will continue doing in the future what it purports to do?* There are two basic components of proof regarding the system's continued proper function: change control and ongoing monitoring of the system. Change control determines how, by whom, and why the system is changed. A detailed log of all changes needs to be kept and reviewed. Without this, there is no way to show that the system is under control.

TIMING OF VALIDATION ACTIVITY

Validation activities fall into one of two categories: prospective validation and retrospective validation. Prospective validation is accomplished while a system is being developed and installed. It is typically completed before a system is accepted for production use. It satisfies the notion that validation is "built in," not added on. Much of the documentation that is required for the validation is produced during the development of the system and should be produced keeping in mind the requirements for validation. It can be very expensive and time-consuming to go back and reproduce some of this documentation after the system is complete. Retrospective validation applies to systems that have been in use for several years (usually) and have strong historical evidence that they are functioning properly (Hambloch 1994).

Importantly, some validation activity relating to particle counting systems will fall into the category of retrospective validation. Evidence that instruments have been functioning properly must be of a tangible nature (e.g., printed reports, calibration files, test results, graphic control charts, quality assurance records, etc.). The validation will then consist of assembling the historical evidence into a validation protocol and filling in any holes that are left. The important point is that much work may be avoided if historic records can be substituted for some portion of time-consuming testing. Although the components of the retrospective approach are more or less widely accepted, the term is not one of the author's favorites. His personal philosophy is to conduct all validations as testing of the instrument in its current state of use, whether or not it has been in service for some time. All of the components in the next section should be considered.

THE COMPONENTS OF VALIDATION

As one searches the literature for information on how to conduct validation of a computer-based instrument system, various listings of the stepwise execution of critical components of the process may be found. Below is one such list (the author's). This list and most others like it are generally applicable both to software validation and that of computer or software-driven analytical systems. With particle counting systems, as with HPLC, gas chromatography (GC), and other instrumental chemical means of analysis, those conducting the validation will find that the demarcation between validation of the software that drives the instrument and the software-governed functions of the instrument becomes blurred; in some cases, as with particle counters, it is not possible to make a clear distinction. With this caveat, the most important components of validation, based on historical perspective and published information from regulatory, compendial, and industry sources are the following 11 items:

1. Risk assessment (also referred to as the validation rationale)
2. Requirements definition (also termed the system specification or spec)
3. Vendor qualification (VQ)
4. Design qualification (DQ)
5. Installation qualification (IQ)
6. Validation test plan development
    - Operational qualification (OQ)
    - Performance qualification (PQ, includes training, SOPs, and other)
7. Test plan execution
8. Maintenance/change control
9. System security, including backup, data archive, and disaster recovery
10. Validation report/certification
11. Validation review/periodic revalidation activity

These items represent general categories of documentation and activity in the validation process; detailed considerations must be made under each of the major headings. The genesis of this list is a historical perspective based on publications, FDA cGMPs, various directives and standards, and the author's limited experience in this area. The following discussions in this chapter provide more specific points as to what should be addressed in each area. Whether particle counting systems or chromatographs (old or new) are involved, the user and those assisting in the validation are advised to formally (i.e., in writing) address each of these components to the extent justified by the operational principles of the system in question, company policy, and logical approach. One good reason for addressing these specific items is that they provide a framework for comprehensive consideration of a given system; another is that they mirror to some extent the expectations of regulatory auditors.

The author would note at this point that the requirements definition, which may be referred to by an investigator as a "specification," is of overwhelming importance. Test plans and qualifications are conducted initially on installations of a system and later based on the significance of system changes. The other components, such as change control, are considered to be ongoing. Again, the manufacturer is best advised to validate all software-based systems and to address each component at an appropriate level of detail, rather than to omit components or spend time justifying why a system is not validated.

Risk Assessment (Validation Rationale)

A risk assessment of a computerized system (particle counting or other) is an evaluation of that system to determine what impact the system has on product quality, product safety, or the data or reports that influence decisions involving product quality or safety. Specific risks to data (including corruption, fraud, misidentification, or loss), risk of system misuse, or risk of instrument failure are also defined and discussed. The degree of risk attributed to a part of the system will determine the extent of validation involved. Importantly, the highest level of validation effort is most appropriately spent in areas that are identified as having the highest levels of associated risk.

Requirements Definition

The requirements definition specifies both a description of the system involved and requirements for operation of the system, including, but not limited to,

- interfacing requirements,
- functional requirements,
- user requirements,
- special requirements, and
- data requirements.

Again, the requirements definition is often considered by auditors to be the most important, single validation component. This document constitutes a detailed description of the system as it currently exists; it can take up to 50 percent of the time required to develop the entire validation package. It is a living document, tied into the change history and user's manual. Any changes to a system after the installation certification must be reflected both in the requirements definition and the change history. Most importantly, the requirements definition must contain a sufficiently detailed description of what the system is required to do both in terms of general functional requirements and specific user requirements so that it can serve as a basis for the test plan and qualification steps.

A user's manual typically lacks the level of detail and functional requirements data that are valuable and useful constituents of the requirements definition. Whereas the requirements definition is a detailed description of how the current version of the system functions, the user's manual must contain user-level information and cannot, in fact, be written at the level of detail required for the requirements definition. Presenting an auditor with a user's manual in lieu of a requirements definition may lead the auditor to evaluate the user's manual according to criteria for the latter document, with tragic consequences.

Functional and user requirements must be listed in the requirements definition. A clear, logically written (general) summary of how the instrument or system functions should be included here; most importantly, specific user requirements (e.g., USP <788> criteria for particle counters) must be referenced and identified alphanumerically so that they may be referenced in the test matrix of the test plan. This cross-reference is the key to ensuring that requirements are met, which is the principle of validation.

Vendor Qualification

Vendors of computerized systems, computerized instruments, or computer software that are used in a function affecting product quality or safety should be selected on the basis of their ability to meet the requirements of that function, including quality requirements. Records of qualified vendors should be maintained. The qualification of vendors will determine the appropriateness of the development process, the adequacy of the code produced, and the extent of the maintenance support available.

Many firms (not, unfortunately, in the author's perspective, including particle counter manufacturers) presently have in place a structured approach to both qualification of the design and manufacture of their product and to the VQ activities of customers; the validation processes of some vendors are much more detailed than those of others. With regard to the purchase of a new system, the manufacturers of particle counters should be subject to this evaluation process; with regard to retrospective validation activities, the process will serve to define both the current and past compliance of the vendor with the quality process.

Without the vendor's help and documents, it is impossible to demonstrate that the computerized system has been validated during its development. The vendor's assistance can speed up the validation tasks usually done by the user, for example, acceptance testing and OQ. In this chapter, some guidelines on how to select and qualify a vendor are given in Huber's books, along with criteria for equipment selection. One procedure for the customer to follow is to send the vendor a checklist with the questions typically asked during an audit. If all of the questions are answered satisfactorily, a formal on-site audit may not be necessary. Huber (1995) suggests the following general categories of questions and explains why they are raised:

- *How well do they know good practice regulations and quality standards?* If the company is not familiar with Good Laboratory Practice (GLP) and quality standards, they cannot provide products and services that help users to meet them.
- *Will they allow an audit?* Suppliers who refuse to be audited should not be considered for future purchases.
- *Do they provide adequate long-term support?* The customer (or potential customer) should ask questions about response time, proper training of staff, and what happens after product obsolescence. Large companies can generally give more reliable long-term support because they are better established in the marketplace.
- *Do they build all the required functions into the product?* Equipment used in manufacturing should be regulated and regularly tested, quality standards should be in place, and errors should be identified and recorded. Built-in functionality designed to meet field requirements helps to reduce laboratory costs.
- *Will they supply source codes to regulatory agencies?* The vendor should give written insurance to prospective users that the source code will be made accessible at the request of regulatory agencies. Source code is also required for complete software validation, which includes structural testing.
- *Do they supply details of algorithms and calculations performed by the instrument system?* Testing of software at the user's site normally includes verification of results, for example, the calculation of amounts from peak areas in chromatography. Mathematical algorithms acquired to do this verification should be included in the user documentation.
- *Do they have an adequate quality system?* There is a growing trend in the pharmaceutical industry toward dealing only with companies whose quality systems are compliant to internationally recognized quality standards, for example, ISO 9001. For vendors not compliant to such a standard, proof should be provided that a similar quality system with documented procedures for product design, development, manufacturing, testing, distribution, and servicing is followed.
- *Will they assist with customer on-site validation?* The instrument should be tested, verified, and qualified at the user's site at installation, after updates, after repair, and after extended use. The vendor should provide validation testing documentation, for example, test procedures, to reduce costs for the user.

Note: The FDA generally does not make a distinction between vendor and customer validations, and the requirements do not change. The requirements definition (or specification) document, for instance, describes why the customer chose one system over another. This logic can often be provided by the vendor. If the vendor will perform some of the validation and provide documentation of the testing, the user may use them. This will not eliminate the need for user validation once the system is installed. It is absolutely essential that the customer review and approve the vendor documentation because the user is ultimately responsible for the validation, even if a vendor has performed it. For in-house testing, the FDA expects to see some evidence that structural or white box testing was performed. This may be impossible to do with vendor-supplied systems if the vendor won't supply the source code (Chapman et al. 1987). Regarding the source code, the user should attempt to get vendor documentation for their source code development and review for files. The user should, at a minimum, make sure that an agreement is in place with vendors such that the source code can be accessed and reviewed when needed (e.g., source code held in escrow). Someone knowledgeable from the user company should review this code during validation to make sure it can be maintained if the company goes out of business. (The actual usefulness to an auditor of a source code written in assembly language has always been questionable.) The source code should be indexed in the event that use should become necessary.

Finally, any material furnished by a vendor must be evaluated for thoroughness and substantial conformance to in-house SOPs. Any off-the-shelf system, any modification to an off-the-shelf version of the software, or any system customized by the user will need IQ and require validation by the user. Integration of a system supplied by the vendor into an existing system will also require work in-house.

Design Qualification

Vendor qualification assesses, among other things, the level of quality (conformance to requirements) built into an instrument before it leaves the factory. Another step in a complete validation or qualification is equally basic; DQ consists of reviewing the final instrument being evaluated for purchase in the context of how well it conforms to the vendor design criteria on which it is based.

For instruments produced more than a decade ago, vendor records of design and engineering specifications and, in fact, final testing required to release an instrument from the factory to the user are typically less than comprehensive, and in a worst case, unavailable. Today, however, due to the widely recognized benefits of ISO 9000 level quality system certification, vendor systems for tracking and documenting the design and development process and testing of instrumentation and software are much more highly developed. Whereas even five years ago many instrument vendors were generally unaware of the drive toward instrument qualification/validation in the highly regulated pharmaceutical industry, many vendors now actively assist customers in developing the validation process. Based on user cooperation, the more proactive validation processes may be conducted, with the initial components (DQ, VQ) begun at the vendor site using vendor information. The design evaluation also serves to provide a foundation of how well the instrument will meet the user's requirements, which will be spelled out in the requirements definition of the user's validation process.

During vendor evaluation activities, DQ provides information in two important ways. First, if the vendor operates according to a structured matrix in software and/or instrument design and execution of that design (e.g., Hewlett Packard, see Huber [1994] and the Hewlett Packard GLP document [1993]), their records of the design-development process will provide a readily followed audit trail that will reveal how closely the instrument or software meets requirements. Second, the customer will be able to use vendor documentation to determine at some level of detail how closely the instrument developed by the vendor matches the user's requirements.

Installation Qualification

Generally, there are three key components of in-house validation activities: the installation of the instrument, execution of the in-house validation, and the ongoing or continuing phase. The latter will typically involve activities such as system annual reviews and change control. IQ involves the setup of the instrument by the vendor or vendor's agent and confirmation that the instrument is operating properly per the vendor's performance criteria (Alford and Kline 1990a). Written certification should be obtained from the vendor that the instrument is installed and performing properly. This ensures that the user can expect that the requirements definition is initially met and that further validation activities can be begun by the user. Rationally, it makes little sense to begin validation activities that have been carefully structured and will be time-consuming on an instrument that is not performing properly.

Validation Plan (Test Plan)

One observation that is needed here relates to the wide variation in the content of the requirements definition and the validation plan between companies. These two documents are without doubt the most important in describing the system, and specific information must be included in one or the other. Of overwhelming importance in the test plan is the sequential outline of testing to be performed, item by item. Each item must reference a specific user requirement as spelled out in the requirements definition. The testing so defined will serve as the foundation for the validation by empirically proving that user requirements are met.

The following subsections of the validation plan contain the most critical information:

- *Purpose:* Describe why this validation is necessary.
- *System description:* Provide a thorough description of the system being validated. Include here a definition of the original data handled by the system, along with the processing or other activity involved with that original data. If the computerized system interacts with other systems, provide a reference or description of that interaction and/or interfacing.
- *Test plan matrix:* Test items precisely cross-referenced to user requirements detail.
- *Scope:* Describe how much of the system is being validated and why.

Some users also employ the test plan to present the principles of OQ, PQ, and SST as they apply to the system. The OQ tests and documents whether each component of the system (i.e., hardware and application software) performs as intended throughout its expected operational ranges. A series of diagnostic and functional tests ensures that each piece of a modular system is operating within acceptable and predetermined performance criteria. Tests and criteria should be described in a suitable protocol or SOP, such that tests are consistently executed and appropriate GMP documentation of the outcome can be obtained. After successful OQ testing of each separate instrument module, the modules are assembled into the complete working instrument. The entire working system is then tested to verify it is operating within acceptable and predetermined performance criteria. This component of OQ testing is mandatory for software-based particle counting systems, just as that for simpler microprocessor-based systems.

PQ tests and documents whether the entire system (i.e., hardware, application software, and associated instruments) performs as intended throughout its expected operational range. The analytical system is expected to perform as required by a specific analysis before the system is released for use in that analysis. Stress or boundary testing may be included in the PQ and involves testing for appropriate functions at specified limits. For particle counters, concentration limits, high and low flow rates, effects of electrical instability vibration, and other factors may also be selected for testing. Qualification timelines are specified in Figure 11.2.

System Suitability Testing

The IQ, OQ, and PQ activities discussed thus far constitute prior-to-use testing except in the case of retrospective validation of "legacy" instruments put into service prior to the widespread application of validation. In addition to this in-depth testing performed prior to system use, time-of-use tests that are performed prior to or during use on a daily or more frequent basis are extremely important. The most widely accepted term for this type of testing is *system suitability testing.* This testing standardizes the integrated instrument system for a specific application. This type of testing serves not only to ensure that the system is operating properly prior to use but also to limit the amount of data that may be generated with instrument performance due to its frequent use.

The author believes that the utility of well-designed SST cannot be overemphasized. The approach applied by most users involves the principle of apical output testing. In a functional overview, an integrated instrument system consists of a pyramid of outputs from its different components (e.g., hardware, software, the instrument itself, and operators). These outputs join at the top of the functional pyramid in a relatively small number of critical performance parameters. According to this concept, outputs are selected for SST that serve as critical assessments of multiple instrument functions that combine to provide data output.

If, in a hypothetical ideal case for a given instrument system, we could find one single test

Figure 11.2. Timing of qualification activities.

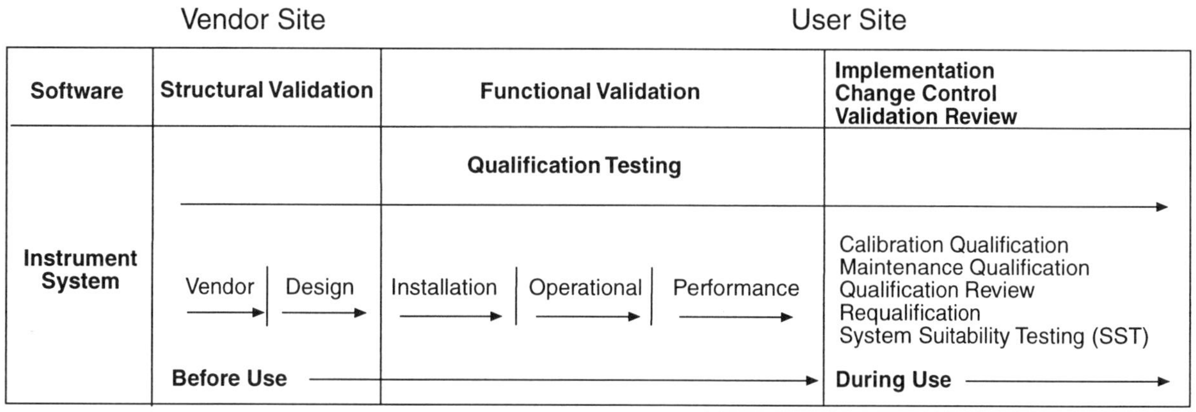

that would serve as a test of resolution, accuracy, precision, electronic function, data handling, and other critical parameters, we could simply run this test prior to using the instrument, and there would be no need for structured validation; all instrument functions would be assured to be operating as intended by this ideal SST. An example (again, hypothetical) might be the rendering of a test spectrum of an NIST standard polystyrene latex (PSL) sphere by Fourier-transform infrared spectroscopy (FT-IR) at such a high level of resolution and with relative peak integrals so closely matching the ideal that absolutely nothing could be amiss. Another example might be a system suitability test micrograph from an electron microscope showing such high resolution that all systems of the instrument would be shown to be working optimally and that no component or subsystem could be out of spec. There are no such ideal tests, of course, but it is very possible to select testing procedures that will provide a high level of assurance that the instrument is operating properly before use. These tests do not constitute a substitute for validation activities, but rather rely on the validation process for their identification and must themselves be validated during PQ.

In the specific case of a particle counter, obtaining the correct number of counts at closely spaced size thresholds for a validated count standard with a continuous particle size distribution serves to confirm the USP <788> operational parameter of flow rate, sample volume accuracy, resolution, and calibration. Critical SST often may be conducted very quickly, but it requires careful planning and test standard design.

Execution of System Testing

Generally, more thought and effort is given to the test plan and its execution than other components of validation. The functional test plan comprises a stepwise evaluation of the function of various activities that are carried out by the software for a software-driven instrument system. The purpose of system testing in the validation plan is to establish a set of tests that exercise the system with respect to each of the requirements identified in the requirements definition, as well as any issues associated with those requirements. Included in this section is a series of discrete tests, each traceable to the requirement it is designed to verify. The following items are included:

- *Cross-reference to requirements:* Reference the actual requirement in the requirements definition for which this test is being carried out.
- *Test procedure:* Describe the specific step-by-step procedures and the exact data input required to carry out this particular test. If printed output is a result of this test, include information required for identifying that printed output, such as "Label the printout as Attachment 4." If electronic files are used for input or generated as output, include

information necessary for identifying those files.

- *Expected results:* Include in text format the expected results or reference here an appropriate document that shows or describes the expected results.
- *Acceptance criteria:* Indicate the criteria for evaluating the actual results, such as actual results must match expected results, or actual results must be within a specific range.
- *Actual results:* Actual results must be documented at the time the test is performed. This documentation may be in a laboratory notebook, or better, on predeveloped and preapproved worksheets, on instrument printouts, or by other types of documentation. All documentation generated for the test must be traceable back to that test. Identify and explain any results that are exceptions to the expected results.

Results may be documented in written format, by screen print, or in checklist form. Examples of test result recording formats for particle counters are provided in later sections of this chapter. A checklist should be detailed enough to allow a nonexpert reviewer to understand exactly what was tested and the result. Stress testing is usually included in the test plan, but it is often poorly understood. As an example, if a system has an operating range of 1 to 50 units, assurance is not obtained by testing all 50 values, but rather just those most likely to provide failure conditions, i.e., 0, 1, 50, and 51, as well as a small number of mid-range values. Similar logic implies that focusing on null entries, negative values (when positive ones are expected), and the entry of numbers into a letter field are more logical approaches than testing every possible letter/number in the expected string.

Both for individual system validations and for company validation programs, responsibilities should be defined at the highest level in the first case. Each review of a test plan should have a specified accountability, i.e., what a review signature means. At the corporate level, sign-off means awareness, accountability, and approval. Another signature item of concern is the electronic signature of data printouts generated by computers. While these are in some contexts acceptable to the FDA, it is necessary that each signature be coded, that a password is necessary for access to document signature blocks, and that the signature block contain a statement regarding what responsibility is being accepted by the individual affixing his or her signature.

Maintenance/Change Control

Change control is performed during the initial validation of a computerized system and is maintained on an ongoing basis throughout the operating life of the system (Alford and Kline 1990b). Change control documentation serves as the requirements definition for the change, and the information in this section should be used to update the requirements definition document for the computerized system. Evaluation of the change should be according to the effect on the operation of the system, resources required to implement the change, regulatory or other requirements, the effect on the quality of the data or product, or other factors.

Configuration management is an essential element of the validation process for a computerized system (see IEEE 1988, 1989, and 1990). It maintains system functionality as the system evolves and provides an audit trail that ensures the system is continually operating in a validated state. With proper implementation of configuration management procedures, it is possible to develop a history of a computerized system from its inception to its current use, which enables maintenance of the system's integrity. Configuration identification is performed when a computerized system is initially validated or when hardware, software, or documentation is added to the system.

Examples of configuration specific items are as follows:

- Requirements definition
- Design specification
- Hardware configuration
- Software configuration
- Hardware manuals
- Software manuals
- User manuals
- SOPs

System Security

During validation, the need for operational control documents (SOPs) for the computerized system must be determined and the necessary documents generated. Manuals and SOPs that document the procedures being performed must be available to the users of the system when necessary. Published literature or vendor documentation may be used as a supplement to the SOP if properly referenced therein. The user should develop operational controls in an SOP for the computerized system. The SOP must minimally contain a system description that provides a general overview of the system and its use, thorough enough that a user can determine what the system consists of and how it is a part of the overall process. The user should also evaluate the following areas for the need for additional operational controls; if any of these controls are applicable, they should be included in the SOPs for the system.

- Backup and archival requirements
- Disaster management
- Security requirements and procedures
- Validation review requirements and procedures
- Training requirements
- Operation procedures
- Maintenance procedures

Disaster Recovery

Disaster recovery planning is required for those computerized systems that support critical business functions if the business function cannot be operated in the event of a disaster affecting the system. Management must evaluate the role the computerized system plays in the operation of critical business functions and determine if a disaster recovery plan is required. Adequate security procedures combined with disaster avoidance procedures may be adequate protection for most computerized systems (Schoenauer and Wherry 1993). This requirement must be taken very seriously; for years, the author personally believed that the likelihood of disaster (e.g., fire, flood, tidal waves, etc.) affecting the company's manufacturing operations was so slight as to be almost nonexistent. This outlook changed with an airplane crash that did in excess of $20 million damage to a manufacturing facility and destroyed the product test data archive.

The essential first step is to identify the criticality of the use of the computerized system in the operation of the business. The threat to the business that would be posed in the event of a disaster must also be assessed. Evaluate the security measures in place for the computerized system. The user should then determine if the following types of measures may be adequate to address the identified threat from a disaster:

- UPS systems (uninterruptible power supply)
- Fire detection and/or suppression
- Independent air handling
- Physical and data security measures
- System redundancies
- Routing system maintenance

Recovery strategies should be planned and may involve the following:

- *Off-site storage:* Determine the contents to be maintained off-site and develop procedures for keeping the off-site contents current.
- *Hardware recovery:* Determine how hardware will be brought back up.
- *Plan activation:* Define the procedure for activating the disaster recovery plan.

Data Security

Data security is based on the following:

- Control and authorization procedures for individuals having access to data at different levels.
- Proper entry of data and proper identification of the individual entering the data.
- Verification of manually or electronically input data.
- Interpretation of error codes, flags, or messages and the corrective action to follow when these occur.
- Proper methods for execution of data changes, including the original data element, the changed data element, identification of the date of change, the individual responsible for the change, and the reason for the change.
- Management responsibility for security of data collection, reduction, and archiving.

There must invariably be a specific definition of the original data acquired, processed, or stored by the computerized system. This definition must be incorporated into an SOP that defines and controls the use of the computerized system. The original data are defined as the first record of an original observation that may be collected on paper or on magnetic, optical, or other media.

If laboratory observations are made so that there is direct entry to magnetic media, the electronic data constitutes the original data. When observations are made initially on paper and are subsequently entered into a computerized system and recorded on magnetic media, the hard-copy forms constitute the original data.

All electronic original data must be capable of being readily retrieved. The individual responsible for direct data input must be identified at the time of the data input. Any change in automated data entries must be made so as not to obscure the original entry, and indicate the reason for change. All changes must be dated, and the responsible person must be identified.

Hard-copy printouts of the electronic original data may be subject to the requirement that the printout be verified as accurate, signed, and dated. The verification may have been established through a previous validation of the computerized system, or it may be performed upon each occasion. The hard copy of the original data may then be permanently stored. A historical file of outdated or modified programs used for generating the electronic original data should be maintained. The files required for maintenance should be those sufficient for adequate retrieval of the electronic original data as well as for processing if required in order to compare with the results contained in a report of that data.

Electronic Records and Electronic Signatures

Based on the Final Rule published by the FDA in the *Federal Register* in 1997, electronic records of original data may be equivalent to hard copy, based on meeting criteria for paper records (e.g., access control, database security review, traceability, signature). The electronic system must be identified to the regulatory agency or its representatives as constituting the original data/database. Two types of electronic archives are recognized: open and closed systems. Open systems are accessible to individuals outside the company and others; access is not adequately controlled to satisfy the requirements for an electronic records archive. Validation of an electronic records system as being a closed system (i.e., access controlled and data security maintained) is necessary before it can be used as an electronic documentation database. The user may elect to use an open electronic database just to enhance user access to data. This data, however, is not considered "original data" under the terms of GLP/cGMP. Remember, an electronic database considered to be original data must be so designated and validated with regard to access control and data security (i.e., data secured and protected from unauthorized access, alteration, corruption, misidentification, or loss).

More recent consideration of the Final Rule by the FDA has gone somewhat beyond what the manufacturer of pharmaceuticals and medical devices originally intended or asked. In deliberations following the issuance of the Final Rule, it became apparent to the FDA that for some types of instruments, the electronic output contained a great deal more information than a hard-copy printout and was, in this context, a more suitable material for archiving than the "hard copy." For example, the resolution of electronic spectral data from an X-ray spectrometer might have higher resolution than the printout, and, if archived, be available for future processing, such as background or spectral subtraction. Thus, there is some indication that for some types of instrumental outputs, the electronic data will be defined as the "original" data; this will mandate the use of the electronic archiving and the electronic signature. One of the key factors to be considered by the user of original data is that these must somehow be applied to the original file to generate new nonoriginal data, and the original data file must be held in an archive where it cannot be changed.

Fortunately, these latter, more complex considerations do not apply to particle count data from LO counters that may be archived in electronic format. The electronic format stores no more data; it simply is a digitized format, which can be printed in analog form.

The concern over data security requires that electronic databases and electronic signatures be considered. The implications for particle count data are the same as for other data relating to product safety, purity or efficacy, or other quality attributes. Per 21 CFR §11.10(e), use of secure, computer-generated, time-stamped audit trails to independently record the data and time of operator entries and actions that create, modify, or delete electronic records is mandated. Record changes should not obscure previously recorded information (see also 21 CFR 11.10, 11.50–11.300).

Specific FDA interests include the following:

- Who is authorized to enter, edit, retrieve, and change data in the system?
- Is there an electronic audit trail that determines what was changed, when it was changed, and who changed it?
- If there is no audit trail, how are data secure from unauthorized access?
- Are procedures used to prevent unauthorized program changes?
- How are data secured from alterations, inadvertent erasures, or loss?
- Who in authorized to alter or have access to programs?
- Is storage of backup disks, tapes, and so on secure?
- Are procedures available in hard copy form as SOPs?

Administrative Security

Administrative security of the instrument system is ensured through

- identification of individuals who have authority to assign or make changes to passwords;
- documented authorization of individuals allowed access to data, functions, and review; and
- timely deletion of accounts and access privileges for terminated employees.

A detailed consideration of security procedures is provided in the PDA reference by Schoenauer and Wherry (1993).

The Validation Report—Certification

The validation report should be comprehensive and in-depth. It should cover all areas of the validation, be written clearly with references to documented evidence to support the contents, and provide meaningful conclusions. The executive summary portion should be a brief overview of the validation. It should summarize the major issues addressed and activities performed during the validation, as outlined in the following 10 items:

1. Identify the system by description and system validation number.
2. Summarize the assessment of the risk involved with using this system.
3. Indicate the scope of the validation.
4. Summarize the validation issues addressed for the system and the activity performed.
5. Include a definition of the original data.
6. Include as applicable an accounting of any configuration changes.
7. Summarize the testing performed. Identify each exception and include its resolution. Provide a statement of the effect the exception has on the validation of the system.
8. Provide a final statement of the result of the validation process.
9. Identify the system's validation documentation and the location of that documentation.
10. Approvals:
    - System coordinator
    - User management
    - Quality assurance

EXTENT OF VALIDATION

Systems to be validated will obviously vary in a number of specifics. It is generally difficult and unwise to place systems in categories based on the level of effort that will be required for validation, but some general considerations are useful at this point in our discussion. Systems are generally of different size and complexity; larger, more complex systems will require more effort than smaller, simpler ones. For example, a particle counter driven by relatively simple software and the assay that it performs are obviously going to take far less effort to validate than a laboratory information management system for a small laboratory. "Canned" or off-the-shelf software systems, especially if vendor information is available will generally be more easily validated than custom or in-house developed packages for critical roles in lab or manufacturing operations. Simpler, lower level of effort approaches may be reasonably selected for some instrument systems (Figure 11.3). These approaches typically leverage vendor validation activities to the maximum extent possible consistent with vendor data quality and the policy of the buyer.

Other systems, based on complexity, cost, criticality of application, company philosophy, or other factors will reasonably be the subject of a more detailed approach, with the individual system components dealt with in terms of rigorous, in-house testing (Figure 11.4). The critical determinant in the decision process regarding how much validation is enough will be what body of information and test results constitutes "adequate" proof that the system functions and will continue to function as intended.

Simpler approaches should meet two criteria: (1) Regardless of the cost or size of the instrument, the validation must achieve the necessary assurance of correct function of the software or system. (2) All of the components of validation addressed in the list above under "Components of Validation" should be addressed at some level of detail. The risk assessment is of overwhelming importance in the pharmaceutical and medical device industries.

HOW MUCH VALIDATION IS ENOUGH?

There is no simple dogma regarding how much testing should be conducted and/or when validation may be stopped. Some considerations in making this decision on a system-specific basis are the following (see Anisfeld 1994; Gold 1996; Vinther 1998):

- Allocate resources (time, detail, and extent of testing) to different components of the system and its operation based on the magnitude of risk assigned. The primary consideration must always be the risk of generating erroneous, mislabeled, or otherwise unacceptable or misleading data output.
- Define and discuss endpoints for validation in each activity (e.g., test plan, data security, training, etc.) prior to beginning the testing and in the final report. Don't make an auditor guess why you stopped where you did.
- Develop an objective rationale for value added (e.g., in terms of data security, confidence in the intended operation) versus extent of validation activity for each validation activity.
- Stop testing when reasonable assurance of as-intended operation is achieved. Vendor DQ and reliability testing may be leveraged to assess likelihood of failure of specific system or software functions; repeating vendor testing generally wastes man-hours without adding security in terms of either soundness or product conformance to requirements.
- Reasonable assurance usually results from a one-time test of functions that will be used; the goal is not testing to failure, and repeated tests of the same function should be avoided.
- Don't validate features or functions not used unless they impact the operation of those that will be used.
- Don't validate beyond the range of use (this does not apply to boundary, stress, or limits testing). The range of operation must be clearly defined and limited by an SOP.
- Don't validate features that are self-flagging if they fail, such as pull-down screens.
- Use SST as part of the rationale for limits chosen for OQ and PQ testing. Its function is to detect errors that will have negative effects on the specific assay performed; in this way, the ongoing suitability testing is a focused and empirical test for system errors that will result in the generation of bad data.

REGULATORY INSPECTIONS

The approach of an FDA inspector to the audit of a computer, software system, or computer-based instrument or machine is based on a relatively simple philosophy (Clark 1988). Some of the more important points of this philosophy are well worth mention. First, the inspector expects the inclusion of key specific components or elements discussed above in any validation.

A specific important observation with regard to inspections is the following: The manufacturer is well advised to have validation packages in place for critical computer-based systems,

Figure 11.3. Validation approach emphasizing vendor information and in-use testing of system operation.

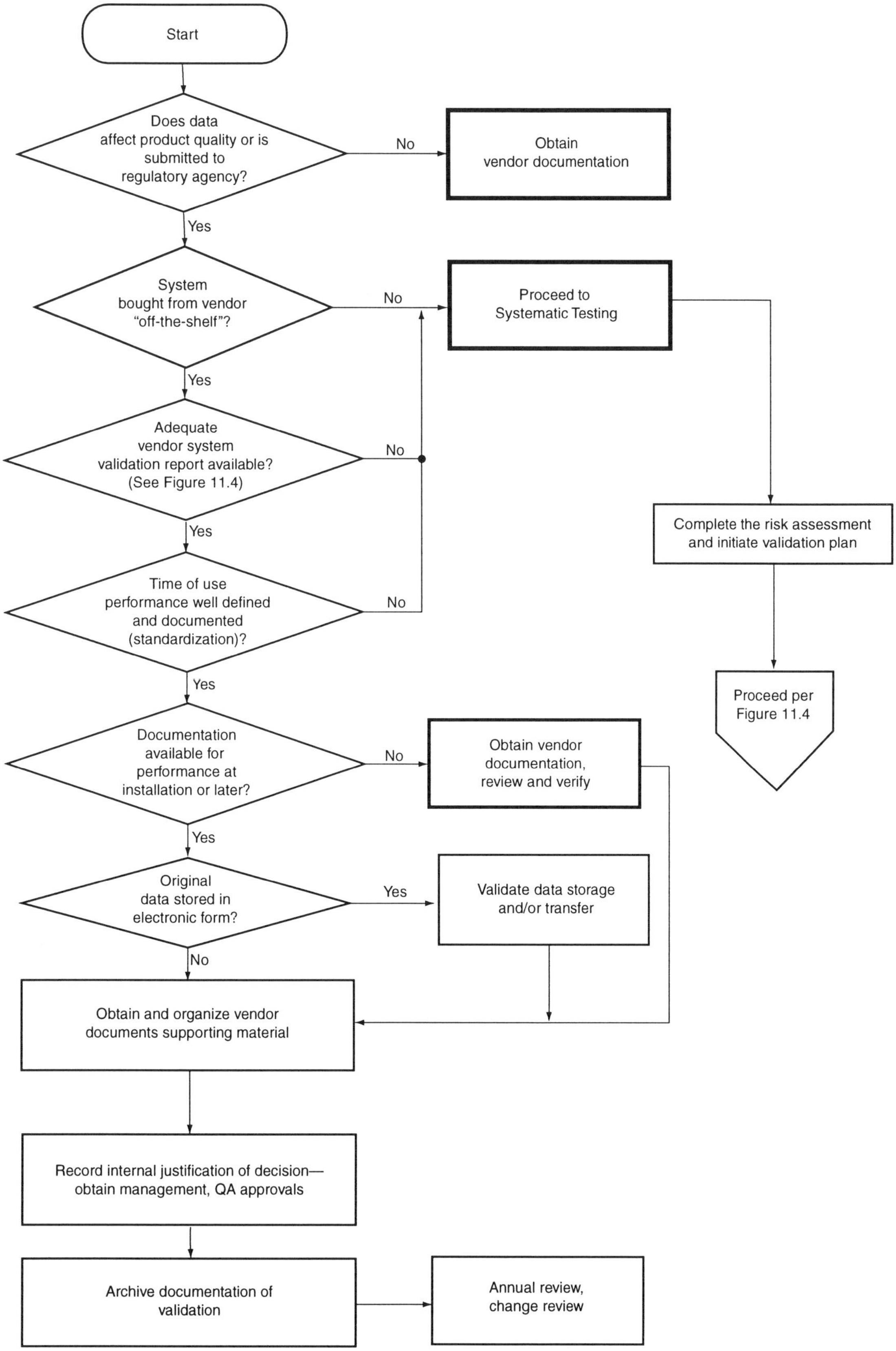

Figure 11.4. Validation approach based on empirical determination by in-house testing.

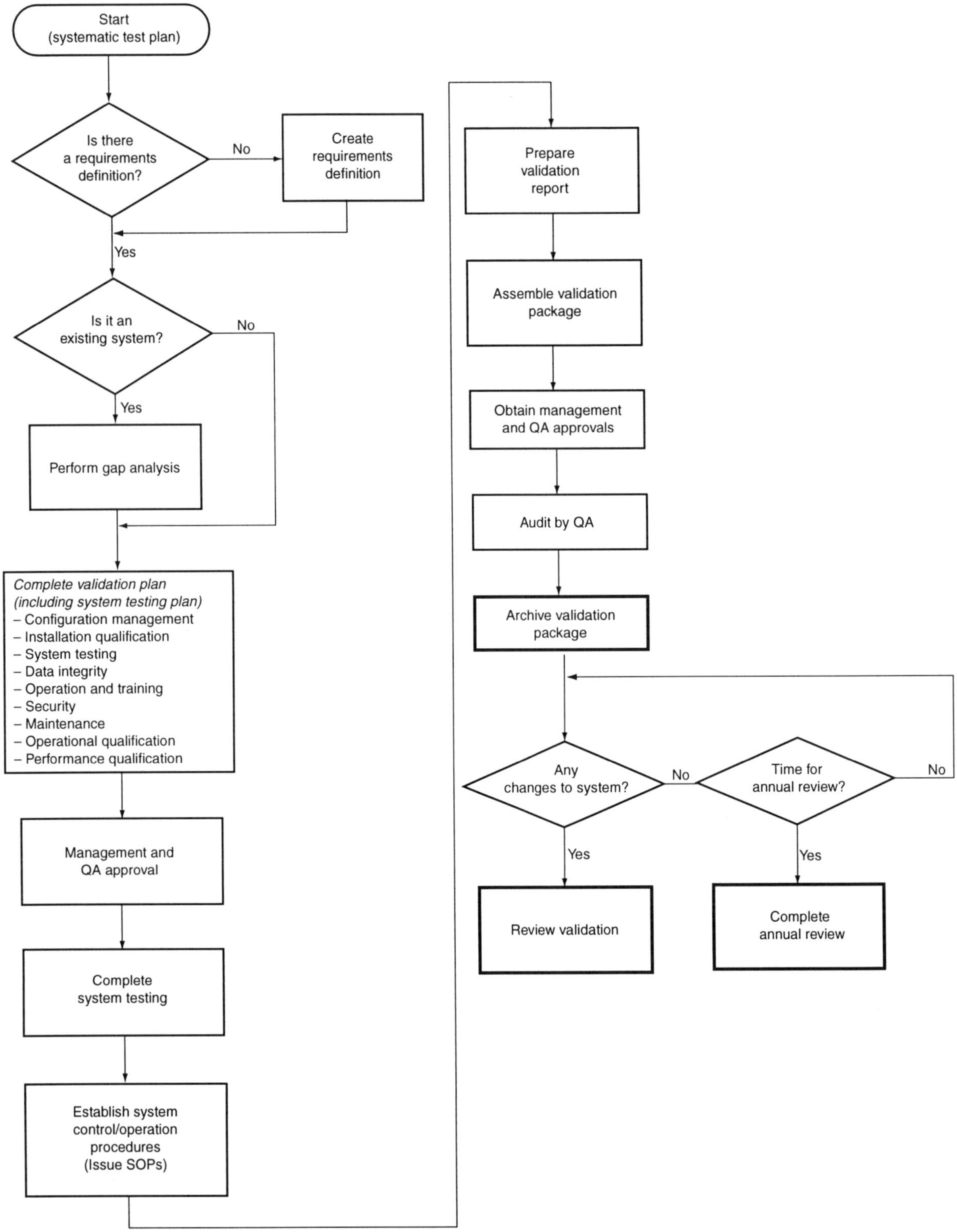

including particle counters, prior to inspection. Packages should be indexed and organized in a format that is easy to follow, with key elements clearly defined. This will result in a field inspector being able to assess the information presented and pass judgment on its suitability. Validation information that is not well organized and presented, or that is presented in complex, user-unfriendly fashion, may result in the field inspector requesting help from an FDA computer specialist. This has proven unfortunate in some cases, leading to an investigation focus on source codes, software, and programming mechanics as well as overall system suitability. A word to the wise is hopefully sufficient.

As in other types of regulatory inspections, serious problems may develop if deviations from SOPs are found without explanations documented, and if undetected errors are found by the inspector. In this regard, simple validation approaches that are uniformly followed are far better than complex, custom approaches tailored to specific systems. The FDA also takes some level of exception to several time-honored approaches that were once widely used in validation by manufacturers:

- *Grandfathering:* The blanket judgment that a system is validated based on its years of prior service is generally unacceptable. On the other hand, the use of historical data and information confirming proper operation as part of a retrospective validation is eminently acceptable and widely practical (Hambloch 1994).
- *Historical data:* While a consideration of historical data has some value, a validation based on historical data will often be more time-consuming than a considered, efficient approach from the ground up. It may also provide less definitive proof of "operation as intended" than a retrospective validation incorporating the steps listed above. Even retrospective validation may also provide items of interest for investigators.
- *In-use testing:* In-use or SST demonstrating no errors with some period of routine use does not serve to test a system beyond the regular range of use and does not substitute for OQ and PQ.
- *Obsolescence:* Interestingly, obsolescence is one reason for the decision not to validate that may be accepted by the FDA. A decision is made not to validate a system based on its planned replacement in a time frame shorter than that required for validation. The cautions here are that once a commitment to replace is made, it must be followed up; the replacement of complex systems requires long-term planning and effort.
- *"Non-GMP" rationales:* Along these lines, the decision not to validate based on an assessment that the system in question is "non-GMP" in application generally does not reflect sound logic.
- *Extensive nonvalidation rationales:* The user should, at all costs, avoid the tendency to develop extensive rationales as to why validation is not necessary. These rationales may take longer to develop than a validation plan. Inspectors will not only be magnetically drawn to unvalidated systems, but they may be interested in the philosophy, logic, and rationale figuring into the decision.

The order of priority of historical validation-related 483 items based on the number of citations is interesting.

- *Noncurrent Requirements Definition:* Generally, the item most often cited is the lack of a current requirements definition or system description, which is expected to be a "living document."
- *Ambiguous SOPs:* Next in popularity are ambiguous SOPs that could lead to unapproved, undocumented changes and inconsistencies in a procedure.
- *Unvalidated systems:* As a general rule, unvalidated systems that are often identified by extensive written rationales as to why validation is unnecessary are targeted.
- *Gaps in testing:* The omission of key elements in the validation documentation (i.e., "gaps" in testing) also receive citations.

No auditor expects a 100 percent level of assurance from validation. This cannot be achieved. Rather, it is expected that the validation documentation will demonstrate a reasonable level of assurance through an objective approach to each component. Far more important is the systematic inclusion at each step of a means of detecting errors that may be made and identifying deviations that may occur.

This will offset in an effective fashion any factor or combination of factors that escaped testing and that may then produce an error.

The following general 483 observations pertaining to validation are instructive:

- There was no validation protocol that detailed how system tests were applied and no documentation to show the results of each test. There was no documentation to show acceptable integration testing of the data acquisition system.
- The current data acquisition system has not been validated and evaluated as to its correctness and effect on program execution. There are no SOPs or systems in place to initiate revalidation when modifications to the software occur or significant hardware changes are made.
- There is no assurance that the analytical chromatographic data, reported as percent purity and potency for each product, are accurate and reliable. An error-prone software program, written in-house, integrated into a "vendor's" software program resulted in an overall error-prone system.
- Significant computational errors were discovered in the software program during the validation process and were ignored. The program was installed and used to acquire and process data. There were no procedures in place to ensure that the data acquisition system was not used when it did not conform to the firm's specifications.
- The firm lacks the necessary controls to ensure the integrity of data generated by its computerized system. Analysts have read/write access to all chromatographic data files stored on the server.
- The firm lacks the controls necessary to the integrity of raw data generated by the computer system. For example, computer software programs that acquire and process analytical data cannot ensure data integrity and traceability. Analysts can overwrite the raw data. Once the results are modified, the changes can be saved to the current file or a newly created data file. Then we dialog boxes in the program to reject analytical data whenever the user considers the results to be out of specification.
- Operations that could affect the integrity of data once acquired and processed by the data acquisition system are not controlled by electronic audit trails that document who, when, and what was changed. There are no records of user transactions when data are deleted, copied, renamed, or purged. Electronic audit trails are not configured for any of the data systems.
- The firm lacks the controls necessary to ensure the integrity of raw data generated by the computer system. For example, operations that delete, corrupt, or modify data before, during, or after data acquisition were available to analysts. Users without the appropriate security level have access to the operating system software and all applications software. Data files *can* be copied, deleted, modified, or rearranged. Users can access dialog boxes that add, delete, and rearrange data in a sequence by appending, moving, inserting, and copying rows.
- There are insufficient security measures in effect to ensure the integrity of chromatographic data housed in the quality control laboratory. There are no security procedures in place to prevent users from accessing all data files. Users have read/write access to the data stored on each hard drive. Employees without passwords have access to data files when analysts leave a terminal without terminating a session.

SPECIFIC CONSIDERATIONS IN LIGHT OBSCURATION COUNTING

Any validation procedure for particle counting systems must begin with a consideration of the technical aspects of the assay. This is particularly important with regard to the compendial LO particle count assay. This test is not a chemical assay, such as a titration, for which standards of accuracy and precision are established and readily available. The instrumental particle counting test is simply intended to indicate whether or not the number of particles in a tested unit are within limits. Specific human factors, such as differences in the way analysts agitate or enter a solution unit, have an effect on the outcome. Typically, in the past, when two identical HIAC 4100 instruments (same sensor and counting electronics) calibrated by the same method are used to count an ideal standard suspension of mixed latex beads, a variation of ±10 percent or greater might be noted between counts from the two instruments. In counting nonideal particles (i.e., extraneous particles in parenteral solutions), this between-instrument variability increases to in excess of 20 percent.

A significant component of variation is the human factor, whereby human analysts perform certain activities of the test with slight differences. With regard to the variability of test articles, between-unit variability of particle counts also increases as unit size decreases, so that a considerable variability component from this source is introduced into the assay result. Although a great deal of effort has been put forth in individual laboratories to minimize the effect of these variables, their impact on the test is still significant in the manual test. In this consideration, the user may find it useful to review USP <1225> and assess the desirability of testing factors such as precision, ruggedness, and robustness of the assay as a basic step preliminary to other validation activities.

Any particle counting methodology will be critically dependent on the instrument used. One of the primary concerns in the validation of particle counting tests according to the philosophy of many users is PQ. If a new instrument is purchased, one goal of the validation is to obtain evidence that the new test, based on a software-driven counter, gives a result equivalent to the test being replaced or phased out. The light obscuration particle counting assay described in USP 23 <788> is a limits test or pass-fail assay. The desired result is a simple prediction of whether a batch or other sample group tested exceeds or meets the established compendial limits. Given the variability of the light obscuration test, it is impractical to test a sufficient number of units to determine actual numbers of particles present in a unit with any reasonable degree of confidence. Therefore, the primary goal of the validation of the assay, whether computer based or not, is a determination of batch suitability that is in agreement between manual and computerized assays (i.e., that the probability of failure of a tested batch was the same for both assays) or old and new assays.

In consideration of all these factors, data obtained by manual and software-driven assays may be tested in four specific areas:

1. Numerical agreement between assay results for two sample groups from the same batch of product, one tested using the older ("manual") method, the other using the software-based system
2. Occurrence of the same count distribution in data from both assays
3. An agreement for mean results of testing across a specific product type, e.g., small-volume injection (SVI) drug units, large-volume injection (LVI) flexible containers, and so on
4. Assessment of equivalence of the two assays with regard to prediction of probability of a batch exceeding or meeting compendial limits.

The last item is considered by some to be the most meaningful test. The procedure involved in testing involves fitting a distribution to the data for five particle sizes counted. This distribution is typically found to be a Poisson or negative binomial distribution. Once the data distribution, mean assay result, and sample size are known, it becomes possible to predict what fraction of all possible sample groups of a specific size from a given population (e.g., batch) would give a failing result. This probability is often extremely low (10^{-6} or greater) and is generally the same for manual and computerized assays. It should be observed in this regard that a difference at the $\geq$ 25 μm size observed frequently results from a mean of 0 counts in one assay, to a fraction of 1 count by the other. While a difference of this type may be proportionally or statistically significant, it is generally not technically significant due to the variables in the assay and the low level of counts with regard to the limits.

Regarding the validation of light obscuration counting procedures, there is one additional, exceedingly important comment. In regard to the application of in-use standards (e.g., a system suitability test), the more frequently a validation of the counting procedure is performed, the less likely the generation of erroneous data. In some chemical analyses (e.g., gas chromatography), a standard may be run with each sample. This approach is not feasible with particle counting, but a generally similar practice can provide enhanced confidence in data. Particle standards generated in the laboratory or purchased can be applied for this purpose. It is relatively simple to generate count standards using the techniques described above and use them on a daily basis for manual counting. Such standards may be monitored for count stability using a counter designated as a standard instrument, on which key electronic and physical parameters are closely monitored; their use constitutes, in effect, ongoing OQ. If an analyst applies standards of this type before and after each day's counting, appropriate functioning of an instrument during a day's work is increased.

Importantly, there is no such thing as counting accuracy (see the paper by Knapp and Abramson [1996]); due to the variable response of particle counters to particles of heterogeneous composition, the accuracy of size data cannot be validated. Every string of count data produced by LO counters from samples that contain unknown particles of differing size, shape, and refractive index simply represents an operational definition made by the instrument. Precision with a given method may be enhanced or ensured, but accuracy applies only to the instrument under consideration. The Knapp and Abramson publication provides a detailed consideration of coincidence or concentration effects. These would seem at best a minor consideration with LO counting per USP <788> due to the cleanliness of solutions tested and the low allowable level of counts generally recorded. Two examples of much more significant errors are threshold drift, which will result in sizing and hence count errors, and undercounting.

System Validation for Particle Counting Systems

Throughout a relatively short life history (since 1969), particle counting systems have evolved from manual models with simple means of sample drawing and Nixie™ tube data presentation to the automated (computer-based) models available today. This evolution is to be expected for any instrument type and is driven by customer needs, advancing technology, and marketing strategy. The reader should be aware of and contemplate this process briefly in choosing a validation strategy.

Two key principles governing the extent of validation or instrument qualification should be kept in mind. First, functions of computer-driven instruments that are not used need not be validated. Accordingly, if a modern computer-driven counter is used in its simplest mode of operation (i.e., simply to collect and display counts that are manually recorded in a laboratory notebook), the extent of validation activity may be significantly decreased. The trade-off here is that much of the capability of the instrument will not be used and time savings may not be realized. Here, validation activity as an investment must be considered.

Secondly, those conducting a validation must heavily weigh the power of applying in-use standards and system suitability checks during the use of the instrument to ensure appropriate performance. Some laboratories use this approach, emphasizing PQ and OQ as the central points of assay assurance rather than the "classical" validation components listed earlier. One large laboratory with which the author is familiar even refers to the process that is applied to ensure instrument function and data validity as "assay qualification" rather than validation, although the major difference in reality is in the term rather than the extent of what was done to ensure an appropriate control. In this approach, more or less time is spent with each use of the instrument to carry out a performance validation, but some of the complexity of the initial validation process may be avoided.

Such ongoing testing of the instrument system has the advantage of ensuring function on a daily basis, which generation of even a comprehensive, monolithic system validation package and annual review do not. Also attractive, and probably the most secure approach, is to combine elements of the system validation approach with some level of use of daily performance checks. The choice is up to the user. Under no conditions should analyst training and instrument use SOPs be neglected. Both of these items must receive due emphasis.

Given these basic considerations, what is a computer-based or computer-driven particle counting system and what are the considerations in its validation? Based on a functional approach incorporating all of the components of such a system, the system may be considered to consist of the components shown in Figure 11.5. The four components of the system are as follows:

1. The particle counter itself. This instrument may be further separated into a sensor, a counting unit (usually a multichannel peak height analyzer), and a sampler that may operate by hydraulic or electrical power to draw the sample.

Figure 11.5. Particle counting system.

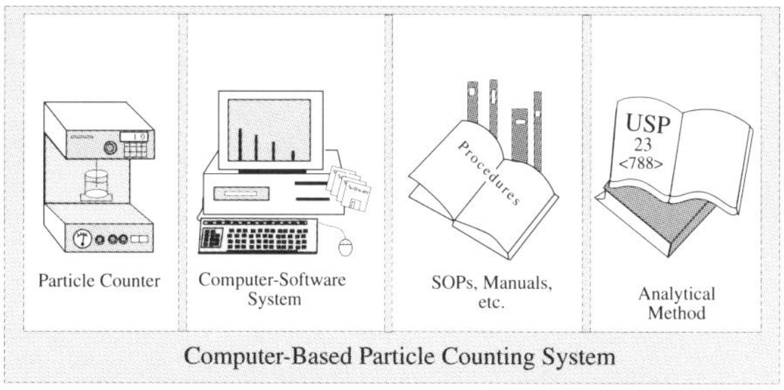

2. The computer system (which includes dedicated software to drive the instrument, perform calibration calculations, and reduce and/or store data.
3. The assay (in the present consideration, this is the USP 23 <788> test for particulate matter in injections).
4. The SOPs, manuals, and specifications that prescribe for the user the methods of use of the instrument. The subjects covered include, but are not limited to, the following:
   - Operating procedures and minor troubleshooting
   - Backup and archival of software and data
   - Disaster recovery plan
   - Security procedures
   - Product sampling and testing
   - Training
   - Data review
   - Out-of-limits data review
   - Retesting

Some considerations of this nature emphasize training and the user to the extent of adding the human analyst to this list. In this regard, no amount of validation is likely to prevent a 483 item if the analyst cannot be shown to have received appropriate training. Interestingly, the computer system and its software is but one element of the four that must be considered, but the consideration of this single element and in particular its interactive functions typically will increase the level of effort for the validation process 5- to 10-fold over a simpler approach to qualification of the manual systems available in the past. Again, obviously, if a computer is not used to drive the system, the other three elements retain their importance and must be dealt with in sufficient detail.

The earlier discussion of steps or components in the validation process is useful in the specific consideration of particle counting systems, in that it provides a format for review that generally applies to any instrument system. As discussed earlier, the distinction of software validation and the validation of interactive functions will become less than clear in some cases. In the following discussion, a significant dependence is placed on the application of in-use standards to verify appropriate function on a frequent basis; the USP <788> operational criteria will also be emphasized. A generalized approach, excluding vendor-related factors, is shown in Figure 11.6. In the figure and in reality, the user must determine what software functions must be validated and how. Validation of instrument function is given a separate overview. Then interactive function is considered, and the two are tied together.

Some counter systems currently available for pharmaceutical applications require the counter to be operated using a microcomputer-based operating software system based on a commercial system, such as Microsoft Windows™.

Figure 11.6. Generalized approach to particle counter validation.

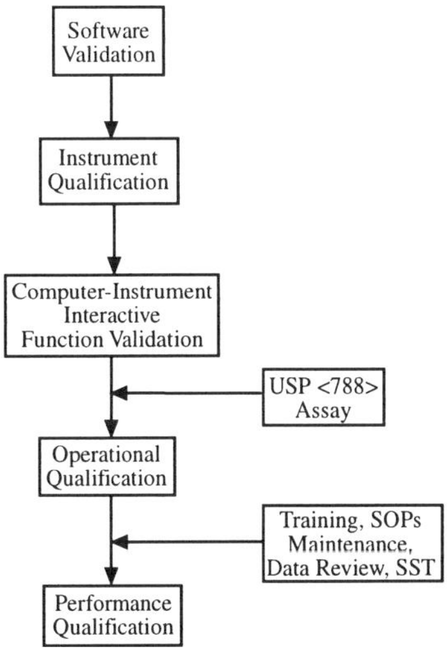

Such a system represents a high level of complexity and may use software routines to store and transmit data, calculate resolution and calibration curves, and produce data from current and historical data. Despite the complexity of this type of system in comparison to the classical stand-alone computer-based systems of 10 years ago, it should be observed that the rigorous format of a formal validation is not arbitrarily required. As discussed earlier, there are several factors that enter our consideration of this type of system and how to validate it in the most time-efficient and cost-efficient fashion.

The user should keep in mind that the basic goal of any validation approach must be minimal effort consistent with proving that the system functions as intended. To be avoided is the *ad absurdum* verification of minor system or software functions and the needless repetition of work previously performed in necessary depth by the vendor. The nature of software used will remain a most important consideration. If the software is vendor generated and sold as a standard package, a significant level of validation activities may have been carried out by the vendor. Repetition of vendor activity in this regard represents wasted effort. Custom software, on the other hand, typically will not have been extensively validated, and the user software validation process described above should be addressed.

Vendor Support for Validation Activities

If validation of the newly purchased or legacy system is to be conducted with minimal effort, documentation of all vendor-conducted validation activities must be available or accessible to an auditor. A very interesting observation here is that the desirability of ISO certification has caused particle counter vendors to pay more attention to documentation of development and quality activities; the user of particle counters may profit. The documentation available from vendors typically includes at least some of the following:

- DQ documents
- IQ documents
- Structural testing
- Functional testing
- Design criteria
- Source code
- Programming review
- Functional specification
- Test plan and execution
- Validation testing report
- Software change control
- System archival/backup procedures
- Training materials, manuals, and SOPs

Extreme diligence is often necessary in acquiring vendor documentation; considerable time may be spent on the telephone in efforts to contact the appropriate individual and then explain what is needed. Once these conditions are met, most vendors are cooperative. Vendor certification of correct performance of the system as installed (IQ) is of significant importance. It provides the link that allows the user to build on the validation activities related both to the instrument and its software that have already been conducted the vendor. The rationale here is as follows: If the vendor has validated the system and its software, it is critical to ensure that the system as installed is functioning properly and is totally consistent with the system validated by the vendor. The basic test here is that any validation activity performed by the vendor on the counter system in question prior to sale will be applicable to a purchased system of the same model provided that the vendor provides appropriate verification of correct function upon installation. Remember, vendor information must be reviewed and meet user standards, or it cannot be used.

Application of In-Use Standards and Operational/Performance Qualification Tests

In the OQ/PQ testing discussed in this section, a battery of tests are described for possible use in the OQ/PQ of LO counters on some periodic basis. The tests serve to validate or qualify the instrument portion of the total system. These tests are used in the laboratory with which I am most familiar; others may be devised and rationalized. The major tenet here is to apply tests that serve to demonstrate compliance with the compendium and/or serve to indicate that the entire system is functioning properly.

Examples of system assurance checks that may be applied are as follows:

- Six-month determinations of size calibration, resolution, count "accuracy," and sample volume accuracy per USP <788> ("prime" calibration)
- Monthly verification of size threshold settings with the same PSL standard spheres used in conducting the USP standardizations in the item above
- Six-month evaluation of electronic reference values for comparators and supply voltages
- Daily use of a count standard (in-use standard or system suitability test)

The first item in the list simply means that the USP performance criteria have been met. This is a mandatory qualification or system suitability test of the counting system, and serves as the cornerstone of the OQ/PQ process. These procedures are performed at not more than 6-month intervals, unless there is some repair or replacement of system components occurs that necessitates more frequent testing. "Drift" of size thresholds is a fairly common ailment of LO counters and can be tested by the simple expedient of testing suspensions of the same PSL calibrant spheres used to satisfy the USP procedure for setting size thresholds. The allowable size variation with respect to that determined when the USP <788> qualification was performed may be set at some realistic value, such as ±10 percent. In addition to these checks, key instrument reference voltages may be monitored electronically when the USP qualification is performed.

While prior validation and documentation are critical, the in-use standardization scheme will ensure collection of valid data on a routine daily basis. One example of such a scheme is as follows: The USP particle count reference standard (PCRS) contains 15 μm monosize PSL latex spheres at a known number concentration in aqueous diluent. This material is generally too expensive for routine use but may be used to generate in-lab standards that are traceable to the PCRS and can be applied on a daily basis to ensure the appropriate function of the count(s) that will be used to assay product per <788>.

The in-use standard is made by gravimetric dilution of standard material containing the same lot of spheres as the PCRS obtained from the same vendor (Duke Scientific). Standard suspensions are prepared by diluting the 15 μm sphere material to a level approximating the PCRS; standard counts are determined on a "master" counter that has been calibrated per the USP <788> requirement and count verified using the PCRS. As an additional step, the number of spheres in the in-use standard must also be quantified microscopically. Daily or twice daily use of this standard to check counters before use ensures proper operation before testing is begun.

Data Integrity

Data integrity is a basic principle that must be reviewed to ensure the necessary proof of "as intended" function is obtained. In the case of data integrity, the user may well decide that the electronic data storage, transfer, and reduction functions of a computer-based counter-system available from the vendor may require the same level of validation as would pertain for a customized system. One approach by which this has been addressed in past is by first flowcharting the course of a block of data from generation to archival, and then verifying that the original data is unchanged by any transfer or storage function. Additionally, software functions that may be used to manipulate the data in storage or that access the storage data directly or indirectly must be annually reviewed.

A separate issue that must be reconciled involves whether or not the data will be subject to hard copy archival. This process may be regarded by some FDA auditors as more secure, but reliance on hard copy denies the user many of the time/cost savings attainable with such computer-based systems. In some instances, both electronic and hard copy archives are used; the latter allows rapid data access and retrieval for statistical overviews and analyses; the former allows hard copy verification as necessary.

When electronic data storage and transfer are used, the issue of data security both from physical (disaster) damage and unauthorized access must be insured. It will also be necessary to comply with electronic signature guidelines if this procedure will be applied in data handling.

Analyst Training

The validation plan should also address the training of system operators. Documentation of this training should be maintained in employee training records. The training requirement specified in <788> of USP 23 must be met and documented. An SOP should also be developed that details operating procedures, minor troubleshooting routines, and other procedures related to the routine use of the system.

Requirements Definition

The requirements definition is a key document in the validation of any system. It defines what the system must do, how it is configured, and will lock in a specific version of software if the counter is computer based. Of extreme importance is that all testing of the system is directed toward ensuring that the system meets the requirements specified.

Specific functional requirements via USP <788> are the following:

- The instrument must be an LO counter.
- It must be capable of counting at ≥ 10 μm and ≥ 25 μm size ranges.
- Sensor resolution must be ≤ 10 percent.
- Calibration must be accomplished with 10 μm, 15 μm, and 25 μm PSL standard spheres.
- Volume accuracy of samples for 5 mL is ±5 percent (must satisfy count accuracy by passing the USP PCRS).

There may also be additional requirements depending on the system, including the following:

- Cumulative and differential counting
- Automatic calibration functions, automated resolution determination, or calibration subjected to "suitable" validation per <788>
- Averaging of counts
- Counting of additional sizes
- Interfacing to laboratory information management systems
- Summary report generation
- Password protection

System Testing

Many of the specifics of validation are dealt with in only summary fashion in this chapter; in other cases, they were omitted due to limited space or in favor of the detailed consideration given in the references cited. System testing with regard to the tests that are applied in validating particle counters may be conducted in several contexts, which should be well understood. Operational testing is closely related to SST, whereby tests are performed that check if function of the instrument within the normally encountered operational conditions is appropriate. An example might be the use of a particle count standard with numbers of particles at the ≥ 10 μm and ≥ 25 μm size ranges within the limits numbers of the USP. This is acceptable in the context of ongoing testing, but at some point, two other types of testing are appropriate: (1) boundary or worst-case testing and (2) stress testing.

These latter types of testing are important due to the fact that instrumental test results will more likely be encountered at some limit (i.e., overconcentration of particles or a zero or blank count). Boundary testing for the counter itself might involve determination of the likelihood of error as particle concentrations approximate the sensor concentration limit; another test case is encountered when the test assay blank or background test is performed. Other test cases are appropriate for software (see Huber 1996). Stress testing is the term describing testing of conditions outside the instrument's specified range. Particle concentrations that significantly exceed the range of a sensor or analyst inputs to a computer with forbidden character types, letters instead of numbers or numbers too high or too low to allow realistic calculation of particle size thresholds are examples of such cases. Such "unrealistic" cases are typically tested in periodic OQ tests, but not in routine operational testing, with system suitability checks.

System-Specific Validation Approaches

The validation of computer-based particle counting systems is made somewhat more complex by virtue of the range of system types in use. Historically, the past 15 years have seen simple, comparator-based counters with

6–12 manually adjustable thresholds supplanted by instruments incorporating microprocessors and pulse height analyzers and, more recently, by instruments based on microcomputers and user-friendly software. Microprocessor-based system assurance constitutes a verification of functional parameters, rather than involved validation understanding. These instruments are discussed briefly in the following section due to their wide use and to the fact that they are part of a logical progression from simple to more complex validation requirements.

To a great extent, the approach followed in validation will depend on the complexity of the system, the degree to which depends on the complexity of the software and vendor documentation. In this regard, a functional hierarchy of counter systems may be established as follows:

- Hardware + simple microprocessor—HIAC Model 4100
- Hardware + interactive-capable microprocessor—HIAC Model 8103
- Hardware (HIAC Models 4100 or 8103) and ancillary software for calibration or standardization
- Software-interactive (APSS 200 and HIAC 9703 with APSS-View™ or Pharm Spec® software, respectively)

The reader should be aware beforehand of two points pertaining to the author's discussion of the validation for specific instruments. First, a comprehensive discussion of how to validate a specific instrument system is beyond the scope of this chapter. The following sections of this chapter thus present some basic considerations for each type of particle counter, which must be considered in the context of the preceding discussion of validation and in view of specific company policies. Secondly, the stated intent of this chapter is to provide an insight into what I believe to be a logical approach. The approach presented may not be in line with the specifics of the validation philosophy of some companies. In this case, the principles described may be useful to the reader, but in every case should be followed only insofar as they conform to company practice. The user must also be aware that although user training is not listed as a component for each system, it is a critical component of instrument use. USP 23 <788> specifically calls out an analyst training requirement.

HIAC Model 4103 System (Figure 11.7)

The HIAC (Pacific Scientific) model 4103 system consists of a model 4100 particle counter mainframe, a model 3000 bottle (syringe) sampler, and a white light or laser diode sensor. This system in a pharmaceutical laboratory will invariably be validated retrospectively, since the counter unit has been obsoleted by the manufacturer and is no longer available for purchase. In this process, consideration of historical data with the instrument and functional details is needed. The counter is based on a microprocessor rather than a computer and has simple, programmable, user adjustable functions. It counts particles in 6 channels with manually set voltage thresholds and must be manually calibrated. This system is used routinely in many labs to count and size particles in injectable solutions. Data obtained with this system will determine whether product is within the USP <788> particle limits at the ≥ 10 μm and ≥ 5 μm size ranges. With regard to risk assessment, count data obtained from this system can be used for New Drug Application (NDA) submissions; development work and data obtained from this system impacts product quality and safety, so it meets the most-used criteria for an instrument that must be validated.

It is important to note that although this instrument is not "computerized," the general

Figure 11.7. HIAC-Pacific Scientific model 4103 system.

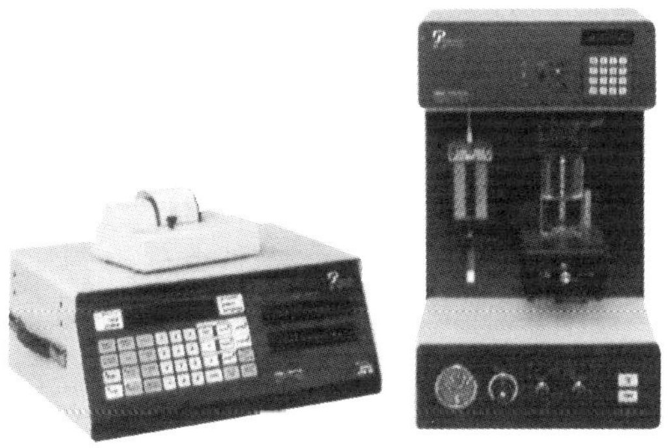

elements of validation specified earlier in this chapter provide a useful matrix of operational factors to be reviewed as this instrument system is validated. For example, a requirements definition or functional specification is as important for this system as for any computerized one; data integrity must be ensured, and so on. The validation should include documentation of SST and the execution of a functional test plan to ensure that the HIAC 4103 particle counting system is accurately sizing and counting particulate matter.

In this simplest type of validation, the procedure is heavily dependent on verifying appropriate function and ensuring the system is standardized per USP <788> methodology. A list of the components of the process might be as follows:

1. Risk assessment
2. Vendor documentation of engineering and development testing of the system
3. Vendor certification of proper performance at installation (IQ)
4. Validation test plan (test of instrument-analyst interactive functions)
5. Execution of test plan
6. OQ (USP <788> system suitability tests on biennial basis or as a result of system repair)
    - Calibration response curve
    - Resolution
    - Sample flow rate determination
    - Sample volume accuracy
    - Sensor concentration limit (vendor specification)
    - Count accuracy test (USP PCRS)
7. PQ
    - Environmental envelope (e.g., HEPA filtration, cleaning procedures)
    - Daily count accuracy test using in-house 15 μm sphere standard
    - Monthly or weekly verification of threshold settings
    - Periodic electronics (comparator) check
    - Verification of sampling procedures
    - Blank or background count testing
    - Sampling plans
    - Data review pathway
    - Responsible individuals
    - Out-of-limits review

    Note: Items 6 and 7 constitute items of the performance assurance plan.
8. Training and SOPs (instrument maintenance and performance records, use and handling of standards, operating procedures, troubleshooting)
9. Change control
10. Disaster recovery
11. Validation report
12. Ongoing validation

The process is shown diagramatically in Figure 11.8. This format represents a minor departure from the principles of validation of software-based systems discussed earlier; since the instrument is not software based, software considerations are not made. (For a software-based system, these elements would simply be added to the flowchart.) Documentation from the vendor is obtained as available. (Vendor documentation of product development for this older type of instrument is often sketchy.) The performance as intended pivots heavily on use of USP <788> performance criteria for OQ and systematic application of in-use standards to ensure function on a daily basis.

For an instrument without software functions, engineering and development test data are secured to some extent as software-related documentation. A specific concern is the degree of design change control during the life cycle of the system (Hokanson 1994). This may be in such parameters as sensor function (e.g., improved revolution, higher concentration limit) or electronics (improved pulse handling decreased count cycle time). These parameters may effect particle sizing and, as a result, count data. The performance assurance plan, which incorporates in-use standards, is also of great importance. This plan for the 4103 system should include all suitability tests and performance standards that are used for the system. Procedure specifications or SOPs must be developed for the six-month size calibration and the monthly size validations.

Figure 11.8. Validation matrix for a nonsoftware-based counter.

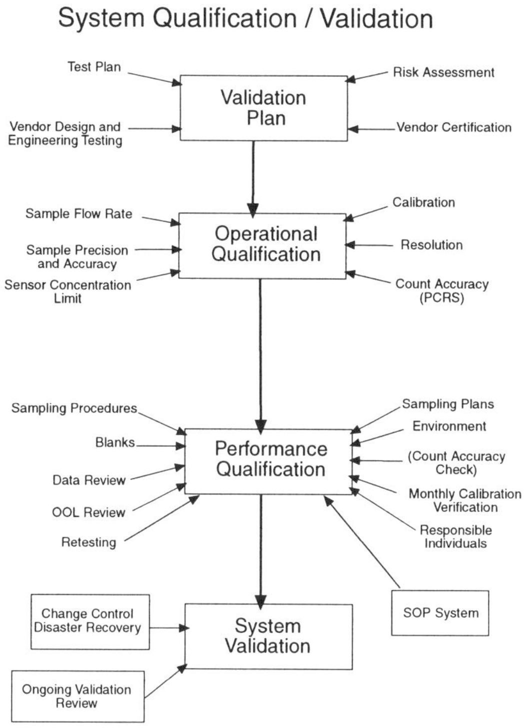

SOPs must also be developed and issued that detail the procedures to be used for such activities as the electronic alignment, preparation, and use of a daily count standard. A functional test plan for the HIAC model 4103 operational parameters (i.e., analyst-instrument interactive functions) must be developed and performed.

Again, continuous validation of this and other particle counting systems is very important. This is a constant confirmation that the overall system is operating properly. Prior to manual performance of the instrumental particle count assay, a particle count standard may be assayed as a system suitability test to ensure that the HIAC instrument is counting and sizing particles properly, as described earlier. A commercially available, multisize standard solution type with known counts at specific sizes (Duke Scientific) is reportedly available and NIST traceable. The counts determined by the test counter are then compared to the known values. If the experimental counts match the expected counts, then the system is considered suitable for use in analysis.

The system test plan will test the operational parameters normally used for daily operation or calibration of the 4103 system and many of the additional instrument functions not routinely used. Daily operation parameters and parameters used during calibration are outlined in procedure specifications. The parameters normally used and parameters occasionally used for the 4103 system will be functionally tested.

The test plan is designed to verify the correct performance of the instrument functions that will be used (functions that will not be used are not validated). The procedure is begun by preparing a test matrix based on vendor description of instrument function and a consideration of which functions of the instrument will be used based on the requirements definition. The functional verification then proceeds by checking each function item against tabularized results. A 15 μm sphere suspension containing approximately 1,000 particles/mL serves as a test medium. A magnetic stirring bar is placed in the suspension on the sample stand of the model 3000 sampler; this sample is mixed constantly throughout the analysis and is used as a test suspension. A suitable printer is connected to the model 4100 mainframe.

Many of the instrument functions will be used for several of the desired operations; they may, however, be mentioned only once in the cross-reference to the requirements definition as proof that the function was tested at least once. Two formats for testing of the microprocessor-controlled parameters for the 4100 mainframe and the model 3000 syringe sampler are shown in Tables 11.1 and 11.2. What is being done here is testing functions that the instrument performs automatically; functions that will not be used are (at least in theory) not required to be verified. This matrix is much simpler than that for more complex systems.

Each of the critical functions of the instrument that will be used must be verified for expected result versus actual result. The acceptance criteria are that the actual results match the expected results. The analyst can use Tables 11.1 and 11.2 to record the actual results. If the actual results match the expected results, the analyst may need only to record a check mark (✓). *Note:* A screen dump is not available, so the checklist is considered acceptable. The author's personal preference is for a written description of the test result rather than a ✓.

Table 11.1. Model 4103 System Functions to Be Verified (partial example)

Item #	Count Control Group	Explanation	Test Result
1.0	SAMPLE MODE	Selects one of three sample modes displayed above this key.	Sample mode selection as indicated.
1.1	MANUAL	START/STOP action of sample controlled manually.	Syringe draw starts and stops according to switch position.
2.0	CHANNEL ADVANCE	Advances and displays next channel data.	Channel advance in sequence as selected.
3.0	START	Initiates a sample run.	Functions properly.

Table 11.2. Model 4103 System Test Plan and Expected Results (partial example)

Test Plan	Expected Result	System Function	Test Result
Set the model 4100 in MANUAL sample mode; set the model 3000 sampler to VOL 1.0; set toggle switch on the model 3000 to MANUAL, set the data mode to CUM; press the start button on the model 4100. After more than 1 mL has been exceeded, press the stop button on the model 4100.	• Counting light on the model 4100 turns on when the start button is pressed and the count light on the model 3000 does not turn on after pressing the start button.	1.1, 3.0	
	• Volume exceeds 1.0 mL and model 3000 still pulls sample.	1.1	
	• The model 3000 cycles off when the stop button is pressed.	4.0	
	• Counts for channels 1–5 are different than the previous run, but count in each channel for this run are the same or decrease with the increased channel number.	21.0	

The principles of the validation performed for this manual system provide a basis on which to validate the more automated instruments that may be used. Importantly, emphasis will continue to be placed on the OQ/PQ areas, with the application of in-use standards and the USP–specified requirements.

HIAC Model 8103 System

The function of more complex counters is simply the sizing and counting of particles, as was the case with the model 4103. Thus, in-use standards remain a powerful means of checking complex internal instrument functions with simple external tests that can be applied by the analyst using the instrument.

The HIAC model 8103 counter (see Figure 8.16) is similar to the earlier model 4100 counter in that it is microprocessor based and has no software interactive functions. The microprocessor of the newer instrument is far more capable, however. In a complex function, the microprocessor automatically determines the calibration curve, once the median voltage peak resulting from calibration particles of known sizes are established manually.

An attendant complexity to the use of this counter involves the capability for using either laser or white light sensors. The white light sensor response is a simple power law relationship of pulse voltage to particle diameter; that for the laser sensor is more complex and must be calculated based on a cubic spline equation. Whichever sensor type is used, the curve fit by the automated function of the instrument must be verified by manual calculation during the validation process. With this exception, the remaining items of the system test plan are similar to those listed above for the model 4100 instrument from the same manufacturer, and a similar system assurance plan must be implemented.

The menus and displays of the model 8103 represent the primary method of communication from the counter to the operator. They are presented in a typical operational sequence, along with additional menus and required or suggested keypad selections. This sequence will continue until all of the operating parameters have been covered or the operations have been completed.

The menus that exist within the system include the following:

- Main function menu
- Miscellaneous function menu
- Counter communications menu
- User defined standard function menu
- Parameter setup function menu
- Sensor calibrations function menu
- Maintenance functions menu
- Calibration function menu
- Additional calibration functions menu
- Display functions menu
- Printer functions menu
- Set calibrations functions menu
- Password functions menu

Both the parameters normally used and parameters occasionally used for the 8103 system must be functionally tested in methodical fashion with a test sequence and expected and actual results recorded and cross checked as was the case for the simpler model 4103 (Table 11.3). In this case, due to the fact that the functions are more complex, the checking will of necessity involve more details and be more time-consuming. Obviously, the last function here (3.03, Set Cal) is a concern; alteration of calibration files is a means of collecting erroneous count data at best and would allow data from a specific sample to be altered. The model 8103 is a much "smarter" instrument than the earlier model 4103. As always, the price for the time saved in routine use through automation features is paid back in some fashion through verification that automated features perform as intended. Of particular concern with the model 8103 is the fact that the instrument not only calculates a calibration curve based on the data obtained when standard particles are run but also stores data from previous calibrations. The calibration curve calculated is different based on whether a white light or laser sensor is used. Calculated curves must be verified manually during the calibration procedure, and retrieved historical data must be subject to criteria for testing data integrity.

Firmware-Based System with Ancillary (Aftermarket) Software

Prior to the use of any particle counter for compendial (USP) testing, critical operational parameters of the instrument must be verified according to the method specified in <788> or by equivalent calibration threshold, resolution, count accuracy, and sample volume accuracy tests. The use of software reduction of data can effect significant time saving at each prime calibration of the instrument, which must occur at six-month intervals or when there is any significant alteration of the system, such as circuit board replacement or sensor repair or replacement.

The basic method for verification of such software includes a comparison of calibration and resolution data generated by the software to

Table 11.3. HIAC Model 8103 Functional Verification

Item #	Execution/Information	Explanation	Result FP
3.0	Cal (Calibration)	Fits calibration curve to data points (check most recent).	FP-calculation check: NB ref.3189, p. 136.
3.01	More	Auto Adj., Quick Adj. Sensor Cal., Maintenance functions.	FP-Functions selected as indicated.
3.02	Show Cal	Shows the current calibrations in the counter.	FP-current Cal listing correct.
3.03	Set Cal	Allows the user to manually enter a calibration or alter an existing one.	FP-Cal #3 (2/5/96) corrected; NB ref.3189, p. 131.

a manual reduction of the same data. Formulas used by the software programs should be consistent with those prescribed in the USP. If data are recorded on an electronic spreadsheet, all validation steps for data transfer and recording must be carried out. In terms of the format of validation required, this is a classical software validation procedure and should be based on a full-blown life cycle approach (as per Hokanson 1994) and comprehensive testing of the software itself, with attention to the components in the list given earlier. An alternative (more attractive) approach is to require the vendor to supply an acceptable software validation package prior to purchase.

Figure 11.10. HIAC model 9703 system.

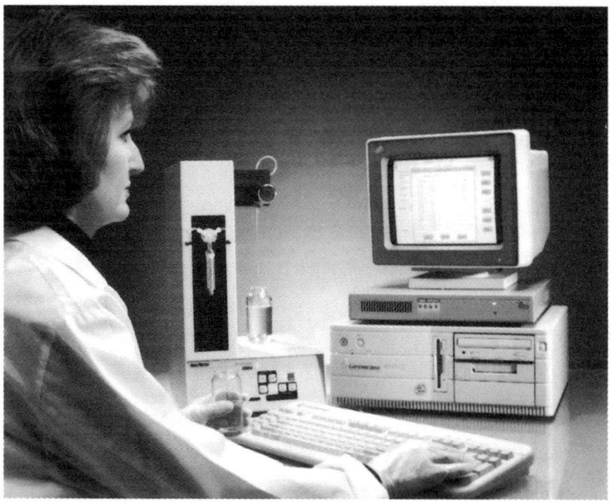

Computer-Based Systems (Particle Measuring Systems Model APSS 200™, PSS Accusizer™, and HIAC 9703 Systems)

Over the past 10 years, technical advances in light obscuration particle counting and an increased user knowledge base related to these instruments has made possible great progress in the application of this mode of particle counting in pharmaceutical labs. The Particle Measurement Systems (PMS) APSS 200™ (see Figure 8.17), the Particle Sizing Systems Accusizer™ (Figure 11.9), and the HIAC 9703 systems (Figure 11.10) are examples of the current level of development of this type of instrument. Use of these instruments to illustrate the principles of validation of this type of counter does not in any way constitute a recommendation for purchase or endorsement for use in fulfilling the USP 23 <788> requirements; all three instruments are representative of approximately the same level of development in instrumentation, and both are applied by pharmaceutical and medical device manufacturers.

Figure 11.9. Particle Sizing Systems Accusizer™ instrument.

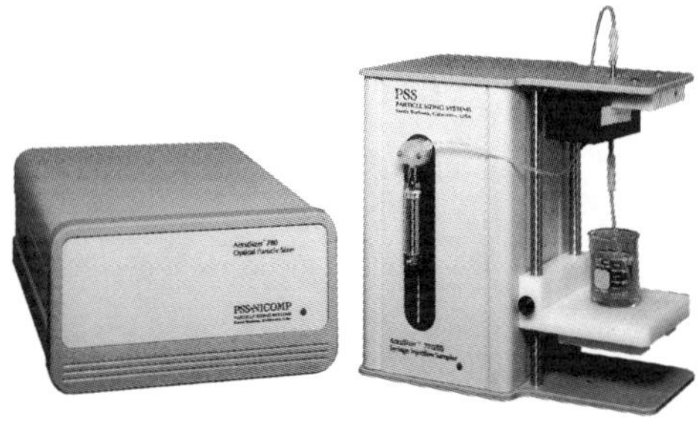

Those contemplating purchase of such a system are best advised to become thoroughly familiar with both the instrument and the vendor and to carefully assess the depth of support available from the vendor. The potential is always present for an FDA audit of the validation, which may require significant assistance from the vendor if this documentation is substituted for in-house generated data.

At this point, a comment regarding validation "packages" available from the vendors of these instruments is needed. The information required for such a package can be divided into two general categories: (1) information that the user may incorporate as furnished by the vendor without modification (e.g., installation documentation or software design specifications) and (2) information that the user must add to or modify in-house based either on the specific requirements of the company involved or the inadequacy of vendor-supplied documentation. The validation packages furnished by PMS and HIAC for their respective instruments show similar strengths and weaknesses with respect to the information provided. The general information on company quality processes, engineering design principles, and system descriptions is adequate and can be leveraged by the user. Specific data related to software or hardware testing is almost completely lacking and may only be available on specific request from the vendor or as a result of specific negotiation. (*Author's note:* The best time to make such requests or to negotiate is prior to purchase.) Examples of test plans in "fill in the blank" format are provided, but results of actual testing that could be used directly by the user are not. Thus, these packages are of limited practical usefulness in the conduct of a validation. Although the information provided is useful, the user is still obliged to develop the majority of the validation components themselves. Particle Sizing Systems is in the process of developing a more rigorous package for the Accusizer™.

The features of these three systems include the following:

- Compact, user-friendly system design
- Integrated sensor design (counter and sensor in one unit)
- Computer-based data processing
- Automated calibration procedures
- Storage of calibration data for interim checks of instrument calibration
- Interactive software functions
- Real-time, multichannel data presentation
- Reporting and data presentation options

To some extent, the ease-of-use features, use functions, and time economies possible with this computer-based system with interactive software will be affected by increased validation requirements. Based on the author's experience, however, the user will find that the benefits resulting from the precision and accuracy of particle counting with this system and the ease of calibration and use outweigh the additional time and resources spent in validation. Benefits that accrue to the use of these newer, software-driven instruments following validation include not only simple, relatively trouble-free operation, but the ability to perform the USP <788> calibration and resolution determination in a fraction of the time originally required manually.

Both the HIAC and PMS systems use a standard personal computer to provide user input and particle count data display, analysis, and storage. The PMS APSS View Windows™-based software features automated computer interactive functions for determining sample volume and flow rate as well as calculating calibration and resolution data. It also allows data storage and retrieval and comparison of a new calibration with previous ones. The Windows™-based Pharm Spec® HIAC software has built-in procedures for all USP <788> testing requirements. Designed to be validatable for the pharmaceutical customer, Pharm Spec® makes parenteral limits testing easy and reduces errors with step-by-step instructions and pass/fail results. The USP <788> calibration includes system performance certification according to the USP <788> standardization test.

User Testing of APSS 200™ and 9703 Systems

The information that readers have obtained thus far in this chapter should enable them to deal with the complexities of assembling the necessary documentation and devising a test plan for a complex system, such as the PMS APSS 200™ or the HIAC 9703. In addition to the considerations shown in Figure 11.8 for a noncomputer-based system, a key consideration of the validation of software and computer-interactive functions is added.

Some of the required documentation for the validation can simply be assembled from vendor information, user manuals, and maintenance documentation. For example, the IQ is obtained directly from the records kept by the manufacturer for each instrument sold and installed. Disaster recovery plans for earthquake, flood, fire, and so on are based on the combination of simple logic and company procedures that pertain to these events. Change control depends simply on identifying a means of tracking any vendor changes to the software or hardware of the system and taking the appropriate actions to ensure that the system in question is upgraded as necessary and that any component of instrument operation affected receives due attention.

In addition to these general considerations, three specific, high-effort areas of the validation activity that will need to be conducted by the user are worthy of consideration: the risk assessment, the requirements definition, and the test plan.

Risk Assessment

In the consideration of most users, these systems will be used to collect data that will ultimately or in a more immediate time frame affect product quality (i.e., USP <788> compliance). Remember, this consideration applies to both instruments used for batch release and for R&D testing, and coordinated validation activity may be conducted to generate a single package. The assessment of risk is, in any case, sufficient to justify validation.

Both of these systems are based on off-the-shelf software that has been adequately validated by the vendor. Documentation of this validation is available. The vendor has established procedures in place for configuration and change control and will provide certification of correct operation upon installation. The system will be used with a performance assurance test and standardization and in-use standards. Prior to use for data reduction, storage, and transfer, these functions will be independently validated by the user or will be the subject of appropriate validation by the vendor. Vendor certification of proper performance at installation is provided (IQ). As a result of this basic consideration of use, validation is required; factors have also been identified that suggest that the validation performed will leverage vendor information and in-use standards or system suitability tests to the maximum extent possible.

Requirements Definition

A basic requirements definition was spelled out in the introductory discussion for particle counting systems. These are simply basic requirements that are defined by USP <788>. Not included are the specific steps that an analyst has to execute to perform a test according to the appropriate SOP or instrument-specific software routines that may be used. The general principle followed here is that the requirements definition will be spelled out by the user of the instrument per the authorizing USP requirement and company SOPs, and the test plan will reference the requirements definition to ensure that the critical instrument functions are tested.

For computer-based systems, examples of requirements, other than those needed to perform the <788> test itself, that must be specified and/or tested include the following:

- Interfacing requirements (to analyst and other systems)
- Processing interactions
- Menu presentations (itemized for all user interactions)
- Threshold setup dialog
- Sample limits dialog
- Save and recall menu
- Graphics data display
- Report selection and setup (effects the critical hard-copy format and report generation)
- Calibration dialog and execution
- Data sharing and transfer/retrieval
- All calculations (calibration and resolution)
- Hardware requirements
- Data requirements
- Input requirements
- Output requirements
- Vendor support requirements
- Data reporting requirements

The OQ/PQ plan for the APSS 200 consists of the same four items listed for the HIAC 4103 system and should include five system suitability tests and performance standards that are used for the systems.

Test Plan for Software-Based Systems

The overall validation process involves addressing the components of validation discussed earlier. VQ and DQ become of great importance since control of the design process and structural testing performed by the vendor will form a foundation for the user's expectations for software or hardware function in the lab environment; the more rigorously the vendor controls design and manufacturing, the more likely the system is to function correctly and reliably in the field.

In the author's experience, this is unlikely to happen. Having performed detailed structural testing to satisfy themselves of appropriate system performance, vendors are reluctant to spend significant amounts of time to validate performance at the level needed by the user. There is a real concern on the part of vendors that the information provided will not be utilized by the user or may fall into the hands of a competitor. Further, many vendors do not understand the regulatory expectations related to use in the pharmaceutical industry and are thus unlikely to conduct their validation under the appropriate format.

The test plan is another matter, if simply in terms of the greater detail required in comparison to the simpler systems discussed earlier. A systematic approach to generating a test plan can be based on the following logic trail:

1. Consider the requirements for the test performed with the instrument:
    - <788> test only?
    - Other?
2. What functions of the instrument will be used?
    - Consider current manual analysis and SOPs (time savings)
    - Leverage automated functions or no change
3. Componentize each function used.
    - Software
    - Hardware
    - Separate approaches
4. Devise a "sub" or mini test plan for each function to be used.
    - Software
    - Hardware
    - Document functions not to be used
5. Generate a limits test or "boundary" test for the range over which each function will be tested.

The generation of a test plan is not as simplistic, but it is the product of a simple, logical thought process. Take as an example the testing of the automated procedure for determination of a calibration or particle size response curve with the APSS 200™ and HIAC 9703 instruments. These instruments are both capable of fitting a curve to the data points generated by running standards at the 10 μm, 15 μm, and 30 μm sizes. The curve generated by the software must be checked against a manual spline curve calculation. A representative page from a test plan for one of these instruments (APSS 200™) is shown in Table 11.4. Note the references to the requirements definition; each item of the requirements definition that is software interactive must be tested.

Validation Report

Upon completion of validation plan activities, a Validation Report summarizing the conduct and results of the study is written and approved. The system will be considered to be validated after

- all requirements are complete and tested,
- there are no known fatal errors, and
- there are no known instances data corruption.

A caution here regards the occurrence of unexpected results. As instrument functional capability increases, the complexity of the test plan increases apace.

In this circumstance, the likelihood that all test steps will be executed exactly as expected becomes less and less. The user is reminded that all unexpected results are not necessarily fatal errors in use of the system or successful completion of a validation. Many unexpected results, in the author's experience, have been found to be relatively minor in terms of impact on use. There are generally three courses of action:

1. Require the vendor to correct the noncompliant condition prior to instrument use.
2. Modify the test plan to further test the unexpected result output and validate its consistent function and lack of significant effect on data.
3. "Work around" the output (given that there is lack of significant effect on data), describing the deviant condition in SOPs for instrument use and in the documentation report.

Such unexpected results are, in fact, empirical proof that the validation process works and a favorable comment on the detail and diligence of the testing procedure. The impact on the use of the instrument ultimately depends on the effect that the error has on the data generated. If this effect is insignificant, the consequence of its occurrence for instrument use may also be insignificant. The flip side of this situation is the discovery of serious errors in testing. Examples of this type of error that the author has observed include failures of software to calculate resolution or calibration curves correctly, failures to pass the USP <788> count accuracy test due to sensor defects, and serious threshold drift occurring when the instrument was left on for a number of days. One model of counter

Table 11.4. Test Plan Example for Software-Based System

Cross Ref. to RD	Procedure/Action	Expected Results	Expected Result Confirmed YES	NO	Comments
3.32.2 3.1.2.1 A5g 3.3.2.14	Turn the power to ft LS-200 (sampler) on	Approximately after 2 minutes sampler status will read "ready" and sensor status will read "Sensor Error."	✓		
3.3.2.2	Turn the power to the Liquilaz (sensor) on.	Approximately after one minute, "Sensor Error" has disappeared on ft status bar. "Ready" is displayed on the status bar.	✓		
3.1.2.110A 3.3.4.1C	Select Logins from Setup Menu; enter ABCDEFGHIJKLMNOP QRSTUVWXY for User ID. Press TAB and enter 123456789012345 for the password. Move the cursor to the privilege box. Click on ADD.	Input may or may not be accepted depending on the pitch of the characters used. Record the actual results in the Comments section.	N/A	N/A	Actual Results: Entire input not is accepted. Letters A–X are displayed for user ID and number 1–9, 0, 1, 2, 3, 4 (14 numbers) are displayed for password. Password ID displayed in the Privilege field. User ID is displayed in the user ID field to the password field.
3.1.2.1A10a 3.3.4.1C	Enter 123456789012345678901234 for User ID. Press TAB and enter ABCDEFGHIJKLMN for password. Tab cursor gets to box. Click on ADD.	Input may or may not be accepted depending on the pitch of the characters used. Record the actual results in the Comments section.	N/A	N/A	Actual Results: Entire input is accepted. Entire input for User ID (24 character) displayed in the User ID field and entire input (14 characters) for password displayed in the password field.
3.1.2.1A10a 3.3.4.1C	Enter Fish for user ID and Farming for password. Press TAB.	The cursor moves to the Privilege field.	✓		
3.1.2.1A.10 3.3.4.1C	Soled Operator for Privilege; Click on Add. Click on OK to accept all changes.	Inputs are accepted and displayed on the System Dialog box. Exit the System Logins dialog.	✓ ✓		
2..2.3.B (all)	Select Threshold Setup from the Setup Menu, enter 5 for the threshold for channel 1, enter 10 for the threshold for channel 2, and leave the remaining channel thresholds blank. Click on OK.	Inputs are accepted and the threshold setup dialog exited. (This indicates the user has supervisory privileges.)	✓		

Table continued on next page.

Table continued from previous page.

Cross Ref. to RD	Procedure/Action	Expected Results	Expected Result Confirmed YES	NO	Comments
	Attempt to soled Logins from the Setup Menu.	Since the user has supervisory privileges, the Logins is right option to select.	✓		
	Exit APSS-View from the Setup Menu. Double click on APSS-View icon. Press Enter to start APSS-View. (Do not enter any name for User Name).	Input is accepted and the APSS-View Main display appears. No name is displayed for User ID.	✓		
	Select Sample Setup from the Setup Menu; enter 10 mL for Sample Volume. Press OK.	Input is entered and the Sample Setup dialog is exited. (This indicates the user has supervisory privileges.)	✓		
Data Recorded By	Date	Reviewed By	Date		

tested proved extremely sensitive to electrical disturbances, and failed criteria for ruggedness by generating spurious counts during electrical storms.

Retesting and Particle Count Data

Some level of consideration of retest procedures is considered to be a part of the PQ by many users of computer or software-based systems. Particle count testing per the USP <788> method is somewhat unusual in the context of tests performed on pharmaceutical products, in that it is a physical rather than a chemical test. Overall, the variability of results for this test will be found to be much greater than chemical methods. The USP does not provide for the use of outlier tests with chemical data, due partly to the theoretical reproducibility of a chemical test for a molecular moiety. It is interesting that the Poisson distribution of a microbial test result, such as the recovery of organisms from surfaces, is considered appropriate for the application of outlier tests. The distribution of particle counts collected from individual injection units will also be found to conform to a Poisson or log normal distribution.

The handling of retests and out-of-limit (OOL) or out-of-specification (OOS) test results is considered by many users of the light obscuration test method to be an important component of the PQ. In this context, a method of handling retests must be decided on before the instrument system goes into use, so that the actions ultimately taken when the counts from an individual unit or a solution pool exceeds limits, appropriately address the high values and comply with sound technical logic and FDA expectations in this area. In the latter regard, the *U.S. v. Barr Laboratories* decision faulted a number of components in Barr Lab's handling of OOS values, including the use of outlier tests and averaging data. The USP <788> test states categorically, however, that the test result is the arithmetic average of the counts from (however many) units tested. The sample pooling procedure for smaller units (< 25 mL volume) in effect provides a physical averaging of the particle counts from the units that are pooled.

Should an individual unit or pool exceed limits, appropriate action to investigate and resolve the failure (i.e., determination of assignable cause) will be required. The investigation to determine cause is made more challenging by the many sources of artifactual counts that can affect light obscuration data (see Chapter 9). A flowchart for investigation/retesting of one type that could be undertaken is shown in Figure 11.11. The user of the method is reminded that outlier tests must only be applied to particle count data based on a diligent investigation and a careful, conscious decision that such a test is appropriate. The user is further warned that the flowchart in Figure 11.11 is intended only as an example of a possible retest scheme; following it does not in any way ensure a result acceptable to the FDA or internal quality functions.

Remember, in investigation, the occurrence of a failing assay result may indicate a bad test, a bad unit, or a bad batch. The follow-up investigation must be objective and based on sound statistical/technical rationale. While the USP allows microscopic retesting, this process may be of greatest value for its investigative value (i.e., Why was the light obscuration result high?). Assignable cause consists of documented evidence of an error related to the analyst, the assay method, or the instrument used in testing. An assumption, especially an undocumented one, or a theory is not an assignable cause. The cause is considered unassignable if no basis for error is found. Those results that are not explained, but merely called into question by successful retesting, must be included in retrospective validation studies and annual reports. Only those test results determined through appropriate failure investigation of drug product to be caused by analyst or operator error can be excluded from retrospective validation studies.

Figure 11.11. Example retest flowchart for LO counting.

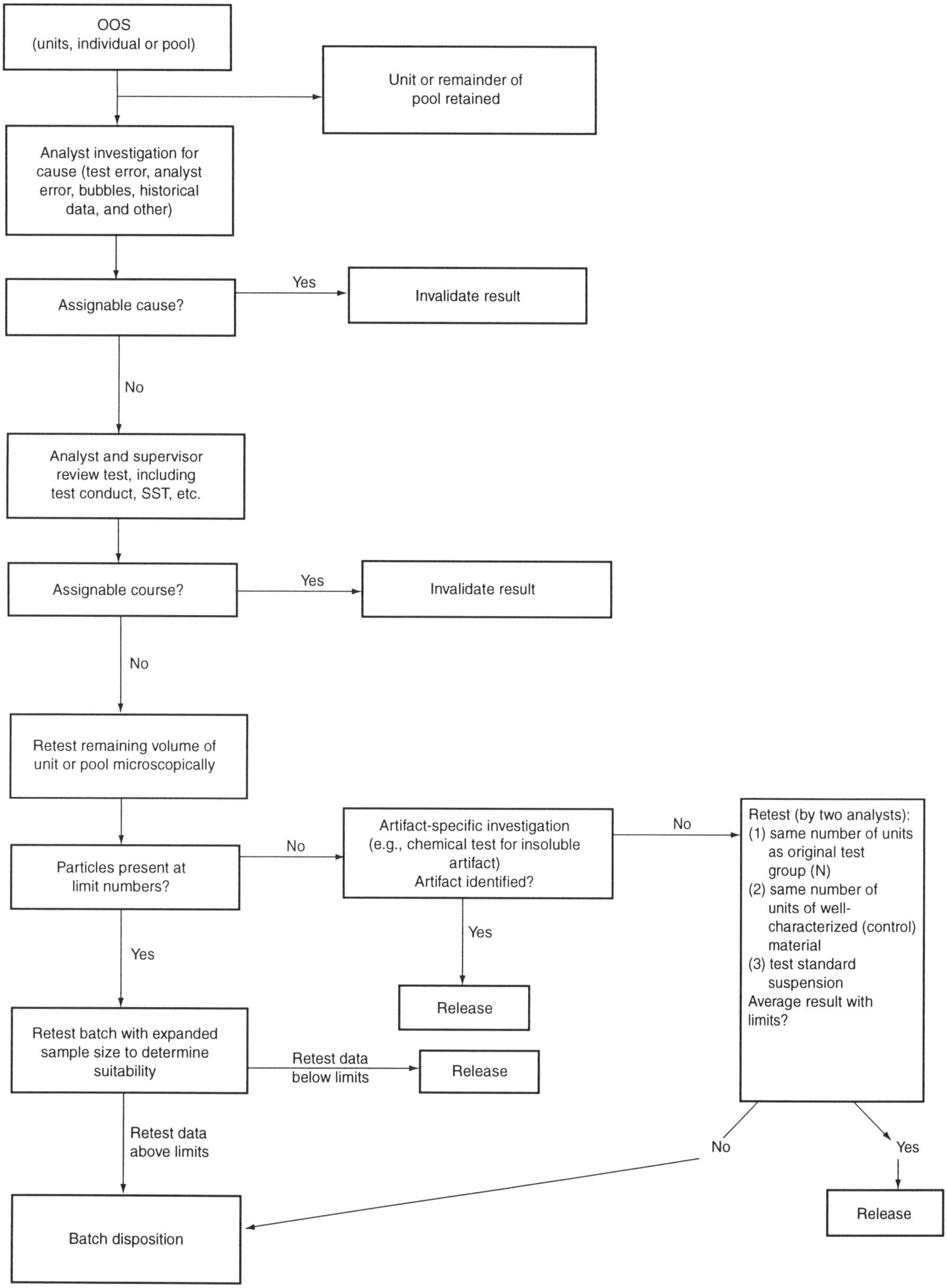

SUMMARY

Validation is an ongoing process of examination, evaluation, and documentation of a computerized system to provide a high degree of assurance that the system will consistently produce a product meeting the predetermined specifications and quality attributes described in the requirements definition. Validation is intended to reduce the risk that product will not conform to requirements or that data will be erroneous. Whether or not a significant reduction of these risks has resulted in the "validation era" is uncertain, based on the number of regulatory observations worldwide. In a simplistic expression of the goal of validation, we may adopt the FDA philosophy of demonstrating that "adequate control" of the use of an analytical system and of critical data exists. For an analytical system, per the USP, data should be as precise and accurate as that of a manual assay.

System validation is an extremely important concern for any analytical system used to collect data related to pharmaceutical products. The conduct of the validation must conform to principles accepted by the FDA, ISO 9000, and other guidelines highly visible throughout the pharmaceutical industry. Although a "high degree" of confidence in the intended function of an instrument system and its data is the goal, complete 100 percent assurance of error-free function cannot be obtained and is not required. A pitfall to be avoided in validation is overkill or excessive testing or documentation, which achieves no increase in the level of practical assurance achieved, and does not increase the value of the product for the customer.

One approach to computer-based system validation that has been followed by some companies is to obtain information from consultants or consulting firms to direct their validation efforts. Different consultants will espouse different routes of validation; in the case of large number of reputable consultants, the different approaches may be equally satisfactory with regard to end results. In the experience of some companies, the amount of work associated with validation has appeared to increase as the square of the number of consultants involved. If consultants are to be used in designing validation plans, the manufacturer is probably best advised to choose one (or at most two) consultants with the necessary credentials and rely on these experts to specify philosophy rather than detail of what is to be done. A consultant cannot possibly be familiar with all of wide range of instruments to be validated or with their data output; thus, their decision on specific items many not be well informed, and reliance may be placed on depth of validation to ensure that no key item is missed.

The validation plan chosen for a specific instrument system must ensure that a rational approach is followed so that errors in system function can be detected and/or avoided. The adequacy of validation will depend on the specifics of the instrumentation involved as well as the nature and use of the data obtained. Application of in-use standards and leverage of vendor documentation are key principles to be followed. Validation in the individual laboratory should always conform to a some "master plan" for the company and operating units of the company to ensure uniformity of approach and consistency in adherence to the elements of validation.

Validation of computer-based analytical systems is clearly an area in which the FDA has allowed the industry significant license in terms of determining policies and procedures. Depending on the course chosen by the manufacturer, this lack of specific FDA directives can make the approach easier or more difficult. The author prefers to be optimistic; in my assessment, the industry is free to pursue a number of approaches to validation that are equally logical and result in equal levels of performance assurance in the operation of computer-based instrumentation.

During inspections of validation documentation, an investigator may raise questions regarding organizational items in documentation, untested functions, and so on. These questions are to be expected and may not be easily dealt with if each system has been addressed in standalone fashion. They may be readily addressed, however, if a uniform rationale has been used for all of the systems addressed. More validation in terms of diverse approaches is not necessarily better validation.

The basic deliverable that the FDA expects is an assurance of control of testing and data based on (1) understanding of instrument systems (computer-based or not) and the data they generate as opposed to blind acceptance; (2) a systematic assurance of the credibility of data generated; and (3) assurance that errors will be avoided to the maximum extent possible and detected if they occur. Thus, in any validation, whether it be of particle counters or chemical analytical instruments, the user is best advised to focus on these issues. Focus on these items and on in-depth understanding of the function and data output of a computer-based system will ensure an adequate validation with minimal costs and effort.

What an inspector wants to find and expects is evidence that the function of a computer-based analytical system has been given careful consideration by the user. There are a number of ways of conducting this careful consideration. Detailed, well-considered approaches taking into account the components of validation listed above are unlikely to be objected to, even if the format of validation is not according to published or widely accepted plans. A careful validation of an analytical technique, including software, instrumentation, and the method itself serves the function of a search for truth; in this case, the "truth" is sound data.

REFERENCES

Agallaco, J. 1987. Validation of existing computer systems. *Pharm. Tech.* (Jan.).

Alford, J. S., and Kline, F. L. 1990a. Computer system validation—Staying current: Installation Qualification. *Pharm. Tech.* (Sept.): 88–104.

Alford, J. S., and Kline, F. L. 1990b. Computer system validation—Staying current: Change control. *Pharm. Tech.* (Sept.): 20–40.

Anisfeld, M. 1990. *International drug GMPs.* Buffalo Grove, Ill., USA: Interpharm Press, Inc., pp. 23–24.

Anisfeld, M. H. 1994. Validation—How much can the world afford? Are we getting value for the money? *J. Pharm. Sci. Tech.* 48:45–49.

Chamberlain, R. 1991. *Computer system validation for the pharmaceutical and medical device industry.* Libertyville, Ill., USA: Alaren Press.

Clark, A. S. 1988. Computer systems validation: An investigator's view. *Pharm. Tech.* (Jan.): 60–66.

Chapman, K. G., J. R. Harris, A. R. Bluhm, and J. J. Errico. 1987. Source code availability and vendor-user relationships. *Pharm. Tech.* (Dec.)

Commission of the European Communities. 1992. EC guide to good manufacturing practices for medicinal products. In *The rules governing medicinal products in the European Community,* vol. IV. Luxembourg: Office of Official Publications for the European Communities.

Drug Information Association. 1988. *Computerized data systems for nonclinical safety assessment.*

EPA. 1990. *Good automated laboratory practices: EPA's recommendations for ensuring data integrity in automated laboratory operations with implementation guidance.* OIRM Good Automated Laboratory Practice draft document. Washington, D.C.: Environmental Protection Agency.

Federal Information Processing Standards. 1976. *Guidelines for documentation of computer programs and automated data systems.* FIPS Publication 38.

Federal Information Processing Standards. 1983. *Guideline for life cycle validation, verification, and testing of computer software.* FIPS Publication 11.

FDA. 1983. *Guide to the inspection of computerized systems in drug processing: Reference materials and training aids for investigators.* Rockville, Md., USA: Food and Drug Administration.

FDA. 1987. *Guideline on general principles of process validation.* Rockville, Md., USA: Food and Drug Administration, Center for Devices and Radiological Health.

FDA. 1988. *Reviewer guidance for computer-controlled medical devices* (draft). Rockville, Md., USA: Food and Drug Administration.

FDA. A course in computerized systems audit and computerized manufacturing. Rockville, Md., USA: Food and Drug Administration, National Center for Toxicological Research.

FDA 91-4179. *Medical device Good Manufacturing Practices manual*—Appendix: Application of the medical device GMPs to computerized devices and manufacturing processes (medical device GMP guidance for FDA investigators). Rockville, Md., USA: Food and Drug Administration.

Furman, W. B., T. P. Layloff, and R. F. Tetzlaff. 1994. Validation of computerized liquid chromatographic systems. Proceedings of the 106th Annual AOAC International Meeting. *J. AOAC Intern.* 77 (5):1314–1318.

Gold, D. H. 1996. Validation: Why, what, when, how much. *J. Pharm. Sci. Tech.* 50:55–60.

Hambloch, H. 1994. Existing computer systems: A practical approach to retrospective evaluation. In *Good computer validation practices: Common sense implementation,* edited by T. Stokes, R. C. Branning, K. G. Chapman, H. Hambloch, and A. J. Trill. Buffalo Grove, Ill., USA: Interpharm Press, Inc.

Hewlett-Packard. Good laboratory practice—Part 1: Vendor-validated computer-controlled HPLC systems. Publ. Number 12-5091-3748E. Waldbronn, Germany: Hewlett-Packard.

Hokanson, G. C. 1994. A life cycle approach to the validation of analytical methods during pharmaceutical product development, Part II: Changes and the need for additional validation. *Pharm. Tech.* (Oct.): 92–100.

Huber, L. 1994. *Good laboratory practice.* Publication number 12-5091-6259E (12-5963-2115E). Waldbronn, Germany, Hewlett-Packard.

Huber, L. 1995. *Validation of computerized analytical systems.* Buffalo Grove, Ill., USA: Interpharm Press, Inc.

Huber, L. 1999. *Validation and qualification in analytical laboratories.* Buffalo Grove, Ill., USA: Interpharm Press, Inc.

IEEE Std 730-1989. *IEEE standard for software quality assurance.*

IEEE Std 828-1990. *IEEE standard for software configuration management plans.*

IEEE Std 983-1986. *IEEE guide for software quality assurance.*

ISO/IEC Guide 25: *General requirements for the competence of calibration and testing laboratories,* 3rd ed. 1990. Geneva: International Organization for Standardization.

ISO 9000-3. 1991. *Guidelines for the Application of ISO 9001 to the Development, Supply, and Maintenance of Software.* Geneva: International Organization for Standardization.

Knapp, J. Z., and L. R. Abramson. 1996. A new coincidence model for single particle counters, Part III. Realization of single particle counting accuracy. *J. Paren. Sci. Tech.* 50:99–122.

Koseisho, M. 1988. *Good laboratory practice attachment: GLP inspection of computer systems.* Tokyo: Ministry of Health and Welfare, Pharmaceutical Affairs Bureau.

Kuzel, N. R. 1985. Fundamentals of computer system validation and documentation in the pharmaceutical industry. *Pharm. Tech.* (Sept.): 251–256.

Kuzel, N. R. 1987. Quality assurance auditing of computer systems. *Pharm. Tech.* (Feb.): 70–76.

Masters, G., and P. Figarioic. 1986. Validation principles for computer systems: FDA's Perspective. *Pharm. Tech.* (Nov.): 181–191.

Matsuda, M. 1994. Guideline on control of computerized systems in drug manufacturing. (Japanese Ministry of Health and Welfare). *J. Pharm. Sci. Tech.* 48:11–15.

PMA's Computer Systems Validation Committee. 1986. Validation concepts for computer systems used in the manufacture of drug products. *Pharm. Tech.* (May): 85–89.

Schoenauer, C. M., and R. J. Wherry. 1983. Computer system validation—Staying current: Security in computerized systems. *Pharm. Tech.* (May): 48–58.

U.S. Pharmacopeia, 23rd rev. Easton, Penn., USA: Mack Publishing Co.

Trill, A. J. 1994. A regulatory perspective. In *Good computer validation practices: Common sense implementation,* edited by T. Stokes, R. C. Branning, K. G. Chapman, H. Hambloch, and A. J. Trill. Buffalo Grove, Ill., USA: Interpharm Press, Inc.

UK Pharmaceutical Industry Computer System Validation Forum (PICSVF). *Good automated manufacturing practice in the pharmaceutical industry (GAMP)*—Draft Guidelines: Validation of automated systems in pharmaceutical manufacture. Randalls Way, Surrey, UK: LOGICA Industry Ltd.

Vinther, A. 1998. Pharmaceutical laboratory compliance—Manufacturer's view. *J. Pharm. Sci Tech.* 52:76–79.

BIBLIOGRAPHY

Altria, K., and D. Rudd. 1995. An overview of method validation and system suitability aspects in capillary electrophoresis. *Chromatographia* 41: 325–331.

Bedson, P., and M. Sargent. 1996. The development and application of guidance on equipment qualification of analytical instruments. *Accred. and Qual. Assur.* 1:265–274.

FDA. 1987. *Guideline for submitting samples and analytical data for methods validation.* Rockville, Md., USA: Food and Drug Administration.

FDA. 1993. *Guide to inspections of pharmaceutical quality control laboratories.* Rockville, Md., USA: Food and Drug Administration.

FDA. 1994. *Reviewer guidance: Validation of chromatographic methods.* Rockville, Md., USA: Food and Drug Administration, Center for Drug Evaluation and Research.

Freeman, M., D. Leng, D. Morrison, and R. P. Munden. 1995. *Pharm. Technol. Eur.* 7 (10):40–46.

Furman, J., T. Layloff, and R. Tetlaff. 1994. Validation of computerized liquid chromatographic systems. *J. AOAC Int.* 77 (5):1314–1318.

Huber, L. 1996. Validation of computerized analytical systems: Parts I, II, and III. *LC-GC* (July, Aug., and Sept.).

Huber, L. 1996. Validation of computerized analytical systems, Part IV: Ensuring ongoing performance. *LC-GC* 14 (10):896–900.

Human Drug cGMP Notes. September 1997, p. 5.

ICH. *Validation of analytical procedures: Methodology.* ICH Harmonized Tripartite Guideline Q213.

Kuwahara, S. 1998. Analytical standards and controls. *BioPharm* (April): 58–59.

Kuwhara, S. 1997. Meeting notes on chromatographic system suitability public workshop. *BioQuality* 2 (7):2–3. (BQ is published by Pocket Guide Press, call 714-380-3204 for information.)

Little, L. 1998. Qualification of legacy equipment in quality control laboratories. *BioPharm* 11 (3): 61–62.

Maxwell, W., and Sweeney, J. 1994. Applying the validation timeline to HPLC system validation. *LC-GC* 12 (9):96–101.

Parriot, D. 1994. Performance verification testing of computerized liquid chromatography equipment. *LC-GC* 12 (2).

Pritchett, T. J. 1998. Qualification of quality control laboratory equipment. *BioPharm* 11 (1):57–58.

Ross, G. 1997. Instrumental validation in capillary electrophoresis and checkpoints for method validation. *Accred. and Qual. Assur.* 2:275–284.

Snyder, L., et al. 1988. *Practical HPLC methods development.* New York: Wiley-Interscience.

XII

SAMPLING AND COLLECTION OF PARTICULATE MATTER FOR ANALYSIS

Before particles can be characterized, identified, or sourced, they must be captured and immobilized. This chapter will provide the pharmaceutical particle analyst with an overview of the collection and isolation of particulate matter. The pharmaceutical particle analyst is most often a generalist and typically does not need in-depth knowledge of specific techniques of particle collection or isolation. Detailed information is readily available from individuals who are specialists or from the literature. Microfiltration is a very complex subject. A great deal of information is available on this topic simply by requesting technical information from filter vendors—Gelman, Pall, Millipore, Sartorius, and others. This information is usually provided free of charge; as an additional benefit, valuable technical contacts may be made with experts in the applications groups of the various manufacturers.

The references for this chapter deal with particle isolation, garments, analytical procedures (isolation techniques for), filtration, particle monitoring, container cleanliness, and air sampling. The Millipore manual (1992) is a very useful aid in selecting particle isolation and collection techniques. The appropriate section of the text by McCrone and Delly (1973c) deals with all aspects of particle collection. The papers by Sokol and Boyd (1968), DeLuca and Bodapatti (1980), and Oles (1978) are useful with regard to the collection of particles from pharmaceutical products and environments. The classic text by Allen (1990) provides fairly detailed insights into a wide range of collection methods. A number of articles that are of specific interest to those involved in the analysis of particles in injectables have been published by some of my associates at Pharmacia-Upjohn. These are well recommended (see Aldrich 1985, 1990; Borchert et al. 1986; Aldrich and Smith 1995). A listing of appropriate standards is also given at the end of the chapter. These are valuable in that they tend to include focused technical information and methodologies.

THEORY OF PARTICLE SAMPLING AND COLLECTION

Before getting into specific techniques and methods, a brief discussion of considerations regarding particulate matter samples in general follows. First, in terms of the analyte, what is a "sample" and what is a "specimen"? These terms are sometimes used interchangeably, but they are properly reserved for the material to be analyzed at different levels of preparation or isolation. The term *sample* is the particulate matter initially presented for analysis; it may be a mixed population containing the specific particles of interest as a component, or a fraction collected based only on gross visual observation. The term *specimen* is typically applied to a fraction of the sample that has been subjected to some variable level of isolation or purification; it often refers to an isolate prepared for observation by a specific instrument.

It is critically important that the sample initially collected for analysis is representative of the problem to be addressed or of the material to be characterized. If the issue at hand involves particles present in an injectable solution, the analyst should ensure that the units received for analysis contain the particles that constitute the problem. Before beginning any analytical work, the analyst must also ensure that particles to be identified have been adequately described by the person submitting the samples or that the specific problem to be addressed is defined in detail. Ideally, the analyst will be able to personally collect the particulate matter sample for analysis. If the problem is point-generated particulate from a filler or component of the process, it is best to have an analyst go to the production area to collect samples. If components of a solution container or device are involved, the parts submitted for analysis should be collected using cleaned forceps and placed in particle-free bags of nylon or some other clean, durable polymer (not polyethylene) that have been tested for cleanliness. Minimally, telephoned instructions on the method of sampling should be given. In the author's experience, samples of particulate matter collected by filtration in another lab and sent for analysis are liable to be contaminated and may be suspect on other grounds. The analyst must never forget that a single touch of a finger on a sample can deposit thousands of skin cells. Particle-free gloves are vital if samples must be handled. Despite precautions, any particle laboratory can expect to receive samples collected with various degrees of care, and the analyst generally learns to make the best of bad situations.

Essential to all investigations is a comprehensive interview with the sample submitter. Conditions of formation, isolation, and so on may hold clues to the identity of the sample particulate matter. Additionally, the submitter may have a "hunch" regarding the source of the particles, often one he or she may have dismissed or ignored, that may be helpful in subsequent work. Coaxing a comprehensive description of the incident or scenario from the requestor is an essential contribution to our success.

The experienced particle analyst thus appreciates the fact that the individuals who submit samples relative to a particle problem can often provide the solution to the problem if the analyst asks the right questions. The location of the particles of interest, circumstances of their occurrence, and the visual appearance often provide the analyst with at least a general idea of the source and identity. An understanding of the process by which an affected solution or device is made is extremely valuable to the analyst. Often, the sample submitter may have a good idea of what the sample particulate matter is, which may greatly decrease the time spent in the analysis in subsequent work. This information source is invaluable if time-critical production processes must be investigated.

No comprehensive particulate matter investigation is complete without a walk-through, or a detailed level of understanding regarding the origin of the sample. Although not always possible (or practical), this part of the investigation may reveal a situation, condition, environment, or other factor that has been taken for granted or otherwise ignored by the sample submitter. In this manner, the submitter and analyst become partners in the foreign material/defect investigation. Consultant labs and off-site investigators are at a distinct disadvantage in this process, and thus must pursue much of this insight from the interview of the submitter, often directing subsequent activities for the submitter to provide supporting information.

The pharmaceutical particle analyst is more often concerned with particle identification than with enumerating particles of different types or sizes in a population. Sampling for identification may require a less meticulous technique (see McCormack et al. 1976). In the simplest case, the entire contents of an affected solution unit can simply be filtered. The analyst can often identify particles on the basis of a rapid light microscopic analysis; if the interest is only the composition of the unknown particles, only a small sample may be necessary. The analyst may identify particles in situ using various instruments, such as a pocket microscope or a hand-held magnifying glass. In the case of an ampoule of drug solution in which visible particles are present, the analyst may simply examine the particles using an inverted microscope. Polarized light may also be used with this instrument, and many commonly occurring extraneous particles in pharmaceuticals (such as paper fibers, cotton, synthetic fibers, glass, metal fragments, rust, skin cells, and rubber) can be identified by this method. The analyst may not, in fact, sample in the usual sense but instead identify the particle in situ using an inverted microscope. For example, assume the sample is an ampoule or vial of injectable drug solution in which a few small particles can be seen in a bright light beam against a black background. If they cannot recognize one or more components, they can, at least, characterize them so that, after removal, they are recognized as the unknown.

If polarized light microscopy is to be performed, the most suitable mounting medium is a viscous liquid with a high refractive index. The Aroclor® series of chlorinated biphenyls or polyphenyls (PCBs) manufactured by Monsanto has fallen out of favor due to environmental concerns. Other mountants that may be substituted include glycerol, Permount®, white Karo® syrup, Canada balsam, or silicone oil. Aroclor® mountants may still be purchased, but their sale is controlled. Some very useful alternative PCB–free mountants are those of the Meltmount® series from Cargille. Available in refractive indices ranging from 1.539 to 1.704, they have the beneficial properties of low coat and low melting point (65°C) and contain no volatile solvents.

SAMPLING GUIDELINES

Sampling for the level of particulate contamination in any system (gases, liquids, or component surfaces) is based on the collection of particles on a membrane filter. Adequate attention must be given to proper sampling techniques in order to ensure valid, reproducible results. Sampling equipment, containers, analytical apparatus, and membrane filters must be clean if subsequent measurements are to be a valid index of the system being tested.

The number of samples, the sample volume, and sampling time depend on both the level of system contamination and the type of measurement being employed. In general, sufficient material must be sampled so that the collected contaminant will be clearly measurable at "dangerous" levels of contamination. Thus, 100 mL would represent an adequate sample from a 1 L solution unit of injectable drug solution, in which the contamination is measured by microscopy (particle counting); 1 to 5 gal (4–20 L) would be a representative sample; and fluid supplies a minimum of 10 ft^3 (280 L) for measurement of the air in a cleanroom environment down to Class 10,000. Proportionally greater volumes are needed for Class 10, 100, and 1,000 cleanrooms.

To be meaningful and reproducible, samples from a distilled water feed line should be as representative as possible of the entire system when the system is operating normally. Samples of viscous fluids should be taken from areas of high turbulence, where particles are mixed throughout the fluid cross-section. When it is necessary to sample a static system, such as a mix tank or part washing kettle, and the system contents cannot be thoroughly mixed, a multilevel sample should be taken. In sampling component surfaces, it is impossible to remove all contaminating particles through the membrane filter. Therefore, it is extremely important to follow the identical sampling (flushing) procedure for any given component or system every time the samples are analyzed. The analytical results from such sampling will not indicate the total extent of surface contamination, but they will yield meaningful and reproducible data on relative contamination levels.

Airborne particulate matter concentration can vary widely (Chapter 9), depending on the operations being carried out. To minimize these variations, sampling times may be extended to a whole shift (8 h), and the results are the average level of particles. Alternatively, sampling times can be short, sometimes only a few minutes, in order to measure how high the exposure is during particle-generating operations such as powder filling. Generally, the sample quantity must be sufficient to reflect variation in the group of units or area being sampled. The key here is to sample across time such that all variations in activity that might affect the airborne particle burden are captured.

Sample Protection

Sample submitters, left to their own devices, will display uncommon ingenuity in the selection of modes of sample collection and transport. Common paths of logic involve the supposition that "cleanroom" wiping material (e.g., cloth wipes or swabs), being clean, are satisfactory for sample collection. Not so. These materials generally present a complex background for analysis and constitute a matrix from which it is almost impossible to remove particles. Another innocent but erroneous conclusion is that containers that contain no readily visible particles (e.g., polybags) are good sample enclosures. Generally, the following hierarchy of suitability is appropriate for sample collection:

- Best
  - On-site: Send particles as they are observed, in original container, if solution borne or on the affected part or other surfaces.
  - Send particulated part wrapped in aluminum foil. Use the center of the roll, and remove from the box before unrolling. Cut with clean scissors.
  - Particulate matter enclosed between clean micro slides with edges taped.
  - Sample collected in precleaned wide mouth glass jar with aluminum foil or Saran Wrap® or Parafilm® (Parafilm® is clean but abrades easily) under screw cap.
- Sometimes Adequate
  - Tape collection (Scotch Magic™ 812)
  - Precleaned nylon bags
  - Snap top plastic vials
  - Gore-Tex (Teflon®) or Tyvek® wipe
- Never
  - Cloth wiping cloth
  - Paper towel or wipe
  - Foam or cotton swab
  - Paper envelope
  - Inside rubber glove
  - Polyethylene bag

Many commercial packaging films are suitable for sample transport and protection. Packaging films commonly in use today include nylon, polyethylene, Mylar®, aluminum foil, and several laminated products. Nylon is often used as a barrier material because it is does not slough off particles readily, and because it can be cleaned and heat sealed to provide complete enclosure of the clean parts. It may have an antistatic compound added during the melting process in order to render it static free. Polyethylene is not recommended as a sample container material because it has a pronounced tendency to release particles of waxes or other additives from the surface and is subject to particle generation caused by abrasive contact with hardware. Polyethylene is excellent, however, as a moisture barrier and is recommended as an overbag for items first sealed in Mylar®/nylon to exclude moisture from the critical hardware. Mylar® and Saran® are relatively nonreactive and generally clean.

The author has had generally good experiences with the use of two other materials for the protection of samples during transport. These are Tyvek® and aluminum foil. The latter material has the advantage that it conforms to the shape of the samples, which may be loose materials such as stoppers or small device components, and forms a tight, durable package. Tyvek® is subject to abrasion, just as polyethylene is, but has low shedding properties and may be used if available. Particles shed by this material are readily identified if encountered. Double-layered wrapping is practiced by some laboratories, with an inner layer of some less durable material, such as Saran®, being overlaid with foil. A caveat with foil is that some lots from some vendors may contain thermally generated particles from the manufacturing process; the best vendor can be identified by making a pouch of the material and rinsing it with filtered water, which is then filtered and tested for particle burden.

A factor that works in the analyst's favor is that the particles that cause a problem are generally monotypic, present in high numbers on or in the article of interest, and are of a composition differing uniquely from any heterogeneous, randomly sourced, extraneous particles that may be associated with the materials used for protection during shipment. If the laboratory is to be involved to any significant extent with samples collected at remote locations and shipped in for analysis, a good plan may be to identify a standard shipping method with materials from known vendors. One vendor of a number of different types of clean packaging materials is Veltek, located in Phoenixsville, Penn.

Sample Container Cleaning

It is often not convenient to collect liquid samples in bottles and send them to the laboratory for analysis. (See Chapter 8 under Cleaning of Glassware.) In lieu of the wrapping materials discussed above, device component parts, stoppers, and the like can be collected into and shipped in clean plastic materials such as Nalgene® plastic bottles. Cleaned glass beakers may be used, with the obvious need to protect them against breakage. Either type of wide-mouthed container may be closed with aluminum foil or a double layer of Saran® or Parafilm® under the foil. Borosilicate glass bottles such as those marketed by the Wheaton Glass Company work very well. Not only the cleanliness of the bottle must be insured, but the cap should be clean and non-shedding, and the cap liner must be clean and inert. Cleanliness of the sample collection vessels is critical. Bottles should be graduated and preferably slightly larger than the sample size.

Before each use, the sample bottles for the collection of liquids may be cleaned as follows:

1. Wash thoroughly with a laboratory detergent solution. For critical work, bottle cleanliness should be checked by filling and shaking with filtered solvent to dislodge any particles. Initially, the solvent should then be filtered and analyzed. If the cleaning procedure is not effective, cleaning with dilute acids or a low surface tension solvent (e.g., Freon®) may be required.
2. Rinse each bottle twice with Millipore® Milli-Q water. In all rinsing operations, a solvent filtering dispenser is especially convenient. By squeezing the bulb on the flask, a stream of solvent is forced from the flask, through a Millipore filter, and out the flexible dispenser tip.

   When many bottles must be cleaned at one time, the Millipore® Filterjet™ solvent dispenser is effective. When connected to a pressurized solvent tank, it provides a strong jet or spray of ultraclean solvent, in a continuous or trigger-controlled action.
3. Rinse the sample bottles with 1.2 μm membrane-filtered distilled or deionized water, Milli-Q water, or isopropyl alcohol (IPA) to remove residual rinse water.
4. If the bottles are to be used to collect oil or apolar liquids, a final rinse with a filtered solvent, such as IPA or Freon® TF may be used at the analyst's discretion.
5. A small square of Saran Wrap® may be placed over the mouth of the sample bottle before replacing the cap to minimize the chances of contaminating the bottle with particles from the screw cap.

METHODS OF COLLECTION/ISOLATION

Basically, the isolation of particles of interest involves the leverage of unique physical or chemical properties of the particle, to separate it from other particles in a population or from a suspending medium. These unique properties include size, density, rigidity, stickiness, shape, electrical charges, and magnetism.

If direct isolation is to be used, shape, color, and transparency serve as descriptions used to select particles of interest. Size is probably the next most common physical property used for isolation, such as in the process of sieving used since ancient times. The most widely used methods not only for particle collection but also for removal is filtration. While the procedure is typically used to isolate solid or rigid particles from fluids, it can also be used to separate apolar liquids from polar ones, and vice versa, based on the affinity of the particle for polar or apolar filter surfaces. Particles of a density greater than a suspending medium will simply settle out if large enough to overcome dispersive forces, such as Brownian motion or turbulence. Charged particles adhere tenaciously to surfaces of opposite electrified charges. Particles may also be collected directly using probes or capillary pipettes.

The complexity of the isolation/collection process generally is directly related to the complexity of the substrate or suspending medium and inversely related to particle size. Specifically, some particles, particularly those found in drug solutions, are "ethereal" in nature and difficult to isolate or collect in a form amenable to analysis. In this category are particles that are amorphous or semiliquid and those that are loose agglomerates of very fine particles. Others may be subject to dissolution on agitation or filtration.

Stated another way, the smaller the particles of interest are, the more sensitive to background and environmental contaminants the analysis will be. Analytical contaminants are pervasive and of a wide range of compositions. For the manipulation and isolation of smaller particles (e.g., $\leq$ 10 μm in size), cleaner air is required, and the analyst is recommended to use the protection advised in the U.S. Pharmacopeia (USP) 23 <788> test (i.e., nonparticle shedding gloves, high efficiency particulate air [HEPA]–filtered directional airflow and water filtered at a retention rating of 0.45 μm or 0.22 μm). The simple goal to be achieved is the same that the analytical chemist pursuing identification of a trace organic unknown strives for—the maximum signal-to-noise ratio and the elimination of spurious peaks (see Delly 1996). Interferences that might confuse the analysis or in a worst case lead to a false conclusion must be avoided. Although airborne particles are some threat to an analysis, contaminants that are associated with objects or materials that approach the analysis closely or are an integral part of the analytical process are the most dangerous. Examples are materials from the garments or gloves of the analyst or material contained in the water or the glassware used in the process.

A few examples of the hazard of extraneous contamination are provided below. Beginning analysts have been confounded many times in their efforts by starch, calcium carbonate, or talc from gloves. Paper fibers are everywhere and may be carried to a membrane by tools or on the hands. Skin cells likewise can come from the hands or from the external surfaces of gloves that were not carefully donned, and cosmetics can leave residues with telltale traces of iron oxide, titanium, or pigments. Sample containers are notorious sources of contaminating particles. Samples sent in polybags are often found to be coated with particles of the waxes used as extrusion aids to produce the tubular film from which bags are made. Nylon bags are typically clean, but must be tested before use to be sure that amide-containing particles are not present. (Amide-containing particles constitute a particularly confusing influence in infrared analyses.) Again, the use of paper sample containers should be avoided at all costs.

The type and size of particle to be identified, characterized, or counted is also a consideration. Large (> 25 μm) black particles on a white plastic part are not likely to be confused with typical contaminants and are readily removed with a probe, so this analysis is relatively insensitive to outside influences. When fibers are to be identified, care must be taken for sources of paper or cotton. A white powder on a blue plastic device part or on a stopper may be readily isolated by a number of means and will probably "outnumber" most contaminants, so precautions may be taken at a less stringent level. When small (< 10 μm) white particles are to be "picked from a spongiform filter, however, the fullest protection may be called for, including HEPA–filtered directional airflow and an awareness that any crumbs of filter medium abraded from the membrane by the probe may resemble the analyte in size and color.

A thought for reflection here is that success in science or technology most often depends on doing things "well enough" and no better. The analyst will learn this approach to particle isolation and sample collection so that the analysis proceeds in a time-effective fashion and, most importantly, provides the correct result.

There is a wide variety of commercial laboratory apparatus available for separating suspended particulate matter from the air and collecting it on some surface or in some container from which it can be recovered for analysis. The choice of equipment will depend on the purpose for which the sample was collected. The collection apparatus may range in complexity from a simple glass bottle to impingers or cascade impactors. It is always important to obtain as complete a history of the sample as possible so that the impact of sampling conditions and procedures on the composition can be evaluated. Methods of isolation/collection used for particulate matter will be pivotally dependent on size and physical properties of the particles and the medium (solid, liquid, gas) in which they are found. Methods of collection range from simple direct isolation through complex chemical isolation procedures that may entail dissolution of the particle followed by crystallization or precipitation. While it is not possible here to discuss possible applications of techniques in detail, Examples have been provided to allow the investigator to further develop specialized techniques.

The most commonly used methods of isolation and collection are the following:

- Direct isolation
- Filtration
- Surface or "fallout plate" sampling
- Adhesion and entrainment
- Sedimentation
- Inertial collection
- Sieving
- Chemical methods

Generally, within each of these categories, there is a range of techniques varying significantly in complexity; the simplest that will serve to isolate the particles of interest for analysis is generally the best. In the following sections of this chapter, the methods are discussed in the order in which they are listed.

Direct Isolation and Manipulation of Particles

Direct isolation is the process of obtaining particles characterized by visual or microscopic observation as desirable for further analysis by manipulation of the individual single particles. The process is indirect in the sense that particles to be dealt with on an individual basis have been initially collected by sedimentation, decantation, or other means. Given that the particles are solid or rigid and of sufficient size, they are simply selected from the matrix visually or aided by the stereomicroscope. A description of the type of tools is included later in this chapter. A suitable probe or similar tool may be used to transfer a solid particle from a surface to a floor slide or support for an infrared microscope or electron microprobe.

A variation of this technique for solutions involves using a micropipette (also called a Pasteur pipette). The tip of the pipette is submerged in the solution with an index finger over the large end and maneuvered near the particle. When finger pressure is released, the solution and the particle will rush into the pipette, and the liquid and the particle can be allowed to flow out onto a filter membrane for further isolation steps. Particles present as a powder on a surface can be collected by gentle scraping with a microspatula, or a wisp of glass wool moistened with solvent.

A variation of the direct technique using a capillary tube may be useful for visible particles observed within a vial or ampoule. Particles are most often visually observed as appearing white (due to diffuse scatter in a transparent injectable solution, but they may be seen as dark colored, if they are within an emulsion or drug suspension or powder. In other circumstances, the analyst may not sample in the usual sense but instead identify the particle in situ (Aldrich and Smith 1995). If the sample is an ampoule of injectable drug solution in which a few small particles are seen, the particles will be difficult to remove with assurance of high collection efficiency and low contamination. Based on the low solution volume, the particles in the ampoule may be studied using an inverted microscope. One can use 100 to 200× magnification and polarized light in observing the particle through the walls of the ampoule. Most contaminants (e.g., paper fibers, cotton, nylon, glass, metal filings, rust, skin, mold, rubber, flyash, etc.) can be quickly and easily identified.

Direct isolation techniques are also, of course, adaptable for the removal of particles from surfaces such as those of production machinery or components (e.g., stoppers). If the particles are present on a hard, smooth surface, it is best to scrape them off with a solvent-cleaned razor blade under a stereomicroscope. The residue on the blade is removed with a tungsten needle and placed on a glass slide for optical examination. Particles on a soft surface can be successfully removed with a polyethylene scraper that will not scratch the surface. To make a polyethylene scraper, a piece of tubing of appropriate diameter is simply cut at a 45° angle.

Particle "picking" involves physical isolation of individual particles (see Teetsov 1977). The microscopist always identifies particles with as little sample treatment as possible. If any particles have to be removed for closer study, only one particle or a few particles at a time are removed. If the membrane filter sample is studied by reflected light under a stereomicroscope, specific particles may be removed by touching each with an Aroclor®-coated, finely pointed, tungsten needle. The needle is coated by touching the tip to a 1 percent benzene or acetone solution of Aroclor® and allowing the solvent to evaporate. This coated tip with a radius approximately that of the particles will pick up one or several specific particles when directed by a steady hand under the stereomicroscope. The particles are then transferred to a small Aroclor® drop on a clean microscope slide. After adding a cover slip, they can be examined in the usual way. Using a similar technique, particles may be isolated for transfer to a carbon or beryllium mount for examination by scanning electron microscopy or X-ray microprobe or to a salt plate for analysis by infrared spectroscopy.

This same sticky needle technique can be used with the inverted microscope to remove specific particles from an opened bottle of solution, an ampoule of injectable drug solution, or from any other container until the tip touches the desired particle. The solution must not dissolve the coating. Karo® syrup or Aquaresin®, both water soluble, can be used in place of Aroclor®. Any water-soluble adhesive should be washed off the particles before they are examined in Aroclor®.

For purposes of identification, some particle samples can be examined directly without further isolation. However, if the particles of interest are embedded in a matrix or must be removed from a retentative mixture on a filter, manipulation will often be required. Techniques for the isolation and collection of individual particles for analysis involves the use of specialized micro instruments to pick up and transfer particles physically from a filter or other collection substrate to a micro slide or sample support for an analytical instrument.

This technique is not new. Louis Pasteur used the microscope to differentiate the optical isomers of tartaric acid. By methodically segregating individual crystals of the two morphological types, he was able to resolve the optically inactive mixtures into the two optically active isomers. This method of isolating individual particles under the microscope using a microneedle, forceps, or another instrument is often essential to the identification of particles of interest. Such "particle picking" requires a hood or other HEPA–filtered enclosure, a clean work area, a steady hand, a stereomicroscope, and the tools shown in Figure 12.1.

Needle probes and glass fiber brushes are used to pick up single particles. The fine forceps is useful for larger particles. Drawn micropipettes can be used to pick up single particles in a drop of fluid. The techniques for isolating single particles are generally simple, but it is only through careful practice that one acquires the skill necessary for capable particle handling down to sizes of 10 μm and less. The techniques, necessary tools, and procedures for particle manipulation are discussed by Teetsov (1977), DeLuca and Bodapatti (1980), and Aldrich and Smith (1995). Tools specifically designed for particle isolation are available from McCrone Associates (Westmont, Ill.).

The tungsten probe most commonly used for direct isolation was developed by Anna Teetsov at the McCrone Research Institute and is still the favorite implement for manual methods. Tungsten wire with a diameter of ≈0.5 mm (available through McCrone Associates) is chemically etched with molten sodium nitrite ($NaNO_2$) to a selected tip diameter, generally in the range of 1 to 200 μm. The tip diameter dictates the isolate size; particles similar in size to the tip diameter will adhere and can be transferred. Some analysts utilize coarse animal fibers for sample preparation for electron microscopy. A cat whisker, thick human eyelash, or similar stiff bristle acts as a reliable probe. Any coarse animal hairs are satisfactory and lend themselves to personalization, which many microscopists favor.

The value of the capability for isolation of particles using these physical techniques cannot be overemphasized. The sensitivity limit of microanalytical instruments allows nanogram quantities of material to be analyzed. If a single crystalline particle can be isolated and mounted on the tip of a 10 μm glass fiber, one can obtain an X-ray diffraction pattern. If a clean 10 μm particle can be mounted on a polished beryllium or carbon specimen support, it can be examined by scanning electron microscopy or analyzed quantitatively with X-ray spectroscopy. Secondary ion mass spectrometry is useful with particles of similar size. Micro Fourier-transform infrared (FT-IR) spectrophotometers can obtain spectra with sharply resolved absorbencies from single particles of 10–20 μm in size. Experts in X-ray diffraction, electron microscopy, or microprobe analysis are not usually adept at handling single particles of a size that approaches the sensitivity of their instruments. They must depend on particle analysts with this technical skill to acquire samples for these instruments.

The following methods for incident and transmitted light observation with a stereomicroscope have proven reliable (Aldrich and Smith 1995). Hold the probe in your dominant hand, steady that hand with the other hand or against the base of the stereomicroscope, and, at

Figure 12.1. Tools for particle isolation (Teetsov 1977).

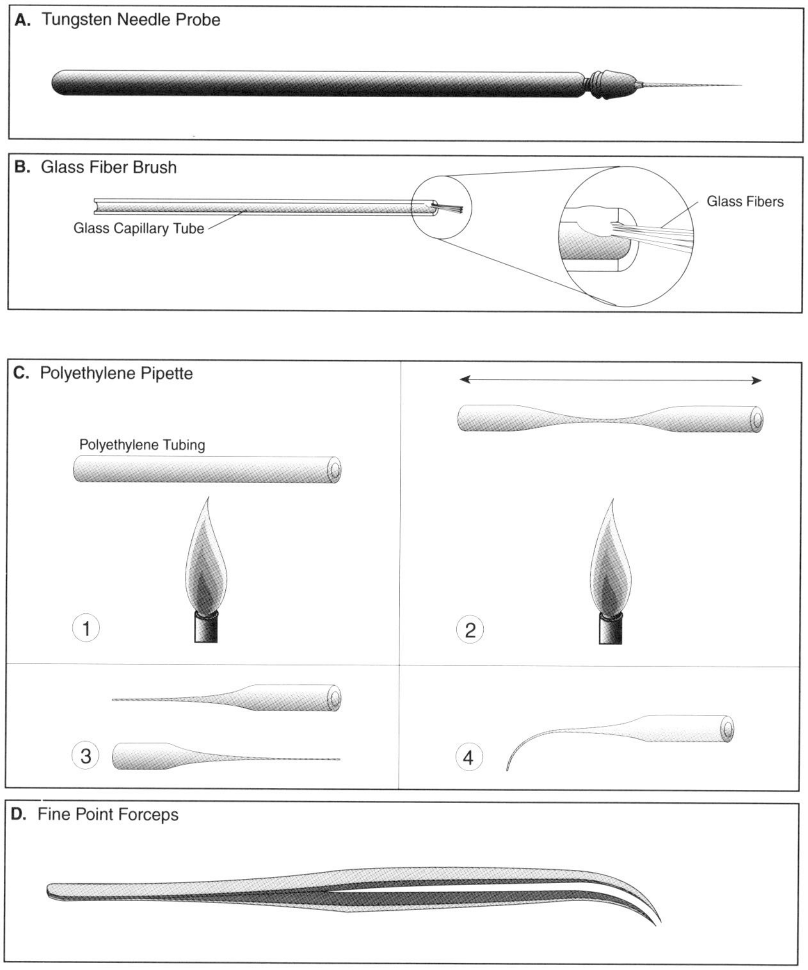

an oblique angle, sweep the selected particle off the substrate. Too much pressure on the particle may cause the particle to exit rapidly, stage left (or right).

Static electrical effects often complicate fine-particle manipulation efforts. One must ensure that the transfer probe is suitably sharp, that plastic containers are removed from the isolation stage, and that the room atmosphere is of reasonable relative humidity (above 30 percent). If still a problem, wet the probe with water or oil, which removes the static and provides some adhering power.

There are general procedures and helpful hints regarding particle isolation for the beginner. These include the following:

- Never take your eyes off the particle during a manipulative operation.
- Remove the particle from the needle only by "washing" it onto a microdrop of liquid in which the particle will be trapped.
- Beware of a static electricity.
- Ensure that the path the needle will take to place the particle on the microscope slide is unobstructed.

Importantly, any particle(s) to be isolated for analysis by instrumental means should be characterized by light microscopy before separation. The mounted particle can then be recognized as the original, not another particle picked up by accident.

Filtration

Filtration is the most commonly used means of collection of particles from liquids, but it may not be appropriate for amorphous or agglomerated particles or particles that are soluble in water or other solvents used to "wash" the membrane following filtration. A variety of commercial filters are available in a wide range of porosities and membrane composition. Depth, screen, and track-etched filters may be employed selectively to suit the type of isolation required; hydrophobic and hydrophilic varieties are even available. The basic equipment common to all sampling techniques includes filters and filter holders for use with vacuum or positive pressure filtration.

The filters used for particle analysis are most frequently one of two types: (1) cast film filters that are spongiform in structure and are made by casting a polymer film containing volatile solvent(s) and pore-forming liquids on a smooth surface (Kesting 1971, 1977), and (2) track-etched or nucleation track filters that are produced by etching a thin polycarbonate film that has been exposed to gamma radiation (Fleischer et al. 1964). The latter type is commonly referred to as "Nuclepore®" filters based on the name of the original vendor. The generic term *membrane,* which may be used in reference to either of the types of filters by different authors, refers to the thinness of these filters and their generally fragile nature. Brock (1983) is an invaluable basic text and discusses the principles of the manufacture of microporous filters of all types, as well as filter function and applications.

The cast film filters often used for particle collection in the lab are open structures, with only a small fraction of the membrane volume occupied by the polymer substrate (Figures 12.2 and 12.3). In a representative filter of this type, 80–85 percent of the membrane is open space (void volume). This openness provides for a relatively high fluid flow rate. In the days before scanning electron microscopy, it was thought that filters of this type contained cylindrical pores of defined size. Examination of a film filter with the scanning electron microscope has revealed that their structure is, in fact, random, being characterized by a polymer network surrounding spherical or oval voids. Depending on the type of cast film filter considered, the polymer structure surrounding the voids of the filter may be almost complete or may consist of fibrous strands. These filters do not trap all particles above their retention rating on the upstream surface; rather, they retain smaller particles within the depth of the medium.

Figure 12.2. Microporous cast film filter structure and function.

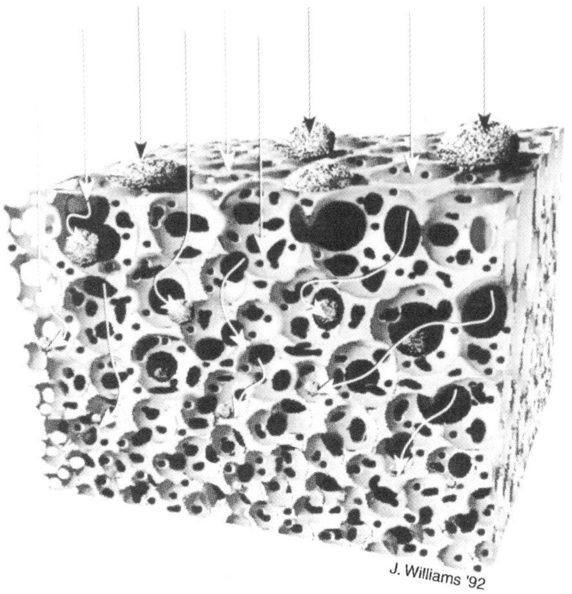

Figure 12.3. Surface structure of a cast film filter.

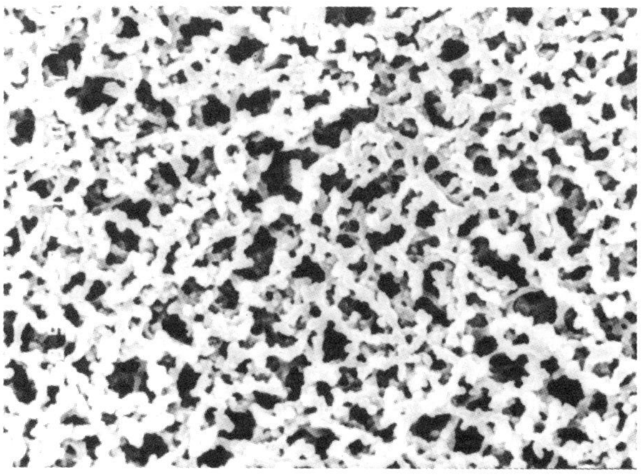

The open-celled, spongiform structure of cast film filters differs significantly from that of nucleation track (Nuclepore®) membranes, which are made by chemically etching polycarbonate films that have been subjected to localized damage by nuclear radiation (Fleischer et al. 1964). A track-etched membrane filter (Nuclepore®) is shown in Figure 12.4. Most characteristic of this filter are the cylindrical pores of uniform diameter that cause the filter to function as a sieve. These filters are suitable for applications in which particles must be trapped on the filter surface. However, as can be seen in the figure, these filters also have less open pore area and, consequently, lower liquid flow rates than the cast film filters; this characteristic makes them unsuitable for many general filter applications. They also tend to "blind" or become blocked more easily than cast film filters, and their extreme thinness (10–15 μm), flexibility, and responsiveness to static electrical charges make handling them difficult. Despite the disadvantages, this type of filter is extremely useful for collecting particles for further analysis; the smooth surface allows particles to be easily removed for transfer to the supports used in analytical instruments, such as scanning microscopes or electron microprobes. Because the surface of the filter is relatively smooth, particles can be picked off or removed for analysis with relative ease; few are lost due to penetration. In addition, a portion of the filter can be cut out, solvent-vapor treated to grip the particles, and vacuum coated for direct scanning electron microscopic observation.

The track-etched filters are manufactured from polymers that are resistant to heat, acids, bases, and organic solvents (e.g., polycarbonate). They are available in controlled pore sizes down to 0.01 μm. A consequence (disadvantage) of the high efficiency of the membrane filters is their high pressure drop and low porosity relative to cast film types; it may take a longer time to filter a given volume using them as compared to a cast film filter. They have the advantage of a nearly complete absence of surface texture (other than the pores themselves) and a single plane for particle collection and examination. Inverting a small piece of filter in a drop of liquid on a microscope slide with repeated lifting and wetting will remove nearly all particles.

Figure 12.4. Track-etched filter medium.

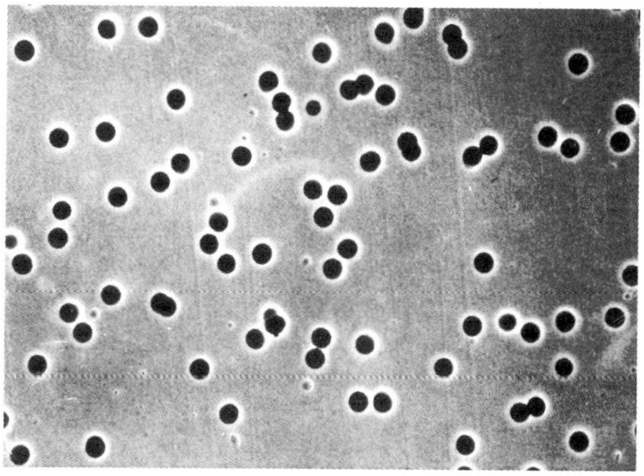

This general discussion of microporous filtration media represents an oversimplification of the real situation. Many types of filter media falling into the general category of cast film filters are made by proprietary processes very different from that used to make cellulose acetate, nitrate, or mixed ester filters. Some filters of this type are made by laminating a fragile medium onto a more durable supporting substrate. These details are not essential to the present discussion. However, the real knowledge of most value to the analyst involves the wide range of physical and chemical properties embodied in the different types of filters. Filters for specific applications in a variety of analytical roles may be selected by consulting with vendors or their application literature.

While the cast film cellulose ester filters and track-etched filters discussed above are useful for the majority of particle isolation procedures undertaken by a pharmaceutical particle analyst, the filtration of solvents, acids, or bases may require the use of appropriately resistant filters. If a question arises regarding the selection of a filter with a specific resistance, the ideal first step is for the analyst to consult the vendor catalogs. These catalogs will generally include a tabular presentation of filter types and their resistance to specific solutions.

Teflon® (polytetrafluoroethylene or PTFE) filters (Figure 12.5) have historically been made by expanding (stretching) Teflon® film over a supporting matrix. Current versions, such as the Millipore® Mitex™ type, have no backing. Polypropylene, polysulfone, and nylon filters are made by a modified cast film process but have wetting properties and chemical resistance that differ from conventional cast film cellulose ester or mixed cellulose ester materials. Polypropylene and untreated nylon filters have a high affinity for apolar liquids (oils) and may be used to remove such materials from aqueous systems by adsorption. Untreated nylon filters have a positive charge and are extremely

Figure 12.5. PTFE expanded mesh filter medium.

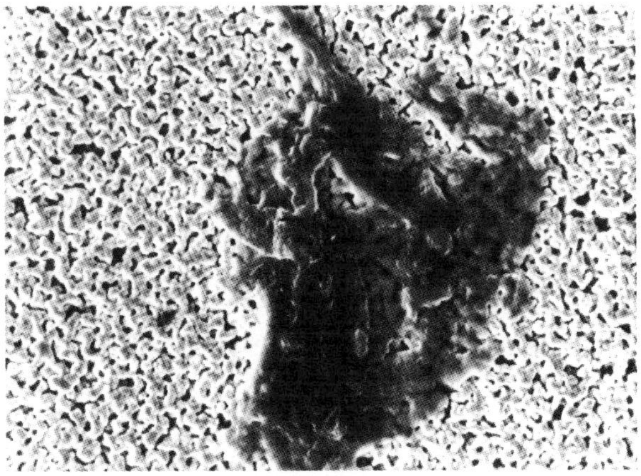

effective at removing charged materials (including biologicals) from solutions and suspensions. Sintered silver filters (Figure 12.6) are produced by the compression of silver particles at high temperature. The collimated holes glass filter (Bares and Lannis 1990) (Figure 12.7) is made by the proprietary technology (etch-resist) used in the manufacture of semiconductor wafers. Each of these filters has properties that may make them useful in special particle isolation endeavors; the analyst is advised to consult the manufacturer's literature for more specific information.

Figure 12.6. Sintered silver filter surface.

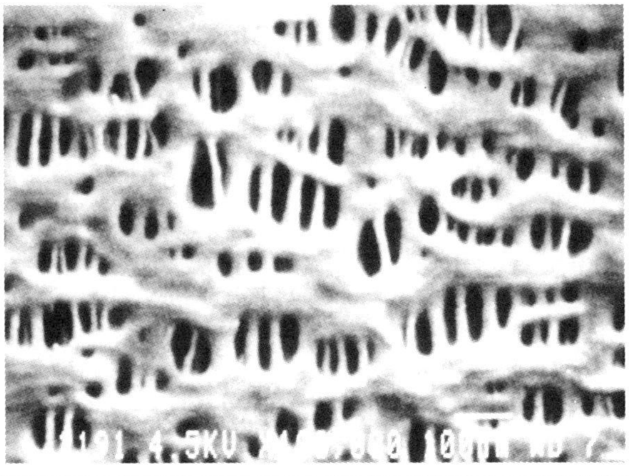

Figure 12.7. Collimated holes glass filter.

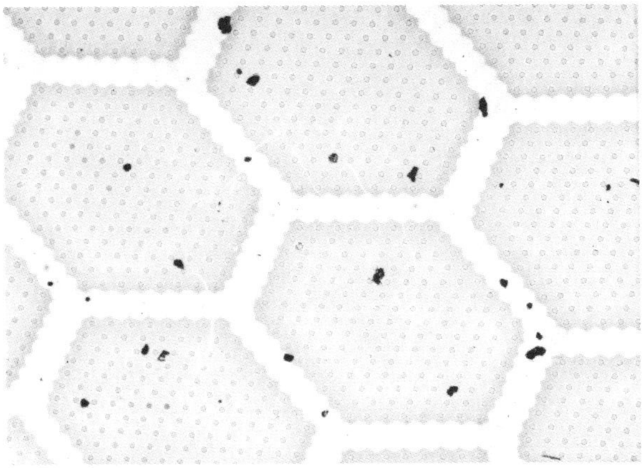

Polysulfone filters are presently offered in both the conventional hydrophobic and inherently hydrophilic types (e.g., Supor®, Gelman Sciences). The latter material, as well as being far stronger than conventional mixed ester membranes, also offers higher flow rates and does not require a surfactant for wetting. Polymeric microporous filters for analytical purposes are also made from acrylic copolymers, polyvinylidine difluoride, PVC, hydrophilic vinyl-acrylic copolymers, and hydroxyl-modified polyamide. "Depth" filters (Figure 12.8) are produced from a mat of fibers, most often paper, glass, or polyester. These filters are thicker than cast film

Figure 12.8. Depth filter (glass). Polyester binding material applied to the surface forms a coating on individual glass fibers.

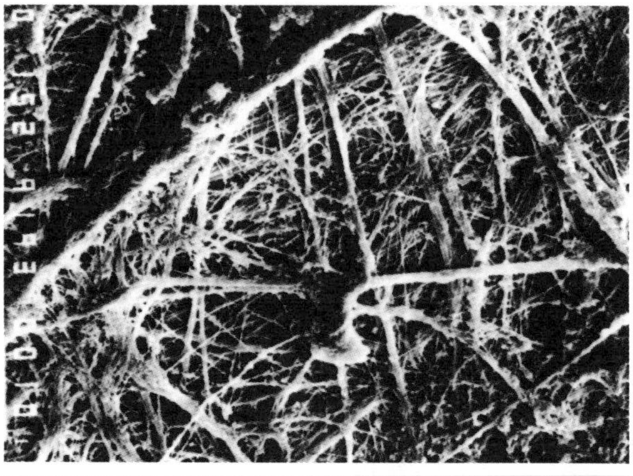

filters and, because of their tendency to entrap particles inside the filter matrix and their surface roughness, are less used for particle identification studies than other types.

Retention Ratings, Pore Size, and Porosity

The analyst must have a basic understanding of critical descriptors applied to filter media of different types: the retention rating, the pore size, and the void volume or porosity. The size of particles that the analyst can expect to be retained on a filter with some high degree of efficiency is expressed as the retention rating. Thus, a filter with a retention rating of 0.8 μm would be expected to retain nearly all particles of this size and greater that are contained in a solution to be filtered.

The difference in structure between cast film filters and track-etched (sieve) filters is dramatically illustrated by their retention characteristics (Figure 12.9). The cutoff or maximum retention is rapidly reached at the retention rating of the track-etched filter; the percent retention of the spongiform cast film filter increases slowly, beginning at a particle size far below the retention rating.

Pore size or pore diameter is directly related to the retention rating and the retention characteristics of sieve-type filters such as Nuclepore® or collimated holes glass filters. Obviously, due to the wide variation of pore sizes in cast film filters and the complex structure of this type of media, the pore size measurement is of less value in predicting performance. It should be emphasized that the pore diameters (typically stated in μm) in the manufacturer's literature for cast film filters are generally not values obtained by direct measurement with an electron microscope. They are typically theoretical values or values based on physical measurements (such as bubble point testing). Pore diameters specified are often mean values, and there is a range of sizes on either side of the mean. (The standard deviation of pore size is typically not specified by the manufacturer.)

Figure 12.9. Particle retention curves for filters of different types.

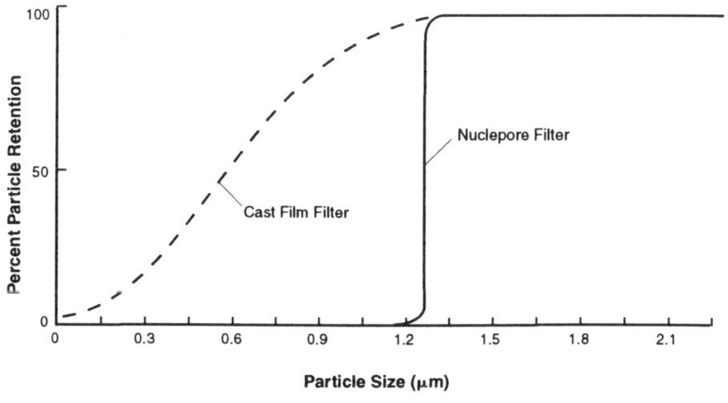

By selecting a filter material soluble in some organic solvent, all of the particles can be isolated for examination by other techniques. Some filters can be rendered transparent simply by the addition of a drop of a clearing liquid having the same refractive index. This is especially useful for rapidly scanning a series of samples, but it restricts the index of the mounting medium to about 1.51. A more satisfactory procedure for clearing a membrane filter is by dissolving it in a few drops of solvent on a microscope slide. On evaporation of the solvent, all of the collected particles are encapsulated in a solid, isotropic film of known refractive index. The particles can be examined directly with a polarizing microscope.

Filtration Apparatus

A wide variety of filter holders and filtration funnels are available for use with microporous filters. These differ in size, material of construction, and intended application. The diameter of the filter used is typically determined by the number of particles to be counted and/or the volume of fluid to be filtered. The most common filter diameters available are 10 μm, 25 μm, and 47 μm. The smaller diameters are desirable for the filtration of small volumes of liquid, but the analyst should remember that the density of particles on a membrane must be appropriate for the counting procedure to be applied. If only small volumes of sample are available and the particle number in the sample is high, dilution of the sample prior to filtration and use of a larger diameter filter membrane may be indicated. Trasen (1968a and 1968b) gives a useful description of filtration techniques.

Filter holders made of metal, glass, or plastic (e.g., polysulfone) are available. The plastic holders are lightweight, have adequate mechanical strength, are low in cost, and are easily cleaned. In applications where filters are to be frequently used and cleaned, a plastic filtration apparatus is often chosen. Whether metal, glass, or plastic filter holders are chosen, the equipment employed must be designed for the intended use (low particle shedding characteristics and smooth surfaces) and should be amenable to easy assembly and disassembly. When only aqueous solutions are being filtered, as is most often the case in the pharmaceutical industry, the filtration apparatus can be adequately cleaned by soaking it in a detergent solution, followed by exhaustive rinsing in a stream of distilled water filtered at a retention rating appropriate to the analysis to be performed.

Numerous filter holders for specialized applications are also available. If pressurized gas or process compressed air is to be tested, high-pressure metal filter holders that securely lock the filter in a housing and are capable of withstanding the pressures of the system to be evaluated are used. If very small airborne particles, such as fine fibers are of interest, conductive carbon-filled plastic holders may be obtained that eliminate the adherence of particles to the material of the holder. Filter holders of all types are available in vendor literature.

Millipore membrane filters for particulate monitoring are generally employed as disks with an underlying porous medium of sufficient strength to support the filter against pressure differentials achieved during filtration. Filter disks are clamped in the filter holders between parallel sealing surfaces or gaskets. Filter holders for use in contamination analysis can generally be divided into two types:

1. Reusable filter holders of stainless steel, glass, or plastic that are commonly used for liquids.
2. Disposable contamination monitors that are transparent, plastic filter holders preassembled with filters in place for air and on-line liquid particulate analysis. All monitors are assembled in a cleanroom environment to minimize background particulate contamination. For many monitors, the average background particle count has been determined and is marked on the outside of each package. Monitors are also available with matched weight membranes to eliminate the need for preweighing test filters in gravimetric analysis.

Filtration may involve hazardous materials (e.g., chemotherapeutic agents), operations, and equipment. It is the responsibility of whoever uses these procedures to establish appropriate safety and health practices and determine the applicability of regulatory limitations prior to their use. For flammable liquids, stainless steel must be used and should be grounded in accordance with the directions provided in the product insert.

Filtration Procedure

The necessary equipment for filtration consists of the following:

- Filter holder, funnel(s) with 100 mL capacity of polysulfone, glass, or other nonshedding material. Complete with flask and rubber stopper. Other sizes may be used when necessary.
- Vacuum source, capable of 25 in. Hg.
- Petri dishes or slides.
- Forceps of stainless steel with unserrated tips.
- Dispensing pressure vessel.
- Filtering jet solvent dispenser, 25 mm.
- Membrane filters, 25 mm or larger diameter, compatible with solvent.
- Cleaned solvents, filtered to a level not to exceed 10 percent of the requirements for the product being tested.

These items (not including the solvents) are available from Gelman and Millipore as well as other sources. By mentioning these vendors, the author does not imply any approval of the companies mentioned nor disapproval of any other firms with whom there has been no experience.

To perform filtration, the base of the funnel is assembled into the filter flask (appropriately shielded against implosion), and a vacuum hose is used to connect the flask to a vacuum pump or other vacuum source. A membrane filter that has previously been cleaned and blanked is placed on the base face up, the funnel is placed on the base, and the locking ring is engaged to clamp the membrane in position. Next, vacuum is applied slowly to the flask to draw the solvent through the membrane. When there is about 0.5 in. of liquid remaining in the funnel, the sides of the funnel are rinsed down to collect all particles. Then vacuum is applied again to draw all of the remaining fluid through the membrane until no liquid is visible on the membrane. Finally, the funnel is removed and placed in a beaker or other container to prevent it from touching the work surface. Using unserrated forceps, the membrane is carefully removed and placed in a precleaned petri dish and dried with the lid ajar in a HEPA airflow. When dry, it is covered with the lid to await analysis.

While filtration is, in theory, a simple technique to apply, there are many "tricks" that are known and applied by the experienced analyst. Careful selection of the membrane type and rinse solvent aids can make the initial isolation more or less selective. One quickly learns that inadequately cleaned glassware will contribute to the isolated material, and examination of the filter may not directly indicate the type of particulate matter that led to the original observation. Often, because of this failing, a novice analyst will erroneously identify extraneous particles as being related to the problem. This concern is addressed by ensuring (through preliminary microscopic examination) that a blank analysis or negative control is sufficiently clean prior to the sample isolation. The same sample glassware and filter type that will be used in the isolation should be used for the blank.

A second type of error may relate to a failure to examine the specimen isolate under the appropriate conditions. When evaluating membranes obtained from an unknown sample, the use of both episcopic (vertical) and oblique (low angle lateral) lighting is recommended as in the current USP 23 <788> test; otherwise, soft, semiliquid, or amorphous particles that become thinned on the membrane surface during filtration can be overlooked. The lateral illumination will allow the analyst to visualize only massive (equant) particles that cast a shadow, an observation that might lead the analyst to conclude that this species was responsible for the visible defect. Episcopic illumination will reveal the semisolid particle species that are closely conformed to the filter contours. Due to the amorphous nature of particles of this type, a large visible particle in the product unit may become a thin coating on the membrane and go undetected if care is not used in the microscopic examination.

Removal/Isolation of Particles from Filters

Once particles have been isolated on a filter, they should be removed manually using the direct particle isolation technique described above then analyzed according to the forensic scheme (light microscopy, thermal analysis, then infrared if warranted) (Aldrich and Smith 1995). Samples may be received on paper, glass fiber, or membrane filters. The particles on the filter should first be studied directly in reflected light; transmitted light may also be helpful in locating embedded particles. If there is a fair amount of sample on a filter, the particles can often be viewed in transmitted light by immersing a portion of the filter in a refractive index medium close to that of the filter (approximately 1.54, 1.52, and 1.51, respectively, for paper, glass, and membrane filters). The filter essentially disappears in the proper medium, and the particles nearly lie in a thin optical plane. It is difficult, however, to isolate individual particles from the oil for further analysis.

Static problems are often encountered with track-etched polycarbonate membranes. Particle removal can more easily be accomplished from these by laying the membrane on a glass slide and wetting the edge with a drop or two of water or appropriate solvent. If only a few particles are present, they may be removed with a tungsten needle to a clean microscope slide for further analysis. Sometimes a particle will not cling to a tungsten needle if it is adhered on a highly charged surface; this is especially true with membrane filters. To avoid this, a drop of collodion solution may be picked up on the tip of the needle. A single particle or a number of particles can be picked up with this drop in the 30 to 60 sec before it dries. If particles are embedded in the coarse meshwork of paper filters, the analyst can use a ball of rubber cement on the tungsten needle as collodion wets the paper fibers.

Importantly, filtration provides not only a means of isolating particles, but it can be used to concentrate particles in quantities sufficient for instrumental analytical techniques, such as crystallography or mass spectrometry. Filtration of an entire large-volume injection (LVI) unit volume or a pooled volume of multiple containers yields the entire particulate matter population greater in size than the membrane retention. If the desired analyte is present in homogeneous populations with few contaminants, this is a powerful technique.

In addition to being isolated on filters in the laboratory, particles are also often received from customers on paper, glass fiber, or membrane filters. The particles on the filter should first be studied directly in reflected light; transmitted light may also be helpful in locating embedded particles. Particles may be removed from filters by picking or scraping with a microtool, but the analyst must keep in mind that the scraping of particles from a filter will not necessarily result in a complete or representative recovery and may result in fragments of the filter material being included in the sample. Isolated particles are transferred to a clean microscope slide for further analysis.

Filtration techniques can also be used to isolate particulate matter to be analyzed by chromatography or other means of analysis requiring a liquid sample. This technique is most useful when the particles to be analyzed constitute a dominant mass component of a population, which can be isolated in large numbers on a filter of glass, Teflon®, polycarbonate, or other substrate in which the particles may be dissolved. Once the differential solubility of the particles is determined (such as a solvent that will dissolve the particles of interest but not others present), the filter bearing the particles is extracted by manual procedures or with the aid of ultrasonic or Soxhlet apparatus.

Particles can be easily removed from soluble membrane filters by dissolving all or a portion of the membrane in a centrifuge tube of solvent, usually acetone, and centrifuging the particles to the bottom of a glass centrifuge tube. The supernatant liquid is then pipetted off, the particles are washed once or twice with fresh solvent, and a drop of the concentrated suspension is evaporated to dryness on a microscope slide. A drop of mounting medium and a cover slip are added to complete the preparation.

Care must be taken in this procedure to avoid dissolving particles of interest or losing them during the decanting of the solvent. The removal of particles from filters may be effected by ultrasonic treatment, but complete removal of particles is not likely to be obtained. Paper filters should generally be avoided in collecting particles for microscopical study. When they must be employed, ultrasonics can sometimes be used to shake particles loose.

Pooled volumes of multiple units of injectables can yield a large mass of particulate matter for analysis. Careful selection of the membrane type and rinse solvent aids subsequent analyses. This is useful for a "clean" separation of all foreign matter and delivers a fully rinsed isolate (analyst's choice of solvent). However, the following precautions must be considered: insufficiently cleaned glassware may contribute to the filter isolate, and examination of the filter may not directly indicate which particulate species led to the defect observation. Often, an experienced microscopist will identify far more than the offending species; even worse, an inexperienced microscopist will identify (and not recognize) artifacts of the filtration procedure. The first of these issues is handled by ensuring that a blank analysis (negative control) is sufficiently clean prior to the sample isolation, using the sample glassware type and same filter type. The second concern requires experience on the part of the analyst; experience comes with time. Quite often, a seemingly large particle in the product fluid becomes a thin sheet on filtration and drying, enhancing the possibility of overlooking its presence. This occurs with amorphous or semisolid particles that flatten into a film or sheet on filtration and must be examined using episcopic rather than lateral top lighting. Static problems are often encountered with track-etched membranes. Particle removal can more easily be accomplished by placing the membrane on a glass slide and wetting the edge with a drop or two of water or appropriate solvent.

Collection of Particles from Environmental Air

Airborne particles may be counted microscopically and with particle counters. Particles collected on a filter may be isolated and identified using the technique discussed above with regard to solution-borne particulate matter. Particles may be collected from air in CEAs by the simple use of a vacuum pump attached to an appropriate filtration funnel. A 47 mm stainless steel filter holder is generally a good choice where large volumes of air or air with a high particle burden are to be sampled. The filter holder should be precleaned by the technique discussed earlier, and the filter precounted for background particles. For ease of use or for critical samples, transparent, disposable, aerosol contamination monitors are available. These monitors have been precleaned during assembly, and the average surface particle background count is supplied with each lot. The Millipore manual (1992) is available free of charge from Millipore Corporation (Bedford, Mass.) and includes particularly useful information on the analysis and enumeration of particles from the air and catalogs of the necessary equipment.

Large quantities of particles removed from the air can be quantitated by gravimetric means. By determining the flow rate of air through the membrane filter and the length of sampling time, the volume of air sampled can be calculated. After counting or weighing the fibers and particles on the membrane filter, the total number counted or weighed is divided by the total sample volume to give counts or weight per unit volume of air. Matched weight monitors, which eliminate the need for preweighing test filters, are available from Millipore for use in gravimetric analysis. Each of these monitors contains two superimposed filters matched in weight to within 0.1 mg.

Vacuum pumps used in aerosol collection should be able to draw enough vacuum to accommodate a range of flow rates from 1 to 150 L/min. Flow regulation, pump weight, and operating time are important considerations. Pumps must either be calibrated by a flowmeter or be used with a flow-limiting orifice. They should be capable of operating without overheating or changing flow rate over the long periods of time often required in air sampling.

Importantly, the flow rate of sampling must be controlled so that the total volume of air sampled can be determined. A simple way to control flow rate is to use flow-limiting orifices such as those available from Millipore. These are inserted into the outlet of either a Millipore 47 mm filter holder or a Millipore aerosol adapter used with a 37 mm contamination monitor. When a specified level of vacuum is applied, air flows through the filter and orifice at a constant rate. The amount of vacuum required to maintain the correct flow rate for each orifice available is listed in Table 12.1. The applied vacuum must be equal to or greater than the specified level.

Table 12.1. Vacuum Requirements for Function of Flow-Limiting Orifices (Courtesy of Millipore Corp.)

Orifice Flow Rate (L/min)	Minimum Required Vacuum (mm Hg)	(in. Hg)
1	300	12
2	300	12
3	300	12
4.9	400	16
10	500	20
14	550	22

As mentioned above, the filtrative collection of particles may be used either for gravimetric analysis or to collect particles for counting, identification, or other analytical methods. As an interesting example of the latter procedure, the particles of an antibiotic powder in the air of an aseptic filling area may be collected on a filter and assayed for antimicrobial activity on the basis of volume of air filtered. Thus, the level of the drug in the air may be related to different airborne count levels, and the numerical variability of drug particles relative to total count variation may be determined. If the process is consistent, the drug component of the total airborne particle count by this means can be shown to represent some constant proportion of the total count so that personnel exposure can be quantitated.

The choice of filters for air monitoring applications is dependent on the purpose for which monitoring is being conducted. Pore size and surface texture of a filter, and the analytical method to be applied, are important considerations in filter choice. In general, microporous cast film filters will efficiently retain airborne particles with a diameter that is much smaller than the stated pore size of the filter in the same fashion as HEPA filters do. For instance, a Millipore Type AA (0.8 μm) mixed ester filter will retain essentially 100 percent of all particles drawn onto it. Particles significantly below the retention rating will be collected within the depth of the filter, however, and will be unavailable for analysis. Nuclepore® membranes, as discussed, have a number of advantages for analytical use. Perhaps most notable is the smooth surface on which all particles above the retention rating may be readily detected by scanning electron microscopy. However, the low porosity of these membranes and resulting low flow rates may pose some difficulties for use in air monitoring.

Collection of Particles from Surfaces

Monitoring of large particles (10 μm) present in low numbers in CEAs may require special techniques. These particles are large enough to settle from air in areas of nonunidirectional flow. Surface contamination with large particles can be measured by light microscopic counting and sizing; for nonaccessible surfaces, one can use tape sampling methods. There are also video-camera-based techniques and scanning laser light scattering devices for this purpose. In electronics manufacturing, these latter devices may be used for evaluation of the product itself, such as silicon wafers or chips. If there are many very small particles < 5 μm on a hard, smooth surface, they may be scraped off with a clean razor or scalpel blade under the stereomicroscope. The residue that may be on the blade is removed with a tungsten needle and placed on a glass slide for optical examination. Small particles (< 5 μm) on a soft surface can sometimes be successfully removed with a polyethylene scraper (Figure 12.10). To make a polyethylene scraper, a piece of 1/4 in. tubing is simply cut at a 45° angle.

If a small number of particles are scattered over a surface and all of them are needed for further analysis, one can use a microvacuum cleaner as suggested by McCrone and Delly (1973a, 1973b) (Figure 12.11) to concentrate the sample onto a 10 mm membrane filter. Small, handheld, battery-powered, portable vacuum cleaners of the type currently marketed may also be used as the basis for a large-scale surface sampling device for larger particles. In this case, a filter holder is inserted in the intake assembly and fitted with a 45 mm filter of sufficiently high retention rating so that a low pressure drop is achieved and adequate vacuum is attained for pickup of particles.

By far, the most practical methods for monitoring larger particles on surfaces, however, is based on microscopic evaluation or instrumental analysis of "witness plates" that have been exposed to the same environment as the product. Although automated devices for reading witness plates are available, the manual light microscopic evaluation of witness plates is the most useful method for pharmaceutical production environments. The technique provides a simple and effective way of measuring large particle fallout and particles generated at process points.

This measurement is sometimes described as a "particle fallout count," but data are also collected for particles that deposit on surfaces in almost any orientation by various mechanisms in additional to gravitational settling. Data for deposited particles are reported in terms of mass or the number of particles deposited per unit surface area per unit time. In some

Figure 12.10. Polyethylene scraper (after McCrone and Delly 1973a).

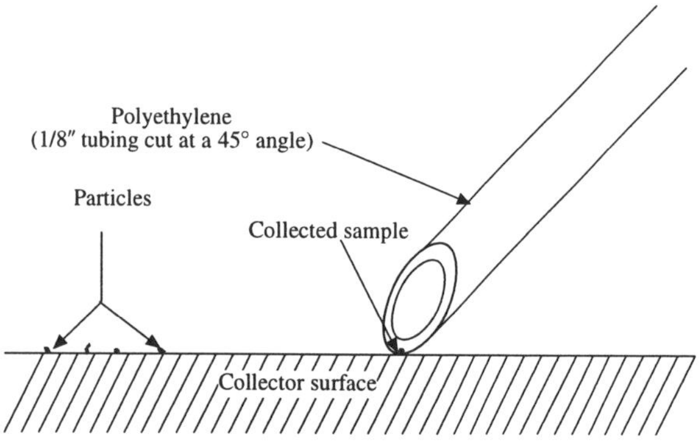

Figure 12.11. Microvacuum device for surface particle collection (McCrone and Delly 1973a, 1973b).

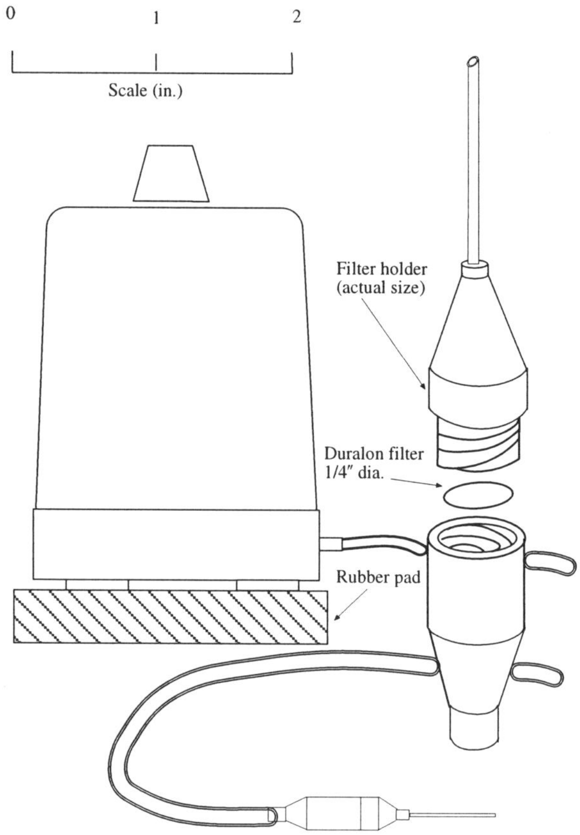

situations, a surface obscuration ratio may be used to define particle deposition.

A partial listing of materials appropriate for use as witness plates, depending on the particle size to be detected and means of measurement, is as follows:

- Microporous membrane filters
- Double-sided adhesive tape
- Petri dishes
- Petri dishes containing a contrasting color (black) polymer, such as polyester resin
- Photographic film (sheet)
- Microscope slides (plain or with evaporated metal film coating)
- Glass or metal mirror plates
- Semiconductor wafer blanks
- Glass photomask substrates

When selecting methods for particle collection and analysis, include the following considerations:

- The surface smoothness of the witness plate must be appropriate to the size of the particles that will be counted so that these particles are easily visible.
- The means of measurement employed must be capable of resolving and measuring the smallest particle to be enumerated. The witness plate material must be consistent with the type of equipment used for counting and sizing particles. Optical microscopes and image analyzers may use glass microscope slides, petri dishes, or transparent tape as suitable witness plate materials. Wafer surface scanners used for particle detection are typically designed to function as semiconductor wafer blanks or photomask substrates. With these detectors, the witness plate surface smoothness should have no defects large enough to give a signal as great as 10 percent of that caused by the minimum particle size to be measured.

For some clean rooms or clean zones, it may be necessary to characterize the environment in terms of particles ≥ 5 mm (macroparticles) that may be deposited on critical surfaces in the cleanroom. These large particles may not be readily measured by optical particle counter procedures. These large particles are most frequently generated by processes occurring at or near the point of emission into the cleanroom environment. They cannot be quantified by optical particle counters due to their adherence to surfaces or to the inability of the counter to entrain them in the sample tube airflow during the short time that they remain airborne.

The material from which the witness plate is constructed should be compatible with the type of equipment used for counting particles on surfaces. Microscopes and image analyzers may use silicon wafers, microscope slides, membrane filters, glass photomasks, or petri dishes as suitable witness plates. Automatic laser-based surface particle detectors are usually designed for silicon wafers but can be adapted for use with other optically flat surfaces. In choosing both the method and the instrument to be used, consideration should be given to its ability to detect particles in the range relevant to the product being manufactured. Other factors that may be taken into consideration include the time required for the analysis and the subjectivity of the method.

Microscopes can be set to detect a minimum particle size of 5 μm on the witness plate. A laser-based surface particle counter is recommended for particles of less than 5 μm, extending into the submicrometer range. A suitable calibration procedure must be established for all equipment to ensure that accurate results are obtained for each test. Manual microscope counts should be performed in accordance with ASTM (American Society for Testing and Materials) F 24-65.

The number and positions of the test points required to test the cleanroom adequately should be determined. Usually, witness plates should be placed near the points of the manufacturing process that are most subject to particulate contamination. Depending on the surface to be tested, the witness plate may be used either vertically or horizontally.

The witness plate should be held in position from its back surface by vacuum pressure. This is the simplest and cleanest method that will ensure surface integrity. Each witness plate should be electrically grounded by, for example, a pressure contact to the lapped (mirror polished) surface of the silicon wafer. If feasible, the witness plate should be recessed into the surface to be evaluated (table, floor, etc.) rather than permitted to lie on top of it.

Witness plate inspection should be performed according to the following steps (see IES-RP-CC018.2 or ASTM E 21.05.AUS-94):

1. Verify that all cleanroom systems are functioning correctly, in accordance with operational requirements.
2. Identify each witness plate and clean it as required to reduce the surface particles to the lowest level possible.
3. Perform a blank reading of the particle density, B, on each plate:

$$B = \frac{N_b}{A_w}$$

where N_b is the number of particles on the witness plate after cleaning but before exposure and A_w is the area of witness plate (cm^2).

4. Maintain 20 percent of the witness plates as controls. These must be handled in the same manner as the other witness plates until the test witness plates are placed in their measurement position. The controls should then be stored in such a manner that additional particles are prevented from settling on their surfaces.
5. Transport all witness plates to the test sites in such a manner as to prevent airborne particles from contaminating their surfaces.
6. Place the test witness plates in their test locations, taking extreme care not to add particles to the test surfaces during transfer from the container. Store the control witness plates in a particle-free environment.
7. Expose the test witness plates for a minimum of 4–48 h, depending on the type of cleanroom, mode of operation, and particle counting apparatus used. The exposure time should be adjusted, if necessary, to obtain a statistically valid reading that satisfies user requirements. It is important to ensure that each witness plate does not carry any significant electrostatic charge.
8. Collect the witness plates by reversing the setup procedure.
9. Perform a particle count on all witness plates, including the controls, N_t.
10. Determine the increase in surface particle density for each witness plate:

$$\frac{N_t - N_b}{A_w}$$

11. Average the values for the increase in particle density for the control plates.
12. Determine the net increase in the density for each witness plate by subtracting the average control density from the witness plate density. Divide the net density by the exposure time. This calculation yields a particle deposition rate in terms of particles per square centimeter per unit time.
13. $$PDR = \frac{(N_t - N_b)test - (\overline{N}_t - \overline{N}_b)control}{A_w t}$$

where PDR is the particle deposition rate, number of particles/cm^2/h; N_t is the number of particles on a witness plate following exposure; $(N_t - N_b)$control is the mean value of change with reference to the particle number on control plates; and t is the time of exposure (h). Report the mean PDR value and its standard deviation.

Adhesion and Entrainment

Another quick sampling procedure for surfaces uses transparent sticky tape (e.g., Scotch Magic™ 812). When applied to a surface and peeled off, loose particles adhere to the adhesive layer of the tape. This tape can then be applied to a clean microscope slide and labeled. The particles, sandwiched between the slide and the tape backing, are protected and can be studied microscopically through the tape backing. If necessary, the particles are removed by lifting the tape; applying a small drop of benzene, acetone, or other solvent; and using a fine needle to make a small ball of adhesive containing the desired particle. A larger portion of particles can also be removed in the same way, placed on a clean slide, and covered with a cover slip for more careful study.

A tape sample may be attached to a clean microscope slide while it is being transported to the laboratory for study. For quick examination, a tape sample may be placed adhesive side down on a microscope slide and observed directly, or a small drop of a refractive index liquid may be placed between the slide and tape in the area to be examined. As with particles collected on filters, particulate matter may be removed from the tape with a solvent. To accomplish this, the adhesive surface of the tape is exposed, and the tape sample is placed in a centrifuge tube with xylene or other appropriate solvent. When the adhesive has dissolved from the tape and has freed the particles, the backing is removed from the tube, and the particles are collected by centrifugation. The supernatant liquid is then pipetted or decanted off and the particles are washed in solvent before drying and mounting. Collections of settled particles in the field are often made by allowing dust to settle onto a thin Aroclor® film on a microscope slide. Particles adhere to the film and cannot be blown away.

A micro adaptation of the tape sampling technique may be practical for the isolation of particles for instrumental analysis by X-ray spectroscopy or other electron microscopic techniques. A 1 percent Formvar® or collodion solution in a volatile solvent is painted onto a group of particles to be collected. On drying, the film is peeled off the substrate with the attached particles and transferred to a carbon or plastic grid for placement in the electron beam instrument. This technique is useful when particles of $< 10\ \mu m$ are handled.

It takes only a few minutes to replicate relatively large surface areas using this technique. The very flexible 10 μm film can be smoothed out into a flat film on a glass slide so that all areas are in focus when examined with high magnification. These replicas show all details of surface irregularity. Most important, all loosely adherent particles are incorporated in the film and picked up with it, including the topographical relationships between surface detail and particles.

The following stepwise procedure is followed for surface replication and particle removal (McCrone and Delly 1973b):

1. The replicating solution consists of a 1–2 percent solution of Formvar® or collodion in ethyl acetate or other solvent. The solution must be thoroughly dissolved before use; it may be heated or left standing for 48 h to ensure dissolution. Some analysts use two solutions: one made up at 0.5 percent, which will give a thinner initial coating, and a more concentrated one to be applied as a backing coat strong enough to readily lift from the surface.
2. A thin layer (usually 1–2 μm) of the collodion solution is applied to the surface to produce cloudy replicas. A second layer of this solution may be necessary to produce a final film thickness of 10 μm. (Thinner films are fragile; thicker films are difficult to flatten and are, therefore, difficult to examine at high magnification.)
3. An incision is made in the film surrounding the area of interest with a cleaned razor blade or sharpened probe.
4. The cutout section of film is then slowly peeled from the surface with fine forceps. The film must be allowed ample time to separate from the surface. Pulling too rapidly will tear the film. A drop of water between the surface and film can be helpful in loosening the replica.
5. The film is now placed replica side down on a microscope slide and smoothed out gently with a membrane filter separator sheet or a piece of clean Saran Wrap® before covering with a coverslip.

This technique works well for polarized light microscopy, for which the film may be dissolved with solvent that is wicked out with a fragment of filter material and replaced with mounting media or refractive index oil. Particles entrained in the film may be transferred to a support for X-ray analysis or placed on a carbon grid for electron diffraction using a transmission election microscope.

SEDIMENTATION TECHNIQUES

Sedimentation allows gravity to do the work of particle collection for the analyst. In a very useful adaptation of this method, a combination of sedimentation and decantation allows particles observed in an LVI container to settle out and be concentrated in a small volume at the container bottom by decantation or by drawing the solution down by suction. Following this initial procedure, the settled material is resuspended, and the settling procedure repeated in smaller vessels until the visible particles are concentrated in a small amount (i.e., a few milliliters) of solution. At this point, the remaining solution with the concentrated particle population may be poured onto a "sponge" comprised of a large diameter cast film filter or a stack of two to three such filters. The filters will soak up the liquid, leaving the particles concentrated in a small area.

Sedimentation under normal gravity conditions is very slow for very small particles, but it can be increased by centrifuging the sedimentation device. Separation by centrifugation has an advantage over gravitational methods in that it increases the rate of particle separation. Placing an aliquot of the product in a clean centrifugation vessel will provide a separation of solid and liquid phases if needed. The technique is generally useful for larger amounts of materials or large pools of near-micrometer-sized solids. The concentrated material can then be evaluated according to the forensic scheme. For some fine particle materials, high speed may be necessary. Some materials can be spun "up" to the surface of a liquid if they are of lower density than the suspending medium, then collected by floating a hydrophilic (e.g., polypropylene) filter on the surface.

The most common commercial sedimentation device is probably the Andreason pipette (Figure 12.12). The pipette consists of a graduated sedimentation vessel (0–20 cm) that holds 500–600 mL when filled to the 20 cm mark. The settling medium is usually water, often with a wetting agent. A two-way stopcock is located just above the ground joint so that particles may be drawn into the bulb and then emptied into another container. To start, the entire system is agitated, and, as settling occurs, samples of the suspension are removed from the bottom of the apparatus at regular intervals.

Figure 12.12. Andreason pipette.

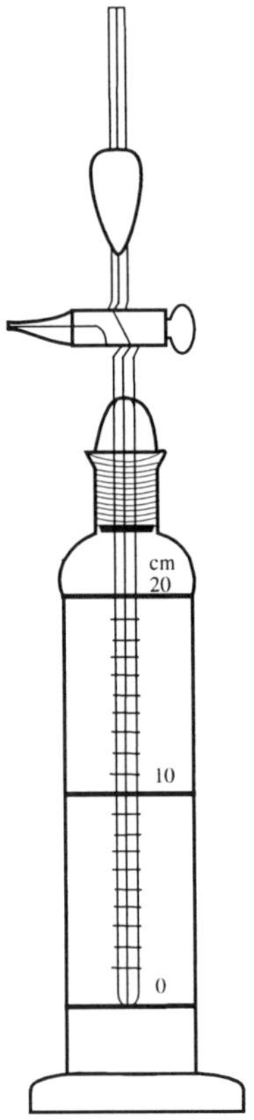

The collection of particles by stagnation techniques simply involves the creation of a zone devoid of air currents that can keep particles suspended (McCrone and Delly 1973a). Stagnation devices simply collect particles as they settle. The simplest stagnation collector consists of an open, wide mouth jar placed where it cannot be tampered with easily. It should be several inches across and deep enough so that wind cannot disturb the collected dust. Distilled water or an IPA:water mixture can be added to prevent reentrainment. Water-soluble particles and those that might react with water or dilute alkaline or acid solutions may be altered. Care should be taken that the jars are not placed too close to walls or furniture that might interfere with wind currents and result in a nonrepresentative sample. Remember that sedimentation methods are useful for collecting large airborne particles, but those in the submicrometer range settle very slowly and are easily reentrained; consequently, sedimentation is not the best way to obtain a sample representative of total suspended particulate matter. To isolate particles, the collected sample is simply vacuum filtered in an aqueous suspension through a membrane filter and the filter air dried.

A dynamic sedimentation technique described by Aldrich and Smith (1995) is termed "panning." Visualization of the defect is aided by the surrounding opaque drug suspension. Often, the material is difficult to locate once the suspension has been agitated, even on the trip to the laboratory. Reasonable success has been achieved by using clean, disposable, plastic, petri dishes, commercially prepared for microbiological cultures (such as the Falcon® 4005 petri dish). The sample is dispersed and poured into a waiting dish in a volume that nearly covers the plastic bottom; a deep fill will not suffice. The pan is then gently rolled about the central axis on a slight angle, 5–10°, that "pans" the suspension into a monolayer of solids and usually reveals the foreign material. The proper background must be chosen for the type of foreign material sought, and two or more differing backgrounds, including transmitted light, may need to be employed. Conducted at low power on a stereomicroscope stage, the isolated particles can be picked out with a tungsten probe, a fiber probe, a fine forceps or withdrawn with a capillary tube. The capillary is often most helpful for isolating the particle within a small amount of the suspension vehicle for later separation on a microscope slide or infrared disk.

Inertial Collection

Impaction utilizes moving particle inertia for collection. Flowing particles possess momentum that causes them to resist changes in direction. The effect is, of course, mass dependent. This property is utilized in two different types of particle-fractionating devices: impactors and cyclones. In both systems, every particle tends to slip toward the outside of a curved path, but particles of highest mass will slip most; small particles more closely follow the gas flow, and this path has the smallest radius of curvature. In both systems, particles of greatest mass, due to the high rate of gas flow and small radius of curvature, impact a solid surface, where they lose their forward momentum. In impactors, the particles remain (hopefully) at the point of impact; in cyclones, the particles fall (hopefully) into a container at the bottom. By causing the airstream to pass through regions of successively higher velocity and successively smaller radii of curvature, particles of successively smaller mass will be collected.

Impingers use momentum and the mass of moving particles to facilitate their collection. The Greenburg-Smith and its modifications are probably the most familiar impinger designs; they are recommended for use in EPA source testing. The Greenburg-Smith impinger (Figure 12.13) has a plate suspended at a fixed distance from the capillary tip. Just as for impactors, the collection efficiency for various size ranges is a function of jet velocity; consequently, the flow rate must be constant if reproducibility is to be achieved. During source testing, however, the flow rate, and hence, the jet velocity, are deliberately varied in order to maintain isokinetic

Figure 12.13. Greenburg-Smith impinger.

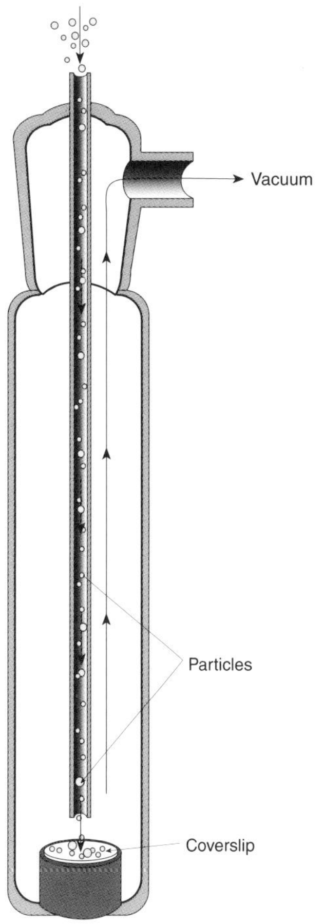

sampling. As it turns out, this is of little consequence, because the impinger follows a highly efficient filter. In this situation, the impingers only cool the hot gases and condense volatile substances.

An impinger can be used to collect particles efficiently down to less than 1 μm. Impingers are especially useful for sampling concentrated aerosols, because there is no significant sample buildup on the collection plate and, consequently, reentrainment is not a serious problem. On the other hand, it would be foolish to use impingers with water if it is suspected that the particles to be examined react with or, worse still, dissolve in water. Bear in mind, too, that gaseous components may form acidic or alkaline solutions that could attack the particles. There is, of course, no reason why water cannot be replaced with another less volatile and/or more inert liquid.

A novel air sampler developed by California Measurements Inc. (Sierra Madre, Calif.) allows small airborne particles to be collected directly onto stub mounts for examination with a scanning electron microscope. The design of this device is based on the inertial (cascade) impactors and classifiers used for environmental sampling (Figure 12.14). The device contains a vertical stack of circular plates with recesses into which the stubs are placed. The plates are placed in a cylindrical enclosure through which air is drawn from top to bottom by a vacuum pump. Based on the geometry of the airflow path over the stubs, smaller or larger particles are collected by inertial impaction. The result, after air is drawn through the device for an appropriate period of time, is a series of stubs bearing size-segregated particles that may be identified and related to a size range in which elevated counts were noted with a particle counter. The use of this device in combination with scanning electron microscopy provides a specialized but effective approach to identifying small particles present in the air of clean areas, such as an aseptic fill complex. The analyst is cautioned that in any clean area where Class 100 conditions are maintained, any sampler of this type will have to be run for prolonged periods of time to collect a representative sample of the particles present.

In general, samples should be taken as close as possible to the point where critical operations are carried out. The vacuum pump should be located away from the sampling area, preferably outside the controlled environmental enclosure itself. In general practice, the filter surface is placed vertically. The number of air samples required for a given area will be based on the floor area, interruptions to the airflow, and the room volume. Samples must be large enough to be measurable and representative. In general, for particle counting analysis with the optical microscope, proportionately more air is required for Class 10, 100, and 1,000 cleanrooms.

The commercially available aerosol analysis monitors discussed earlier are also very useful for the collection of aerosol particles and probably operate more according to inspection principles than filtrative ones. These typically consist of a clear polycarbonate cassette with a removable cover containing a 25 mm or 37 mm cellulose acetate filter. Ports are provided at the top and bottom of the housing for the application of a vacuum and the entrance of sample air. On obtaining elevated counts of airborne particles, monitors may be placed in the affected cleanroom or production area and used with a vacuum pump to collect particles for analysis for some specified period of time. As is the case with filters used to collect particles from solution, the filters may be examined directly with a light microscope or sections may be cut from the filter and used as samples for scanning electron microscopy or X-ray spectroscopy.

Figure 12.14. Cascade impactor for collection of fine particles (California Measurement, Inc.).

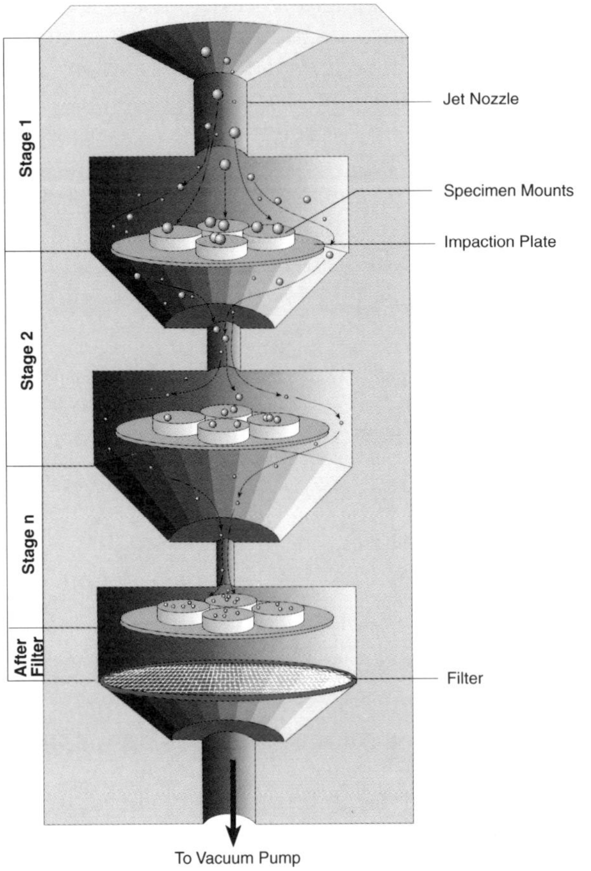

Sieving

Sieving is probably the oldest method of particle size separation, when adequate sample is available. It reportedly was originated by the ancient Egyptians. Sieving, using woven-wire sieves, is limited to about 74 μm dry and to about 30 μm wet. Smaller mesh, woven-wire sieves vary considerably in opening size. Electrolytically formed micromesh sieves are made to cover the mesh size range from 5 to 90 μm. The problem of clogging of the openings is solved by wet sieving and, for the smallest sizes, ultrasonic vibration in the liquid suspension during sieving. Ultrasonic frequencies of 40 kc/sec do not damage the sieve, and the openings do not clog. These finer sieves have a low throughput because the area of the openings is only 1.2 percent for the 5 μm sieve, 3.5 percent for the 10 μm sieve, 11 percent for the 20 μm sieve, and 25 percent for the 30 μm sieve. The micromesh sieves have, however, much more uniform openings than woven-wire sieves, ±2 μm over a sieve area of 20 × 20 cm and ±1 μm over smaller areas. They also have a flat upper surface. Micromesh® sieves (Buckbee-Mears Co., Minneapolis, Minn.) have been tested against other size fractionation procedures using both glass spheres and quartz.

The use of sieving for both particle isolation and sorting has decreased in recent years, due in large part to particle population analyzers. It remains a valuable separation technique for the particle analyst when visible-sized particles are of interest in difficult to filter materials such as viscous pharmaceutical dosage forms and plasma fractions. Allen (1990) and USP 23 (1995) include fairly detailed descriptions (see also Table 12.2) of sieving analysis.

Sieving, as a discipline distinct from filtration, is too often overlooked as a means of particle isolation. It has historically been used for quantitative analysis rather than characterizing or identifying particles. One key area of application involves the isolation of large particles (e.g., > 25 μm) from viscous solutions, such as concentrated carbohydrates, drug "mother liquors" or plasma, or plasma fractions. In some instances, particles shed by a blood collection device must be separated from blood, which is in itself a particulate suspension. (Handling of blood or blood components requires special restrictions regarding the area in which the material is handled and biohazard training for the personnel involved.) By judicious choice of the sieve size to be used and the sieving apparatus setup, the analyst can build a high-flow, reusable isolation mechanism tailored to a specific application. The Micromesh® sieves are ideal for such an application, as they are of stainless steel, with the screen durably sealed in a retainer ring; they are suitable for the exhaustive cleanup and repeated reuse that may be required in such applications.

Table 12.2. USP Sieve Sizes

Nominal Designation No.	Sieve Opening
2[1]	9.5 mm
3.5	5.6 mm
4	4.75 mm
8	2.36 mm
10	2.00 mm
14	1.40 mm
16	1.18 mm
18	1.00 mm
20	850 μm
25	710 μm
30	600 μm
35	500 μm
40	425 μm
45	355 μm
50	300 μm
60	250 μm
70	212 μm
80	180 μm
100	150 μm
120	125 μm
200	75 μm
230	63 μm
270	53 μm
325	45 μm
400	38 μm

[1]Designated as 3/8 in. in ASTM Specification E 11-70.

Chemical Isolation Technique

The principles of chemical isolation, based on differential solubilities, have been well known to chemists for over 100 years. The reader is directed to Aldrich (1990) and Aldrich and Smith (1995) for detailed discussions of novel and routine chemical techniques of particle isolation. The application window for chemical methods involves the occurrence of a material not amenable to isolation in sufficient quantity for analysis by physical means or dictated by the method of choice, such as X-ray crystallography, whereby particles must often be collected, concentrated, redissolved, and recrystallized. The material must exhibit a high solubility in an organic solvent immiscible with the liquid suspending the particles. Classical extraction of volumes of the product solution may be performed to concentrate the analyte. As an example, silicone or solid organic particles may be extracted into methylene chloride, which is then treated to allow the materials to crystallize or evaporated to leave the material is a concentrated residue. Solvent treatment may also be used to remove materials occurring as "frosts" on plastic or rubber parts. One application is solvent-soaked wisps of glass wool used to concentrate the particulate matter for further purification by dissolution and crystallization.

Included in this category are several techniques that are more closely related to chemical techniques than to the physical techniques historically used by particle analysts. Despite the term I have used, these techniques are not frequently applied by chemists. Conventional extraction techniques can be utilized for particle isolation/collection in cases where an immiscible liquid is present, as determined by light microscopic examination. Extraction techniques may be used for confirmation of the results of an instrumental analysis and also in analysis of trace contaminants that have been detected during a particle investigation. Material present as solid particles can be dissolved into an extracting phase, then recrystallized to give samples for X-ray crystallography. The extracting liquid must in all cases be clean and free of extraneous particles.

Chemical extraction is based on the affinity of a particulate material for one solvent over another and/or differential solubility. Physical affinity techniques can also be applied based on the affinity of some apolar liquid chemicals that exist as microdroplet dispersions in aqueous systems. A classical example is provided by polydimethyl siloxane oils used for lubricants, sealants, and coatings in medical devices and containers. If the presence of these or similar materials is suspected in an injectable, the material may be filtered through a membrane filter of a composition that encourages the preferential adsorption of the dispersed material. Teflon® filters such as Millipore's Mitex™ or nylon membranes will serve quite well to concentrate and collect the oil from the drug. The material may then be dissolved off the filter in a solvent, the solvent blown down, and the residue analyzed.

Conventional extraction techniques (carried out with rigorous attention to cleanliness) can also be utilized in cases where an immiscible liquid is present in an aqueous solution, as determined by light microscopic examination. In such cases, filtration of a solution giving high levels of LO (light obscuration) counts at > 2 μm and > 5 μm will result in the detection of levels of particles far too low to account for the high instrumental counts. Extraction techniques may be used as confirmation of the microspectroscopic evaluations and also in the evaluation of product components when searching for the source of a trace contaminant that has been identified in a particle investigation. Material present as solid particles can be dissolved into an extracting phase then recrystallized. The extracting liquid must in all cases be clean and free of extraneous particles. Aldrich and Smith (1995) describe a variation of this method called "zone drying" used to address "defect" incidents that may be due to immiscible oils and particulate aggregates, often consisting of multiple components.

This mode of separation may be the only successful direct method of analysis. Solutions that appear to be hazy may contain immiscible oils; this can usually be confirmed by the observation of suspended, tiny droplets with Brownian motion by direct light microscopic examination (at 400× magnification) of the hazy solution. The method for zone drying is as follows: The sample is placed on a waiting substrate (again, direct placement onto a salt plate for infrared analysis eliminates the need for a secondary transfer of the separated material) and then is slowly dried while viewing under the stereomicroscope. The drying process may be aided by introducing a nitrogen stream or even the analyst's directed breath onto the plate. Separation and, usually, coalescence of immiscible and relatively insoluble phases will occur. In an example described to the author, silicone oil had been observed on careful inspection of the vial as small immiscible fluid droplets, which to the unaided eye appeared as particles.

SAMPLING OF DEVICE PARTS OR CONTAINER COMPONENTS

In addition to collecting particles for analysis from solutions, the air, and process points, it may be necessary to collect particles from various physical objects, such as the components of medical devices, closures, or parts of process machinery. The method used for this purpose depends primarily on the size of the object to be analyzed and the presence or absence of any surface coatings in which particles might be adhered or entrapped. If the part is large enough to be readily grasped or handled, it may be held in the analyst's gloved hand and rinsed over a filtration funnel with a forceful jet of water from a gun-type solvent dispenser with an integral filter. Particle-free vinyl gloves are well suited to this operation, but other types of gloves (e.g., nitrile) may be used if care is taken to ensure that, after washing in filtered water, they add no significant numbers of background particles to the analysis. Larger parts may be rinsed by immersion or by use of a solvent jet from a pressure vessel while enclosed in a clean plastic bag of nylon or other material or in a cleaned plastic or glass beaker that serves to collect the rinse solution for filtration.

Although handling and collecting device components, closures, or other discrete articles would seem to be a very simple matter, it is worthy of some reflection. If the hands are to be used, appropriate gloves are required. Forceps of various types are often ideal; they must be cleanable, be composed of nonshedding material, and have a tip configuration such that the parts collected are not abraded or cut. Forceps used should be reserved for this purpose and not exposed to contamination. As with any tool used by human workers, contact transfer and touch contamination are concerns. Cleanroom and microelectronics trade magazines list vendors of many different types of forceps, swabs, and probes useful for clean sample handling.

If the parts to be tested are small, they may simply be submerged in a suitably clean wash fluid in an ultraclean glass beaker. Particles may then be removed by sonication or gentle manual agitation and isolated by pouring the wash solution into a filter funnel. It is important to remember that when particles are dislodged, they will settle rapidly and may be lost at the bottom of the collection vessel unless the solution is swirled to resuspend them and then poured off fairly rapidly. A second rinse of the collection vessel may also be used to effect a more complete collection of particles. Small objects may be rinsed in groups to collect larger numbers of particles. If particles are adhered to the test articles by oil or grease, aqueous washing may be ineffective in their removal. In this case, a detergent solution or a suitable solvent may be used to enhance removal. Freon® is a useful solvent if silicon oil or grease is involved. With the current restrictions on the use of fluorinated solvents, however, methylene chloride may be substituted.

In sampling the surfaces of parts or components, it is practically impossible to remove contaminating particles quantitatively for filtration through a membrane filter. Thus, it is important to follow an identical rinsing or flushing procedure for any given component or system each time the samples are analyzed. The analytical results for such sampling will not yield absolute counts of surface adherent particles, but they will yield meaningful and reproducible data on relative contamination levels.

The liquid flushing technique depends on fluids being passed through or over a part, component, or system in such a manner as to contact all critical surfaces. The effluent fluid is collected and passed through a membrane filter, impinging the particles on the membrane. The particulate contamination on the membrane is then subjected to microscopic analysis to size and count particles when specified, or to identify particulate matter using various microscopic techniques. In some companies, the collected fluid is not filtered but is introduced into an LO particle counter for sizing and counting. The disadvantage of this technique is not being able to identify particulate matter as needed to define problem areas during manufacturing or during the failure analysis of malfunctioning components.

If filtration is used to collect particles washed from parts, the procedure is basically the same as described earlier. In practice, the tested part is washed in the fluid in the filtration funnel or jet washed over the funnel. The part to be tested is held inside the filter funnel with nonmarring forceps and is flushed with a stream of solvent from a pressure vessel. After all critical surfaces have been contacted, the part is removed from the funnel and placed on a freshly cleaned surface to await results of the analysis.

The so-called "water-break" tests have been used for years by manufacturers of optical equipment, lenses, and other items. This test is mainly used to determine the presence of organic films (e.g., hydrocarbons) but can also locate particles on a surface if they are 5 μm or larger. The technique is subjective in that it depends almost solely on the judgment and experience of the operator and the purity of the water. The analysis is performed by positioning the item to be tested at a 45° angle in a clean tray. Starting at the top, slowly pour water over the entire surface. If all of the water drains off the surface without leaving any spots or streaks, the surface is considered to be clean of particles or films. If particles or oils are present, the water film will break in spots.

SAMPLING OF CLEANROOM GARMENTS

There is an understandably high level of interest in the testing of cleanroom garments and wiping materials. The pharmaceutical manufacturer most often has an interest either in validating the low particle shedding characteristics of garments purchased or verifying the effectiveness of washing when nondisposable woven fabric garments are used. While a brief discussion of testing methodology and particle collection from this source is included here, the pharmaceutical manufacturer should consider requiring the vendor of cleanroom materials and garments or washing sources to perform particle testing rather than performing these analyses in-house. This opinion is based primarily on the specialized and time-consuming nature of this analysis. A number of appropriate test methodologies are given in ASTM and IES (Institute of Environmental Sciences) standards.

Generally, fabric materials may be tested either wet or dry. Logically, wiping materials are more frequently tested using a solvent, and garments are most often tested dry. The simplest dry test consists of clamping sections of the material to be tested across the opening of a filtration funnel so that clean air may be drawn through the fabric and a filter by means of a vacuum pump. Per ASTM F 51-68, 5 μm filtered air is drawn through 5 designated 0.01 ft^2 areas of a single thickness of the garment fabric at a rate of 14 L/min for 1 min per area. Loose particulate contaminants on or in the garment are impinged on the surface of the filter that is then examined microscopically to determine the number of particles (> 5 μm) removed from the garment. For garment monitoring per this standard, it is customary to count and tabulate particles in only two categories:

1. All particles with the major dimension greater than 5 μm
2. Fibers (longer than 100 μm with a length-to-width [aspect] ratio exceeding 10 to 1)

Other methods are also available that test the garment or fabric of interest under dynamic conditions. For dry testing, the Helmke drum test is applied by many vendors of cleanroom gloves, masks, wipers, and garments (Figure 12.15). Samples to be tested are placed in a rotating drum and are tumbled to release particles. The air in the drum is drawn through a sampling probe, and the particles ≥ 0.5 μm released from the samples are counted by a light-scattering particle counter. A Helmke drum apparatus or equivalent (as specified in IES-RP-CC003.2 [1997]) is placed in a HEPA–protected hood, and the speed of the drum rotation is adjusted to 10 rpm ± 0.1 rpm. Alternatively, the air from the drum may be drawn through a filter cassette placed in-line between the drum and vacuum pump and analyzed microscopically.

Yet another method of particle collection from fabric items in a dry state is based on flexing the test article. In this procedure, the stress is imparted to the wiper by a flexing device, the design of which is based on the Gelbo Flex Tester (as described in IES-RP-CC004.2 [1997]). The unit consists of a metal and Plexiglas® enclosure, housing two opposed, cylindrical heads, one stationary and the other attached to a rotating and reciprocating motor-driven arm. When activated, the movable head approaches the stationary one and rotates approximately 180° in the course of its full stroke. A return of the head to its resting position completes 1 cycle. The rate of flexing is 1 Hz.

A sample is mounted in the flexor by wrapping and clamping the ends of the wiper around the heads that are in their extreme open position. Optimally, the width of the section of fabric tested will correspond to the circumference of the flexor heads, yielding a cylinder of the cloth (the edges of which just abut) between the heads. The distance between the heads may be altered by sliding the stationary one on its supporting shaft in the axial direction and locking it

Figure 12.15. Helmke drum garment test.

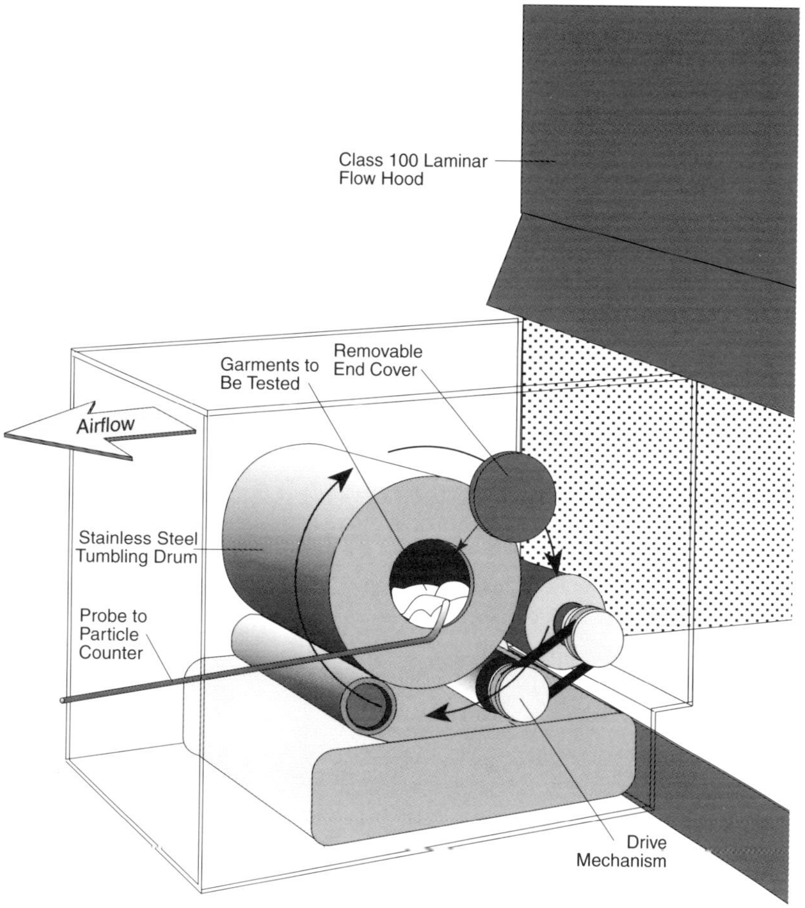

in place by means of a set screw. The particles generated by the flexed wiper are collected by an airborne count sampling probe located beneath the wiper. The counter should sample air at the rate of 1 CFM (ft^3/min) from the chamber in which the wiper is stressed. Thus, if the sample is stressed for 1 min (60 cycles), 1 ft^3 of air will have been drawn from the chamber, and the particles it contains counted. The duration of flexing of the sample may be adjusted depending on the physical characteristics of the wiper, its intended use, and the magnitude of the stress associated with that use.

It has been the author's experience that dry testing significantly understates the number of particles that might actually be released when a wiper is used in a wet condition. The existence of cohesive forces and static charge, present in all wipers but to different extents, holds particles in place and allows only a portion of them to be removed in the dry state. It has been estimated that approximately one million times more energy would have to be imparted to a dry wiper to remove the same burden of particles as are removed by testing in water. When testing is performed on a wetted sample, these forces are dissipated, and the release of particles from a wiper more closely approximates what will occur during actual use.

In the simplest application of wet testing, the fabric to be tested is clamped in a filtration funnel apparatus, and a specified volume of water is drawn through the cloth by a vacuum and collected in an ultracleaned vacuum flask. The solution containing particles from the sample is then filtered for microscopic analysis. Another method makes use of a biaxial shaker of the type commonly employed to mix paint in hardware stores per IES-RP-CC004.2 (1997). The test fabric is placed in a 4 L jar with 600 mL of filtered water. The jar and its contents are shaken for 5 min; then the particles may be isolated by filtration. A similar method (IES-RP-CC005.2 [1997]) is applied to gloves.

Since the magnitude of the cohesive forces adhering particles to the bulk fabric varies with the composition of the material, and since these forces are eliminated by the characteristics of water in a wet test, one would expect that different particle generations from the wipers would result in the wet and dry tests. This, in fact, is the case. Typically, fabric with high percentages of synthetic constituents yield relatively low dry particle counts. On wet testing, however, these wipers produce comparatively high counts. On the other hand, those wipers containing cellulosic components give relatively high particle generation in the dry test and a low level of particles in the wet test. This is only an empirical rule of thumb, as the particle generation on any material will be contingent on other factors as well, including finish of cut edge, handling, and the presence of binders, surfactants, and so on.

ON–LINE SAMPLING

Process liquids may contain contaminants, such as resin particles or microorganisms, picked up in the purification process. Particles can also be picked up from components such as valves and from jointing compounds. Boiler feed and sterilized cooling water may contain contamination arising from particulate matter in makeup water and/or soils present in the system prior to startup. Other sources of contamination are corrosion and mechanical wear.

As is the case in sampling compressed and other gases from the transport lines, liquid sampling must be conducted with three concerns: (1) obtaining a representative sample of the materials in the lines based on fluid flow and particle sedimentation characteristics, (2) being certain that the sample point valve does not add or subtract particles to or from the sample, and (3) use of appropriate clean and nonshedding collection vessels.

Direct sampling, where samples are filtered directly from a fuel or water line, is, outwardly, a simple procedure. Direct sampling can be used to eliminate the possibility of introducing extraneous contamination from the sample containers. (See the 1992 Millipore handbook for a range of applicable methods.) If the liquid is under pressure in excess of 10 psi, samples may be filtered directly at the sampling point by means of a disposable contamination monitor. The sampler/monitor assembly is simply plugged into a quick-release sample valve installed in the system. The line operating pressure forces liquid through the assembly. The initial volume is directed through a bypass to flush the sample coupling. The sample volume then is passed through the monitor into a graduated container attached to the sampler base. After sampling, the test filter may be sent to the laboratory for particle counting or gravimetric analysis. For gravimetric analysis, the control filter method must be used. This method involves placing two preweighed filters, one above the other, in the holder through which the test sample is passed. The lower filter serves as the control.

The best sampling valves provide a straight fluid path when open, such as ball or plug valves. They should be flushed clean when first installed. If connected at a tee in the system, the valve should be oriented upward or horizontally so that any sediment in the fluid stream will not settle into the tee and valve.

In dynamic systems, sampling valves should be located at points where bends or changes in direction or pipe dimensions create turbulence. If possible, they should be installed just before the most contamination-sensitive components in the system.

In static systems, such as storage tanks, the sampling valve should be located at the outlet connection. In the case of drums and similar fluid containers, the conventional stopcock or drain valve will usually suffice for sampling purposes.

Precleaned bottles or other containers cleaned as described earlier in this chapter can also be used for sampling fluid lines. These systems should be operated for several minutes before taking the sample. This ensures that contaminants are evenly distributed throughout the system. The following procedure is used:

1. With the system operating, open the sampling valve and allow sufficient liquid to flow into a waste container to flush out the valve. The sample must never be taken while the sampling valve is first being opened.

2. Remove the cap from the sample bottle and hold it in a free hand.
3. Immediately place the bottle into the liquid stream and collect the desired volume. Do not "rinse" the container in the sample fluid prior to collection.
4. Remove the container and replace the plastic film and cap.
5. Label the container.
6. Turn off the sampling valve after removing the bottle.
7. Avoid wiping the sampling valve or the neck of the bottle with a cloth or paper towel, since this may introduce fibers into the sample.
8. The sample bottle should be returned to the laboratory promptly for filtration and analysis.

POWDER SAMPLING

Bulk pharmaceutical chemicals (bulk drug raw materials) are currently coming under a higher level of inspectional scrutiny than in previous years; much of the material used in the United States to manufacture sterile drug products is produced abroad, calling for an increased level of inspectional effort and enhanced vigilance on the part of the Food and Drug Administration (FDA).

From the perspective of an FDA review chemist, impurities in drug substances present a series of problems that an applicant should have addressed in the New Drug Application (NDA) submission: finding, quantitating, isolating, and identifying impurities. This sequence leads to proposals for impurity limits by an NDA applicant, and FDA decisions on the specifications—both methods and limits—for impurities in the drug substance. FDA decisions on limits are reached jointly with the pharmacology reviewer and may involve the medical review staff as well.

Impurities may be present as particles as well as in soluble chemical form. Field audits indicate that this concern over impurities is now extended to include particulate matter in both crystallized and lyophilized bulk drug. A specific example of particulate impurities is provided by the presence of amorphous particles. These materials are specifically called out in USP <788> as showing an indistinct, gelatinous, or semisolid appearance. In a bulk drug material, these materials may represent an impurity such as a low polymer of the drug indicating that degradation has occurred due to the presence of moisture or residual solvent. Worse yet, they may be indicative of ongoing degradation that might continue after aseptic filling or mixing of the final product. This material may be detected in some cases by microscopy at levels below the sensitivity of chemical analytical technique.

Hence, the microscopic assay can serve as an important descriptor of drug quality with regard both to degradants, impurities, and extraneous particles. In the latter case, the presence of large numbers of extraneous contaminants may lead to questions regarding the chemical purity of the drug. There is also the possibility of impurities in particulate form, which are chemical variants of the drug substance (e.g., enantiomers, diastereoisomers, geometric isomers, etc.). Sampling of bulk drugs and other pharmaceutical powders calls for special techniques; the analyst must always be mindful of the hazards of potent or potentially toxic materials. In no case should a material be sampled or tested without first becoming familiar with the Material Safety Data Sheet (MSDS) information for the product.

Bulk powders present special problems in sample preparation. The essential principle to be observed in collecting samples of a bulk powder material is that of ensuring that the sample contains particles of all sizes in the correct concentration proportions. The analyst working with powders must realize the difficulties involved and understand the ways in which the segregation of different particle sizes in a powder is likely to occur. The sampling procedure must then be designed to minimize the effects of segregation. Sampling of powders is described in some detail by Allen (1990).

Here is a reminder to the analyst regarding the cleanliness of the sampling devices and equipment. Most powder sampling devices were originally designed to obtain representative samples of the powder itself. When pressed into service to collect contaminant particles for analysis, their cleanliness must be ensured by rigorous cleaning. The analyst must also be aware of the extraneous particle burden that can build up in containment hoods used to handle drug powders. The best procedure, if possible, is to perform only the initial collection of powder in this containment area and to dissolve the powder in a clean vessel so that it can be removed from the isolation area for further processing, such as filtration.

Adequate sampling is particularly important when a 50–100 kg mass of process material must be reduced stepwise to a gross sample of only a few kg, a lab sample of 50–100 g, and a measurement sample of a few milligrams. Generally, the number of fines will decrease with the extent to which air is mixed into the sample. Particles segregate when powder is poured into a heap or into a collection vessel.

In another commonly occurring scenario, the analyst will be asked to collect samples from a drug powder batch that may consist of a number of 50 kg (or larger) drums. This is not a simple task. The powder in the drums has most certainly segregated by size during shipment. A way must be found to sample throughout the volume of powder without introducing contamination. Maintenance of sterility is an overriding consideration when sterile bulk drug powders are to be sampled. For this reason, samples for particle analysis may be collected by the powder manufacturer and shipped separately to the laboratory. This practice is, in general, acceptable, but assurance must be obtained that the sample supplied is representative both with regard to particle size distribution and levels of extraneous particles.

Scoop sampling (the simplest method) consists of plunging a scoop or other collection tool into the bulk material and withdrawing a sample. This method is likely to result in errors since the whole volume of the sample is accessible to the sampling device, and the sample is typically taken from the surface where it may not be typical of the mass. Large particles that have segregated at the top of a container during shipment will be disproportionately represented in the scoop sample. Similarly, sampling at the lower levels of a large container may result in a collection of more fine particles than would be present if the container were thoroughly mixed. Thus, sampling by means of a scoop or spatula is generally unsuitable for collecting a representative sample. A better method is the use of a "thief" to sample from various depths within a large container. A laboratory scale version of this device is shown in Figure 12.16. Larger models are available for use on drums of powder.

The coning and quartering method of sample dividing (Figure 12.17) consists of pouring the material into a conical heap and relying on its radial symmetry to give four identical samples when the heap is flattened and quartered with a metal cutter. This method will give reliable results if the heap is symmetrical about a vertical axis, and if the two cutting planes coincide with the axis of the heap. In practice, the heap may not be symmetrical, and symmetry of

Figure 12.16. Miniature powder "thief" or sampling spear.

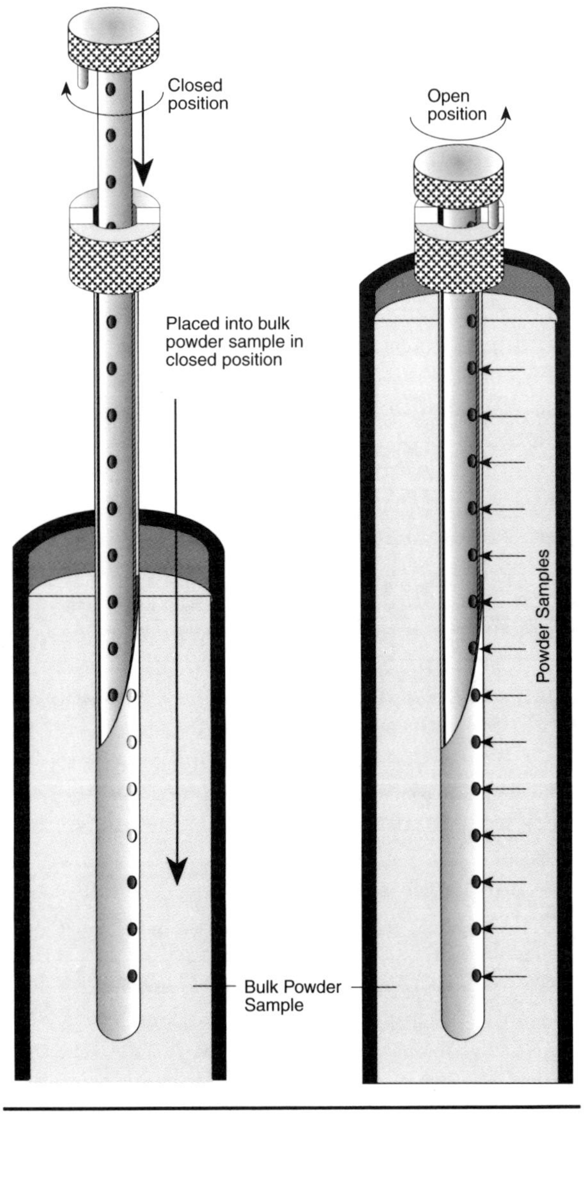

Figure 12.17. Cone and quarter method of sample collection.

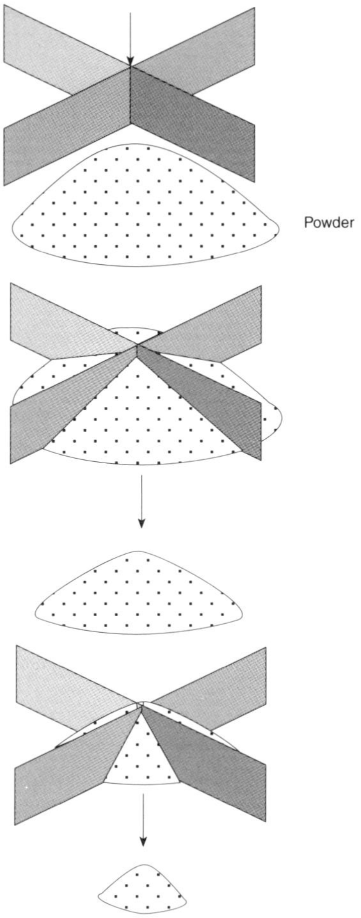

cutting is difficult to achieve. Vertical size segregation occurs in forming the heap, and any departure from symmetry in the cutting will lead to differences in the size of the four portions into which the heap is cut. The method is very dependent on the skill of the operator. If coning and quartering is possible, the quantity of material available is usually such that it can be separated using a device (such as a chute splitter shown in Figure 12.18), in which the sample is split into half volumes of decreasing size until the working sample is isolated.

The thief, chute splitter, and cone and quarter methods can be used in applications where small numbers of samples are to be analyzed, and the workload is insufficient to justify purchase of a dedicated powder sampling apparatus. If samples of powder material are to be routinely analyzed, specialized, motor-driven sample mixing devices (such as a spinning riffler, a rotary sample divider, or an oscillating sample divider) may be used to obtain uniform and reproducible sampling (Allen 1990).

The standard deviation (SD) for the number of particles counted in specific size ranges varies for the different methods of sampling. Generally, variability decreases with regard to two primary rules of sampling powders:

1. Most reproducible sampling is obtained when the mass of powder to be sampled is in motion.
2. The best sample for analysis is comprised of many samples of smaller mass taken from the total volume of the bulk material.

Figure 12.18. Chute splitter.

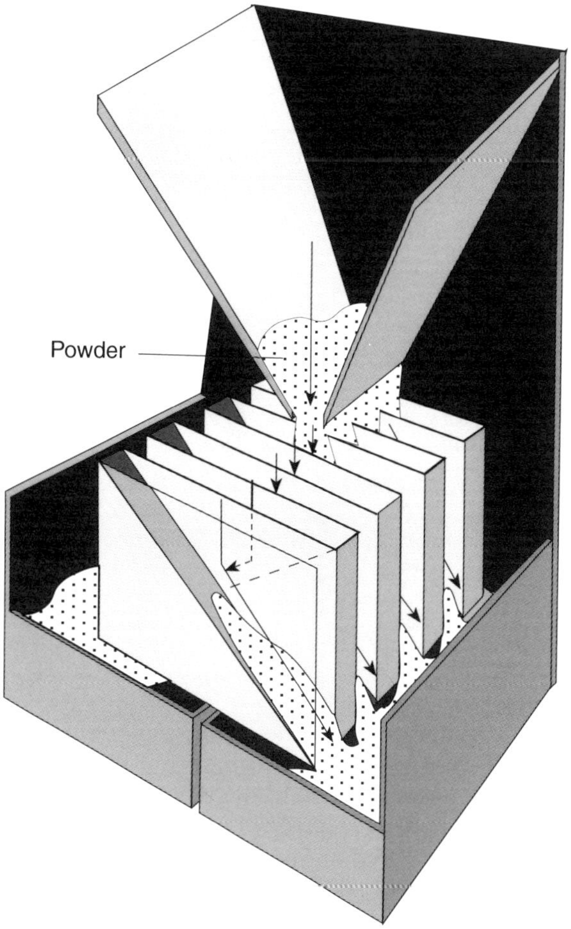

Measurements based on cone and quarter or scoop sampling approximates 10 percent. A chute splitter can result in as low as 1–2 percent variation (SD). The spinning riffler can result in samples for which a 0.1 to 0.3 percent SD is attainable.

Once the powder sample is reduced to a workable size, the second difficult problem facing microscopic examination is the preparation of a slide containing a uniformly dispersed, representative sample of the powder. Several methods of accomplishing this are available, but the final result often depends more on the skill and technique of the analyst than the procedure used. If a permanent slide is not required, a useful procedure is simply to place a small sample of the powder on a microscope slide and add a few drops of some dispersing fluid. Some analysts mix the powder in the fluid with a flexible spatula; others roll it with a glass rod. Both procedures may produce fracture of the particles. A preferable alternative is to use a small camel-hair brush. Further dispersing fluid may then be added until the concentration is satisfactory. A drop of the suspension is then placed on another slide with the brush, and a cover slip carefully placed so as to exclude air bubbles. For a semipermanent slide, the cover slip may be sealed with amyl acetate or glue. Silicone oil and glycerol are two other satisfactory dispersing fluids. Some powders may require the addition of a dispersing agent or surfactant to eliminate aggregation.

Another acceptable method for the production of slide mounts is to place a small representative sample of the powder to be analyzed in a 10 mL beaker, add 2 or 3 mL of a solution containing 1–2 percent collodion in amyl acetate, stir vigorously, and place a drop of the resulting suspension on the still surface of distilled water in a large beaker. The film produced as this drop spreads and evaporates may then be picked up on a clean microscope slide for examination. A powder material may also be dispersed in a volatile liquid in which it is insoluble, and a drop of the resulting suspension transferred to a slide. When the particulate sample is dry, a drop of mounting medium is placed over the particles that may be further dispersed as necessary by sliding a cover slip over the drop of mountant in a circular pattern.

SUMMARY

A wide variety of particle isolation and collection methods is available to the pharmaceutical particle analyst. There is also a wide variety of commercial equipment available for separating suspended particulate matter from the air and collecting it on some surface or in a container from which it can be recovered for analysis. The choice of equipment should depend on the purpose for which the sample was collected. Frequently, however, samples come to the laboratory for quantity or size identity. It is always important to obtain as complete a history of the sample as possible so that the impact of sampling conditions and procedures on the composition can be evaluated.

The complexity of the isolation/collection process generally is directly related to the complexity of the substrate or suspending medium and inversely related to particle size. Specifically, some particles, particularly those found in drug solutions are "etherial" in nature and difficult to isolate or collect in a form amendable to analysis. In this category are particles that are amorphous or semiliquid and those that are loose agglomerates of fine particles or subject to dissolution on agitation or filtration.

The most commonly used analytical path involves the isolation of particles from solution by filtration onto a microporous filter followed by isolation of the particle(s) of interest using a fine-pointed transfer probe. Once the particle is isolated, it may be analyzed by polarized light microscopy, X-ray spectrometry, or other instrumental means. Specialized methods are also available for the collection of particles from surfaces, containers, process points, garments, and the atmosphere.

A key critical issue is to obtain a particulate matter sample representative of the problem to be solved. Sample integrity and traceability are essential. Further, the question being asked by the project submitter is essential: Is enumeration, characterization, or identification required? Clean transport of samples must also be ensured. The best process here is to test and standardize on a single type of appropriate sample container—one for dry samples, another for liquid.

The experienced particle analyst appreciates the fact that the individuals who submit samples relative to a particle problem can often provide the solution to the problem if the analyst asks the right questions. The location of the particles of interest, circumstances of their occurrence, and their visual appearance often provide the analyst with at least a general idea of the source and identity. An understanding of the process by which an affected solution or device is made is essential to the analyst in this process.

REFERENCES

Aldrich, D. S. 1985. Particulate formation as a result of packaging. In *Proceedings of the European Conference on Visible and Subvisible Particles in Parenteral Products,* pp. 261–279.

Aldrich, D. S. 1990. Identification of parenteral particles. In *Proceedings of the PDA International Conference on Particle Detection, Metrology and Control,* pp. 269–282.

Aldrich, D. S., and M. A. Smith. 1995. Pharmaceutical application of infrared microspectroscopy. In *Practical guide to infrared microspectroscopy,* edited by H. J. Humecki. New York: Marcel Dekker.

Allen, T. 1990. *Particle size measurement,* 4th ed. New York: Chapman and Hall.

Bares, D., and M. Lannis. 1990. The potential of sieve filters for particulate determination. In *Proceedings of the International Conference on Particle Detection, Metrology, and Control,* pp. 467–501.

Beeson, R. D., and E. R. Crutcher. 1983. Hardware cleaning and sampling for cleanliness verification and contamination control microscopy. Technical Bulletin 72-4. Mt. Prospect, Ill., USA: Institute of Environmental Sciences.

Borchert, S. J., Abe, A., Aldrich, S. D., Fox, L. E., Freeman, J. E., and White, R. D. 1986. Particulate matter in parenteral products: a review. *J. Paren. Sci. and Technol.* 40:212–239.

Brock, T. D. 1983. *Membrane filtration: A user's guide and reference manual.* Madison, Wis., USA: Science Technology Institute.

DeLuca, P. P., and S. Bodapatti. 1980. Guidelines for the identification of particles in parenterals. *FDA Guidelines* (July) No. 3.

Delly, J. G. (1996) Microscopy: The setup and operation of the polarized-light microscopy lab for particle identification. In *Liquid and surface-borne particle measurement handbook,* edited by J. Z. Knapp, T. A. Barber, and A. Lieberman. New York: Marcel Dekker.

Elford, W. J. 1933. The principles governing the preparation of membranes having graded porosities. *Transactions of the Faraday Soc.* 33: 1094–1106.

Fleischer, R. L., P. B. Price, and E. M. Symes. 1964. Novel filter for biological materials. *Science* 143:249–250.

Kesting, R. E. 1971. *Synthetic polymeric membranes.* New York: McGraw-Hill.

Kesting, R. E. 1977. Asymmetric cellulose acetate membranes. In *Reverse osmosis and synthetic membranes,* edited by S. Sourirajan. Publication No. 15627. Ottawa, Canada: National Research Council of Canada, pp. 89–111.

Kesting, R. E., A. Murray, K. Jackson, and J. Newman. 1981. Highly anisotropic microfiltration membranes. *Pharm. Tech.* 4:53–60.

Marshall, J. C., and T. H. Meltzer. 1976. Certain porosity aspects of membrane filters: Their pore distribution and anisotrophy. *Bull. Parent. Drug Assoc.* 30:214–225.

McCormack, J., J. E. C. Harris, and H. J. Sullivan. 1976. Single particle characterization by optical microscopy and associated techniques. *Proc. Analyst. Div. Chem.* Soc. 13:344–348.

McCrone, W. C., and J. G. Delly. 1973a. *Principles and techniques, Vol. I: The particle atlas,* 2nd ed. Ann Arbor, Mich., USA: Ann Arbor Science Publishers.

McCrone, W. C., and J. G. Delly. 1973b. *The particle analyst's handbook, Vol. IV: The particle atlas,* 2nd ed. Ann Arbor, Mich., USA: Ann Arbor Science Publishers.

McCrone, W. C., and J. G. Delly. 1973c. *The particle atlas: Light microscopy,* vol. II, ed. II. Ann Arbor, Mich., USA: Ann Arbor Science Publishers.

Michaels, A. S. 1976. Synthetic polymeric membranes: Practical applications: past, present and future. *Pure and App. Chem.* 46:193–204.

Millipore. 1992. *Detection and analysis of particulate contamination.* Bedford, Mass., USA: Millipore Corp.

Oles, P. J. 1978. Particle analysis and identification in the pharmaceutical industry. *Microscopy* 26: 41–48.

Schmidt, W. H. 1964. Control and analysis of particulate matter by membrane filtration. *Bull. Paren. Drug Assoc.* 18 (6):25–31.

Sladek, K. J., and T. J. Leahy. 1981. Retention of bacteria by membrane filters. Annual Meeting, Society of Food and Dairy Microbiologists. 4–9 October, in Montreal, Canada.

Sokol, M., and J. Boyd. 1968. Sampling techniques in analysis for particulate matter. *Bull. Paren. Drug Assoc.* 22:9–12.

Teetsov, A. S. 1977. Techniques of small particle manipulation. *Microscope* 25:103–113.

Trasen, B. 1968a. Detection and reduction of particulate matter in pharmaceuticals. *Bull. Paren. Drug Assoc.* 22:1–8.

Trasen, B. 1968b. Membrane filtration technique in analysis for particulate matter. *Bull. Paren. Drug Assoc.* 22:1–8.

Zierdt, C. H. 1979. Adherence of bacteria, yeast, blood cells, and latex spheres to large-porosity membrane filters. *Applied and Environ. Microb.* 38:1166–1172.

Specifications and System Standards

Air Force Technical Order 00-25-203. 1992. *Contamination control of aerospace facilities.* Hdqtrs, Wright Patterson AFB, Ohio.

APHA. 1977. *Methods of air sampling and analysis,* 2nd ed. Washington, D.C.: American Public Health Association.

ASTM F 24-65. 1983. *Standard method for measuring and counting particulate contamination on surfaces.* Philadelphia: American Society of Testing and Materials.

ASTM F 318-78. 1996. *Sampling airborne particulate contamination in clean rooms for handling aerospace fluids.* Philadelphia: American Society for Testing and Materials.

ASTM F 312-69. 1992. *Methods for microscopical sizing and counting particles from aerospace fluids on membrane filters.* Philadelphia: American Society of Testing and Materials.

ASTM F 51-68. 1989. *Standard test method for sizing and counting particulate contaminants in and on clean room garments.* Philadelphia: American Society of Testing and Materials.

ASTM E 20-68. 1987. *Analysis by microscopical methods for particle-size distribution of particulate substances of subsieve sizes.* Philadelphia: American Society for Testing and Materials.

ASTM F 25-68. 1995. *Sizing and counting airborne particulate contamination in clean rooms and other dust-controlled areas designed for electronic and similar applications.* (Equivalent to SAE ARP-743 except only two particle size ranges are counted.) Philadelphia: American Society of Testing and Materials.

ASTM F 71-68. 1988. *Using the morphological key for the rapid identification of fibers for contamination control in electron devices and microelectronics.* Philadelphia: American Society of Testing and Materials.

ASTM F 154-76. 1994. *Identification of structures and contaminants seen on specular silicon surfaces.* Philadelphia: American Society of Testing and Materials.

ASTM F 311-78. 1992. *Processing aerospace liquid samples for particulate contamination analysis using membrane filters.* Philadelphia: American Society of Testing and Materials.

ASTM F 314-70. 1989. *Identification of metallic and fibrous contaminants in aerospace fluids.* Philadelphia: American Society of Testing and Materials.

ASTM F 490-77. 1989. *Microscopical sizing and counting particles on membrane filters using image shear.* Philadelphia: American Society of Testing and Materials

Federal Test Method Standard No. 791a. 1989. Method 3009-T: *Lubricants, liquid fuels, and related products.*

Federal Standard 209E. 1992. *Cleanroom and work station requirements, controlled environment.* Washington, D.C.: General Services Administration.

IES-RP-CC018.2. 1997. *Cleanroom housekeeping—operating and monitoring procedures.* Mt. Prospect, Ill., USA: Institute of Environmental Sciences.

IES-RP-CC003.2. 1997. *Garments required in clean rooms and controlled environmental areas.* Mt. Prospect, Ill.: Institute of Environmental Sciences.

IES-RP-CC004.2. 1997. *Wipers used in clean rooms and controlled environments.* Mt. Prospect, Ill.: Institute of Environmental Sciences.

IES-RP-CC005.2. 1997. *Recommended practice of clean room gloves and finger cots.* Mt. Prospect, Ill.: Institute of Environmental Sciences.

National Aerospace Standard, NAAS 1638. 1994. *Cleanliness requirements of parts used in hydraulic systems.* National Aerospace Standards Committee, Aerospace Industries Association of America.

MIL-STD-1246C. 1994. *Product cleanliness levels and contamination control program.* Philadelphia: Dept. of Defense Standardization Documents Order Desk.

SAE ARP-598. 1991. *Determination of particulate contamination in liquids by the particle count method.* Aerospace Recommended Practice. Warrendale, Penn., USA: Society of Automotive Engineers.

SAE ARP-743. 1991. *Procedure for the determination of particulate contamination of air in dust controlled spaces by the particle count method.* Aerospace Recommended Practice. Warrendale, Penn., USA: Society of Automotive Engineers.

XIII

APPLICATION AND IN–USE TESTING OF HEPA FILTERS

Air filters and filtration devices of other types are critical to controlling particulate matter levels in manufacturing and in product; they are a key component of a process that has been validated for acceptably low levels of particulate matter and to provide adequate sterility assurance levels. Attaining cleanliness in pharmaceutical manufacturing areas through the removal of particles from the air demands air filtration at a high level of efficiency. This is accomplished using a variety of types of air filters with varying particle removal efficiencies. The filters are typically used in a linear arrangement with "coarser" or lower retention rating filters placed upstream of filters with a higher retention rating. In this arrangement, the air becomes progressively cleaner as it passes through a series of filters on its way from the blowers of the ventilation system to the process area.

The final filters used in the ventilation system supplying air to a pharmaceutical or device process area are periodically tested for their ability to retain particles. These tests are referred to as penetration and/or leak tests. From the time a filter is manufactured until a cleanroom is certified, a filter may be tested multiple times. The testing adds value for end user requirements, specification, and Good Manufacturing Practice (GMP) compliance. Test data are also used to improve the manufacturing process, to control contaminants from people and raw materials, and ultimately to ensure product cleanliness.

The types of air filters most commonly used and test criteria most often used are described as below, in order of increasing retention rating:

- *ASHRAE* (American Society of Heating, Refrigerating, and Air-Conditioning Engineers) *medium efficiency (85 percent) filters:* These filters are challenge certified using 72 percent standardized test dust (fine), Molocco black (carbon), and 5 percent #7 cotton linters (fibers). The particle size distribution of the test dust is as follows:
  - 0–5 μm: 39 ± 2 percent
  - 5–10 μm: 18 ± 3 percent
  - 10–20 μm: 18 ± 3 percent
  - 20–40 μm: 18 ± 3 percent
  - 40–60 μm: 9 ± 3 percent

  In testing, 70 mg/m^3 of the test aerosol is dispersed into the inlet section of the test fixture. Dust passing through the test filter is collected on a final filter for gravimetric analysis. The amount of dust fed and collected on the final filter is determined by weighing, and the penetration of the test filter is calculated as the ratio of dust collected to dust fed.

- *95 percent efficiency DOP (dioctyl phthalate) filters:* These filters are characterized by a lower initial pressure drop than high efficiency particulate air (HEPA) filters, a larger dust-holding capacity, and a minimum efficiency of 95 percent on 0.3 μm particle size. They are sometimes referred to generically as "hospital grade" filters. They are often used in hospital operating rooms, food-handling operations, bioventilation systems, and in pharmaceutical and microelectronics facilities for less critical applications in higher classification areas.
- *HEPA filters:* The HEPA filter is defined in IES-RP-CC001.3 as a replaceable, extended-media, dry-type filter in a rigid frame having a minimum particle collection efficiency of 99.97 percent for 0.3 μm thermally generated DOP (or specified alternative aerosol) particles, and a maximum clean-filter pressure drop of 2.54 cm (1.0 in.) WG (water gauge) when tested at rated airflow capacity.
- *ULPA (ultralow particulate air) filters:* These are single-use, extended-medium, dry-type filters in a rigid frame having a minimum particle-collection efficiency of 99.999 percent (that is, a maximum particle penetration of 0.001 percent) for particles in the size range of 0.1 to 0.2 μm, when tested in accordance with the methods of IES-RP-CC007.1.
- *Super ULPA filter:* An extended-media, dry-type filter in a rigid frame made with filter media that have a minimum particle collection efficiency of >99.9999 percent (that is, a maximum particle penetration of < 0.001 percent) for particles of the most penetrating particle size (MPPS) when tested at the same velocity as the average media velocity in the filter in accordance with the methods of IES-RP-CC021.

The ULPA filter is rare in the medical/pharmaceutical products industry. The level of air cleanliness attained with this type of filter is generally not required, even in isolator systems. The HEPA filter in process areas is capable of supporting much better than Class 100 conditions; at this level of air cleanliness, process-generated particulate matter becomes the overwhelming consideration. This source of particulate matter is not amenable to control by air filtration. The 95 percent DOP and ASHRAE filters are widely used in higher (i.e., less clean) classification (e.g., Class 10,000 to Class 100,000) where adequate control of airborne particulate matter may be achieved at a much reduced cost for testing, initial purchase, and power consumption (a consideration sometimes overlooked) than with HEPA filters. ASHRAE filters are often applied as prefilters, with HEPA filters downstream.

At the outset, the reader is alerted to the evolving nature of the technology for the application and testing of high efficiency air filters. In the United States, the Institute of Environmental Sciences (IES) standards (Recommended Practices) such as IES-RP-CC001.3 in this area are under revision, ISO (International Organization for Standardization) 14644-3 contains revised information on filter testing, and work has gone forward in Europe to change classification methods for air filters and eliminate the historical designations (HEPA, ULPA, etc.) (see Wepfer 1986, 1995). This evolution of technology, coupled with the changing regulatory environment makes it necessary for those involved with HEPA filter application and testing to keep close track of the developing standards.

The reader is also warned regarding the limited scope of this chapter. Only the testing of and application of HEPA filters is dealt with in any detail in this chapter. This is appropriate due to the almost exclusive application of this type in the pharmaceutical and device industries. Whereas in other industries, particularly microelectronics, the increasing requirements for cleanliness have driven the development and application of filters of ever-increasing efficiency, pharmaceutical products manufacturers have retained the use of HEPA filters and simply advanced the test and use technologies related to this type. The result of these refinements in technology has been increased air and environmental cleanliness without significant changes in the design of the filters used. This approach is eminently practical. Whereas in microelectronics removal of particles of ever-decreasing size is the concern, the particles of most critical interest in the manufacture of pharmaceuticals are viables; these particles have not changed significantly in size during the history of pharmaceutical manufacture. The emphasis in this industry both on the part of regulators and manufacturers is thus appropriate.

The focus regarding control and removal of particles in environmental air is on process control, the total protection afforded the product by pressure gradients, airflow directionality, and air filtration with HEPA filters, rather than emphasis on filtration efficiency.

As a further note, the text of this chapter emphasizes the testing of HEPA filters using an oil aerosol challenge and aerosol photometer detection of leaks and quantitation of aerosol. Other methods have seen limited use in pharmaceutical manufacturing controlled environment areas (CEAs). This is understandable based both on Food and Drug Administration (FDA) expectations in the United States and the historical perspective and large database for this method. Other methods are available, most notably the use of polystyrene latex (PSL) test particles applied in combination with a laser optical particle counter (OPC) for detection. This alternate is generally more intricate in application and time-demanding than the photometer method.

HEPA filters are used extensively in commercial applications where it is necessary to remove airborne particles of 0.01 μm to 20 μm in size. Applications usually can be classed as product protection, personnel protection, or biohazard containment. HEPA filters are manufactured in an assortment of sizes and airflow capacities and can be made from a variety of materials. The efficiency of HEPA filters (99.97 percent) is for the "most penetrating" particle size. This latter elusive performance descriptor is influenced by the composition of the filter medium, the velocity of the air passing through the medium, and other factors. It is necessary to monitor the operating parameters of air handling systems that contain HEPA filters, due to the dynamic nature of such systems. As filters load with particulate matter, their efficiency increases. Eventually, the resistance to airflow (pressure drop) will increase to a point where it is necessary to replace the filter. The filters must be tested periodically for leakage due to damage to the medium or due to air bypassing the filter seal, which is the subject of this chapter.

In the pharmaceutical industry, HEPA filters are most commonly used in what is referred to as a "terminal" installation (i.e., the filter is located at the point where the clean air enters the cleanroom or CEA. Critical production (e.g., fillers) may be surrounded by individual Plexiglas® enclosures that are supplied with air through a dedicated filter in close proximity to the process point to be protected.

For some other applications, HEPA filters are installed in a housing or filter bank in a remote air handler or section of ductwork. These filters are known as in-line or ducted HEPA filters. For pharmaceutical applications, the standard filter size is approximately 24 in. × 48 in. and have a rated capacity of 5,000 to 10,000 CFM (ft^3/min). The filters have a corrugated, pleated, or folded paper filter medium. The depth of the filter "pack" may range from 2 in. to 6 in. The pleats in the medium may be separated by aluminum separators, or the filter may be of the minipleat design that uses strips of medium or creases in the filter pack to separate the pleats. This construction allows for an extended filter surface within a compact filter frame. An 8 ft^3 filter may have 300–600 ft^3 or more of filter surface area; because of the corrugated construction, an airflow velocity through the filter medium of 2–5 FPM (ft/min) will result in an air velocity at the face of the filter in the range 90 to 100 FPM. The terminal HEPA filters used in pharmaceutical manufacturing areas are often installed in a 2 ft ceiling grid and connected to the room supply air system with a flexible duct.

Periodic testing of HEPA filters in critical applications for integrity (absence of leaks) is generally believed by the FDA and accepted by manufacturers to be essential to verify their function (LaRocca 1985). In the United States, the most widely used in situ challenge (leak test) procedure involves the use of a cold DOP aerosol generated upstream of the filter and an aerosol photometer to scan the filter face for leaks. Problems with the cold DOP method include nonuniform application of the challenge, questionable physiologic properties of the material used, complexity of the test procedure, and DOP's tendency to degrade filters and decrease performance. This test method is more than 35 years old, and developments in HEPA testing technology within the last decade offer possible alternate approaches to testing filters in pharmaceutical applications.

The reader should keep in mind an important philosophical point regarding the application of in-use testing of HEPA filters. Leak tests, which have some value for filters used in laminar flow cleanrooms, are of far less usefulness if filters are used in turbulent flow installations, such as high classification production areas. The dogma relating to field testing for leaks suggests that a point leak in a filter used in a laminar (directional) airflow application can result in a focused, concentrated stream of contaminant particles being carried for some distance downstream of the filter face. This particle stream is, hypothetically, a source of potential contamination for critical process components, such as open vials about to be filled in a low classification (e.g., Class 100) area. While the author is skeptical of this depiction, it has wide acceptance. In nonunidirectional airflow areas, the air is constantly mixed by the turbulent nature of the circulation, and any particles passing into the area through a point leak will be of little consequence due to the mixing and dilution effect.

Periodic testing is not widely applied in other industries, such as aerospace and microelectronics, that require more stringent air cleanliness levels than does the manufacture of injectables and other pharmaceutical materials. This suggests that the pharmaceutical manufacturer may be pursuing a cost-without-benefit procedure due to regulatory and/or internal requirements that lack sound technical justification. This chapter provides (1) a consideration of HEPA filter function; (2) a review of currently available test methods; (3) a discussion of their applicability in pharmaceutical manufacturing environments; and (4) an examination of possible alternates to the DOP challenge that may be more suitable for application in the pharmaceutical and medical device industries. The references are divided into two sections: literature references cited in the text and standards and sources.

PRINCIPLES OF HEPA FILTRATION

The term *high efficiency particulate air*, although broadly applied, is actually a misleading, simplistic description. As we have already seen, the term includes a wide variety of filter types manufactured to comply with a generic set of performance criteria. Importantly, current deliberations of ISO Working Groups and national standards organizations may result in new definitions of air filter performance and revised nomenclature (see Wepfer 1995). The discussion in this chapter deals with HEPA filters in typical form, and the reader is referred to standards (some of which are listed at the end of this chapter) and vendors of filters and media for more specific information.

The conventional HEPA filter medium is made of nonwoven glass fibers of small diameter oriented randomly throughout the depth of the filter medium. The filter medium is produced by generating a suspension of glass filters in an aqueous suspending medium, then "wet laying" this suspension onto a metal support (usually a fine mesh belt) on which it can be pressed out into a thin layer and air dried. The fibers are commonly held in place by a small amount of a binder material of acrylic or other polymer.

The morphology of HEPA filter media may differ significantly based on the manufacturer and grade of filter involved (Figures 13.1a and 13.1b) (Liu et al. 1983); in the author's experience, it is difficult to relate the apparent density of the filter medium to performance, since depth or thickness is an important consideration. A mat with more "fine" (submicrometer) fibers will sometimes show a higher retention efficiency. The function of the assembled filter is governed not only by media morphology as shown in the figure but also by physical construction, such as the depth of pleats in the filter, the density of the pleats, and so on. The HEPA filter medium functions as a depth filter and does not have a controlled pore size; spaces between fibers are typically much larger than the particles captured. As air moves through the filter, entrained particles contact the fibers or adhere to other particles that have already stuck to the fibers. When a particle contacts either a fiber or a previously adhered particle, electrostatic or Van der Waals force attractions are established between the fiber and the captured particle (Tillery 1987). A HEPA filter has a minimum average efficiency of 99.97 percent when challenged with a thermally generated, DOP aerosol with a mean particle size of 0.3 μm at a concentration of 10 μg to 100 μg of DOP per liter of air.

Three primary mechanisms—impaction, diffusion, and interception—cause small particles to bump into a capture point (Liu et al. 1985; Liu and Lee 1982), as shown in Figure 13.2. In the process of capture by impaction, particles that have sufficient momentum (i.e., enough mass) proceed in a straight line as the air flows around fibers and are impacted onto the fiber surface. In the process of capture by diffusion, smaller particles that move about randomly touch capture points. The third mechanism of capture occurs when a particle strikes a fiber on a grazing trajectory (interception). A fourth (less important)

Figure 13.1. Morphology of HEPA filter material based on measurement of media in sheet form:

a. Medium with 99.99 percent retention efficiency for 0.3 μm DOP particles.

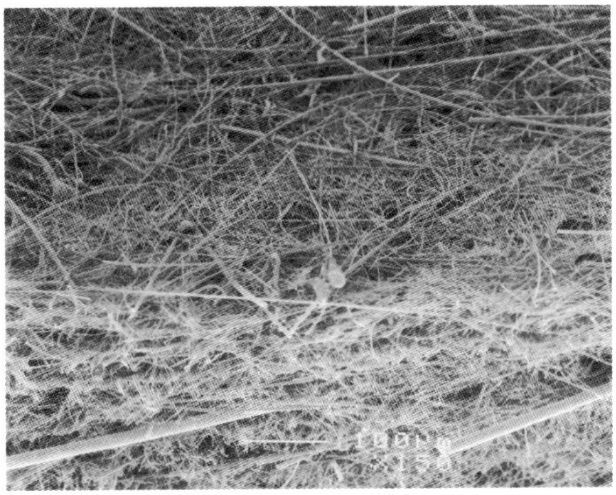

b. Medium with 99.97 percent retention efficiency for 0.3 μm DOP particles.

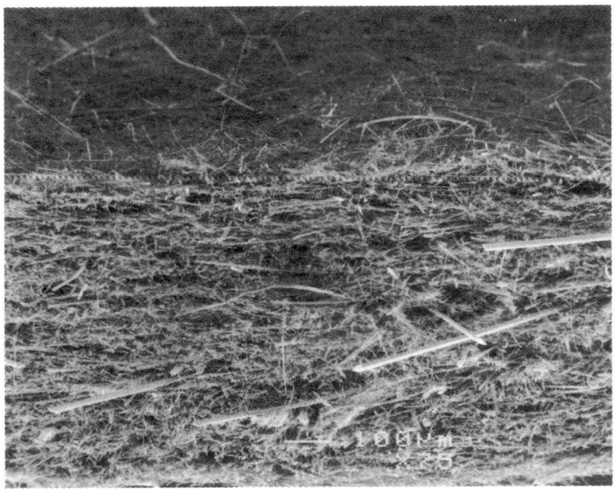

Figure 13.2. Capture of particles by HEPA filter medium.

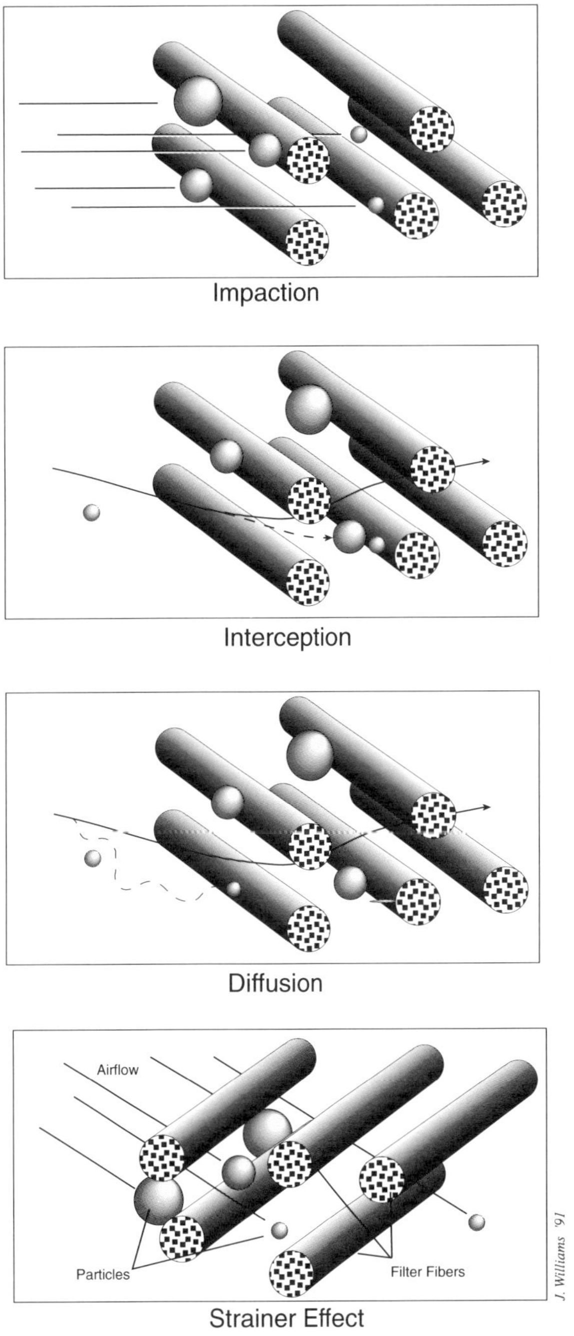

mechanism of filtration—screening or straining—can also occur in a HEPA filter. Straining implies that the spaces between the fibers are smaller than the particles that are captured.

Because larger particles are captured by impaction and smaller particles by diffusion, and because the total capture efficiency for a given particle size is the sum of the effects of diffusion and impaction for that specific size, there is (theoretically) one particle size that is most difficult to capture. Twenty-five years ago, it was believed that HEPA filters were least efficient for the 0.5 μm particle size. As more sensitive equipment for measuring particles was developed, the theoretical, most penetrating particle size decreased. Some 10 to 12 years ago, a particle measuring 0.3 μm was thought to be most

Figure 13.3. Penetration of HEPA filter media by particles of different sizes.

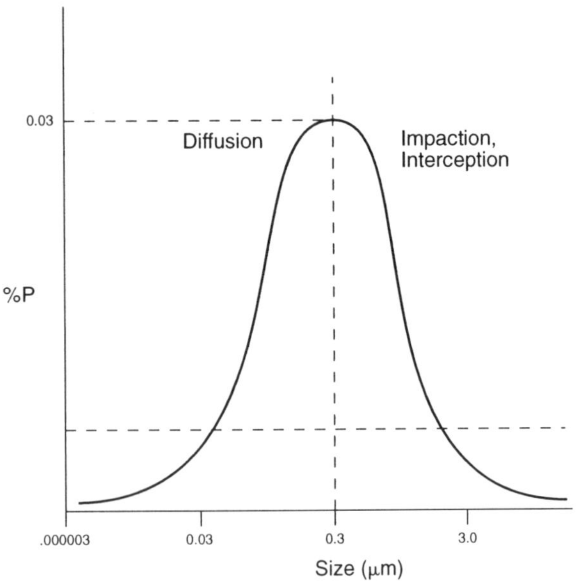

difficult to capture (Figure 13.3). The estimate was subsequently decreased to 0.2 μm; most recently, laser particle counters suggest 0.12 μm to 0.15 μm to be the most penetrating size (see Johnson et al. 1990a and 1990b; Liu et al. 1993; Bergman et al. 1989; Lee and Liu 1982; Xiaowei et al. 1990).

The filter medium or "paper" is a random arrangement of fiber glass fibers. These fibers range in size from 0.1 μm to 10 μm in diameter. The density of the medium is only about 10 percent fibers and 90 percent air and the thickness or depth of the medium is 0.010 in. to 0.025 in. All filters require energy to move air through the medium. The greater the resistance to airflow, the greater the energy cost to operate the fan or blower. The pressure drop across new or clean HEPA filters can range from 0.4 in. to 1.0 in. WG at the rated flow. As the filter loads with particles trapped from the air, the resistance increases. This increase in resistance is not linear to the amount of loading. As the filter fibers collect particles, additional particles will actually attach to the previously collected particles. These form projections or dendrites that serve the same effect as additional filter medium. With this additional surface area, particles have a greater chance of coming in contact with a fiber or dendrite. As the filters continue to load, a point will be reached where the blower is no longer capable of delivering the rated volume of air at the increased resistance. When the maximum fan capacity is achieved, the pressure drop will remain constant, and the air volume will show a decrease. One should be aware that the pressure drop across a filter is meaningless without knowing the volume of air being delivered through the filter.

Most HEPA filters are constructed with the medium arranged in an array of narrow, parallel pleats. A 2 ft × 4 ft × 6 in. HEPA filter will typically contain 150–200 pleats 3–5 in. deep. The pleats serve to increase the filter surface area and align air or gas as it flows through the filter. The resistance of HEPA filter paper to gas flow is reasonably uniform, so that approximately equal volumes of gas move simultaneously through each pleat of the filter. These two factors, the uniform resistance of the medium and the large number of parallel pleats, cause the air to assume laminar flow characteristics 6–12 in. downstream of the HEPA filter face. Recent HEPA designs (Figure 13.4) use "dimples" or ridges impressed into the medium to effect separation of the pleats rather than the thin separators of aluminum used in older designs. The use of HEPA filters does not guarantee laminar flow, since parameters such as flow velocity and pressure drop also effect the flow pattern. Some HEPA filters are not intended for laminar flow applications.

A comment regarding airflow regimes should be made here. The term *laminar,* once almost universally used to describe airflow issuing from the face of a HEPA filter, has fallen into disfavor today. The term was originally defined as movement of air in parallel streamlines. Since this condition implies equal velocity between adjacent streamlines (which we know does not exist), the term *unidirectional* is used in many standards today. In this chapter the author will continue to use the earlier term with this caveat.

Nominally, a HEPA filter has a minimum average efficiency of 99.97 percent when challenged with a thermally generated, DOP aerosol having a mean particle size of 0.3 μm at a concentration of 100 μg DOP per liter of air. This acceptance test method was standardized years ago in MIL-STD-282 when the 0.3 μm particle size was believed to be the most penetrating. This test of average efficiency is performed

Figure 13.4. Pleated HEPA filter media (courtesy Flanders Filters, Inc.).

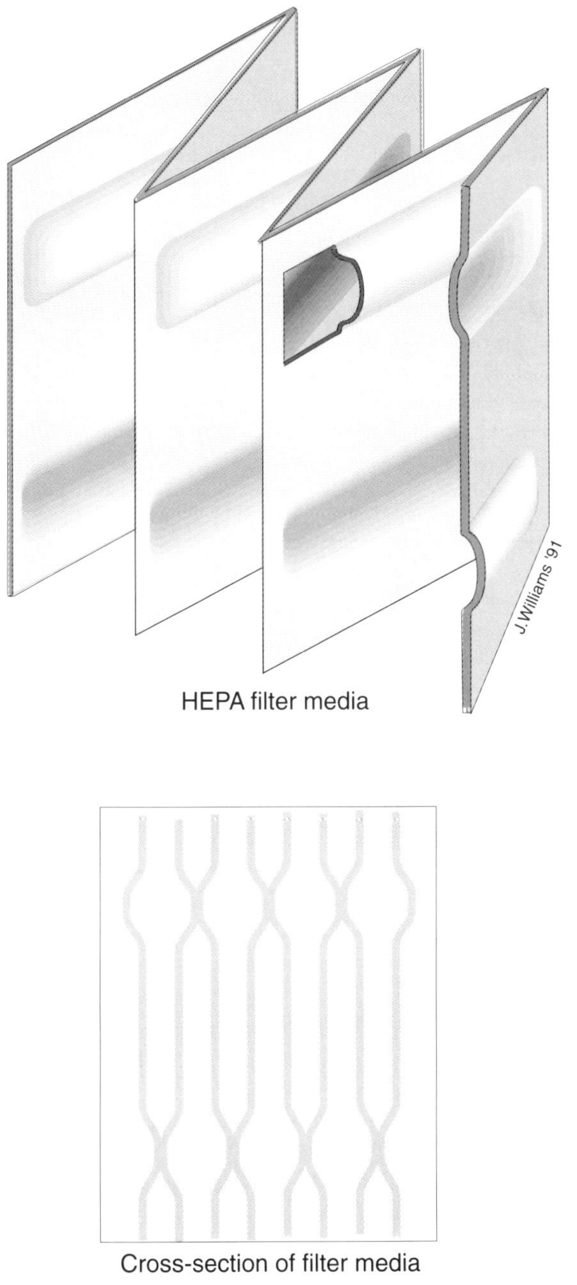

using a specialized test apparatus (Q-107 Penetrometer) generating a hot DOP fog (Hinds et al. 1978). This device (Figure 13.5) uses a thermal aerosol to test the average efficiency of the filter. This does not constitute a leak test, but rather specifies the average efficiency as downstream aerosol concentration versus upstream challenge concentration.

Local areas of greater penetration (either in the edge sealant between the filter element and the filter's integral frame or in the element itself) may be present without seriously affecting the outcome of this test, since particles passing through the leaks are diluted by the greater amount of clean air passing through the filter. In some cases, these areas may be detected as leaks by a postinstallation challenge test (Cadwell 1985). In the certification test, leaks can be tolerated as long as the overall penetration through the filter does not exceed .03 percent. In the now obsolete BS (British Standard) 5295: Part 2 Appendix D (1989), the guidance on air filtration systems D2(a) referenced efficiency tests in accordance with BS 3928 (sodium flame) and discussed the maximum permitted concentration for HEPA installation leaks as 0.001 percent for Class C, D, E, and F and 0.01 percent for all other classes. This is called "scan testing." The British Standard did not address "in place" total leakage tests as does IES-RP-CC006.2. When purchasing HEPA filters, it is generally to the user's benefit to specify compliance with MIL-STD-282.

The replacement of worn-out HEPA filters with newer, less expensive filters that do not meet the appropriate specifications in service may result in a general increase in airborne particle counts. Filters whose downstream face has been "scanned" by the supplier for any detectable leaks are available at extra cost. In these units, leaks are sealed before the filter is "certified" and shipped. Many years ago, HEPA filter manufacturers, confronted with the prospect of failing a field test that could locate defects escaping detection in the overall efficiency test with the penetrometer, began to offer factory-probed filters destined for laminar flow cleanrooms. The Type C filter, per IES-RP-CC001.3, is "one that has been tested for overall penetration and, in addition, has been leak tested using air generated DOP smoke. . . ."

As mentioned earlier, most air cleanliness requirements in the pharmaceutical industry, with the possible exception of containment applications, are satisfied by achieving average filter efficiencies of 99.95 percent or greater; areas of greater penetration ("pinhole leaks") may be tolerated as long as the overall penetration does not exceed 0.03 percent. This is not the case in critical laminar flow systems (clean

Figure 13.5. Q-107 certification test apparatus for HEPA filters (courtesy of Flanders Filters, Inc., Technical Bulletin 581-D).

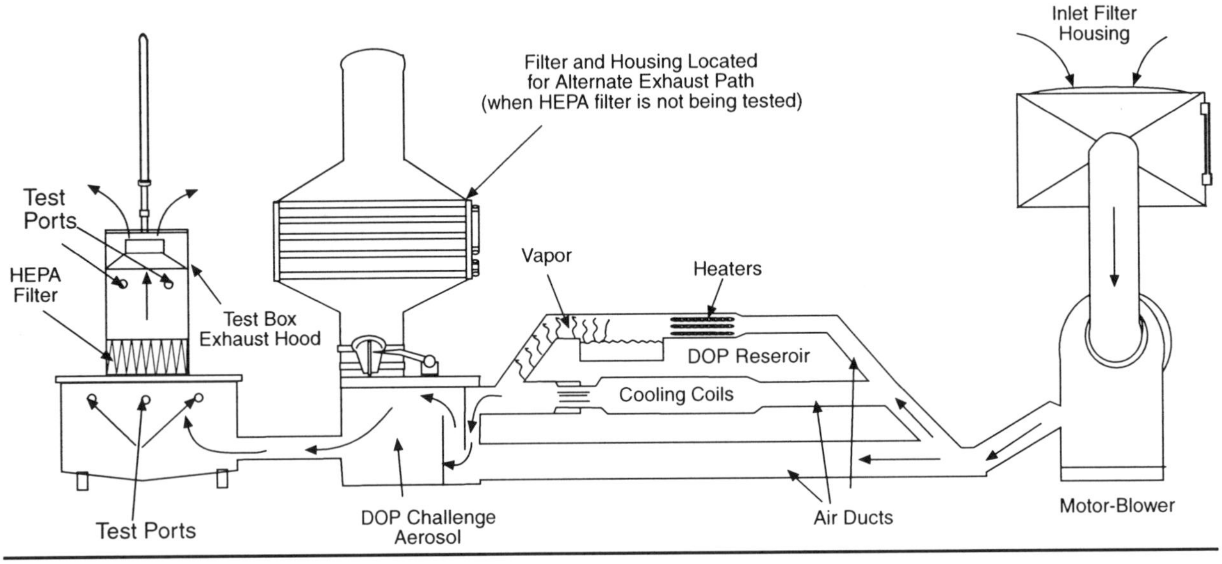

workstations, cleanrooms, downflow hoods), where HEPA filters are located at the point where the air supply enters the cleanroom or work area. In this application, particles passing through a pinhole leak are not diluted by the volume of clean air passing through the filter for some distance downstream of the filter face, and it is possible that the product or process requiring particle-free air could be located directly downstream of a pinhole leak.

Filters produced with similar vendor specifications may show very different performance characteristics (Moelter et al. 1992; Moelter 1988); fibrous filter media produced with similar fiber codes, thickness, and area mass may show significantly different values of penetration and pressure drop. The depth of the pleats, as one critical parameter, governs the surface area of the filters in a given frame size. Measurement of the local area mass distribution of similar filter media has shown that the quality of the media decreases with decreasing SD of the local area mass. The inhomogeneities responsible obviously affect the penetration and the pressure drop of filter media. Loughborough (1991) reported highly variable results regarding particle retention versus pressure drop characteristics of four filter designs. Other factors, such as unit cost, volume flow rate, and ease of installation, obviously must be taken into account when considering the optimum choice for such pharmaceutical applications.

Since HEPA filters are expensive to replace, most users try to prolong their life. Aside from physical damage associated with in-use testing, the biggest factor influencing the life span of HEPA filters is their becoming loaded with particles (e.g., fibers, dust, etc.) with subsequently increased resistance to airflow. When resistance across the filter exceeds the capability of the motor/blower and airflow drops below the required level, the filters must be replaced. All HEPA installations should incorporate a prefilter to accumulate coarse debris to minimize damage to the filter and to extend its useful life. The life of a HEPA filter may be extended up to 400 percent simply by using a 85 percent prefilter (Lieberman 1994). Users should select a prefilter that maximizes the trade-off between efficiency and loading capacity (Bresson 1986). Prefilters must be changed frequently, whenever buildup of contamination is observed. Some installations incorporate an "intermediate" filter between the prefilter and the HEPA filter to further protect and prolong the life of the HEPA filter; they are often of pleated construction and are more efficient than prefilters.

A new, properly manufactured HEPA filter is designed to handle the specified flow rate and air volume while not exceeding a 1 in. WG pressure drop. This should be verified by acceptance testing by the end user. Pressure drop is a significant issue because filters with fewer pleats of filter material in the frame may pass MIL-STD-282 but exceed the maximum initial filter resistance of 1.0 in. WG. In use, as the filter is loaded with dirt and contaminants, the pressure drop will increase, and the motor/blower unit will be required to work harder to deliver the required airflow performance. Thus, if a new filter's resistance is greater than this value, its service life may be significantly reduced. Blowers specified for HEPA systems generally have adequate horsepower capacity to push the "advertised" volume of air up to about 1.5 in. to 2.0 in. WG. Above that level, it is desirable to change the HEPA filter.

DEFINITIONS OF A "LEAK"

The definition of a leak is the critical factor in challenge methodology. Based on accepted U.S. standards and historical perspective, a leak in a HEPA filter is represented by the passage of particles in excess of a 10^{-4} fraction (0.01 percent) of the upstream DOP challenge aerosol. This empirical definition may be reinforced by characterization of a leak in terms of the actual "hole" size in a filter that will pass a 10^{-4} concentration of a cold DOP aerosol. This opening size can be mechanistically defined using capillary tubes or apertures of known diameter inserted in the medium of a test filter (Sadjadi et al. 1990; NEBB 1988). This rationale results in the definition of a leak as a minute opening in the filter medium that has no retention of particles of any size in the aerosol challenge. Figure 13.6 shows "leak" areas from HEPA filters; while the damage or "hole" in the medium is obvious in Figures 13.6a and 13.6b, Figure 13.6c shows no evidence of damage. The leak in the latter case may be due to a thickness variation or subsurface void.

Probably the most useful statement the author can make here is to warn the reader about variable definitions for the word *leak*. Historically, U.S. standards have dwelt on the 0.01 percent figure. The NEBB method will result in a large amount of a DOP aerosol being passed through a filter. The same may be said for the procedure in DIN-24184. In fact, based on technical considerations and the structure of more recent international standards (e.g., ISO

Figure 13.6. Scanning micrograph of leak areas detected by a DOP test.

a. "Pinhole" or surface void in leak area.

b. Pleat damage leak area.

c. Leak area with no surface defect.

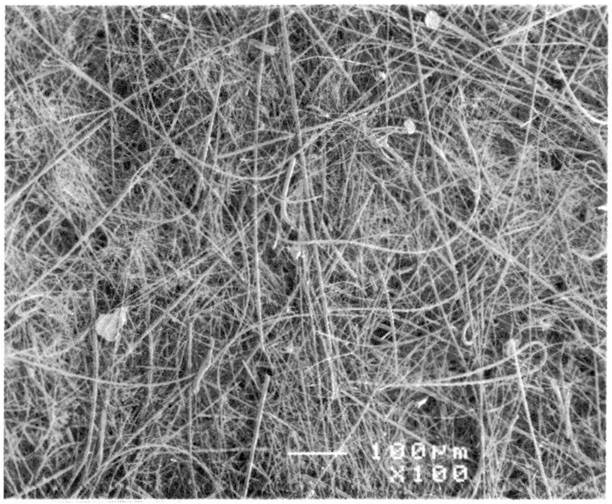

14644-3), the definition of a leak may be specified by a user for a specific installation and agreed upon by the user and those testing the facility. Ultimately, with regard to the U.S. regulatory structure, a leak could be considered a penetration of a HEPA filter below that necessary to compromise the classification of the CEA in question. This approach becomes complex, and most users will be satisfied to test against the 0.01 percent status quo.

This 10^{-4} definition has been the historically preferred and empirically correct one for HEPA filters. A leak can also be specified as a local area of the filter with decreased retention capability. In fact, the entire methodology pivots around what we define to be a leak; there is considerable opinion at present that more critical definitions of a leak may be necessary. For many years, leaks were defined based simply on the detection of fractional penetration. More recently, the extremely critical relationship of the rate at which a photometer or particle counter probe is moved across the filter face and the response characteristics of the instrument to the probability of leak detection has been examined and incorporated into test methods. The ISO 14644 series of documents have resulted in significant discussions of how multiple factors relating to detection capabilities of the method used are critical in the definition of a leak. In the future, in critical applications of HEPA filters, it seems likely that specifics of leak definition will be subject to even more scrutiny. This will affect both the pharmaceutical manufacturer and the aerospace and electronics industries.

In general terminology, the smallest leak that must be detected and repaired is called a "threshold leak." Specifications for HEPA filters designate the threshold leak size when scanning is required. Leaks equal to and larger than the threshold leak size are defined to be unacceptable by specification. Leaks smaller than the threshold size are considered to be insignificant. A criterion of 0.01 percent for threshold leaks for HEPA filters is thus established in the standards and recommended practices that use DOP, the Laskin nozzle, and photometer scanning methods. The size of a leak is defined as the photometer reading when the probe is stationary over the leak, expressed as a percentage of the photometer reading upstream of the filter. Threshold leak sizes less than 0.01 percent are appropriate for filters that are more efficient than HEPA filters.

IN–USE TESTING OF HEPA FILTERS

Field testing of HEPA filters to check for damage and to verify leak-free installation is accomplished by "leak testing" rather than by overall efficiency testing, as described above. These two types of testing are sometimes confused by users, and efficiency ratings are incorrectly assigned to installed filters in air handling systems. Acceptance criteria for leak testing do not directly correlate with those used in efficiency testing. In the United States, leak testing in the field is most often performed with portable DOP aerosol generation and detection equipment, which is a proven (though dated) technology. This challenge aerosol is polydisperse and is typically used in lower concentrations than in efficiency testing performed by the filter manufacturer. A large number of leak test methods are available, and those used in Europe differ somewhat from those applied in the United States. Commonly used test methods are summarized in Tables 13.1 and 13.2.

TESTING WITH POLYDISPERSE DIOCTYL PHTHALATE AND OTHER OILS

The basis for applying this HEPA integrity test procedure can be traced to the requirements listed in the now obsolete Federal Standard, FS-209B. It is worthy of note that the following excerpt does not specify periodic testing and does not quantify the level of upstream aerosol. In section 50.1.a of the document, the following requirement is given:

> An in-place filter test should be made to determine whether the HEPA filter bank has any significant leaks. Test should be made to determine leaks (1) in the filter media itself, (2) in the bond between the filter media and the interior of the filter frame, (3) between the filter frame gasket and the filter bank supporting frames, and (4) between the supporting frames and the walls or ceilings. Leak test should be made by introducing a high concentration of smoke or fog (for example, cold generated DOP fog) into the plenum upstream of the HEPA filters. The mass concentration should be of the order of 10^4 above the minimum

Table 13.1. Methods for HEPA Filter Efficiency Measurement

Method	Aerosol Generation & Property			Aerosol Detection	
Standard Test	Material	Generation Dispersity	Technique	% Measured Detection	Penetration
DOP (U.S. Standard)	DOP	Monodisperse	Vaporization/condensation	Photometer	0.001 and up
Sodium Flame (British Std.)	NaCl	Polydisperse	Atomizing 2%aqueous soln	Flame photometer	0.001 to 0.01
Methylene Blue (Alternate British Std.)	Methylene blue	Polydisperse	Atomizing 1% aqueous soln	Stain density	0.01 to 50
Uranine (French Std.)	Uranine	Polydisperse	Atomizing 1% aqueous soln	Fluorometer	0.001 and up
Monodisperse Aerosol (Electrical Detection)	DOP	Monodisperse	Vaporization/condensation	Electrical aerosol detector	0.001 and up
Polydisperse Aerosol	Ambient aerosol	Polydisperse	Ambient counter (CNC)	Condensation nucleus	0.1 and up
CNC Detection	Gold & silver atomizer	Polydisperse	Vaporization/condensation & spray drying	CNC	0.001 and up
Monodisperse & Polydisperse Aerosol (Optical Detection)	Latex spheres	Monodisperse	Ultrasonic nebulization	Photometer	0.01 and up
	Amb. aerosol	Polydisperse	Ambient	Optical counter	0.01 and up
	PSL, oleic acid, uranine	Both	Spray drying & vaporization/condensation	Photometer	0.01 and up
Airborne Radioactivity (Alpha Detection)	Ambient aerosol with decay product	Polydisperse	Ambient	Alpha detector	1.0 and up

Table 13.2. International Standards Related to HEPA– and ULPA–Filter Testing and Classification

Standard or Guideline	Country	Aerosol[1] and Its Detection			Efficiency	Flow + Resistance	Classification	Design Specs	Leak Limits
		Material	Avg. Size	Detection					
BS 3928, Sodium Flame	UK	NaCl salt	0.60 μm	Mass relat.	x	x	—	—	—
AFNOR X44-013	F	NaCl salt	0.60 μm	Mass relat.	x	x	—	—	—
AFNOR NF X44-011, Uranine	F	Uran. salt	0.15 μm	Mass relat.	x	x	—	—	—
DIN 24184	D	Paraf. oil	~0.45 μm	Area relat.[2]	x	x	x	—	x
M 7605	A	NaCl salt	0.60 μm	Mass relat.	x	x	—	—	—
SWKI 84-2	CH	NaCl salt	0.60 μm	Mass relat.	x	x	x	—	2
MIL-STD-282 (DOP)[3]	USA	DOP oil	0.30 μm	Area/qty rel.	x	x	—	—	—
Mil. Spec. F-51068 F[3]	USA	DOP oil	0.30 μm	Area/qty rel.	x	x	—	x	—
Mil. Spec. F-51477[3]	USA	DOP oil	0.30 μm	Area/qty rel.	x	x	—	x	x
IES-RP-CC001.3[3]	USA	DOP oil	0.30 μm	Area/qty rel.	x	x	—	x	x
IES-RP-CC007.1[3]	USA	DOS oil	~0.18 μm	Quantity rel.	x	x	—	—	—
prEN 1822.1 (Jan. 1995)	Europe	DEHS oil	MPPS	Quantity rel.	x	x	x		x

[1]Mass-related average diameter.

[2]Reference to DIN 24184—detection of "oil thread" by visible light scattering.

[3]All of these test methods use the same test aerosol and the same aerosol detection technique.

sensitivity of the photometer used as the detector.

Importantly, the test described is based on a simple proportionality of leak penetration to the challenge aerosol; the 10^{-4} fraction of the challenge aerosol, which will be defined as a leak, equates to 0.01 percent. A concentration of DOP aerosol of 10 μg/L is generally accepted as being the lowest effective dose in the type of testing prescribed in FS-209B:

- FS-209B section 50.1.a gives an example of 80–100 μg/L, which has been taken literally by some users.
- IES-RP-CC001.3 states that the test aerosol should have a concentration of 10–20 μg/L.
- IES-RP-CC006.2 indicates that a concentration of approximately 10 μg/L of air is adequate.

Over time, the leak test requirement has become an industry standard for filters used in critical pharmaceutical applications. Thus, based on the extensive precedent established by testing with DOP, the FDA will expect any alternative method of integrity testing HEPA filters to show equivalency to the cold generated DOP method.

In application of the probe test, the cold DOP challenge aerosol with a particle size range from 0.1 μm to 3.0 μm (polydispersed) is generated and introduced upstream of the filter bank while the system is in operation. The downstream side is probed with a portable forward light-scattering photometer (Figure 13.7). Pinhole leaks and filter-to-frame leaks are identified and patched.

The apparatus almost universally used to generate a polydisperse challenge "fog" of DOP contains a device designated a Laskin nozzle for its inventor (Echols and Young 1963). A single Laskin nozzle is illustrated in Figure 13.8.

There are two sets of holes in the Laskin nozzle. One set of four holes is located directly beneath the collar around the bottom of the tube, and the second set of four holes is located in the collar. Each hole is positioned directly above the corresponding hole at the base of the tube. The air flowing out of the holes in the tube causes the DOP oil to be drawn through the holes in the collar, shearing the liquid into an aerosol. Unlike the more homogeneous, "monodispersed" particles generated by the hot DOP test, the cold DOP aerosol is polydisperse, having a "light-scattering mean droplet size distribution" specified generally in IES, American Association for Contamination Control (AACC), and other standard documents as follows:

- 99 percent less than 3.0 μm
- 95 percent less than 1.5 μm
- 92 percent less than 1.0 μm
- 50 percent less than 0.72 μm

Figure 13.7. Leak scan test using aerosol photometer and DOP aerosol.

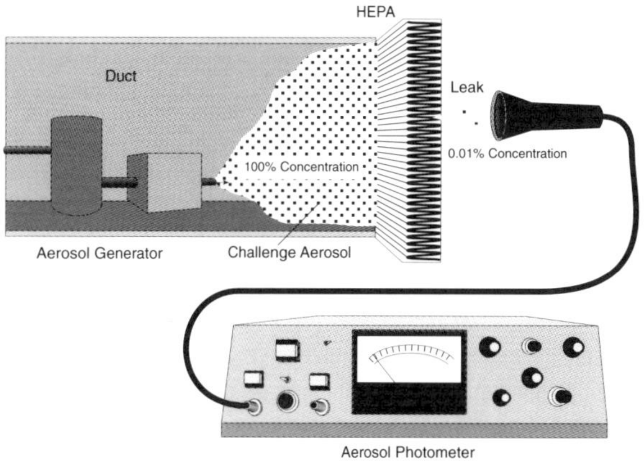

Figure 13.8. Laskin nozzle apparatus for ambient (cold) DOP aerosol generation (reprinted from IES-RP-CC006.2).

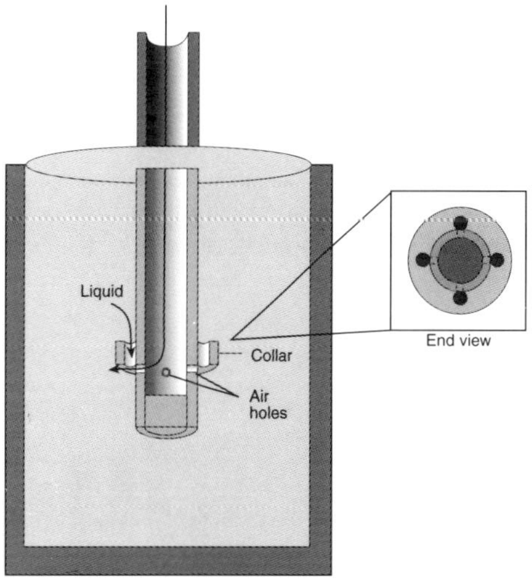

- 25 percent less than 0.45 μm
- 10 percent less than 0.35 μm

On a number basis, the figures describe a generally log normal aerosol particle population, with much larger numbers at the smaller sizes and billions of particles per cubic foot.

Given the importance of challenge aerosol characteristics in testing, this historical presentation is somewhat unfortunate. Light scattering in this context relates generally to total volumes of particles at all sizes in the aerosol, but the meaning of "light-scattering mean droplet size" is unclear. Light-scattering methods do not measure particle size directly, and any test based on light scattering will be heavily dependent on the extent to which calibrant particles approximate the test material in refractive index, shape, and transparency. The size distribution of test aerosols is discussed in Gerbig and Keady (1985), Ettinger et al. (1969), and Yan et al. (1990). Personal communications from Dr. Ron Wolff (Eli Lilly and Company, Toxicology Research Laboratories) and Dr. Werner Bergman (Lawrence Livermore National Laboratory, Special Projects Division) indicate that characterization of the Laskin test aerosol using aerodynamic particle sizing results in a distribution more similar to that shown in Figure 13.9. This distribution is consistent with a count median diameter of approximately 0.7 μm diameter. The caveat above with regard to the refractive index dependency of light-scattering measurements applies particularly to particle counters, since these methods are typically calibrated with PSL spheres (see Chapter 9). The aerodynamic particle size measurement is based heavily on mass, so that the measurements made by this means should be less dependent on refractive index than those from single beam particle counters.

One alternative to using the Laskin nozzle for fog generation is found in the TDA-5A aerosol generator offered by the ATI Corporation of Owings Mills, Maryland. In this device, an aerosol of consistent particle size distribution is created by discharging a regulated quantity of DOP liquid onto a large heated area

Figure 13.9. Laskin nozzle-generated aerosol particle size distribution by aerodynamic particle size (courtesy of Dr. Werner Bergman).

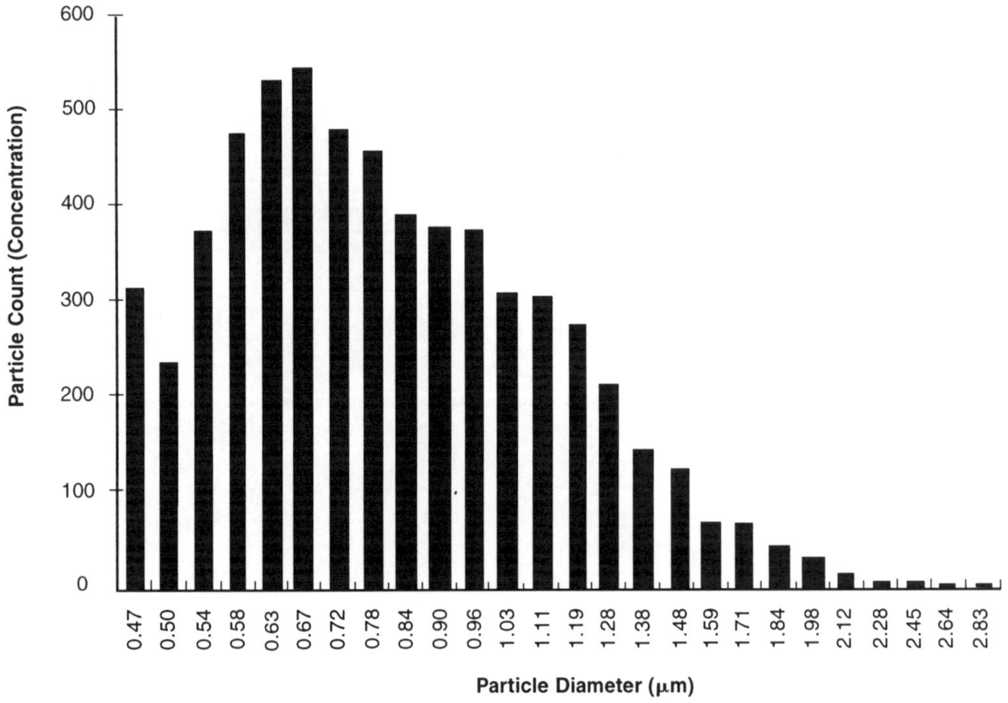

(Figure 13.10). The DOP is vaporized and reconstituted into a polydispersed aerosol by a small amount of nitrogen delivered at only 5 pounds of pressure. The efficiency at which the aerosol is produced enables the TDA-5A generator, when used in conjunction with TDA-2E particle detection units, to provide enough aerosol to leak test filtration systems from 500 to 30,000 CFM (Echols and Young 1963).

Systems of more than 7,500 CFM may be tested for leaks at significantly less than 100 μg/L when TDA-2E series photometers with their increased sensitivity are used. The particle size distribution generated is, for all practical purposes, identical to that produced by a Laskin nozzle. The instrument is lightweight and produces volumes of fog equivalent to a six-nozzle Laskin-type generator.

PHOTOMETER TECHNOLOGY

The device typically used to test for leaks downstream of a filter is a white light aerosol photometer. In older photometers, the operator simply set the needle of the meter to read at the zero point, while holding the probe at the filter face and sampling the effluent from the filter at a scan rate of not more than 10 FPM. (Current photometers contain an internal filter for setting the zero reading.) Before testing, a reading is taken through an orifice in the plenum upstream of the filter. This reading must comprise either a 100 percent concentration value or a value 10^4 times the minimum sensitivity of the instrument. If the challenge is insufficient, an adjustment is made by increasing the air pressure into the generator and checking the upstream concentration reading until the correct limit is attained.

The photometer probe is connected by flexible tubing to the intake of the light-scattering chamber of the photometer. To test the filter, the operator scans the perimeter of the pack and then, using slightly overlapping strokes, probes the entire face of the filter. Most photometers sample at 1.0 CFM. Air is drawn through the sensing chamber, and any entrained particles present in the sampled air deflect the light source onto the sensitive area of a photomultiplier tube. This causes the needle on the meter to move, indicating the size of the leak by the meter reading. A reading of greater than .01 percent indicates a leak that is unacceptable, and the affected area must be repaired. Thus, a leak may not pass the aerosol fog in greater proportion than 1:10,000 relative to the clean air that surrounds it. Probe-tested filters are frequently and erroneously described as "zero probe" or "99.99" filters, inferring that they have a higher efficiency rating than the minimum efficiency

Figure 13.10. ATI TDA-5A aerosol generator schematic.

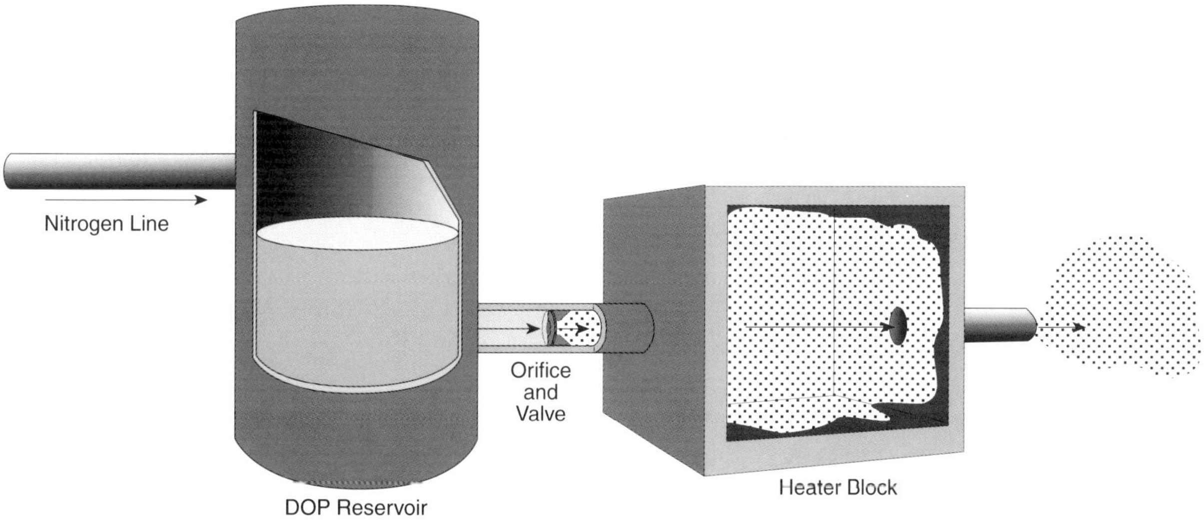

of 99.97 percent required for HEPA filters at manufacture.

A very important consideration regards the sensitivity of the photometer and whether linear or log scale measurements are used. The advantages of more sensitive photometer designs are intuitively obvious. If the minimum sensitivity limit of the photometer is, for example, .0001 μg/L of DOP aerosol rather than .001 μg/L, then a significantly lower concentration of fog may be used as a challenge upstream of the filter. This is based on the simple nature of leak detection, whereby a value of .01 percent of the challenge concentration will be considered a leak. The user of any photometer should carefully evaluate calibration curves supplied by the vendor on a periodic basis. If the calibration curve suggests that the level of challenge aerosol applied is significantly greater than the minimally appropriate 10 μg/L, or that there are advantages in using a linear scale rather than a log scale measurement in the test, long-term advantages may be gained by modifying the test procedure or upgrading the instrumentation.

Photometer designs currently in widest use are based on white light and use collection of the near-forward scattered illumination to quantitate the test aerosol (Figure 13.11). An example of the current development of this type of instrument is provided by the model manufactured by ATI Inc.

A more recent photometer design is the TSI (Minneapolis, Minn.) Model 8110 (Figure 13.12), which incorporates a higher sensitivity, laser illumination, and proprietary collection optics. This optical design is not currently available for leak testing of HEPA filters by the probe method.

Figure 13.11. ATI Model TDA-2E photometer sensor schematic.

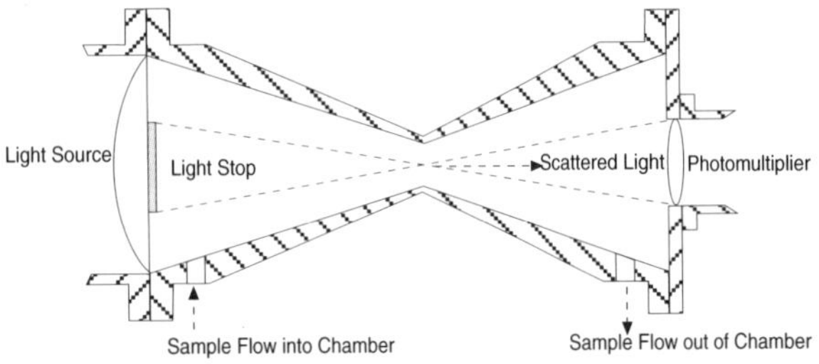

Figure 13.12. TSI Model 8110 sensor schematic.

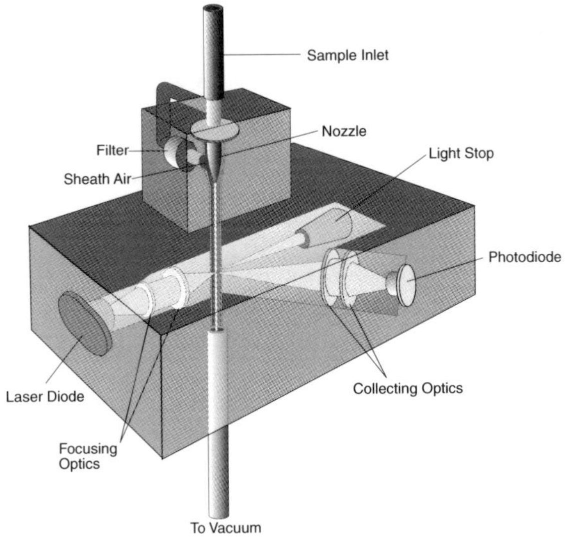

Hopefully, more sensitive photometer designs of this type will become available for use in leak testing large HEPA filters in the future.

ADDITIONAL CONSIDERATIONS IN THE USE OF COLD DOP CHALLENGE

Despite the wide range of tests for HEPA filters as shown in Table 13.1, no alternative to challenging HEPA filters with DOP or other oils currently exists that is acceptable to both pharmaceutical manufacturers and regulatory authorities. Two key issues that must be dealt with when selecting any test method are the nature of the material to be used as a challenge aerosol and the certainty of leak detection with a minimum amount of time expended in testing. There is little doubt that the application of a conventional, cold DOP challenge at a concentration of 80–100 μg/L upstream of the tested filter represents "overkill" in terms of the certainty of leak detection and shortens the useful life of HEPA filters. While this high concentration of aerosol may have been required for the photometers available when FS-209B was written, some current model instruments will effectively detect leaks with as low as 10 μg/L of aerosol upstream of the filter. It must be kept in mind when using lower aerosol concentrations, however, that any challenge method is based on an even dispersion of aerosol at the back face of the filter. Turbulent airflow may have the effect of dispersing the challenge aerosol below the 100 percent level of the photometer; thus, a leak might pass less aerosol than could be detected downstream in the area being probed.

Many of the problems resulting from the in-use testing of filters in the pharmaceutical industry relate to filter installation. Older type HEPA installations in some production facilities often involve large fan prefilter units that drive air to HEPA filters at the ends of plenums or ducts. The ducts often run for a significant distance from the blower, and one blower may feed several HEPA filters through a manifold arrangement. With this layout, obtaining a uniform concentration of test aerosol at the upstream face of HEPA filters becomes difficult, unless the challenge can be introduced a short distance upstream of the filter (Kirk et al. 1988). This may prove impossible, and the user often resorts to a high concentration of aerosol from multiple nozzle generators introduced at a location some distance from the filter(s) to be tested. Since the fog follows the airflow downstream and is diluted, the filter most remote from the generator receives a reduced concentration of challenge; those closest may be overchallenged and loaded with DOP.

One way to address this issue is to replace older filter housings with those of current manufacture that incorporate ports on the front face through which challenge aerosol can be injected, as illustrated in Figure 13.13 (Cadwell 1985). These designs incorporate a manifold upstream of the filter to spread the challenge fog and also have a second port through which a photometer sample tube may be inserted to measure upstream concentration. Filters in this type of housing may be challenged in a fraction of the time required for the conventional test with an aerosol generator at a remote location. A caveat with this type of challenge is that the user needs to ensure that the aerosol injection method used results in an even, sufficiently concentrated, challenge fog uniformly distributed across the upstream surface of the filter.

Generally, the pharmaceutical manufacturer is not only concerned with testing fragile HEPA filters with excessive concentrations of DOP but also with the frequency of the test. One of the principal factors driving a search for new ways to challenge HEPA filters is the fact that tests of filters with DOP, incurring either physical damage by test personnel or through loading of filters, may deteriorate filter performance and shorten filter life. In other words, testing by this methodology actually increases the likelihood that HEPA filters will fail between tests. In the electronics industry, where HEPA filters are most often not tested unless room air counts exceed limits, HEPA filters have a life span of up to 10 years (Sem 1986). The life span is generally shorter in the pharmaceutical industry, where DOP testing is periodically conducted. A specific significant difficulty with DOP testing relates to the high concentration of fog that may be necessary to challenge remote filters or large filter banks (Franklin et al. 1987).

Figure 13.13. HEPA filter housing of current design with ports for challenge aerosol introduction and measurement (courtesy of Flanders Filters, Inc.).

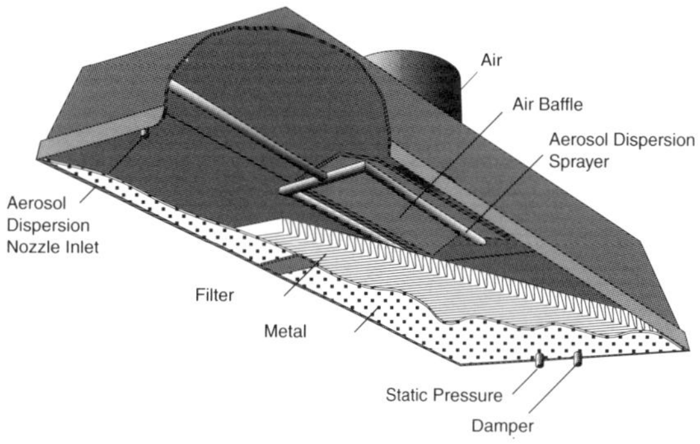

Under normal operating conditions, the "efficiency" of HEPA filters does not degrade very quickly. In fact, barring any physical damage, HEPA filters will become more efficient with use. During normal use, loading with dirt or contaminants inevitably clogs some of the flow channels in the media. Eventually, this loading will significantly increase resistance to airflow and require filter replacement, since airflow cannot be maintained. Gupta et al. (1993) found this to occur when the increase in pressure drop across a HEPA filter was measured as a function of solid particle mass loading, using three materials with different particle morphologies. In this work, NaCl, NH_4Cl, and Al_2O_3 particles in several different size distributions were generated as challenge aerosols. In all cases, the specific resistance of the filter cake increased as the mass median particle diameter decreased. The advantages of this approach over other methods are the use of a more convenient and easily measured particle diameter; the described approach does not require the measurement or prior knowledge of the particle cake porosity. This correlation can be used to predict either the final pressure difference across a loaded HEPA filter or the maximum mass that can be loaded onto a filter for a specified pressure difference (Letourneau et al. 1988; Novick et al. 1992).

This work verified the findings of Leibold and Wilhelm (1991). In this report, the evaluation of HEPA filters showed the importance of particle size on filtration characteristics during high particle loading. For particle diameters < 0.4 μm, homogeneous dust structures on the surface of filter media are built up by smaller particles after particle loading of various durations. A progressive increase in pressure drop at the most penetrating particle size was observed for all of the filtration velocities investigated.

Smith et al. (1991) investigated loading at various relative humidity conditions. Tests were conducted at both high and low relative humidity. For both high and low relative humidity, the filter resistance increased progressively with filter weight gain. For a given weight increase, the filter resistance was greater for the high humidity case. Filter efficiency was essentially unchanged for the low humidity case and was always better than 99.99 percent. Salt loading with high humidity indicated a slight decrease in efficiency with values no lower than 99.98 percent at the highest loading.

The loading of HEPA filters by the DOP aerosol used in on-site testing has been known for many years to have a detrimental effect on filter life. Interestingly, loading by testing may also result in decreased filter efficiency. Payet et al. (1992) found this to be the case when they modeled the effects of mass loading of a HEPA filter in the filtration of submicrometer liquid aerosol particles. Penetration of the test medium increased during clogging by a liquid aerosol, irrespective of particle size within the range of interest (0.02–0.5 μm). This occurred in a fashion not critically dependent on the physicochemical properties of the test aerosol. Investigation showed that the increase in penetration could be explained in part by an increase in interstitial velocity and a decrease in the number of fibers available for particle capture.

A key difficulty with DOP testing relates to the fact that excessive levels are often applied. The detrimental effects of DOP on filter life are less pronounced if the challenge is properly applied. Filters in place for 6 years, which were tested on a semiannual basis, were evaluated by Crosby (1992). The initial pressure drop through these HEPA filters averaged .5–55 in. WG at approximately 100 FPM velocity. After 6 years (or 12 DOP tests), the average pressure drop was .65–7 in. WG, and the air velocity average was 94–96 FPM. The particle counts within the clean area downstream averaged 25–75 particles 0.5 μm and greater per ft^3 of air with the room at rest. During production periods, the particle counts averaged from 25 to 150 particles/ft^3 of air. The following experiment was also performed: A ducted HEPA filter module, 2 ft by 4 ft, was subjected to airflows at approximately 105 FPM. The initial pressure drop was .352 in. WG. This same module was additionally subjected to approximately 20 μg DOP per liter of air for a period of 16 h. At this time, there was no visible downstream evidence of DOP. Air baffles in the duct adjacent to the filter did collect enough DOP to form droplets. The pressure drop through the filter at 16 h was .413 in. WG or a net gain of .061 in. WG.

Potential Safety Hazards of DOP

The actual hazard of DOP use remains a matter for individual investigation. There was a time when DOP was a suspected carcinogen and on the Occupational Safety and Health Administration's (OSHA) list of hazardous materials (Guiney 1982; NIOSH 1988; NTP 1980). It has since been removed from the list and is now given minimal safety protection requirements. The NIOSH and National Toxicology Program (NTP) technical bulletins in the reference list provide considerations related to the toxicology of DOP. The actual risk is probably extremely low (Tancrede et al. 1987). There is a high level of concern about the deposition of DOP onto products such as high resolution optics or microelectronic circuits due to vaporization occurring long after the filters have been challenged. The determination to use or not use DOP as the challenge aerosol is still a matter for individual interpretation, but the FDA at present only accepts the DOP challenge test. Currently, the only accepted alternative to DOP testing in other industries is the ambient particle challenge, as outlined in IES-RP-CC006.2 or ASTM (American

Society for Testing and Materials) F 91-70. This method has some advantages as well as disadvantages, which will be covered later in this discussion.

Test Aerosol Concentration

To optimize the testing of HEPA filters using DOP, the user must know the amount of upstream challenge aerosol, the particle size distribution, and the relationship between these items and the instrument used to detect the aerosol. The particulars regarding the use of Laskin nozzle generators are discussed in IES-RP-CC013 and by Weckerly (1990). The correct amount of upstream challenge aerosol is the mass that, when related to surface area times air speed, will result in a known minimum amount of aerosol penetrating through the filter's media and subsequently detected by a photometer. The challenge aerosol must be of a sufficient concentration to cause penetration should a flaw or leak actually exist. With too low a concentration, the photometer will not be able to sense legitimate leaks and flaws due to the dilution of the concentration, and there will be a lower probability of locating a defect. This will give the user a false sense of security, and most likely will lead to the sporadic appearance of unwanted particles in the clean area as time passes. The user must always bear in mind that a significant portion of the aerosol is precipitated to the upstream surface of the filter before ever entering the filter medium. A concentration that is too high will reliably detect flaws and leaks, but it can also lead to premature filter failure, the deterioration of photometer optics, and the potential challenge agent outgassing from the filters at high levels.

The sensitivity of a photometer is the least amount of the aerosol detectable, such that the resulting aerosol signal is greater than the noise signal of the instrument (signal-to-noise ratio). The correct concentration (and in many cases the minimal concentration) of a challenge aerosol such as DOP for HEPA and ULPA grade filters, is between 10 μg/L and 20 μg/L. This amount will allow for sufficient media penetration and photometer detection but will not cause filter or photometer contamination or appreciable outgassing to occur (more than 1 μg/ft^3 at a linear flow rate of 90 FPM). Photometer sensitivity relates to the upstream aerosol as shown in Table 13.3.

Table 13.3. Sensitivity of Photometer at 25 μg/L of DOP Aerosol

Sensitivity Range	% Reading	Concentration (μg/L)
100	100	25
10	100	2.5
1	100	.25
.1	1,000	.025
.01	100	.0025
.01	10	.00025

A nomograph chart, specifically designed for use with the generator and photometer and coupled to the filter system under test, will result in reliable and repeatable test information. The nomograph will allow the challenge aerosol to be adjusted within the appropriate range before testing begins. The example of a nomograph in Figure 13.14 (Weckerly 1990) displays the filter face areas on the left side of the graduated scale and airflow on the right side. Between these columns is a column listing various air pressure settings. A line drawn from one known to the other known, passes through a point in the middle (air pressure) column. This point gives the needed setting for the air pressure gauge on the aerosol generator for a sample plotted for a 2 ft by 4 ft filter (effective filtering area of approximately 7 ft^2) with a 90 FPM airflow rate. The pressure setting is 17 psi. The user knows that in this example, or in any actual case using a certified nomograph, the right amount of challenge aerosol is being used.

VARIABLES IN LEAK DETECTION (McDonald 1993, 1994a, 1994b)

In addition to aerosol concentration and photometer sensitivity, variables associated with the methodology include the following:

- Penetration that defines leak (typically as a fraction of the upstream concentration
- Particle size distribution differences before and after passing through the filter
- Length of probe in direction of travel
- Width of probe

Figure 13.14. Nomograph chart for estimate of aerosol generating operational parameters.

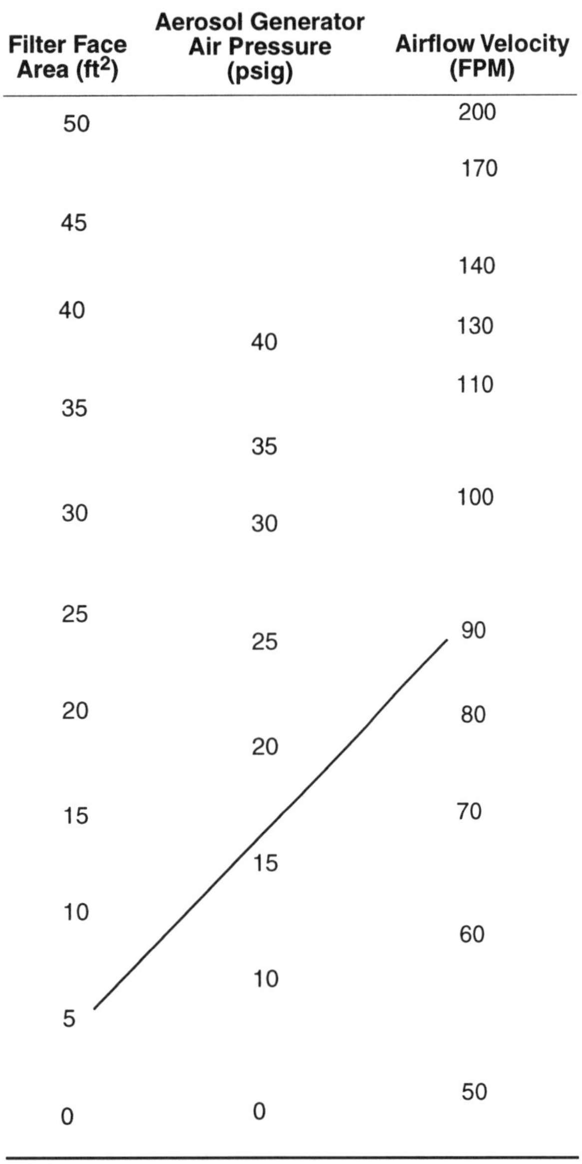

- Shape of probe
- Response time of photometer
- Area scanned per unit time by system
- Signal-to-noise ratio of photometer
- Velocity of airflow

These variables are factors in determining the scan rate used in testing. The smallest leak that must be detected and repaired is called the threshold leak. Leaks equal to and larger than the threshold leak size are defined as unacceptable by specification. Leaks that are smaller than the threshold size are considered to be insignificant. A criterion for threshold leaks for HEPA filters is established in the standards and recommended practices that use the DOP, Laskin nozzle, and photometer scanning methods (i.e., 0.01 percent of upstream concentration). This definition is widely accepted and is reasonably transferable between test sites because the method and instrumentation are standardized.

A critical problem with historical test procedures is found in the highly variable nature of leak detection based on the response characteristics of the photometers used and the spontaneous, subjective manner in which they were (and are) used. Simply stated, the probe of the photometer must remain over a leak long enough so that the aerosol from the leak has a chance to fill the view volume of the sensor, plus the time required for the electronics to register the change in light scatter associated with the leak. This consideration becomes more complex as we begin to consider other factors, such as flow rate of the instrument and probe shape. With regard to probe shape, a rectangular or square probe obviously picks up aerosol differently than a circular one; overlapping or adjacent scan paths of a circular probe will result in the leak being under the probe for different periods of time, dependent on whether it passes under the central or peripheral area of the probe. Since standards governing the testing of HEPA filters published prior to the current version of IES-RP-CC006.2 did not include a consideration of sampling instrument flow rate, the statistical reliability of detecting leaks based on the relationship between the photometer sampling time and the time that the probe is over a leak was not dealt with as a critical parameter. Other prior work considered the effect of the counter flow rate but not the motion (i.e., rate of movement of the probe). The current version of IES-RP-CC006.2 provides a consideration of scan rates applicable to both single particle counters and photometers (McDonald 1994a). The actual mathematics required for relating probe shape and size, flow rate, photometer response, scan rate, and leak rate are complex but are reduced to a simple form in the standard.

The photometer scanning method remains widely used, particularly by the manufacturers of injectables and devices, and there was not a documented basis for the scanning speed when using that method until the present version of IES-RP-CC006.2. The previous revision of IES-RP-CC006.2 gave a single, maximum scan speed regardless of the probe geometry. That scan speed was established when round probes were frequently used. Later, square probes were used because the sensitivity to leaks varies across the width of a circular probe. More recently, rectangular probes have come into use. Identical scan speed with different probe geometry means that the probe is over a leak for differing amounts of time. The photometer response time and the time that the probe is over a leak affect the magnitude of the meter response. The IES Working Group revised the Recommended Practice by requiring that the maximum area scan rate be the same for rectangular probes as it was at the original maximum speed when a square probe was used. The latest version of IES-RP-CC006.2 requires the user to account for scan rate and photometer response and sensitivity.

Measurements using artificial leaks clearly show there is a relationship between scanning speed and photometer response. Increased scanning speed reduces the sensitivity to leaks. The photometer response is modeled as a linear, first-order system with an exponential response. The photometer response in the moving case is related to the stationary case, the probe geometry, the scanning speed, and a response time. The model is a reasonable representation of the experimental measurements. Response times are calculated from the measurements. The response time depends on the instrument settings.

Key assumptions used to derive the equations for scan rate (McDonald 1993) are as follows:

- The probe used to scan the filter is rectangular.
- The flow into the probe is isokinetic (i.e., the velocity in the entrance of the probe is the same as the surrounding flow exiting the downstream face of the filter.
- The leak is a point source, which is small compared to the probe.
- The linear scan rate and the probe dimension in the scan direction are such that the probe is over the leak for a period of time that is equal to or greater than that necessary for the detection of a leak at the 0.01 percent penetration rate.
- There are no background counts downstream of the filter, except those due to particles that penetrate undamaged media.
- The filter is of sufficient efficiency that the number of particles penetrating the media is significantly less than the number of particles penetrating the smallest leak detected.

The "penetration" of a leak is defined as the ratio of downstream counts when the probe is stationary over the leak to upstream counts (or meter readings when using a photometer). There must be no significant contribution to the downstream reading due to particles penetrating undamaged media. A leak may also be defined in terms of the leak flow. The historical leak criterion is a downstream reading equal to 0.01 percent of the upstream reading using DOP and a 1 CFM photometer. This means that the threshold leak is 0.0001 CFM at the test condition of 90 FPM filter face velocity. Likewise, a leak penetration of 0.001 percent is a leak flow of 0.00001 CFM.

In testing, an audible alarm threshold is often used to signal the existence of a leak. Manual scanning methods are employed in many cases; in some instances, the operator can watch a penetration meter built into the probe that is used to scan the filter. The operator using manual scanning can use experience, visual evidence of damage, and signals that are below the threshold leak size to improve his or her ability to find small leaks. Nonetheless, the reduction of sensitivity with increasing scan speed affects the response of the system. The impact of reduced sensitivity during scanning becomes more important as the threshold leak size decreases.

The photometer used in these measurements detects particles in the sample airstream by measuring the intensity of the light scattered from many particles in a scattering chamber. When the scan probe passes over a leak, the flow into the scattering chamber changes abruptly from particle-free air to air laden with the particles penetrating the leak. After the probe passes over the leak, the flow into the scattering chambers is basically particle free. If the flow into the chamber uniformly displaces the air in the chamber, then the concentration in the sensing zone is zero until the first half of the chamber is filled. Then there is a step change to the maximum concentration. In this case, the photometer response is a delay while the first-half chamber fills, followed by a very rapid rise to the maximum value. After the probe passes the leak, there would be a rapid drop back to the near-zero level over the intact filter.

The time required to fill the view volume of the photometer is the volume V divided by the flow rate Q: V/Q. A representative view volume is approximately 6 cm (2.5 in.) in diameter at the base with an overall length of about 16 cm (6.5 in.). The volume of half of the chamber is 80 cm^3 (5 $in.^3$) and the flow rate is 28 L/min (1 ft^3/min) for a filling time of 0.17 sec.

The flow geometry suggests that uniform displacement flow is unlikely. The velocity of the jet entering the view volume is on the order of 20 m/sec (66 ft/min). At a Reynolds number of approximately 7,000, the jet is turbulent. The Reynolds number for the flow through the neck of the chamber is on the order of 4,000. Given these conditions, a good alternative model is that the flow into the view volume is always fully mixed with the air in the sensing zone. These considerations dictate that the response of the photometer increases with time over the leak up to a maximum (Figure 13.15). If the probe remains stationary, and the challenge of aerosol upstream of the filter is inserted in pulsed fashion, the curve of Figure 13.15a will result, with the slope of the detection curve affected only by the time required for the aerosol to reach the detection level and the response of the photometer, due to electronics rise time and the time necessary for the sensor volume to be filled with a detectable level of aerosol. Figure 13.15a assumes the time required for both of these events is negligible, so a vertical response curve to detection level results.

In Figure 13.15b, the probe is moving. The "sawtooth" or "shark fin" response profile represents the curve of 13.15a impacted by the motion of the probe over the leak and movement off the leak area as a function of scan rate. In this case, the response and decay of the response of the photometer are elliptical, and the photometer does not indicate a leak. As detailed in the current version of IES-RP-CC006.2 and McDonald (1994b), the scan rate must be such that the maximum response is reached for uniform leak detection (Figure 13.16).

The particles filling the photometer view volume can be modeled by a first-order linear model with an exponential response. For this analysis, it is assumed that the response of the entire system, including the sampling probe, the tubing connecting it to the photometer, the photometer view volume, and the photometer electronics can also be modeled as a first-order linear system. It is further assumed that the time constant of the exponential response is independent of the pulse height (size of the leak) and pulse duration (the scanning speed). Consideration of these factors results in an interaction of scan rate on photometer response of the type shown in Figure 13.17.

In IES-RP-CC006.2, the linear scan rate is controlled by a requirement that the maximum area scan rate should not exceed 0.093 m^2/min (1 ft^2/min). The area scan rate Ar is given by the probe width W times the scan rate Sr. The test procedure sets the probe sample flow rate Q at 28.3 L/min (1 CFM) and the maximum area scan rate. For a specified airflow velocity V_f, the minimum time the probe is over a point is

Figure 13.15. Particle counter–photometer response to leaks in a stationary and moving mode.

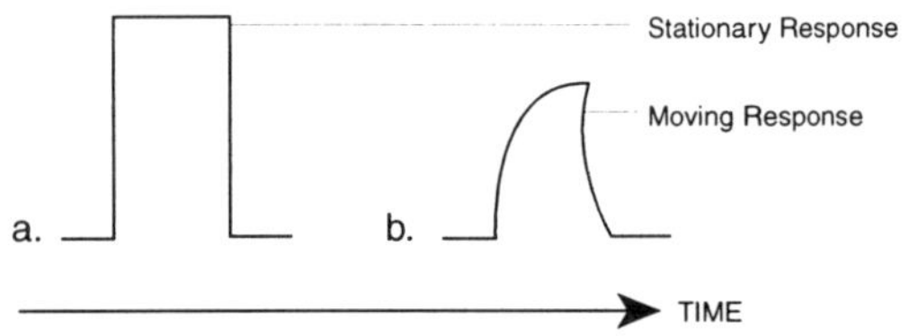

Figure 13.16. Photometer maximum response with time over leak (courtesy of McDonald 1993). Probe length in scan direction is 0.5 in.

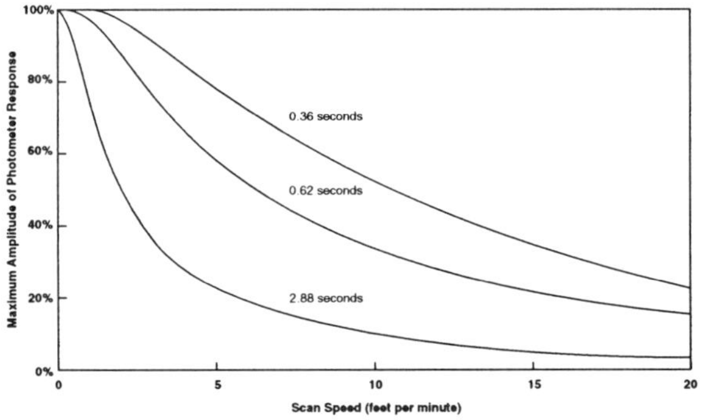

Figure 13.17. Regression of seconds/inch (scan rate) on photometer.

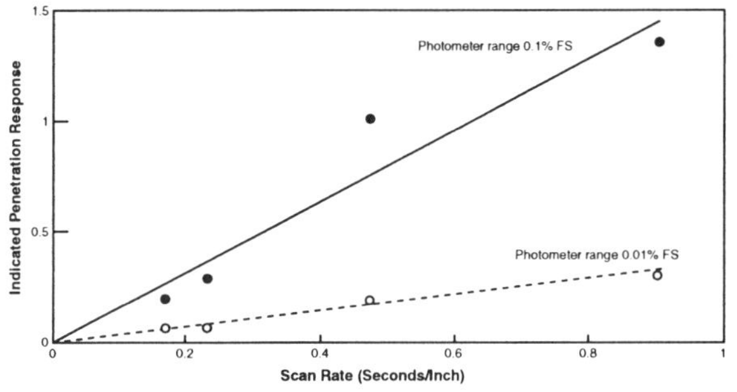

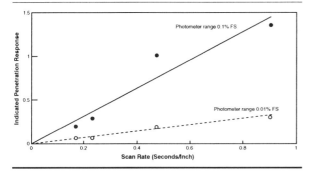

Figure 13.18. Modeled versus actual photometer response (McDonald 1993).

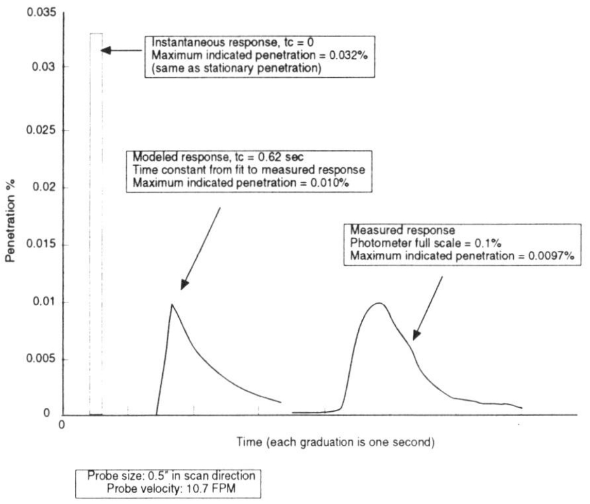

established by the maximum *Ar*. For a common filter exit velocity of 0.5 m/sec (100 FPM), the minimum time the probe is over a leak is 0.6 sec. The constant *Ar* guarantees that a rectangular probe is over a point leak for the same time regardless of the aspect ratio of the rectangular inlet if the V_f is always the same. The constant *Ar* is not a sufficient restriction to ensure that a rectangular probe is over a point leak for the same time if differing filter airflow velocities are involved. Most importantly, for a photometer test, the indicated value of penetration obtained with a moving probe will be less than that obtained with the photometer stationary over the leak. This is due primarily to the fact that the probe may remain over the leak for less than one counting interval.

Thus, the response from a 10^{-4} leak may be actually less than 0.01 percent if too fast a scan rate is used. This response is also dependent on which scale of the photometer is used. This seriously affects the sensitivity of leak detection (Table 13.4 [from McDonald 1994b]; Figure 13.18).

Table 13.4. Scan Rate Effect on Leak Detection

Probe Motion	Photometer Range	Alarm Setting	Smallest Leak Detected
Stationary	0.1%	0.01% (fs/10)	0.01%
Stationary	0.01%	0.01% (fs)	0.01%
10 FPM	0.1%	0.01% (fs/10)	0.0303%
10 FPM	0.01%	0.01% (fs)	0.125%

The implications of the data shown in Table 13.4 and Figure 13.18 are that the scanning of leaks is less sensitive than a stationary reading; at a 10 FPM scan speed with a 0.5 in. probe, the maximum penetration response is 0.33 of the stationary response when using the 0.1 percent range and 0.08 of the stationary response when using the 0.01 percent range. Thus, a leak that measures 0.01 percent with a stationary probe will register a peak response of 0.0033 percent when scanned at 10 FPM with 0.5 in. probe with the photometer set on 0.1 percent full scale. The same leak will only register a peak response of 0.0008 percent when the photometer is on the 0.01 percent fs. Using the more sensitive scale actually decreases the magnitude of the response. However, if the alarm is set at 5 divisions on the meter, then a 0.01 percent leak will be detected on the 0.001 percent range and will not be detected on the 0.1 percent range.

ALTERNATE MEANS OF FILTER TESTING IN THE PHARMACEUTICAL INDUSTRY

The information presented in Table 13.1 and the review of recent work pertaining to in situ testing of HEPA filters suggest a number of alternatives to the conventional cold DOP test. A number of methodologies have been suggested as viable alternatives for the pharmaceutical manufacturer wishing to avoid using the 80–100 μg/L concentration sometimes used with conventional Laskin nozzle generators. In addition to the application of more sensitive laser light-based photometers discussed earlier, these methods involve the following:

- Laser particle counters instead of photometers for leak testing
- Substitute oil materials in place of DOP
- Monodisperse latex aerosols
- CNCs (condensation nucleus counters) or high-sensitivity particle counters with ambient aerosols
- Thermal DOP generators that produce a fog containing larger numbers of particles at the most penetrating size
- NaCl particles (Murrow and Nelson 1973)
- Silica test dust
- Substitute "average penetration" tests for a probe test
- Pulsed aerosol generation and testing
- Simultaneous upstream and downstream measurement of the test aerosol concentration
- Continuous multipoint monitoring (Sem 1986)
- Uranine test aerosols

Laser Submicrometer Particle Counters in Leak Testing

A laser particle counter may be used in place of a photometer for leak testing. These instruments are constrained in application by their low particle concentration limit, but their sensitivity compensation for this factor requires specialized test methods. These instruments also measure particle size. When used with a computer, particle size distributions both upstream and downstream can be generated from the filter initially used to measure penetration at 0.3 μm; they can also be used to measure penetration versus particle size down to 0.1 μm. The first work in this method was done at the University of Minnesota (see Liu et al. 1993; Liu and Lee 1982; Lee and Liu 1980). That work confirmed that there is a particle size is most penetrating. This size was not 0.3 μm but was in the range 0.1 to 0.2 μm. This information became useful as semiconductor design dimensions shrank to the point that there was a need to know and control penetration at 0.1 μm.

Laser particle counters and atmospheric aerosols are now being widely used in leak tests, just as they are in the penetration test. There are drawbacks in this application for HEPA filters. Consider a leak-free ULPA filter with penetration of .001 percent that is 10 times less than the .01 percent defined as a leak for the HEPA. The low penetration of the filter allows this type of test. With a HEPA filter, the entire face would show high particle counts in this size range because of the relatively higher penetration through "good" media; filtered penetration is large and not distinguishable from a leak.

When using an atmospheric–laser counter leak test on ULPA filters in a cleanroom, the advantages noted for the penetration test are still present (lower overall exposure to the filter), but some technical problems arise:

- Upstream concentration is usually too low immediately upstream of the filter. Compensation is made by finding a way to introduce unfiltered air, removing prefilters, cracking the access door in the air supply, and so on.
- The stability of upstream concentration is often unsuitable. With temporary methods of raising concentration, particle counts can be highly variable (e.g., activity outside the CEA). Compensation is made by constantly monitoring upstream and adjusting the acceptable downstream count as necessary.
- Particle counters overload (i.e., exceed their concentration limit) at "high" upstream concentrations. How high varies with the model. A diluter on the upstream sample is usually necessary.

Substitute Oil Challenge Materials

Within the past 10 years, a number of challenge materials have been evaluated by filter manufacturers and users as possible replacements for DOP. These alternate liquids may be used in Laskin nozzle generators to produce aerosols for in-place testing of HEPA filters in exactly the same manner as DOP. In studies done by the

Army Chemical Center (Carlon et al. 1989; Carlon and Guelta 1990), as well as in industry (Steffen and Girres 1992), Emery 3004, which is available from Henkel Chemical Corporation in Cincinnati, Ohio, has been shown to function almost identically to DOP with regard to aerosolization properties and photometer detection. The general properties of DOP and these substitute materials are shown in Table 13.5. Properties of substitute materials are discussed by Hinds et al. (1981, 1982, 1983). A range of quantitative data provided by Crosby is duplicated in Tables 13.5, 13.6, and 13.7. The data in Table 13.8 are from Moore et al. (1993).

One concern with the use of an alternate liquid is that a photometer calibrated using a specific concentration of polydispersed DOP aerosol from a Laskin generator might not detect and quantitate the alternate accurately, and thus be less effective in detecting a leak. This proved to be the case in the data presented by Crosby (1990) (Table 13.6).

It is apparent from the data that the different substitute liquids do give concentrations that vary over a wide range, as would be expected based on their different physical properties; the droplet generation process is affected by both liquid viscosity and surface tension. More importantly, photometric reactions to the various substitute liquids do not seem to correspond to their concentrations in μg/L. The photometer used to collect the data in Table 13.7 was calibrated against a DOP aerosol. Aerosol photometers utilize the forward light-scattering principle as discussed earlier. The light scattered forward is a function of the aerosol diameter, and the aerosol of these substitute liquids may differ in aerosol mass median aerodynamic diameter from DOP. Thus, the aerosol photometer readings would not be expected to relate accurately with the actual gravimetric reading unless the photometer was calibrated initially against the particular aerosol. Refractive index differences between the oils shown in Table 13.5 are not great enough to cause significant differences in sizing. When using a substitute liquid for DOP, the variation in concentration must be considered. Also, the photometer and generator may be calibrated together using the actual substitute liquid to ensure accurate test results. Interestingly, Emery 3004 appears to elicit a stronger response from a photometer than DOP with 85 μg/L, giving a 125 percent response. This suggests that smaller quantities of Emery material may be used in testing.

Based on the variable sensitivity of photometers for the different oils, ATI suggests the corrections shown in Table 13.7 (Crosby 1993).

The correction may be simplified by eliminating one of these variables. If the photometer is calibrated by the manufacturer using the substitute aerosol, the internal reference factor from Table 13.7 may be disregarded. In this case, the generator pressure factor must still be taken into account, since the same pressure on the same nozzle with a substitute does not consistently yield the same concentration of aerosol.

As an example of utilizing a substitute liquid for DOP (in this case Emery 3004), the generator pressure was increased by a factor of 1.15 based on Table 13.7. The normal 20 psig is multiplied by 1.15, which gives an operating

Table 13.5. Some Properties of DOP and Substitute Materials (Crosby 1990)

Material	Specific Gravity	Viscosity (CP)	Boiling Point (°C)	Refractive Index
DOP	0.983	82 @ 20°C	350	1.485
Dioctyl sebacate (DOS)	0.915	17.4 @ 25°C	240	1.448
Mineral oil, heavy	0.845–0.905	>125 @ 0°C	360	1.471
Mineral oil, light	.853	37 @ 20°C	—	1.467
Corn oil	.918	—	—	1.464
Emery 3004	.819	16.9 @ 40°C	401	1.4556
Polyethylene glycol (PEC)	1.128	105 @ 25°C	—	1.465
Paraffin oil (BP ENERPAR M-002)	.848 @ 15°C	14.8 @ 40°C	301	1.466

Table 13.6. Aerosol Output of Alternate Oil Material (Crosby 1990)

Generator Liquid Level	Gravimetric Concentration (µg/L)	Photometric Concentration Reading (%)
DOP	102.0	100
DOS	77.7	82
Mineral oil, heavy	90.0	100
Mineral oil, light	66.0	87
Corn oil	83.3	95
Emery 3004	85.0	112
PEG 400	67.1	56
Paraffin oil	79.4	90

Table 13.7. Correction Factors for Use of Alternate Oils (Crosby 1993)

Liquid Substitutes for DOP	Generator Pressure Factors	Photometer Adjustment Factors
DOP	1.00	1.00
DOS	1.22	0.96
Mineral oil, heavy	1.10	0.90
Mineral oil, light	1.34	0.79
Corn oil	1.17	0.88
Emery 3004	1.15	0.73
PEG 400	1.33	1.11
Paraffin oil	1.21	0.89

pressure of 23 psig for the generator. Other challenge aerosol agents with their own specifically designed generators require their own calculated nomographs. Emery 3004 is officially approved by the U.S. Army Surgeon General as a replacement for DOP.

Presently, the use of substitute oils seems to be the most promising first alternate method for oil aerosol testing. A major pharmaceutical manufacturer (Eli Lilly and Company) has completed an extensive comparison of Emery 3004 with extremely promising results (Moore and Marshall 1993; Moore et al. 1994). These data were reported at International Society of Pharmaceutical Engineers (ISPE) conferences and shared with FDA personnel at national, regional, and local levels. Based on the favorable results of comparison testing, Eli Lilly and Company is in the process of converting from DOP to Emery 3004 for all in-place testing of HEPA filters, and the material has been accepted by the FDA as an acceptable substitute for DOP in pharmaceutical manufacturing. Other companies are expected to follow. Representative results from the data of these authors are shown in Table 13.8. Based on these data, testing with the two materials may be expected to yield closely consistent results, with the Emery 3004 yielding somewhat more sensitive leak detection.

The high degree of similarity shown in the test results is borne out by the extremely similar nature of the particle size distribution of Laskin-generated aerosols from these materials (Figure 13.19). For both the Emery 3004 and DOP, the mass median particle diameter is approximately 1.3 µm, and the count median diameter is approximately 0.7 µm. This similarity between the materials indicates that one can be substituted for the other in Laskin-type generators and produce particle size distributions that are nearly identical. Since particle size is effectively the same for DOP and Emery 3004, these materials would be expected to yield closely similar results in filter integrity testing.

Monodisperse Latex Aerosols

The first widely acceptable alternate to oils such as DOP, PSL spheres, is now available. A suitable system for dispersing PSL spheres in air has been devised and offered by Clean Room Sciences in Phoenix, Arizona. This company offers the CRS Model µS-25K-1 aerosol generator and various filter scan test solutions. Among the latter is the TS-µ265 material, in which the spheres used are 0.26 µm in size, which approximates the most penetrating particle size for most HEPA filters and is counted with high efficiency by laser-based, light-scattering particle counters. The generator is constructed in a lightweight, compact carrying case of nonconductive, high-impact plastic. It utilizes ultrasonic nebulization of the specially prepared particle-carrying solutions, in which the surface tension is reduced by the addition of surfactants. At its highest setting, the generator will produce up to 10^6 particles/ft^3 in 25,000 ft^3 of air. A particle counter with a 0.1 µm or 0.2 µm sensitivity limit

Table 13.8. Comparative Data Collected Using Emery 3004 and DOP (Moore et al. 1993)[1]

Testing Series	DOP Greater Than Emery 3004	DOP Equal to Emery 3004	DOP Less Than Emery 3004	Total
I	6 (15%)	11 (27.5%)	23 (57.5%)	40
II	7 (11.7%)	7 (11.7%)	40 (76.6%)	60
III	1 (6.3%)	4 (25%)	11 (68.7%)	16
TOTAL	14 (12%)	22 (19%)	80 (69%)	116

1. Penetration readings for DOP greater than/equal to/less than those for Emery 3004.

Figure 13.19. Number distribution of aerosol droplets by aerodynamic particle sizing (courtesy of Dr. Ron Wolff).

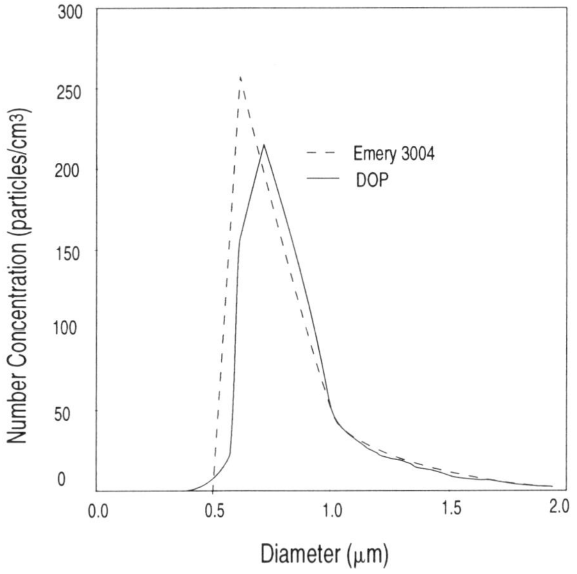

is used to quantitate aerosol upstream and search for leaks.

The filter test solutions are compounded, beginning with a clean (i.e., low particle burden) demineralized water base. A bacterial growth inhibitor is added, along with special surfactants to control the surface tension of finished solutions. PSL spheres are added to the solution at the specific size(s) required for the optimum challenge, depending on the filter efficiency. The resulting TS-H305 solution for HEPA filter testing and TS-μ265 solution for ULPA filter testing are provided in 1 gal lots or 4 gal to a case. A gallon of solution will last approximately 10–12 h in a CRS generator set at full capacity.

Counter factors are also an issue with this method. If a counter has a saturation limit of 10^6 particles/ft^3, it will require that a challenge aerosol of $>10^6$/ft^3 be diluted for accurate counting. This dilution can be accomplished in the necessarily precise fashion, but the operation is technically nontrivial. Once generators under development by the vendor are available, which produce up to $1 \times 10^{6-7}$ particles/ft^3, any major objections to the method should be reevaluated.

The μS-25K-1 generator and TS solutions combine to make a testing system that is exceptionally efficient. Depending on the application, these systems are able to detect leaks in HEPA and ULPA filters with an accuracy equal to or better than DOP methods. The setup of test conditions is easier, cleaner, and faster than DOP or any other leak test method. Although these systems have some limitations, it is likely that they will become the future systems of choice for filter scanning and filter efficiency testing. Critics of this alternate methodology raise the same objections as are voiced with regard to ambient challenge (i.e., the number of particles upstream that are marginal or too low for time-effective, reproducible detection of leaks). Under some conditions of use, surfactant aggregates may be created, with resultant confusion of upstream counts. In this regard, it should be noted that the PSL challenge consists of spheres at or near the most penetrating size, whereas a large proportion of the particles in a cold DOP aerosol are not functional in testing, since they are too large and simply coat the back of the filter medium. Thus, the method of PSL challenge is more efficient in application. Despite its perceived drawbacks regarding attainable upstream concentration, the advantages of selecting the most penetrating sphere size and precise quantitation of leak counts based on known penetration characteristics of that sphere size are highly significant to the pharmaceutical manufacturer. Further evaluation of possible correlation between DOP and PSL testing will be required by regulatory organizations before replacement of DOP with PSL spheres can be seriously considered.

Ambient Aerosol Challenge Method

The ambient aerosol challenge method utilizes a CNC or higher sensitivity particle counter with a lower sensitivity limit of 0.05 μm (Biermann and Bergman 1988; Gogins et.al. 1987). The CNC method is described in IES-RP-CC006.2, Section 5.2, and in ASTM F 91-70 (1976). The basic principle of the method involves comparing the number of particles penetrating the filter as it is scanned to the upstream particle count. The ASTM procedure involves using a 50 mm diameter glass laboratory funnel as the sampling probe and a CNC as the particle detector. Atmospheric aerosols from the ambient air are used as the test aerosol, and the particle concentration is generally above 1,000 particles/cm^3. If the concentration is below 1,000 particles/cm^3, a double-pole single throw (DPST) relay is used to increase the particle concentration. A specific wiring arrangement of the DPST allows for a cyclic or buzzer-like action of the relay, and the arcing action at the contacts allow for generation of a large number of nuclei that can be used as a source of particles for leak scanning. The traversing velocity of the probe is specified to be between 2 and 4 in./sec. Finally, particle counts at least 10 times the background are to be used as evidence of a leak. Obviously, if only 1 CNC is used in testing, variations in the challenge aerosol may go undetected, leading to variability in leak detection. The advantage of using a CNC is that extremely small particles (e.g., 0.01 μm) present in large numbers upstream of the filter serve as the challenge aerosol.

Particle counters may also be used to test HEPA filters with ambient aerosol. In one application of this method, two OPCs are used, having a minimum sampling flow rate of 1.0 CFM (0.5 L/sec), ideally with a minimum sensitivity of 0.1 μm or greater. By definition, the particle size discrimination capability should be at least 0.3 μm when testing 99.99 percent for HEPA filters, and at least 0.12 μm when testing 99.999 percent for ULPA filters. The individual conducting the test first verifies that the design airflow velocity has been adjusted to 90 FPM ± 20 percent. With both particle counters in place, the particle count in the plenum upstream of the filter installation to be tested is then established. The upstream concentration should be on the order of 3×10^5 particles for the single size, corresponding to the most penetrating particle size for the type of filter installation being tested. Readings of both particle counters should be within 2 percent of each other a minimum of 8 times during 10 consecutive 1 min samples taken from the upstream plenum. The probe of 1 particle counter is then returned to a point immediately downstream of the filter face center line, and the counter is allowed to clean down and stabilize by collecting counts until readings are below 1.0×10^{-4} the upstream average particle concentration/ft^3.

During testing, the upstream counter is continuously monitored by one qualified technician, while another scans the downstream side of the filter installation. The upstream concentrations should not deviate more than 10 percent from the predetermined upstream average during any time period while scanning is being conducted. For scanning, the probe inlet of the particle counter must be of sufficient cross-sectional area to maintain the probe inlet velocity at the test airflow rate through the filter (i.e., isokinetic). The filter and installation are then scanned at a rate of 5 FPM. Any detected particle count is cause for rescanning that area at a rate not to exceed 1 FPM.

With 300,000 particles/ft^3 upstream challenge, the indicated count representing a significant leak of 0.01 percent would be any count greater than 3 per 5 sec period. All leaks greater than 10^{-4} times the upstream concentration are reported. An unacceptable leak is defined as a sustained particle count greater than 0.01 percent of the measured upstream challenge by a quantity of particles at a specified size, at a sample distance of 1–2 in. from the face of the filters. Periodic surges are not necessarily indicative of leaks but could be burst releases from crevices and so on. In such cases, suspected areas must be rechecked to verify the presence of a leak. The CNC counter method may be applied in arrays as well as with single counters (Holänder et al. 1990). It must be kept in mind that with the CNC method, higher upstream concentration beyond a certain point does not equate to more sensitive leak detection (Sadjadi et al. 1990). Since CNC counters can detect particles of 0.01 μm in size, the increased overall penetration of the filter with excessive aerosol upstream results in a detrimental increase in background counts.

The objections raised by many (including the FDA) to the ambient aerosol challenge method include the variable nature of the upstream "challenge" concentration and the fact that this challenge level is, at best, consistent only with the detection of gross leaks. Scan times required with low challenge levels may be protracted. This is particularly critical if a particle counter rather than a CNC is used because of the larger particles that must be detected. Typically, the air for clean areas is made up of a significant recirculated volume of air that has already been filtered so that counts in excess of 300/ft^3 are extremely rare; at this level of counts upstream, reproducible detection of a leak might require 2–5 sec sampling at the leak site; thus, for this method to be time effective, a level of counts upstream of the filter 5 times higher than the norm would be required. The statistics and sampling rates for using particle counters for leak detection are discussed by Greiner (1990) and are well covered by McDonald (1993); the counts and times necessary are also shown in the nomograph of Figure 13.20.

One of the key issues in defining new methods for use in challenging HEPA filters in the pharmaceutical industry is the need for comparing between different test methods. Bierman and Bergman (1988) conducted both theoretical and experimental comparisons of two instruments—a CNC and an aerosol light-scattering photometer—to determine how they measured the penetration of HEPA filters. In some experiments, leaks were simulated around the test filter to ascertain how such leaks affected the two instruments' penetration measurements. The results of these experiments indicate that differences in the measurements between the two instruments might vary widely (i.e., from no difference to a factor of 10). These variations were found to be related to certain functions that depend on particle size: the response function of the photometer, the filter penetration, and the distribution of the test aerosol. The use of appropriate aerosol diagnostics enabled the authors to measure these functions and compare numerical calculations to experimental results.

Figure 13.20. Scan rate nomograph for ambient particle challenge and particle counter or CNC (Greiner 1990).

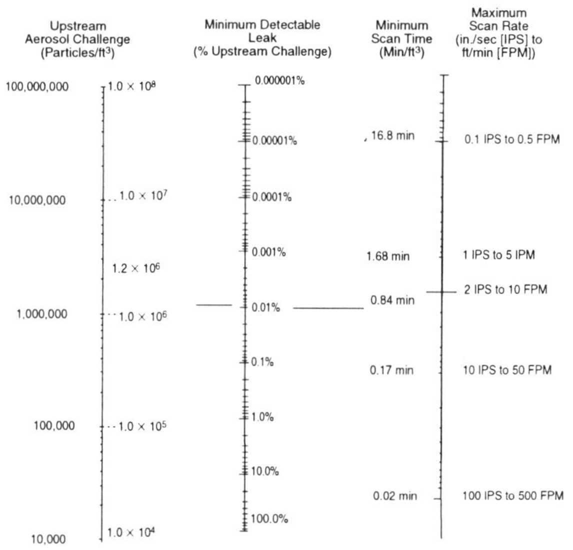

Smaller Particle Test Aerosols

As mentioned earlier, a basic difficulty in present applications of leak testing of HEPA filters relates to the questionable suitability of the test aerosols used. A representative distribution of DOP aerosol particles from a Laskin nozzle generator is shown in Figure 13.21.

A significant fraction of this aerosol by volume resides in particles too large to be effective in leak detection, which may simply be impacted against the upstream surface of the filter and proceed no further. The 0.3 μm particle size specified as most penetrating in historical studies is based on HEPA filters less efficient than those currently manufactured. Whereas earlier studies defined the most penetrating particle size as 0.3 μm, more recent studies using the currently available high-resolution, high-sensitivity particle spectrometer (for example, LAS-X Model, Particle Measuring Systems, Boulder, Colorado) indicates a size in the range of 0.12 μm (Ortiz et al. 1991; Osaki and Kanagora 1990; Payet et al. 1992). Figure 13.22 illustrates a generalized efficiency curve of the type obtained by these authors at standard HEPA operating conditions and pressure drops.

Aerosols of a smaller mean particle size than a Laskin nozzle output may be produced by several methods, including an electrostatic

Figure 13.21. Particle size distribution of aerosol from a Laskin nozzle generator determined by particle counter (data courtesy of Werner Bergman).

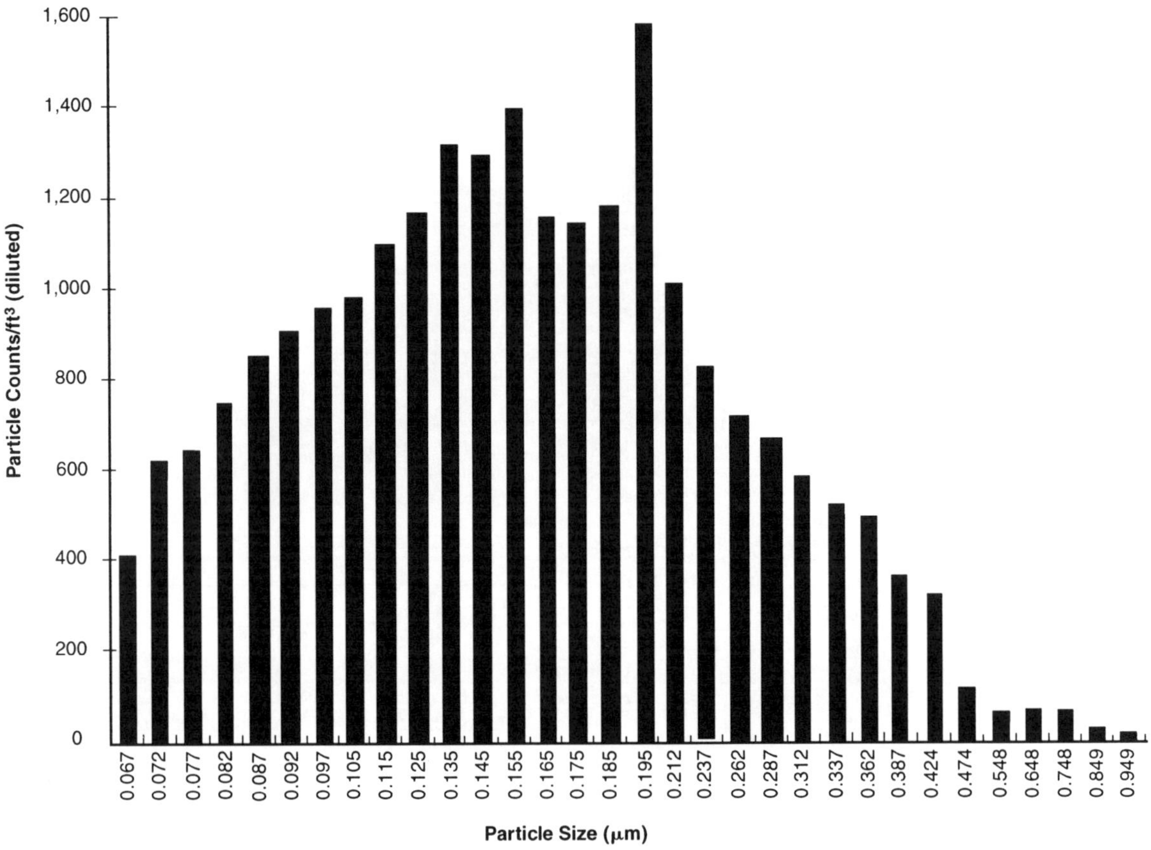

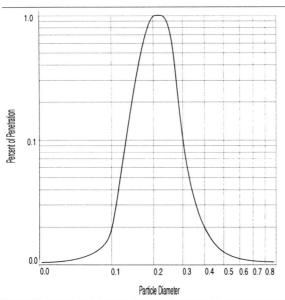

Figure 13.22. Variation in HEPA penetration based on particle size.

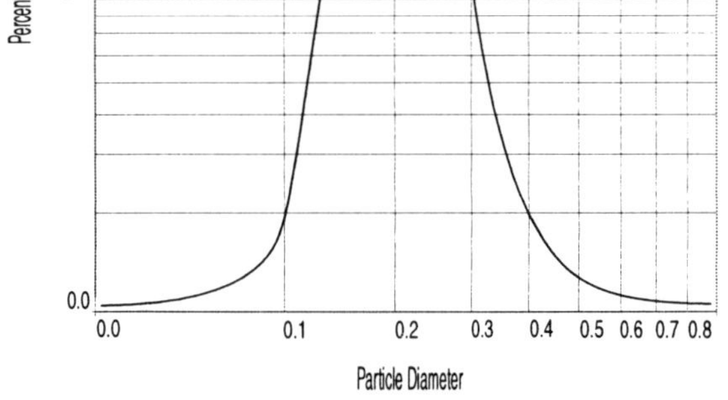

(electrical mobility) classifier in combination with a nebulizer, which produces a monodisperse fog (Payet 1992) or the conditioned aerosol method of Johnson et al. (1990a and b) shown in Figure 13.23.

The advantages of the apparatus shown in Figure 13.23 include its simplicity and its capability of providing an aerosol concentration greater than 1.0×10^6 particles/cm^3. This device produces a stable, predictable output size distribution with a geometric mean particle diameter of approximately 0.18 μm and a geometric SD of 1.6. In Figure 13.24, the output of this device is compared to that of a Laskin nozzle. The inherent stability of this distribution results in extremely reproducible filter efficiency measurements.

An oil challenge fog, although most commonly used in combination with an aerosol photometer, may also be used with CNCs. Johnson et al. (1990a) described an automated filter tester

Figure 13.23. Aerosol generator for producing "conditioned" DOP aerosol (Johnson et al. 1990a) (courtesy of TSI Inc.).

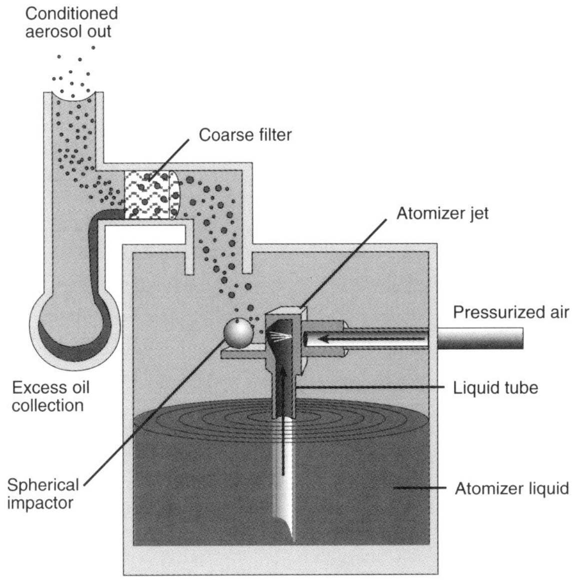

Figure 13.24. Conditioned aerosol particle size distribution (after Johnson et al. 1990a).

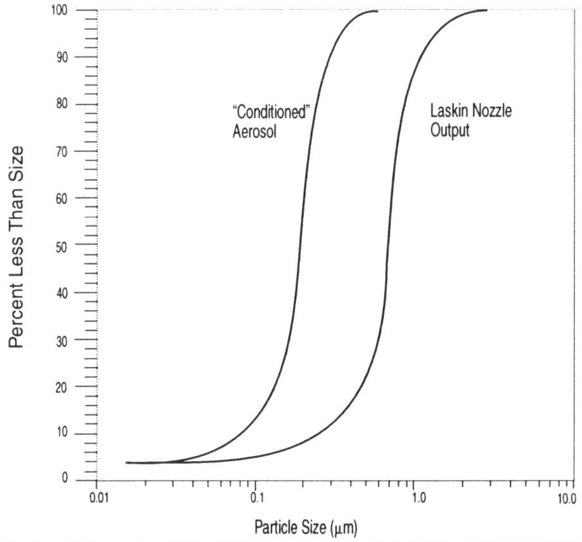

for production testing of HEPA and ULPA filter media in protective masks with penetrations as low as 0.00001 percent (efficiencies of 99.99999 percent). The filter tester can perform testing with various liquid aerosols, including DOP and paraffin oil aerosols. The automated filter tester employs the generator shown in Figure 13.23. Real-time filter penetration data are generated using two CNCs, one measuring upstream and the other measuring downstream of the filter. The two CNCs, along with a special purge air system, enable the tester to measure filter efficiencies to 99.9999 percent in 10 sec. The filter test automatically measures filter penetration, pressure drop, and flow rate. This application for testing HEPA cartridges is obviously quite different from that of testing framed HEPA filters in CEAs, but it raises the possibility of using smaller size aerosols in scanning applications in combination with more sensitive photometers. A comparison of the TSI automated filter tester and the commonly used Q-127 hot DOP test fixture showed a correlation coefficient better than 0.999.

Ortiz et al. (1991) discussed the application of aerosol counters in combination with an oil fog challenge to measure the smaller particles in a cold DOP aerosol. In this study, the penetration of a 2-stage HEPA filter system was measured by several laboratories using laser particle counters. The purpose of this filter testing was to evaluate a test method for determining the performance of 2-stage HEPA filter systems. The test method is being developed by the ASTM F 21 Working Group on filtration (ASTM F 1471-93) and involves challenge of the filters using an oil mist aerosol and subsequent measurement of aerosol penetration using a laser particle spectrometer at 0.1 μm to 0.2 μm. The current MIL-STD-282 applicable for single-stage filter systems measures the filter penetration using a photometer, which has decreased sensitivity for the particles < 0.3 μm in size that penetrate leaks in high numbers; it also requires that the challenge aerosol mean diameter be 0.3. This method yields little information regarding the particle size penetration.

Monodisperse Sodium Chloride Solid Aerosols

If a dilute aqueous solution of NaCl or other solute is passed through a nebulizer, evaporation of the aerosolized solution droplets leaves behind a spherical solid, NaCl crystallite. This means of aerosol generation may be used to produce aerosols of suitable concentration for HEPA testing based on monodisperse aerosols of the most penetrating particle size. The TSI Corporation offers two generators of this type capable of generating aerosols of a concentration of 10^7 particles/cm^3. TSI Models 3470 and 3472 are condensation-type aerosol generators capable of producing highly monodisperse solid or liquid particles at high concentrations. Users make coarse adjustments to particle size using a precise digital temperature control, and then use the instrument's unique gas-flow bypass valve for exceptionally fine adjustments. The Model 3470 offers a particle size range from 0.1 μm to 8 μm; the Model 3472 provides a range from 0.1 μm to 3 μm but operates at a much higher flow rate. These instruments nebulize either a colloidal suspension or dilute solution, such as NaCl in water. This nebulized material is dried to produce nuclei-forming particles that pass through a thermostat-controlled bubbler that contains the substance to be vaporized. From here, both the condensation nuclei and the vaporized substance pass into a cooling zone, where the substance condenses onto the nuclei in a uniform manner, producing a highly monodisperse aerosol with potential for use with either photometers or particle counters.

Silica Test Dust

A method using ultrafine, super hydrophobic Dri-Test silica was described by Bauman (1987). While this method is not currently used to any appreciable extent, it is worthy of mention as a novel approach. The vendors claimed that particle concentrations of 10^5 higher than DOP aerosols were attained with the silica, thus reducing the testing time required for the integrity test for HEPA filters. However, the silica particle load was not believed significant on the back of the HEPA filters. Theoretically, this increased the life of the tested HEPA filter. The silica was dyed with a fluorescent agent so that under ultraviolet (UV) light a leak in the filter could be detected without an aerosol photometer. The particle size distribution of the silica was advertised to be tighter than a DOP aerosol, and the particles were of smaller average size.

The solid silica material was prepared from thermal SiO_2 by modifications that rendered its surface hydrophobic and fluorescent (to UV). The fluorescent tag enabled the identification of tested filters. A portable aerosol generator at a delivery rate of 1.24 ft^3/min delivered 10 in. per ft^3. HEPA filters reportedly showed no blinding (blockage) with the Dri-Test powder. The delivered aerosol was stated to have a bimodal size distribution, with peaks near .08 μm and .150 μm. It was unaffected by ambient moisture and was classified as harmless by its manufacturer.

The Dri-Test aerosol generator consisted of a gravity feed-type dispensing system that was aided by compressed air. The silica was stored and dispensed from a funnel, which incorporated rotational mixing vanes to maintain a homogeneous mixture, into a rotor assembly with the aid of gravity and compressed air. The rotor consisted of a solid ring with a series of small pockets along its circumference that dispersed the silica at a variable rate into a venturi. Aerosol concentration readings were taken using the aerosol photometers currently employed with the cold generated DOP method.

Toxicity testing of the silica indicated the material to be nontoxic. The following toxicity tests were performed:

- Acute oral toxicity
- Primary dermal irritation
- Primary eye irritation
- Inhalation

A notable drawback of this material is the difficulty of handling and containing it and its tendency to "float" and leave coatings on surfaces, devices, and instrumentation within a wide radius of the point at which it is handled.

Average Downstream In-Place Testing

An interesting approach to a containment method for leak testing is described by Bergman et al. (1989). A test procedure is described for evaluating filter penetrations as low as 10^{-9} at 0.1 μm particle diameter. In comparison, the present U.S. nuclear filter certification test has a lower penetration limit of 10^{-5}. This new test procedure is unique not only in its high sensitivity, but also in avoiding the undesirable effect of filter clogging. The procedure consists of 2 steps: (1) challenging the test filter with a very high concentration of heterodisperse aerosol for a short period of time while passing all or a significant portion of the filtered exhaust into an inflatable bag, and (2) measuring the aerosol concentration in the bag using a laser particle counter sensitive to 0.07 μm diameter particles. The ratio of particle concentration in the bag to the concentration challenging the filter gives the filter penetration as a function of particle diameter. The bag functions as a particle accumulator for subsequent analyses to minimize the filter exposure time. Particle losses in the bag over time are reported to be negligible when the measurements are taken within 1 h. Filter penetration measurements taken in the conventional direct-sampling method compared to that of the indirect sampling method were in excellent agreement.

Pulsed Aerosol

Testing with a pulsed, rather than a continuous, aerosol has many advantages, including decreased loading of filters. Parker et al. (1993) describe a technique developed at the Harwell Laboratory for the in situ testing of HEPA filters using multiple pulses of test aerosol; this technique significantly decreases the DOP loading of tested filters. The pulse test apparatus consists of a modified forward light-scattering photometer coupled to a portable microcomputer fitted with an internal data acquisition and control card. The microcomputer switches an aerosol generator on and off via an external relay driver unit. Using this apparatus, the filter bank is challenged by a small number of equal length, constant concentration pulses of aerosol at timed intervals. The aerosol concentration data upstream of the filter bank are logged to a disk by the computer. The process is then repeated for downstream concentration, with increased photometer gain to give maximum sensitivity. The collected data are analyzed using a computer spreadsheet package; recorded aerosol pulses are combined and integrated, and the background data subtracted. The downstream data are then divided by upstream pulse data to give the filter penetration. Using this technique, the sensitivity of the in situ filter test has been greatly improved. Penetrations approaching 10^{-5} can be measured, allowing HEPA filters mounted in series to be successfully tested.

Uranine Fluorescent Dye Method

Challenge with uranine dye particles is commonly practiced in Europe but is rarely used in the United States. The method is reportedly applied by the National Aeronautics and Space Administration (NASA) to some extent. Jalon (1987) described testing HEPA filters with uranine at NASA's Goddard Space Flight Center, including the test equipment, the upstream concentration, the particle size distribution of the test aerosol, the penetration of different size particles, and the leak detection sensitivity of the test. The mass concentration of a uranine test aerosol may be measured by different methods. Mulcey et al. (1990) introduced a technique for measuring the mass concentration of a fluorescent aerosol. The study focused on the detection of a uranine (fluorescein sodium salt) aerosol, implemented during standardized efficiency tests of HEPA filters. The results indicate that application of the method provides a much quicker response than the conventional uranine method, but that the response differs too much from the standard DOP test to be routinely used in determining HEPA filter efficiency.

SOURCES OF LEAKS

A number of factors can cause leakage or bypass in HEPA filters in pharmaceutical applications. Prominent sources of leaks in HEPA filters and HEPA filter–containing enclosures include the following:

- Poor design of the filter housing
- Poor workmanship/inadequate quality control by the manufacturer
- The filter media itself
- Adhesive bond failure between filter medium and frame
- Poorly fitted filter frame and housing frame seals
- Faulty fit of the housing sealing frame and the housing
- Damage incurred in-transit or due to handling
- Misalignment or improper assembly of the HEPA filter into its housing
- Inhomogeneities in media
- Breaks in media caused by flexion or vibration

- Handling/installation damage
- Room seam/penetration induction leakage

Parameters affecting the detection of small leaks in HEPA filters are discussed by Sadjadi et al. (1990), Bhoda (1993), Sharaf and Troutman (1988), and Osaki (1990). Based on the data of Sadjadi et al. (1990), it is likely that leaks no smaller than 50 μm in size will be detected by a polydisperse aerosol challenge with DOP. The National Environmental Balancing Bureau (NEBB) standard (1988) for testing cleanrooms suggests that verification of photometer performance (i.e., detection of 0.01 percent of challenge level) can be obtained using a 24-gauge hypodermic syringe needle siliconed in place in the filter. The lumen of a 24-gauge needle approximates a 300 μm diameter; this dimension would seem to be much larger than that of a "pinhole" leak.

All HEPA filters are fragile, and the glass filter medium is liable to be broken or torn by any mishandling of the grille work by plant personnel when the filter grille is removed to perform a challenge. Also, filters are not infrequently damaged by contact with the photometer probe used to scan the filter face. Unfortunately, in many cases, the personnel performing the challenge testing are not well versed regarding the delicate nature of the filters and need further training in proper procedure. When testing, it must be kept in mind that a leak in a HEPA filter cannot be assumed to be a uniform size. The amount of challenge material to which the leak channel is exposed will depend critically on the average concentration of challenge material upstream and the extent to which turbulence behind the filter causes the concentration at a leak to vary in size. In the absence of turbulence, the random distribution of test particles according to fluid mechanics may cause a significant temperature variation in the challenge material to which a leak is exposed during the critical interval of time that the leak area is being probed.

In some situations, leaks can be eliminated by tightening or realigning the HEPA filter or the cabinet/housing components. A common cause for leakage is debris lodged between the filter housing and the filter seal or mounting channel. Therefore, it is advisable to clean these surfaces carefully and thoroughly when installing new HEPA filters. Damaged or worn seals should also be replaced. Some leaks located in the pleats of a filter cannot be reached from the surface. To seal a hole of this type, the filter must first be removed. Then a liquid compound (such as Rulabon No. 2227 available from Rubber Latex Company of America, Clifton, N.J.), may be poured down through the fold from each side to harden and seal the area.

The maximum allowable size of patches has been a subject of some debate over the years. Historically, a generally accepted figure was a maximum of 5 percent of the filter face area with no dimension of the patch greater than 38 mm (1.5 in.). There are a number of pertinent considerations in regard to patch size. First, personnel applying patches should be trained and adept at the procedure. The leaks detected, if the filter face has not suffered gross damage as a result of being installed or tested, can generally be localized to an area of 1 mm or less. Thus, a drop of sealant of 5 mm in diameter should be sufficient to seal a leak if it has been sufficiently well isolated. Frequently, large excesses of sealant are applied because of a failure to identify the exact area of the leak.

Patches have the effect of making areas of the filter impermeable to airflow, so that flow volume at a constant upstream pressure is decreased, resulting in an increase of the pressure drop across the filter. The patch also has the effect of disrupting laminarity of airflow for some distance downstream and coupling any mechanical vibration from the HVAC fans to the filter medium; thus, the probability increases that the medium will tear in the area of the patch. The likelihood of damage in the patch area is also increased by the increased weight of the patch and its resistance to airflow.

Despite these negative factors, it has not been documented that patches larger than 5 percent on HEPA filters in noncritical applications, such as nonlaminar flow applications in high classification areas, will result in any significantly enhanced rate of filter leak development or failure. In such applications, patches with a total area of up to 20 percent of the filter face with no single patch having a dimension of greater than 125 mm may be allowed if pressure drops across the filter at a flow rate of 90 FPM ± 15 FPM are acceptable.

PATCHING OF LEAKS

In leak testing of HEPA filter systems, not only the downstream face of the filter but also the peripheries must be scanned, while paying particular attention to the bond between the filter media and its frame, the seal between the filter frame and its housing, and any other potential leak sites (seams, welds, bolt holes, rivets, screws, electrical cord penetrations, etc.). The number and variety of leak sites identified by thorough DOP testing and photometer scanning is often high. Any leaks detected must be sealed. This is almost invariably conducted using silicone room temperature vulcanizing (RTV) sealant (e.g., G. E. RTV Silicone Sealer Number 108 or Dow Corning 732). These materials have the consistency of a thick paste, when applied from a cartridge using a caulking gun, and dry into a clear elastomer. Seams where the filter meets the holding frame and where the Plexiglas® panels, table surface, and top panel meet the unit are areas of frequent leaks. A bead of sealer applied along these areas on the inside is generally effective in stopping leaks. Taping or sealing these outside seams with duct tape will give added assurance that particles will not be pulled in and along with the airflow through seam cracks too small to be readily seen.

Minimal amounts of sealant should always be used, and it is always best to try to pinpoint the leak visually to locate the hole, tear, or crack and mark it, so that it can be sealed completely. Leak sites may migrate, especially during attempts to seal a leak between the filter housing or frame and the filter media (Thomas and Crane 1963). Generally, the blower should be turned off before attempting to seal a leak. By doing this, the repair sealant can be pushed into the leak site more effectively. Then allow the sealant to set before it is retested with the blower in the "on" position. In all cases, repaired areas must be retested. When searching for pinhole leaks in filter pleats, a 3 × 5 in. index card is often useful in isolating suspected leak areas. The card is placed first to one side of the suspect leak area, then to the other side while photometer readings are taken; by this procedure the leak is isolated within a small area that may be patched in a precise manner, thereby avoiding excessive use of sealant.

It is of note that ISO 14644-3 (which is scheduled for issue in 1999) limits the patch size to 0.5 percent of the filter face area, with a maximum dimension of 3.0 cm^2.

CURRENT DEVELOPMENTS IN HEPA TESTING

As the author remarked earlier, the use and testing of HEPA filters is an evolving technology, both in respect to standards development and in the actual methods of testing applied in the manufacture of filters and in testing installed filters. With regard to standards that are being developed, both ISO and IES documents reflect changed photometer test methods with regard to those currently applied by U.S. manufacturers based on IES-RP-CC001.3.

ISO 14644-3 (Metrology) notably gives more attention to scan rate and probe size than the current IES procedure and also contains information regarding measurement of homogeneity of the upstream aerosol. This document is still in the form of a committee draft, but this section seems not likely to change. Excerpts of interest are as follows:

> **B6.2.1 Choice of upstream aerosol**
>
> A pneumatically or thermally generated polydisperse aerosol (e.g., DOP or *materials* with equivalent properties) must be added to the natural upstream aerosol to reach the required challenge concentration. The *MMD* for this production method will typically be between 0.5 to 0.7 μm with a geometric standard deviation of up to 1.7.
>
> **B6.2.2 Concentration of upstream aerosol challenge and its verification**
>
> The concentration of the aerosol challenge upstream of the filter should be between 10 mg/m^3 and 100 mg/m^3. A concentration lower then 20 mg/m^3 would reduce the sensitivity for leak detection.
>
> Appropriate measures should be taken, for the verification of the homogenous mixing of the added aerosol to the supply airflow. The first time a system is tested, it should be validated that sufficient aerosol mixing is taking place. For such validation all injection and sampling points have to be defined and recorded.

The ratio of all concentration measurements taken immediately upstream of the filters shall, whether at different times or locations, not exceed the value of 2 and the average concentration shall be used as upstream aerosol concentration.

B6.2.3 Determination of probe size

The sample probe inlet size should be calculated from consideration of the sample flow rate of the photometer, and the filter exit airflow velocity so that the probe inlet air velocity approximates the filter exit airflow velocity. The sampling probe should be of square or rectangular configuration. In case of a rectangular probe configuration, W_p/D_p should be between 6 and 1 and:

$$D_p = \frac{F_a}{\left(V \times W_p\right)}$$

where D_p is the probe dimension parallel to the scan direction [cm], F_a is the actual sample flow rate of the photometer [cm^3/s], V is the filter exit airflow velocity [cm/s], and W_p is the probe dimension perpendicular to the scan direction [cm]

B6.2.4 Determination of scan rate

The probe traverse scan rate S_r when using a 3 × 3 cm probe should not exceed *5 cm/s.* With a rectangular probe, the maximum area scan rate should not exceed *15* cm^2/s.

B6.2.5 Test procedure for installed filter system leakage scan test

The test is performed by introducing the specific challenge aerosol upstream of the filter(s) and searching for leaks by scanning the downstream side of the filter(s) and the grid or mounting frame system with the photometers probe as follows:

- Measurements of the aerosol upstream of the filters according to section B6.2.2 are to be taken first to verify the aerosol concentration and also its homogenous distribution.
- The probe should then be traversed at a scan rate not exceeding the value S_r, stated in section B6.2.4, using slightly overlapping strokes. The probe is thereby held in a distance of approximately 5cm from the downstream filter face or the frame structure.

Indications are that the IES future document specifically directed at leak testing (HEPA and ULPA Filter Leak Tests IES-RP-CC034.1) will also provide significant detail in the included method for photometer testing. McDonald (1994b) and Moore et al. (1994) discuss comparative test methods and selection of photometer scan rates, probe sizes, and other test parameters. One early drafter of the document contains a very interesting synopsis of summary test considerations:

> The smallest leak to be detected during a leak test is called the designated leak size. The designated leak size establishes an upper limit for the size of leaks that are not detected during scanning and remain in the filter system after the scanning and repairs are completed. The choice of the size of the designated leak depends on the efficiency of the filters, on the type of test (factory, field bench, or post installation) and on the intended use of the clean air device or cleanroom. When HEPA filters are scanned with the photometer test method and aerosol generated by a Laskin nozzle, it is customary to define the designated leak to have a standard leak penetration of 0.0001 (0.01 percent). For ULPA filters it is appropriate to use smaller values to define the designated leak. See Table 1 for other values. Using smaller designated leak sizes will cause the scan test to take longer or will require higher challenge aerosol concentrations. The buyer needs to balance the added time, cost, and challenge aerosol deposited on the filters with the benefit of finding very small leaks.

Any industry will have some level of concern for the time (man-hour) requirements for filter testing. Of some (increased) interest currently are automated scanning devices of the type shown in Figure 13.25. These may be constructed in a form appropriate for field testing use for point leaks. Features of interest are the mechanical drive of the probe of the filter at a predetermined rate, with a vertical separation of scan paths ensuring overlap, and polymer brushing and bearing surfaces providing smooth, uniform probe movement.

SUMMARY

No alternative to the use of a cold DOP challenge for in-place testing of HEPA filters currently exists that will be acceptable to both manufacturers and regulatory authorities. The two primary concerns to be dealt with when selecting any test method are the nature of the material to be used as a challenge aerosol, and the certainty of leak detection with a minimum amount of time expended in testing. The application of a conventional cold DOP challenge at a concentration of 20–80 μg/L upstream of the tested filter represents some level of "overkill" in terms of certainty of leak detection and also results in damage to filters. Ambient challenge with particles 0.1 μm to 0.3 μm in size is generally time-consuming and is highly dependent on the upstream concentration of particles. An ambient challenge approach relying on the higher numbers of 0.05 μm particles in air upstream of the tested filter with CNC is better, but it does not test HEPA filters at a most penetrating size and is also time-consuming.

One certain improvement to the conventional DOP challenge method is simply to perform a DOP challenge in the most efficient way (i.e., optimize the present method). This would include the use of the most sensitive aerosol photometers currently available. These instruments will potentially provide the same certainty of leak detection as the present test, with a maximum of 10 μg/L challenge aerosol concentration upstream, and avoid the use of the

Figure 13.25. Precision scan automated scanning fixture (Precision Scan System courtesy of Flanders Filters).

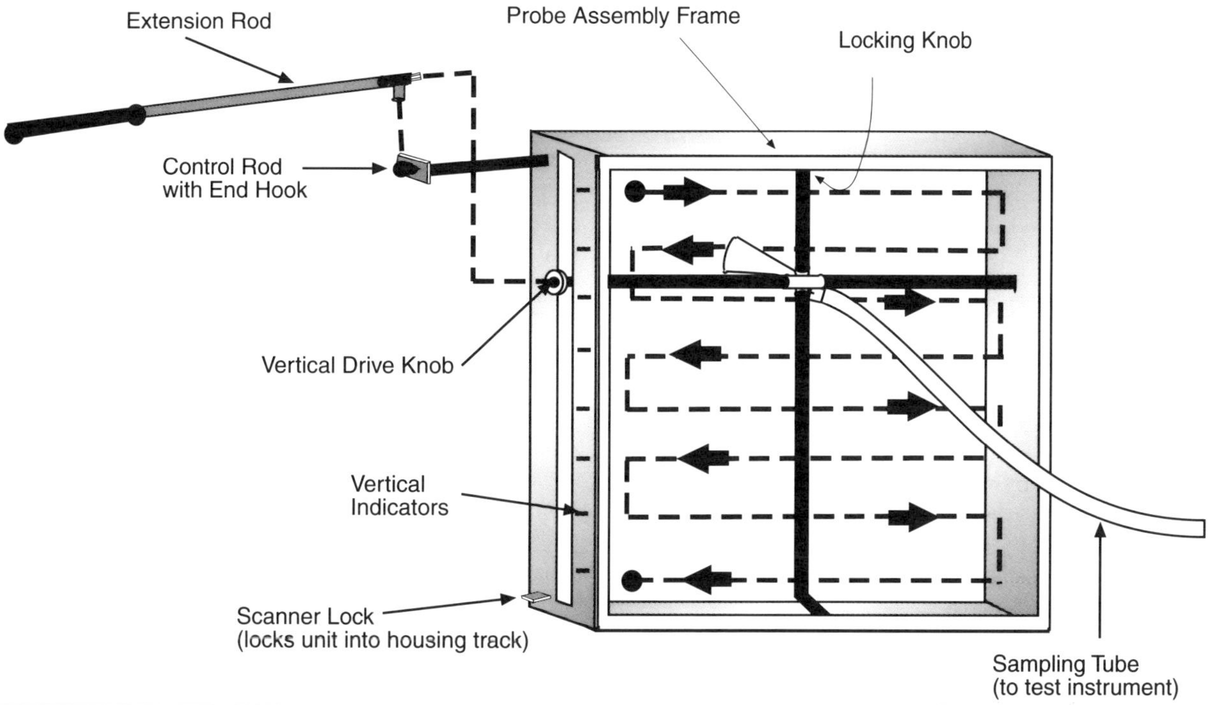

80–100 μg/L concentration often applied with conventional Laskin nozzle generators. The fact that Emery 3004 behaves in a fashion almost identical to DOP in testing and has decided advantages over DOP was sufficient for its acceptance by the FDA based on appropriate testing.

The use of monodisperse (0.2 μm to 0.3 μm) PSL aerosols generated by ultrasound appears to be a promising method for future use. This procedure serves the purpose of avoiding DOP use, with consequent lower filter loading and decreased human exposure. The use of significantly different techniques (such as the monosize PSL challenge test) will likely require careful validation of the methodology, including specifics of leak sizes detectable with the alternate before regulatory approval is considered. Importantly, the FDA maintains an open mind regarding alternate methods, simply requesting that "adequate validation" be performed. Validation can be expected to require significant man-hours and planning, so a manufacturer must carefully weigh costs and benefits before undertaking the task. In the final analysis, the FDA will simply require that alternate methodologies be proven to be equivalent to the current method in efficiency of leak detection. Validation of alternate methods using oils other than DOP is reportedly being planned by several pharmaceutical manufacturers.

While the execution of testing to validate equivalency of an alternate test method as meeting regulatory requirements in the United States promises to be a significant undertaking, the test plan for such a study is relatively simplistic. The definition of a leak is the critical factor in challenge methodology. Based on accepted standards and historical perspective, a leak in a HEPA filter is represented by passage of particles in excess of a 10^{-4} fraction (0.01 percent) of the upstream DOP challenge aerosol. This empirical definition may be reinforced by characterization of a leak in terms of the actual "hole" size in a filter that will pass a 10^{-4} concentration of a cold DOP aerosol. This opening size can be mechanistically defined using capillary tubes or apertures of known diameter inserted in the medium of a test filter (Sadjadi et al. 1990). While a capillary opening that passes 10^{-4} fraction of a cold DOP aerosol may differ in size from that which passes a similar concentration of another challenge material (such as monosize .26 μm PSL spheres), this experimental method will allow the proportionality of the new challenge aerosol to leak to be experimentally determined. Given that counting statistics, detection sensitivity, and aerosol uniformity are taken into consideration, this method should be applicable with a range of challenge materials and counting or photometric methods. Once the proportionality is defined experimentally, the validation procedure might then be based on a systematic verification that any real leak in a HEPA filter detected using a well-mixed, precisely quantitated DOP aerosol at 10 μg/L upstream and a suitably sensitive photometer downstream could be detected using the candidate replacement material at a predetermined upstream concentration with a suitable means of detection downstream. This latter component of the validation must remain the empirical test for the proposed methodology.

Several of the methods outlined in this chapter show promise for replacing the DOP test that is presently so widely used. Before the FDA accepts these methods, such as the monosize PSL challenge, more detailed definitions of what constitutes a leak will have to be determined. Until the anatomy of leaks in HEPA filters and the mechanics of leaks with regard to the precise particle sizes most effective in detecting leaks are resolved, it may prove impossible to validate alternate methods with the level of certainty necessary to satisfy regulatory requirements.

As discussed in Chapter 5, the ISO 14644 series will require testing of HEPA filters at periodic intervals (per 14644-2) and will provide test methods using laser particle counters and aerosol photometers (per 14644-3). Based on the author's involvement with the development of this series, there appears to be a widespread opinion in Europe (and to a lesser extent, Japan) that more carefully structured test methods are needed with a well-defined probability of leak detection (see Wepfer 1995). New, more stringent test methods have been proposed. This sentiment appears to mirror the concerns earlier expressed in the United States (McDonald 1993) and reflected in the latest version of IES-RP-CC006.2.

ISO 14644-3 calls for determining the extent of mixing of the upstream aerosol and precise calculation of the scan rate. These two additions will probably necessitate the change of some company SOPs for compliance as will compliance with the new IES document.

ACKNOWLEDGMENTS

The author wishes to express his gratitude for the editorial and technical assistance related to this chapter received from a number of individuals knowledgeable in HEPA filter testing. These include Mr. Alvin Lieberman (Particle Measuring Systems), Mr. Roger Jones (Lydall Corp.), Mr. Mark Cutler (Flanders Filters, Inc.), Mr. Dave Crosby (ATI Inc.), Mr. Gary Rolf (Clean Room Sciences), Mr. Don Moore (Eli Lilly), Dr. Werner Bergman (Lawrence Livermore Laboratory), Dr. Ron Wolff (Eli Lilly), Mr. Keith Reals (ATI Inc.), Mr. Bruce McDonald (Donaldson Corporation), and Mr. Tyler Beck (TSI Inc.).

The author is particularly in the debt of Mr. Bruce McDonald for his provision of general information for this chapter and in particular for considerations in scanning a filter for leaks.

REFERENCES

Bauman, A. J. 1987. Particulate silica test agents for HEPA filters. *Proceedings of the 33rd Meeting of the Institute of Environmental Sciences,* pp. 452–453.

Bergman, W. 1994. Personal communication.

Bergman, W., L. Foiles, and C. Mariner. 1989. New filter efficiency test for future nuclear grade HEPA filters. *Proceedings of the 20th DOE/NRC Nuclear Air Cleaning Conference,* pp. 1189–1207.

Bhoda, V. K. 1993. Designing and constructing the next generation of HEPA filters. *Microcontamination* (December): 31–35.

Biermann, A. H., and W. Bergman. 1988. Filter penetration measurements using a condensation nuclei counter and an aerosol photometer. *J. Aerosol Sci.* 19 (4):471–483.

Bresson, J. F. 1986. HEPA filter test activities at the Department of Energy. *Proceedings of the 19th DOE/NRC Nuclear Air Cleaning Conference,* pp. 852–862.

Cadwell, G. H. 1985. *Update on DOP as an aerosol challenge for testing of HEPA filters used in clean rooms.* Washington, N.C., USA: Flanders Filters, Inc.

Carlon, H. R., and M. A. Guelta. 1990. Safe replacement materials for DOP in "hot smoke" aerosol penetrometer machines. *Proceedings of the 21st DOE/NRC Nuclear Air Cleaning Conference,* pp. 154–173.

Carlon, H. R., M. A. Guelta, and B. V. Gerber. 1989. *A study of candidate replacement materials for DOP in filter-testing penetrometer machines.* Technical Report CRDEC-TR-053. Aberdeen Proving Ground, Md., USA: U.S. Army Chemical Research, Development and Engineering Center.

Crosby, D. W. 1990. Concentration produced by a Laskin nozzle generator: A comparison of substitute materials and DOP. *Proceedings of the 21st DOE/NRC Nuclear Air Cleaning Conference,* pp. 1–7.

Crosby, D. W. 1992. Personal communication.

Crosby, D. W. 1993. Photometer adjustments when using DOP substitutes. *Performance Review* (Winter): 9–10.

Echols, W., and H. J. A. Young. 1963. *Studies of portable air-operated aerosol generators.* NRL Report 5929. Washington, D.C.: Naval Research Lab.

Ettinger, H. J., J. D. DeField, D. A. Bevis, and R. N. Mitchell. 1969. HEPA filter efficiencies using thermal and air-jet generated dioctyl phthalate. *American Industrial Hygiene Association Journal* 30:87–94.

Flanders. 1994. *HEPA filters and filter testing,* 3rd ed. 1994. Bulletin No. 581D. Washington, N.C., USA: Flanders Filters, Inc.

Franklin, B., M. Pasha, and C. A. Bronger. 1987. Remote aerosol testing of large size HEPA filter banks. *Waste Manage.* 3:493–499.

Gerbig, F. T., and P. B. Keady. 1985. Size distributions of test aerosols from a Laskin nozzle. *Microcontamination* (July): 56–61.

Gogins, M., B. McDonald, R. Nicholson, and R. Cardinal. 1987. Design and operation of optimized high efficiency filter element test systems. *Proceedings of the Institute of Environmental Sciences,* pp. 124–133.

Greiner, J. 1990. HEPA filter leaks testing using the particle counter leak method. *Clean Rooms* (October): 37–39.

Guiney, P. D. 1982. Acute toxicity assessment of polyalphaolefin (PAO) synthetic fluids. *Proceedings of the Symposium on Synthetic and Petroleum Based Lubricants,* pp. 381–389.

Gupta, A., V. J. Novick, P. Biswas, and P. R. Monson. 1993. Effect of humidity and particle hygroscopicity on the mass loading capacity of high efficiency particulate air (HEPA) filters. *Aerosol Science and Technology* 19 (1):94.

Hinds, W., J. Macher, and M. W. First. 1981. Size distribution of aerosols produced from substitute materials by the Laskin cold DOP aerosol generators. *Proceedings of the 16th DOE Nuclear Air Cleaning Conference,* pp. 125–138.

Hinds, W., J. Macher, and M. First. 1982. Size distributions of test aerosols produced from materials other than DOP, *J. Environ. Sci.* 25:20–21.

Hinds, W. C., J. M. Macher, and M. W. First. 1983. Size distribution of aerosols produced by the Laskin aerosol generator using substitute materials for DOP. *American Industrial Hygiene Association Journal* 44 (7):495–500.

Hinds, W., M. First, D. Gibson, and D. Leith. 1978. Size distribution of "hot DOP" aerosol produced by ATI Q-127 aerosol generator. *Proceedings of the 15th DOE Nuclear Air Cleaning Conference,* pp. 1130–1144.

Holänder, W., W. Dunkhorst, and H. Lödding 1990. A condensation nucleus counter array for clean room applications. *Proceedings of the International Symposium,* pp. 66–77.

Jalon, S. 1987. Testing HEPA filters with uranine. *Proceedings of the 33rd Meeting of the Institute of Environmental Sciences,* pp. 436–438.

Johnson, E. M., B. R. Johnson, M. Huang, and S. K. Herweyer. 1990a. New CNC based automated filter tester for fast penetration testing of HEPA and ULPA filters and filter media. *Proceedings of 36th Annual Technical Meeting of the Institute of Environmental Sciences,* pp. 128–133.

Johnson, E. M., B. R. Johnson, S. K. Herweyer, and M. Huang. 1990b. An automated CNC based filter test for fast penetration testing of HEPA and ULPA media. *Solid State Technology,* pp. 113–117.

Kirk, G. Q., C. E. Childress, and D. D. Schmoyer. 1988. Methods of testing HEPA filters with short upstream approaches. *Proceedings of the 20th DOE/NRC Nuclear Air Cleaning Conference,* pp. 1210–1215.

LaRocca, P. T. 1985. Testing requirements for HEPA filters and clean work stations. *Pharmaceutical Manufacturing* (March): 47–68.

Lee, K. W., and B. Y. H. Liu. 1980. On the minimum efficiency and the most penetrating particle size for fibrous filters. *Air Poll. Control Assoc. J.* 30:377–381.

Leibold, H., and J. G. Wilhelm. 1991. Investigations into the penetration and pressure drop of HEPA filter media during loading with submicron particle aerosols at high concentrations. *J. Aerosol. Sci.* 22 (1):S773–S776.

Letourneau, P., P. Mulcey, and J. Vendel. 1988. Prediction of HEPA filter pressure drop and removal efficiency during dust loading. *Proceedings of the 20th DOE/NRC Nuclear Air Cleaning Conference,* pp. 984–993.

Lieberman, A. 1994. Personal communication.

Liu, B. Y. H., J.-K. Lee, H. Mullins, and S. G. Danisch. 1993. Respirator leak detection by ultrafine aerosols: A predictive model and experimental study. *Aerosol Science and Technology* 19 (1): 15–26.

Liu, B. Y. H., and K. W. Lee. 1982. Experimental study of aerosol filtration by fibrous filters. *Aerosol Sci. and Technol.* 1:35–46.

Liu, B. Y. H., K. L. Rubow, and D. Y. H. Pui. 1985. On the performance of HEPA and ULPA filters. *Proceedings of the 31st Annual Technical Meeting of the Institute of Environmental Sciences,* pp. 25–28.

Loughborough, D. 1991. The dust holding capacity of HEPA filters. *Proceedings of the Nuclear Air Cleanliness Conference,* 1:155–172.

McDonald, B. R. 1993. Scanning high efficiency air filters for leaks using particle counting methods. *J. of the IES* (Sept./Oct.): 28–36.

McDonald, B. 1994a. Personal communication.

McDonald, B. 1994b. Scanning HEPA filters with photometers. *J. of the IES* (Sept./Oct.): 32–38.

Moelter, W. 1988. Fast, automated testing of HEPA and ULPA filter media. *Filtr. Sep.* 25 (6):417–418.

Moelter, W., W. Schmitz, and H. Fissan. 1992. Future practical contamination control. *Mech. Eng Publ.* pp. 575–579.

Moore, D. R., and J. G. Marshall. 1993. An alternative to the use of dioctyl phthalate (DOP) for HEPA filter testing in the pharmaceutical industry. *Proceedings of the ISPE Seminar,* pp. 27–41.

Moore, D. R., J. G. Marshall, and M. A. Kennedy. 1994. Comparative testing of challenge aerosols in HEPA filters with controlled defects. *Pharm. Eng.* 14 (2):78–89.

Mulcey, P., P. Pybot, and J. Vendel. 1990. Real time detection of a fluorescent aerosol application to the efficiency measurements of HEPA filters. *Proceedings of the 21st DOE/NRC Nuclear Air Cleaning Conference,* 1:95–108.

Murrow, J. L., and G. O. Nelson 1973. HEPA-filter testing: Comparison of DOP and NaCl aerosols. *Proceedings of the 12th AEC Air Cleaning Conference,* 12:808–816.

NIOSH. 1988. *Occupational safety and health guideline for di-2-ethyhexyl phthalate (DEHP) potential human carcinogen.* Washington, D.C.: U.S. Dept. Health and Human Services.

Novick, V. J., P. R. Monson, and P. E. Ellison. 1992. The effect of solid particle mass loading on the pressure drop of HEPA filters. *J. Aerosol Sci.* 23 (6):657–665.

NTP. 1980. *Technical report on the carcinogenesis bioassay of di(2-ethylhexyl)phthalate.* CAS No. 117-81-7. Research Triangle Park, N.C., USA: National Toxicology Program.

Ortiz, J. P., A. H. Biermann, and R. M. Nicholson. 1991. Preliminary test results of a round robin test program to evaluate a multi-state HEPA filter system using single-particle size counters. *Adv. Filtr. Sep. Technol.* 4:213–217.

Osaki, M., and A. Kanagora. 1990. Performance of high efficiency particulate air filters. *J. Nuclear Sci. and Technol.* 27:875–882.

Parker, R. C., M. Marshall, and R. B. Bosley. 1993. New method for in-situ filter testing using pulses of aerosol. *Environ. Eng.* 6 (2):9–12.

Payet, S., D. Boulaud, G. Madelaine, and A. Renoux. 1992. Penetration and pressure drop of a HEPA filter during loading with submicron liquid particles. *J. Aerosol Sci.* 23 (7):723–735.

Sadjadi, R., S. M. Liu, and Y. H. Benjamin. 1990. Characteristics of pin-hole leaks in HEPA and ULPA filters. *Proceedings of the 36th Annual Technical Meeting of the Institute of Environmental Sciences,* pp. 23–27.

Sem, G. J. 1986. A case for continuous multipoint particle monitoring in semiconductor clean rooms. *IES Proceedings,* pp. 432–438.

Sharaf, M. A., and S. J. Troutman. 1988. Comparative size distribution and filter penetration measurements of DOP and corn oil aerosols. *Particulate Science and Technology* 5:207–217.

Smith, P. R., I. H. Leslie, E. C. Hensel, T. M. Schultheis, J. R. Walls, and W. S. Gregory. 1991. Investigation of salt loaded HEPA filters. *Proceedings of the 21st DOE/NRC Nuclear Air Cleaning Conference,* 1:366–75.

Steffen, D. H., and C. K. Girres. 1992. *Emery 3004 as a challenge aerosol: Operational and experience at Westinghouse Hanford Company.* Technical Report WHESA-1509-FP. Riceland, Wash., USA: Westinghouse Hanford Company.

Tancrede, M., R. Wilson, L. Zeise, and E. A. C. Crouch. 1987. The carcinogenic risk of some organic vapors indoors: A theoretical survey. *Atmos. Environ.* 21 (10):2187–2205.

Thomas, J. W., and G. D. Crane. 1963. Aerosol penetration through 9 mil HV-70 filter paper with and without pinholes. *Proceedings of the DOE/NRC Nuclear Air Cleaning Conference,* pp. 189–199.

Tillery, M. I. 1987. Applications of high efficiency air filtration. In *Filtration principles and practices,* edited by M. J. Matteson and C. Orr. New York: Marcel Dekker.

Weckerly, J. 1990. The missing link in filter testing: Challenge aerosol concentration. *Clean Rooms* (October): 20–21.

Wepfer, R. 1986. Production control of ULPA Filters. *Proceedings of the 8th ICCCS Symposium.* 154–169.

Wepfer, R. 1995. Characterization of HEPA and ULPA filters by proposed new european test methods. *Filtration and Separation* (June): 29–35.

Wolff, R. K. 1994. Personal communication.

Xiaowei, Y., M. W. First, and S. N. Rudnick. 1990. Characteristics of Laskin nozzle generated aerosols. *Proceedings of the 21st DOE/NRC Nuclear Air Cleaning Conference,* pp. 131–140.

Yan, X., M. First, and S. Rudnick. 1990. Characteristics of Laskin nozzle generated aerosols. *Proceedings of the 21st DOE/NRC Nuclear Air Cleaning Conference,* pp. 167–171.

STANDARDS AND SOURCES

AFNOR standard NF X44-101. 1983. *Definitions, classification, procedure for testing of controlled dust content chambers.* Paris.

AS1807.6. 1989. *Cleanrooms, workstations and safety cabinets: Determination of integrity of terminally mounted HEPA filter installations.* North Sydney, New South Wales: Standard Association of Australia.

ASHRAE standard 52–76. 1975. *Method of testing and cleaning devices used in general ventilation for removing particulate matter.* Atlanta: American Society of Heating, Refrigerating, and Air-Conditioning Engineers.

ASME N510. 1989. *Testing of nuclear air treatment systems.* Fairfield, N.J., USA: American Society of Mechanical Engineers.

ASME N509. 1980. *Standard for nuclear power plant air cleaning units and components.* Fairfield, N.J., USA: American Society of Mechanical Engineers.

ASTM F 91-70. 1976. *Testing for leaks in the filters associated with laminar flow clean rooms and clean work stations by use of a condensation nuclei detector.* Philadelphia: American Society of Testing and Materials.

ASTM F 1471-93. 1993. *Test method for air cleaning performance of a high-efficiency particulate air cleaning system.* Philadelphia: American Society for Testing and Materials.

BS 3928. 1965. *Method of test for low penetration air filters.* London: British Standards Institution.

BS 5295. 1989. *Environmental cleanliness in enclosed spaces: Specification for clean rooms and clean air devices.* London: British Standards Institution.

DIN 24184. 1974. *Type testing of high efficiency submicron particulate air filters.* Berlin: Deutsche Institut fur Normierunge.

FS-209D. 1988. *Clean room and work station requirements, controlled environment.* Mt. Prospect Ill.: Institute of Environmental Sciences.

FS-209E. 1992. *Airborne particulate cleanliness classes in cleanrooms and clean zones.* Mt.

Prospect, Ill., USA: Institute of Environmental Sciences.

IES-RP-CC001.3. 1993. *HEPA and ULPA filters.* Mt. Prospect, Ill., USA: Institute of Environmental Sciences.

IES-RP-CC002. 1986. *Laminar flow clean air devices.* Mt. Prospect, Ill., USA: Institute of Environmental Sciences.

IES-RP-CC006.2. 1993. *Testing clean rooms.* Mt. Prospect, Ill., USA: Institute of Environmental Sciences.

IES-RP-CC007.1. 1992. *Testing ULPA filters.* Mt. Prospect, Ill., USA: Institute of Environmental Sciences.

IES-RP-CC013. 1986. *Equipment calibration or validation procedures.* Mt. Prospect, Ill., USA: Institute of Environmental Sciences.

IES-RP-CC034.1. 1998. HEPA and ULPA filter leak tests. Mt. Prospect, Ill., USA: Institute of Environmental Sciences.

JACA10C. 1978. *Standard of test method for air cleaning devices.* Tokyo: Japan Air Cleaning Association.

MIL Spec. F-51068. 1989. *Filters, particulate, high efficiency, fire resistant.* Philadelphia: Standardization Document Order Desk.

MIL Spec. F-51079E. 1988. *Filter medium, fire resistant, high efficiency.* Philadelphia: Standardization Document Order Desk.

MIL-F-51477. 1991. *Military specification for filters, particulate, high efficiency, fire resistant, biological use, general specification for HEPA filters.* Philadelphia: Standardization Document Order Desk.

MIL-STD-282. 1985. *Filter units, protective clothing, gas-mask components and related products: Performance test methods.* Philadelphia: Standardization Document Order Desk.

NEBB. 1996. *Procedural standards for certified testing of cleanrooms,* 1st ed. Vienna, Va., USA: National Environmental Balancing Bureau.

NEF3-43. *Quality assurance testing of HEPA filters.* Oak Ridge Tenn., USA: Department of Energy, Office of Scientific and Technical Information.

NSF. 1987. *NSF standard 49 for class II (laminar flow) biohazard cabinetry.* Ann Arbor, Mich., USA: National Sanitation Foundation.

prEN 1822.1. 1995. *High efficiency particulate air filters (HEPA and ULPA).* Luxembourg: Commission of the European Communities.

prEN 1822.4. 1997. *Testing air filter elements for leaks.* Luxembourg: Commission of the European Communities.

QJ2214-91. 1991. *Cleanliness classifications and verification for clean rooms/clean zones.* Beijing: Chinese Aerospace Ministry.

SWKI Guideline 84-2. 1984. *Classification and utilization of air filters.* Bern, Switzerland: Swiss Association of Heating and Air Conditioning Engineers.

Appendix 1

SOURCES OF STANDARDS AND DOCUMENTS

ANSI: American National Standards Institute

11 West 42nd Street
New York, NY 10036
212-642-4900
212-302-1286 fax

ASTM: American Society for Testing and Materials

100 Barr Harbor Drive
West Conshohocen, PA 19428
610-832-9500
610-832-9555 fax
610-832-9585 publications

BSI: British Standards Institution

389 Chiswick High Road
London W4 4AL, United Kingdom
01 81 996 9000
01 81 996 7400 fax

Global Engineering Documents

7730 Carondelet Avenue, Suite 407
Clayton, MO 63105
800-854-7179
800-726-6418 fax

IES: Institute of Environmental Sciences

940 East Northwest Highway
Mt. Prospect, IL 60065
847-255-1561
847-255-1699 fax

JSA: Japan Standards Association

1-24 Akasaka 4, Minato-ku
Tokyo 107, Japan

MOD: Ministry of Defence

Directorate of Standardization
First Avenue House
High Holborn
London WCIV 6HE

NFPA: National Fluid Power Association

3333 North Mayfair Road
Milwaukee, WI 54222-3219
414-778-3344
414-778-3361 fax
414-778-3363 orders

NIST: National Institute of Standards and Technology

SRM Program
Building 202, Room 204
Gaithersburgh, MD 20899
301-975-6776
301-948-3730 fax

SAE: Society of Automotive Engineers, Inc.

400 Commonwealth Dr.
Warrendale, PA 15096-0001
412-776-4841
412-776-5760 fax

USP: United States Pharmacopeia

12601 Twinbrook Parkway
Rockville, MD 20852
800-227-8772

Appendix 2

TRADEMARKS

Following is a list of trademarked items, processes, and products mentioned in some context in this book. The author has made every effort to acknowledge appropriately any trademarks included in the text. It is possible, however, due to the various citation formats used in references, books, promotional literature, and professional literature, that some may have been inadvertently omitted. Should any doubt arise as to whether a trademark is appropriate where one is not included in the list, the reader is requested to investigate further. Inquiries may be made of the vendor of a product or process in question, or the web site database for the U.S. Patent Office (http://www.uspto.gov/tmdb/index.html) may be consulted.

AAMI®	Association for the Advancement of Medical Instrumentation, Arlington, VA 22201
ALP®	Automated Liquid Packaging, Inc., Woodstock, IL 60173
Analyslide®	Filter examination and storage container, Gelman Sciences, Ann Arbor, MI 48106
Aroclor®	Mounting media, Monsanto Co., St. Louis, MO 63167
Bev-A-Line®	Hytrel®-lined flexible PVC tubing, Thermoplastic Processes, Inc., Stirling, NJ 07980
Bon-Ami®	Polishing cleanser, Faultless Starch/Bon Ami Co., Kansas City, MO 64101
BSI™	British Standards Institute, London, England
CA-Cricket® Graph™	Graph Computer Associates International, Inc., San Jose, CA 95131
Climet®	Climet Instruments, 1320 W. Cotton Ave., Redlands, CA 92374
Coulter Counter®	Electronic particle counter, Coulter Electronics, Inc. Hialeah, FL 33010
Coulter®	Coulter Electronics, Inc., Hialeah, FL 33010
Dacron®	Polyester fiber, DuPont Chemicals, Wilmington, DE 19898
Darvon®	Darvon-N, propoxyphene napsylate USP, Eli Lilly, Indianapolis, IN 46285

Dimple Pleat™	Pleated high efficiency air filtration medium, Flanders Filters Inc., Washington, NC 27889
Donaldson®	Air filters, Donaldson Company Inc., Minneapolis, MN 55440
Duco®	Plastic cement, DuPont Chemicals, Wilmington, DE 19898
Easy-Lab®	Lab automation software, Zymark Corp., Hopkinton, MA 01748
EISAI®	Eisai Co., Ltd., Tokyo, Japan
Ektachrome®	Photographic color film, Eastman Kodak Company, Rochester, NY 14650
Elzone®	Particle measurement system, Particle Data, Inc., Elmhurst, IL 60126
EXECUSTAT®	Data analysis system, Strategy Plus, Inc., Princeton, NJ 08540
Extar®	Polyester-based polymer, Eastman Kodak Co., Kingsport, TN 37662
Flanders®	Filter, Flanders Filters Inc., Washington, NC 27889
Formvar®	Polyvinyl formal plastic, Monsanto Co., St. Louis, MO 63167
Freon®	Precision cleaning agent, E. I. DuPont De Nemours & Co., Wilmington, DE 19898
Gelman®	Membrane Cellulose Filters, Gelman Instrument Co., Ann Arbor, MI 48103
Gore-Tex®	Expanded PTFE, W. L. Gore and Associates Inc., 555 Paper Mill Road, Newark, DE 19714
Hewlett-Packard®	Hewlett-Packard Co., Palo Alto, CA 94304
HEPA®	Air filters, HEPA Corporation, Anaheim, CA 92806
HIAC/Royco®	Division of Pacific Scientific Company, Silver Spring, MD 20904
Hytrel®	Polyester elastomer, DuPont Chemicals, Wilmington, DE 19898
IEST™	Institute of Environmental Sciences and Technology, Mount Prospect, IL 60056
Lasair®	Aerosol particle counters, Particle Measuring Systems, Inc., Boulder, CO 80301
Leitz®	E. Leitz Inc., Rockleigh, NY 07647
Lotus®	Software, Lotus Development Corp., Cambridge, MA 02142
Luminaire™	Electronically operated backlit display illuminator, Clearr Corp., Minnetonka, MN 55345
Meltmount®	Mounting media, Cargille Laboratories, Cedar Grove, NJ 07009
Metricell®	Microporous polypropylene media, Gelman Sciences, Ann Arbor, MI 48106
Micromeretics®	Micromeretics Corp., Norcross, GA 30093
Microsoft®	Microsoft Corp., One Microsoft Way, Redmond, WA 98052
Microsoft® Word	Word processing program, Microsoft Corporation, Redmond, WA 98052-6399
Microsoft® Excel	Worksheet analysis, exchanging data, customizing, automating, Microsoft Corporation, Redmond, WA 98052-6399
Milli-Q®	Water purification system, Millipore Corp., Bedford, MA 01730
Millipore®	Millipore Corporation, Bedford, MA 01730
Mylar®	Flexible film for packaging purposes, E.I. DuPont De Nemours and Co., Wilmington, DE 19898
NIST®	National Institute of Standards and Technology, U.S. Department of Commerce, Gaithersburgh, MD 20899
Nuclepore®	Track-etched membranes, Costar Corp., Cambridge, MA 02140
NYLAflo®	Nylon media capsule filter, Gelman Sciences Inc., Ann Arbor, MI 48103

Olmpus®	Olmpus Corporation, Lake Success, NY 11042
OPTIPHOT®	Nikon Microscopes and Parts, Nippon Kogaku K.K., Tokyo, Japan
Orlon®	Acrylic fiber, DuPont Chemicals, Wilmington, DE 19898
Pall®	Pall Corporation, Long Island, NY 11542
Particle Measuring Systems®	Particle Measuring Systems Inc., 1855 S. 57th Court, Boulder, CO 80301
Permount®	Microscope slide mounting medium, Fisher Scientific, Pittsburgh, PA 15219
Petrislide®	Membrane filter viewing container, Millipore Products, Bedford, MA 01730
Plexiglas®	Acrylic sheet, Rohm and Haas Co., Atlanta, GA 30338
Polaroid®	Instant film, Polaroid Corporation, Cambridge, MA 02139
Rion®	Rion Co., Ltd., Tokyo, Japan
SAE®	Society of Automotive Engineers, Inc., Warrendale, PA 15096
Scotch® Magic Tape	Transparent tape, 3M, St. Paul, MN 55101
Spectrex®	Spectrex Corp., Redwood City, CA 94063
STA-GRAPHICS®	Statistical graphics system, STSC, Inc., Rockville, NM 20852
Supor®	Inherently hydrophilic polysulfone membrane, Gelman Corporation, Ann Arbor, MI 48106
Tacky Mat®	Entrance mat for cleaning shoe soles, Liberty Industries, Inc., East Berlin, CT 06023
Teflon®	Fluorocarbon resin, DuPont Chemicals, Wilmington, DE 19898
Triton® X-100	Detergent, Rohn and Haas Co., Philadelphia, PA 19105
TSI®	TSI Incorporated, St. Paul, MN 55164
Tween® 20	Detergent, ICI America, Inc., Wilmington, DE 19897
Tygon®	Plastic tubing, Nortar Performance Plastics, Wayne, NJ 07470
Tyvek®	Fabrics of man-made fibers suitable for apparel, E.I. DuPont De Nemours and Co., Wilmington, DE 19898
Ultipor®	Filters, Pall Corp., Glen Cove, NY 11542
Vanox®	Microscopes, Olmpus Corp., New York, NY 11042
Windows®	Software information and data management system, Microsoft Corp., Redmond, WA 98052
Zymate®	Robotic system, Zymark Corp., Hopkinton, MA 01748

INDEX